Lecture Notes in Computer Science 10746

Commenced Publication in 1973
Founding and Former Series Editors:
Gerhard Goos, Juris Hartmanis, and Jan van Leeuwen

Editorial Board

More information about this series at http://www.springer.com/series/7412

Marcello Pelillo · Edwin Hancock (Eds.)

Energy Minimization Methods in Computer Vision and Pattern Recognition

11th International Conference, EMMCVPR 2017
Venice, Italy, October 30 – November 1, 2017
Revised Selected Papers

 Springer

Editors
Marcello Pelillo (iD)
Ca' Foscari University of Venice
Venice
Italy

Edwin Hancock (iD)
University of York
York
UK

ISSN 0302-9743   ISSN 1611-3349   (electronic)
Lecture Notes in Computer Science
ISBN 978-3-319-78198-3   ISBN 978-3-319-78199-0   (eBook)
https://doi.org/10.1007/978-3-319-78199-0

Library of Congress Control Number: 2018937378

LNCS Sublibrary: SL6 – Image Processing, Computer Vision, Pattern Recognition, and Graphics

Printed on acid-free paper

This Springer imprint is published by the registered company Springer International Publishing AG
part of Springer Nature
The registered company address is: Gewerbestrasse 11, 6330 Cham, Switzerland

Preface

This volume contains the papers presented at the 11th International Conference on Energy Minimization Methods in Computer Vision and Pattern Recognition (EMMVCPR). The year 2017 marked the 20th anniversary of the inaugural meeting of the conference series, which took place in Venice in May 1997, and we wanted to celebrate by bringing the event back to Venice. The conference was held from October 30 to November 1, 2017, in the unique setting of Palazzo Franchetti, a historical Venetian palace facing the Grand Canal. The conference was co-located with ICCV 2017, the 16th International Conference on Computer Vision, which was held in Venice Lido the week before EMMCVPR.

The call for papers produced 51 submissions, resulting in the 37 papers appearing in this volume, nine of which were presented as long talks, and the remainder as short talks. We make no distinction between these two types of paper in this book.

The accepted papers cover a wide range of problems and perspectives including clustering and quantum computing, image processing and video analysis, shading and light, as well as optimization methods, and address both theoretical issues as well as real-world problems. In addition to the contributed papers, the conference featured two invited talks by Joachim Buhmann, from ETH Zurich (Switzerland), and Paolo Frasconi, from the University of Florence (Italy).

We would like to take this opportunity to express our gratitude to all those who helped to organize the conference. First of all, thanks are due to the members of the Steering and the Scientific Committees and to all the additional reviewers. Special thanks are due to the members of the Organizing Committee, Sebastiano Vascon and Ismail Elezi, together with Cristiana Fiandra and her team at "The Office" for managing the conference organization, and also Roberta D'Argenio and Agnese Boscarol, from the European Center for Living Technology (ECLT), for their administrative support. We gratefully acknowledge generous financial support from the Bertinoro International Center for Informatics (BICI) and the Department of Environmental Sciences, Informatics and Statistics (DAIS) of Ca' Foscari University of Venice.

We also offer our appreciation to the editorial staff at Springer in producing this book, and for supporting the event through publication in the LNCS series. We thank all the authors and the invited speakers for helping to make this event a success, and producing a high-quality publication to document the event. We warmly acknowledge the help of Sebastiano Vascon in assembling the final proceedings volume.

February 2018

Marcello Pelillo
Edwin Hancock

Organization

Program Committee

Ognjen Arandjelovic	University of St. Andrews, UK
Xiang Bai	Huazhong University of Science and Technology, China
Michael Bronstein	University of Lugano, Switzerland
Andres Bruhn	University of Stuttgart, Germany
Sotirios Chatzis	CUT, Cyprus
Xilin Chen	Institute of Computing Technology, China
Jim Clark	McGill University, Canada
Daniel Cremers	Technical University of Munich, Germany
Gianfranco Doretto	West Virginia University, USA
Francisco Escolano	University of Alicante, Spain
Mario Figueiredo	Instituto de Telecomunicações, Instituto Superior Técnico, Portugal
Alexander Fix	Oculus Research, USA
Boris Flach	Czech Technical University in Prague, Czech Republic
Davi Geiger	NYU, USA
Edwin Hancock	University of York, UK
Hiroshi Ishikawa	Waseda University, Japan
Ian Jermyn	Durham University, UK
Fredrik Kahl	Lund University, Sweden
Zoltan Kato	University of Szeged, Hungary
Anton Konushin	Lomonosov Moscow State University, Russia
Hamid Krim	North Carolina State University, USA
Longin Jan Latecki	Temple University, USA
Hongdong Li	ANU, Australia
Stan Li	CBSR, CASIA, China
Stephen Maybank	Birkbeck College, UK
Marcello Pelillo	University of Venice, Italy
Thomas Pock	Graz University of Technology, Austria
Antonio Robles-Kelly	CSIRO, Australia
Emanuele Rodola	Technical University of Munich, Germany
Samuel Rota Bulò	Fondazione Bruno Kessler, Italy
Christoph Schnoerr	Heidelberg University, Germany
Fiorella Sgallari	University of Bologna, Italy
Ali Shokoufandeh	Drexel University, USA
Hugues Talbot	Université Paris Est, France
Wenbing Tao	Huazhong University of Science and Technology, China

Andrea Torsello   Università Ca Foscari, Italy
Carole Twining   ISBE, University of Manchester, UK
Olga Veksler   University of Western Ontario, Canada
Richard Wilson   University of York, UK
Josiane Zerubia   Inria, France

Additional Reviewers

Chiotellis, John   Piccirilli, Marco
Haefner, Bjoern   Pidhorskyi, Stanislav
Jones, Quinn   Rossi, Luca
Laude, Emanuel   Stoll, Michael
Lei, Zhen   Sun, Kun
Liao, Shengcai   Wan, Jun
Loewenhauser, Benedikt   Wang, Xiaobo
Ma, Zhanyu   Welk, Martin
Marin, Dmitrii   Xia, Guisong
Maurer, Daniel   Xiao, Yang
Moellenhoff, Thomas   Xu, Yongchao
Motiian, Saeid   Yang, Yang
Möllenhoff, Thomas   Zhu, Xiangyu

Contents

Propagation and Time-Evolution

Inference, Labeling and Relaxation

Clustering and Quantum Methods

Ising Models for Binary Clustering
via Adiabatic Quantum Computing

Christian Bauckhage[1,2](✉) ⓘ, E. Brito[1,2], K. Cvejoski[1,2], C. Ojeda[1,2],
Rafet Sifa[1,2], and S. Wrobel[1,2]

[1] Fraunhofer IAIS, Sankt Augustin, Germany
christian.bauckhage@iais.fraunhofer.de
[2] B-IT, University of Bonn, Bonn, Germany

Abstract. Existing adiabatic quantum computers are tailored towards minimizing the energies of Ising models. The quest for implementations of pattern recognition or machine learning algorithms on such devices can thus be seen as the quest for Ising model (re-)formulations of their objective functions. In this paper, we present Ising models for the tasks of binary clustering of numerical and relational data and discuss how to set up corresponding quantum registers and Hamiltonian operators. In simulation experiments, we numerically solve the respective Schrödinger equations and observe our approaches to yield convincing results.

1 Introduction

After decades of being a mainly theoretical endeavor, quantum computing now appears to become practical. First working quantum computers are being sold, major industrial players invest into corresponding research and development, and a growing number of voices predicts rapid technological progress [1–3]. All these developments will likely impact machine learning and pattern recognition because quantum computers promise fast solutions to the kind of optimization problems dealt with in these fields [4–8].

In this paper, we are therefore concerned with the feasibility of quantum computing for pattern recognition. In particular, we consider binary clustering of numerical or relational data and show how to approach these problems using *adiabatic quantum computing* [9].

Our interest in adiabatic quantum computing stems from the fact that recent technological progress has happened mainly in this area [10,11]. Available devices are geared towards finding low energy states of Ising models

$$H(\boldsymbol{s}) = \sum_{i,j=1}^{n} Q_{ij}\, s_i\, s_j + \sum_{i=1}^{n} q_i\, s_i \tag{1}$$

where H is a Hamiltonian or energy operator, $\boldsymbol{s} \in \{-1,1\}^n$ is a vector that represents the state of a system of n bipolar entities s_i (say, magnetic spins),

© Springer International Publishing AG, part of Springer Nature 2018
M. Pelillo and E. Hancock (Eds.): EMMCVPR 2017, LNCS 10746, pp. 3–17, 2018.
https://doi.org/10.1007/978-3-319-78199-0_1

$Q \in \mathbb{R}^{n \times n}$ is called the coupling matrix, and $q \in \mathbb{R}^n$ models external influences to the system.

Adiabatic quantum computing relies on the adiabatic theorem [12] which states that if a quantum mechanical system in a low energy configuration or ground state of a Hamiltonian is subjected to gradual changes, it will end up in the ground state of the resulting Hamiltonian. To harness this principle for problem solving, one prepares a system in the ground state of a simple, problem independent Hamiltonian and adiabatically evolves it to a Hamiltonian whose ground state represents a solution to the problem at hand. One of the general challenges of adiabatic quantum computing is thus to devise suitable problem Hamiltonians. However, if the problem at hand can be expressed as an Ising energy minimization problem, these Hamiltionians are readily constructed. In this sense, the quest for adiabatic quantum implementations of pattern recognition algorithms becomes the quest for Ising model (re-)formulations of their objective functions.

Our presentation in this paper is therefore structured as follows: First of all, we present Ising models for the problems of computing $k = 2$ means clustering and balanced graph cuts.

Second of all, we discuss technical details of adiabatic quantum computing and show how to set up a system of qubits along with two appropriate Hamiltonians to perform quantum adiabatic binary clustering.

Third of all, we present simple practical examples that demonstrate the feasibility of adiabatic quantum clustering and illustrate how appropriately prepared systems of qubits evolve towards a solution.

Finally, we conclude with a discussion of related work and a summary of our contributions.

2   An Ising Model for Numerical Binary Clustering

In this section, we address the problem of binary clustering of numerical data and derive a heuristic for $k = 2$ means clustering that leads to an Ising model. Throughout, we let $X \subset \mathbb{R}^m$ denote a sample of $n = |X|$ data points and assume w.l.o.g. that these data are normalized such that the sample mean $\bar{x} = 0$.

Recall that hard k-means clustering is a prototype-based clustering technique that partitions the data in X into k disjoint subsets

$$X_i = \left\{ x \in X \ \middle| \ \left\| x - \bar{x}_i \right\|^2 < \left\| x - \bar{x}_j \right\|^2 \ \forall \, i \neq j \right\} \qquad (2)$$

of size $n_i = |X_i|$ whose centroids are given by

$$\bar{x}_i = \frac{1}{n_i} \sum_{x \in X_i} x. \qquad (3)$$

2.1 The "Standard" k-Means Objective Function

Given its definition of a cluster in (2), the basic problem of k-means clustering is to find appropriate cluster prototypes $\bar{x}_i$. Well known algorithms such as those of Lloyd [13], Hartigan [14], or MacQueen [15] accomplish this by minimizing the *within cluster scatter*

$$S_W(k) = \sum_{i=1}^{k} \sum_{x \in X_i} \left\| x - \bar{x}_i \right\|^2 \tag{4}$$

with respect to the $\bar{x}_1, \dots, \bar{x}_k$.

Since the minimization objective in (4) involves the distances $\|x - \bar{x}_i\|$ that occur in the definition of a cluster, it formalizes an intuitive idea. Yet, despite its seeming simplicity, k-means clustering proves to be NP hard [16]. Algorithms such as those in [13–15] are therefore but heuristics for which there is no guarantee that they will find the global minimum of (4).

2.2 An Alternative k-Means Objective Function

A rarely used fact in the context of k-means clustering is that the problem of minimizing (4) is equivalent to the problem of maximizing the *between cluster scatter*

$$S_B(k) = \sum_{i,j=1}^{k} n_i n_j \left\| \bar{x}_i - \bar{x}_j \right\|^2. \tag{5}$$

This can be easily seen using Fisher's analysis of variance [17]. It establishes that the *total scatter* of a sample of data points can be written as

$$S_T = \sum_{x \in X} \left\| x - \bar{x} \right\|^2 = S_W(k) + \frac{1}{2\,n} S_B(k) \tag{6}$$

which—since S_T and n are constants—is to say that any decrease of S_W in (4) implies an increase of S_B in (5).

2.3 An Ising Model for $k = 2$ Means Clustering

For the case where $k = 2$, the maximization objective in (5) simplifies to

$$S_B(2) = 2\,n_1 n_2 \left\| \bar{x}_1 - \bar{x}_2 \right\|^2. \tag{7}$$

Interestingly, this expression provides an intuition for the fact that k-means clustering is agnostic of cluster shapes and distances and thus tends to produce clusters of about equal size even in situations where this seems unreasonable [18]. In order for S_B in (7) to be large, both the distance $\|\bar{x}_1 - \bar{x}_2\|$ between the two cluster centers and the product $n_1 n_2$ of the two cluster sizes have to be large. However, since the sum $n_1 + n_2 = n$ is fixed, the product of the sizes will be maximal if $n_1 = n_2 = \frac{n}{2}$.

This observation provides us with a heuristic handle to rewrite the objective in (7) which will then allow for setting up an Ising model for $k = 2$ means clustering.

Assuming that at a solution produced by conventional k-means clustering we will likely have $n_1 \approx n_2 \approx \frac{n}{2}$ allows for the following approximation or reformulation of the objective function

$$2\,n_1 n_2 \left\|\bar{\boldsymbol{x}}_1 - \bar{\boldsymbol{x}}_2\right\|^2 \approx 2\,\frac{n^2}{4}\left\|\bar{\boldsymbol{x}}_1 - \bar{\boldsymbol{x}}_2\right\|^2$$
$$= 2\left\|\frac{n}{2}\left(\bar{\boldsymbol{x}}_1 - \bar{\boldsymbol{x}}_2\right)\right\|^2 = 2\left\|n_1\,\bar{\boldsymbol{x}}_1 - n_2\,\bar{\boldsymbol{x}}_2\right\|^2 \tag{8}$$

and, using this heuristic, the problem of $k = 2$ means clustering becomes to solve

$$\underset{\bar{\boldsymbol{x}}_1, \bar{\boldsymbol{x}}_2}{\operatorname{argmax}} \left\|n_1\,\bar{\boldsymbol{x}}_1 - n_2\,\bar{\boldsymbol{x}}_2\right\|^2. \tag{9}$$

Next, we express the norm in (9) in a form that does not explicitly depend on the cluster means $\bar{\boldsymbol{x}}_i$. To this end, we gather the given data in a data matrix $\boldsymbol{X} = [\boldsymbol{x}_1, \ldots, \boldsymbol{x}_n] \in \mathbb{R}^{m \times n}$ and introduce two binary vectors $\boldsymbol{z}_1, \boldsymbol{z}_2 \in \{0,1\}^n$ which indicate cluster memberships in the sense that entry l of $\boldsymbol{z}_i$ is 1 if $\boldsymbol{x}_l \in X_i$ and 0 otherwise. This way, we can write $n_1\,\bar{\boldsymbol{x}}_1 = \boldsymbol{X}\boldsymbol{z}_1$ as well as $n_2\,\bar{\boldsymbol{x}}_2 = \boldsymbol{X}\boldsymbol{z}_2$ and thus

$$\left\|n_1\,\bar{\boldsymbol{x}}_1 - n_2\,\bar{\boldsymbol{x}}_2\right\|^2 = \left\|\boldsymbol{X}\left(\boldsymbol{z}_1 - \boldsymbol{z}_2\right)\right\|^2 = \left\|\boldsymbol{X}\boldsymbol{s}\right\|^2. \tag{10}$$

Here, $\boldsymbol{s}$ is guaranteed to be a bipolar vector because, in hard k-means clustering, every given data point is assigned to one and only one cluster so that

$$\boldsymbol{z}_1 - \boldsymbol{z}_2 = \boldsymbol{z}_1 - (\boldsymbol{1} - \boldsymbol{z}_1) = 2\,\boldsymbol{z}_1 - \boldsymbol{1} = \boldsymbol{s} \in \{-1,1\}^n. \tag{11}$$

This, however, establishes that there is an Ising model formulation for the problem of $k = 2$ means clustering of zero mean data. On the one hand, zero mean data implies

$$\bar{\boldsymbol{x}} = \frac{1}{n}\left(n_1\,\bar{\boldsymbol{x}}_1 + n_2\,\bar{\boldsymbol{x}}_2\right) = \boldsymbol{0} \Leftrightarrow n_1\,\bar{\boldsymbol{x}}_1 = -n_2\,\bar{\boldsymbol{x}}_2 \tag{12}$$

so that the two means will always be of opposite sign. On the other hand, since $\left\|\boldsymbol{X}\boldsymbol{s}\right\|^2 = \boldsymbol{s}^T \boldsymbol{X}^T \boldsymbol{X}\boldsymbol{s}$, the problem in (9) is equivalent to solving

$$\underset{\boldsymbol{s} \in \{-1,1\}^n}{\operatorname{argmax}} \boldsymbol{s}^T \boldsymbol{X}^T \boldsymbol{X}\boldsymbol{s} \quad \Leftrightarrow \quad \underset{\boldsymbol{s} \in \{-1,1\}^n}{\operatorname{argmin}} -\boldsymbol{s}^T \boldsymbol{X}^T \boldsymbol{X}\boldsymbol{s}. \tag{13}$$

Because of (12) this will necessarily produce a vector $\boldsymbol{s}$ whose entries are not all equal and thus induce a clustering of the data in $\boldsymbol{X}$. If we furthermore substitute $\boldsymbol{Q} = \boldsymbol{X}^T \boldsymbol{X}$, the problem in (23) can also be written as

$$\underset{\boldsymbol{s} \in \{-1,1\}^n}{\operatorname{argmin}} -\sum_{i,j=1}^{n} Q_{ij}\,s_i\,s_j \tag{14}$$

and is now recognizable as an instance of an Ising energy minimization problem.

Looking at (14), two final remarks appear to be in order. First of all, the coupling matrix $Q = X^T X$ is a Gram matrix. The clustering heuristic derived in this section therefore allows for invoking the kernel trick and is thus applicable to a wide variety of practical problems.

Second of all, (14) exposes the "hardness" of $k = 2$ means clustering for it reveals it to be an integer programming problem. That is, it shows that binary clustering is to find a suitable label vector $s \in \{-1, 1\}^n$ whose entries assign data points to clusters. A naïve approach to $k = 2$ means clustering would therefore be to evaluate (14) for each of the 2^n possible assignments of n data points to 2 clusters. On a conventional digital computer this is of course impractical if n is large. On a quantum computer, on the other hand, we could prepare a system of n qubits that is in a superposition of 2^n states each of which reflects a possible solution. The trick is then to evolve the system such that, when its wave function is collapsed, it is more likely to be found in a state that corresponds to a good solution. Below, we discuss how to accomplish this.

3 An Ising Model for Relational Binary Clustering

Next, we address the problem of binary clustering of relational data. In particular, we revisit the idea of graph clustering via minimum cuts, focus on normalized cuts as popularized in [19], and analyze their properties in order to derive a novel cut criterion that leads to an Ising model.

Throughout, we restrict our discussion to undirected and unweighted graphs $G = (V, E)$ consisting of vertices $V = \{v_1, \ldots, v_n\}$ and edges $E \subseteq V \times V$. For these, we recall several basic properties:

The adjacency matrix A of an unweighted graph is a binary $n \times n$ matrix where $A_{ij} = 1$ if $(v_i, v_j) \in E$ and $A_{ij} = 0$ otherwise. Moreover, if the graph is undirected, A will be symmetric.

The degree vector d of G is an n-dimensional vector whose entries

$$d_i = \sum_{j=1}^{n} A_{ij}. \tag{15}$$

count the degrees of the vertices v_i. The degree matrix of G is a diagonal matrix $D = \text{diag}(d)$ and it is easily verified that both the degree matrix as well as the outer product matrix dd^T are symmetric and positive semidefinite.

The Laplacian matrix of G is given by $L = D - A$. Again, if G is undirected, its Laplacian is a symmetric and positive semidefinite matrix.

Finally, the volume $\text{vol}(G)$ of G amounts to the sum of all its vertex degrees

$$\text{vol}(G) = \sum_{i=1}^{n} d_i = \mathbf{1}^T d \tag{16}$$

and, for brevity, we will henceforth write it as $\eta \equiv \text{vol}(G)$.

3.1   Normalized Cuts for Balanced Graph Clustering

The problem of clustering a given graph G into two disjoint subgraphs G_1 and G_2 such that $V_1 \cup V_2 = V$, $V_1 \cap V_2 = \emptyset$, and $E_1 \cap E_2 = \emptyset$ is often formalized as the problem of finding an appropriate graph cut of minimum cost

$$\text{cut}(V_1, V_2) = \text{cut}(V_2, V_1) = \sum_{\substack{v_i \in V_1 \\ v_j \in V_2}} A_{ij}. \tag{17}$$

Interestingly, the costs of cutting G can be expressed in terms of its graph Laplacian. Assuming a bipolar indicator vector $s \in \{-1, 1\}^n$ whose entries

$$s_i = \begin{cases} +1, & \text{if } v_i \in V_1 \\ -1, & \text{if } v_i \in V_2 \end{cases} \tag{18}$$

indicate which of the clusters a vertex will belong to after the cut, straightforward algebra [20] reveals that

$$\text{cut}(V_1, V_2) = \tfrac{1}{4} s^T L s. \tag{19}$$

However, since naïvely minimizing (19) might lead to unbalanced partitions, various modifications have been proposed. A particularly popular heuristic is to consider normalized cuts

$$\text{Ncut}(V_1, V_2) = \frac{s^T L s}{\eta_1} + \frac{s^T L s}{\eta_2} = s^T L s \cdot \frac{\eta_1 + \eta_2}{\eta_1 \eta_2} \tag{20}$$

which trade off the size of cuts and the volumes of the resulting subgraphs [19].

Finally, it is easy to see that the problem of minimizing (20) with respect to $s \in \{-1, 1\}^n$ is an integer programming problem and thus NP hard. The Ncut problem is therefore usually relaxed to spectral clustering [19, 21, 22]. However, on an adiabatic quantum computer, we could again prepare a system of n qubits which are in a superposition of the 2^n possible solutions and evolve the system such that the optimal solution is more likely to be measured. This, however, requires an optimization objective that can be expressed as an Ising model. Since the Ncut objective in (20) does not meet this requirement, we next derive a balanced cut criterion that corresponds to an Ising model.

3.2   An Ising Model for Balanced Graph Clustering

Looking at (20), we realize that, in order for Ncut to be small, the quadratic form $s^T L s$ has to be small and the product $\eta_1 \eta_2$ has to be large.

However, since $\eta_1 + \eta_2 = \eta$ is a constant, the product $\eta_1 \eta_2$ will be maximal if $\eta_1 = \eta_2 = \frac{\eta}{2}$. In other words, the Ncut criterion implicitly enforces $\eta_1 \approx \eta_2$. We may therefore just as well require the squared difference $(\eta_1 - \eta_2)^2$ to be small. Regarding this idea, we note that

$$\eta_1 - \eta_2 = \sum_{v_i \in V_1} d_i - \sum_{v_j \in V_2} d_j = s^T d \tag{21}$$

so that the squared difference of the volumes of the subgraphs that result from a graph cut can be written as

$$(\eta_1 - \eta_2)^2 = s^T dd^T s. \tag{22}$$

This observation immediately leads to yet another heuristic for balanced minimum cuts: combining (19) and (22), we can formalize graph clustering as the task of solving the following equality constrained quadratic binary optimization problem

$$s^* = \underset{s \in \{-1,1\}^n}{\operatorname{argmin}} \; s^T L s \quad \text{s.t.} \quad s^T dd^T s = 0 \tag{23}$$

whose Lagrangian is given by

$$\mathcal{L}(s, \lambda) = s^T L s + \lambda s^T dd^T s. \tag{24}$$

Contrary to the Ncut formulation, our formulation is now an instance of an Ising energy minimization problem. However, in order to minimize (24) on an adiabatic quantum computer, we need to specify the Lagrange multiplier λ so that the problem indeed only depends on the bipolar indicator vector s. To this end, we propose to simply choose $\lambda = \frac{2}{\eta}$. We then have $0 \leq \frac{2}{\eta} s^T dd^T s \leq 2\eta$ and, since we also always have $0 \leq s^T L s < 2\eta$, neither term on the right hand side of (24) will dominate the minimization objective.

In summary, our derivation leads to the following graph cut criterion, which, for want of a better name, we will call an Ising cut

$$\begin{aligned} \text{Icut}(V_1, V_2) &= s^T L s + \frac{2}{\eta} s^T dd^T s \\ &= s^T \left(L + \frac{2}{\eta} dd^T \right) s = s^T Q s = \sum_{i,j=1}^{n} s_i \, Q_{ij} \, s_j. \end{aligned} \tag{25}$$

4   Adiabatic Quantum Binary Clustering

To perform adiabatic quantum binary clustering of n entities, we consider a system of n qubits that is in a superposition of 2^n basis states

$$|\psi(t)\rangle = \sum_{i=0}^{2^n - 1} a_i(t) |\psi_i\rangle \tag{26}$$

whose time dependent amplitudes $a_i \in \mathbb{C}$ obey $\sum_i |a_i|^2 = 1$. We understand each of the different basis states

$$|\psi_0\rangle = |000\ldots000\rangle \tag{27}$$
$$|\psi_1\rangle = |000\ldots001\rangle \tag{28}$$
$$|\psi_2\rangle = |000\ldots010\rangle \tag{29}$$
$$|\psi_3\rangle = |000\ldots011\rangle \tag{30}$$

$$\vdots$$

as an indicator vector that represents one of the 2^n possible assignment of n entities to 2 clusters and use the common shorthand to express tensor products, for instance

$$|\psi_1\rangle = |000\ldots001\rangle = |0\rangle \otimes |0\rangle \otimes \ldots \otimes |1\rangle. \tag{31}$$

If a system such as in (26) evolves under the influence of a time-dependent Hamiltonian $H(t)$, its behavior is governed by the Schrödinger equation

$$\frac{\partial}{\partial t}|\psi(t)\rangle = -i\,H(t)\,|\psi(t)\rangle \tag{32}$$

where we have set $\hbar = 1$. In adiabatic quantum computing, we consider periods ranging from $t = 0$ to $t = \tau$ and assume the Hamiltonian at time t to be given as a convex combination of two static Hamiltonians, namely

$$H(t) = \left(1 - \tfrac{t}{\tau}\right) H_B + \tfrac{t}{\tau} H_P. \tag{33}$$

H_B is called the *beginning Hamiltonian* whose ground state is easy to construct and H_P is the *problem Hamiltonian* whose ground state encodes the solution to the problem at hand.

For simple Ising models such as the ones in (14) and (25), there are by now standard suggestions for how to set up a suitable problem Hamiltonian [9]. In particular, we may define

$$H_P = \sum_{i,j=1}^{n} Q_{ij}\, \sigma_z^i \sigma_z^j \tag{34}$$

where σ_z^i denotes the Pauli spin matrix σ_z acting on the ith qubit, that is

$$\sigma_z^i = \underbrace{I \otimes I \otimes \ldots \otimes I}_{i-1\,\text{terms}} \otimes \sigma_z \otimes \underbrace{I \otimes I \ldots \otimes I}_{n-i\,\text{terms}}. \tag{35}$$

The beginning Hamiltonian is then typically chosen to be orthogonal to the problem Hamiltonian, for instance

$$H_B = \sum_{i=1}^{n} \sigma_x^i \tag{36}$$

where σ_x^i is defined as above, this time with respect to the Pauli spin matrix σ_x.

To compute a clustering, we then let $|\psi(t)\rangle$ evolve from $|\psi(0)\rangle$ to $|\psi(\tau)\rangle$ where $|\psi(0)\rangle$ is chosen to be the ground state of H_B. That is, if λ denotes the smallest eigenvalue of H_B, the initial state $|\psi(0)\rangle$ of the system corresponds to the solution of

$$H_B|\psi(0)\rangle = \lambda\,|\psi(0)\rangle. \tag{37}$$

Finally, upon termination of its evolution, a measurement is performed on the n qubit system. This will cause the wave function $|\psi(\tau)\rangle$ to collapse to a particular basis state and the probability for this state to be $|\psi_i\rangle$ is given by the amplitude $|a_i(\tau)|^2$. However, since the adiabatic evolution was steered towards the problem Hamiltonian H_P, basis states that correspond to ground states of H_P are more likely to be found.

On an adiabatic quantum computer, this process be carried out physically. On a digital computer, we may simulate it by numerically solving

$$\left|\psi(\tau)\right\rangle = -i \int_0^\tau H(t) \left|\psi(t)\right\rangle dt. \tag{38}$$

5   Practical Examples

In this section, we present several examples which demonstrate the practical feasibility of the above ideas. Our examples are of didactic nature and mainly intended to illustrate the adiabatic evolution of n qubit systems. We thus restrict ourselves to rather small $n = n_1 + n_2$ in order to be able to comprehensibly visualize how the amplitudes of 2^n basis states evolve over time.

In each experiment, we simulate quantum adiabatic evolutions on a digital computer. To this end, we set up the corresponding problem Hamiltonian H_P, the beginning Hamiltonian H_B, and its ground state $\left|\psi(0)\right\rangle$ as discussed above and use the *Python* quantum computing toolbox *QuTiP* [23] to numerically solve (38) for $t \in [0, \tau = 50]$.

5.1   Adiabatic Quantum $k = 2$ Means Clustering

Next, we consider two examples of adiabatic quantum clustering of numerical data. In both cases, we consider a data matrix $\boldsymbol{X} = [\boldsymbol{X}_1, \boldsymbol{X}_2] \in \mathbb{R}^{2 \times n}$ whose column vectors form two clusters of size n_1 and n_2, respectively, and let an n qubit system evolve in order to uncover these clusters.

Figure 1(a) shows a zero mean sample of $n = 8$ data points. In order to quantum adiabatically cluster these data into two clusters, we consider the Ising model in (14) using $\boldsymbol{Q} = \boldsymbol{X}^T \boldsymbol{X}$. Figure 1(c) illustrates the temporal evolution of the amplitudes $|a_i(t)|^2$ of the $2^8 = 256$ basis states $\left|\psi_i\right\rangle$ the quantum system $\left|\psi(t)\right\rangle$ can be in. At $t = 0$, all states are equally likely but over time their amplitudes begin to increase or decrease. At $t = \tau$, two of the basis states have considerably higher amplitudes than the others so that a measurement will likely cause the system to collapse to either of these more probable states.

The top 8 most likely basis states for the system to be found in at the end of this experiment are listed in Table 1. Looking at the table, we recognize that the two most likely states are indeed $\left|00111111\right\rangle$ and $\left|11000000\right\rangle$ which, when understood as cluster indicator vectors, both produce the result in Fig. 1(b).

For the six next most likely states, we observe that they will produce clusters of sizes 3 and 5 and 4 and 4, respectively. In other words, although our heuristic approximation of the k-means objective in (9) and the resulting Ising model in (14) were based on the assumption of clusters of about equal size, this practical result shows that the model can nevertheless produce clusters of imbalanced sizes if the data are distributed in a manner that would require it.

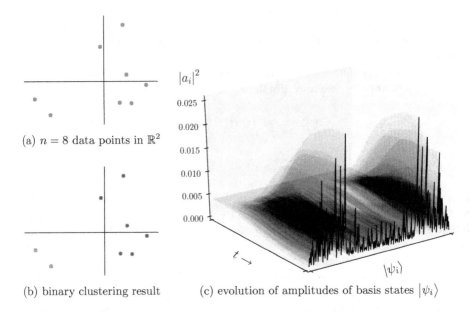

(a) $n = 8$ data points in $\mathbb{R}^2$

(b) binary clustering result

(c) evolution of amplitudes of basis states $|\psi_i\rangle$

Fig. 1. Example of adiabatic quantum $k = 2$ means clustering using the Ising model in (14). (a) zero mean sample of 8 data points; (b) corresponding clustering result; (c) adiabatic evolution of a system of 8 qubits. During its evolution over time t, the system is in a superposition of $2^8 = 256$ basis states $|\psi_i\rangle$ each representing a possible binary clustering. Initially, it is equally likely to find the system in any of these states. At the end, two basis states have noticeably higher amplitudes $|a_i|^2$ than the others and are therefore more likely to be measured; these are $|00111111\rangle$ and $|11000000\rangle$ and they both induce the result in (b).

Table 1. The 8 most likely states for the system in Fig. 1 to be in at $t = \tau$.

$|\psi_i\rangle$	$|a_i|^2$
$|00111111\rangle$	0.02547562
$|11000000\rangle$	0.02547562
$|11010000\rangle$	0.02104650
$|00101111\rangle$	0.02104650
$|00110111\rangle$	0.01929660
$|11001000\rangle$	0.01929660
$|11010001\rangle$	0.01812229
$|00101110\rangle$	0.01812229

Figure 2(a) shows $n = 16$ data points arranged in a pattern that suggests the use of kernel k-means clustering. Accordingly, we choose the coupling matrix for the Ising model in (14) to be a centered kernel matrix

$$Q_{ij} = k(\boldsymbol{x}_i, \boldsymbol{x}_j) - \frac{1}{n} \sum_l k(\boldsymbol{x}_i, \boldsymbol{x}_l) - \frac{1}{n} \sum_k k(\boldsymbol{x}_k, \boldsymbol{x}_j) + \frac{1}{n^2} \sum_{k,l} k(\boldsymbol{x}_k, \boldsymbol{x}_l) \quad (39)$$

where $k(\boldsymbol{x}_i, \boldsymbol{x}_j) = \exp\left(-\|\boldsymbol{x}_i - \boldsymbol{x}_j\|^2 / 2\sigma^2\right)$ and $\sigma = 0.25$. During its quantum adiabatic evolution, the corresponding 16 qubit system $|\psi(t)\rangle$ is in a super-position of $2^{16} = 65536$ basis states but evolves to a configuration in which two of these basis states have much higher amplitudes than the others (see Fig. 2(c)). Here, the two most likely basis states are $|1111111100000000\rangle$ and $|0000000011111111\rangle$ and they cluster the data in a manner a human observer would expect (see Fig. 2(b)).

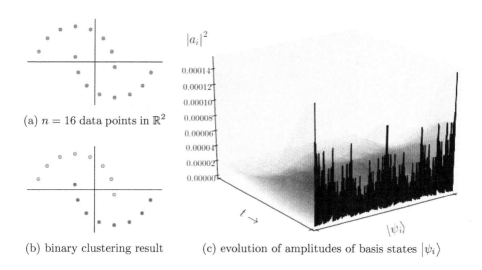

(a) $n = 16$ data points in $\mathbb{R}^2$

(b) binary clustering result

(c) evolution of amplitudes of basis states $|\psi_i\rangle$

Fig. 2. Example of adiabatic quantum kernel $k = 2$-means clustering. (a) sample of 16 data points; (b) corresponding clustering result; (c) adiabatic evolution of a system of 16 qubits. Throughout, the system is in a superposition of $2^{16} = 65536$ basis states. Upon termination of the evolution, the two most likely basis states for the system to be found in are $|1111111100000000\rangle$ and $|0000000011111111\rangle$.

5.2   Adiabatic Quantum Graph Clustering

Next, we present an example of adiabatic quantum graph clustering using the Ising model in (25).

Figure 3(a) shows a graph composed of two recognizable communities. To produce it, we created two Erdős-Rényi random graphs of 6 and 10 vertices, respectively, where the edge probability was set to $p = 0.5$ in both cases. To connect these subgraphs, we then randomly drew two edges between them.

Figure 3(c) shows the temporal evolution of the amplitudes $|a_i(t)|^2$ of the $2^{16} = 65336$ basis states $|\psi_i\rangle$ the quantum system $|\psi(t)\rangle$ used to cluster this

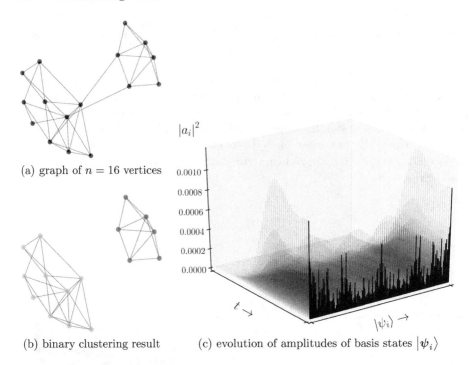

(a) graph of $n = 16$ vertices

(b) binary clustering result

(c) evolution of amplitudes of basis states $|\psi_i\rangle$

Fig. 3. Example of adiabatic quantum clustering of a graph composed of two Erdős-Rényi graphs. Over time, the corresponding qubit system is in a superposition of $2^{16} = 65536$ basis states $|\psi_i\rangle$ representing possible clusterings. Initially, it is equally likely to find the system in any of these states. At the end, two states have noticeably higher amplitudes $|a_i|^2$ than the others and are therefore more likely to be measured; these are $|0000000000111111\rangle$ and $|1111111111000000\rangle$ and they both induce the clusters in (b).

graph can be in. At $t = 0$, all basis states are equally likely but over time their amplitudes begin to increase or decrease. At $t = \tau$, two of the basis states have considerably higher amplitudes than the others so that a measurement of the quantum system at this time will likely cause it to collapse to either one of these more probable states.

Table 2 shows the top 8 most likely basis states for the qubit system of this experiment to be found in at the end of its adiabatic evolution. Looking at this table, we recognize that the two most likely basis states are $|0000000000111111\rangle$ and $|1111111111000000\rangle$ which, due to obvious symmetries, both produce the result shown in Fig. 3(b).

Table 2. The 8 most likely states for the system in Fig. 3 to be in at $t = \tau$.

$|\psi_i\rangle$	$|a_i|^2$
$|0000000000111111\rangle$	0.00094453
$|1111111111000000\rangle$	0.00094453
$|1111111111010000\rangle$	0.00058813
$|0000000000101111\rangle$	0.00058813
$|1111111111100000\rangle$	0.00058813
$|0000000000011111\rangle$	0.00058813
$|0000000000110111\rangle$	0.00056642
$|1111111111001000\rangle$	0.00056642

6 Related Work

Interest in quantum computing for machine learning and pattern recognition is steadily increasing [5,7,8,24–28]. Indeed, many algorithms in these fields deal with demanding optimization problems for which quantum computing is expected to yield considerable speed-up.

Unsurprisingly, quantum computing approaches to problems in the context of binary clustering have been considered before. However, prior work reported in [5,6,8] follows the paradigm of *quantum gate computing*. Though equivalent to adiabatic quantum computing, quantum gate computing algorithms do not attempt to optimize Ising models and are, as of this writing, not yet practically feasible for lack of corresponding hardware and, crucially, the availability of quantum oracles.

Lloyd et al. [7] present an adiabatic quantum implementation of Lloyd's heuristic algorithm [13] for k-means clustering which has a run time complexity of $O\big(k \log(mn)\big)$ rather than $O\big(\text{poly}(mn)\big)$ where m denotes data dimensionality. However, their approach hinges on the availability of a quantum random access memory (qRAM), a theoretical concept which, as of this writing, is not yet technically feasible. While the approach we discussed in this paper is admittedly more restricted in that it only allows for $k = 2$ means clustering, its Hamiltonians are simple and easy to work work with, and it does not require a qRAM.

Graph clustering, too, has been previously studied from the point of view of quantum computing. In a very recent contribution, Ushijima-Mwesigwa et al. [29] actually perform graph clustering experiments on D-Wave computers [10,11]. Rather than an Ising model derived from the idea of normalized cuts as proposed above, they consider an Ising model based on Newman's modularity measure [30]. Their experiments show this approach to equal or even outperform current state-of-the art methods.

7 Summary

After decades of being a mainly theoretical concept, quantum computing is now becoming practical. Companies like IBM, Google, or Microsoft invest increasing resources into corresponding research and development [1] and further rapid progress is expected [2,3]. All of this will likely impact statistical learning and pattern recognition, because quantum computers have the potential to accelerate the kind of optimization procedures that are at the heart of many algorithms in these fields.

Since at the time of this writing, adiabatic quantum computers appear to be most advanced, we were concerned with adiabatic quantum computing for data analysis, especially for data clustering. From an abstract point of view, the problem of setting up pattern recognition algorithms for adiabatic quantum computing can be seen as the problem of expressing their objective functions in terms of Ising models because adiabatic quantum computers are tailored towards minimizing Ising energies. We thus derived Ising models for (kernel) $k = 2$-means clustering and balanced graph cuts. Simulated examples demonstrated that the Schrödinger equations governing the dynamics of corresponding qubit systems can indeed be used to cluster data.

References

1. Technology Quarterly: Quantum Devices. The Economist, March 2017
2. Castelvecchi, D.: Quantum computers ready to leap out of the lab in 2017. Nature **541**(7635), 9–10 (2017). https://doi.org/10.1038/541009a
3. Gibney, E.: D-Wave upgrade: how scientists are using the world's most controversial quantum computer. Nature **541**(7638), 447–448 (2017). https://doi.org/10.1038/541447b
4. Grover, L.: From Schrödinger's equation to the quantum search algorithm. J. of Phys. **56**(2), 333–348 (2001). https://doi.org/10.1119/1.1359518
5. Aïmeur, E., Brassard, G., Gambs, S.: Quantum clustering algorithms. In: Proceedings ICML (2007)
6. Aïmeur, E., Brassard, G., Gambs, S.: Quantum speed-up for unsupervised learning. Mach. Learn. **90**(2), 261–287 (2013). https://doi.org/10.1007/s10994-012-531305
7. Lloyd, S., Mohseni, M., Rebentrost, P.: Quantum algorithms for supervised and unsupervised machine learning. arXiv:1307.0411 [quant-ph] (2013)
8. Wiebe, N., Kapoor, A., Svore, K.: Quantum algorithms for nearest-neighbor methods for supervised and unsupervised learning. Quantum Inf. Comput. **15**(3–4), 316–356 (2015)
9. Albash, T., Lidar, D.: Adiabatic quantum computing. arXiv:1611.04471 [quant-ph] (2016)
10. Bian, Z., Chudak, F., Macready, W., Rose, G.: The Ising model: teaching an old problem new tricks. Technical report, D-Wave Systems (2010)
11. Johnson, M., Amin, M., Gildert, S., Lanting, T., Hamze, F., Dickson, N., Harris, R., Berkley, A., Johansson, J., Bunyk, P., Chapple, E., Enderud, C., Hilton, J., Karimi, K., Ladizinsky, E., Ladizinsky, N., Oh, T., Perminov, I., Rich, C., Thom, M., Tolkacheva, E., Truncik, C., Uchaikin, S., Wang, J., Wilson, B., Rose, G.: Quantum annealing with manufactured spins. Nature **473**(7346), 194–198 (2011). https://doi.org/10.1038/nature10012

12. Born, M., Fock, V.: Beweis des Adiabatensatzes. Zeitschrift für Physik **51**(3–4), 165–180 (1928). https://doi.org/10.1007/BF01343193
13. Lloyd, S.: Least squares quantization in PCM. IEEE Trans. Inf. Theory **28**(2), 129–137 (1982). https://doi.org/10.1109/TIT.1982.1056489
14. Hartigan, J., Wong, M.: Algorithm AS 136: a k-means clustering algorithm. J. Roy. Stat. Soc. C **28**(1), 100–108 (1979). https://doi.org/10.2307/2346830
15. MacQueen, J.: Some methods for classification and analysis of multivariate observations. In: Proceedings Berkeley Symposium on Mathematical Statistics and Probability (1967)
16. Aloise, D., Deshapande, A., Hansen, P., Popat, P.: NP-hardness of euclidean sum-of-squares clustering. Mach. Learn. **75**(2), 245–248 (2009). https://doi.org/10.1007/s10994-009-5103-0
17. Fisher, R.: On the probable error of a coefficient correlation deduced from a small sample. Metron **1**, 3–32 (1921)
18. MacKay, D.: Information Theory, Inference, and Learning Algorithms. Cambridge University Press, Cambridge (2003)
19. Shi, J., Malik, J.: Normalized cuts and image segmentation. IEEE Trans. PAMI **22**(8), 888–905 (2000). https://doi.org/10.1109/34.868688
20. Dhillon, I.: Co-clustering documents and words using bipartite spectral graph partitioning. In: Proceedings KDD (2001)
21. von Luxburg, U.: A tutorial on spectral clustering. Stat. Comput. **17**(4), 395–416 (2007). https://doi.org/10.1007/s11222-007-9033-z
22. Fiedler, M.: A property of eigenvectors of nonnegative symmetric matrices and its application to graph theory. Czech. Math. J. **25**(4), 619–633 (1975)
23. Johansson, J., Nation, P., Nori, F.: QuTiP 2: a python framework for the dynamics of open quantum systems. Comput. Phys. Commun. **184**(4), 1234–1240 (2013). https://doi.org/10.1016/j.cpc.2012.11.019
24. Biamonte, J., Wittek, P., Pancotti, N., Rebentrost, P., Wiebe, N., Lloyd, S.: Quantum machine learning. arXiv:1611.09347 [quant-ph] (2016)
25. Dunjiko, V., Taylor, J., Briegel, H.: Quantum-enhanced machine learning. Phys. Rev. Lett. **117**(13), 130501 (2016). https://doi.org/10.1103/PhysRevLett.117.130501
26. Schuld, M., Sinayskiy, I., Petruccione, F.: An introduction to quantum machine learning. Contemp. Phys. **56**(2), 172–185 (2014). https://doi.org/10.1080/00107514.2014.964942
27. Wiebe, N., Kapoor, A., Svore, K.: Quantum perceptron models. In: Proceedings NIPS (2016)
28. Wittek, P.: Quantum Machine Learning. Academic Press, London (2014)
29. Ushijima-Mwesigwa, H., Negre, C., Mniszewski, S.: Graph partitioning using quantum annealing on the D-Wave system. arXiv:1705.03082 [quant-ph] (2017)
30. Newman, M.: Modularity and community structure in networks. PNAS **103**(23), 8577–8582 (2006). https://doi.org/10.1073/pnas.0601602103

Quantum Interference and Shape Detection

Davi Geiger and Zvi M. Kedem[✉]

Department of Computer Science, Courant Institute of Mathematical Sciences,
New York University, New York, USA
{dg1,kedem}@nyu.edu

Abstract. We address the problem of shape detection in settings where large shape deformations and occlusions occur with clutter noise present. We propose a quantum model for shapes by applying the quantum path integral formulation to an existing energy model for shapes (a Bayesian-derived cost function). We show that the classical statistical method derived from the quantum method, via the Wick rotation technique, is a voting scheme similar to the Hough transform. The quantum phenomenon of interference drives the quantum method for shape detection to excel, compared to the corresponding classical statistical method or the statistical Bayesian (energy optimization) method. To empirically demonstrate our approach, we focus on simple shapes: circles and ellipses.

Keywords: Shape · Hough transform · Interference
Statistical methods · Energy minimization · Wick rotation

1 Introduction

We are given an image $\mathcal{I}$ containing multiple shapes, some of which are possibly occluded and/or deformed, and there is possible clutter noise data. To detect the shapes in $\mathcal{I}$, we propose a new method, which is rooted in quantum theory. The main advantage of the quantum method over existing methods is its high accuracy and resistance to deformations thanks to the quantum interference phenomenon as described, e.g., by Feynman and Hibbs [1] and Feynman [2].

The input data is specified by a set of N feature points $X = \{x_1, x_2, ..., x_N\}$ in $\mathbb{R}^D$. Here we focus on 2-dimensional data, so $D = 2$. An image $\mathcal{I}$ may be specified by the set of points, as shown for example in Fig. 1. Otherwise, feature points can be extracted from $\mathcal{I}$, using, for example SIFT features, HOG features, or maximum of wavelet responses, etc.

1.1 Our Approach

We start with an energy-minimization view of shape detection. We consider a standard error criterion of the energy model for fitting a shape to the data using

© Springer International Publishing AG, part of Springer Nature 2018
M. Pelillo and E. Hancock (Eds.): EMMCVPR 2017, LNCS 10746, pp. 18–33, 2018.
https://doi.org/10.1007/978-3-319-78199-0_2

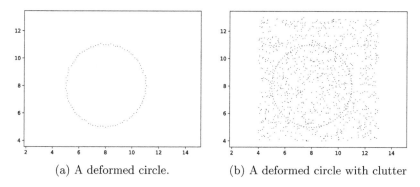

(a) A deformed circle.   (b) A deformed circle with clutter

Fig. 1. (a) A circle centered at $(8, 8)$, radius 3, and with 100 points, is deformed: for each point x_i a random value drawn from a normal distribution with standard deviation $\eta_i = 0.05$ is added along the radius. (b) 1000 points of clutter are added by sampling from a uniform distribution inside a box formed by corners at points: $(4, 4)$, $(4, 13)$, $(13, 4)$, $(13, 13)$ with area $B = 81$. Thus the density is $n_c = \frac{1000}{81} = 12.35$. We work with spatial precision of 0.01.

Bayesian modeling. In such modeling, deformations are modeled as *noise* distorting the shape. In contrast, applying the *quantum path integral* method [1] allows us to interpret the shape's distortion as evidence for multiple evolution paths. The key component of the quantum method, not present in the classical statistical method, is *interference*. The total quantum probability contains probability terms that *cancel* each other.

In physics, the quantum mechanics phenomenon of particle interference [2] has no counterpart in classical mechanics. Due to quantum interference, events near events of maximum probability may themselves have zero probability. In contrast, in the classical approach, events close to maximum-probability events generally also have high probability, making maximum-probability events less salient.

The quantum method offers, in addition to interference, a technique known as Wick rotation [3] to derive a corresponding classical statistical method. We show that this classical statistical method is a voting scheme similar to the Hough transform [4] (generalized to any shape energy model).

When addressing shape-detection problems in the classical setting, solutions close to the maximum-probability solutions have probabilities almost as high, making it difficult to discern the best solution. In contrast, in the quantum setting, thanks to interference, solutions near the maximum-probability solutions end up with low probability. We demonstrate this *saliency effect*, by examining shapes with large deformations, where shape detection is particularly difficult. We show that the quantum method preserves the sharpness of the optimal solution much better than the classical method, which assigns high probabilities to many more suboptimal solutions.

1.2   Previous Work

Shape detection was an early subject of study in computer vision, with one of its best methods developed in 1959 by Hough [4]. As our paper is not a comprehensive review paper, we focus on only a small set of references, most closely related to our work.

The work of Yuille [5] is representative of early approaches for using energy based models in computer vision, including for shape detection. We use such a view to describe classical shapes in an energy-minimization setting.

Quantum methods have already had an impact on computer vision research. One example addresses graph matching and quantum walks as studied by Emms et al. [6,7]. The matching problem was abstracted using an auxiliary graph that connects pairs of vertices from the graphs. A discrete (or continuous)-time quantum walk was simulated on this auxiliary graph and the quantum interference on the auxiliary vertices indicates possible matches. While they reported significantly reduction to the space of possible permutation matchings, the performance did not improve over the best existing/classical techniques. Rangrajan and Gurumoorthy [8] and Gurumoorhty et al. [9] applied the Schrödinger equation to shape detection by noting and validating a relation between the Schrödinger equation and distance transforms. That relation, however, cannot be expanded to other optimization methods. Moreover, there was no demonstration of the quantum method outperforming a classical method.

The Schrödinger equation is obtained from the modeling of behaviors of physical systems, which is not our setting. Different than in "Quantum Mechanics" our optimization criteria comes from computer vision problem of shape modeling (without any physical modeling as it is the case of Schrödinger equation). Because our method only relies on mathematical techniques from quantum theory, it is very general and *offers an approach to other optimization problems*.

Our work extends the approach of Cicconet et al. [10], of which one of us is a co-author, where it proposes a complex representation for circles, based on distances, to outperform a non-complex one. Such an idea is very specific for distances and is not derived as a quantum framework. The work presented here has a broader objective, which is to provide a general quantum method for the detection of all shapes. Even when our method is applied to circle detection, it differs from the work in [10], which is based only on distances.

1.3   Paper Organization

Section 2 presents a general description of shapes, which is easily adapted to any optimization method. Section 3 presents a dynamic model of shapes. Section 4 lays out our main proposal of using quantum theory to address the shape detection problem. Our approach also leads naturally to a classical method that behaves like a voting scheme and we also establish a connection to Hough transforms. Section 5 presents experiments with circle and ellipses to show that the quantum method outperforms classical statistical methods. Section 6 presents theoretical and empirical analyses of the quantum method for shape detection and a comparison with the Hough method. We demonstrate that for larger deformations the quantum method outperforms the Hough method. Section 7 concludes the paper.

2   Energy Model for Shape Specification

A shape S may be defined by the set of points x satisfying $S_\Theta(x) = 0$, where Θ is a set of parameters specifying S. Let μ be a shape's center, which in our setting is (μ_x, μ_y). The choice of μ is in general arbitrary, though frequently there is a natural choice for μ for a given shape, such as its "center of mass," i.e., the mean position of its points. We consider all translations of $S_\Theta(x)$ as representing a single shape, so that a shape is translation-invariant. It is then convenient to describe the shapes as $S_\Theta(x - \mu)$, with the parameters Θ not including the parameters of μ. Thus, we specify a shape by the set of points X such that

$$S_\Theta(x - \mu) = 0 \quad \text{for all } x \in X.$$

The more complex a shape is, the larger is the set of parameters required to describe it. For example, to describe a circle, we use three parameters (μ_x, μ_y, r) listing its center and radius, i.e., $S_{\Theta=(r)}(x - \mu) = 1 - \frac{(x-\mu)^2}{r^2}$ (see Fig. 1a) To specify a square with sides L we can write $S_{\Theta=(L)}(x - \mu) = \frac{1}{L^2}\left(\left|\,|x - \mu_x| - \frac{L}{2}\right| + \text{ReLu}\left(|y - \mu_y| - \frac{L}{2}\right)\right) \times \left(\text{ReLu}\left(|x - \mu_x| - \frac{L}{2}\right) + \left|\,|y - \mu_y| - \frac{L}{2}\right|\right)$, where $\text{ReLu}(x) = \max(0, x)$. An ellipse can be specified by $S_{\Theta=(\Sigma)}(x - \mu) = 1 - (x - \mu)^\mathsf{T}\, \Sigma^{-1}\,(x - \mu)$, where Σ is the covariance matrix, specified by three independent parameters.

We also require that a shape representation be such that if all the values of the parameters in Θ are 0, then the set of points that belong to the shape "collapses" to just $X = \{\mu\}$. This is the case for the parameterizations of the examples above.

Shape Complexity: A large class of shapes can be described by polynomials and a larger class by ratios of polynomials, each specified uniquely and unitlessly. The higher-degree polynomials require more parameters and thus are more general. We also allow for non-linear functions to be utilized to specify a shape, as for the square shape above.

Energy Model: Given a shape model, we can create an *energy model* per data point x as

$$\mathrm{E}_x(S_\Theta, \mu) = |S_\Theta(x - \mu)|, \tag{1}$$

where the L_1 norm could be generalized to an L_p norm, but in this paper we work with L_1 because of its simplicity and robust properties. The smaller $\mathrm{E}_x(\Theta, \mu)$, the more it is likely that Θ, μ are the true parameters of the shape.

2.1   Deformations

To address realistic scenarios we must study the detection of shapes under deformations. When deformations are present, the energy (1) is no longer zero for
deformed points associated with the shape $\mathcal{S}_\Theta(\boldsymbol{x} - \boldsymbol{\mu})$. Let us describe deformations from an ideal shape data point $\boldsymbol{x}_i^S$ as a perturbation to it, i.e., by adding
$\boldsymbol{\eta}_i$ to it. Thus, $\boldsymbol{x}_i = \boldsymbol{x}_i^S + \boldsymbol{\eta}_i$. Then the points of the deformed shape, $\boldsymbol{x}_i$, satisfy

$$0 = |\mathcal{S}_\Theta(\boldsymbol{x}_i^S - \boldsymbol{\mu})| = |\mathcal{S}_\Theta(\boldsymbol{x}_i - \boldsymbol{\eta}_i - \boldsymbol{\mu})|.$$

We may assume that the deformations of a shape occur in the directions perpen-
dicular to the ideal shape's tangents, i.e., along the direction of $\nabla_{\boldsymbol{x}}\mathcal{S}_\Theta(\boldsymbol{x}-\boldsymbol{\mu})\big|_{\boldsymbol{x}_i}$,
where $\nabla_{\boldsymbol{x}}$ is the gradient operator. Thus, we can write $\boldsymbol{\eta}_i = \eta_i\,\hat{\boldsymbol{n}}_i$

$$\hat{\boldsymbol{n}}_i = \frac{\nabla_{\boldsymbol{x}}\mathcal{S}_\Theta(\boldsymbol{x} - \boldsymbol{\mu})\big|_{\boldsymbol{x}_i}}{\left|\nabla_{\boldsymbol{x}}\mathcal{S}_\Theta(\boldsymbol{x} - \boldsymbol{\mu})\big|_{\boldsymbol{x}_i}\right|}, \tag{2}$$

where η_i can be positive or negative depending on whether the deformation
is outwards or inwards with respect to the shape. For example, for a circle
$\mathcal{S}_{\Theta=(r)}(\boldsymbol{x} - \boldsymbol{\mu}) = 1 - \frac{(\boldsymbol{x}-\boldsymbol{\mu})^2}{r^2}$, and therefore $\nabla_{\boldsymbol{x}}\mathcal{S}_r(\boldsymbol{x} - \boldsymbol{\mu})\big|_{\boldsymbol{x}_i} = 2\frac{(\boldsymbol{x}_i-\boldsymbol{\mu})}{r^2} \propto \hat{\boldsymbol{r}}_i$,
where $\hat{\boldsymbol{r}}_i$ is a unit vector pointing outwards in the radius direction at point $\boldsymbol{x}_i$.
Thus, $\boldsymbol{\eta}_i = \eta_i\,\hat{\boldsymbol{r}}_i$.

2.2   A Bayesian Model

We consider a Bayesian model with one hyper-parameter (λ). Each data point
x_i either belongs to the shape or to the clutter and so produces a likelihood

$$P_\lambda(\boldsymbol{x}_i, \boldsymbol{M}_i|\Theta, \mu) = \frac{1}{C}e^{-[M_i|\mathcal{S}_\Theta(x_i-\mu)|+\lambda(1-M_i)]},$$

where $\boldsymbol{M}_i \in \{0, 1\}$. The marginal probability over the assignment $\boldsymbol{M}$ is

$$P_\lambda(\boldsymbol{x}_i|\Theta, \mu) = \sum_{M_i=0}^{1} P_\lambda(\boldsymbol{x}_i, \boldsymbol{M}_i|\Theta, \mu) = \frac{1}{C}\left(e^{-|\mathcal{S}_\Theta(x_i-\mu)|} + e^{-\lambda}\right).$$

Each point $\boldsymbol{x}$ and deformation is conditionally independent given the shape
specification parameters. Thus, the final probability for all points is given by

$$P_\lambda(\boldsymbol{X}|\Theta, \mu) = \frac{1}{C}\prod_{i=1}^{N} P_\lambda(\boldsymbol{x}_i|\Theta, \mu) = \frac{1}{C}\prod_{i=1}^{N}\left(e^{-|\mathcal{S}_\Theta(x_i-\mu)|} + e^{-\lambda}\right). \tag{3}$$

3   Shape Dynamics

Our task is to detect *static* shapes from a single image $\mathcal{I}$. Quantum theory, however, was developed for systems of particles that evolve over time. We invoke a *hidden time parameter*, such an idea has been used in computer vision before. For example, Witkin [11] and Lindeberg [12] proposed to use scale space to describe shape evolution to better compare different shapes. Siddiqi and Kimia [13] described a time evolution equation to produce shape-skeletons.

We view a shape as the result of an evolution from an initial collapsed state $\boldsymbol{X}_0 = \boldsymbol{\mu}$ at time $t = 0$ to its final state $\boldsymbol{X}_T$ at time $t = T$. The set $\boldsymbol{X}_T$ is the set of points satisfying $\mathcal{S}_{\Theta_T}(\boldsymbol{x} - \boldsymbol{\mu}) = 0$, and $\boldsymbol{X}_0 = \boldsymbol{\mu}$ is the single point satisfying $\mathcal{S}_{\Theta_T=0}(\boldsymbol{x} - \boldsymbol{\mu}) = 0$. Reversely, we can consider the collapsed state as the final state and the given shape as the initial state. The dynamics is invariant by time reversal.

Parameterizing the evolution with $t \in [0, T]$, we have the variables $\boldsymbol{x}(t)$, $\frac{\mathrm{d}\boldsymbol{x}(t)}{\mathrm{d}t}$, $\Theta(t)$, and $\frac{\mathrm{d}\Theta(t)}{\mathrm{d}t}$. Here $\Theta(t)$ represents all other shape parameters and is, in general, a multidimensional variable. The mathematical formalism used in classical mechanics to derive evolution equations assigns to a problem a Lagrangian, which is a function $\mathcal{L}(\boldsymbol{x}, \boldsymbol{v}, \Theta, \boldsymbol{v}_\Theta, t)$, where $\boldsymbol{v} = \frac{\mathrm{d}\boldsymbol{x}(t)}{\mathrm{d}t}$ and $\boldsymbol{v}_\Theta = \frac{\mathrm{d}\Theta(t)}{\mathrm{d}t}$. From the Lagrangian, an action is constructed in the time interval $[0, T]$ as $\mathcal{A}_0^T = \int_0^T \mathcal{L}(\boldsymbol{x}, \boldsymbol{v}, \Theta, \boldsymbol{v}_\Theta, t)\mathrm{d}t$.

Different *trajectories* or *paths* $\mathcal{P}_0^T = (\boldsymbol{x}(t), \Theta(t)) \mid t \in [0, T])$ yield different values of the action, so $\mathcal{A}_0^T = \mathcal{A}_0^T(\mathcal{P}_0^T)$ is a function of the path. The optimal path $\mathcal{P}_0^{*T} = (\boldsymbol{x}^*(t), \Theta^*(t)) \mid t \in [0, T])$ is an extremum of the action. We may write $\mathcal{A}_{\Theta(0)\to 0}^{\boldsymbol{x}(0)\to\boldsymbol{\mu}}(\mathcal{P}_0^T) = \mathcal{A}_0^T(\mathcal{P}_0^T)$ to explicitly state the initial and final states of the path. As stated before, the action is invariant to a time reversed path.

In order to specify the action for the shape detection problem we use the optimization criterion from (1) and scale it by the time interval T, i.e.,

$$\mathcal{A}_{\Theta(0)\to 0}^{\boldsymbol{x}(0)\to\boldsymbol{\mu}}(\mathcal{P}_0^T) = |\mathcal{S}_{\Theta(0)}(\boldsymbol{x}(0) - \boldsymbol{\mu})|\, T. \tag{4}$$

Note that we disassociate the path $\boldsymbol{x}(t)$ from the parameter evolution $\Theta(t)$. The paths are perpendicular to the shape gradient, and following (2)

$$\boldsymbol{x}(t) = \boldsymbol{\mu} + \int_{\tau=0}^{t} \frac{\mathrm{d}\boldsymbol{x}(\tau)}{\mathrm{d}\tau}\mathrm{d}\tau = \boldsymbol{\mu} + \int_{\tau=0}^{t} v\,\hat{\boldsymbol{n}}(\tau)\,\mathrm{d}\tau, \tag{5}$$

where $v = \left|\frac{\mathrm{d}\boldsymbol{x}(t)}{\mathrm{d}t}\right|$ is the speed of the evolution from the center outwards (assumed to be a free parameter that does not vary over time). The unit vector $\hat{\boldsymbol{n}}(t)$ is along the shape's gradient throughout the evolution. Of course, the set of parameters $\Theta(t)$ also evolves and impacts on the evolution of $\hat{\boldsymbol{n}}(t)$. For example, a circle of radius r and with a (hidden) evolution of time interval T yields a time evolution equation $\boldsymbol{x}(t) = \boldsymbol{\mu} + vt\,\hat{\boldsymbol{r}}$ where $\hat{\boldsymbol{r}} = \hat{\boldsymbol{n}}(t)$ and the optimal speed is $v = r/T$. In this case, $\hat{\boldsymbol{r}} = \hat{\boldsymbol{n}}(t)$ is constant throughout the evolution. When multiple paths at different $\hat{\boldsymbol{r}}$ directions are started at time 0 with the same rate $v = r/T$, an expanding (non deformed) circle is obtained. If we consider different speeds v, possibly with non-optimal values, deformed circles will be created.

An optimal path, $\mathcal{P}_0^{*T}$, causes the action in (4) to take the minimum value, zero. So the evolution of a given ideal shape yields an action that is zero throughout such evolution. However, for non-optimal paths $\mathcal{P}_0^T$, the action will not be zero.

4   Quantum Shape Detection and Interference

Following the path integral point of view of quantum theory [1], we consider the wave propagation to evolve by the integral

$$\psi_0(\boldsymbol{\mu}) = \int d\mathcal{P}_0^T \, \mathcal{K}(\mathcal{P}_0^T) \, \psi_\Theta(\boldsymbol{x}), \tag{6}$$

where $\mathcal{P}_0^T$ is a path from initial state $(\boldsymbol{x}(0), \Theta(0)) = (\boldsymbol{x}, \Theta)$ to final state $(\boldsymbol{x}(T), \Theta(T)) = (\boldsymbol{\mu}, 0)$. The integral is over all possible paths. The Kernel $\mathcal{K}$ is of the form

$$\mathcal{K}(\mathcal{P}_0^T) = \frac{1}{C} e^{i\frac{1}{\hbar} \mathcal{A}_{\Theta \to 0}^{x \to \mu}(\mathcal{P}_0^T)} = \frac{1}{C} e^{i\frac{T}{\hbar} |S_\Theta(x-\mu)|}, \tag{7}$$

where a hyper-parameter, $\hbar$, is introduced. It has the notation used in quantum mechanics for the reduced Planck's constant, but here it will have its own interpretation and value, as we discuss subsequently.

Optimal paths will have zero action while longer or shorter paths, by considering different speeds v in (5), account for deformations. In this way, the sum over all paths, described by (6), is a sum over all possible values of v (the degree of freedom to construct paths). In a more general setting, analogous with the speed v, we have associated degrees of freedom to the evolution of the $\Theta(t)$ shape parameters. So, the sum over all paths corresponds to the sum over all degrees of freedom.

We consider an empirical estimation of $\psi_\Theta(\boldsymbol{x})$ to be given by a set of impulses at the given data set $\boldsymbol{X}$, i.e., $\psi_\Theta(\boldsymbol{x}) = \frac{1}{\sqrt{N}} \sum_{i=1}^{N} \delta(\boldsymbol{x} - \boldsymbol{x}_i)$, where $\delta(x)$ is the Dirac delta function. The normalization ensures that the probability is one when integrated everywhere. Note that $\psi_\Theta(\boldsymbol{x})$ is a pure state, that is, a superposition of impulses. Then, substituting this state (6) with the kernel provided by (7) yields the evolution (contraction) of the probability amplitude from the shape data to the shape center

$$\psi_\Theta(\boldsymbol{\mu}) \approx \sum_{i=1}^{N} \int d\boldsymbol{x} \, \frac{e^{i\frac{T}{\hbar} |S_\Theta(x-\mu)|}}{C} \frac{1}{\sqrt{N}} \delta(\boldsymbol{x} - \boldsymbol{x}_i) = \frac{1}{C\sqrt{N}} \sum_{i=1}^{N} e^{i\frac{T}{\hbar} |S_\Theta(x_i-\mu)|} \tag{8}$$

Thus, points of the deformed shape, $\boldsymbol{x}_i$, are interpreted as evidence of different quantum paths (different speed parameters v). For an ideal shape there is only optimal classical paths. The quantum probability associated with this probability amplitude (a pure state) is given by $P(\boldsymbol{\Theta}) = |\psi_{\boldsymbol{\Theta}}(\boldsymbol{\mu})|^2$, i.e.,

$$P_{\boldsymbol{\Theta}}(\boldsymbol{\mu}) = \psi_{\boldsymbol{\Theta}}(\boldsymbol{\mu})\,\psi_{\boldsymbol{\Theta}}^*(\boldsymbol{\mu}) = \frac{1}{C^2 N}\left(\sum_{i=1}^{N} e^{i\frac{T}{\hbar}|\mathcal{S}_{\boldsymbol{\Theta}}(\boldsymbol{x}_i - \boldsymbol{\mu})|}\right) \cdot \left(\sum_{j=1}^{N} e^{-i\frac{T}{\hbar}|\mathcal{S}_{\boldsymbol{\Theta}}(\boldsymbol{x}_j - \boldsymbol{\mu})|}\right),$$

which can also be expanded as

$$P_{\boldsymbol{\Theta}}(\boldsymbol{\mu}) = \frac{1}{C^2 N}\sum_{i=1}^{N}\left[1 + 2\sum_{j>i}^{N}\cos\left[\frac{T}{\hbar}\left(|\mathcal{S}_{\boldsymbol{\Theta}}(\boldsymbol{x}_i - \boldsymbol{\mu})| - |\mathcal{S}_{\boldsymbol{\Theta}}(\boldsymbol{x}_j - \boldsymbol{\mu})|\right)\right]\right]. \quad (9)$$

It is convenient to define the phase $\phi_{ij} = \frac{T}{\hbar}\left(|\mathcal{S}_{\boldsymbol{\Theta}}(\boldsymbol{x}_i - \boldsymbol{\mu})| - |\mathcal{S}_{\boldsymbol{\Theta}}(\boldsymbol{x}_j - \boldsymbol{\mu})|\right)$.

4.1 Interference

We now elaborate on the interference phenomenon arising from the cosine terms in the probability (9), the driving force to the *saliency effect*. Consider any two points $\boldsymbol{x}_i, \boldsymbol{x}_j \in \boldsymbol{X}$. If the two points belong to the shape, then $|\mathcal{S}_{\boldsymbol{\Theta}}(\boldsymbol{x}_i - \boldsymbol{\mu})| \approx |\mathcal{S}_{\boldsymbol{\Theta}}(\boldsymbol{x}_j - \boldsymbol{\mu})|$ and $|\phi_{ij}| \ll 1$, resulting in a large cosine term, $\cos\phi_{ij} \approx 1$. If the two points belong to the clutter, $|\mathcal{S}_{\boldsymbol{\Theta}}(\boldsymbol{x}_i - \boldsymbol{\mu})|$ and $|\mathcal{S}_{\boldsymbol{\Theta}}(\boldsymbol{x}_j - \boldsymbol{\mu})|$ are likely to be quite different, and small values of $\hbar$ will result in larger values of $|\phi_{ij}|$. Pairs of clutter data points not belonging to the shape, with large and varying phase differences, will produce varied cosine terms, some positive and some negative. In summary, points of the shape "strengthen" each other, whereas the points of the clutter could "weaken" each other.

Similarly, if the shape is evaluated at misplaced center location or misplaced set of parameters, even points belong to the shape will behave like clutter, canceling each other and not supporting such a parameter choice.

4.2 Linear-Complexity Computation in the Size of the Data Set

Note that even though the probability reflects a pair-wise computation as seen in (9), we evaluate it by taking the magnitude square of the probability amplitude, given by (8), which is computed as a sum of N complex numbers. Thus, the complexity of the computations is linear in the data set size, N.

4.3 A Note on Bayesian Theory

In this work we proposed a simple Bayesian probabilistic model described by (3). By no means did we imply that this is the only possible Bayesian model. In fact, the probability model we just obtained by the quantum method, that is (9), could be derived as a Bayesian model since, in principle, any "good/correct" probabilistic model can be Bayesian-derived. In this sense, we do not regard the quantum method as being in opposition to Bayesian theory. However, we assert that the quantum method, with novel interpretation of the information content of the data, will automatically produce new probability models from classical optimization models. In particular it produces pair-wise interactions with negative terms (with interference). At the same time, computationally, the method is still linear in the number of data points. So, what we are proposing is to add the quantum method to the set of tools for deriving probabilistic models from classical optimization ones.

4.4   A Classical Statistical Version of the Quantum Criterion

We derive a classical probability from the quantum probability amplitude via the Wick rotation [3]. This mathematical technique frequently employed in physics, transforms quantum physical systems into statistical physical systems and vice-versa. It consists of replacing the term $i\frac{T}{\hbar}$ by a real hyper-parameter α in the probability amplitude. (In statistical physics, α is related to the time evolution of the particle ensemble.) Considering the probability amplitude (8), the Wick rotation yields

$$P_\Theta(\boldsymbol{\mu}) = \frac{1}{Z}\sum_{i=1}^{N} e^{-\alpha|\mathcal{S}_\Theta(\boldsymbol{x}_i - \boldsymbol{\mu})|}. \tag{10}$$

We can interpret this as follows. Each data point $\boldsymbol{x}_i$ produces a vote $v(\Theta, \boldsymbol{\mu}|\boldsymbol{x}_i) = e^{-\alpha|\mathcal{S}_\Theta(\boldsymbol{x}_i - \boldsymbol{\mu})|}$, with values between 0 and 1. The hyper-parameter α controls the weight decay of the vote. Clutter data with larger values for the shape error will cast negligible votes, and therefore this classical measure is resistant to clutter and occlusions.

Interestingly, this probability model resembles a Hough transform with each vote $v(\Theta, \boldsymbol{\mu}|\boldsymbol{x}_i) = e^{-\alpha|\mathcal{S}_\Theta(\boldsymbol{x}_i - \boldsymbol{\mu})|}$ being approximated by the binary vote

$$v(\Theta, \boldsymbol{\mu}|\boldsymbol{x}_i) = u\left(\frac{1}{\alpha} - |\mathcal{S}_\Theta(\boldsymbol{x}_i - \boldsymbol{\mu})|\right), \tag{11}$$

where $u(x)$ is the Heaviside step function, $u(x) = 1$ if $x \geq 0$ and zero otherwise, i.e., $u = 1$ if $|\mathcal{S}_\Theta(\boldsymbol{x}_i - \boldsymbol{\mu})| \leq \frac{1}{\alpha}$ and $u = 0$ otherwise. The parameter $\frac{1}{\alpha}$ clearly defines the error tolerance for a data point $\boldsymbol{x}_i$ to belong to the shape $\mathcal{S}_\Theta(\boldsymbol{x} - \boldsymbol{\mu})$. Hough transforms, in classical shape detection, are resistant to occlusions and clutter. Still, these statistical models of shape do not exhibit the interference phenomenon. We will study and show how such a difference affects the detection of shapes in images in the presence of deformations.

4.5   Evaluation of $\hbar$ and α for Detection of the Center μ

The amplitude probability given by Eq. (8) has the hyper-parameters $\hbar$ and T where the ratio $\frac{T}{\hbar}$ scales up the magnitude of the shape values. Without loss of generality we can set $T = 1$ and focus solely on the hyper-parameter $\hbar$. The smaller is $\hbar$, the more a pair of inputs $(\boldsymbol{x}_i, \boldsymbol{x}_j)$ will map to random phase values $\left(\varphi_i = \mod\left(\frac{1}{\hbar}|\mathcal{S}_\Theta(\boldsymbol{x}_i - \boldsymbol{\mu})|, 2\pi\right), \varphi_j = \mod\left(\frac{1}{\hbar}|\mathcal{S}_\Theta(\boldsymbol{x}_j - \boldsymbol{\mu})|, 2\pi\right)\right)$ in the unit circle. The larger $\hbar$ is, the smaller the phase φ_i is.

A large parameter $\hbar$ can help in aligning shape points to similar phases. That suggests large values for $\hbar$. At the same time, $\hbar$ should help in misaligning pair of points where at least one of them does not belong to the shape. That suggests small values of $\hbar$. We then choose $\hbar$ so that

$$\hbar = \frac{1}{\pi} \max_i |\mathcal{S}_{\Theta^*}(\boldsymbol{x}_i - \boldsymbol{\mu}^*)|. \tag{12}$$

For choosing the parameter α, we note that for $\frac{1}{\alpha} \geq \max_i |\mathcal{S}_{\Theta^*}(\boldsymbol{x}_i - \boldsymbol{\mu}^*)|$ all shape points will vote for the true center. Thus, choosing $\frac{1}{\alpha} = \max_i |\mathcal{S}_{\Theta^*}(\boldsymbol{x}_i - \boldsymbol{\mu}^*)|$ will guarantee all votes for the true shape parameters and give a lower vote for shape points evaluated from the displaced parameters. We also did search for lower values of $\frac{1}{\alpha}$ and in our experiments such changes did not improve the Hough transform performance.

In summary, we choose $\pi \hbar = \frac{1}{\alpha} = \max_i |\mathcal{S}_{\Theta^*}(\boldsymbol{x}_i - \boldsymbol{\mu}^*)|$.

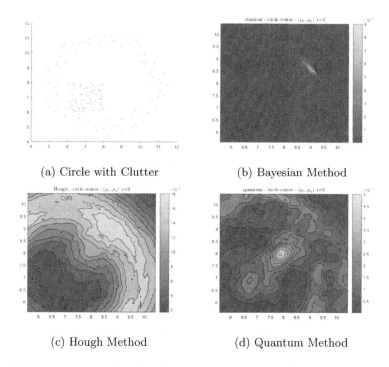

(a) Circle with Clutter   (b) Bayesian Method

(c) Hough Method   (d) Quantum Method

Fig. 2. (a) A circle centered at $(8, 8)$, radius 3, and with 100 points, is deformed with $\eta = 0.5$. 100 clutter noise points are added uniformly to a box with diagonal corners at $(6, 6)$ and $(8, 8)$. The radius value is fed to the methods. The figures have the probabilities normalized across the range of center hypothesis depicted (where the two diagonal corners range from $(5.5, 5.5)$ to $(10.5, 10.5)$). (b) Bayesian method applied to the data is in general more sharply distributed due to the probability being the result of multiplications per data point. However, the clutter has a strong effect on pushing the center away from its true location. There is no choice of λ to revert such scenarios. (c) Hough method with $\frac{1}{\alpha} = 0.37$ estimated to include all circle points that have been deformed. It responds better to clutter noise. (d) Quantum method applied with $\hbar = 0.12$ outperform both methods due to the ability to have probability cancellations.

5   Experiments with Center Detection

We devised experiments with circles and with ellipses to demonstrate how the
quantum method (9) can estimate shape parameters in the presence of noise
and large deformation in comparison with the classical statistical methods (3)
and (11). We focused on center detection accuracy, providing the other true shape
parameters Θ. Note that all the models are "one hyper-parameter models" and
have the same variables (the shape parameters and center). For uniform clutter
noise and low deformations, the Bayesian model works quite well. We did not
show such results as it is evident from the first experiments that the Bayesian
method cannot handle large deformation or clutter noise that is not uniform.

The first set of experiments, shown in Fig. 2, compares the three methods (3), (11), and (9) for large deformations ($\eta = 0.5$). The results suggest weak

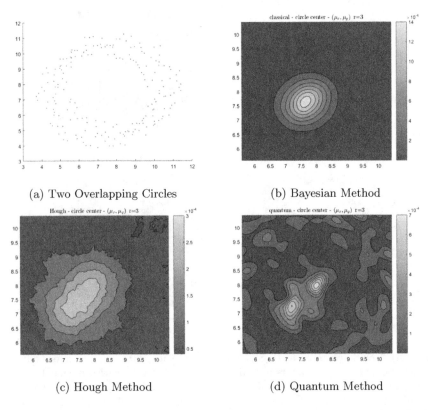

(a) Two Overlapping Circles   (b) Bayesian Method

(c) Hough Method   (d) Quantum Method

Fig. 3. (a) Two overlapping circles deformed with $\eta = 0.5$, radius $r = 3$ and with
100 points each. The circle centers are at $(8, 8)$ and $(7.2, 7.2)$. The radius is fed to the
method. (b) The Bayesian method. The method mixes all data yielding the highest
probability approximately in the "middle" of both centers and no suggestion of two
peaks/circles/centers exists. (c) The Hough method with $\frac{1}{\alpha} = 0.37$. The method yields
a probability that is more diluted and includes the correct centers, but does not suggest
two peaks. (d) The quantum criterion with $\hbar = 0.12$. The *saliency effect* of the quantum
method leads to two peak detections at nearby solutions.

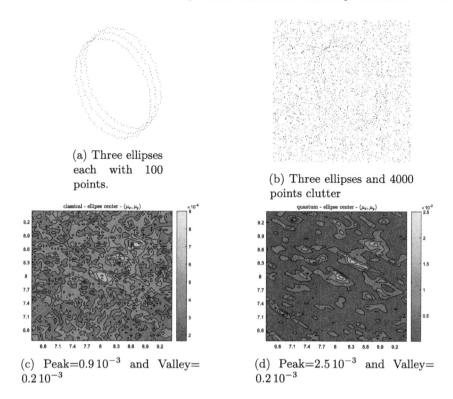

(a) Three ellipses each with 100 points.

(b) Three ellipses and 4000 points clutter

(c) Peak=0.9 10^{-3} and Valley= 0.2 10^{-3}

(d) Peak=2.5 10^{-3} and Valley= 0.2 10^{-3}

Fig. 4. Deformed ellipses, each with large axis $a = 3$, small axis $b = 2$, angle $\theta = 2$, and deformation $\eta = 0.05$. Centers at $\mu = (8.0, 8.0); (8.3, 8.3); (8.6, 8.6)$. We used the classical statistical method (10) with $\frac{1}{\alpha} = 0.04$ and the quantum criterion with $\hbar = 0.03$. In all of these experiments, the *saliency effect* of the quantum method leads to greater contrast between peak solutions and nearby solutions.

accuracy for the (one hyper-parameter) Bayesian method since the method can not ignore clutter noise (every data point is "explained" by the Bayesian method with a positive probability contribution). The Hough method has the capacity of completely ignoring data and is therefore able to perform better in the presence of clutter. The quantum method shows the *saliency effect* as probability cancellation offers a better method to discredit false centers near the true center solution (an analysis of this phenomena is provided in the next section).

The second set of experiments for two circles in the presence of large deformations ($\eta = 0.5$), shown in Fig. 3, compares the three methods (3), (11), and (9). The results suggest weak accuracy for the (one hyper-parameter) Bayesian method. The method has not modeled two circles and thus finds the best "one circle" center description. The Hough method, does have the capacity to detect several centers. However, its limitation is evident, creating a probability distribution that includes both centers but does not distinguish them as two centers. The quantum method exhibit the *saliency effect* as the probability cancellation offers a better way to discredit false centers near the true center solution with the true center solutions emerging from the probability distribution.

The last set of experiments, shown in Fig. 4, uses three overlapping ellipses and compares the two methods (10) and (9). Note that the deformations are ten times smaller then previous experiments on circles ($\eta = 0.05$) and therefore the overlapping is not as extensive. The experiments suggests that both methods are able to "peak" the three centers in the presence of noise. The peak of the center detection vs. the valley of the center detection is more pronounced for the quantum method.

We next present an analysis for the circle shape (the analysis method can be extended to any shape) to demonstrate that the main conclusions obtained from the experiments can also be derived theoretically.

6   Analysis of the Circle Shape Center Detection

We analyze the quantum method described by the probability (9) and compare it with the Hough method (11). This analysis is carried for the simple example of the circle.

We consider the circle shape, $\mathcal{S}_r(\boldsymbol{x} - \boldsymbol{\mu}) = 1 - \frac{(x-\mu)^2}{r^2}$ of true radius r and its evaluation not only at the true center $\boldsymbol{\mu}^*$ but also at small displacements from it $\boldsymbol{\mu} = \boldsymbol{\mu}^* + \delta\boldsymbol{\mu}$ where $\frac{\delta\mu}{r} < 1$ with $\delta\mu = |\delta\boldsymbol{\mu}|$. So

$$
\begin{aligned}
\mathcal{S}_i = \mathcal{S}_r(\boldsymbol{x}_i - \boldsymbol{\mu}) &= 1 - \frac{(x_i - \mu)^2}{(r)^2} = 1 - \frac{((x_i^S + \boldsymbol{\eta}_i) - (\boldsymbol{\mu}^* + \delta\boldsymbol{\mu}))^2}{(r)^2} \\
&= -2\frac{\eta_i - \hat{r}_i \cdot \delta\boldsymbol{\mu}}{r} - \left(\frac{\eta_i^2 + (\delta\mu)^2 - 2\eta_i(\hat{r}_i \cdot \delta\boldsymbol{\mu})}{r^2} \right) \\
&= -[2(a_i - b_i) + (a_i^2 + b^2 - 2a_i b_i)],
\end{aligned}
\tag{13}
$$

where $a_i = \frac{\eta_i}{r}$ and $b_i = \frac{\hat{r}_i \cdot \delta\mu}{r}$, $b = \frac{\delta\mu}{r} = \max_i b_i$, and $a = \frac{\eta}{r} = \max_i a_i$ and we applied to the ideal circle data points $\boldsymbol{x}_i^S = r\hat{r}_i$ a uniformly distributed deformation $\boldsymbol{\eta}_i = \eta_i \hat{r}_i$, where $\eta_i \in (-\eta, \eta)$. We are considering deformations within $a < 1$ and displacements from the true center within $b < 1$. We interpret $|\mathcal{S}_i|$ as a random variable and we assume that the sampling of the circle is such that the variation of b_i is uniform. This assumption simplifies our calculations, but the result is robust to other reasonable samplings. So $P_b(b_i) = \frac{1}{2b} = \frac{r}{2\delta\mu}$ and with the deformations uniformly distributed we also have $P_a(a_i) = \frac{1}{2a} = \frac{r}{2\eta}$. Finally, since the deformations are independent of the shape's position, a_i and b_i are independent, and $P_{\mathcal{S}}(|\mathcal{S}_i|) = P_a(a_i)P_b(b_i) = \frac{1}{4ab} = \frac{r^2}{4\eta\,\delta\mu}$.

Quantum Case: Inserting shape Eq. (13) into the quantum probability amplitude of (8), for N_C circle points, results in

$$\psi_r(\mu^* + \delta\mu) \approx \frac{1}{C\sqrt{N_C}} \sum_{i=1}^{N_C} e^{i\frac{1}{\hbar}|S_r(x_i - \mu)|}$$

$$\approx \frac{1}{C\sqrt{N_C}} N_C \int_{-a}^{a} da_i \int_{-b}^{b} db_i \, e^{i\frac{1}{\hbar}|2(a_i - b_i) + (a_i^2 + b^2 - 2a_i b_i)|} P_a(a_i) P_b(b_i)$$

$$= \frac{\sqrt{N_C}}{C} \mathcal{I}_{ab}(e),$$

where $\mathcal{I}_{ab} = \frac{1}{4ab} \int_{-a}^{a} da_i \int_{-b}^{b} db_i \, e^{i\frac{1}{\hbar}|2(a_i - b_i) + (a_i^2 + b^2 - 2a_i b_i)|}$ and at the true center

$$\psi_r(\mu^*) \approx \frac{\sqrt{N_C}}{C} \mathcal{I}_a,$$

where $\mathcal{I}_a = \mathcal{I}_{ab=0} = \frac{1}{2a} \int_{-a}^{a} da_i \, e^{i\frac{1}{\hbar}|2a_i + a_i^2|}$. Note that the quantum probability $P_r(\mu) = |\psi_r(\mu)|^2$ at any of the considered locations μ scales linearly with N_C. The ratio of the probabilities at a displaced center and true center is then

$$Q_C(a, b, \hbar) = \frac{P_r(\mu^*)}{P_r(\mu^* + \delta\mu)} \approx \frac{|\mathcal{I}_a|^2}{|\mathcal{I}_{ab}|^2}.$$

These integrals, $\mathcal{I}_{ab}$ and $\mathcal{I}_a$, can be evaluated numerically, or via integration of a Taylor series expansions, followed by numerical evaluation of the expansions.

Hough Case: Inserting shape Eq. (13) into the vote for the Hough transform given by (11) results in

$$v(r^*, \mu^* + \delta\mu|x_i) = u\left(\frac{1}{\alpha} - |2(a_i - b_i) + (a_i^2 + b^2 - 2a_i b_i)|\right)$$

and interpreting the Hough total vote, $V_r^H(\mu^*) = \sum_{i=1}^{N_C} v(r, \mu^*|x_i)$, as an average over a function of the random variable $|S_i|$ multiplied by the number of votes, we get

$$V_r^H(\mu^* + \delta\mu) \approx N_C \int_{-a}^{a} da_i \int_{-b}^{b} db_i \, u\left(\frac{1}{\alpha} - |2(a_i - b_i) + (a_i^2 + b^2 - 2a_i b_i)|\right)$$
$$\cdot P_a(a_i) P_b(b_i).$$

The ratio of the votes at the displaced center and the true center is then

$$H(a, b, \alpha) = \frac{V_{r*}^H(\mu^*)}{V_{r*}^H(\mu^* + \delta\mu)}.$$

Figure 5 show these two ratios, $H(a, b, \alpha)$ and $Q(a, b, \alpha)$, and demonstrates the better accuracy of the quantum method due to the *saliency effect*.

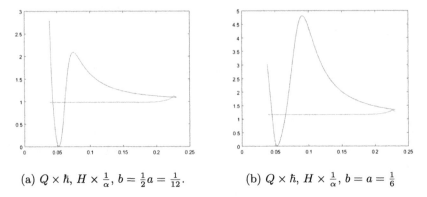

(a) $Q \times \hbar$, $H \times \frac{1}{\alpha}$, $b = \frac{1}{2}a = \frac{1}{12}$.   (b) $Q \times \hbar$, $H \times \frac{1}{\alpha}$, $b = a = \frac{1}{6}$

Fig. 5. $Q(a,b,\hbar) \times \hbar$ (blue) for $\hbar \in (0.047, 0.2802)$ and $H(a,b,\alpha) \times \frac{1}{\alpha}$ (red) for $\frac{1}{\alpha} \in (0.044, 0.37)$ for circles with radius $r = 3$ and deformations $\eta = 0.5$, i.e., $a = \frac{1}{6}$. Both plots have 200 points, with uniform steps in their respective ranges. For (a) $b = \frac{1}{2}a$ and for (b) $b = a$ (or center displacements $\delta\mu = 0.25, 0.5$, respectively.) Clearly, the quantum method probability for $\hbar = 0.12$ (the solution to (12)), decreases much faster than the Hough method probability (for any $\frac{1}{\alpha}$). These plots illustrate the *saliency effect* and the experimental observations of the previous section. (Color figure online)

7   Conclusion

We proposed a quantum method for shape detection. Our approach starts from classical shape modeling and then applies a mathematical method, namely the path integral method, to obtain a quantum formulation of the same classical problem. The path integral method was invoked to account for each data point using our insight to interpret a shape deformation data as evidence of an evolution of a path (where the optimal path yields no deformation data). From the quantum model obtained, we derived a statistical classical model, via the Wick rotation, and we also established a theoretical connection to general Hough transform. The main advantage of the quantum method over existing classical statistical methods is its high resistance to large amounts of deformation. Quantum interference has the ability to cancel the contributions from the wrong shape hypothesis centers (or any other parameter), while enhancing the contributions from the shapes at the true parameters. The computations are very efficient as they are linear in the number of data points. The method is quite general for shapes and can be extended to other energy optimization problems.

It is worth stressing that nowhere in our development we relied on physical concepts, such as forces or energies, as the usage of the Schrödinger equation would imply. Our only model was a computer vision model of shapes. Quantum theory was an alternative statistical method, a purely mathematical construct applicable to a variety of different fields and models, if it is beneficial to do so.

Acknowledgments. The first author thanks the National Science Foundation for the Award Number 1422021, which partially supported this research. Both authors wish to thank Dan Pinkel for numerous interesting conversations about these ideas and methods and the anonymous reviewers for valued feedback.

References

1. Feynman, R., Hibbs, A.: Quantum Mechanics and Path Integrals: Emended by D. F. Steyer. Dover Publications, New York (2012)
2. Feynman, R.: The Feynman Lectures on Physics, vol. 3. Addison Wesley, Boston (1971)
3. Wick, G.C.: Properties of Bethe-Salpeter wave functions. Phys. Rev. **96**, 1124–1134 (1954)
4. Hough, P.V.C.: Machine analysis of bubble chamber pictures. In: 2nd International Conference on High-Energy Accelerators and Instrumentation, Proceedings Geneva, Switzerland, pp. 554–558 (1959)
5. Yuille, A.L.: Energy functions for early vision and analog networks. Biol. Cybern. **61**(2), 115–123 (1989)
6. Emms, D., Wilson, R., Hancock, E.: Graph matching using the interference of continuous-time quantum walks. Pattern Recogn. **42**, 985–1002 (2009)
7. Emms, D., Wilson, R., Hancock, E.: Graph matching using the interference of discrete-time quantum walks. Image Vis. Comput. **27**, 934–949 (2009)
8. Rangarajan, A., Gurumoorthy, K.S.: A Schrödinger wave equation approach to the eikonal equation: application to image analysis. In: Cremers, D., Boykov, Y., Blake, A., Schmidt, F.R. (eds.) EMMCVPR 2009. LNCS, vol. 5681, pp. 140–153. Springer, Heidelberg (2009). https://doi.org/10.1007/978-3-642-03641-5_11
9. Gurumoorthy, K.S., Rangarajan, A., Banerjee, A.: The complex wave representation of distance transforms. In: Boykov, Y., Kahl, F., Lempitsky, V., Schmidt, F.R. (eds.) EMMCVPR 2011. LNCS, vol. 6819, pp. 413–427. Springer, Heidelberg (2011). https://doi.org/10.1007/978-3-642-23094-3_30
10. Cicconet, M., Geiger, D., Werman, M.: Complex-valued Hough transforms for circles. In: Proceedings, 2015 IEEE International Conference on Image Processing (ICIP), pp. 2801–2804, September 2015
11. Witkin, A.P.: Scale-space filtering. In: Proceedings of the Eighth International Joint Conference on Artificial Intelligence, pp. 1019–1022 (1983)
12. Lindeberg, T.: Scale-space theory: a basic tool for analysing structures at different scales. J. Appl. Stat. **2**(21), 224–270 (1994). (Supplement on Advances in Applied Statistics: Statistics and Images: 2)
13. Siddiqi, K., Kimia, B.B.: A shock grammar for recognition. In: Computer Vision and Pattern Recognition, Proceedings CVPR 1996 (1996)

Structured Output Prediction and Learning for Deep Monocular 3D Human Pose Estimation

Stefan Kinauer[1](✉)(iD), Riza Alp Güler[1](✉)(iD), Siddhartha Chandra[1],
and Iasonas Kokkinos[2,3]

[1] CentraleSupélec, INRIA Saclay, Gif-sur-Yvette, France
{stefan.kinauer,riza.guler}@inria.fr
[2] Facebook AI Research, Paris, France
[3] University Collage London, London, UK

Abstract. In this work we address the problem of estimating 3D human pose from a single RGB image by blending a feed-forward CNN with a graphical model that couples the 3D positions of parts. The CNN populates a volumetric output space that represents the possible positions of 3D human joints, and also regresses the estimated displacements between pairs of parts. These constitute the 'unary' and 'pairwise' terms of the energy of a graphical model that resides in a 3D label space and delivers an optimal 3D pose configuration at its output. The CNN is trained on the 3D human pose dataset 3.6M, the graphical model is trained jointly with the CNN in an end-to-end manner, allowing us to exploit both the discriminative power of CNNs and the top-down information pertaining to human pose. We introduce (a) memory efficient methods for getting accurate voxel estimates for parts by blending quantization with regression (b) employ efficient structured prediction algorithms for 3D pose estimation using branch-and-bound and (c) develop a framework for qualitative and quantitative comparison of competing graphical models. We evaluate our work on the Human3.6M dataset, demonstrating that exploiting the structure of the human pose in 3D yields systematic gains.

1 Introduction

Human pose estimation has made rapid progress thanks to deep learning, as witnessed by the improvements reported on large-scale benchmarks [1–7]. In this work we focus on the more challenging task of 3D human pose estimation from a single monocular image, which can have many applications in human-computer interaction, augmented reality, and can eventually lead to addressing generic 3D object pose estimation.

Recovering 3D information from a single 2D image is clearly ill-posed, given that different 3D scenes can project to the same 2D image. However, exploiting task-specific prior knowledge can increase the probability of the more plausible scenes. All leading approaches to 3D human pose estimation, such as [8–11],

---

S. Kinauer and R. A. Guler have contributed equally to this work.

ⓒ Springer International Publishing AG, part of Springer Nature 2018
M. Pelillo and E. Hancock (Eds.): EMMCVPR 2017, LNCS 10746, pp. 34–48, 2018.
https://doi.org/10.1007/978-3-319-78199-0_3

rely on incorporating prior knowledge about the structure of the 3D human body. Two-stage approaches, e.g. [9,10,12], firstly detect joint positions in 2D and subsequently lift joints into 3D by relying on prior knowledge about the 3D human pose. The advantage of such approaches is that they can exploit large datasets constructed for the prediction of 2D landmarks - the disadvantage is that errors in the 2D stage can propagate to the 3D predictions and can often not be recovered from. Inherently 3D approaches [13,14] discretize the depth variable and train a CNN to score every possible combination of position and depth with respect to the presence of a joint - one can understand that the CNN learns to use the scale of the joint to guess its depth. This was recently shown in [13] to deliver results that are largely superior over previous 2-stage approaches. More recent works have delivered further improvements by fusing the 2D and 3D streams [15].

In directly regressing the pose from the input image, the aforementioned approaches do not explicitly impose constraints that exploit the dependencies between the human joints. In our understanding, what is missing from existing works is a method to exploit the structure of the human pose in a natively 3D setup. Authors in [16] acknowledge this deficiency of contemporary methods, and propose to use a stacked denoising auto-encoder to learn these dependencies implicitly. Other approaches to combining structured prediction with deep learning have recently been successfully pursued in 2D human pose estimation e.g. [17,18], while current approaches to incorporating structure in feedforward CNNs for pose estimation rely on cascading, or stacking the outputs of CNNs in 2D [5,6], which can become prohibitive when done in 3D, due to the increased memory and computation load. In this work we develop novel techniques that allow us to 'explicitly' capture the dependencies between human joints via an energy function that consists of unary and pairwise terms, and thereby pursue this direction in the arguably harder 3D setting.

Our contribution consists in showing that one can combine a volumetric representation with a structured model that imposes constraints between the relative positions of parts. Rather than relying exclusively on a feed-forward architecture, we show that one can append a structured prediction algorithm that propagates information on a graph that represents the part positions, and still remain computationally efficient. In particular, we train a CNN to not only populate the 'unary' 3D score maps for each part, but also to provide estimates of the relative positions of parts in 3D. This coupling of the part positions results in an optimization problem, that we treat as yet-another layer of a deep network, and thereby add functionality to the network (Fig. 1).

We can describe our contributions as (i) allowing for a high resolution in the estimation of the pose without increasing the computation/memory budget (Sect. 2.1) (ii) constructing the coupling term so as to make optimization tractable while working with a high resolution in the depth coordinates (Sect. 2.2) (iii) exploiting rich connectivity, i.e. allowing the edges in the underlying graph of the model to form loops, by using fast approximate inference that combines Branch-and-Bound with ADMM (Sect. 2.3) and (iv) using Deeply Supervised

Network (DSN) [19] training to accelerate training, by forcing the 2D and 3D inputs to structured prediction to reflect part of the solution, while also leveraging on larger 2D datasets during training (Sect. 2.4).

We evaluate our approach on the Human3.6M dataset. Our results indicate that blending local information about the 3D pose of parts with information about the 3D displacements of parts yields systematic improvements over approaches that rely on either of these two cues alone, while incurring a marginal computational overhead, in the order of a fraction of a second.

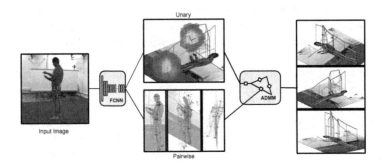

Fig. 1. We consider the task of 3D human pose estimation from a single RGB image. Our approach involves a fully convolutional neural network that provides a 'bottom-up' estimate of the 3D positions of parts and their relative displacements, and a structured prediction layer that combines them into a coherent estimate of the pose. The whole architecture is trained end-to-end, allowing us to optimize the CNN outputs with the respect to the subsequent pose estimation algorithm.

2   Method

We start by formulating our approach in terms of a structured prediction problem, and then provide the details about the individual components of our proposed approach. We represent the pose Φ in terms of the concatenation of the 3D coordinates of N individual parts ϕ_i

$$\Phi = \{\phi_1, \ldots, \phi_N\}. \tag{1}$$

Given an image I, we score a candidate pose in terms of a graphical model that considers individual properties of parts, as well as properties of some of their pairwise combinations:

$$S_I(\Phi) = \sum_{i=1}^{N} \mathcal{U}_i(\phi_i) + \sum_{i,j \in \mathcal{E}} \mathcal{P}_{i,j}(\phi_i, \phi_j), \tag{2}$$

where $\mathcal{U}$ stands for unary and $\mathcal{P}$ for pairwise potentials, and $\mathcal{E}$ is the set of edges used in our graphical model. The unary and pairwise terms are delivered by the CNN, while the structured prediction layer couples the parts through the optimization of Eq. 2. If we consider a generic cost function, this can be challenging even for simple cases, let alone for the 3D pose space we are working with. Our main technical contributions aim at making the construction and optimization of Eq. 2 tractable while still exploiting the structure of the output space.

2.1   Quantized Regression for Depth Estimation

One of the main challenges in constructing a volumetric CNN is that the amount of memory and computation scales linearly in the granularity of the depth quantization, requiring to tradeoff accuracy for speed/memory. The root of the problem is that the underlying quantity is continuous, but plain regression-based models may be neither sufficiently accurate, nor expressive enough to capture the uncertainty and multimodality of the depth value caused by depth ambiguity, or occlusion.

Instead, we follow recent successful developments in object detection [20, 21], dense correspondence estimation [22], and pose estimation [23] where a combination of classification and regression is used to attack the image-based regression problem. We use a first classification stage to associate a confidence value with a set of non-overlapping depth intervals, corresponding to a coarse quantization of the depth value. If we have N classes and a depth range of, say D units, the k-th class is associated with a quantized depth of $q_k = k\frac{D}{N}$. This however may be at a very coarse depth resolution. We refine this coarse estimate by combining it with the results of a regression layer that aims at recovering the residual between the ground-truth depth values and their quantized depth estimates.

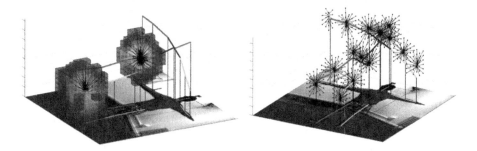

Fig. 2. Unary 3D coordinates via quantized regression. To efficiently regress the unary 3D coordinates, we use a divide and conquer strategy. We begin by quantizing the 3D space into voxels. We estimate the score of each joint belonging to each of these voxels using a classifier. Finally we regress a residual vector per voxel which indicates the offset between the center of the voxel and the continuous 3D position of each joint. *Left:* Sigmoid function on classified voxels and regressed residual vectors (in black) for two joints. *Right:* Regressed residual vectors for all joints.

As shown in Fig. 2 this strategy allows us to 'retarget' the voxels to 3D positions that lie closer to the actual part positions, without requiring the exhaustive sampling of the 3D space. In particular a voxel v lying at the k-th depth interval will become associated with a novel 3D position of part i, $p_i^v = k\frac{D}{N} + r_i(v)$, where $r_i(v)$ is the residual regressed by our network for the i-th part type at voxel v.

The value of the associated unary terms, $U_i(p_i^v)$, is obtained in terms of the inner product between a joint-specific weight vector, w_i and a feature vector extracted from the CNN's output at the 2D position associated with voxel v.

2.2   Efficient Optimization with Quadratic Pairwise Terms

Having described how the unary terms are constructed in our model, we now turn to the pairwise terms and the resulting optimization problems. The expression for the pairwise term in Eq. 2, $\mathcal{P}_{i,j}(\phi_i, \phi_j, I)$ would suggest constructing a six-dimensional function. Instead, as in the Deformable Part Model paradigm [24], we use a pairwise term that penalizes deviations from a nominal displacement $\mu_{i,j}$:

$$\mathcal{P}_{i,j}(\phi_i, \phi_j, I) = -\sum_{d=1}^{3} c_d(\phi_{d,i} - \phi_{d,j} - \mu_{d,i,j})^2, \tag{3}$$

where the c_d parameters allow us to calibrate the importance of the different dimensions. These parameters are forced to be positive, while the expression in (3) corresponds to the log-probability under an axis-aligned Gaussian model, centered at the predicted part position. We note that as in [25,26], $\mu_{i,j}$ is image-dependent, and in our case is the output of a sub-network which is trained end-to-end. This enables us to capture dependencies between parts, where an estimate of their ideal displacement is combined with the local evidence provided by their unary terms.

One important advantage of the pairwise terms is that since they encode the relative position of parts they are often easier to model, since e.g. the distance between human joints is much more predictable than the actual positions of the joints. As such they can simplify the overall problem.

Another crucial advantage of the particular form of the pairwise term is that by virtue of being in the form of a quadratic cost function, it can easily be bounded from above and below using interval arithmetic - in particular, we rely on the 3D Branch-and-Bound algorithm introduced recently in [27] to efficiently search over optimal combinations of parts in 3D. A brute-force, dynamic programming-type algorithm for solving this task would require a quadratic number of operations, since it would need to compare pairs of points. Our implementation has a low-constant linear complexity for the construction of per-part KD-tree data structures, and logarithmic best-case complexity for the subsequent optimization. In practice optimization requires less than a tenth of a second on a CPU, while further accelerations could be obtained through GPU-based implementations.

2.3   Network Connectivity: From Star-Shaped to Loopy Graphs

The Branch-and-Bound (BB) algorithm we use for efficient inference only accommodates a star-shaped graph topology. This can be problematic if one wants to model human pose in terms of a tree-structured graph, or introduce loops to capture more constraints. For this we employ master-slave type approximate inference techniques that allow us to use BB for slave problems and coordinate them through a master. In particular we rely on the Alternating Direction Method of Multipliers (ADMM) [28–30] which matches the continuous nature of the pose estimation problem [30]. The approach to subdivide difficult problems into smaller and easier ones has before been seen in [31]. The authors introduced Dual Decomposition to optimize MRF-type energies, outperforming former state of the art of "tree-reweighted message passing" algorithms. Later works on ADMM like [30] borrow from developments outlined in [28] to reach convergence in a lower number of iterations. Loopy graphs are subdivided into easier to handle trees and coordinated via a master problem, which turns out to be updating the dual variables.

   The method we outline below uses approximate inference to obtain solutions in $\omega(T \log N)$ operations, where T is a low constant in the order of tens, N is the number of voxels, and $\log N$ is the cost of re-solving the slave subproblems. The ω (best-case) notation relates to the (exact) Branch-and-Bound algorithm, which also empirically has typically this performance. Even though the ADMM-based results are now only approximately optimal, the cost function being optimized reflects more accurately the problem structure, which can positively affect accuracy.

   We consider the case where the set of graph edges in Eq. 2 corresponds to a graph with loops. Denoting by $R \subset 1 \ldots K$ the subset of point indices belonging to more than one star graph, our optimization problem can equivalently be rewritten as follows:

$$\max S(\varPhi) = \sum_{i=1}^{N} S_i(\varPhi_i) \quad \text{s.t.} \quad \varPhi_i(r) = u(r) \quad \forall r \in R, \tag{4}$$

where S_i is a set of loop-free subproblems, defined so that $S(\varPhi) = \sum_{i=1}^{N} S_i(\varPhi_i)$ for a common solution $\varPhi$. The consistency is enforced by the 'master', to whom the 'slave' subproblems S_i deliver their solutions $\varPhi_i$ - obtained through Branch-and-Bound. In particular a relaxation to the constraints is updated and used to reset the problem solved by the slaves - at each step the relaxation becomes tighter and at convergence consistency is guaranteed. Dual Decomposition relaxes the constraints in Eq. 4 by introducing a Lagrange Multiplier $\lambda_i(r)$ for each agreement constraint. ADMM augments this with a quadratic constraint violation penalty resulting in an *augmented Lagrangian* function:

$$\mathcal{A}(\varPhi, u, \lambda) = \sum_{i=1}^{N}(S_i(\varPhi_i) + \sum_{r \in R} \langle \lambda_i(r), \varPhi_i(r) \rangle) - \sum_{r \in R}((\sum_{i=1}^{N} \lambda_i(r))u(r) - \frac{\rho}{2} \sum_{i=1}^{N}(\varPhi_i(r) - u(r))^2)$$

$$\tag{5}$$

where ρ is a positive parameter that controls the intensity of the augmenting penalty. The quadratic term ensures rapid convergence by acting like a regularizer of the solutions found across different iterations. To maximize the augmented Lagrangian, ADMM iteratively performs the following steps:

$$\Phi_i^{t+1} = \mathrm{argmax}_{\Phi_i} \mathcal{A}(\Phi_i, u^t, \lambda^t) \tag{6}$$

$$u^{t+1} = \mathrm{argmax}_u \mathcal{A}(\{\Phi_i^{t+1}\}, u, \lambda^t) \tag{7}$$

$$\lambda_i^{t+1}(r) = \lambda_i^t(r) - \rho(\Phi_i^{t+1}(r) - u^{t+1}(r)) \tag{8}$$

In words, the slaves efficiently solve their sub-problems and update the master about Φ_i, then the master sets $u^{t+1}(r)$, and the current multipliers $\lambda_i^{t+1}(r)$, and communicates them back to the slaves for the next iteration. Unlike [30] who used dynamic programming to efficiently solve the slave problems, here we combine ADMM with the Branch-and-Bound algorithm. Interestingly, both of the additional terms contributed by the master problem to the slave problems, $\lambda_i(r)u(r)$, $(\Phi_i(r) - u(r))^2$ can be easily bounded using interval arithmetic, allowing for a straightforward incorporation into the original Branch-and-Bound method. With these changes we have observed similar convergence behavior as the one reported in [30]; In typically 15–20 (sometimes even less) ADMM iterations the slaves converge to a consistent pose estimate.

2.4   Deeply Supervised 2D- and 3D-Learning

We have observed substantial simplifications in the learning procedure by employing Deeply Supervised Network (DSN) [19] training. In particular we use loss functions that directly operate on the unary and pairwise terms, before these are integrated through structured prediction. We empirically observed that this substantially accelerates and robustifies learning, by helping the network come up with good 'proposals' to the subsequent combination stage.

As discussed in Sect. 2.1, the unary coordinates are obtained by adding the quantization and regression signals. Rather than expect this result to be correctly obtained only by back-propagation from the last layer, we also associate a classification and regression problem with each 2D image position.

We associate every pixel with a set of discrete labels corresponding to quantized depth values. For each joint we learn a different classification function; we consider a voxel as being positive if the respective joint is within certain proximity to the center of the 3D volume. We train this classifier using the cross-entropy loss. We also regress residual vectors between voxel centers and groundtruth joints using an L1 loss which is only active when a voxel is close enough to 3D landmarks.

For the pairwise terms, we regress vectors that point from each 3d joint to others. Similar to the unary coordinates, we regress these quantities in a fully-convolutional manner. The smooth L1 loss for the pairwise offsets between a specific joint and the rest of the joints is only active on pixels within certain proximity to the specific joint.

2.5   Training with a Structured Loss Function

Having outlined our cost function and our optimization algorithm, we now turn to parameter estimation. Our graphical model is defined in Eq. 2, and the pairwise terms are described in Eq. 3. As outlined in the preceding sub-sections, our network generates the unary terms $U_i(p_i^v)$, the nominal displacements $\mu_{i,j}$ and the 3D coordinates ϕ_i. In this section we describe training of all these parameters, as well as the calibration parameters c in Eq. 3, using a structured loss function [30, 32, 33] that reflects the geometric nature of the problem we want to address. Once our loss function is defined, back-propagation can be used to update all of the underlying network parameters.

While authors in [34] use an Intersection-over-Union (IoU) based structured loss for the task of detection, given that in this setup we have access to continuous ground truth values that naturally capture the underlying geometry of the problem, we opt for simplicity and use a more straightforward structured loss function.

Given that Φ denotes the 3D coordinates for a candidate configuration of parts (Eq. 1), and $\hat{\Phi}$ denotes the groundtruth 3D coordinates, we use the Mean Euclidean Distance, $\Delta(\hat{\Phi}, \Phi) = \frac{1}{P} \sum_{p=1}^{P} \|\phi_p - \hat{\phi}_p\|_2$ as a loss for our learning task, penalizing the 3D displacement of our estimated landmarks from their ground truth positions. As in standard structured output prediction, we use this loss to induce a set of constraints in pose space:

$$S(\hat{\Phi}) > S(\Phi) + \Delta(\hat{\Phi}, \Phi) \ \forall \Phi, \tag{9}$$

requiring that the score of the ground truth configuration should be greater than the score of any other configuration by a margin depending on how far the particular configuration is from the ground truth.

Since this cannot hold in general, we introduce slack variables $\xi : \xi(\Phi) = max(S(\Phi) + \Delta(\hat{\Phi}, \Phi) - S(\hat{\Phi}), 0)$. Thus, the slack variables represent the violations of the constraints in Eq. 9, and our goal here is to learn the model parameters that minimize the slack variables.

Standard training of structural SVMs [30, 32, 33] typically finds the most violated configuration given by $\Phi^* = \text{argmax}_\Phi(S(\Phi) + \Delta(\hat{\Phi}, \Phi) - S(\hat{\Phi}))$ and tries to reduce the violation of this configuration by updating the model parameters appropriately via the cutting-planes or Franke-Wolfe algorithm. In this work we use the standard stochastic gradient algorithm to minimize these slack variables. We do so by first finding K most violated configurations for each input sample (K is a hyper-parameter which affects the convergence speed; we set $K = 20$ based on experiments on a validation set). We then compute the sub-gradients of the model parameters with respect to each of these violated constraints and back-propagate them through the network.

3   Experimental Evaluation

3.1   Network Architecture

In our experiments we use a fully-convolutional 151 layer ResNet, with weights initialised from a model pre-trained on MPII for 2D body pose estimation [3]. Both 3D and 2D branches of our network are implemented as single-level convolution layers branching from the last layer of the ResNet. The input images to the system are cropped and rescaled to a fixed size of 320×320; the downsampling factor of our network is 16, leading to a cube of $20 \times 20 \times 20$ dimensions for 3D unary detection and residual regression branches and 20×20 spatial dimensions for the 2D branches.

3.2   Dataset

We use the largest available 3D human pose dataset Human3.6M (29) to train and evaluate our approach. The dataset consists of 3.6 million video frames of daily life activities performed by actors whose 3D joint locations are recorded by motion capture systems. Following the recent works in the literature, we have used frames from subjects S1, S5, S6, S7 and S8 for training and S9 and S11 for testing. We have used frames from all 4 cameras and all 15 actions in our training and testing in an action-agnostic manner. We have sub-sampled the videos at 10 frames per second. Several videos that suffer from *drift* of the groundtruth joints are removed from the dataset.

Due to the projective geometry, it is not possible to obtain "groundtruth data-cubes" from 3D poses. In particular, we cannot assume 3D points project to 2D points according to an orthogonal projection model. To cope with this issue, Pavlakos et al. [13] create a data-cube using image coordinates for x and y dimensions and real-world coordinates for the z dimension (distances relative to the root node). At test time the depth of the root node and the intrinsic camera parameters are used to obtain 3D pose estimates.

Unlike their approach, which requires knowledge of the root node's z-coordinate at test time, we estimate z coordinates such that the ratio of standard deviations of real-world and projected coordinates in x, y dimensions is preserved in the z dimension. This approximation naturally introduces some reconstruction error, but leads to a system that estimates pose up-to a similarity transform agnostic to the distance of the person to the camera and the intrinsic camera parameters.

3.3   Joint Training with 2D Pose

Our network is initialized with ResNet parameters obtained by training for 2D joint localization on the MPII dataset, but we observe that including samples from MPII as training samples increases performance - apparently not doing so results in the network forgetting about 2D joint localization. As in [35] we modify the labelled joints of the Human3.6M dataset in order to be able to utilize the

2D data. In particular we include a joint of "thorax" between shoulders that is connected to the "neck" and discarding "chin" and "abdomen" joints. The resulting skeleton structure is identical to the one of MPII. We have verified that two identical networks trained with baseline and MPII-type label structures lead to equivalent evaluation scores, thus it is fair to compare to existing methods. The active losses for an MPII sample are 2D detection and X and Y pairwise offset values, while the 3D position estimates are ignored.

3.4   Results

Since our groundtruth comes in the form of projected coordinates, we can obtain the 3D pose only up-to a similarity transform. We report "reconstruction error", which is measured as the mean euclidean distance to the ground truth, after applying Procrustes analysis.

Table 1. Comparison of average reconstruction errors for different graph topologies.

	Directions	Discussion	Eating	Greeting	Phoning	Photo	Posing	Purchases
UNARY alone	49.69	49.45	47.77	50.69	54.80	57.35	43.76	44.11
Center star	49.41	49.26	47.35	49.93	50.97	56.12	43.62	**43.43**
Stick figure	49.13	49.19	47.15	49.70	50.50	55.57	43.53	43.59
Extended stick figure	49.16	49.07	47.35	49.82	50.67	55.45	43.60	43.57
2-hop	**48.89**	**48.75**	**47.07**	**49.40**	**49.82**	**55.31**	**43.30**	43.47
	Sitting	Sit. down	Smoking	Waiting	Walk dog	Walking	Walk tog.	Average
UNARY alone	65.39	95.76	53.53	46.27	51.53	41.59	49.52	53.48
Center star	61.50	**78.09**	52.51	45.88	50.63	41.08	49.41	51.42
Stick figure	60.14	79.46	51.52	45.74	50.59	40.73	49.33	51.12
Extended stick figure	**59.94**	78.51	**51.42**	46.01	50.39	40.89	49.32	51.08
2-hop	60.48	78.20	51.69	**45.63**	**50.16**	**40.74**	**49.17**	**50.87**

We experimented with a number of graph topologies and notice that performance depends on the graph structure: **center star** describes the graph topology where all joints are connected to one central root node at the human's torso. It performs better than "unary only", indicating that the body center "knows" something about the other body parts.

Stick figure is a graph that directly corresponds to the human skeleton, i.e. the wrist is connected to the elbow, the elbow is connected to the shoulder, and so on. Clearly the shoulder knows better where the elbow has to be than the root node in the torso. This structure clearly performs better than "center star".

Extended stick figure is an extension to "stick figure", containing all its edges plus additional connections between the elbows of left and right arm, left and right knee, head to shoulders and torso to knees. This shows that additional loops boost performance, stabilizing against outliers or false evidence.

2-hop follows the human skeleton like "stick figure" and adds connections from every joint to its indirect (2-)neighbours in the skeleton. This connects, for example, hand with shoulder and ankle to hips and left to right knee. "2-hop" performs best, helping to resolve occlusions and improving accuracy.

Our experiments, reported in Table 1 clearly indicate that the 2-hop graph topology outperforms all of the other structures that we experimented with. This indeed justifies using approximate inference (ADMM), since these results require employing a loopy graph.

Table 2. A comparison of our approach to methods that report reconstruction error in literature.

	Average error
Yasin et al. [36]	108.3
Rogez and Schmid [37]	88.1
Tome et al. [8]	70.7
Pavlakos et al. [13][a]	53.2
(Ours) Unary	53.48
(Ours) ADMM	50.87

[a]Camera calibration parameters and real world z coordinate of root node assumed to be known at test time.

Table 3. Reconstruction errors for videos for specific cameras and test subjects in the Human3.6M dataset.

	Cam1	Cam2	Cam3	Cam4	Average
S-9	55.62	51.24	56.10	55.22	54.54
S-11	51.14	42.86	47.83	41.90	45.91
Average	53.72	47.64	52.59	49.57	

In Table 2 we compare the performance of our method to existing methods. Our results indicate that (a) our quantizatoin + regression-based unary network already delivers excellent results, at the level of the current state of the art. (b) Structured prediction yields an additional, quite substantial boost.

We note that there are some methods that only use a single camera or only S-11 frames as test samples and the rest of the videos for training. In order to compare our approach to such works, we present our results per camera and per

Fig. 3. Exemplar pose estimates by ADMM inference: Blue indicates the ground truth pose, whereas red and green is the solution obtained from "unaries alone" and ADMM respectively. (Color figure online)

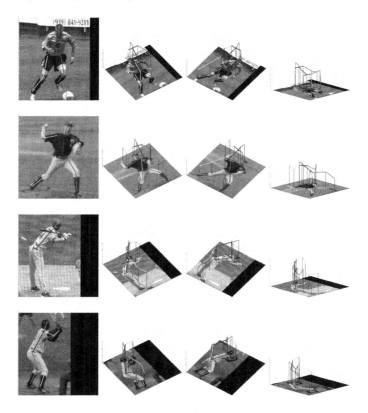

Fig. 4. Monocular 3D pose estimation results on LSP dataset.

subject in Table 3. Our results show that we are also outperforming the very recent work of [35], who uses only S-11 as test set and obtains 48.3, which is inferior with respect to our S-11 result (45.91), even though we have not used S-9 for training.

We provide qualitative results in Fig. 3, demonstrating cases where the ADMM inference clearly increases the pose estimation performance. Figure 4 shows some example images from the LSP dataset in the left column, augmented with the inferred body skeleton. The other three columns illustrate the plausible 3D structure as inferred by our approach.

4   Conclusion

In this work we have introduced an efficient method for 3D human pose estimation from 2D images. To this end, we augment the functionality of existing deep learning networks by adding a final layer that optimizes an energy function with variables in three dimensions. We have shown our method to deliver state-of-the-art 3D human pose estimation results, and intend to explore ways of extending our method to other tasks, such as general 3D object pose estimation. Another avenue of investigation is to further expand our optimization approach to handle even more general pairwise potentials. Furthermore we intend to make our optimization module software publicly available where it can be easily adopted for other applications in the computer vision community.

Acknowledgements. This work has been funded by the European Horizon 2020 programme under grant agreement no. 643666 (I-Support).

References

1. Toshev, A., Szegedy, C.: DeepPose: human pose estimation via deep neural networks. In: Proceedings of the IEEE Conference on Computer Vision and Pattern Recognition, pp. 1653–1660 (2014)
2. Pishchulin, L., Insafutdinov, E., Tang, S., Andres, B., Andriluka, M., Gehler, P.V., Schiele, B.: DeepCut: joint subset partition and labeling for multi person pose estimation. In: Proceedings of the IEEE Conference on Computer Vision and Pattern Recognition, pp. 4929–4937 (2016)
3. Insafutdinov, E., Pishchulin, L., Andres, B., Andriluka, M., Schiele, B.: DeeperCut: a deeper, stronger, and faster multi-person pose estimation model. In: Leibe, B., Matas, J., Sebe, N., Welling, M. (eds.) ECCV 2016. LNCS, vol. 9910, pp. 34–50. Springer, Cham (2016). https://doi.org/10.1007/978-3-319-46466-4_3
4. Cao, Z., Simon, T., Wei, S.E., Sheikh, Y.: Realtime multi-person 2D pose estimation using part affinity fields. arXiv preprint arXiv:1611.08050 (2016)
5. Newell, A., Yang, K., Deng, J.: Stacked hourglass networks for human pose estimation. In: Leibe, B., Matas, J., Sebe, N., Welling, M. (eds.) ECCV 2016. LNCS, vol. 9912, pp. 483–499. Springer, Cham (2016). https://doi.org/10.1007/978-3-319-46484-8_29
6. Wei, S.E., Ramakrishna, V., Kanade, T., Sheikh, Y.: Convolutional pose machines. In: Proceedings of the IEEE Conference on Computer Vision and Pattern Recognition, pp. 4724–4732 (2016)
7. He, K., Gkioxari, G., Dollár, P., Girshick, R.: Mask R-CNN. arXiv preprint arXiv:1703.06870 (2017)
8. Tome, D., Russell, C., Agapito, L.: Lifting from the deep: convolutional 3D pose estimation from a single image. arXiv preprint arXiv:1701.00295 (2017)
9. Chen, C.H., Ramanan, D.: 3D human pose estimation = 2D pose estimation + matching. arXiv preprint arXiv:1612.06524 (2016)
10. Bogo, F., Kanazawa, A., Lassner, C., Gehler, P., Romero, J., Black, M.J.: Keep it SMPL: automatic estimation of 3D human pose and shape from a single image. In: Leibe, B., Matas, J., Sebe, N., Welling, M. (eds.) ECCV 2016. LNCS, vol. 9909, pp. 561–578. Springer, Cham (2016). https://doi.org/10.1007/978-3-319-46454-1_34

11. Mehta, D., Rhodin, H., Casas, D., Sotnychenko, O., Xu, W., Theobalt, C.: Monocular 3D human pose estimation in the wild using improved CNN supervision. arXiv preprint arXiv:1611.09813v3 (2017)

12. Guler, A., et al.: Human joint angle estimation and gesture recognition for assistive robotic vision. In: Hua, G., Jégou, H. (eds.) ECCV 2016. LNCS, vol. 9914, pp. 415–431. Springer, Cham (2016). https://doi.org/10.1007/978-3-319-48881-3_29

13. Pavlakos, G., Zhou, X., Derpanis, K.G., Daniilidis, K.: Coarse-to-fine volumetric prediction for single-image 3D human pose. arXiv preprint arXiv:1611.07828 (2016)

14. Burenius, M., Sullivan, J., Carlsson, S.: 3D pictorial structures for multiple view articulated pose estimation. In: Proceedings of the IEEE Conference on Computer Vision and Pattern Recognition, pp. 3618–3625 (2013)

15. Tekin, B., Márquez-Neila, P., Salzmann, M., Fua, P.: Fusing 2D uncertainty and 3D cues for monocular body pose estimation. arXiv preprint arXiv:1611.05708 (2016)

16. Tekin, B., Katircioglu, I., Salzmann, M., Lepetit, V., Fua, P.: Structured prediction of 3D human pose with deep neural networks. CoRR abs/1605.05180 (2016)

17. Tompson, J.J., Jain, A., LeCun, Y., Bregler, C.: Joint training of a convolutional network and a graphical model for human pose estimation. In: NIPS (2014)

18. Yang, W., Ouyang, W., Li, H., Wang, X.: End-to-end learning of deformable mixture of parts and deep convolutional neural networks for human pose estimation. In: Proceedings of the IEEE Conference on Computer Vision and Pattern Recognition, pp. 3073–3082 (2016)

19. Lee, C., Xie, S., Gallagher, P.W., Zhang, Z., Tu, Z.: Deeply-supervised nets. In: AISTATS (2015)

20. Girshick, R.: Fast R-CNN. In: Proceedings of the IEEE International Conference on Computer Vision, pp. 1440–1448 (2015)

21. Ren, S., He, K., Girshick, R., Sun, J.: Faster R-CNN: towards real-time object detection with region proposal networks. In: Advances in Neural Information Processing Systems, pp. 91–99 (2015)

22. Guler, R.A., Trigeorgis, G., Antonakos, E., Snape, P., Zafeiriou, S., Kokkinos, I.: DenseReg: fully convolutional dense shape regression in-the-wild. In: The IEEE Conference on Computer Vision and Pattern Recognition (CVPR), July 2017

23. Papandreou, G., Zhu, T., Kanazawa, N., Toshev, A., Tompson, J., Bregler, C., Murphy, K.P.: Towards accurate multi-person pose estimation in the wild. CoRR abs/1701.01779 (2017)

24. Felzenszwalb, P.F., Huttenlocher, D.P.: Pictorial structures for object recognition. Int. J. Comput. Vis. **61**(1), 55–79 (2005)

25. Chen, X., Yuille, A.L.: Articulated pose estimation by a graphical model with image dependent pairwise relations. In: Advances in Neural Information Processing Systems, pp. 1736–1744 (2014)

26. Sapp, B., Toshev, A., Taskar, B.: Cascaded models for articulated pose estimation. In: Daniilidis, K., Maragos, P., Paragios, N. (eds.) ECCV 2010. LNCS, vol. 6312, pp. 406–420. Springer, Heidelberg (2010). https://doi.org/10.1007/978-3-642-15552-9_30

27. Kinauer, S., Berman, M., Kokkinos, I.: Monocular surface reconstruction using 3D deformable part models. In: Hua, G., Jégou, H. (eds.) ECCV 2016. LNCS, vol. 9915, pp. 296–308. Springer, Cham (2016). https://doi.org/10.1007/978-3-319-49409-8_24

28. Boyd, S., Parikh, N., Chu, E., Peleato, B., Eckstein, J.: Distributed optimization and statistical learning via the alternating direction method of multipliers. Found. Trends® Mach. Learn. **3**(1), 1–122 (2011)

29. Martins, A.F., Smith, N.A., Aguiar, P.M., Figueiredo, M.A.: Dual decomposition with many overlapping components. In: Proceedings of the Conference on Empirical Methods in Natural Language Processing, pp. 238–249. Association for Computational Linguistics (2011)

30. Boussaid, H., Kokkinos, I.: Fast and exact: ADMM-based discriminative shape segmentation with loopy part models. In: Proceedings of the IEEE Conference on Computer Vision and Pattern Recognition, pp. 4058–4065 (2014)

31. Komodakis, N., Paragios, N., Tziritas, G.: MRF optimization via dual decomposition: message-passing revisited. In: IEEE 11th International Conference on Computer Vision, ICCV 2007, pp. 1–8. IEEE (2007)

32. Joachims, T., Finley, T., Yu, C.N.J.: Cutting-plane training of structural SVMs. Mach. Learn. **77**(1), 27–59 (2009)

33. Pepik, B., Stark, M., Gehler, P.V., Schiele, B.: Multi-view and 3D deformable part models. IEEE Trans. Pattern Anal. Mach. Intell. **37**(11), 2232–2245 (2015)

34. Zhang, Y., Sohn, K., Villegas, R., Pan, G., Lee, H.: Improving object detection with deep convolutional networks via Bayesian optimization and structured prediction, pp. 249–258 (2015)

35. Sun, X., Shang, J., Liang, S., Wei, Y.: Compositional human pose regression. arXiv preprint arXiv:1704.00159 (2017)

36. Yasin, H., Iqbal, U., Kruger, B., Weber, A., Gall, J.: A dual-source approach for 3D pose estimation from a single image. In: Proceedings of the IEEE Conference on Computer Vision and Pattern Recognition, pp. 4948–4956 (2016)

37. Rogez, G., Schmid, C.: MoCap-guided data augmentation for 3D pose estimation in the wild. In: Advances in Neural Information Processing Systems, pp. 3108–3116 (2016)

Dominant Set Biclustering

Matteo Denitto[1]([✉]), Manuele Bicego[1], Alessandro Farinelli[1],
and Marcello Pelillo[2]

[1] Department of Computer Science, University of Verona, Verona, Italy
matteo.denitto@univr.it
[2] ECLT - University of Venice, Venice, Italy

Abstract. Biclustering, which can be defined as the simultaneous clustering of rows and columns in a data matrix, has received increasing attention in recent years, being applied in many scientific scenarios (e.g. bioinformatics, text analysis, computer vision). This paper proposes a novel biclustering approach, which extends the *dominant-set clustering* algorithm to the biclustering case. In particular, we propose a new way of representing the problem, encoded as a graph, which allows to exploit dominant set to analyse both rows and columns simultaneously. The proposed approach has been tested by using a well known synthetic microarray benchmark, with encouraging results.

1 Introduction

Biclustering, also widely known as co-clustering, can be defined as the simultaneous clustering of both rows and columns of a given data matrix [5,12,17]. With respect to clustering, the main differences of biclustering consist in the exploitation of *local information* (instead of global) *to retrieve subsets of rows sharing a "similar" behaviour in a subsets of columns, and vice versa* (instead of subsets of rows sharing a similar behaviour among *whole* the columns). Although bi-clustering was born and mainly applied to analyse gene expression microarray data [5,22], it has been recently exploited in a more various range of applications from clickstream data [18], passing by recommender systems [19], to different Computer Vision scenarios (such as facial expression recognition [16], motion and plane estimation [8]).

Different biclustering techniques have been proposed in the past – for a comprehensive review please refer to [12,17,22,23] – each one characterized by different features, such as computational complexity, effectiveness, interpretability and optimization criterion. Various of such previous approaches are based on the idea of adapting a given clustering technique to the biclustering problem, for example by repeatedly performing rows and columns clustering [10,14].

This paper follows the above-described research trend, and proposes a novel biclustering algorithm, which extends and adapts to the biclustering scenario the *dominant-set* based clustering. The concept of dominant set can be depicted from various points of view, since it involves optimization theory, graph theory,

M. Pelillo and E. Hancock (Eds.): EMMCVPR 2017, LNCS 10746, pp. 49–61, 2018.
https://doi.org/10.1007/978-3-319-78199-0_4

game theory and pattern recognition. Approaching it from a clustering perspective, given a set of objects V to group, a dominant set $C \subseteq V$ is a subset of objects with two well-defined properties: (i) all elements belonging to C should be highly similar to each other, and (ii) no larger cluster should contain C as a proper subset. Thus, C can be informally expressed as a *maximally coherent* set of data items [4,21]. Practically, a dominant set C is represented by a characteristic vector $\mathbf{x}$ where an entry x_i represents how likely the object v_i belongs to the retrieved cluster. In [4,21] authors provide a clustering algorithm based on dominant sets which is theoretically solid and supported by several experimental evaluations. Differently from classical clustering approaches, dominant-set clustering does not provide a partition of the data, and it can be successfully exploited in highly noisy contexts or scenarios with outliers. Moreover, dominant-set clustering can be exploited also in cases where the similarity matrix between objects is asymmetric. These two last considerations provide a solid link between dominant-set and classical biclustering approaches which, in most of the cases, do not provide a partition of the data matrix, and deal with non-squared matrices (hence, not symmetric) [17].

A first step toward the usage of dominant sets in the biclustering scenario has been presented in [26]. In this case authors propose to retrieve the bicluster by iteratively sorting and shifting the rows/columns of the given data matrix. Such sorting and shifting is made according to the dominant-set characteristic vector. However this algorithm do not exploit the potential of dominant set to group both rows and columns simultaneously, which is the core of classic biclustering algorithms and of this paper. Another approach similar in spirit is the one presented in [9]. In fact, authors of [9] tackle biclustering by retrieving *bi-cliques* on a bipartite-graph adjacency matrix. Although the technique derived shows some similarities with the one proposed in this paper, the resulting technique does not provide dominant sets as results and for this reason we do not present its details in this manuscript.

For this reason we decided to investigate how the concept of dominant set can be extended to the biclustering scenario. Among the different biclustering algorithm typologies, one branch involves the representation of the problem as a edge-cutting problem in a weighted bipartite graph, where one set of nodes represents rows and the other represents the columns [1,13]. However, a dominant set from a graph theory perspective is equivalent to *maximal clique* [4], and thus it cannot be applied directly on a bipartite graph (since a maximal clique in a bipartite graph is composed by only two nodes). We thus provide a novel simple graph representation for the biclustering problem. This involves the exploitation of data matrix entries as a *similarity* measure between rows-columns couples. The intuition behind this usage of the data matrix comes from the consideration that in various biclustering scenarios (such as recommender systems, gene expression analysis, clickstream data) the information encoded in the data matrix represents how much a row "is important" for that particular column. Moreover, to obtain a theoretically appropriate dominant set, we modified the bipartite graph adjacency matrix following the baselines provided in [21,29].

We evaluate the performance of the proposed approach on both synthetic and real datasets, favourably comparing with the current the state-of-the-art.

The remainder of paper is organized as follows: Sect. 2 introduces the dominant-set clustering approach, clearing the connections with other fields; our algorithm is then described in Sect. 3, whereas the experimental evaluation is given in Sect. 4; finally Sect. 5 concludes the paper.

2   Dominant Set Clustering

In this section we summarize the contributions provided in [4], providing the background knowledge concerning the dominant-set clustering algorithm.

Clustering is the problem of organizing a set of data elements into groups in a way that each group satisfies an *internal homogeneity* and *external inhomogeneity* property. Differently from classical clustering approaches the dominant-set algorithm faces the problem from a game theory perspective, instantiating a *non-cooperative clustering game* where the notion of a cluster turns out to be equivalent to a classical equilibrium concept from (evolutionary) game theory, as the latter reflects both the internal and external cluster conditions [4]. As discussed in Sect. 1, the internal condition asserts that elements belonging to the cluster should have high mutual similarities, whereas the external property claims that a cluster cannot be further extended by introducing external elements.

Formally, let $G = (V, E, \omega)$ be a weighted graph representing a clustering problem instance, where $V = \{1, \ldots, n\}$ is a finite set of vertices (representing the objects to group), $E \subseteq V \times V$ and $\omega : E \rightarrow \mathbb{R}$. We adopt $A_{ij} = \omega(i, j)$ to denote the graph adjacency matrix, representing the objects similarities. Given a non-empty subset of objects $C \subseteq V$, we define the *average weighted in-degree* of $i \in V$ with respect to C as:

$$\mathrm{awindeg}_C(i) = \frac{1}{|C|} \sum_{j \in C} A_{ij},$$

where $|C|$ is the cardinality of C. Also, if $j \in C$ we define

$$\phi_C(i, j) = A_{ij} - \mathrm{awindeg}_C(j),$$

which is a measure of the relative similarity of object i with object j with respect to the average similarity of object j with elements in C. Let us define the weight of an object i with respect to a set C:

$$W_C(i) = \begin{cases} 1, & \text{if } |C| = 1 \\ \sum_{j \in C \setminus \{i\}} \phi_{C \setminus \{i\}}(i, j) W_{C \setminus \{i\}}(j), & \text{otherwise.} \end{cases}$$

Please note that such definition is inductive and not circular, since in the sum j takes belongs to C excluding the interested point i. Now, we define the total weight of C as:

$$W(C) = \sum_{i \in C} W_C(i).$$

The weight $W_C(j)$ captures the strength of the coupling between vertex j and the other elements in the set relative to the overall coupling among the vertices. Such properties of the weighting are the basis for the formalizations of the notion of dominant set as a notion of a cluster.

Definition 1. *A non-empty subset of objects $C \subseteq V$ such that $W(T) > 0$ for any non-empty $T \subseteq C$, is said to be a* dominant set *if*

1. $W_C > 0$, *for all $i \in C$,*
2. $W_{C \cup i}(i) < 0$, *for all $i \notin C$.*

This definition provides conditions that correspond to the two main properties of a cluster: the internal and external conditions mentioned above.

Such formulation of clustering has connections with other scientific fields (namely optimization theory, graph theory and game theory), this links have been theoretically proven in [4]. The understanding of these links are needed to perceive how authors in [4] decide to compute a dominant set/cluster from the similarity matrix A. Moreover, the basis of how we decided to extend such approach to biclustering are theoretically founded in the theorems lying behind such connections.

Summarizing, in [4] they show that:

- with respect to *optimization theory*: a dominant set can be can be characterized in terms of local solutions pf the following standard quadratic program

$$
\begin{aligned}
\text{maximize} \qquad & f(\mathbf{x}) = \mathbf{x}^T A \mathbf{x} \\
\text{subject to} \qquad & \mathbf{x} \in \Delta \subset \mathbb{R}^n,
\end{aligned}
\tag{1}
$$

where

$$
\Delta = \left\{ \mathbf{x} \in \mathbb{R}^n : \sum_{j \in V} x_j = 1 \text{ and } x_j \geq 0 \text{ for all } j \in V \right\}.
$$

And, particularly, they show that if C is a dominant set of A, then its characteristic vector $\mathbf{x}^C$ is a strict local solution to (1). Conversely, if $\mathbf{x}^*$ is a strict local solution to (1) then its support $\sigma = \sigma(\mathbf{x}^*)$ is a dominant set of A. Where $\sigma(\mathbf{x})$ is defined as the index set of the positive components in $\mathbf{x}$.
- with respect to *graph theory*: a dominant set of A corresponds to a *maximal clique* in the corresponding graph. This means that the nodes corresponding to the support vector of $\mathbf{x}^C$ are a maximal clique, and thus they are all connected to each other and such clique cannot be expanded.
- with respect to *game theory*: if C a dominant set of A, then its characteristic vector $\mathbf{x}^C$ is an *Evolutionary Stable Strategy* of the corresponding *clustering game*[1].

---

[1] The idea is to setup a symmetric, non-cooperative game, called clustering game, between two players. Data points V are the strategies available to the players and the similarity matrix A encodes their payoff matrix.

Please for all the details and demonstration concerning these connections refer to [4].

Finally, once the problem has been theoretically instantiated, the authors of [4] provide two different approaches to retrieve a dominant set from A. Both strategies have roots in the game theory domain and it should not be surprising that methods developed in this context can be used to find dominant sets, given the tight relation that exists between dominant sets and the game-theoretic notion of equilibrium. The first involves *replicator dynamics* [24], whereas the other concerns *infection and immunization dynamics* [3].

3 Biclustering with Dominant Set

In this Section we provide our formulation of the biclustering problem, including the details of how dominant set are extended to such scenario.

As mentioned in Sect. 1, biclustering aims at the simultaneous clustering of rows and columns of a given data matrix. Formally, we denote as $D \in \mathbb{R}^{n \times m}$ the given data matrix, and let $R = \{1, \ldots, n\}$ and $K = \{1, \ldots, m\}$ be the set of row and column indices. We adopt D_{TL}, where $T \subseteq R$ and $L \subseteq K$, to represent the submatrix with the subset of rows in T and the subset of columns in L. Given this notation, we can define a *bicluster* as a submatrix D_{TL}, such that the subset of rows of D with indices in T exhibits a "coherent behavior" (in some sense) across the set of columns with indices in L, and vice versa. The choice of coherence criterion defines the type of biclusters to be retrieved (for a comprehensive survey of biclustering criteria, see [17,20]).

In this paper we propose to tackle biclustering exploiting the principles of dominant set definition. Although a preliminary approach toward this direction has already be presented in literature, authors of [26] present an iterative rows/columns clustering algorithm which does not fully exploit dominant set potentials. Specifically, the technique proposed in [26] defines a weighted correlation measure adopted to build a similarity matrix between the rows of the given data matrix. On such matrix authors apply dominant-set clustering and they exploit the characteristic vector $\mathbf{x}^C$ to sort the data matrix rows. This result in a data matrix where rows belonging to the bicluster are shifted to the bottom. At this point they compute a similarity matrix for the columns adopting $\mathbf{x}^C$ as weight for the correlation, giving more importance to the rows belonging to the biclusters. At this point dominant-set clustering is applied on the columns similarity matrix. The idea is that weighting the columns correlation with respect to the characteristic vector (computed on the rows) should help in retrieving a subset of columns acting similarly in that particular subset of rows. Columns are then shifted according to their characteristic vector and such operations is iteratively repeated twice for the rows and twice for the columns [26]. The resulting data matrix now contains the bicluster in the bottom-right position. To retrieve the actual bicluster authors compute the correlation between consecutive rows (starting from the bottom), and they stop when such correlation is below a certain threshold (same procedure applies for retrieving bicluster columns).

Hence, in [26], authors exploit the result of dominant-set clustering to iteratively order rows and columns to obtain a data matrix where the bicluster is isolated in the bottom-right portion of the matrix.

One branch of techniques exploits the weighted bipartite graph representation to face biclustering. In this context the common choice is to represent with two distinct sets of nodes the rows R and the columns K of the data matrix. Then, connect with edges only nodes belonging to different sets, and the weights are none other than the data matrix entries. Given this graph, the problem is thus formulated as an edge cutting problem where the surviving edges define the rows and the columns belonging to the bicluster [1]. Such cutting is obviously guided by a pre-defined objective function.

Alternatively to what previously presented, what we propose in this paper is to adopt a graph where rows and columns are represented by a unique set of nodes. Hence, given a data matrix D, we instantiate the biclustering problem as a graph $G = (V, E, \omega)$ where the vertices $V = \{1, \ldots, n+m\}$ represent the rows ($\{v_1, \ldots, v_n\}$) and columns ($\{v_{n+1}, \ldots, v_{n+m}\}$) of D. With this representation, we can easily encode the bipartite graph mentioned above. In fact, this can be obtained by introducing the data matrix D in correspondence of the positions connecting rows and columns in the adjacency matrix A. Practically, $A([1, \ldots, n], [n+1, \ldots, n+m]) = D$ and, to obtain a consistent adjacency matrix, we also define $A([n+1, \ldots, n+m], [1, \ldots, n]) = D^T$. The other portions of A, representing row-row and column-column similarities, are set to 0 (resulting in no edges connecting such vertices).

Once the bipartite graph is represented through a squared similarity matrix, we can now exploit dominant set definitions (usually applied in clustering context) to obtain a bicluster. In fact, since the built adjacency matrix contains high values only in row-columns positions (assuming a positive data matrix D), we expect the dominant set to be a group of rows presenting high similarities in a subset of columns (and vice versa). However, recalling the link between dominant set and graph theory (presented in Sect. 2), a dominant set of A is equivalent to a maximal clique in the correspondent graph. Hence, since the graph is bipartite (no row-row/column-column edges), a maximal clique is composed by just two nodes: one row and one column. Particularly, by adding another node we cannot have a clique since the subset of nodes will lack of one edge.

To overcome this, we add a negative value $-\alpha$ (where $\alpha \geq 0$) on the main diagonal of the similarity matrix A. This has been proved to be equivalent to solve a standard quadratic problem where the values of the off-diagonal entries of A are increased by α, and the main diagonal is set to 0 [21,29]. Note that adding α on the off-diagonal entries introduces edges between row-row and column-column nodes, resulting in a classic graph (not bipartite). Thus applying dominant set on this latter version of A results in an actual maximal clique, where a subset of rows will be selected simultaneously with a subset of columns. We depicted how the similarity matrix A is built in Fig. 1.

Intuitively, this is obtained because, independently by value adopted for α, the actual information is still contained in the rows-columns portions of A

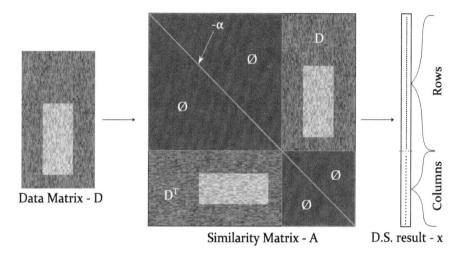

Fig. 1. Scheme of the algorithm

(in fact the value of α is the same for all the entries, and hence it is not informative). It has also been theoretically proved that increasing the value of α will increase the dimension of the resulting clique [4, 21, 29], as shown in Fig. 2.

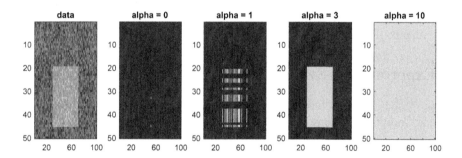

Fig. 2. Different results varying alphas

Summarizing, given a data matrix D with n rows and m columns we represent the biclustering problem as a graph having $n + m$ vertices, where the first n represent the rows and the remaining m represent the columns. Rows and columns are connected to each other with weights corresponding to the entries of D, whereas row-row and column-column edges have weights equal to α. A dominant set in such graph is a maximal clique isolating group of rows presenting high similarities in a group of columns, hence a bicluster.

To obtain such bicluster/dominant set we resort to *replicator dynamics*. The replicator dynamics are deterministic game dynamics that have been developed in evolutionary game theory. It considers an idealized scenario whereby individuals are repeatedly drawn at random from a large, ideally infinite, population to play a two-player game. In contrast to classical game theory, here players are not supposed to behave rationally or to have complete knowledge of the details of the game. They act instead according to an inherited behavioral pattern, or pure strategy, and it is supposed that some evolutionary selection process operates over time on the distribution of behaviors. Particularly, we adopt the iterative discrete-time replicator dynamics, which are given by

$$x_i(t+1) = x_i(t)\frac{(A\mathbf{x}(t))_i}{\mathbf{x}(t)^T A\mathbf{x}(t)}, \tag{2}$$

for $i \in V$. For further details concerning the theoretical basis relying under the connections between replicator dynamics and dominant set, we refer interested readers to the recent summary [4].

The resulting algorithm – called *Dominant Set Biclustering (DSB)* – is thus parsimonious in terms of space $\mathcal{O}(n+m)$ and efficient in terms of time of execution $\mathcal{O}(n+m)$. The parameters of such approach are: (i) the off-diagonal value α, (ii) the convergence of the replicator dynamics (which can be defined with a maximum number of iterations or with a threshold between consecutive changing of $\mathbf{x}$.

Please note that, although the method recovers one bicluster at a time (as widely exploited in literature [2,6,7]), there exists different heuristic to "mask" the obtained bicluster and to look for the next one. Specifically, to mask the retrieved bicluster we put zeros in the corresponding positions inside the adjacency matrix A.

4   Experimental Evaluation

The proposed approach has been evaluated using two sets of synthetic datasets and one Computer Vision dataset divided in two problems.

4.1   Synthetic Experiments

The two synthetic benchmarks are created to simulate gene expression matrices containing a single bicluster. In the first dataset the implanted biclusters are constant valued bicluster (we call this "Constant Bicluster Benchmark"), while in the second dataset additively coherent biclusters were used (we call this "Evolutionary Bicluster Benchmark").

In both cases, each matrix has been generated using the following procedure: (i) we generate a 50×50 matrix containing random values uniformly distributed between 0 and 1; (ii) we insert a constant valued (or additively coherent valued) bicluster, whose dimension was 25% of the matrix size, the bicluster was inserted in a random position; (iii) finally, the entire matrix has been perturbed with Gaussian noise. The standard deviation of the Gaussian noise is a percentage of the difference between the mean of the entries belonging to the bicluster and the mean of the background. 5 different noise levels (i.e. percentages) were used, ranging from 0 (no noise) to 0.2 (high noise). For each noise level, 30 matrices have been generated, resulting in a total of 75 matrices.

The quality of the retrieved biclusters have been assessed using two standard indices, also employed in [28]: (i) *purity*: percentage of points retrieved by the algorithms which actually belong to the real bicluster; (ii) *inverse purity*: percentage of points belonging to the true bicluster which have been retrieved by the algorithms. Calling C the bicluster found by the algorithm and L the ground truth, the indices are calculated as follows:

$$\text{Purity} = \frac{|C \cap L|}{|C|}, \qquad \text{Inverse Purity} = \frac{|L \cap C|}{|L|}.$$

The proposed approach has been compared with four other biclustering algorithms, including the preliminary one adopting dominant set (mentioned in Sect. 1). The results for the OOB, EBG and BAP algorithms have been taken from [8], whereas for WCC (the first approach resorting to dominant set) we implemented the code following the indications presented in [26] and adopting the suggested values for the parameters.

The results for the Constant and Evolutionary Bicluster benchmarks are shown in Fig. 3, where purity (a, c) and inverse purity (b, d) are displayed for the different methods, while varying the noise level. Each point represents the average over the 30 runs of the given noise level. In the plot, a full marker indicates that the difference between the considered method and the proposed approach is statistically significant[2].

The results evidently show that the proposed approach significantly outperforms the current state-of-the-art, especially when the level of noise increases, thus confirming the potentials of dominant sets in complex highly noisy situations. Concerning WCC, it is expectable that the performances on the constant bicluster benchmark are influenced by the exploitation of the weighted correlation coefficient. In fact, supposing to correctly select the columns involved in the bicluster, the behaviour of the bicluster in the selected columns and the one of the background is similar (since the value of the bicluster is constant). Thus, it is difficult for the method to differentiate between these two situations. This is also confirmed by the better performance of WCC in the evolutionary bicluster benchmark, where background and bicluster have different behaviours (background is constant and the bicluster evolves). However, in both cases the proposed approach provides better quality results, demonstrating that the exploitation of a more solid framework involving dominant set is sound.

Multiple Structure Recovery Dataset. *Multiple structure recovery* (MSR) concerns the extraction of multiple models from noisy or outlier-contaminated data. MSR is an important and challenging problem, which emerges in many computer vision applications [11,15,27]. In general, an instance of an MSR

---

[2] We performed a t-test for each noise level (on the result of the 30 matrices), we set the significance level to 5%.

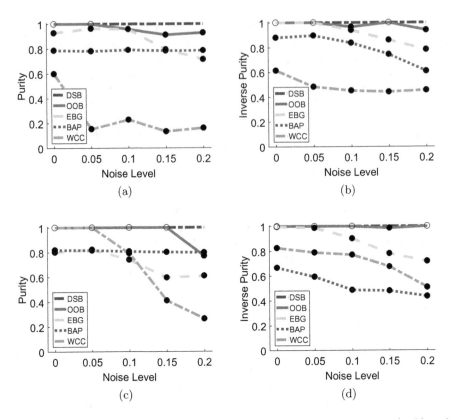

Fig. 3. Purity (a, c) and Inverse Purity (b, d) for matrices with constant (a, b) and additive coherent (c, d) biclusters

problem is represented by a *preference matrix* containing, in one dimension, the points under analysis, and in the other, the hypotheses/structures to which points should belong. The entry (i, j) in this matrix indicates how well a certain point i is represented by the given hypothesis/structure j.

The Adelaide dataset, which has been already exploited for assessing quality of biclustering algorithms [8], involves two type of MSR problems: motion and plane estimation. Given two different images of the same scene, where several objects move independently, motion segmentation aims at recovering subsets of point matches that undergo the same motion. Given two uncalibrated views of a scene, plane segmentation consists in retrieving the multi-planar structures by fitting homographies to point correspondences. The AdelaideRMF dataset[3] is composed of 38 image pairs (19 for motion segmentation and 19 for plane segmentation), with matching points contaminated by strong outliers. The ground-truth segmentations are also available. As in [8,25], we adopt the misclassification errors to assess the results.

---

[3] https://cs.adelaide.edu.au/~hwong/doku.php?id=data.

Table 1 presents the results. We report two different results for DSB; we run the algorithm varying the parameters, and on the basis of the considered results the performances can slightly vary. The last columns of Table 1 (*DSB best*) shows the results for DSB where we consider for each different matrix the best performance with respect to the misclassification error (varying the parameters). The results in the sixth column (*DSB best set*), which are slightly worse than the previous, are obtained by selecting the best set of parameters values minimizing the misclassification error (one for the motion segmentation and one for the plane estimation). Please note that two columns slightly differs, demonstrating that dominant sets strongly resist to both noise and the massive presence of outliers. With respect to other techniques, Table 1 shows that the proposed approach improves the results of the state-of-the-art in the plane segmentation dataset, and that we also provide comparable result on the motion segmentation dataset.

Table 1. Misclassification error (ME %) for motion segmentation (above) and planar segmentation (bottom). k is the number of models and % out is the percentage of outliers.

	k	%out	T-lnkg	RCMSA	RPA	DSB best set	DSB best
biscuitbookbox	3	37.21	3.10	16.92	3.88	10.42	6.17
breadcartoychips	4	35.20	14.29	25.69	7.50	5.48	5.48
breadcubechips	3	35.22	3.48	8.12	5.07	5.21	5.21
breadtoycar	3	34.15	9.15	18.29	7.52	11.44	11.44
carchipscube	3	36.59	4.27	18.90	6.50	4.24	4.24
cubebreadtoychips	4	28.03	9.24	13.27	4.99	9.48	9.48
dinobooks	3	44.54	20.94	23.50	15.14	14.16	14.16
toycubecar	3	36.36	15.66	13.81	9.43	16.00	16.00
biscuit	1	57.68	16.93	14.00	1.15	16.36	16.36
biscuitbook	2	47.51	3.23	8.41	3.23	2.63	2.63
boardgame	1	42.48	21.43	19.80	11.65	8.96	8.96
book	1	44.32	3.24	4.32	2.88	10.69	10.69
breadcube	2	32.19	19.31	9.87	4.58	11.57	9.50
breadtoy	2	37.41	5.40	3.96	2.76	3.12	3.12
cube	1	69.49	7.80	8.14	3.28	3.31	3.31
cubetoy	2	41.42	3.77	5.86	4.04	4.81	4.81
game	1	73.48	1.30	5.07	3.62	1.71	1.71
gamebiscuit	2	51.54	9.26	9.37	2.57	4.57	4.57
cubechips	2	51.62	6.14	7.70	4.57	7.04	7.04
mean			9.36	12.37	5.49	7.96	7.62
median			7.80	9.87	4.57	7.04	6.17

	k	%out	T-lnkg	RCMSA	RPA	DSB best set	DSB best
unionhouse	5	18.78	48.99	2.64	10.87	25.00	25.00
bonython	1	75.13	11.92	17.79	15.89	4.04	4.04
physics	1	46.60	29.13	48.87	0.00	2.83	0.94
elderhalla	2	60.75	10.75	29.28	0.93	5.14	2.80
ladysymon	2	33.48	24.67	39.50	24.67	10.54	10.54
library	2	56.13	24.53	40.72	31.29	13.95	13.95
nese	2	30.29	7.05	46.34	0.83	0	0
sene	2	44.49	7.63	20.20	0.42	0.40	0
napiera	2	64.73	28.08	31.16	9.25	13.24	13.24
hartley	2	62.22	21.90	37.78	17.78	3.12	1.56
oldclassicswing	2	32.23	20.66	21.30	25.25	8.44	8.44
barrsmith	2	69.79	49.79	20.14	36.31	51.03	51.03
neem	3	37.83	25.65	41.45	19.86	25.72	15.76
elderhallb	3	49.80	31.02	35.78	17.82	25.88	18.82
napierb	3	37.13	13.50	29.40	31.22	20.84	20.84
johnsona	4	21.25	34.28	36.73	10.76	20.37	20.37
johnsonb	7	12.02	24.04	16.46	26.76	19.87	19.87
unihouse	5	18.78	33.13	2.56	5.21	3.69	3.69
bonhall	6	6.43	21.84	19.69	41.67	38.76	38.76
mean			24.66	28.30	17.20	15.41	14.19
median			23.38	29.40	17.53	13.24	13.24

5 Conclusions

In this paper we proposed a novel algorithm facing the biclustering problem. Such algorithm extends the definition of dominant sets (already exploited for clustering) to the biclustering scenario. It involves a novel paradigm to represent the problem, coupled with solid theoretical basis. Dominant sets representing the bicluster are efficiently computed resorting to discrete-time replicator dynamics. The algorithm performances have been assessed on both synthetic and real datasets, providing better quality solutions when compared with the state-of-the-art.

References

1. Ahmad, W., Khokhar, A.: cHawk: an efficient biclustering algorithm based on bipartite graph crossing minimization. In: VLDB Workshop on Data Mining in Bioinformatics (2007)
2. Ben-Dor, A., Chor, B., Karp, R., Yakhini, Z.: Discovering local structure in gene expression data: the order-preserving submatrix problem. J. Comput. Biol. **10**(3–4), 373–384 (2003)
3. Bulò, S.R., Bomze, I.M.: Infection and immunization: a new class of evolutionary game dynamics. Games Econ. Behav. **71**(1), 193–211 (2011)
4. Bulò, S.R., Pelillo, M.: Dominant-set clustering: a review. Eur. J. Oper. Res. **262**(1), 1–13 (2017)
5. Cheng, Y., Church, G.: Biclustering of expression data. In: Proceeding of Eighth International Conference on Intelligent Systems for Molecular Biology (ISMB00), pp. 93–103 (2000)
6. Cheng, Y., Church, G.M.: Biclustering of expression data. In: ISMB, vol. 8, pp. 93–103 (2000)
7. Denitto, M., Farinelli, A., Figueiredo, M.A., Bicego, M.: A biclustering approach based on factor graphs and the max-sum algorithm. Pattern Recogn. **62**, 114–124 (2017)
8. Denitto, M., Magri, L., Farinelli, A., Fusiello, A., Bicego, M.: Multiple structure recovery via probabilistic biclustering. In: Robles-Kelly, A., Loog, M., Biggio, B., Escolano, F., Wilson, R. (eds.) S+SSPR 2016. LNCS, vol. 10029, pp. 274–284. Springer, Cham (2016). https://doi.org/10.1007/978-3-319-49055-7_25
9. Ding, C., Zhang, Y., Li, T., Holbrook, S.R.: Biclustering protein complex interactions with a biclique finding algorithm. In: 2006 Sixth International Conference on Data Mining, ICDM 2006, pp. 178–187. IEEE (2006)
10. Farinelli, A., Denitto, M., Bicego, M.: Biclustering of expression microarray data using affinity propagation. In: Loog, M., Wessels, L., Reinders, M., Ridder, D. (eds.) Pattern Recognition in Bioinformatics. Lecture Notes in Computer Science, vol. 7036, pp. 13–24. Springer, Heidelberg (2011). https://doi.org/10.1007/978-3-642-24855-9_2
11. Fitzgibbon, A.W., Zisserman, A.: Multibody structure and motion: 3-D reconstruction of independently moving objects. In: Vernon, D. (ed.) ECCV 2000. LNCS, vol. 1842, pp. 891–906. Springer, Heidelberg (2000). https://doi.org/10.1007/3-540-45054-8_58
12. Flores, J.L., Inza, I., Larraaga, P., Calvo, B.: A new measure for gene expression biclustering based on non-parametric correlation. Comput. Methods Programs Biomed. **112**(3), 367–397 (2013). http://www.sciencedirect.com/science/article/pii/S0169260713002605
13. Gao, B., Liu, T.Y., Zheng, X., Cheng, Q.S., Ma, W.Y.: Consistent bipartite graph co-partitioning for star-structured high-order heterogeneous data co-clustering. In: Proceedings of the Eleventh ACM SIGKDD International Conference on Knowledge Discovery in Data Mining, pp. 41–50. ACM (2005)
14. Getz, G., Levine, E., Domany, E.: Coupled two-way clustering analysis of gene microarray data. Proc. Nat. Acad. Sci. USA **97**(22), 12079–12084 (2000)
15. Häne, C., Zach, C., Zeisl, B., Pollefeys, M.: A patch prior for dense 3D reconstruction in man-made environments. In: 2012 Second International Conference on 3D Imaging, Modeling, Processing, Visualization and Transmission (3DIMPVT), pp. 563–570. IEEE (2012)

16. Khan, S., Chen, L., Zhe, X., Yan, H.: Feature selection based on co-clustering for effective facial expression recognition. In: 2016 International Conference on Machine Learning and Cybernetics (ICMLC), pp. 48–53. IEEE (2016)
17. Madeira, S., Oliveira, A.: Biclustering algorithms for biological data analysis: a survey. IEEE Trans. Comput. Biol. Bioinf. **1**, 24–44 (2004)
18. Melnykov, V.: Model-based biclustering of clickstream data. Comput. Stat. Data Anal. **93**, 31–45 (2016)
19. Mukhopadhyay, A., Maulik, U., Bandyopadhyay, S., Coello, C.A.C.: Survey of multiobjective evolutionary algorithms for data mining: part ii. IEEE Trans. Evol. Comput. **18**(1), 20–35 (2014)
20. Oghabian, A., Kilpinen, S., Hautaniemi, S., Czeizler, E.: Biclustering methods: biological relevance and application in gene expression analysis. PloS One **9**(3), e90801 (2014)
21. Pavan, M., Pelillo, M.: Dominant sets and hierarchical clustering, p. 362. IEEE (2003)
22. Pontes, B., Giráldez, R., Aguilar-Ruiz, J.S.: Biclustering on expression data: a review. J. Biomed. Inf. **57**, 163–180 (2015)
23. Prelic, A., Bleuler, S., Zimmermann, P., Wille, A., Bhlmann, P., Gruissem, W., Hennig, L., Thiele, L., Zitzler, E.: Comparison of biclustering methods: a systematic comparison and evaluation of biclustering methods for gene expression data. Bioinformatics **22**(9), 1122–1129 (2006)
24. Smith, J.M.: Evolution and the theory of games. In: Smith, J.M. (ed.) Did Darwin Get It Right?, pp. 202–215. Springer, Boston (1988). https://doi.org/10.1007/978-1-4684-7862-4_22
25. Soltanolkotabi, M., Elhamifar, E., Candès, E.J.: Robust subspace clustering. Ann. Stat. **42**(2), 669–699 (2014)
26. Teng, L., Chan, L.: Discovering biclusters by iteratively sorting with weighted correlation coefficient in gene expression data. J. Sig. Process. Syst. **50**(3), 267–280 (2008)
27. Toldo, R., Fusiello, A.: Image-consistent patches from unstructured points with J-linkage. Image Vis. Comput. **31**(10), 756–770 (2013)
28. Tu, K., Ouyang, X., Han, D., Honavar, V.: Exemplar-based robust coherent biclustering. In: SDM, pp. 884–895. SIAM (2011)
29. Zemene, E., Pelillo, M.: Interactive image segmentation using constrained dominant sets. In: Leibe, B., Matas, J., Sebe, N., Welling, M. (eds.) ECCV 2016. LNCS, vol. 9912, pp. 278–294. Springer, Cham (2016). https://doi.org/10.1007/978-3-319-46484-8_17

Bragg Diffraction Patterns as Graph Characteristics

Francisco Escolano[1]([✉]) and Edwin R. Hancock[2]

[1] Department of Computer Science and AI,
University of Alicante, 03690 Alicante, Spain
sco@dccia.ua.es
[2] Department of Computer Science, University of York,
York YO10 5GH, UK
erh@cs.york.ac.uk

Abstract. In this paper we establish a link between diffraction theory and graph characterization through the Schrödinger operator. This provides a natural way of characterizing wave propagation on a graph. In order to do so, we compute the spatio-temporal Fourier transform of the operator and then pack its spherical representation in a point of a Stiefel manifold. We show that when the temporal interval of analysis is set according to quantum efficiency principles the proposed approach outperforms the alternatives in graph discrimination.

Keywords: Diffraction · Schrödinger operator · Stiefel manifolds

1 Introduction

Graph characterization aims to provide a succinct way of representing graph structure that can be used distinguish or compare different types of graph, without applying graph or subgraph isomorphism (procedures that are known to be NP-complete). Popular and effective methods include random walks [1], the Ihara zeta function [2] and the spectral radius [3].

Of particular interest to us here is recent work based around the analysis of the heat kernel of a graph. The heat kernel is the solution of the heat equation on a graph, with the Laplacian matrix playing the role of conductivity matrix, i.e. controlling the flow of heat along edges with time. If the eigenvalues and eigenvectors of the Laplacian are known, then the heat kernel can be found by exponentiating the Laplacian eigensystem with time. The heat kernel determines the time evolution of a continuous time random walk on a graph, and this leads to several possible characterizations or signatures for graphs. Bai and Hancock [6] show that the moments of the heat kernel trace (i.e. its Mellin transform) are linked to the Riemann zeta function. Sun, Ovsjanikov and Guibas [11] histogram the elements of the heat kernel trace to compute the heat kernel signature, and use this for shape recognition. In a recent paper, Escolano et al. [4] introduced an alternative technique based on the analysis of the heat flow on a graph.

© Springer International Publishing AG, part of Springer Nature 2018
M. Pelillo and E. Hancock (Eds.): EMMCVPR 2017, LNCS 10746, pp. 62–75, 2018.
https://doi.org/10.1007/978-3-319-78199-0_5

Heat flow is derived from the heat kernel, which is the solution of the heat diffusion equation. It provides a method to represent the heat transfer between the nodes of a graph over time.

Closely related to this work on the heat kernel is the wave kernel signature (WKS) [7]. This involves histogramming the elements of the wave kernel, which is the solution of the complex wave equation or Schrödinger equation associated with the graph's Laplacian matrix. While the heat equation describes how heat is transferred in a system, the Schrödinger equation characterizes the dynamics of a particle in a quantum system. In fact, the continuous time quantum walk on a graph is the solution to the Schrödinger equation, with the normalised Laplacian playing the role of a Hamiltonian. In this setting, the quantum nature of the Schrödinger equation and its complex-valued solutions give rise to many interesting non-classical effects, including quantum interferences. These interferences have proved to be useful in several applications, including the detection of symmetric motifs in graphs via continuous-time quantum walks [8] and graph embedding by means of quantum commute times [9].

One difference between the approach in [7] and ours is that in the WKS, the time variable is not considered. In order to do so, the limiting average time behaviour (in the infinity) is computed. However, Rossi et al. [8] show that this choice is sub-optimal when used for measuring the similarity between two graphs. Alternatively, our approach relies on choosing proper finite limiting times. Our process is data driven (validated by the experiments) but herein we argue in favor of relating these limiting times with the *transport efficiency* of the quantum walk. Therefore, the long-term objective of this line of research is to choose the limiting times that *maximize transport efficiency.*

Another difference between WKS and our representation is that we do not compare the wave signatures between two nodes in the graph, but consider simultaneously all of them as forming a time-parameterized wave.

It must be stressed though that we do not claim that our method is in any sense a quantum algorithm. So we do not consider the issue of whether the characterization developed is observable or not. We are primarily interested in the complex nature of the characterization provided by the Schrödinger equation and the resulting non-stationarity and non-ergodicity of the dynamic system associated with it. Since the dynamic system is non-stationary and non-ergodic, it makes sense neither to characterize it using its steady state behaviour (since this does exist) nor its phase transitions (as is the case in the heat flow method). Instead we turn to the Fourier transform as a natural way of providing a frequency domain characterization of the time evolution of the complex wave equation, and use this instead of the heat flow trace [4].

The resulting representation relates frequency and graph structure by establishing a link between structural pattern analysis and diffraction theory. Diffraction theory is the basis of methods such as X-ray diffraction which allows molecular structure to be recovered from diffraction patterns. The seminal achievement here was the determination of the structure of DNA from Rosalind Franklin's skilfull diffraction imagery by Crick and Watson [12]. The diffraction pattern is a spatial pattern, and so it is an embedding of a graph on the 2D plane. The symmetry planes of the graph-structure are manifest as sets of geometrically regular frequency peaks. Thus we transform the characterzation of graph structure into a problem of searching for geometric regularity in a set of points.

In Sect. 2 we compare the Schrödinger operator with the heat kernel (both are governed by the eigensystem of the graph Laplacian). In Sect. 3 we analyze the operator and build a parametric representation from its spatio-temporal power spectrum. Such representation is transferred to a point in a suitable Stiefel manifold and principal angles are used for graph comparison. In Sect. 4 we show that the proposed approach outperforms state-of-the-art graph matching methods. Finally, in Sect. 5 we present our conclusions and future work.

2   Heat Kernel Vs. Schrödinger Operator

2.1   Heat Kernel

Let $G = (V, E)$ be an undirected graph where V is its set of nodes and $E \subseteq V \times V$ is its set of edges. The Laplacian matrix $L = D - A$ is constructed from the $n \times n$ adjacency matrix A with $n = |V|$, in which the element $A(u, v) = 1$ if $(u, v) \in E$ and 0 otherwise, where the elements of the diagonal $n \times n$ degree matrix are $D(u, u) = \sum_{v \in V} A(u, v)$. The $n \times n$ heat kernel matrix K_t is the fundamental solution of the heat equation

$$\frac{\partial K_t}{\partial t} = -LK_t, \tag{1}$$

and depends on the Laplacian matrix L and time t. The form of the heat kernel matrix is $K_t = e^{-Lt}$. The continuous time random walk starting at $p_0 \in \mathbb{R}^n$ evolves as $p_t = K_t p_0$, where p_t is the state of the random walk at time t. The spectral decomposition of the Laplacian is $L = \Phi \Lambda \Phi^T$, where $\Phi = [\phi_1 | \phi_2 | \ldots | \phi_n]$ is the $n \times n$ matrix of ordered eigenvectors according to the corresponding eigenvalues $0 = \lambda_1 \leq \lambda_2 \leq \ldots \leq \lambda_n$, and $\Lambda = diag(\lambda_1 \ \lambda_2 \ \ldots \ \lambda_n)$. Therefore, the spectral decomposition of the heat kernel is $K_t = \Phi e^{-\Lambda t} \Phi^T$ where $e^{-\Lambda t} = diag(e^{-\lambda_1 t} \ e^{-\lambda_2 t} \ \ldots \ e^{-\lambda_n t})$, that is, the heat kernel and the Laplacian share their eigenfunctions, which are contained in Φ. Both the columns and the rows of Φ define orthonormal basis: $\phi_i^T \phi_j = \delta_{ij}$.

2.2   Schrödinger Operator

The Schrödinger equation describes how the complex state vector $|\psi_t\rangle \in \mathbb{C}^n$ of a continuous-time quantum walk varies with time [10]:

$$\frac{\partial |\psi_t\rangle}{\partial t} = -iL|\psi_t\rangle. \tag{2}$$

Given an initial state $|\psi_0\rangle$ the latter equation can be solved to give $|\psi_t\rangle = \Psi_t |\psi_0\rangle$, where $\Psi_t = e^{-iLt}$ is a complex $n \times n$ *unitary matrix*. In this paper we refer to Ψ_t as the *Schrödinger operator* and we focus our attention on the operator itself and not on the quantum walk process. In this regard, Stone's theorem [13] establishes a one-to-one correspondence between a time parameterized unitary matrix U_t and a self-adjoint (Hermitian) operator $H = H^*$ such that there is a unique Hermitian operator satisfying $U = e^{itH}$. Such an operator H is the *Hamiltonian*. In the case of graphs we may set $H = -L$ and then we have that $\Psi_t = e^{-itL}$ is a unitary matrix for $t \in \mathbb{R}$.

Unitary matrices play a fundamental role in *characterizing complex wave equations* in a manner analogous to that performed by doubly stochastic matrices in characterizing diffusion processes. A $n \times n$ complex matrix U is unitary if $U^\dagger U = U U^\dagger = I_n$, where $U^\dagger$ is the conjugate transpose, that is $(A^\dagger)_{ij} = \overline{A_{ji}}$. Therefore, both the rows and columns of U form a orthonormal basis in $\mathbb{C}^n$. U is diagonalizable via the factorization $U = V \Lambda V^\dagger$ where Λ contains the complex eigenvalues of U and V is unitary and its columns contain the eigenvectors of U. Combining the latter diagonalization with the property $|det(U)| = 1$ we have that all the complex eigenvalues of U must lie on the unit Argand circle. They must have either the form $e^{i\theta}$ or $e^{-i\theta}$ where θ is a rotation angle. More precisely, for Ψ_t we obtain the spectral decomposition $\Psi_t = \Phi e^{-it\Lambda} \Phi^T$, where Φ contains the eigenvectors of L and $e^{-i\Lambda t} = diag(e^{-i\lambda_1 t} e^{-i\lambda_2 t} \ldots e^{-i\lambda_n t})$, the complex eigenvalues of Ψ_t, rely on the ones of the Laplacian.

Therefore, the Laplacian controls the dynamics of both the heat kernel and the Schrödinger operator according to the similarity between Eqs. 1 and 2. However, this similarity is misleading since Ψ_t is complex valued. The physical dynamics induced by the Schrödinger equation is therefore totally different from that of the heat equation, due to the existence of oscillations and interferences.

In this paper we address the question of whether the Schrödinger operator may be used to characterize the structure of a graph. Empirical analysis on different graph structures shows that both the heat kernel and the Schrödinger operator evolve with time in a manner which strongly depends on graph structure.[1] However, the underlying physics and the resulting dynamics are quite different (see Fig. 1 where for the heat kernel we represent $K_t(u, v)$ and for the Schrödinger operator we show the squared magnitude $|\Psi_t(u, v)|^2$). In the case of heat flow, heat diffuses between nodes through the edges, eventually creating transitive links (allowing effective energy exchange between nodes that are not directly connected by an edge), until reaching a stationary equilibrium state. The Schrödinger operator defines a wave which yields a faster energy propagation through the system (e.g. for a 100 nodes line graph, it takes $t = 50$ time steps for the Schrödinger operator to reach every possible position on the graph, taking more than twice this time in the case of the heat kernel [4]). Moreover, due to the negative components of the complex amplitudes, interferences are created, producing energy waves [5]. The main difference is that because of its wave nature the Schrödinger operator never reaches an equilibrium state. In other words, it is non-ergodic. Graph connectivity imposes constraints on the

---

[1] Videos showing the evolution of both heat kernel and Schrödinger operator are available at http://www.dccia.ua.es/~pablo/downloads/schrodinger_operator.zip.

distribution of energy. In the case of the heat kernel, a larger number of energy distribution constraints implies the creation of more transitive links with time [4]. This is true in the case of the Schrödinger operator, for which higher frequency and more symmetrical energy distribution patterns are also observed.

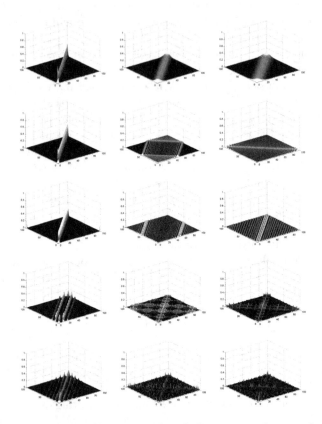

Fig. 1. Evolution with time ($t = 1, 25$ and 100). From top to bottom: heat kernel for a 100 node line graph, Schrödinger operator for a 100 node line graph, Schrödinger operator for a 100 node circle graph, Schrödinger operator for a 10×10 grid graph with 4 neighbour connectivity and Schrödinger operator for a 10×10 grid graph with 8 neighbour connectivity. (Courtesy of Pablo Suau)

3   Analysis of the Schrödinger Operator

3.1   Non-ergodicity

In order to explore the ergodicity of the Schrödinger operator we consider both its spectral decomposition $\Psi_t = \Phi e^{-it\Lambda}\Phi^T$ and that of the heat kernel $K_t = \Phi e^{-t\Lambda}\Phi^T$, that is

$$\Psi_t = \sum_{k=1}^{n} e^{-it\lambda_k} \phi_k \phi_k^T \text{ and } K_t = \sum_{k=1}^{n} e^{-t\lambda_k} \phi_k \phi_k^T, \qquad (3)$$

where λ_k is the k-th eigenvalue of the Laplacian L and ϕ_k its corresponding eigenvector. Therefore, both operators are specified by the eigenfunctions of the Laplacian but in a very different way. The spectral decomposition of the heat kernel demonstrates that it is dominated by the lowest eigenvalues, due to the fact that $\lim_{t\to\infty} e^{-t\lambda_k} = 0$. However, the limit of $e^{-it\lambda_k} = \cos(t\lambda_k) - i\sin(t\lambda_k)$ when t tends to infinity is undefined. Thus, there are two important differences with the heat kernel. Firstly, the Schrödinger operator never converges (it is non-ergodic), and secondly, it is not dominated by any particular eigenvalue. This is consistent with the well known physics of waves since the Schrödinger operator is a linear combination of waves.

3.2   Regimes Dynamics of Wave Propagation

The behavior of the Schrödinger operator at small and large times responds to different aspects of graph structure. At low t, the edge constraints contained in the Laplacian dominate (see left column in Fig. 1). At high t, on the other hand, it is the path structure that dominates (see the rest of the columns in Fig. 1).

In addition, the two regimes can be explained by the fact that the largest amplitudes occur at low frequencies. More precisely, each entry $\Psi_t(u,v)$ is described by a *linear combination of complex rotations*:

$$\Psi_t(u,v) = \begin{cases} \sum_{k=1}^{n} e^{-i\lambda_k t}\phi_k(u)\phi_k(v) & \text{if } u \neq v \\ \sum_{k=1}^{n} e^{-i\lambda_k t}\phi_k^2(u) & \text{otherwise.} \end{cases} \tag{4}$$

We let $z_k(u,v) = \phi_k(u)\phi_k(v)$ if $u \neq v$ and $z_k(u,v) = \phi_k(u)^2$ otherwise. In this case $z_k(u,v) \in \mathbb{R}$ for each value relies on the $k-$th eigenvector ϕ_k of the Laplacian. Since

$$|\Psi_t(u,v)|^2 = \sum_{k=1}^{n}\sum_{l=k}^{n} z_k(u,v)z_l(u,v)2\cos(t(\lambda_l - \lambda_k)) \tag{5}$$

we have that

$$\lim_{t\to 0} |\Psi_t(u,v)|^2 = 2\sum_{k=1}^{n}\sum_{l=k}^{n} z_k(u,v)z_l(u,v) \tag{6}$$

yields the maximal amplitude at (u,v) since $z_k(u,v)$ and $z_l(u,v)$ are time independent. As t increases the differences $\lambda_l - \lambda_k$, which are also time independent, become significant. They define lower or equal amplitudes and the characteristic frequency content of the wave emerges as expected. Low amplitudes dominate due to the ordering of the eigenvectors of the Laplacian $0 = \lambda_1 \leq \lambda_2 \leq \ldots \leq \lambda_n$ although there are always n terms where $\lambda_l = \lambda_k$. The latter property (dominance) is preserved as much as the graph is connected.

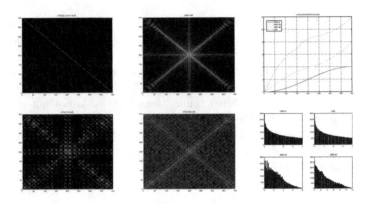

Fig. 2. Power spectra of the Schrödinger operator for different graphs of 400 nodes at $t = 25$: circle (loop) graph (top-left), line graph (top-middle), 20×20 grid graph with 4 neighbor connectivity (bottom-left) and 20×20 grid graph with 8 neighbor connectivity. For the latter graphs we also show their spectra (top-right) and the multiplicity of each value of Δ_{kl} (bottom-right).

3.3 Expressiveness of the Schrödinger Power Spectra

The discrete Fourier transform (DFT) of the squared magnitude of the Schrödinger Operator Ψ_t is

$$
\mathcal{F}_t(\omega_u, \omega_v) = \sum_{u,v=1}^{n} |\Psi_t(u,v)|^2 e^{-i(\omega_u u + \omega_v v)}
$$

$$
= \sum_{u,v=1}^{n} \left(\sum_{k=1,l=k}^{n} Z_{kl} \delta(t\Delta_{kl} - (\omega_u u + \omega_v v)) \right.
$$

$$
\left. + \sum_{k=1,l=k}^{n} Z_{kl} \delta(t\Delta_{kl} + (\omega_u u + \omega_v v)) \right), \tag{7}
$$

where ω_u and ω_v are the angular frequencies, $\Delta_{kl} = \lambda_l - \lambda_k \geq 0$, $Z_{kl} = z_k(u,v)z_l(u,v)$, and $\delta(.)$ is the Dirac delta function resulting from the Fourier transforms of $2\cos(t(\Delta_{kl})) = e^{it\Delta_{kl}} + e^{-it\Delta_{kl}}$ (see Eq. 5) for $k = 1, \ldots, n$, $l = k, \ldots, n$. After shifting we have that the amplitude $A_t(\omega_u, \omega_v) = |\mathcal{F}_t(\omega_u, \omega_v)|$ is given by pooling the values relying on $\sum_{k=1,l=k}^{n} Z_{kl}$ at all points (u,v) belonging to the lines $t\Delta_{kl} = \omega_u u + \omega_v v$ ($t\Delta_{kl}$ gives the distance to the origin and the vector $[\omega_u, \omega_v]^T$ is perpendicular to the direction of the line). Therefore the energy (power) distribution is determined by both the spectrum of the Laplacian, which defines the gaps Δ_{kl}, and its eigenvectors which define the values of $\sum_{k=1,l=k}^{n} Z_{kl}$.

In Fig. 2 we show the power spectra of the Schrödinger operators for scaled versions of the graphs analyzed in Fig. 1. These images resemble responses to diffraction gratings (interference patterns). In diffraction theory, interference patterns emerge when waves are bent around edges or slits. Constructive and destructive interferences occur producing alternate bright and dark fringes (see for instance the Young's experiment) which fade away from the center. The distribution of the so called Bragg's peaks (associated to constructive interferences) relies both on wavelengths and the number and spacings between the slits, as well as it also depends on the incidence angle. Fringes become sharper, for instance, as the number of slits is increased but, in this case, they are characterized by less and less significant maxima of intensity. In X-ray crystallography, the interdependence between the spatial distribution (e.g. a lattice) of the atoms, the properties of incident light and diffraction patterns is exploited to infer the tridimensional density of electrons in a crystal as well as to solve the structure of organic molecules like proteins. When applying this ideas to characterize pure topological structures like graphs, we realized that the Schrödinger operator provides a natural way of encoding the latter interdependences: the complex exponentiation of the Hamiltonian (the negative Laplacian) produces a wave equation completely determined by the spectrum and eigenvectors of such Hamiltonian. In addition, there is a correspondence between interference patterns and Fourier transforms. Actually, the Fourier transform in Eq. 7 has the same form of an aperture used in Fraunhofer diffraction: $a[\delta(x - S/2) + \delta(x + S/2)]$ where S is the distance between two slits. This gives us an interpretation of $A[\delta(t\Delta_{kl} - (\omega_u u + \omega_v v)) + \delta(t\Delta_{kl} + (\omega_u u + \omega_v v))]$ where $A = \sum_{u,v=1}^{n} \sum_{k=1,l=k}^{n} Z_{kl}$ in terms of the topological constraints that must be satisfied in order to produce Bragg's peaks. In our case, the role of the slits is played by the spectra (more precisely by the gaps Δ_{kl}) and the eigenvectors of the Laplacian. They determine what frequencies (energies in the power spectra) can be seen in the diffraction pattern. For instance, in circle (ring) graphs Fig. 2 (top-left) shows that the energy distribution may be constrained to lie at $0 = u + v$. For a line (path) graph (top-middle) we have a richer energy distribution although the line and circle graphs are quasi iso-spectral. Grid graphs are endowed with even richer diffraction patterns (larger range of eigenvalues).

The above rationale can be summarized as follows. Graphs produce diffraction patterns in the power spectra imposed by their Laplacians. This intuition is again borrowed from physics. Therefore, the analysis of the the Schrödinger operator can be posed in terms of analyzing its power spectra in order to explain the distribution of the different frequency amplitudes and their meaning. So far we have provided a geometric interpretation. In this regard, it is key to find the relation, if any, between the anisotropies in the power spectra and the lack of regularity in the structure. Such anisotropy is poorly contemplated in well known models of holistic image characterization [14,15]. However, despite existing models in image characterization are not directly applicable to describe graph-based diffraction patterns, the underlying methodology (including PCA/SVD eigenspaces) can be extended to incorporate anisotropy. In order to do that we will exploit the spatio-temporal nature of the Schrödinger operator.

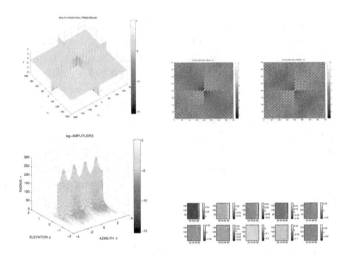

Fig. 3. Spatio-temporal power spectra of the Schrödinger operator for a 20×20 grid graph with 4 neighbor connectivity. Top-left: planes $\omega_u = 0$, $\omega_v = 0$, $\omega_{t'} = 0$. Top-center/right: detail of $\omega_{t'} = 0$ and $\omega_{t'} = 3$ showing parallel high pooled lines. Bottom-left: spherical coordinates of log-amplitudes. Bottom-right: 10 principal eigenvectors of the $\theta - \phi$ space.

3.4 Characterization of the Spatio-Temporal Schrödinger Power Spectra

Let $\mathcal{F}(\omega_u, \omega_v, \omega_t)$ be the spatio-temporal DFT of $|\Psi(u,v,t)|^2$. It is straightforward to extend Eq. 7 to include time variation. As expected, after shifting the transform we have that the amplitudes $A(\omega_u, \omega_v, \omega_t) = |f(\omega_u, \omega_v, \omega_t)|$ are given by pooling the values relying on $\sum_{k=1,l=k}^{n} Z_{kl}$ at all points (u,v,t) belonging to the planes $t\Delta_{kl} = \omega_u u + \omega_v v + \omega_t t$. Furthermore, since the gaps Δ_{kl} are time independent, scaled temporal frequencies $\omega_t t$ can be seen as offsets in the spatial constraints $t\Delta_{kl} = \omega_u u + \omega_v v$. Such offsets are needed to explain the spatio-temporal behavior of the Schrödinger operator. More precisely, for $t > 0$ and $\omega_t \neq 0$ only the contributions $\sum_{k=1,l=k}^{n} Z_{kl}$ at (u,v,t) where (u,v) do not satisfy $t\Delta_{kl} = \omega_u u + \omega_v v$ are taken into account for computing the amplitudes $A(\omega_u, \omega_v, \omega_t)$.

An interesting particular case of the latter rationale is to pool amplitudes from (u,v) satisfying constraints which are orthogonal to the spatial ones. For instance, in Fig. 3 (top-left) we show the spatio-temporal log-amplitudes for the planes $w_u = 0$, $w_v = 0$ and $w_{t'} = 0$ where $t' = t - T/2$ being $[0,T]$ the temporal interval of analysis. The graph analyzed is the 20×20 grid with 4 neighbors connectivity. In Fig. 3 (top-center/right) we show respectively the planes $\omega_{t'} = 0$ and $\omega_{t'} = 3$. Both of them are characterized by high log-amplitudes at lines $w_v = w_u \pm k$, with $k \geq 0$, which are orthogonal to those which have a similar degree of pooling at a particular t' (see Fig. 2 (bottom-left)). The highest pooling is obtained at $k = 0$, that is $w_u = w_v$, and it decreases as $|k|$ increases.

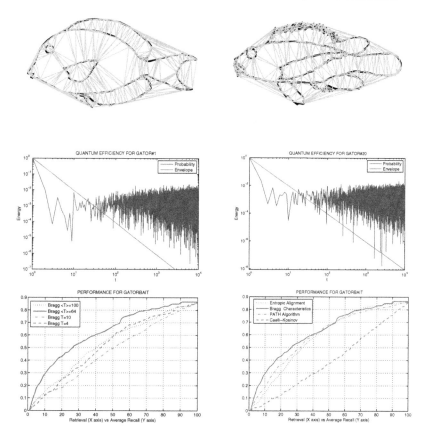

Fig. 4. Experimental results. Top: delaunay graphs of Gatorbait - Gator#1 (left) vsGator#20 (right). Middle: quantum efficiency for Gator#1 (left) vs QE for Gator#20 (right); The intercept for the first one is smaller almost one order of magnitude that that of the second. Bottom: average Recall vs Retrieval curves. Best performance is achieved with $T = 64$ on average, even when larger values of T are assumed (left). The result outperforms state-of-the-art graph matching algorithms (right). In all cases classes with only one element are not considered for measuring the performance.

This happens for $sign(\omega_u) = sign(\omega_v)$. Otherwise we have the inverse case: log-amplitudes increase with $|k|$ (at $\omega_{t'} = 0$ such increase is more spatially constrained than at $\omega_{t'} = 3$).

Once the role of temporal frequencies is clarified, it is convenient to change the coordinate system in order to better visualize the angular asymmetries in the spatio-temporal domain (anisotropy). Given $(\omega_u, \omega_v, \omega_t)$ its spherical coordinates are given by (r, θ, ϕ) where $r = \sqrt{\omega_u^2 + \omega_v^2 + \omega_t^2}$ is the radius, $\theta = \tan^{-1}(\frac{\omega_v}{\omega_u})$, $-\pi \leq \theta \leq \pi$ is the azimuthal angle in the $\omega_u - \omega_v$ plane and $\phi = \cos^{-1}(\frac{\omega_t}{r})$, $-\frac{\pi}{2} \leq \phi \leq \frac{\pi}{2}$ is the elevation angle. Therefore, r encodes the magnitude of the spatio-temporal frequencies, θ refers to the relation between spatial frequencies and ϕ gives the relative importance of temporal frequencies. In addition, for a given pair $\alpha_s = (\theta, \phi)$ the power spectrum $A(\alpha_s)^2$ decays with r and such decay does not follows, in general, a power law. In addition, for $\alpha_{s+\Delta} = (\theta + \Delta, \phi + \Delta)$, with $|\Delta| > 0$ as small as possible, we have that $A(\alpha_{s+\Delta})^2$ differs significantly from $A(\alpha_s)^2$ in the general case (directional anisotropy).

In Fig. 3 (bottom-left) we plot the $r - \theta - \phi$ space for the log-amplitudes. The representation is symmetric with respect to the elevation axis $\theta = 0$ and it is periodic with respect to the azimuthal axis $\phi = 0$. Therefore, for the sake of computational efficiency we can define a discrete $\theta - \phi$ *elevation-azimuth space* by setting: $\theta \in [0, \pi/2]$, $\phi \in [0, \pi]$. Such space relates spatial and temporal frequencies. In addition, for each discrete radius $r \in [0, r_{max}]$, where $r_{max} = n/2$, we define a *sample space* X_r as the log-amplitudes $\log A(r, \alpha_s)$ at all coordinates of the $\theta - \phi$ parametric space. Performing SVD/PCA analysis on the set of sample spaces $\mathcal{S} = \{X_r\}$ the principal eigenvalues $\lambda_1 \geq \lambda_2 \geq \ldots \geq \lambda_p$ with $p \ll d$, where $d = \delta_\theta \times \delta_\phi$ is the number of cells, encode the degree of directional anisotropy. Their associated d–dimensional eigenvectors $u_1, u_2, \ldots, u_p$ define a subspace where we compress all the spatio-temporal information of the operator.

The latter representation allows us to map a graph to a multi-layered parametric space (one layer per radial samples). Then, such space is encoded by a set of eigenvectors as it is done when analyzing image sequences. In this regard, it becomes very useful to consider each set of eigenvectors (subspace) as a point in a given manifold in order to exploit the geodesics defined in it. The natural choice is to consider that $U = [u_1 \, u_2 \, \ldots \, u_p]$ is a point in the Stiefel manifold $St(p, d) = \{U \in \mathbb{R}^{d \times p} : U^T U = I_p\}$, that is, the set of $d \times p$ matrices with orthonormal columns [16]. In Fig. 3 (bottom-right) we show the first $p = 10$ eigenvectors which define the Stiefel point associated to the 20×20 4N grid graph. Given the spatial structure of log-amplitudes in spherical coordinates, global details appear close to $\pi/2$ in the azimuthal axis whereas local details are highlighted at lower values.

Stiefel manifolds are endowed with a Riemannian mathematical structure and therefore it is more convenient to define dissimilarities beyond the Frobenius or Euclidean distances, that is, which account for the curvature of the manifold where these subspaces lie [19, 20]. One of these dissimilarities relies on the concept of principal angle [17, 18]. Given two points U and V in $St(p, d)$, the principal angles $0 \leq \theta_1 \leq \theta_2 \leq \ldots \leq \theta_p \leq \pi/2$ between the subspaces $Span(U)$ and $Span(V)$ satisfy that $\cos(\theta_i)$ are the singular values of $U^T V$ and the geodesic distance between U and V is given by $||\Theta||$ where $\Theta = [\theta_1 \, \theta_2 \, \ldots \, \theta_p]$. In this paper, given two graphs $G_X = (V_X, E_X)$ and $G_Y = (V_Y, E_V)$ and the Stiefel points U_X and U_Y derived from the corresponding spatio-temporal Schrödinger power spectra, we will quantify the dissimilarity between the two graphs in terms of the principal angles.

4   Experiments

4.1   GatorBait Database and Quantum Efficiency

In order to test our graph characterization method we use the *GatorBait_100* ichthyology database. GatorBait has 100 shapes representing fishes from 30 different classes. Shapes are discretized and then Delaunay triangulation graphs (included in the publicly accessible UA Graph Database[2]) are retained for testing graph comparison/matching algorithms [21].

Gaps distribution is very important since it is known that the transport efficiency of quantum walks relies on it [23]. More precisely,

$$|\bar{a}(t)|^2 = |(1/n) \sum_{k=1}^{n} e^{-it\lambda_k}|^2$$

is the probability that a continuous-time quantum walk returns to the origin at t. Such quantity is usually characterized by two regimes: for low-mid values of t it decreases; at higher values, quantum oscillations around the long-term average dominate. This happens if the probability density function over the Δ_{kn} (the larger gaps) follows a power-law distribution, which is a mild assumption for Delaunay graphs. If so, the temporal range where quantum oscillations vanish is bounded by the intercept of $t^{-2(1+\nu)}$ (the so called *envelope* of the process) where ν is the power exponent. The smaller the intercept the higher the efficiency. This allows us to set the optimal value for T within the range of the envelope. The intercept for each graph induces a partial order that can be used for scaling T (see Fig. 4 (top) where vertical axes are fixed according to the minimal values of $|\bar{a}(t)|^2$). It also explains why too-low or too-high uniform values of T produce less discriminative characterizations that setting $T = 64$ *on average* (Fig. 4 (bottom-left)).

4.2   Comparison with Graph Matching Algorithms

In Fig. 4 we compare the discriminability of or characterization with state-of-the-art graph matching algorithm like Entropic Manifold Alignment [21,22] (which outperforms many others), the PATH algorithm [24] and the Caelli-Kosinov spectral algorithm [25], not evaluated in previous experiments. Their cost functions or associated kernels are used for estimating similarity after alignment. In Fig. 4 (bottom-right) we show that our approach (setting $T = 64$ on average) outperforms the alternatives in terms of average recall in the part of the curve where the number of retrievals is small or medium. Only when a high number of retrievals is allowed (usually avoided in practice) the alternatives slightly improve our characterization.

---

[2] http://www.rvg.ua.es/graphs/dataset01.html.

5   Conclusion

In this paper we have proposed the use of Bragg diffraction patterns to characterize graphs. The representation of the spatio-temporal Fourier transform of the Schrödinger operator in terms of a Stiefel points produces high discrimination rates provided that quantum efficiency is considered. Future works include the formulation of finding the optimal T that maximizes quantum transport efficiency.

Acknowledgements. F. Escolano is funded by the project TIN2015-69077-P of the Spanish Government.

References

1. Aziz, F., Wilson, R.C., Hancock, E.R.: Graph characterization via backtrackless paths. In: Pelillo, M., Hancock, E.R. (eds.) SIMBAD 2011. LNCS, vol. 7005, pp. 149–162. Springer, Heidelberg (2011). https://doi.org/10.1007/978-3-642-24471-1_11
2. Peng, R., Wilson, R., Hancock, E.: Graph characterization via ihara coefficients. IEEE Trans. Neural Netw. **22**(2), 233–245 (2011)
3. Das, K.C.: Extremal graph characterization from the bounds of the spectral radius of weighted graphs. Appl. Math. Comput. **217**(18), 7420–7426 (2011)
4. Escolano, F., Hancock, E., Lozano, M.A.: Heat diffusion: thermodynamic depth complexity of networks. Phys. Rev. E **85**(3), 036206(15) (2012)
5. Rossi, L., Torsello, A., Hancock, E.R., Wilson, R.C.: Characterizing graph symmetries through quantum Jensen-Shannon divergence. Phys. Rev. E **88**(3), 032806(9) (2013)
6. Xiao, B., Hancock, E., Wilson, R.: Graph characteristics from the heat kernel trace. Pattern Reogn. **42**(11), 2589–2606 (2009)
7. Aubry M., Schlickewei, U., Cremers, D.: The wave kernel signature: a quantum mechanical approach to shape analysis. In: IEEE International Conference on Computer Vision (ICCV), Workshop on Dynamic Shape Capture and Analysis (4DMOD) (2011)
8. Rossi, L., Torsello, A., Hancock, E.: Approximate axial symmetries from continuous time quantum walks. In: Gimel'farb, G. (ed.) SSPR/SPR 2012. LNCS, pp. 144–152. Springer, Heidelberg (2012). https://doi.org/10.1007/978-3-642-34166-3_16
9. Emms, D., Wilson, R.C., Hancock, E.: Graph embedding using quantum commute times. In: Escolano, F., Vento, M. (eds.) GbRPR 2007. LNCS, vol. 4538, pp. 371–382. Springer, Heidelberg (2007). https://doi.org/10.1007/978-3-540-72903-7_34
10. Farhi, E., Gutmann, S.: Quantum computation and decision trees. Phys. Rev. A **58**, 915–928 (1998)
11. Sun, J., Ovsjanikov, M., Guibas, L.J.: A concise and provably informative multiscale signature based on heat diffusion. Comput. Graph. Forum **28**(5), 1383–1392 (2009)
12. Watson, J.D., Crick, F.H.C.: A structure for deoxyribose nucleic acid. Nature **171**(4356), 737–738 (1953)
13. Stone, M.H.: On one-parameter unitary groups in Hilbert space. Ann. Math. **33**(3), 643–648 (1932)

14. Oliva, A., Torralba, A.: Modeling the shape of a scene: a holistic representation of the spatial envelope. Int. J. Comput. Vis. **42**(3), 145–175 (2001)
15. Torralba, A., Oliva, A.: Statistics of natural image categories. Network **14**, 391–412 (2003)
16. Edelman, A., Arias, T.A., Smith, S.T.: The geometry of algorithms with orthogonality constraints. SIAM J. Matrix Anal. Appl. **20**(2), 303–353 (1999)
17. Kim, T.-K., Kittler, J., Cipolla, R.: Discriminative learning and recognition of image set classes using canonical correlations. IEEE Trans. Pattern Anal. Mach. Intell. **29**(6), 1005–1018 (2007)
18. Absil, P.-A., Mahony, R., Sepulchre, R.: Optimization Algorithms on Matrix Manifolds. Princeton University Press, Princeton (2008)
19. Turaga, P.K., Veeraraghavan, A., Srivastava, A., Chellappa, R.: Statistical computations on Grassmann and Stiefel manifolds for image and video-based recognition. IEEE Trans. Pattern Anal. Mach. Intell. **33**(11), 2273–2286 (2011)
20. Harandi, M.T., Sanderson, C., Shirazi, S.A., Lovell, B.C.: Graph embedding discriminant analysis on Grassmannian manifolds for improved image set matching. In: CVPR 2011, pp. 2705–2712 (2011)
21. Escolano, F., Hancock, E.R., Lozano, M.A.: Graph matching through entropic manifold alignment. In: CVPR 2011, pp. 2417–2424 (2011)
22. Escolano, F., Hancock, E.R., Lozano, M.A.: Graph similarity through entropic manifold alignment. SIAM J. Imaging Sci. **10**(2), 942–978 (2017)
23. Mülken, O., Blumen, A.: Continuous-time quantum walks: models for coherent transport on complex networks. Phys. Rep. **502**(2–3), 37–87 (2011)
24. Zaslavskiy, M., Bach, F., Vert, J.-P.: A path following algorithm for the graph matching problem. IEEE Trans. PAMI **31**(12), 2227–2242 (2009)
25. Caelli, T., Kosinov, S.: An eigenspace projection clustering method for inexact graph matching. IEEE Trans. PAMI **26**(4), 515–519 (2004)

Motion and Tracking

Variational Large Displacement Optical Flow Without Feature Matches

Michael Stoll[(✉)], Daniel Maurer, and Andrés Bruhn

University of Stuttgart, 70569 Stuttgart, Germany
{stoll,maurer,bruhn}@vis.uni-stuttgart.de
http://cvis.visus.uni-stuttgart.de/index.shtml

Abstract. The optical flow within a scene can be an arbitrarily complex composition of motion patterns that typically differ regarding their scale. Hence, using a single algorithm with a single set of parameters is often not sufficient to capture the variety of these motion patterns. In particular, the estimation of large displacements of small objects poses a problem. In order to cope with this problem, many recent methods estimate the optical flow by a fusion of flow candidates obtained either from different algorithms or from the same algorithm using different parameters. This, however, typically results in a pipeline of methods for estimating and fusing the candidate flows, each requiring an individual model with a dedicated solution strategy. In this paper, we investigate what results can be achieved with a pure variational approach based on a standard coarse-to-fine optimization. To this end, we propose a novel variational method for the simultaneous estimation and fusion of flow candidates. By jointly using multiple smoothness weights within a single energy functional, we are able to capture different motion patterns and hence to estimate large displacements even without additional feature matches. In the same functional, an intrinsic model-based fusion allows to integrate all these candidates into a single flow field, combining sufficiently smooth overall motion with locally large displacements. Experiments on large displacement sequences and the Sintel benchmark demonstrate the feasibility of our approach and show improved results compared to a single-smoothness baseline method.

1 Introduction

The estimation of optical flow has been a core problem in computer vision for decades. Many successful methods for solving this task belong to the class of variational approaches. Based on the minimization of a continuous energy functional consisting of a data and a smoothness term, such methods offer dense and sub-pixel accurate results as well as a transparent modelling. Since the pioneering work of Horn and Schunck [14], a lot of progress has been made on both the modelling and the optimization side. On the modelling side, modern smoothness priors allow the estimation of flow fields with both gradual transitions [6,11] and sharp motion discontinuities [16,27], while modern data terms cope with noise [4], outliers and varying illumination [11]. On the optimization side, coarse-to-fine schemes [17] have been proposed that allow to handle large displacements

M. Pelillo and E. Hancock (Eds.): EMMCVPR 2017, LNCS 10746, pp. 79–92, 2018.
https://doi.org/10.1007/978-3-319-78199-0_6

of relatively large objects. Fast motions of small objects, however, are still hard to capture.

One widely used approach to alleviate this problem is the integration of descriptor matches [8, 21, 25]. While such methods allow to handle arbitrarily large motion, they heavily rely on the uniqueness of the underlying descriptors. Hence, in case of weakly textured regions or repetitive patterns, such methods are likely to produce false matches which deteriorate the final optical flow estimation. Recent approaches face this problem by applying a-posteriori regularization to a set of descriptor matches in order to improve its quality [12].

While there are scenarios where the large displacement problem of small objects is intrinsically unsolvable – e.g. in the presence of multiple non-unique instances – a surprisingly large share of large displacement cases can actually be solved even *with* a-priori regularization, i.e. regularization during the estimation of the matches. In order to understand in which cases large displacements can still be recovered correctly, we have to distinguish two scenarios: (i) One problem is that small objects may not be present on that coarse-to-fine level that is necessary for the estimation of the displacement [8]. This case cannot be handled by standard coarse-to-fine optimization without further data transformations [18]. (ii) Another problem – which, however, has hardly been adressed in the literature so far – is the influence of the balance between data and smoothness term on the estimation of large displacements. For small objects that undergo large displacements, it is typically cheaper to violate the constancy assumptions in the data term (due to the small spatial extent) than to violate the regularity assumptions in the smoothness term (due to the large motion gradient). This particularly holds for large values of the smoothness parameter that are typically required to obtain noise-free flow fields. So even if there is sufficient data on the appropriate coarse-to-fine level, the smoothness term will suppress the estimation of the corresponding large displacement.

In this context, Brox and Malik [8] made the observation that fast motion of high-contrast objects is more likely to be accurately estimated than the motion of low-contrast objects. This is related to the fact that there is an implicit weighting of the constancy assumptions with the corresponding image gradient as observed in [27]. In view of the data costs, mismatches of high-contrast objects are thus more expensive than those of low-contrast objects. This, in turn, suggests to use constraint normalization as in [27] when estimating large displacements.

Contributions. In this work, we address the aforementioned problem of that the appropriate smoothness weight may depend on the local motion pattern. By proposing a variational method that jointly estimates and fuses candidate flows with different smoothness weights into a final flow field, we show that many large displacement scenarios can actually be resolved without using additional feature matches. In contrast to related work from the literature that typically relies on a one-way pipeline based on a discrete fusion of pre-computed flows, we model the entire approach as a single minimization problem based on standard coarse-to-fine optimization. Moreover, we demonstrate the benefit of constraint normalization when estimating large displacements. Please note that we do not focus on designing an overall top-performing method but rather on pushing the limits of pure variational approaches w.r.t. large displacements.

Related Work. To handle large displacements, Brox and Malik [8] proposed to integrate descriptor matches into variational methods by means of a similarity term. While Stoll *et al.* [21] improved the sensitivity of this strategy w.r.t. to outliers by restricting the integration of such matches to promising locations, Weinzaepfel *et al.* [25] investigated the use of improved descriptors. In contrast, Xu *et al.* [26] refrained from using a similarity term, and proposed to enhance the upsampled flow initialization by integrating SIFT-matches at each level of the coarse-to-fine optimization. In contrast to our work, all these methods rely on feature descriptors to estimate large displacements.

Tu *et al.* [23] used a similar strategy as [26] but they considered proposals generated by PatchMatch [3] and by varying the smoothness weight of a variational method. Similarly, Lempitsky *et al.* [15] considered flows obtained by different methods and different parameter sets in a discrete fusion approach. Both works [15,23], however, did not investigate the benefit of varying the smoothness weight for large displacement optical flow.

In all cases, descriptor matching and match integration are separate steps.

2   Baseline Method

Let us start by introducing our baseline optical flow method which is the Complementary Optic Flow method [27]. It is a variational approach where the optical flow $\mathbf{w} = (u, v)^\top$ between two input color images $\mathbf{f}_1 = (f_1^1, f_1^2, f_1^1)^\top$ and $\mathbf{f}_2 = (f_2^1, f_2^2, f_2^3)^\top$ is computed as the minimizer of the following energy:

$$\mathcal{E}_{base}(\mathbf{w})_\alpha = \int_\Omega \mathcal{E}_D(\mathbf{w}) + \alpha\, \mathcal{E}_S(\mathbf{w})\, d\mathbf{x}\,. \tag{1}$$

Here, $\mathcal{E}_D$ is the data term, $\mathcal{E}_S$ is the smoothness term, $\alpha > 0$ is a balancing weight and $\mathbf{x} = (x, y)^\top \in \Omega$ is the location within the image domain $\Omega \subset \mathbb{R}^2$.

Data Term. The data term relates the two input images via the optical flow and is given by [27]

$$\mathcal{E}_D(\mathbf{w}) = \delta\, \Psi_D\left(\sum_{c=1}^3 \left(\sqrt{\theta^c} \cdot (f_2^c(\mathbf{x}+\mathbf{w}) - f_1^c(\mathbf{x}))\right)^2\right)$$

$$+ \gamma\, \Psi_D\left(\sum_{c=1}^3 \left|\begin{pmatrix} \sqrt{\theta_x^c} & 0 \\ 0 & \sqrt{\theta_y^c} \end{pmatrix} \cdot (\nabla f_2^c(\mathbf{x}+\mathbf{w}) - \nabla f_1^c(\mathbf{x}))\right|^2\right)\,. \tag{2}$$

It comprises the brightness constancy and the gradient constancy assumption in order to allow for illumination robust flow estimation [7]. Moreover, to reduce the influence of large gradients, constraint normalization [20] is applied via the weights $\theta^c = 1/(|\nabla f_2^c|^2 + \zeta^2)$ and $\theta_*^c = 1/(|\nabla f_{2,*}^c|^2 + \zeta^2)$ (with $* \in \{x, y\}$), where ζ is a regularization parameter that prevents divisions by zero. Finally, both assumptions are rendered robust under noise by applying a sub-quadratic penalizer [4] – here given by the Charbonnier function [10] $\Psi_D(s^2) = 2\epsilon_D^2\sqrt{1 + s^2/\epsilon_D^2}$ with contrast parameter ϵ_D. The non-negative weights δ, γ serve as balancing factors.

Smoothness Term. As smoothness term, we consider the anisotropic complementary smoothness term [27]

$$\mathcal{E}_S(\mathbf{w}) = \sum_{i=1}^{2} \Psi_{S_i}\left(|\mathcal{J}\mathbf{w}\cdot\mathbf{r}_i|^2\right), \tag{3}$$

that penalizes the directional derivatives of the flow by projecting the Jacobian $\mathcal{J}$ onto the local directions $\mathbf{r}_1$, $\mathbf{r}_2$ of maximum and minimum information contrast. In this context, the directions $\mathbf{r}_1$ and $\mathbf{r}_2$ are the eigenvectors of the so-called *regularization tensor* [27] which reads

$$R_\rho = K_\rho * \sum_{c=1}^{3}\left[\delta\,\nabla f_1^c\nabla f_1^{c\top} + \gamma\left(\nabla f_{1_x}^c\nabla f_{1_x}^{c\top} + \nabla f_{1_y}^c\nabla f_{1_y}^{c\top}\right)\right], \tag{4}$$

where $*$ denotes convolution with a Gaussian K_ρ of standard deviation ρ.

Following [24], we apply the edge-enhancing Perona-Malik penalizer [5] given by $\Psi_S(s^2) = \epsilon_{S_1}^2\log\left(1 + s^2/\epsilon_{S_1}^2\right)$ in $\mathbf{r}_1$-direction and the edge-preserving Charbonnier penalizer [10] in $\mathbf{r}_2$-direction; the former with contrast parameter ϵ_{S_1} and the latter with contrast parameter ϵ_{S_2}.

3   Joint Estimation and Fusion Model

After we have discussed the baseline method in the previous section, we are now in the position to describe our joint estimation and fusion model. Similar to methods from the literature that include descriptor matches [8,21,25], we want to estimate an optical flow $\mathbf{w}_f$ using the baseline method $\mathcal{E}_{base}$ and some similarity term $\mathcal{E}_{sim}$ that feeds N candidate flows $\mathbf{w} = \{\mathbf{w}_1,\ldots,\mathbf{w}_N\}$ from the candidate model $\mathcal{E}_{cand}$ into the solution. To this end, we propose the joint variational model

$$\mathcal{E}(\mathbf{w}, \mathbf{w}_f) = \mathcal{E}_{base}(\mathbf{w}_f)_{\alpha_f} + \mathcal{E}_{sim}(\mathbf{w}, \mathbf{w}_f) + \mathcal{E}_{cand}(\mathbf{w}), \tag{5}$$

that consists of three terms. On the one hand, as baseline model, we use the approach from the previous section with smoothness weight α_f. On the other hand, as candidate model, we consider multiple instances of the baseline model $\mathcal{E}_{base}(\mathbf{w})_\alpha$ with different smoothness weights α_i that estimate the corresponding candidate optical flows $\mathbf{w}_i$. It is given by

$$\mathcal{E}_{cand}(\mathbf{w}) = \lambda_C \cdot \sum_{i=1}^{N} \mathcal{E}_{base}(\mathbf{w}_i)_{\alpha_i} \tag{6}$$

Due to the different smoothness weights, the single instances can capture different levels of motion details, i.e. displacement scales. Finally, in order to couple the candidate flows $\mathbf{w}_i$ and the final optical flow $\mathbf{w}_f$, we introduce a similarity term $\mathcal{E}_C$ for each of these instances weighted by a parameter β_i. The combined similarity term reads

$$\mathcal{E}_{sim}(\mathbf{w}, \mathbf{w}_f) = \sum_{i=1}^{N} \beta_i \, \mathcal{E}_C(\mathbf{w}, \mathbf{w}_f)_i \, , \tag{7}$$

where the distinct similarity terms are defined as

$$\mathcal{E}_C(\mathbf{w}, \mathbf{w}_f)_i = \int_{\Omega} c_i(\mathbf{x}, \mathbf{w}) \cdot \Psi_C \left(|\mathbf{w}_i - \mathbf{w}_f|^2 \right) d\mathbf{x} \, . \tag{8}$$

Here, c_i is a local confidence function for the candidate flow $\mathbf{w}_i$ and Ψ_C is the Charbonnier penalizer [10] that makes the estimation more robust against outliers in the candidate flows. In Sect. 4, we will define appropriate confidence functions c_i that steer the local influence of each instance flow $\mathbf{w}_i$ on the final flow $\mathbf{w}_f$. The overall weight λ_C balances $\mathcal{E}_{cand}(\mathbf{w})$ and $\mathcal{E}_{base}(\mathbf{w}_f)$ by steering the direction of information flow between the candidate flows and the final flow. The higher it is, the more remains the estimation of the candidates $\mathbf{w}$ unaffected by the similarity term and the information only flows from $\mathbf{w}$ to $\mathbf{w}_f$ via $\mathcal{E}_{sim}$ while backward information flow is suppressed.

4 Smoothness Weights and Confidence Functions

Since we desire candidate flows at different smoothness scales, the questions arise how to choose the global smoothness weights of these flows and how to locally decide which flow candidate is the most appropriate. Let us discuss these two issues in the following sections.

4.1 Smoothness Weights

First of all, we define a maximum smoothness weight α_1 which is intended to be appropriate at most locations. Moreover, we consider smoothness weights that are significantly smaller in order to be able to capture large displacement motions. Our choice for the smoothness weights α_i of the flow candidates $\mathbf{w}_i$ is an exponential decrease w.r.t. α_1:

$$\alpha_i := \frac{\alpha_1}{2^{i-1}} \, . \tag{9}$$

With this choice, we can cover a wide range of different smoothness scales with only a low number of candidate flows. By the example of the Tennis sequence [8] depicted in Fig. 1 (top row), one can see at which smoothness scale the different motion patterns appear. While the first, smoothest flow covers the background motion and the overall motion of the Tennis player smoothly, the second flow covers the motion of the racket and the arm well, the third flow covers the motion of the hand and the right foot while the fifth flow covers the motion of the ball.

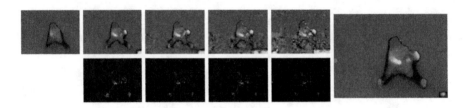

Fig. 1. Top row: Candidate flows with isotropic regularisation. **Bottom row:** normalized visualizations of the local confidence functions $c_1, \ldots, c_5$. **Right:** Final flow.

4.2 Assumptions on Local Confidences

Given a set of candidate flows $\mathbf{w}_i$ with different smoothness scales, we take into account the considerations from the introduction to state the local assumptions on how to integrate these flows in the estimation of the final flow $\mathbf{w}_f$:

1. A less smooth flow is likely to fulfill the data term better than a smoother flow, independently from being reliable or unreliable. Hence, a less smooth flow shall only have influence if it provides significantly less data costs than both the next smoother flow candidate and the smoothest flow candidate (similar to considerations in [21]).
2. The less smooth a flow is, the more texture is necessary in order to achieve meaningful flow vectors (similar to [8]). Otherwise, we might likely get trapped into the aperture problem.
3. A less smooth flow should not be considered if the data is unreliable (i.e. in over- or undersaturated regions).

In order to integrate those assumptions in our local confidence functions c_i, we need measures for the data cost and for the local structure. While the data costs are simply given by evaluating the data term, we compute the structure tensor [13] to measure structureness [8], both on local patches to increase robustness.

4.3 Composition of the Local Confidence Function c_i

Following the assumptions from the last section, we model the local confidence function c_i (where i is the index of the candidate flow) as the product of three weights which will be defined in the following.

Structureness Weight. Let $s(\mathbf{x})$ be the smaller eigenvalue of the structure tensor (integrated over a 7×7 neighborhood) of the reference frame f_1, let $\bar{s}$ be its average value over the whole image and let $r_i = \frac{\alpha_1}{\alpha_i}$. The structureness weight is then defined as

$$w_i^s(\mathbf{x}) := \left(\frac{s(\mathbf{x})}{\bar{s}} \right)^{\kappa_s \cdot \log(r_i)}, \qquad (10)$$

where the exponent κ_s is a free parameter. Here, the structureness weight is more pronounced for less smooth candidate flows (i.e. if r_i is bigger).

Cost Reduction Weight. Let $\mathcal{E}_D$ be the data costs and let $\rho_{L \times L}(g, \mathbf{x})$ be a functional that averages the function g in a $L \times L$ neighborhood around $\mathbf{x}$. The following two functions describe the patch-wise energy improvement of flow $\mathbf{w}_i$ compared to the previous, smoother flow $\mathbf{w}_{i-1}$ and the first and smoothest flow $\mathbf{w}_1$, respectively:

$$\delta_{\mathrm{prev},L}(\mathbf{x}, \mathbf{w}, i) := \rho_{L \times L}(\mathcal{E}_D(\mathbf{w}_{i-1}), \mathbf{x}) - \rho_{L \times L}(\mathcal{E}_D(\mathbf{w}_i), \mathbf{x}) \,,$$
$$\delta_{\mathrm{first},L}(\mathbf{x}, \mathbf{w}, i) := \rho_{L \times L}(\mathcal{E}_D(\mathbf{w}_1 \ ), \mathbf{x}) - \rho_{L \times L}(\mathcal{E}_D(\mathbf{w}_i), \mathbf{x}).$$

The cost reduction weight is then defined as

$$w_i^d(\mathbf{x}) := \log\left(1 + e^{\kappa_d(\delta_{\mathrm{prev},L}(\mathbf{x}, \mathbf{w}, i) + \delta_{\mathrm{first},L}(\mathbf{x}, \mathbf{w}, i))}\right), \tag{11}$$

where κ_d is a free parameter. Please note that this function resembles a linear one for large arguments of the exponential while it approaches zero for decreasing (negative) arguments.

Data Reliability Weight. We define $\chi_I(\mathbf{x})$ as an indicator function that excludes under- and oversaturated regions. It reads

$$\chi_I(\mathbf{x}) = \begin{cases} 1 & \text{if } f_1^c(\mathbf{x}) > \tau \text{ and } f_1^c(\mathbf{x}) < 255 - \tau \quad \forall c \in \{1, 2, 3\} \\ 0 & \text{else} \end{cases}, \tag{12}$$

where $\tau = 1$ is a robustness threshold.

Overall Confidence Function. The overall confidence functions $c_1, \ldots, c_N$ are then defined as follows

$$\hat{c}_i(\mathbf{x}, \mathbf{w}) := w_i^d(\mathbf{x}) \cdot w_i^s(\mathbf{x}) \cdot \chi_I(\mathbf{x}) \qquad (i > 1). \tag{13}$$

In order to be numerically robust, they are bounded from above via

$$c_i(\mathbf{x}, \mathbf{w}) := \min\left(\hat{c}_i(\mathbf{x}, \mathbf{w}), 1000\right). \tag{14}$$

Since the smoothest flow $\mathbf{w}_1$ serves as reference, it should be used everywhere except for those locations where a less smooth flow could improve the result. Hence, we define the confidence c_1 of the smoothest flow as

$$c_1(\mathbf{x}, \mathbf{w}) := 1, \tag{15}$$

which corresponds to the confidence of the other flows at average structured areas with only a small energy reduction.

Exemplary visualizations of these local confidence functions c_i for the Tennis sequence are shown in Fig. 1 (bottom row) where brighter values indicate higher confidence. As one can see, for each large displacement we have a high confidence in the smoothest candidate flow that is able to capture it.

5   Minimization

The whole variational model is minimized in a standard coarse-to-fine setting with warping and incremental computations [17]. Due to the nonlinearity of the penalizer functions, we additionally apply the lagged nonlinearity method in order to transform the nonlinear subproblems into series of linear equation systems. These linear equation systems are then solved using a multicolor variant of the successive overrelaxation (SOR) method [1].

Please note that in Eq. 8 the flow **w** is apparent in both the confidence functions and the coupling term. In order to avoid multiplications of unknowns during the minimization, in each coarse-to-fine level we compute the confidence functions based on the flow from the previous level. This can also be seen as a lagged nonlinearity method regarding the computation of the confidences.

6   Evaluation

In order to evaluate the performance of our method, we conducted several experiments. These include a qualitative comparison against LDOF [8] that investigates the large displacement capabilities of our method, an experiment that analyzes the effect of constraint normalization in this context, an experiment that evaluates the effect of different types of data costs and a quantitative experiment on the MPI Sintel benchmark [9] that shows improvements compared to the baseline method. In all experiments, we optimized only the following parameters: the number N of candidates, the data weights δ and γ and the smoothness weight α_1. To this end, we used the downhill simplex method as implemented in [22]. The remaining parameters are kept fixed throughout all experiments. They are given by $\beta_i = \alpha_f = \alpha_1$, $L = 5$, $\lambda_C = 1000$, $\kappa_s = 0.3$, $\kappa_d = 5$, $\epsilon_D = 0.01$, $\zeta = 0.01$, $\epsilon_{S_1} = 0.02$, $\epsilon_{S_2} = 0.03$.

6.1   Large Displacement Sequences

In our first experiment, we evaluate the performance of our method in the context of large displacements. To this end, we consider various challenging large displacement sequences from the literature and compare our results to those of the method of Brox and Malik (LDOF) [8] which has introduced descriptor matching in variational methods for large displacement optical flow. The parameters for all sequences are $\delta = \gamma = 0.5$, $\alpha_1 = 2$ and $N = 7$ candidate flows.

In Figs. 2 and 3 we show the results of both the publicly available implementation of LDOF and our novel variational method for large displacement optical flow. As one can see, our method correctly estimates the large displacements that LDOF is able to estimate – and even some more (see e.g. Tennis sequence 496). This particularly includes the displacements of the tennis balls that evidently extent their sizes. The extremely challenging Bird sequence [26] shows the limitations of both methods as none of them could capture the motion of the bird's head. In order to demonstrate that the correct estimation of large displacements

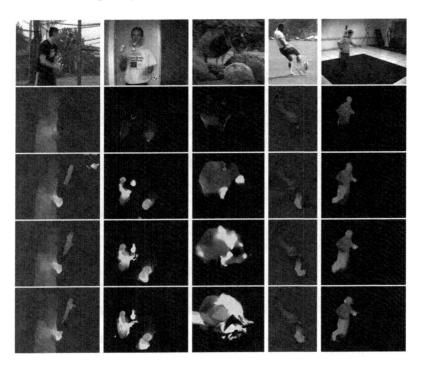

Fig. 2. Left to right: Baseball sequence [26], Beanbags sequence [2], Bird sequence, Football sequence [26], Human Eva sequence [19]. **Top to down:** Overlayed frames, baseline result, LDOF, our result (isotropic), our result (anisotropic).

does not depend on the anisotropic regularizer, we also added results for our method with an isotropic smoothness term (which is also used in LDOF).

While we have chosen the number of candidate flows fixed for all sequences, one may actually improve the results further by choosing it according to the extent of large displacements. For the beanbags sequences, already a value of $N = 3$ is sufficient, while we need a value of $N = 7$ in order to capture the motion of the tennis ball in the Tennis sequence 577.

6.2   Constraint Normalization

In our second experiment we show that constraint normalization [27] is helpful in the context of large displacements. To this end, we estimated flow fields without normalization and with normalization for different values of the normalization parameter ζ. While the general benefits of the constraint normalization have already been shown in [27], Fig. 4 shows the results on two large displacement sequences. As one can see particularly at hand of the tennis balls, both the deactivation of the constraint normalization and a too high value of ζ inhibit the estimation of large displacements. A too low value for ζ, in contrast, leads to noisier results. Using constraint normalization with a value between 0.001 and 0.01 (our standard value) for ζ provides the best results.

Fig. 3. Left to right: Tennis sequences 496, 502, 538, 577 [8]. **Top to down:** Over-layed frames, baseline result, LDOF, our result (isotropic), our result (anisotropic).

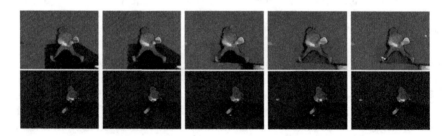

Fig. 4. From left to right: No constraint normalization, $\zeta = 1$, $\zeta = 0.1$, $\zeta = 0.001$, $\zeta = 0.00001$. **From top to bottom:** Tennis sequences 496 and 577.

6.3   Influence of the Data Constancy Assumptions

In our third experiment, we analyze the two types of data terms we used in our model w.r.t. their data costs and their influence on the fusion scheme. While the Brightness Constancy Assumption (BCA) can produce high costs at any part of a mismatched object, the Gradient Constancy Assumption (GCA) can only produce data costs where edges are involved. It is hence a lot sparser (see Fig. 5, top row). As can be seen from the bottom row of Fig. 5, the fusion using only the GCA data term is by far inferior to the results of using BCA or combining both data terms. The data costs of a pure GCA data term for incorrect matches are too low and hence it cannot compete with the smoothness term which prevents

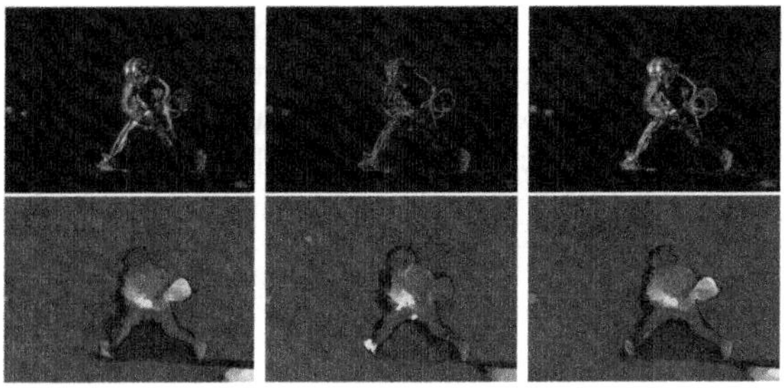

Fig. 5. From left to right: Brightness Constancy Assumption (BCA), Gradient Constancy Assumption (GCA) and both combined. **From top to bottom:** Data costs of the baseline flow (brighter grey values indicate larger energies), final result.

the motion discontinuity of a large displacement. In contrast, when including the BCA, the denser data costs make the misestimation of large displacements more expensive and thus increase the probability to estimate large displacements correctly. This shows that data costs with dense coverage for mismatched objects are important for our fusion scheme.

6.4   MPI Sintel Benchmark

In our fourth experiment, we compare our strategy with the baseline method (Complementary Optical Flow [27]) on the MPI Sintel benchmark [9]. To this end, we use our method with the first order complementary regularizer and computed results both for the training and the evaluation data.

Regarding the training data, Table 1 shows a clear improvement over the baseline ($N = 0$). The average endpoint error (AEE) decreases from 4.273 down to 3.974 (by 7%). This behavior is confirmed by the results for the evaluation data sets that are listed on the MPI Sintel webpage where our method is denoted as *ContFusion* and the baseline is denoted as *COF*. Here, the error decreases from 6.496 to 6.263 (by 3.6%) for the clean pass and from 8.204 to 7.857 (by 4.2%) for the final pass. This shows that the our novel strategy of simultaneous estimation and fusion of motion candidates is also beneficial in a quantitative sense.

Table 1. Quantitative results on the clean training data of the MPI Sintel benchmark.

N	0	1	2	3	4	5	6
AEE	4.273	4.191	4.136	3.974	3.984	4.134	4.316

6.5   Limitations

The behavior at occlusions is a limitation of our method. This can be seen both visually at the large displacement sequences (in Figs. 2 and 3) and quantitatively at the unmatched EPE in the MPI Sintel benchmark (that increases compared to the baseline). Additionally to regions with mismatched objects, occluded regions produce potentially high data costs. Since our confidence function heavily relies on data costs, correct smooth flows are replaced by less smooth candidate flows that lead to a smaller local data energy but are often meaningless.

7 Conclusion

In this work, we pushed the limits of variational approaches that are minimized using a standard coarse-to-fine scheme a little bit further w.r.t. large displacements. We have shown that many large displacement cases from the literature can be estimated without the need for descriptor matches. The weaknesses of prior variational methods in these cases are not due to weak data representations on coarse resolutions but due to a weight balancing of data term and smoothness term that is inappropriate for large displacement optical flow estimation. With multiple instances of the baseline model and appropriate choices of weighted similarity terms, we can estimate different scales of motions within a single variational model that simultaneously estimates and fuses candidate flows with different smoothness weights. The findings were confirmed by the evaluation which showed a good performance for large displacements and an improvement over its baseline method.

Limitations include the behavior at occluded regions where advanced occlusion handling would be necessary. Future work includes the handling of severe illumination changes where the BCA is not applicable at all and the GCA alone cannot help to estimate large displacements correctly, as well as the inclusion of second order smoothness terms for non-fronto-parallel motion patterns.

Acknowledgements. We thank the German Research Foundation (DFG) for financial support within project B04 of SFB/Transregio 161.

References

1. Adams, L., Ortega, J.: A multi-color sor method for parallel computation. In: Proceedings of International Conference on Parallel Processing, pp. 53–56 (1982)
2. Baker, S., Scharstein, D., Lewis, J.P., Roth, S., Black, M.J., Szeliski, R.: A database and evaluation methodology for optical flow. Int. J. Comput. Vis. **92**(1), 1–31 (2010)
3. Barnes, C., Shechtman, E., Goldman, D.B., Finkelstein, A.: The generalized Patch-Match correspondence algorithm. In: Daniilidis, K., Maragos, P., Paragios, N. (eds.) ECCV 2010. LNCS, vol. 6313, pp. 29–43. Springer, Heidelberg (2010). https://doi.org/10.1007/978-3-642-15558-1_3
4. Black, M.J., Anandan, P.: Robust dynamic motion estimation over time. In: Proceedings of IEEE Computer Society Conference on Computer Vision and Pattern Recognition, pp. 292–302 (1991)
5. Black, M.J., Anandan, P.: The robust estimation of multiple motions: parametric and piecewise smooth flow fields. Comput. Vis. Image Underst. **63**(1), 75–104 (1996)

6. Bredies, K., Kunisch, K., Pock, T.: Total generalized variation. SIAM J. Imaging Sci. **3**(3), 492–526 (2010)
7. Brox, T., Bruhn, A., Papenberg, N., Weickert, J.: High accuracy optical flow estimation based on a theory for warping. In: Pajdla, T., Matas, J. (eds.) ECCV 2004. LNCS, vol. 3024, pp. 25–36. Springer, Heidelberg (2004). https://doi.org/10.1007/978-3-540-24673-2_3
8. Brox, T., Malik, J.: Large displacement optical flow: descriptor matching in variational motion estimation. IEEE Trans. Pattern Anal. Mach. Intell. **33**(3), 500–513 (2011)
9. Butler, D.J., Wulff, J., Stanley, G.B., Black, M.J.: A naturalistic open source movie for optical flow evaluation. In: Fitzgibbon, A., Lazebnik, S., Perona, P., Sato, Y., Schmid, C. (eds.) ECCV 2012. LNCS, vol. 7577, pp. 611–625. Springer, Heidelberg (2012). https://doi.org/10.1007/978-3-642-33783-3_44
10. Charbonnier, P., Blanc-Féraud, L., Aubert, G., Barlaud, M.: Two deterministic half-quadratic regularization algorithms for computed imaging. In: Proceedings of IEEE International Conference on Image Processing, pp. 168–172 (1994)
11. Demetz, O., Stoll, M., Volz, S., Weickert, J., Bruhn, A.: Learning brightness transfer functions for the joint recovery of illumination changes and optical flow. In: Fleet, D., Pajdla, T., Schiele, B., Tuytelaars, T. (eds.) ECCV 2014. LNCS, vol. 8689, pp. 455–471. Springer, Cham (2014). https://doi.org/10.1007/978-3-319-10590-1_30
12. Drayer, B., Brox, T.: Combinatorial regularization of descriptor matching for optical flow estimation. In: Proceedings of British Machine Vision Conference, pp. 42.1–42.12 (2015)
13. Förstner, W., Gülch, E.: A fast operator for detection and precise location of distinct points, corners and centres of circular features. In: Proceedings of ISPRS Intercommission Conference on Fast Processing of Photogrammetric Data, pp. 281–305 (1987)
14. Horn, B., Schunck, B.: Determining optical flow. Artif. Intell. **17**, 185–203 (1981)
15. Lempitsky, V., Roth, S., Rother, C.: FusionFlow: discrete-continuous optimization for optical flow estimation. In: Proceedings of IEEE Computer Society Conference on Computer Vision and Pattern Recognition, pp. 1–8 (2008)
16. Nagel, H.H., Enkelmann, W.: An investigation of smoothness constraints for the estimation of displacement vector fields from image sequences. IEEE Trans. Pattern Anal. Mach. Intell. **8**, 565–593 (1986)
17. Papenberg, N., Bruhn, A., Brox, T., Didas, S., Weickert, J.: Highly accurate optic flow computation with theoretically justified warping. Int. J. Comput. Vis. **67**(2), 141–158 (2006)
18. Sevilla-Lara, L., Sun, D., Learned-Miller, E.G., Black, M.J.: Optical flow estimation with channel constancy. In: Fleet, D., Pajdla, T., Schiele, B., Tuytelaars, T. (eds.) ECCV 2014. LNCS, vol. 8689, pp. 423–438. Springer, Cham (2014). https://doi.org/10.1007/978-3-319-10590-1_28
19. Sigal, L., Balan, A.O., Black, M.J.: HumanEva: synchronized video and motion capture dataset for evaluation of articulated human motion. Int. J. Comput. Vis. **87**(1–2), 4–27 (2010)
20. Simoncelli, E.P., Adelson, E.H., Heeger, D.J.: Probability distributions of optical flow. In: Proceedings of IEEE Computer Society Conference on Computer Vision and Pattern Recognition, pp. 310–315 (1991)

21. Stoll, M., Volz, S., Bruhn, A.: Adaptive integration of feature matches into variational optical flow methods. In: Lee, K.M., Matsushita, Y., Rehg, J.M., Hu, Z. (eds.) ACCV 2012. LNCS, vol. 7726, pp. 1–14. Springer, Heidelberg (2013). https://doi.org/10.1007/978-3-642-37431-9_1

22. Stoll, M., Volz, S., Maurer, D., Bruhn, A.: A time-efficient optimisation framework for parameters of optical flow methods. In: Sharma, P., Bianchi, F.M. (eds.) SCIA 2017. LNCS, vol. 10269, pp. 41–53. Springer, Cham (2017). https://doi.org/10.1007/978-3-319-59126-1_4

23. Tu, Z., Poppe, R., Veltkamp, R.C.: Weighted local intensity fusion method for variational optical flow estimation. Pattern Recogn. **50**, 223–232 (2016)

24. Volz, S., Bruhn, A., Valgaerts, L., Zimmer, H.: Modeling temporal coherence for optical flow. In: Proceedings of International Conference on Computer Vision, pp. 1116–1123 (2011)

25. Weinzaepfel, P., Revaud, J., Harchaoui, Z., Schmid, C.: Deepflow: large displacement optical flow with deep matching. In: Proceedings of International Conference on Computer Vision, pp. 1385–1392 (2013)

26. Xu, L., Jia, J., Matsushita, Y.: Motion detail preserving optical flow estimation. IEEE Trans. Pattern Anal. Mach. Intell. **34**, 1744–1757 (2012)

27. Zimmer, H., Bruhn, A., Weickert, J., Valgaerts, L., Salgado, A., Rosenhahn, B., Seidel, H.-P.: Complementary optic flow. In: Cremers, D., Boykov, Y., Blake, A., Schmidt, F.R. (eds.) EMMCVPR 2009. LNCS, vol. 5681, pp. 207–220. Springer, Heidelberg (2009). https://doi.org/10.1007/978-3-642-03641-5_16

Temporal Semantic Motion Segmentation Using Spatio Temporal Optimization

Nazrul Haque[1](✉), N. Dinesh Reddy[2], and Madhava Krishna[1]

[1] International Institute of Information Technology, Hyderabad, India
nazrulatharhaque@gmail.com
[2] Robotic Institute, Carnegie Mellon University, Pittsburgh, USA

Abstract. Segmenting moving objects in a video sequence has been a challenging problem and critical to outdoor robotic navigation. While recent literature has laid focus on regularizing object labels over a sequence of frames, exploiting the spatio-temporal features for motion segmentation has been scarce. Particularly in real world dynamic scenes, existing approaches fail to exploit temporal consistency in segmenting moving objects with large camera motion.

In this paper, we present an approach for exploiting semantic information and temporal constraints in a joint framework for motion segmentation in a video. We propose a formulation for inferring per-frame joint semantic and motion labels using semantic potentials from dilated CNN framework and motion potentials from depth and geometric constraints. We integrate the potentials obtained into a 3D (space-time) fully connected CRF framework with overlapping/connected blocks. We solve for a feature space embedding in the spatio-temporal space by enforcing temporal constraints using optical flow and long term tracks as a least-squares problem. We evaluate our approach on outdoor driving benchmarks - KITTI and Cityscapes dataset.

1 Introduction

Understanding scene dynamics has always been a crucial component in outdoor robotic navigation. In outdoor scenes, scene understanding is facilitated by predicting spatially separated bounding boxes [19,20] on objects or associating a *label* with each pixel [1,31], in an image. For a holistic perception of the scene, the prediction unfolds in assigning a *semantic* or *motion* label. However, both the cues provide complementary information about a scene and are highly interrelated. Joint information about an object such as Moving Car or Stationary pedestrians significantly aids in path planning for an autonomous vehicle. Semantic property of an object can help infer the motion label of the pixel and vice versa.

In both static and dynamic environments, convolutional neural networks have gained enormous success in accuracy for predicting semantic labels in image space. On the other hand, motion segmentation poses many challenges, particularly in scenes where the camera is found to be in motion. Indeed, in the presence

M. Pelillo and E. Hancock (Eds.): EMMCVPR 2017, LNCS 10746, pp. 93–108, 2018.
https://doi.org/10.1007/978-3-319-78199-0_7

of multiple moving objects, generating and tracking various prospective motions becomes challenging. While epipolar geometry constraints work well with moving object detection, they tend to fail in degenerate cases where both the moving object and camera lie in the same subspace. This relative configuration between the camera and the object in motion is common in on-road scenes. Traditional motion segmentation algorithms formulate the problem as clustering the trajectories into affine subspaces. This often leads to a sparse segmentation resulting in different clusters, each representing a motion model in the scene. Further, supervoxel projections on the trajectory clusters give rise to a dense segmentation. In most cases, the projections do not respect the object boundaries and hence, this is followed by a graph based probabilistic model such as Markov Random Field (MRF) to enforce appearance constraints.

Recent literature [17,18,27,30] have leveraged semantic property into a probabilistic framework, generating per-frame moving object proposals with dense predictions. Camera motion often leads to discontinuities in the flow magnitude. Optical flow magnitude for nearby stationary objects may be found to have a larger magnitude than far away stationary objects. This, in turn, effects the motion likelihood that is obtained using depth information in a similar fashion. Thus to eliminate failure cases semantic property comes into role. The intuition behind such a reasoning is that the likelihood of a moving wall or road is less as compared to a moving car or pedestrian.

In this paper, we focus on the problem of joint semantic and motion segmentation. Our method takes a sequence of stereo sequences as input and generates per frame motion probabilities using stereo pairs. For semantics, we use a dilated convolution neural network for predicting per-pixel semantic class. We also learn the correlation between the semantic label and motion likelihood. Further, We propose a novel joint probability formulation in a discrete label space consisting of joint labels. Under this, each image pixel is labeled with both semantic and motion class jointly. Motion property of an object is better perceived by the object tracks over the temporal space. To infer motion probabilities and enforce long term correspondences in image space, we use a dense fully connected CRF (Conditional Random Field) with time as an additional dimension in its feature space. The trajectory constraints are enforced by solving a linear least squares equation for optimizing position features in pairwise constraints in CRF.

The temporal constraints are enforced using dense point trajectories and optical flow. In addition, the spatial properties are preserved by including a second order regularization term in the least squares formulation. The spatial term exploits appearance similarity and edge maps for minimizing the distance between corresponding points in the sequence. For inference, we use an extension of the mean-field based algorithm. The inference is carried out on overlapping connected components in CRF.

In summary, following are the key contributions of our work.

– We present an *end-to-end probabilistic framework* that performs joint semantic and motion segmentation for a sequence of stereo frames.

– We provide a method for integrating semantic constraints with dense point object trajectories to obtain motion segmentation.
– We present results on several sequences on outdoor driving benchmarks.

We evaluate our approach on challenging KITTI on-road dataset. We are able to achieve 4.71% and 17.91% improvement in IOU accuracy on our annotated test dataset for Moving Car and Moving Pedestrian detection, respectively over M-CRF [18], while in comparison to STMOP [5], we show an improvement of 8.04% than M-CRF [18] in moving objects detection.

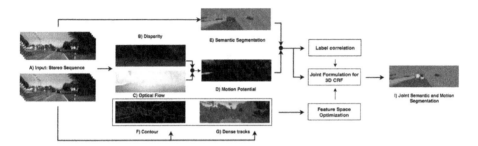

Fig. 1. Illustration of the proposed pipeline: Our framework takes a long sequence of stereo images as input (A). We compute motion potential (D) using depth (B) and Optical Flow (C). Semantic Segmentation (E) is carried out on the input images. Further, we calculate label compatibility between and object and motion class. We compute edge maps (H) and dense point trajectories (G) and solve a least-squares optimization for joint label CRF feature space enforcing temporal consistency and spatial constraints. Using the optimized feature space, we propose a joint CRF formulation in space-time volume. This leads to a temporally consistent joint semantic and motion segmentation (H).

2   Related Work

Scene understanding has fair amount of literature in both semantic and motion aspects. Traditional semantic segmentation approaches have tackled the problem as a multi label classification problem. Classifiers are trained with descriptive features as input for dense pixel labeling [22], followed by a maximum a posteriori inference (MAP) in a conditional random field (CRF) [12,18]. With the advent of convolutional neural networks (CNN), the dense pixel predictions have made significant progress on the accuracy [24]. Fully Convolution Networks [15] have made it possible for the architectures to handle inputs with arbitrary size. The outputs from the Convnets are upsampled by learning a Deconvolution layer [16], resulting in pixel wise predictions. The literature also includes architectures where Convnets followed by a CRF formulation [32] attain significant improvement in accuracy. Koltun [31] proposed an architecture for dense pixel predictions with dilated convolutions and presents state-of-the-art results in semantic segmentation. In recent work, Kundu [14] has shown results for temporally consistent semantic segmentation over a video sequence using long term tracks into a 3D CRF formulation. Existing architectures suffer loss in resolution due to pooling layers, while dilated convolutions sustain exponential expansion of the receptive field without loss in the coverage area. This also leads to a higher resolution output.

In outdoor scenes, motion segmentation has been extensively studied. Traditional algorithms based on epipolar geometry [28] are bound to fail in degenerate cases. The problem is tackled with good precision using geometric constraints [13], frame depth and camera ego-motion. In [13], degeneracy is handled by enforcing flow vector constraints using the camera trajectory obtained from visual SLAM (VSLAM). While the approaches demonstrate good accuracy on the on-road benchmark, they fail to exploit object trajectories for a consistent motion segmentation. In contrast, our approach uses dense point trajectories and semantic constraints in a joint framework for moving objects detection.

In other seminal works [11,26], the problem is formulated as affine clustering of the trajectories into corresponding subspaces. [11,33] do not scale well on on-road datasets where both camera and the moving object lie in the same subspace. [26] used in-frame shear constraints for merging perspective affine models and has shown benchmark results on outdoor scenes. The output, however, is a sparse collection of points belonging to different subspaces based on the motion property exhibited. [5] used uncalibrated frame sequences for generating per frame moving object proposals. The proposals are refined by training dual pathway CNN with both the RGB image and optical flow as input. Further, labels are propagated using random walkers on motion affinities of long term tracks. While the approach works well with video segmentation benchmarks, they do not scale well on outdoor scenes. Our approach likewise exploits the long term tracks, while preserving semantic properties, and uses flow bound constraints for generating per-frame motion likelihood.

In recent literature, semantic properties have been exploited for motion segmentation. [2] uses contextual descriptors for object level motion detection, while semantics has been incorporated in a convolutional neural network architecture [30] for analyzing pedestrians behavior. With the advent of CNNs, efforts for learning joint labels in an end-to-end architecture has been studied in [8] using feature amplification, exploiting short term consistency. The idea has also been complemented by the work by Valada et al. [27] where joint learning is performed using two streams, each tasked to learn semantic and motion attributes. The two parallel streams are fused for joint learning and probability maps thus obtained are subsequently upsampled to obtain joint dense predictions. [8,18,27] fail to establish long term correspondences over a sequence for motion detection, whereas our approach uses long term tracks for establishing correspondences between the frames and enforce spatial constraint using appearance and edge features using the formulation.

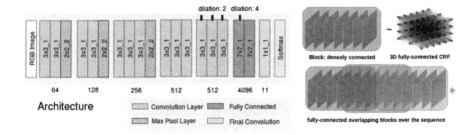

Fig. 2. (Left) Dilated CNN architecture for semantic segmentation - $w \times h\_s$: Layer with kernels of width w, height h, and stride s. Numbers on the top and bottom of each layer depict the number of channels in the output and dilation factor, respectively. (right) 3D CRF structure: The long sequence falls under overlapping blocks. For each block, a dense CRF is formulated and feature optimization is carried out for each set.

3   Dynamic Video Joint Segmentation

In this section, we present our joint labeling framework for outdoor sequences using a stereo camera. We describe our procedure for obtaining per-frame semantic potentials and motion initialization using stereo vision. We present our joint formulation for dense CRF in space-time volume with overlapping connected blocks in the succeeding section. Further, we describe the least squares formulation in the joint label space for long term consistency, using dense point trajectories and spatial constraints. We also provide an illustration of our approach in Fig. 1.

3.1   Semantic Segmentation

For semantic class segmentation, we use a deep learning architecture specifically engineered for dense predictions. The architecture is adapted from fully convolutional VGG 16 net [15,23] and modifications applied from work by Yu and Koltun [31] using dilated convolutions. The last two pooling layers in the VGG architecture were detached and following convolutions are dilated with a factor of *2* for each pooling layer abducted. The dilated architecture benefits dense predictions by generating higher resolution output without losing global context. The network architecture used is shown in Fig. 2. The network takes full size color images as input and the output from the softmax layer is upsampled using a learned Deconvolution layer.

3.2   Motion Potential

We calculate per-frame motion likelihood using stereo pairs. The motion likelihood of a pixel is initialized as the difference between the predicted flow and optical flow vector. The camera extrinsics are calculated using libviso [7]. Given a stereo pair at consecutive time instants, the method estimates visual odometry by minimizing the sum of the reprojection errors on the both the images (left and right). The predicted flow vector of a pixel is stated as:

$$F' = (KRK^{-1}X + \frac{KT}{z}) - X \tag{1}$$

where R and T are the rotation and translation of the camera respectively, K is the camera Intrinsic matrix, X is the pixel coordinate, z is the depth of the given pixel from the camera and F' is the displacement of the pixel under the camera motion. The difference between the predicted and optical flow gives the motion potential for the pixel.

4   Joint Labeling in Space-Time Volume

In this section, we propose a space-time CRF formulation for joint semantic and motion labeling on a sequence of stereo frames. We also incorporate long term tracks in the joint feature space.

4.1   Spatio-Temporal CRF

Given a sequence of frames, we divide the video sequence into overlapping blocks and formulate a Fully Connected dense CRF on each block (Fig. 2). We extend the 3D CRF formulation of [14] to joint label space. In [14], position features of the CRF model are optimized using dense trajectories for generating temporally consistent semantic segmentation. Since motion segmentation is perceived better with tracking objects over a sequence, we provide an extension of the framework for joint label space. We introduce the terms used in this paper. Each pixel in the video volume is located by the vector $\mathbf{p} = (n, t, i) \in \mathbb{R}^3$. Here, n is the block number, t denotes time dimension inside the block n, representative of the frame number relative to the block and i is the index of the pixel in the image. $\mathbf{P}$ represents a group of pixels in the volume. The location of a pixel $\mathbf{p}$ is given by $\bar{x}_p$ in the image space. Also, the RGB color vector of a pixel $\mathbf{p}$ is represented by $\mathbf{C}_p$.

For a given block in the video volume, we define the problem of joint semantic and motion segmentation as finding a minimal cost labeling on a the joint label space $\mathcal{L} = \{l_1, l_2, \ldots, l_k\}$ for a set of random variables $\mathcal{X}_p = \{x_1, x_2, \ldots, x_N\}$. We denote s as the total number of semantic class labels. Each random variable can take a single label from joint label space $\mathcal{L}$, where $k = s * 2$, as each object class label can be associated with two motion classes (moving or stationary). For instance, l_i could be a moving car, stationary road, etc. The energy cost term for a label assignment x is defined as:

$$E(x|\mathbf{P}) = \sum_i \psi_i(x_i) + \sum_{(i,j)\mathcal{N}} \psi_{i,j}(x_i, x_j) \tag{2}$$

where $\mathcal{N}$ is the neighborhood constitution of the random field defined on the pairs of variables. In the random field, a clique covers each block and each pixel is covered by two overlapping blocks. Subsequently, each variable falls under two fully connected subgraphs.

The unary term $\psi_i(x_i)$, represents the cost of assigning label x_i to pixel $\mathbf{i}$. For joint labeling, we propose the unary cost formulation as:

$$\psi_i(x_i) = -\log\left(\bar{\phi}_i(x_i)\right) \tag{3}$$

$$\phi_i(x_i) = \underbrace{\phi_{i,s}(x_i)}_{Object} \cdot \underbrace{\phi_{i,m}(x_i)}_{Motion} \cdot \underbrace{\phi_{i,s,m}(x_i)}_{Correlation} \tag{4}$$

Here, $\bar{\phi}_i(x_i)$ is obtained after normalizing the joint probability distribution $\phi_i(x_i)$ in the range 0−1. $\phi_{i,s}(x_i)$ is the probability of the pixel belonging to the object class s corresponding to the joint label x_i and inferred from the probabilities obtained from the softmax layer in our trained dilated ConvNet described in Sect. 3.1. We express the motion term as:

$$\phi_{i,m}(x_i) = \begin{cases} \lambda & if\ \bar{l}(m) = 0 \\ ||F'(\bar{x}_i) - F(\bar{x}_i)|| & if\ \bar{l}(m) = 1 \end{cases} \tag{5}$$

where, m is the motion attribute in the joint label. The function $\bar{l}(m)$ returns 1 if the motion label space is moving and 0 otherwise. For instance, $\bar{l}(m)$ will return 1 in case of 'Moving Car'. F' is the predicted flow (Sect. 3.2) and F, the optical flow vector of the pixel i. $||F'(\bar{x}_p) - F(\bar{x}_p)||$ represents the normalized motion potential corresponding to the pixel. λ is a learned term calculated using RANSAC algorithm over a small set of annotated images.

$\phi_{i,s,m}(x_i)$ represents the object class label motion compatibility and is given as:

$$\phi_{i,s,m}(x_i) = c(s, m) \tag{6}$$

where, $c(s, m) \in [0, 1]$. Here $c(s, m)$ represents the correlation between the semantic class s and motion attribute m. In other words, this can be seen as the motion compatibility for semantic class s. We follow the work of [18] for calculating label correlation between the two classes using MAHR algorithm [10].

The pairwise term $\psi_{i,j}(x_i, x_j)$ stimulates similar pixels to have same labeling. With Gaussian kernels, the pairwise term [12] given as:

$$\psi_{i,j}(x_i, x_j) = \mu(x_i, x_j) \sum_{z=1}^{Z} \omega^z \kappa^z(\mathbf{f}_i, \mathbf{f}_j) \tag{7}$$

Here, $\mathbf{f}_i$ and $\mathbf{f}_j$ are features analogous to $\mathbf{i}$ and $\mathbf{j}$ pixel respectively. $\mu(x_i, x_j)$ is a label compatibility term and ω^z are the weights. The kernels κ^z is given as:

$$\kappa^z(\mathbf{f}_i, \mathbf{f}_j) = \exp\left(-\frac{||\mathbf{f}_i - \mathbf{f}_j||}{\sigma_z^2}\right) \tag{8}$$

where, σ_z is the model parameter learned using grid search on a subset of annotated training set. The feature space $\mathbf{f}_i$ is a six dimensional vector $\in \mathbb{R}^6$ that consists of position, color and time corresponding to the pixel $\mathbf{i}$, i.e., $\mathbf{f}_i = (\bar{x}_i, \mathbf{C}_i, t_i)$.

4.2 Feature Space Optimization

The six dimensional feature space do not scale well in dynamic outdoor scenes as the pixels tend to displace under the camera and object motion. Thus, time as an additional term in the feature space does not model the pixel correspondences in the space-time volume. We use the formulation proposed in [14] for optimizing position features ($\bar{x}_i$) to reduce the Euclidean distance between the corresponding points in the volume while enforcing spatial constraints for preserving object shapes. In [14], the position features were optimized for temporally consistent semantic segmentation using long term tracks and edge maps. However, we use the underlying formulation for enforcing temporal consistency in joint semantic and motion label space. The dense point trajectories and spatial constraints enforce label similarity with motion potential in unary space. Thus, for the feature space ($\bar{x}_i, \mathbf{C}_i, t_i$), the position features ($\bar{\mathbf{x}}_i$) are optimized using a least squares formulation. The optimized feature space is represented as ($\mathbf{x}_i, \mathbf{C}_i, t_i$) which is obtained after minimizing the proposed energy term.

We use the linear least squares formulation in [14] and is given as:

$$E^{\mathcal{SM}}(x) = E_d^{\mathcal{SM}}(x) + E_s^{\mathcal{SM}}(x) + E_t^{\mathcal{SM}}(x) \tag{9}$$

where x are the position features in the block n, consisting of R frames, with N pixels in each frame. We now explain each of the terms in the energy Eq. 9. A single pixel is denoted by (n, t, i) as described in Sect. 4.1, with n as the block number.

Data $E_d^{\mathcal{SM}}(x)$: Let $r = \lfloor R/2 \rfloor$ be the reference frame. The energy term prevents the points in the reference from drifting far from their original position in the volume. If P^r is the set of pixels in the reference frame and $\bar{x}_p$ be the original feature space, the term is given as:

$$E_d^{\mathcal{SM}}(x) = \sum_{p \in P^a} (\mathbf{x}_p - \bar{x}_p)^2 \tag{10}$$

Spatial $E_s^{\mathcal{SM}}(x)$: This spatial energy term ensures that object shapes are preserved with color and boundary constraints. This is formulated over the 4-connected pixel grid and given as:

$$E_s^{\mathcal{SM}}(x) = \sum_{t=1}^{R} \sum_{i=1}^{N} \left(\mathbf{x}_{(n,t,i)} - \sum_{j \in \mathcal{N}_)} \omega_{ij} \mathbf{x}_{(n,t,i)} \right)^2 \tag{11}$$

where, $\mathcal{N}_i$ are the neighbors of the point (n, t, i) The weights ω_{ij} reduces the regularization effect at boundaries for preserving object shapes and given as:

$$\omega_{ij} = \exp \left(-\frac{||C_{(n,t,i)} - C_{(n,t,j)}||}{\sigma_1} \right) \exp \left(-\frac{q_p}{\sigma_2} \right) \tag{12}$$

Here, q_p is the contour strength of the pixel $\mathbf{p}$ - (n, t, i). Hence, the second term in Eq. 12 is related to the contour strength at the pixel, while the initial term is based on the color difference between the pixels (n, t, i) and (n, t, j). $q_p \in [0, 1]$, where higher value indicates presence of an edge at the pixel. For calculating contour strength q_p, we use Structured Forests Edge Detection [4] implementation.

Temporal $E_t^{SM}(x)$: The energy term uses correspondences obtained from the dense point trajectories and optical flow for bringing corresponding points closer in the feature space and is given as:

$$E_t^{SM}(x) = \sum_{(p,q)\in\mathcal{Y}} (\mathbf{x}_p - \mathbf{x}_q)^2 \tag{13}$$

where, $\mathcal{Y}$ is the super set of corresponding points in the frames of the block n. Here, points p and q belong to different frames. The energy term ensures that the tracked points over the frames are assigned the same label over the joint label space. This also enforces label compatibility for pixels exhibiting similar motion behavior over the frames. We use the implementation by Sundaram et al. [25] for calculating long term tracks.

4.3   Inference

Inference has been a challenging problem for dense CRFs and becomes even more challenging with a set of overlapping Fully Connected blocks. We follow the work of Kundu et al. [14], an extension of mean-field inference algorithm by Koltun [12]. Since a pixel is covered by two overlapping blocks, let $\mathcal{N}_i^1$ and $\mathcal{N}_j^2$ represent the two set of neighbors of the pixel $\mathbf{i}$. We define an alternative distribution over the random variables of the CRF, $Q_i(z_i)$, and define Q as $Q(z) = \prod_i Q_i(z_i)$. Here $Q_i(z_i)$ is a multi class distribution over the joint semantic and motion label space. The mean-field approach minimizes the distance between the Q and the true distribution P. The inference for joint label space is given as:

$$Q_i(x_i = l) = \frac{1}{Z_i} \exp\left(-\psi_i(x_i) - \sum_{l'\in\mathcal{L}}\sum_{j\in\mathcal{N}_i^1} Q_j(x_j = l')\psi_{ij}(l, l')\right.$$
$$\left. - \sum_{l'\in\mathcal{L}}\sum_{j\in\mathcal{N}_i^2} Q_j(x_j = l')\psi_{ij}(l, l')\right) \tag{14}$$

where, $\psi_i(x_i) = \psi_{i,s}(x_i) . \psi_{i,m}(x_i) . \psi_{i,s,m}(x_i)$ and Z_i is the normalization factor. As proposed in [12] we can efficiently solve the pairwise summations, given as Potts model, using Gaussian convolutions. In cases where blocks do not fit into the memory, inference is carried out in chunks and predictions are scaled across the divisions using the heuristic from the work by [14].

5   Evaluations and Results

For evaluation of our approach, we use two renowned on-road datasets.

KITTI: We use the KITTI-Tracking benchmark dataset [6] which consists of diverse on-road sequences taken by a stereo camera, mounted on a driving car. For quantitative evaluation, we use the largest publicly available annotated dataset (200 images) by [8]. The images were annotated with 11 semantic classes, i.e., *Building, Vegetation, Sky, Car, Sign, Road, Pedestrian, Fence, Pole, Sidewalk* and *Cyclist*. Further, each image was also annotated with moving and non-moving labels. We use the results provided by [18] on 200 images for both quantitative and qualitative evaluation, which is a subset of frames provided by [8] and we manually annotate the remaining images. Thus, we form our KITTI-Test dataset consisting of 200 images from five different sequences of KITTI Tracking dataset. Existing semantic segmentation annotations for KITTI dataset is insufficient to train a dilated CNN from scratch. Hence for fine-tuning our network, we manually annotate 56 images from KITTI sequences and ensure no overlap of sequences/images with our KITTI-Test dataset. The annotated images together with 146 KITTI Odometry images labeled by Ros et al. [21] forms our KITTI-Training dataset.

Cityscapes: The Cityscapes dataset [3] is relatively new and consists of challenging urban sequences from over 50 cities with varying dynamic objects and weather conditions. Pixel-wise ground truth semantic annotations are publicly provided for 2975 training and 500 validation images, for a single image in each video snippet. However, we use the semantic annotations for training our dilated CNN for the task of semantic segmentation. We show qualitative results obtained using our approach for joint segmentation on the video sequences provided in the validation set.

Training: We train our network for 20,000 iterations on Cityscapes training annotated set, with learning rate as 10^{-5}. Thereafter, the network was fine-tuned for 10,000 iterations on our KITTI-Training dataset, with learning rate and momentum as 10^{-4} and 0.9 respectively. We use this trained network for obtaining semantic segmentation for KITTI sequences. For obtaining semantic prediction for Cityscapes video sequences, we use the pre-trained dilated CNN network provided by [31]. The model was trained on Cityscapes training annotated benchmark with 19 semantic classes.

For computing disparity, we use Semi Global block matching algorithm [9]. We use state-of-the-art DeepFlow [29] for Optical Flow computation. The motion compatibility term is learned using our manually annotated training dataset. In the experiments carried out, the block-size used for KITTI is 10, while for Cityscapes we consider 25 frames in a single block. The lower block size for KITTI is attributed to the high frame rate, where beyond a given block size we lose tracks for moving objects. In the following section, we show an extensive qualitative and quantitative evaluation of our approach.

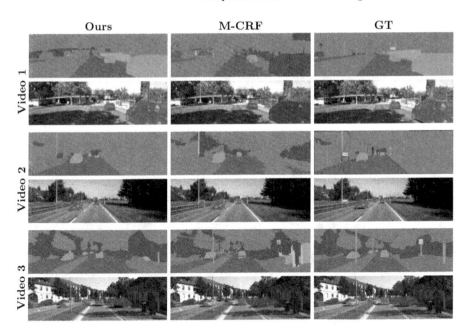

Fig. 3. Qualitative evaluation of our approach against M-CRF [18] and Ground Truth Labeling on our KITTI test dataset. The color convention of our joint labeling can be referred from Table 1. Left To Right: (1) Joint segmentation results from our approach on KITTI sequences. We also show overlay image for better visualization, where moving and stationary cars are overlayed by *green* and *blue* colour respectively. (2) Output from Multi Layer CRF [18] (3) Ground Truth annotations. (Color figure online)

Qualitative Evaluation: We qualitatively evaluate our approach with respect to Ground Truth and M-CRF [18]. Figure 3 shows qualitative comparisons on video sequences for our KITTI-test dataset. The label color spectrum is consistent with the label descriptions given in Tables 1 and 3. Particularly, *blue* and *green* colors denote static and moving car respectively. In Fig. 3, video sequence 1, we able to segment moving car approaching from behind while M-CRF [18] gives the stationary label to the object. M-CRF tends to rely on optical flow and in cases where a new object appears, optical flow is found to be inconsistent. With the incorporated temporal consistency in our approach, we are able to identify motion attributes for incoming objects in the scene. This is also evident in the video sequences 2 and 3 with multiple moving cars. In the sequences, M-CRF misses out on moving objects which are relatively far from the camera due to the inconsistent disparity in those regions. We are able to identify motion relatively farther from the camera which reiterates the role of a temporally consistent framework.

Temporal optimization in a joint label space proves advantageous in many ways. M-CRF relies on consistency constraints enforced through pairwise potentials in CRF and is found to have incorrectly labeled patches on the moving cars. With temporal optimization in joint label space, our approach enforces spatial

Fig. 4. Qualitative Evaluation on KITTI dataset. The results show proficiency of our approach across diverse scenes with challenging conditions. (Color figure online)

Fig. 5. Qualitative Evaluation on the CityScapes dataset. The joint segmentation obtained depict robustness of our approach with various dynamic objects. (Color figure online)

coherence. There is also clear demarcation between the boundaries of the moving object and its surroundings in contrast to M-CRF [18]. This is attributed to the spatial constraints enforced in the least-squares formulation in combination with semantic segmentation. While M-CRF emphasizes a strong motion prior with a separate layer for semantics, our approach enforces consistent segmentation with long term tracks and integration with semantics in unary space. Also, the strong semantic prior obtained from the dilated CNN produces better object labeling.

Results across diverse scenes: To showcase the proficiency of our approach in diverse scenes and different moving objects, we show results on various KITTI-Tracking and Cityscapes sequences. We show both dense joint segmentation results and overlay-ed images with vehicle and pedestrian classes. The joint results are shown in Figs. 4, 5 and on complete sequences in the supplementary video https://youtu.be/6kq8_FgwYFA. Figure 4(a) and (c) show highway scenes which consist of multiple cars moving with high speed. Our approach is correctly able to segment fleet of cars moving with high velocities (a). In Fig. 4(c) moving cars at the turns are clearly distinguished from nearby static cars in the scene, owing to the spatial constraints imposed while performing joint optimization. Due to the limited amount of annotated data for fine-tuning dilated CNN for KITTI dataset, the pedestrian classes are not identified with good precision across all sequences of KITTI. However, in sequences where semantic prediction for pedestrians is accurate, as shown in Fig. 4(c), we are able to capture the motion behavior accurately using our approach.

Results with different moving agents: To show performance on different dynamic objects such as pedestrians, bikes, etc., we show qualitative results on Cityscapes dataset. The Cityscapes dataset proves beneficial in this regard due to its large semantic segmentation training dataset. The color codes in our output are consistent with the label spectrum given in Table 3 and Cityscapes official color codes [3] for the remaining static classes. For better visualization, all vehicles are clubbed under a single class where moving and static vehicles are shown with *green* and *blue* color respectively. Similar policy is followed for 'human' classes where moving and static person is shown with *dark yellow* and *red* color respectively. Also the void or unlabeled regions belonging to the camera mounted car are taken as road. Figure 5(c) and (d) showcase our results on urban crowded scenes. We are able to segment moving pedestrians from the stationary ones with high precision and accurate boundaries which showcase the utility of our joint temporal optimization along with the spatial boundary constraints. The accurate differentiation can also be seen with the diverse vehicle classes present in the figure. The results are complemented in Fig. 5(a) which portrays a more common on-road urban street scene with human obstacles. In Fig. 5(a) our approach is correctly able to differentiate between on-road moving and stationary pedestrian as well as identify the motion attributes of pedestrians at a relatively larger distance from the camera. Figure 5(d) shows a person rested on a moving bike and it can be seen that our framework is able to categorize both of the agents as moving with precise boundaries. The results show the effectiveness of our approach in handling various classes across diverse scenes.

Table 1. (Left) Joint Motion Segmentation evaluation. We compare our method with Multi layer CRF [18]

Model	Moving		Stationary	
	Car	Pedestrian	Car	Pedestrian
M-CRF	68.87	19.27	28.72	16.08
Ours	**73.58**	**37.18**	**45.57**	**28.30**

Table 2. (Right) Motion Segmentation evaluation on two annotated highway sequences. We compare our method with STMOP [5] and Multi layer CRF [18]

Model	Stationary	Moving
STMOP	86.58	51.53
M-CRF	88.89	81.53
Ours-M	**96.94**	**89.17**

Quantitative Evaluation. Quantitative evaluations are carried out with respect to M-CRF [18] and STMOP [5] which have been evaluated on video motion segmentation. For semantic labeling, we show our evaluation for semantic label space w.r.t to M-CRF [18], semantic CNN [31] and T-CRF [14].

Semantics: For quantitative evaluation of our semantic segmentation obtained after joint formulation, we compare our results with existing M-CRF, dilated CNN [31] and temporal semantic CRF [14]. Evaluation is staged by cross-verifying each pixel with the corresponding Ground Truth label. The evaluation metric used is intersection over union, defined as $TP/(TP + FP + FN)$, where TP represents True Positive, FP False Positive and FN False Negative over each pixel in the image. We use the KITTI-test dataset for quantitative evaluation of our approach. Table 1 shows quantitative evaluation in joint label space. Although we cannot see huge improvements over the semantic temporal CRF to our method, this can be attributed to the fact that label transfer using this method on dynamic and static object perform the same way. Our method shows an improved segmentation of the moving objects.

Motion: We show quantitative evaluation of our motion segmentation with respect to existing approaches in both stereo(M-CRF) and monocular setup STMOP. For a fair comparison with STMOP, we use a subset of our annotated test sequences consisting of relatively fewer moving cars. Also, we use the best supervoxel projection in the proposals generated by STMOP. Table 2 shows quantitative evaluation of our approach. The main contribution of the proposed method can be seen in the Table 1. We observe a clear improvement in motion segmentation compared to the older methods which combined semantics and motion as a joint problem and geometric based methods. Most of the previous methods have attempted motion segmentation using geometric cues but could not get high accuracy because of the constraints in geometry. We show that combining learning based methods to geometric constraints can boost the accuracy of motion segmentation.

Table 3. Quantitative evaluation semantic segmentation with respect to M-CRF [18], dilated [31] and [14] on our annotated KITTI dataset.

Method	Building ■	Vegetation ■	Sky ■	Car ■■	Sign ■	Road ■	Pedestrian ■■	Fence ■	Pole ■	Sidewalk ■
Multi Layer CRF [17]	43.56	65.41	70.01	71.17	2.06	59.29	39.2	50.40	9.71	31.96
Dilated [31]	59.38	83.16	**91.41**	82.17	**12.59**	83.72	**65.74**	52.69	**42.01**	41.29
Kundu et al. [14]	60.95	**83.41**	91.11	**82.63**	6.96	84.63	64.81	**53.03**	20.54	42.22
Ours	**61.20**	83.18	91.38	80.87	3.11	**84.69**	64.32	53.01	17.81	**43.87**

6   Conclusion

We presented an end-to-end framework for joint semantic motion segmentation on a video. The proposed method integrates semantic constraints with dense point correspondences. We show results on multiple sequences of KITTI and release our annotations for comparisons.

We look at the problem of dynamic scene understanding and approach the problem using graphical models as end-to-end neural networks for these tasks need stronger cues. Looking at the end-to-end model for complete dynamic scene understanding is still an open problem. We believe that the dataset released and the current approach can form a basis for such models.

References

1. Badrinarayanan, V., Handa, A., Cipolla, R.: Segnet: a deep convolutional encoder-decoder architecture for robust semantic pixel-wise labelling. arXiv preprint arXiv:1505.07293 (2015)
2. Chen, T., Lu, S.: Object-level motion detection from moving cameras. IEEE Trans. Circ. Syst. Video Technol. **27**, 2333–2343 (2016)
3. Cordts, M., Omran, M., Ramos, S., Rehfeld, T., Enzweiler, M., Benenson, R., Franke, U., Roth, S., Schiele, B.: The cityscapes dataset for semantic urban scene understanding. In: CVPR (2016)
4. Dollár, P., Zitnick, C.L.: Fast edge detection using structured forests. PAMI **37**, 1558–1570 (2015)
5. Fragkiadaki, K., Arbeláez, P., Felsen, P., Malik, J.: Learning to segment moving objects in videos. In: CVPR. IEEE (2015)
6. Geiger, A., Lenz, P., Urtasun, R.: Are we ready for autonomous driving? the KITTI vision benchmark suite. In: CVPR (2012)
7. Geiger, A., Ziegler, J., Stiller, C.: Stereoscan: Dense 3D reconstruction in real-time. In: Intelligent Vehicles Symposium (IV) (2011)
8. Haque, N., Reddy, D., Krishna, M.: Joint semantic and motion segmentation for dynamic scenes using deep convolutional networks. In: VISAPP (2017)
9. Hirschmuller, H.: Stereo processing by semiglobal matching and mutual information. IEEE Trans. PAMI **30**, 328–341 (2008)
10. Huang, S.J., Yu, Y., Zhou, Z.H.: Multi-label hypothesis reuse. In: KDD. ACM (2012)
11. Jain, S., Madhav Govindu, V.: Efficient higher-order clustering on the grassmann manifold. In: ICCV, pp. 3511–3518 (2013)
12. Koltun, V.: Efficient inference in fully connected CRFS with Gaussian edge potentials. In: NIPS (2011)
13. Kundu, A., Krishna, K., Sivaswamy, J.: Moving object detection by multi-view geometric techniques from a single camera mounted robot. In: IROS (2009)
14. Kundu, A., Vineet, V., Koltun, V.: Feature space optimization for semantic video segmentation. In: CVPR (2016)
15. Long, J., Shelhamer, E., Darrell, T.: Fully convolutional networks for semantic segmentation. In: ICCV, pp. 3431–3440 (2015)
16. Noh, H., Hong, S., Han, B.: Learning deconvolution network for semantic segmentation. In: ICCV, pp. 1520–1528 (2015)
17. Reddy, N.D., Singhal, P., Chari, V., Krishna, K.M.: Dynamic body VSLAM with semantic constraints. In: IROS (2015)
18. Reddy, N.D., Singhal, P., Krishna, K.K.: Semantic motion segmentation using dense CRF formulation. In: ICVGIP (2014)

19. Redmon, J., Divvala, S., Girshick, R., Farhadi, A.: You only look once: Unified, real-time object detection. In: CVPR (2016)
20. Ren, S., He, K., Girshick, R., Sun, J.: Faster R-CNN: towards real-time object detection with region proposal networks. In: NIPS (2015)
21. Ros, G., Ramos, S., Granados, M., Bakhtiary, A., Vazquez, D., Lopez, A.: Vision-based offline-online perception paradigm for autonomous driving. In: WACV (2015)
22. Shotton, J., Johnson, M., Cipolla, R.: Semantic texton forests for image categorization and segmentation. In: CVPR. IEEE (2008)
23. Simonyan, K., Zisserman, A.: Two-stream convolutional networks for action recognition in videos. In: NIPS, pp. 568–576 (2014)
24. Simonyan, K., Zisserman, A.: Very deep convolutional networks for large-scale image recognition. arXiv:1409.1556 (2014)
25. Sundaram, N., Brox, T., Keutzer, K.: Dense point trajectories by GPU-accelerated large displacement optical flow. In: Daniilidis, K., Maragos, P., Paragios, N. (eds.) ECCV 2010. LNCS, vol. 6311, pp. 438–451. Springer, Heidelberg (2010). https://doi.org/10.1007/978-3-642-15549-9_32
26. Tourani, S., Krishna, K.M.: Using in-frame shear constraints for monocular motion segmentation of rigid bodies. JIRS **82**(2), 237–255 (2016)
27. Vertens, J., Valada, A., Burgard, W.: SMSnet: semantic motion segmentation using deep convolutional neural networks. In: IROS (2017)
28. Vidal, R., Sastry, S.: Optimal segmentation of dynamic scenes from two perspective views. In: CVPR, vol. 2 (2003)
29. Weinzaepfel, P., Revaud, J., Harchaoui, Z., Schmid, C.: Deepflow: large displacement optical flow with deep matching. In: ICCV (2013)
30. Yi, S., Li, H., Wang, X.: Pedestrian behavior understanding and prediction with deep neural networks. In: Leibe, B., Matas, J., Sebe, N., Welling, M. (eds.) ECCV 2016. LNCS, vol. 9905, pp. 263–279. Springer, Cham (2016). https://doi.org/10.1007/978-3-319-46448-0_16
31. Yu, F., Koltun, V.: Multi-scale context aggregation by dilated convolutions. arXiv preprint arXiv:1511.07122 (2015)
32. Zheng, S., Jayasumana, S., Romera-Paredes, B., Vineet, V., Su, Z., Du, D., Huang, C., Torr, P.H.: Conditional random fields as recurrent neural networks. In: ICCV, pp. 1529–1537 (2015)
33. Zografos, V., Nordberg, K.: Fast and accurate motion segmentation using linear combination of views. In: BMVC (2011)

Depth-Adaptive Computational Policies
for Efficient Visual Tracking

Chris Ying[1][(✉)] and Katerina Fragkiadaki[2][(✉)]

[1] Google Brain, Mountain View, CA, USA
chrisying@google.com
[2] Machine Learning Department, CMU, Pittsburgh, PA, USA
katef@cs.cmu.edu

Abstract. Current convolutional neural networks algorithms for video object tracking spend the same amount of computation for each object and video frame [3]. However, it is harder to track an object in some frames than others, due to the varying amount of clutter, scene complexity, amount of motion, and object's distinctiveness against its background. We propose a depth-adaptive convolutional siamese network that performs video tracking adaptively at multiple neural network depths. Parametric gating functions are trained to control the depth of the convolutional feature extractor by minimizing a joint loss of computational cost and tracking error. Our network achieves accuracy comparable to the state-of-the-art on the VOT2016 benchmark. Furthermore, our adaptive depth computation achieves higher accuracy for a given computational cost than traditional fixed-structure neural networks. The presented framework extends to other tasks that use convolutional neural networks and enables trading speed for accuracy at runtime.

Keywords: Visual tracking · Metric learning
Conditional computation · Deep learning

1 Introduction

Multilayer neural networks are the defacto standard machine learning tools for many tasks in computer vision, including visual tracking [3]. Current visual trackers use a fixed amount of computation for every object and video frame [3,14,20–22]. However, different video scenes have varying amount of complexity, background clutter, object motion, camera motion, or frame rate. Fixed compute architectures do not adapt to the difficulty of the input and are can be suboptimal computation-wise.

In this work, we propose neural architectures whose computation adapts to the difficulty of the task from frame to frame, rather than being fixed at runtime. We focus on the task of visual tracking in videos. We present learning

---

C. Ying—Work done as student at the Machine Learning Department, CMU.

M. Pelillo and E. Hancock (Eds.): EMMCVPR 2017, LNCS 10746, pp. 109–122, 2018.
https://doi.org/10.1007/978-3-319-78199-0_8

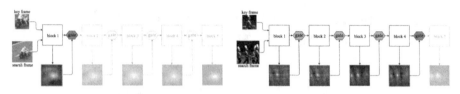

Fig. 1. Adaptive neural computational policies for visual tracking. At each frame, we match the key frame depicting the object of interest to a search region cropped around the location of the detected object in the previous frame. Our controllers (gates) decide how many blocks of the VGG net [18] (divided into 5 blocks of layers) to compute before computing a cross-correlation map and determining the target location. In general, deeper layers yield more accurate predictions but also require more computational power. Our depth-adaptive model picks the first depth for the uncluttered scene on the left and the fourth for the cluttered scene on the right. Red gate color denotes halting of computation at that gate. (Color figure online)

algorithms for training computational policies that control the depth of Siamese convolutional networks [9] for video tracking. Siamese convolutional networks track by computing cross-correlation maps of deep features, between a *key frame*, where the object is labeled, and a *search frame*, where the object needs to be localized, as shown in Fig. 2. The peak of the cross-correlation map denotes the presence of the target object. We observe that in many search frames, the tracked object is similar to the object in the key frame and distinct from its background, and it would be computationally wasteful to compute elaborate features for its detection. To address this, we propose *conditional* computation controlled by gating functions that dynamically determines how many layers of our convolutional feature extractor should be computed before computing the cross-correlation map and thus the target's location. In Fig. 1, our model uses only the first block of convolutional layers to find the motorcyclist against road, but uses 4 convolutional blocks to find the correct drummer among visually-similar peers. Our gate controllers are trained in an end-to-end differentiable framework without the need for sample-intense reinforcement learning. We test our model on the challenging VOT2016 dataset [11] and demonstrate that we perform video tracking with close to state-of-the-art accuracy at real-time speeds.

2   Related Works

2.1   Metric Learning for Visual Tracking

Metric learning approaches for visual tracking learn an appearance distance function between image box crops, so that the distance is large between image crops depicting different objects, and the distance is small between image crops depicting deformations of the same object instance. An accurate distance function then can be used to localize an object by computing the distances between the key frame and various crops within the search frame. Such a distance function can be learned using (a) Siamese networks [9], which use the same neural network weights to extract features from a pair of images before using a single fully connected layer to predict the distance, they are trained using contrastive loss function that minimizes distance between same instance examples and requires distances to be above a certain margin for dissimilar examples. (b) Triplet networks, [8] trained with a ranking loss that ensures distance of positive pairs is lower than the distance of negative pairs, and obviates the need of a margin hyper-parameter.

2.2 Conditional Computation

Conditional computation refers to activating different network components depending on the input and serves as a promising way to reduce computational cost without sacrificing representational power. In [2], conditional computation is implemented by selectively activating different weights in each layer and is trained via reinforcement learning. [17] uses a sparse gating function to determine which sub-networks to execute (each of which are "experts" for different inputs), and shows that it is possible to train the gating and network weights jointly via back-propagation. Graves [5] proposed an adaptive computation model for Recurrent Neural Networks (RNNs), that determines (depending on the input) the number of computational (pondering) steps required before producing an accurate output. Recent work [4], adapts this model to convolutional networks for object detection in static images, where the network is trained to learn the number of convolutional layers to be evaluated per image location, e.g., "easy" image regions (e.g., sky) should require less computation than more feature-rich ones (e.g., a car). Our work differs in that (a) we use conditional computation in videos, rather than static images, and (b) the input we predicate computational decisions on is the quality of a cross-correlation tracking map, as opposed to image classification accuracy.

2.3 Estimating or Back-Propagating Gradients

A central question in all works that learn adaptive computation policies is how to train discrete *gates/controllers*, the discrete elements that determine how computation should be scheduled. Researchers have typically used non-differentiable score function estimators (a.k.a. REINFORCE [23]) for estimating the gradient with respect to such binary thresholds. REINFORCE has been shown to yield gradients with very high variance and requires too many samples for informative gradients to be estimated. High sample complexity is an attribute of many other model-free RL methods, e.g., Q-learning [15]. Indeed, recent works that use such RL techniques for neural net architectural search [25] or conditional computation [13], only scale to small networks and datasets [13], or use large computational resources for training, e.g., in recent work 800 GPUs were used concurrently [25], as opposed to a single GPU in our case. Alternatively, researchers have used soft, differentiable gates with carefully designed differentiable architectures for image generation (DRAW [7]), accessing an external memory (Neural Turing Machines [6]), deciding halting of a recurrent networks [5], etc. Our training scheme, which similarly uses soft and differentiable gates during training to provide meaningful gradients, allows us to scale our policies to controlling deep neural architectures.

3 Depth-Adaptive Fully-Convolutional Siamese Networks

Our model builds upon fully-convolutional Siamese networks from [3], a state-of-the-art model for visual object tracking, which uses the same convolutional neural network to extract deep features from the key and search frames. The model then uses 2D cross-correlation to efficiently calculate the similarity score of the object in the key frame to every spatial location in the search frame, as shown in Fig. 2. This implicitly implements a triplet network-like loss by penalizing all the negative locations and increasing the similarity at the true location. Unlike [3], which uses an AlexNet-like [12] architecture and trains from scratch using ImageNet Video dataset [16], we use a VGG feature extractor [18] pretrained from the ImageNet static image classification dataset.

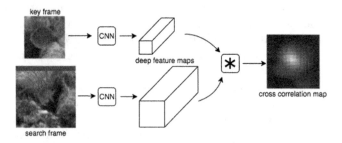

Fig. 2. Siamese network with 2D cross-correlation for key-search frame pairs. The deep feature maps for the key and search frames are extracted by the same convolutional neural network. ∗ denotes 2D cross-correlation.

We extend the fully-convolutional Siamese network for depth-adaptive computation by first dividing the convolutional layers into 5 "blocks" of convolutions and adding intermediate cross-correlations after every convolutional block. To finetune the convolutional weights, we calculate the softmax cross-entropy loss $\mathcal{L}_i$ between each of the computed cross-correlation maps c_i and a ground-truth map G for $i = 1, \ldots, 5$ in our tracking training set. The ground-truth map is a 2D Gaussian centered at the true location of the object in the search frame.

$$\mathcal{L}_i = \texttt{softmax-cross-entropy}(c_i, G) \tag{1}$$

Furthermore, we introduce parametric gating functions between each of the convolutional blocks, which act as controllers for the depth of the VGG feature extractor at runtime. These gating functions take as input the cross-correlation map computed using the deep features at the current depth, and output a *confidence score for halting computation* at that particular depth. In theory, we could use a convolutional neural network to extract the confidence score from each cross-correlation map. However, we would like our gating functions to be computationally inexpensive, so instead we use a small set of intuitive features to capture the "quality" (certainty) of each cross-correlation map, such as, kurtosis (measures "peakiness"), entropy, top-5 max peak values, and the first 5 moments. Let f denote the shallow feature extractor that given a cross-correlation map outputs the features above. We then learn a linear predictor, parameterized by ϕ_i to output the confidence score $\mathbf{g}_i$ for the gate at each depth, re-scaled to $(0, 1)$ via a sigmoid function, as follows:

$$\mathbf{g}_i(c_i; \phi_i) = \text{sigm}(f(c_i)^T \cdot \phi_i) \in (0, 1). \tag{2}$$

Our full depth-adaptive model is depicted in Fig. 3. At training time, we use soft gates in order to use back-propagation for learning the gate weights, and at test time, we use hard gate thresholding, to halt computation at a particular network depth.

To train the model effectively and achieve a satisfactory trade-off of tracking accuracy and computational savings, we found the following two design choices to be crucial:

1. **Intermediate supervision:** Rather than training using the loss at the deepest layer only, like [3], we use a sum of tracking losses at all layers. This introduces intermediate supervision, which has been found to be useful in non-adaptive computational architectures such as [19, 24].
2. **Budgeted gating:** We found that directly using the confidence scores $\mathbf{g}_i$ is insufficient for learning a depth-adaptive policy since each score does not affect the scores at other depths, which leads to polarized policies (either always use the shallowest depth or always use the deepest depth). Instead, we use a "budgeted" confidence score $\mathbf{g}_i^*$, in Eq. 3, where the scores sum to 1.0 and we have the desired behavior that a higher confidence score in a shallower depth corresponds to less need for deeper layers and vice-versa.

$$\mathbf{g}_i^*(c_i; \phi_i) = \begin{cases} (1 - \sum_{j=1}^{i-1} \mathbf{g}_j^*(c_j; \phi_j)) \mathbf{g}_i(c_i; \phi_i), & i \in [1, 4] \\ 1 - \sum_{j=1}^{4} \mathbf{g}_j^*(c_j; \phi_j), & i = 5. \end{cases} \tag{3}$$

We cannot train the gate parameters $\phi_i, i = 1 \ldots 5$ and VGG weights jointly, since the gate feature extractor f is non-differentiable. Thus, we train in two phases. In the first phase, we finetune the VGG weights by minimizing the non-gated loss at all depths $\mathcal{L}^{\text{conv}}$:

$$\mathcal{L}^{\text{conv}} = \sum_{i=1}^{5} \mathcal{L}_i \tag{4}$$

In the second phase, we fix the VGG weights and train the gate parameters by minimizing a loss function $\mathcal{L}^{\text{gate}}$ that combines tracking loss and computational cost in all depths:

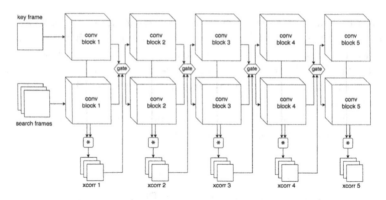

Fig. 3. Depth adaptive Siamese convolutional networks. Convolutional weights are shared between the two network stacks. Each conv block includes 2–4 convolutions with ReLU activation. The feature maps of the key and search frames at the end of each block are cross-correlated to yield 5 cross-correlation (xcorr) maps. A gating function is added at the end of each convolutional block that controls whether the network stops at this layer, or continues computation to a higher depth.

$$\mathcal{L}^{\text{gate}} = \underbrace{\sum_{i=1}^{5} \mathbf{g}_i^* \mathcal{L}_i}_{\text{tracking loss}} + \lambda \underbrace{\sum_{i=1}^{5} p_i \mathbf{g}_i^*}_{\text{computational cost}} , \tag{5}$$

where the hyper-parameter λ trades off tracking accuracy and computational efficiency. We found that $\lambda \in [0.5, 1.0]$ resulted in a diverse set of depths of computation (greater or less than that range generally led to "polarized" results, either all deepest layer or all shallowest layer). The parameter p_i encodes the relative incremental computational cost of each successive layer. For our experiments, we set each p_i to the incremental additional cost as reported in Table 1 with the $p_1 = 1.0$. For example, $p_2 = 2.43 - p_1 = 1.43$.

Much like [17], our training method provides balanced updates to all gates, meaningful gradients, and requires less training data. Though soft gates are used during training to enable back-propagation, at runtime, we threshold the budgeted confidence score and halt computation at a gate if the score exceeds some tune-able value. The score is a value in $[0, 1]$ and in our experiments we set the threshold at 0.25, 0.5, or 0.75 for increasing degrees of strictness (i.e. higher threshold means we are less likely to accept the tracking result at a shallower layer).

3.1   Implementation Details

The base architecture we use for the convolutional layers is the 19-layer VGG architecture [18]. We remove the fully connected layers of the architecture and treat the remaining convolutional and max-pool layers as the feature extractor.

The VGG architecture is divided into 5 blocks of convolutions with 2–4 convolutional layers each, each ending in a max-pool layer. We remove the last max-pool layer in order to keep the deep feature maps as large as possible (i.e. 16×16 in the last layer). To improve training, we normalize the key and search feature maps via batch normalization and rescale the output cross-correlation map to $[0, 1]$. Cross-correlation is an expensive operation so to keep the computational costs low, we downsample the feature maps to 16×16 before cross-correlation.

Since training is performed with a single key frame and a batch of search frames, cross-correlation can be efficiently implemented on GPU by performing 2D convolution on the search feature maps with the key feature maps as the filter, treating the feature channels as the input channel size (the output channel size is 1).

Our model is implemented in TensorFlow v1.0.0 [1] using pretrained VGG network weights on ImageNet [16] for image classification. All training and evaluation was performed on a single NVIDIA TITAN X GPU, an Intel Xeon E5-2630 v3 CPU, and 16 GB of RAM.

To efficiently implement hard-gating, we use TensorFlow's control flow operators (`tf.cond`). Hard-gating is only fully efficient when the batch size of the search frames is 1 since the computation is bottlenecked by the deepest cross-correlation map that is required by a sample in a batch. In practice, this is not as much of an issue since consecutive frames tend to use similar depths for prediction.

4   Experiments

We train and test our model on the Visual Object Tracking dataset VOT2016 [11]. The dataset consists of 60 videos with a total of 21455 frames of various resolutions. Each frame is labelled with the box corner coordinates of a bounding box that corresponds to a single object being tracked in the video. The videos have noisy backgrounds, the object can change shape or orientation, and there is occlusion in some frames. Since the VOT2016 dataset does not include a train-validation split, we randomly pick 25% of the videos to hold out as the test set. Note that the VOT dataset is designed for an evaluation-only competition so our results are not directly comparable to existing benchmarks. Our goal is not necessarily to beat state-of-the-art methods, but rather to present a useful technique for fast video tracking which can improve nearly any convolutional model.

We preprocess the videos by selecting a key frame every 10 frames and the subsequent up-to-100 frames as the search frames. We resize and crop the key frames to 128×128 centered at the tracked object such that there is at least 25% padding around the bounding box. Each of the search frames are resized with the same scale and cropped to 256×256 such that the frame is centered at the object at the previous frame. If the cropped search frame extends beyond the edge of the image, we pad the extra pixels with the mean RGB value of the dataset. The predicted object box is found using the position of the maximum

value in the cross-correlation map as the offset and the bounding box dimensions
are the same as the key frame reference box.

4.1   Evaluation Metrics

We measure tracking accuracy using Intersection-Over-Union (IOU) between the
predicted object box b^{pred} and the ground truth box b^{gt}:

$$IOU(b^{pred}, b^{gt}) = \frac{|b^{pred} \cap b^{gt}|}{|b^{pred} \cup b^{gt}|}. \tag{6}$$

We measure IOU at up-to 1, 5, and 25 frames ahead of the key frame, e.g., for
IOU@25, we take the average IOU of the tracker with key frame t and search
frames $t+1, t+2, \ldots, t+25$. The larger the frame gap between key frame and
search frame, the more the tracking target deforms and the harder it is to track.

   We measure computational cost by computing the number of floating point
operations (FLOPs) required to perform tracking on a batch of 25 search frames.
Experimentally, we find that FLOPs is a good proxy for true computational cost
as measured in frames-per-second (FPS). The reason FLOPs is the preferable
metric is that FPS is heavily tied to hardware and software constraints, which
may prevent the architecture from achieving the theoretical speedup.

4.2   Siamese Tracker Performance

We finetune our VGG feature extractor starting from pretrained weights using
the tracking loss of Eq. 4. The tracking performance during finetuning is shown
in Fig. 4. Using pretrained weights allows the model to reach peak performance
after only a few epochs. For this evaluation we use the full network depth.

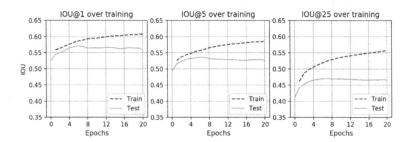

Fig. 4. Finetuning. Training and testing IOU curves during metric learning for different
frame gaps. Starting with weights pretrained on ImageNet image classification, our
VGG feature extractor fast reaches top performance.

   Our implementation is comparable to the top submissions to the VOT2016
competition [10] in IOU over sequence length, as seen in Fig. 5. Though it is not

as accurate as the best trackers at small sequence lengths, it is competitive with many trackers at around 100 frames. Note also that the top four trackers from VOT2016 run at under 1 FPS while our system runs at over 54 FPS using the shallowest layer (`xcorr1`) and over 37 FPS using the deepest layer (`xcorr5`). Note that our code was not optimized so FPS is not a very good metric for computational cost. Additional software engineering (e.g. multithreaded inputs, better GPU utilization) should bring the `xcorr1` FPS to well over 100.

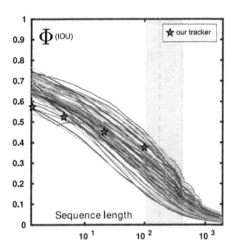

Fig. 5. VOT2016. Comparison of our full-computation model against top submissions to VOT2016 [10]. The stars represent our tracker's accuracy at selected sequence lengths.

4.3   Effect of Intermediate Supervision

We compare the performance of the tracker when trained with loss only at the deepest layer against the tracker when trained with losses added in all depths. The IOU@25 comparison is shown in Fig. 6.

Training with all intermediate losses yields increased accuracy as depth increases, while training with only the deepest yields a big jump in accuracy at depth 5 but lower accuracy at shallower depths. If computational cost is not a factor, using only the deepest loss gives around 0.01 IOU benefit over using all depths. Since our goal is to use depth-adaptive feature extraction, intermediate supervision is essential.

4.4   Depth-Adaptive Computational Policies

The computational policies we compare are:

- Fixed-depth: always use the cross-correlation map at a fixed depth (i.e. `xcorr1`, ..., `xcorr5`).

– Soft-gating: use a sum of the cross-correlation maps at all the depths, weighted by the budgeted confidence score. This policy does not save computational cost but serves as a baseline that also utilizes the gating functions.
– Hard-gating: we halt computation if the budgeted confidence score exceeds a tune-able threshold, which is a model hyper-parameter. We report performance while varying hyper-parameters in order to obtain accuracy/computation trade offs across the whole spectrum.

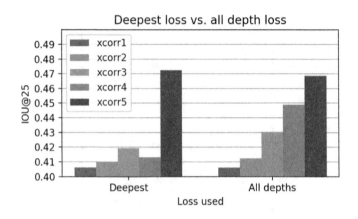

Fig. 6. Intermediate supervision. Supervision at intermediate layers (as opposed to the top layer only) increases the accuracy of the intermediate cross-correlation maps, and makes depth-adaptation at runtime worthwhile.

Table 1. Theoretical FLOPs for varying network depth. xcorr$i$ denotes network evaluation up until the i-th cross-correlation map. Soft-gating uses all five cross-correlation maps weighted by the gating confidence. The gating feature computation is negligible in comparison to convolutional feature extraction.

Gating policy	FLOPs ($\times 10^9$)	Relative to xcorr1
xcorr1	2.78	1.00×
xcorr2	67.70	2.43×
xcorr3	160.75	5.78×
xcorr4	253.79	9.12×
xcorr5	280.37	10.07×
Soft-gating	280.53	10.08×
Hard-gating	Varies	Varies

Table 1 shows the theoretical FLOPs required to compute the cross-correlation maps for a single key-search batch of 1 key frame and 25 search frames. For simplicity of calculation, the values only include the floating point multiplication operations, which comprise the overwhelming majority of the computation. As mentioned earlier, FLOPs is a better metric than FPS because it is independent of implementation details of the algorithm.

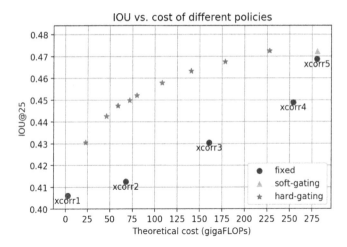

Fig. 7. Accuracy versus computation curves for our model and baselines. We generated the curve for our model (*hard gating*) by varying the relative weight λ of computational cost and tracking accuracy. For our fixed depth baseline, we obtain five points by varying the number of convolutional blocks from 1 to 5. Top left of the diagram is more desirable. Our model clearly outperforms the non-learned fixed-depth policies.

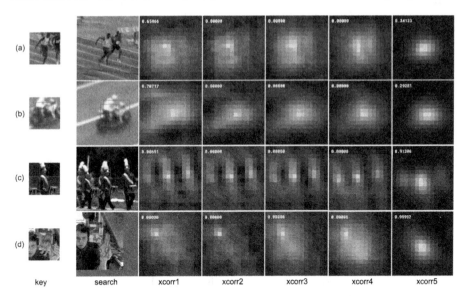

Fig. 8. Tracking results. Green box is ground truth, red box is prediction, red numbers are confidence weights. In (a), the tracker learns that `xcorr1` is sufficient for tracking. In (b), the tracker learns that it needs to compute `xcorr5` in order to confidently track the object. (Color figure online)

Figure 7 compares the accuracy and computational cost of each of the policies. Both soft and hard-gating can achieve accuracy values that exceed any fixed-depth policy and furthermore hard-gating uses significantly less computational cost to achieve the same or better accuracy. By varying hyper-parameter λ, we achieve different trade-offs of accuracy and computational cost depending on the requirements of the task.

The cross-correlation maps for selected video frames can be viewed in Fig. 8.

5   Discussion

Our experimental results show that our proposed depth-adaptive fully convolutional siamese network successfully tracks at accuracies comparable to state-of-the-art submissions to VOT2016. We demonstrate that our learned depth-adaptive policies can outperform fixed-depth networks while using significantly less computational power. Furthermore, we show that we can easily trade accuracy for computational cost as necessary by changing the model hyper-parameters.

Our work is limited by the following factors:

- Our model is finetuned purely on VOT2016 data unlike most other submissions which use ImageNet VID, or other tracking datasets for training the Siamese network. VOT2016 dataset is considerably smaller than ImageNet VID and the videos are considered more difficult to track.
- Fully-convolutional Siamese networks do not support updating the tracking model (i.e. key frame object appearance) since it does not naively support bounding box rescaling.
- Our final cross-correlation map resolution is 16×16, which is too coarse for fine-grained tracking. We experimented with higher resolution cross-correlation maps and obtained 15% improvement to IOU. However, this method is incompatible with hard-gating since it requires cross-correlation at multiple depths.

6   Conclusion

We have presented a conditional computation model for visual tracking, where computational depth is allocated based on a frame's tracking difficulty. Our model is comprised of continuous weight filter variables and discrete learned controllers that dynamically manage the depth of the network at runtime, and balance accuracy with computational cost. The proposed model saves computation on "easy" frames without sacrificing representational power on difficult frames that require deeper features to track. We show our model outperforms naive non-adaptive policies by a significant margin, as measured by accuracy at a various computational costs. Paths for future work include multi-scale tracking and more complex gating features, potentially using the history of the cross-correlation maps to determine computation. Though this work investigates policies that control the depth of the network, other promising "actions" for conditional computation in visual tracking include adaptive spatial computation e.g., by using motion to focus attention to the moving parts of the scene, or frame skipping e.g., by using only a subset of frames to localize the target without sacrificing tracking accuracy.

The methods presented in this work can be extended to any task which uses deep neural networks. By treating neural networks as a series of composable feature extractors, we have the ability to select feature embeddings at various degrees of complexity, which can both reduce computational cost and potentially improve performance.

References

1. Abadi, M., Agarwal, A., Barham, P., Brevdo, E., Chen, Z., Citro, C., Corrado, G.S., Davis, A., Dean, J., Devin, M., Ghemawat, S., Goodfellow, I., Harp, A., Irving, G., Isard, M., Jia, Y., Jozefowicz, R., Kaiser, L., Kudlur, M., Levenberg, J., Mané, D., Monga, R., Moore, S., Murray, D., Olah, C., Schuster, M., Shlens, J., Steiner, B., Sutskever, I., Talwar, K., Tucker, P., Vanhoucke, V., Vasudevan, V., Viégas, F., Vinyals, O., Warden, P., Wattenberg, M., Wicke, M., Yu, Y., Zheng, X.: TensorFlow: large-scale machine learning on heterogeneous systems. Software: tensorflow.org (2015)
2. Bengio, E., Bacon, P., Pineau, J., Precup, D.: Conditional computation in neural networks for faster models. CoRR, abs/1511.06297 (2015)
3. Bertinetto, L., Valmadre, J., Henriques, J.F., Vedaldi, A., Torr, P.H.S.: Fully-convolutional siamese networks for object tracking. In: Hua, G., Jégou, H. (eds.) ECCV 2016. LNCS, vol. 9914, pp. 850–865. Springer, Cham (2016). https://doi.org/10.1007/978-3-319-48881-3_56
4. Figurnov, M., Collins, M.D., Zhu, Y., Zhang, L., Huang, J., Vetrov, D.P., Salakhutdinov, R.: Spatially adaptive computation time for residual networks. In: CVPR (2017)
5. Graves, A.: Adaptive computation time for recurrent neural networks. CoRR, abs/1603.08983 (2016)
6. Graves, A., Wayne, G., Danihelka, I.: Neural turing machines. CoRR, abs/1410.5401 (2014)
7. Gregor, K., Danihelka, I., Graves, A., Rezende, D.J., Wierstra, D.: DRAW: a recurrent neural network for image generation. In: ICML, pp. 1462–1471 (2015)
8. Hoffer, E., Ailon, N.: Deep metric learning using triplet network. CoRR, abs/1412.6622 (2014)
9. Koch, G.: Siamese neural networks for one-shot image recognition. Ph.D. thesis, University of Toronto (2015)
10. Kristan, M., et al.: The visual object tracking VOT2016 challenge results. In: Hua, G., Jégou, H. (eds.) ECCV 2016. LNCS, vol. 9914, pp. 777–823. Springer, Cham (2016). https://doi.org/10.1007/978-3-319-48881-3_54
11. Kristan, M., Matas, J., Leonardis, A., Felsberg, M., Cehovin, L., Fernandez, G., Vojir, T., Hager, G., Nebehay, G., Pflugfelder, R.: The visual object tracking VOT2015 challenge results. In: ICCV, pp. 1–23 (2015)

12. Krizhevsky, A., Sutskever, I., Hinton, G.E.: Imagenet classification with deep convolutional neural networks. In: Advances in neural information processing systems, pp. 1097–1105 (2012)
13. Liu, L., Deng, J.: Dynamic deep neural networks: optimizing accuracy-efficiency trade-offs by selective execution. arXiv:1701.00299 (2017)
14. Ma, C., Huang, J.-B., Yang, X., Yang, M.-H.: Hierarchical convolutional features for visual tracking. In: ICCV, pp. 3074–3082 (2015)
15. Mnih, V., Kavukcuoglu, K., Silver, D., Graves, A., Antonoglou, I., Wierstra, D., Riedmiller, M.A.: Playing atari with deep reinforcement learning. arXiv:1312.5602 (2013)
16. Russakovsky, O., Deng, J., Su, H., Krause, J., Satheesh, S., Ma, S., Huang, Z., Karpathy, A., Khosla, A., Bernstein, M., Berg, A.C., Fei-Fei, L.: Imagenet large scale visual recognition challenge. IJCV 115(3), 211–252 (2015)
17. Shazeer, N., Mirhoseini, A., Maziarz, K., Davis, A., Le, Q.V., Hinton, G.E., Dean, J.: Outrageously large neural networks: the sparsely-gated mixture-of-experts layer. CoRR, abs/1701.06538 (2017)
18. Simonyan, K., Zisserman, A.: Very deep convolutional networks for large-scale image recognition. CoRR, abs/1409.1556 (2014)
19. Szegedy, C., Liu, W., Jia, Y., Sermanet, P., Reed, S., Anguelov, D., Erhan, D., Vanhoucke, V., Rabinovich, A.: Going deeper with convolutions. In: CVPR (2015)
20. Wang, L., Ouyang, W., Wang, X., Lu, H.: Visual tracking with fully convolutional networks. In: 2015 IEEE International Conference on Computer Vision (ICCV), pp. 3119–3127, December 2015
21. Wang, N., Yeung, D.-Y.: Learning a deep compact image representation for visual tracking. In: Burges, C., Bottou, L., Welling, M., Ghahramani, Z., Weinberger, K. (eds.), Advances in Neural Information Processing Systems, vol. 26, pp. 809–817 (2013)
22. Weng, S.-K., Kuo, C.-M., Tu, S.-K.: Video object tracking using adaptive kalman filter. J. Vis. Commun. Image Represent. 17(6), 1190–1208 (2006)
23. Williams, R.J.: Simple statistical gradient-following algorithms for connectionist reinforcement learning. Mach. Learn. 8(3), 229–256 (1992)
24. Xie, S., Tu, Z.: Holistically-nested edge detection. CoRR, abs/1504.06375 (2015)
25. Zoph, B., Le, Q.V.: Neural architecture search with reinforcement learning. arXiv:1611.01578 (2016)

Multiframe Motion Coupling for Video Super Resolution

Jonas Geiping[1(✉)], Hendrik Dirks[2], Daniel Cremers[3], and Michael Moeller[1]

[1] University of Siegen, Siegen, Germany
{jonas.geiping,michael.moeller}@uni-siegen.de
[2] University of Münster, Münster, Germany
mail@hendrik-dirks.de
[3] Technical University of Munich, Munich, Germany
cremers@tum.de

Abstract. The idea of video super resolution is to use different view points of a single scene to enhance the overall resolution and quality. Classical energy minimization approaches first establish a correspondence of the current frame to all its neighbors in some radius and then use this temporal information for enhancement. In this paper, we propose the first variational super resolution approach that computes several super resolved frames in one batch optimization procedure by incorporating motion information between the high-resolution image frames themselves. As a consequence, the number of motion estimation problems grows linearly in the number of frames, opposed to a quadratic growth of classical methods and temporal consistency is enforced naturally.

We use infimal convolution regularization as well as an automatic parameter balancing scheme to automatically determine the reliability of the motion information and reweight the regularization locally. We demonstrate that our approach yields state-of-the-art results and even is competitive with machine learning approaches.

1 Introduction

The technique of video super resolution combines the spatial information from several low resolution frames of the same scene to produce a high resolution video. A classical way of solving the super resolution problem is to estimate the motion from the current frame to its neighboring frames, model the data formation process via warping, blur, and downsampling, and use a suitable regularization to suppress possible artifacts arising from the ill-posedness of the underlying problem. The final goal is to produce an enhanced, visually pleasing high resolution video in a reasonable runtime. However the number of flow computations in this approach increases quadratically with the number of frames. Moreover, due to the strategy of super resolving each frame separately, temporal consistency cannot be enforced explicitly. Yet the latter is a key feature of a

© Springer International Publishing AG, part of Springer Nature 2018
M. Pelillo and E. Hancock (Eds.): EMMCVPR 2017, LNCS 10746, pp. 123–138, 2018.
https://doi.org/10.1007/978-3-319-78199-0_9

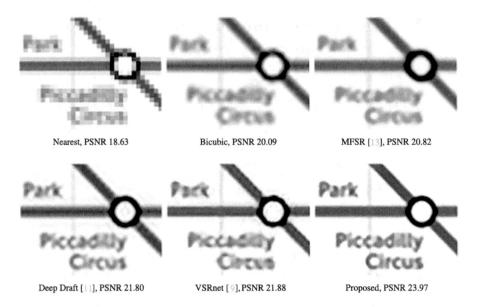

Nearest, PSNR 18.63   Bicubic, PSNR 20.09   MFSR [13], PSNR 20.82

Deep Draft [11], PSNR 21.80   VSRnet [9], PSNR 21.88   Proposed, PSNR 23.97

Fig. 1. Results for super resolving a set of 13 images of a London tube map by a factor of 4. Due to the idea of jointly super resolving multiple frames, our approach behaves superior to the competing variational approach [13]. While approaches based on learning [9,11] are sharp, they sometimes have difficulties resolving structures they were not trained on.

visually pleasing video: Even if a method generates a sequence of high quality high resolution frames, temporal inconsistencies will be visible as a disturbing flickering.

In addition, choosing the right strength of the regularization is a delicate issue. While a small regularization allows significant improvements in areas where the motion estimation is precise, it can lead to heavy oscillations and ringing artifacts in areas of quick motion and occlusions. A large regularization on the other hand avoids these artifacts but quickly oversmoothes the image and hence also suppresses the desirable super resolution effect.

Contributions of this work. We propose a method that jointly solves for all frames of the super resolved video and couples the high resolution frames directly. Such an approach tackles the drawbacks mentioned above: Because only neighboring frames are coupled *explicitly*, the number of required motion estimations grows linearly with the number of frames. However by introducing this coupling on the unknown high resolution images directly, all frames are still coupled *implicitly* and information is exchanged over the entire sequence.

Furthermore, we tackle the problem of choosing the right strength of spatial regularity by proposing to use the *infimal convolution* between a strong spatial and a strong temporal regularization term. The latter allows our framework to automatically select the right type of regularization locally in a single convex optimization approach that can be minimized globally. To make this approach robust we devise a parameter choice heuristic that allows us to process very different videos.

As illustrated in Fig. 1 our approach yields state-of-the-art results. While Fig. 1 is a synthetic test consisting of planar motion only, we demonstrate the performance of the proposed approach on several real world videos in Sect. 4.

The literature on super resolution techniques is vast and it goes beyond the scope of the paper to present a complete overview. An extensive survey of super resolution techniques published before 2012 can be found in [18]. We will focus on recalling some recent approaches based on energy minimization and deep learning techniques.

Variational Video Reconstruction. A classical variational super resolution technique was presented in [25] in which the authors propose to determine a high resolution version of the i-th frame via

$$\min_{u^i} \|D(b * u^i) - f^i\|_{H^{\epsilon_d}} + \lambda \|\nabla u^i\|_{H^{\epsilon_r}} + \sum_{j \neq i} \|D(b * W^{j,i} u^i) - f^j\|_{H^{\epsilon_d}}, \quad (1)$$

where $\|\cdot\|_{H^{\epsilon_d}}$ denotes the Huber loss, D a downsampling operator, b a blur kernel, λ a regularization parameter, and $W^{j,i}$ a warping operator that compensates the motion from the j-th to the i-th frame and is computed by an optical flow estimation in a first processing step. The temporal consistency term is based on $D(b * W^{j,i} u^i) - f^j$ and hence compares each frame to multiple low resolution frames. Figure 2a shows all couplings $W^{j,i}$ needed to use this approach for a sequence of frames.

Mitzel et al. [15] use a similar minimization, albeit with the l^1 norm instead of Huber loss. In comparison to [25] they do not compute all needed couplings but approximate them from the flows between neighboring frames, which allows for a trade-off between speed and accuracy.

Liu and Sun [12] proposed to incorporate different (global) weights $\theta_{j,i}$ for each of the temporal consistency terms in Eq. (1), and additionally estimate the blur kernel b as well as the warping operators $W^{j,i}$ by applying alternating minimization. In [13], Ma et al. extended the work [12] for the case of some of the low resolution frames being particularly blurry. Similar to (1) the energies proposed in [12,13] do not enforce regularity between the high resolution frames u^i directly and require quadratically many motion estimations. Furthermore both works focus on a simplified downsampling procedure that is easier to invert than our more realistic model.

In a recent work [3] on time continuous variational models, the authors proposed to use an optical flow penalty $\|\nabla u \cdot \boldsymbol{v} + u_t\|_1$ as a temporal regularization for joint image and motion reconstruction. While the optical flow term is exact in the temporally continuous setting, it would require small motions of less than one pixel to be a good approximation in a temporally discrete video.

Learning based approaches. With the recent breakthroughs of deep learning and convolutional neural networks, researchers have promoted learning-based methods for super resolution [9–11,22,30]. The focus of [22,30] is the development of a real-time capable super resolution technique, such that we will concentrate our comparison to [9–11], which focus on high image quality rather than computational efficiency.

Note that [9,11] work with motion correction and require optical flow estimations. Similar to the classical variational techniques they register multiple neighboring frames to the current frame and hence also require quadratically many flow estimations. The very deep convolutional network VDSR [10] is a conceptually different approach that does not use any temporal information, but solely relies on the training data.

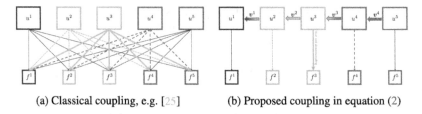

(a) Classical coupling, e.g. [25]   (b) Proposed coupling in equation (2)

Fig. 2. Illustrating different kinds of temporal couplings for an examples sequence of 5 frames: (a) shows how classical methods couple the estimated high resolution frames with all input data, (b) illustrates the proposed coupling. Each frame is merely coupled to its corresponding low resolution version.

2   Proposed Method

For a sequence $f = f^1, \ldots, f^n$ of low-resolution input images we propose a multi-frame super resolution model based on motion coupling between subsequent frames. Opposed to any of the variational approaches summarized in the previous section, the energy we propose directly couples all (unknown) high resolution frames $u = u^1, \ldots, u^n$. Our method jointly computes the super resolved versions of n video frames at once via the following minimization problem,

$$\min_{u} \sum_{i=1}^{n} \|D(b * u^i) - f^i\|_1 + \alpha \inf_{u=w+z} R_{\text{temp}}(w) + R_{\text{spat}}(z). \qquad (2)$$

The first term is a standard data fidelity term similar to (1). The key novelty of our approach is twofold and lies in the way we incorporate and utilize the motion information as well as the way we combine the temporal information with a spatial regularity assumption. The latter combines an extension of a spatio-temporal infimal convolution technique proposed by Holler and Kunisch in [8] with an automatic parameter balancing scheme.

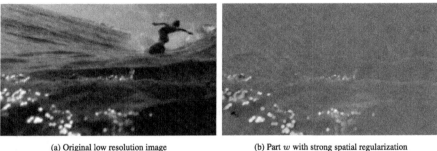

(a) Original low resolution image   (b) Part w with strong spatial regularization

(c) Super resolution result u (d) Part $u - w$ with strong temporal regularization

Fig. 3. Illustrating the behavior of infimal convolution regularization: The super resolution result u of the low resolution input data from (a) is given in Subfigure (c). Subfigures (b) and (c) illustrate the division of u into the two parts w and $z = u - w$ determined by the infimal convolution regularization (3).

2.1 Spatio-Temporal Infimal Convolution and Parameter Balancing

The second term in (2) denotes the *infimal convolution* [5] between a term R_{temp}, which is mostly focused on introducing temporal information, and a term R_{spat}, which is mostly focused on enforcing spatial regularity on u. The infimal convolution between the two terms is defined as

$$(R_{\text{temp}}\Box R_{\text{spat}})(u) := \inf_{u=w+z} R_{\text{temp}}(w) + R_{\text{spat}}(z). \tag{3}$$

It can be understood as a convex approximation to a logical OR connection and allows to optimally divide the input u into two parts, one of which is preferable in terms of the costs R_{temp} and the other one in terms of the costs R_{spat}. The respective costs are defined as

$$R_{\text{spat}}(u) = \sum_{i=1}^{n} \left\| \sqrt{(u_x^i)^2 + (u_y^i)^2 + (\kappa W(u^i, u^{i+1}))^2} \right\|_1, \tag{4}$$

$$R_{\text{temp}}(u) = \sum_{i=1}^{n} \left\| \sqrt{(\kappa u_x^i)^2 + (\kappa u_y^i)^2 + (W(u^i, u^{i+1}))^2} \right\|_1, \tag{5}$$

for $\kappa < 1$, where the subscripts x and y denote the x- and y-derivatives, and W denotes the photoconsistency

$$W(u^i, u^{i+1})(x) = \frac{u^i(x) - u^{i+1}(x + v^i(x))}{h} \tag{6}$$

given a motion field v, see Sect. 3.1. The parameter h encodes the scaling of time and space dimensions and is estimated automatically as the ratio of warp energy to gradient energy on a bicubic estimate u_0:

$$h = \frac{\|\mathcal{W}u_0\|_1}{\|\partial_x u_0\|_1 + \|\partial_y u_0\|_1}, \tag{7}$$

where $\mathcal{W}u_0$ denotes the vector-valued image obtained by stacking all $W(u^i, u^{i+1})$, and $\partial_x u_0, \partial_y u_0$ denote the stacked x- and y-derivatives of all frames of the sequence.

Since the warp operator is multiplied with h^{-1} this provides an image-adaptive way to make sure that the spatial and temporal regularity terms are in the same order of magnitude. Note that such a term also makes sense from a physical point of view: Since u_x and u_y measure change in space and $W(u^i, u^{i+1})$ measures change per time, a normalization factor with units 'time over space' is necessary to make these physical quantities comparable. A related discussion can be found in [8, Sect. 4].

The idea for using the infimal convolution approach originates from [8] in which the authors used a similar term with a time derivative instead of the operator W for video denoising and decompression. The infimal convolution automatically selects a regularization focusing either on space or time at each point. At points in the image where the warp energy $W(u^i, u^{i+1})$ is high, our approach automatically uses strong total variation (TV) regularization. In this sense it is a convex way of replacing the EM-based local parameter estimation from [13] by a joint and fully automatic regularization method with similar effects: It can handle inconsistencies in the motion field v by deciding to determine such locations by R_{spat}. On the other hand introducing strong spatial regularity can suppress details to be introduced by the temporal coupling. The infimal convolution approach allows favoring the optical flow information without over-regularizing those parts of the image, where the flow estimation seems to be faithful.

Figure 3 demonstrates the behavior of the infimal convolution by illustrating the division of one frame into the two parts w and $z = u - w$ of (3). Areas in which the optical flow estimation is problematic are visible in the w variable and hence mostly regularized spatially. All other areas are dominated by strong temporal regularization.

2.2   Multiframe Motion Coupling

A key aspect of our approach is the temporal coupling of the (unknown) *high resolution frames* u. It is based on color constancy assumptions and couples the entire sequence in a spatio-temporal manner using only linearly many flow fields v^i. Figure 2 illustrates the difference of the temporal coupling of previous energy minimization techniques and the proposed method. Besides only requiring linearly many flow fields, the high resolution frames are estimated jointly such that temporal consistency is enforced directly. Note that the energies (1), or the ones of [12, 13, 15] decouple and solve for each high resolution frame separately with the temporal conformance only given by the consistency of the low resolution frames f^i, so that inconsistent flickering in high resolution components is not accounted for.

3   Optimization

The optimization is performed in a two-step procedure: We compute the optical flow on the low resolution input frames and upsample the flow to the desired resolution using bicubic interpolation. Then we solve the super resolution problem (2). We experimented extensively with an alternating scheme, c.f. [12], however the effective resolution increase through this recurring optical flow computation is marginal as we will discuss in Sect. 4.4. An alternative approach shown by the authors of [25] would be to compute the high resolution optical flow on a bicubic video estimate. However our experiments showed that our approach was as precise while being much more efficient.

3.1   Optical Flow Estimation

The optical flow v on low resolution input frames f^i is calculated via

$$
\begin{aligned}
v = \arg\min_{v} \ & \sum_{i=1}^{n-1} \int_{\Omega} \|\nabla f^i(x) - \nabla f^{i+1}(x + v^i(x))\|_1 \; dx \\
& + \int_{\Omega} |f^i(x) - f^{i+1}(x + v^i(x))| \; dx + \beta \sum_{j=1}^{2} \|\nabla v_j^i\|_{H^\epsilon}.
\end{aligned}
\tag{8}
$$

It consists of two data terms, one that models *brightness constancy* and one that models *gradient constancy*, as well as a Huber penalty ($\epsilon = 0.01$) that is enforcing the regularity of the flow field. Note that (8) describes a series of $n-1$ time-independent problems. To solve each of these problems we follow well-established methods [24, 27, 29] and first linearize the brightness- and gradient constancy terms using a first order Taylor expansion with respect to the current estimate $\tilde{v}^i$ of the flow field resulting in a convex energy minimization problem for each linearization. We exploit the well-known iterative coarse-to-fine approach [1, 2] with median filtering. A detailed evaluation of this strategy can be found in [24]. We use a primal-dual algorithm with preconditioning [6, 19] to solve the convex subproblems within the coarse-to-fine pyramid using the CUDA module of the FlexBox framework [7].

3.2 Super Resolution

Unlike previous approaches, the super resolution problem (2) does not simplify to a series of time-independent problems, since individual frames are correlated by the flow. Consequently, the problem is solved in the whole space/time domain. First, we want to deduce that (2) can be rewritten in the form

$$\arg\min_{u,w} \|\mathcal{A}u - f\|_1 + \alpha \left\| \begin{pmatrix} \nabla w \\ \kappa \mathcal{W}w \end{pmatrix} \right\|_{2,1} + \alpha \left\| \begin{pmatrix} \kappa \nabla(u-w) \\ \mathcal{W}(u-w) \end{pmatrix} \right\|_{2,1} \tag{9}$$

where $u = (u^1, \ldots, u^n)$, $f = (f^1, \ldots, f^n)$, and $\mathcal{A} = \mathrm{diag}(DB, \ldots, DB)$ denotes a linear operator, i.e. a matrix in the discrete case after vectorization of the images u^i, that contains the downsampling and blur operators. We use an averaging approach for the downsampling, e.g. [25] and choose the subsequent blur operator as Gaussian blur with variance dependent on the magnification factor, e.g. $\sigma^2 = 0.6$ for a factor of 4. Similarly, the gradients on w and $u - w$ are block-diagonal operators consisting of the gradient operators of the single frames along the diagonal. The operator $\mathcal{W}$ is also linear and can be seen as a motion-corrected time derivative. The notation $\|\cdot\|_{2,1}$ is used to denote the sum of the ℓ^2 norms of the vector formed by two entries from the gradient and one entry from the warping operator $\mathcal{W}$.

Based on the flow fields v from the first step, we write the functions of the form $u^i(x + v^{i-1}(x))$ as $W^{i-1}u^i$, where the W^i are bicubic interpolation operators, such that $u^i(x + v^{i-1}(x)) \approx W^{i-1}u^i$. The final linear operator $\mathcal{W}u$ consists of $n - 1$ entries of the form $u^i - W^i u^{i+1}$ and one final block of zeros, acting as zero Neumann boundary conditions in time.

Similar to the flow problem, we used an implementation of the primal-dual algorithm in the PROST [16] framework but also provide an optional binding to Flexbox [7]. Our code is publicly available on Github[1] for the sake of reproducibility.

4   Numerical Results

We choose static parameters $\alpha = 0.01, \beta = 0.2$ and $\kappa = 0.25$ across all of our different datasets and figures as we found them to yield a good and robust trade off for arbitrary video sequences for a magnification factor of 4.

To be able to super resolve color videos we follow a common approach [9,10,28] and transform the image sequence into a YCbCr color space and only super resolve the luminance channel Y with our variational method. The chrominance channels Cr and Cb are upsampled using bicubic interpolation. Since almost all detail information is concentrated in the luminance channel, this simplification yields almost exactly the same peak signal-to-noise ratio (PSNR) as super resolving each channel separately.

---

[1] https://github.com/HendrikMuenster/superResolution.

To process longer videos, we use our method with frame batches in the size of a desired temporal radius and use the last computed frame from each batch as boundary value for the next batch to ensure temporal consistency.

We evaluate the presented algorithm on several scenes with very different complexity and resolution. Included in our test set is one simple synthetic scene consisting of a planar motion of the London subway map (*tube*), shown in Fig. 1, four common test videos [9,12,21] (*calendar, city, foliage, walk*), three sequences from [11,21] (*foreman, temple, penguins*), and four sequences from a realistic and modern UHD video sequence (*sheets, wave, surfer, dog*) [23] subsampled to 720p, that contain large non-linear motion and complex scene geometries.

For the sake of this comparison we focused on an upsampling factor of 4, although our variational approach is able to handle arbitrary positive real upsampling factors in a straightforward fashion

We evaluate nearest neighbor (NN) and bicubic interpolation (Bic), Video Enhancer [20] (a commercial upsampling software), the variational approach [13] (MFSR), as well as the learning based techniques Deep Draft [11], VSRnet [9], and VDSR [10] using code provided by the respective authors along with our proposed method and reimplementations of the variational methods [15,25] (with $\alpha = 0.1$). For the sake of fairness in comparison of [15] to [25] we computed all necessary optical flows directly instead of approximating them. We consider 13 frames of the *tube, city, calendar, foliage, walk* and *foreman* sets and 5 frames of the larger *temple, penguins, sheets, surfer, wave* and *dog* sets. The PSNR and structural similarity index measure (SSIM) [26] were determined for the central image of each sequence after cropping 20 pixels at each boundary. This was done so that the classical coupling methods [11,13,15,25] are properly evaluated at the frame with maximal information in each direction for a given batch of frames.

4.1 Evaluation of Proposed Improvements

We present several incremental steps in this work. To delineate the contributions of each, we will consider the average PSNR value score over our data sets in the bar plot to the right. The baseline is given by nearest neighbors interpolation with 25.61 dB. Bicubic interpolation yields an improvement to 27.28 dB and total variation upsampling, e.g. [14], adds further 0.31 dB. As a next step we consider our model

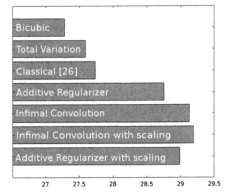

of coupling frames directly, but without the infimal convolution. Instead we consider a simpler additive regularizer first,

$$R(u) = \alpha||\mathcal{W}u||_1 + \alpha||\nabla u||_{2,1} \tag{10}$$

using this regularizer results in an average PSNR of 28.75 dB. It turns out that this method is already 1.01 dB better than the classical coupling of [25], due to failure cases in several fast-moving sets. In these cases, computing the optical flow between frames that are further apart is too error-prone, whereas the flow between neighboring frames is still reasonable to compute.

Next we consider our robustness improvements: Coupling spatial and temporal regularizers via the proposed infimal convolution (3) increases the PSNR value by 0.38 dB to 29.13 dB for fixed $h = 1$. Adapting the spatio-temporal scaling h with the heuristic (7) finally adds 0.06 dB. Note this choice of h can also be used directly for the additive regularizer, yielding 28.99 dB. A memory constrained implementation of the proposed method might want to rely just on that.

We report run times of 24 s per frame (~40% optical flow, ~60% super resolution) for our medium sized datasets (13 frames) on a NVIDIA Titan GPU. Although these results are on a modern GPU, the flow and the super resolution problem are implemented in a general purpose framework without direct communication and with linear operators in explicit matrix notation. Further increase in speed could be obtained by porting to a specialized framework avoiding matrix representations. For comparison, our implementation of classical coupling, e.g. [25] with the same framework needs 126 (~86% optical flow, ~14% super resolution) seconds per frame.

4.2   Choice of Forward Model

During our comparison to other approaches we found out that there was a significant disparity in the choice of operators for the forward model and subsequent data generation. Whereas our approach follows the works [15, 25] and uses a bicubic downsampling process, other works [11–13] use a Gaussian kernel followed by an asymmetric 'striding' operation, which keeps every n-th pixel in each direction for a downsampling factor of n. The Gaussian kernel in [12] is further chosen to be the theoretically optimal kernel. It turns out that this forward model is firstly easier to invert and secondly favors different strategies. Using it with our infimal convolution approach yields sharper results, significantly improving the PSNR values, e.g. the *city* dataset. However the direct use of the additive regularizer, Eq. (10), is the optimal choice, outperforming infimal convolution and results of [12] with up to 2.5 dB. This is a direct consequence of the perfect match of data simulation and construction as discussed in detail in [17, Chap. 2].

To have a proper evaluation, we generate data by using Matlab's bicubic image rescaling in our experiments, including color dithering and an anti-aliasing filter, followed by a clipping to obtain image values in $[0, 1]$. We explicitly do not use this operator in our reconstruction, c.f. Eq. (9). Note that these shortfalls are not limited to variational methods: Neural networks equally benefit from training on exactly the same data formation process that is later used for testing.

4.3   Comparison to Other Methods

The results for all test sequences and algorithms are shown in Table 1. We structured the methods into three categories; simple interpolation based methods, variational super resolution approaches that utilize temporal information but do not require any training data, and deep learning methods. We indicate the three categories by vertical lines in the tables.

Our method consistently outperforms simple interpolation techniques and also improves upon competing variational approaches, especially for complex motions like *walk* or *surfer*. Comparing to the learning based methods, our model based technique seems to be superior on those sequences that contain reasonable motion or a high frame rate. On sequences with particularly large motion and strong occlusions, e.g. *penguins* or *foreman*, the very deep convolutional neural network [10] performs very well, possibly because it does not rely on any motion information but produces high quality results purely based on learned information.

Besides the fact that our approach remains competitive even for the aforementioned challenging data sets in terms of the PSNR values, we want to stress the importance of temporal consistency: Consistency of successive frames is required for a visually pleasing video perception and the lack thereof in other methods immediately yields a disturbing flickering effect. Demo videos showcasing this effect can be found on our supplementary web page[2], including a comparison of the consistency of our approach to the VSRnet and VDSR methods.

For a visual inspection of single frames, we present the super resolution results obtained by various methods on a selection of four data sets in Fig. 4.

4.4   Numerical Analysis

In light of the results of [12], where alternating the optical flow (OF) estimation and super resolution was beneficial for a simplified and controlled data generation, we experimented with its application to our data model and more sophisticated regularization. However, as mentioned, applying the alternating procedure does not increase the video quality. The authors of [13] (who extend the model of [12] to include motion blur) report a similar behavior ([13], Fig. 9, $\beta = \infty$).

---

[2] http://www.vsa.informatik.uni-siegen.de/en/superResolution.

Fig. 4. Super resolution by a factor of 4, zoom into datasets *calendar, walk, foreman, wave*. PSNR values computed as described in Sect. 4. One can see the effective resolution increase of our method for the writing in *calendar*, faces in *walk* and wave front in *wave* as well as the robustness of the approach for the challenging *foreman* sequence.

Table 1. SSIM and PSNR values (4x upsampling) from left to right: nearest neighbor, bicubic, commercial VideoEnhancer software [20], Mitzel et al. [15] adapted for accuracy, see Sect. 4, Unger et al. [25], Multi-Frame Super resolution [13], MMC (our approach), DeepDraft ensemble learning [11], VSRnet [9], VDSR [10].

SSIM	NN	Bic	[20]	[15]	[25]	[13]	MMC	[11]	[9]	[10]
Tube	0.800	0.846	0.898	0.943	0.937	0.877	**0.945**	0.883	0.901	0.918
City	0.596	0.634	0.702	0.760	0.745	0.653	**0.762**	0.726	0.680	0.688
Calendar	0.621	0.652	0.706	**0.778**	0.764	0.686	0.772	0.738	0.705	0.726
Foliage	0.760	0.797	0.809	0.859	0.857	0.809	**0.873**	0.852	0.831	0.836
Walk	0.776	0.833	0.858	0.855	0.853	0.825	**0.894**	0.841	0.875	0.886
Foreman	0.880	0.918	0.924	0.939	0.938	0.923	0.949	0.923	0.941	**0.953**
Temple	0.835	0.874	0.893	0.910	0.909	0.878	0.924	0.820	0.916	**0.927**
Penguins	0.939	0.962	0.966	0.970	0.967	0.965	0.969	0.951	0.976	**0.979**
Sheets	0.948	0.971	0.978	0.978	0.978	0.972	**0.981**	0.974	0.979	0.979
Surfer	0.967	0.980	0.979	0.952	0.954	0.945	0.983	0.934	0.985	**0.986**
Wave	0.941	0.956	0.964	0.963	0.964	0.955	**0.971**	0.961	0.964	0.966
Dog	0.955	0.971	0.974	0.971	0.972	0.970	0.975	0.970	**0.977**	**0.977**
Average	0.835	0.866	0.888	0.906	0.903	0.872	**0.917**	0.881	0.894	0.902
PSNR										
Tube	18.63	20.09	21.73	23.57	23.11	20.82	**23.97**	21.80	21.88	22.36
City	23.35	23.95	24.75	25.38	25.14	24.23	**25.57**	24.92	24.45	24.60
Calendar	18.07	18.71	19.49	20.45	20.13	19.20	**20.51**	19.91	19.36	19.63
Foliage	21.21	22.21	23.19	23.41	23.38	22.40	**24.25**	23.45	23.00	23.16
Walk	22.74	24.37	25.37	24.29	24.13	23.98	**26.81**	25.00	25.95	26.40
Foreman	26.40	28.66	29.31	29.51	29.13	28.39	31.62	28.95	31.02	**32.54**
Temple	24.15	25.47	26.29	26.79	26.76	25.84	27.66	25.35	27.39	**27.90**
Penguins	29.17	31.77	32.82	32.55	32.78	32.54	32.91	30.56	34.63	**35.00**
Sheets	29.68	32.76	33.73	33.49	33.55	32.27	**34.23**	33.01	33.86	33.85
Surfer	30.59	32.91	33.29	26.52	27.15	29.11	34.42	30.48	30.45	**34.96**
Wave	30.73	31.96	32.82	32.81	32.81	31.85	**33.77**	32.43	33.03	33.33
Dog	32.58	34.48	35.07	34.54	34.77	34.15	35.18	34.09	35.63	**35.71**
Average	25.61	27.28	28.16	27.77	27.74	27.07	**29.19**	27.50	28.39	29.13

We investigate this further by running our approach with samples from the Sintel MPI dataset [4], which contains ground truth OF and several levels of realism denoted by 'albedo', 'clean' and 'final', respectively. We compared PSNR values by running our method with estimated optical flow and running our method with the ground truth OF for all three realism settings, c.f. Table 2.

Table 2. PSNR values computed on Sintel [4] dataset bandage_1

Rendering	GT flow	Our OF
Albedo	32.53	31.91
Clean	27.88	27.68
Final	33.31	34.65

Interestingly, we do not profit from the ground truth OF on realistic data. Our super resolution warping operator W penalizes changes in the brightness of the current pixel to the corresponding pixels in neighboring frames, i.e. brightness constancy as does our OF. It turns out that the estimated OF yields matchings that are well suited for super resolution despite not being the physically correct ones.

In light of this discussion the effectiveness of an alternating scheme is questionable. Even if the repeated OF computations converged to the GT OF, performance would not necessarily improve. The performance can only improve if the new OF would yield a refined pixel matching. [13] report for the case of heavy motion blur that recognizing and eliminating particularly blurry frames can refine their matchings in an alternating minimization. However it remains unclear how this translates into a generalized strategy, when all frames are equally low on details.

5   Conclusions

We have proposed a variational super resolution technique based on a multiframe motion coupling of the unknown high resolution frames. The latter enforces temporal consistency of the super resolved video directly and requires only as $N - 1$ optical flow estimations for N frames. By combining spatial regularity and temporal information with an infimal convolution and estimating their relative weight automatically, our method adapts the strength of spatial and temporal smoothing autonomously without a change of parameters. We provided an extensive numerical comparison which demonstrates that the proposed method outperforms interpolation approaches as well as competing variational super resolution methods, while being competitive to state-of-the-art learning approaches. For small motions or sufficiently high frame rate, our results are temporally consistent and avoid flickering effects.

Acknowledgements. J.G. and M.M. acknowledge the support of the German Research Foundation (DFG) via the research training group GRK 1564 Imaging New Modalities. D.C. was partially funded by the ERC Consolidator grant 3D Reloaded.

References

1. Black, M.J., Anandan, P.: The robust estimation of multiple motions: parametric and piecewise-smooth flow fields. Comput. Vis. Image Underst. **63**(1), 75–104 (1996)
2. Brox, T., Bruhn, A., Papenberg, N., Weickert, J.: High accuracy optical flow estimation based on a theory for warping. In: Pajdla, T., Matas, J. (eds.) ECCV 2004. LNCS, vol. 3024, pp. 25–36. Springer, Heidelberg (2004). https://doi.org/10.1007/978-3-540-24673-2_3
3. Burger, M., Dirks, H., Schönlieb, C.B.: A variational model for joint motion estimation and image reconstruction. arXiv preprint arXiv:1607.03255 (2016)

4. Butler, D.J., Wulff, J., Stanley, G.B., Black, M.J.: A naturalistic open source movie for optical flow evaluation. In: Fitzgibbon, A., Lazebnik, S., Perona, P., Sato, Y., Schmid, C. (eds.) ECCV 2012. LNCS, vol. 7577, pp. 611–625. Springer, Heidelberg (2012). https://doi.org/10.1007/978-3-642-33783-3_44

5. Chambolle, A., Lions, P.L.: Image recovery via total variation minimization and related problems. Numer. Math. **76**(2), 167–188 (1997)

6. Chambolle, A., Pock, T.: A first-order primal-dual algorithm for convex problems with applications to imaging. J. Math. Imaging Vis. **40**(1), 120–145 (2011)

7. Dirks, H.: A flexible primal-dual toolbox. arXiv preprint (2016). http://www.flexbox.im

8. Holler, M., Kunisch, K.: On infimal convolution of TV-type functionals and applications to video and image reconstruction. SIAM J. Imaging Sci. **7**(4), 2258–2300 (2014)

9. Kappeler, A., Yoo, S., Dai, Q., Katsaggelos, A.K.: Video super-resolution with convolutional neural networks. IEEE Trans. Comput. Imaging **2**(2), 109–122 (2016)

10. Kim, J., Kwon Lee, J., Mu Lee, K.: Accurate image super-resolution using very deep convolutional networks. In: CVPR, June 2016

11. Liao, R., Tao, X., Li, R., Ma, Z., Jia, J.: Video super-resolution via deep draft-ensemble learning. In: ICCV, pp. 531–539 (2015)

12. Liu, C., Sun, D.: On Bayesian adaptive video super resolution. IEEE Trans. Pattern Anal. Mach. Intell. **36**(2), 346–360 (2014)

13. Ma, Z., Liao, R., Tao, X., Xu, L., Jia, J., Wu, E.: Handling motion blur in multi-frame super-resolution. In: CVPR, pp. 5224–5232 (2015)

14. Marquina, A., Osher, S.J.: Image super-resolution by TV-regularization and breg-man iteration. J. Sci. Comput. **37**(3), 367–382 (2008)

15. Mitzel, D., Pock, T., Schoenemann, T., Cremers, D.: Video super resolution using duality based TV-L1 optical flow. In: Pattern Recognition, pp. 432–441 (2009)

16. Möllenhoff, T., Laude, E., Moeller, M., Lellmann, J., Cremers, D.: Sublabel-accurate relaxation of nonconvex energies. In: CVPR, June 2016. https://github.com/tum-vision/prost

17. Mueller, J., Siltanen, S.: Linear and Nonlinear Inverse Problems with Practical Applications. Society for Industrial and Applied Mathematics, Philadelphia (2012)

18. Nasrollahi, K., Moeslund, T.B.: Super-resolution: a comprehensive survey. Mach. Vis. Appl. **25**(6), 1423–1468 (2014)

19. Pock, T., Cremers, D., Bischof, H., Chambolle, A.: An algorithm for minimizing the Mumford-Shah functional. In: ICCV, pp. 1133–1140. IEEE (2009)

20. Infognition Co., Ltd: Videoenhancer 2 software, version 2.1. http://www.infognition.com/videoenhancer/

21. Xiph.org. Redistributable Video Test Media Collection. https://media.xiph.org/video/derf/

22. Shi, W., Caballero, J., Huszár, F., Totz, J., Aitken, A.P., Bishop, R., Rueckert, D., Wang, Z.: Real-time single image and video super-resolution using an efficient sub-pixel convolutional neural network. In: CVPR, pp. 1874–1883 (2016)

23. Sony Corporation: Sony 4K UHD surfing screen test demo. CC-BY License

24. Sun, D., Roth, S., Black, M.J.: A quantitative analysis of current practices in optical flow estimation and the principles behind them. IJCV **106**(2), 115–137 (2014)

25. Unger, M., Pock, T., Werlberger, M., Bischof, H.: A convex approach for variational super-resolution. In: Goesele, M., Roth, S., Kuijper, A., Schiele, B., Schindler, K. (eds.) DAGM 2010. LNCS, vol. 6376, pp. 313–322. Springer, Heidelberg (2010). https://doi.org/10.1007/978-3-642-15986-2_32

26. Wang, Z., Bovik, A.C., Sheikh, H.R., Simoncelli, E.P.: Image quality assessment: from error visibility to structural similarity. IEEE Trans. Image Process. **13**(4), 600–612 (2004)
27. Wedel, A., Pock, T., Zach, C., Bischof, H., Cremers, D.: An improved algorithm for TV-L^1 optical flow. In: Cremers, D., Rosenhahn, B., Yuille, A.L., Schmidt, F.R. (eds.) Statistical and Geometrical Approaches to Visual Motion Analysis. LNCS, vol. 5604, pp. 23–45. Springer, Heidelberg (2009). https://doi.org/10.1007/978-3-642-03061-1_2
28. Yang, J., Wright, J., Huang, T.S., Ma, Y.: Image super-resolution via sparse representation. IEEE Trans. Image Process. **19**(11), 2861–2873 (2010)
29. Zach, C., Pock, T., Bischof, H.: A duality based approach for realtime TV-L^1 optical flow. In: Hamprecht, F.A., Schnörr, C., Jähne, B. (eds.) DAGM 2007. LNCS, vol. 4713, pp. 214–223. Springer, Heidelberg (2007). https://doi.org/10.1007/978-3-540-74936-3_22
30. Zhang, Z., Sze, V.: Fast: free adaptive super-resolution via transfer for compressed videos. ArXiv https://arxiv.org/abs/1603.08968 (2016)

Illumination-Aware
Large Displacement Optical Flow

Michael Stoll$^{(\boxtimes)}$, Daniel Maurer, Sebastian Volz, and Andrés Bruhn

University of Stuttgart, 70569 Stuttgart, Germany
{stoll,maurer,volz,bruhn}@vis.uni-stuttgart.de
http://cvis.visus.uni-stuttgart.de/

Abstract. The integration of feature matches for handling large displacements is one of the key concepts of recent variational optical flow methods. In this context, many existing approaches rely on confidence measures to identify locations where a poor initial match can potentially be improved by adaptively integrating flow proposals. One very intuitive confidence measure to identify such locations is the matching cost of the data term. Problems arise, however, in the presence of illumination changes, since brightness constancy does not hold and invariant constancy assumptions typically discard too much information for an identification of poor matches. In this paper, we suggest a pipeline approach that addresses the aforementioned problem in two ways. First, we propose a novel confidence measure based on the illumination-compensated brightness constancy assumption. By estimating illumination changes from a pre-computed flow this measure allows us to reliably identify poor matches even in the presence of varying illumination. Secondly, in contrast to many existing pipeline approaches, we propose to integrate only feature matches that have been obtained from dense variational methods. This in turn not only provides robust matches due to the inherent regularization, it also demonstrates that in many cases sparse descriptor matches are not needed for large displacement optical flow. Experiments on the Sintel benchmark and on common large displacement sequences demonstrate the benefits of our strategy. They show a clear improvement over the baseline method and a comparable performance as similar methods from the literature based on sparse feature matches.

1 Introduction

Optical flow estimation is a key problem in computer vision. For solving this task, variational methods belong to the most successful techniques as they combine a transparent modeling with dense and accurate results. Since the seminal work of Horn and Schunck [17] three decades ago there has been an enormous progress in the field regarding both modeling and optimization. While advanced data terms have been developed that are robust against outliers [5] and varying illumination [14], adaptive smoothness terms have been suggested that are able to estimate flow fields with gradual transitions [7,14,22] and sharp edges [1,21,30,33].

© Springer International Publishing AG, part of Springer Nature 2018
M. Pelillo and E. Hancock (Eds.): EMMCVPR 2017, LNCS 10746, pp. 139–154, 2018.
https://doi.org/10.1007/978-3-319-78199-0_10

Moreover, coarse-to-fine warping schemes [8] have been proposed to address the intrinsic incapability of linearized data terms to compute large displacements. Nevertheless, if objects move significantly further than their own extent, such schemes still have difficulties to capture the underlying motion.

In contrast to dense variational methods, sparse descriptor matching techniques do not suffer from these difficulties. Hence, Brox and Malik [9] proposed to incorporate feature matches from such approaches into a variational model using a dedicated similarity term. While this improved the large displacement capability of variational methods, the underlying strategy turned out to be rather sensitive to outliers. To alleviate this problem, Stoll *et al.* [26] proposed to restrict the integration to carefully selected locations. Alternatively, Weinzaepfel *et al.* [31] considered the use of more robust matches. A conceptually different approach to integrate feature matches has been suggested by Xu *et al.* [32]. Instead of using a similarity term, they fuse the up-sampled flow vectors at each coarse-to-fine level with precomputed SIFT proposals. Similarly, Tu *et al.* [28] make use of PatchMatch [3] and a variational approach to create such proposals.

Although some of the aforementioned approaches based on similarity terms or candidate fusion rely on confidence measures when integrating matches, these measures are either purely image-based [9] or they do not explicitly consider illumination changes when using flow information [26,28,31]. Hence, in the presence of illumination changes, such methods have typically difficulties to reliably identify locations where such matches are particularly needed.

Moreover, the actual benefit of integrating feature matches strongly depends on the underlying scene. If there are multiple objects with equal appearance "jumping" through a scene, neither the smoothness assumption of variational methods nor the uniqueness constraint of feature matching techniques are fulfilled. In fact, in those cases the large displacement problem cannot be solved at all. In most other cases, however, the problem can at least be alleviated by integrating feature matches. Surprisingly, many of these cases can even be handled with explicit regularization if the amount of smoothness is sufficiently reduced. Such a de-regularization has for instance been proposed in for similarity terms by Stoll *et al.* [25] and for candidate fusion by Lempitsky *et al.* [18] and Tu *et al.* [28]. However, apart from the work of Stoll *et al.* [25], there has been no evaluation of the benefit of this strategy for large displacements so far. Moreover, also in the latter case, no illumination changes have been considered.

Contributions. Although our approach is closely related to pipeline approaches that seek to adaptively integrate feature matches [9,26,31], our work sticks out in two important points: Firstly, we create our large displacement proposals by altering the smoothness weight of more sophisticated, dense variational methods. Secondly, and even more importantly, we propose a novel intuitive confidence measure based on the illumination-compensated brightness constancy assumption that allows to reliably select candidate flows even in the presence of illumination changes. In that sense, our work also significantly differs from [25] that makes use of feature matches from a variational setting but proposes a simple non-pipelined approach that does not consider varying illumination.

Organization. In Sect. 2 we start by discussing the variational method that serves as baseline for integrating feature matches. In Sect. 3 we then present the different steps of our novel pipeline approach that generates these feature matches: the illumination estimation, the identification of candidate regions, the computation of candidate matches as well as the adaptive selection of flow proposals. In this context, we also discuss the integration of the proposals into the baseline model. Finally, we present an experimental evaluation of our method in Sect. 4 and conclude the paper with a summary in Sect. 5.

2   Baseline Method

Let us start by discussing our baseline optical flow model. It is used to compute an initial flow field that is later on refined by integrating feature matches. As baseline we have chosen the approach of Zimmer *et al.* [33], which has already shown good results in the context of large displacement optical flow [26]. It computes the optical flow $\mathbf{w} = (u, v)^\top$ as minimizer of the following energy

$$\mathcal{E}(\mathbf{w}) = \int_\Omega \mathcal{E}_\mathrm{D}(\mathbf{w}) + \alpha \, \mathcal{E}_\mathrm{S}(\mathbf{w}) \, d\mathbf{x}, \tag{1}$$

where $\mathbf{x} = (x, y)^\top \in \Omega \subset \mathbb{R}^2$ denotes a location, $\mathcal{E}_\mathrm{D}$ and $\mathcal{E}_\mathrm{S}$ are the data and the smoothness term, respectively, and α steers the amount of regularization.

Data Term. The data term $\mathcal{E}_\mathrm{D}$ provides a certain degree of illumination-robustness by combining a brightness (BCA) and a gradient constancy assumption (GCA) with separate robustification [10]. It is given by

$$\mathcal{E}_\mathrm{D}(\mathbf{w}) = \Psi_\mathrm{D}\left((f_2(\mathbf{x} + \mathbf{w}) - f_1(\mathbf{x}))^2\right) + \lambda \, \Psi_\mathrm{D}\left(|\nabla f_2(\mathbf{x} + \mathbf{w}) - \nabla f_1(\mathbf{x})|^2\right), \tag{2}$$

where, f_1, f_2 are the input images, $\Psi_\mathrm{D}(s^2) = 2\varepsilon_\mathrm{D}^2\sqrt{1 + s^2/\varepsilon_\mathrm{D}^2}$ is the Charbonnier penalizer [11] with contrast parameter ε_D, and λ is a balancing weight. Moreover, to reduce the influence of large gradients, constraint normalization [33] is applied.

Smoothness Term. The smoothness term $\mathcal{E}_\mathrm{S}$ consists of the anisotropic first-order regularizer

$$\mathcal{E}_\mathrm{S}(\mathbf{w}) = \sum_{i=1}^{2} \Psi_{\mathrm{S}_i}\left(|\mathcal{J}\mathbf{w} \cdot \mathbf{r}_i|^2\right). \tag{3}$$

It penalizes the projections of the Jacobian $\mathcal{J}$ of the flow $\mathbf{w}$ in the local directions $\mathbf{r}_1, \mathbf{r}_2$ of maximum and minimum information contrast. These directions are computed as the eigenvectors of the regularization tensor [33], Regarding the penalization in $\mathbf{r}_1$- and $\mathbf{r}_2$-direction, we follow [26,29] and employ the edge-enhancing Perona-Malik penalizer [6] $\Psi_{\mathrm{S}_1}(s^2) = \varepsilon_{\mathrm{S}_1}^2 \log\left(1 + s^2/\varepsilon_{\mathrm{S}_1}^2\right)$ and the edge-preserving Charbonnier penalizer [11] with contrast parameters $\varepsilon_{\mathrm{S}_1}$ and $\varepsilon_{\mathrm{S}_2}$.

Similarity Term. When incorporating additional flow candidates $\mathbf{w}_\mathrm{p}$ to refine the initial flow field later on, we extend the model (1) with a similarity term [9]

$$\mathcal{E}_\mathrm{P}(\mathbf{w}) = \beta \cdot \chi \cdot \Psi_\mathrm{P}\left(|\mathbf{w} - \mathbf{w}_\mathrm{p}|^2\right), \tag{4}$$

In this context, β is a weighting parameter, χ is a binary activation mask and Ψ_P is the Charbonnier penalizer with contrast parameter ε_P. While the extended functional shares similarities with the approaches in [9, 26, 31], the mask χ and the candidate flows $\mathbf{w}_p$ are constructed considerably different. Moreover, compared to [26], a local confidence weight is not required and β can be chosen quite large, since, in general, our feature matches hardly contain outliers.

3   Integration of Reliable Matches

Before detailing on the construction and integration of candidate matches $\mathbf{w}_P$, let us first give an overview of the overall pipeline approach; see Fig. 1. After computing an initial flow $\mathbf{w}_{\text{base}}$ with our baseline, we use this flow to explicitly estimate the illumination changes between both frames. Compensating the first frame by these illumination changes then allows us to rely on illumination-compensated image data in the remaining pipeline. This turns out to be useful for the following three steps: (i) Firstly, we can use the registration error based on the illumination-compensated brightness constancy as an intuitive confidence measure to identify candidate regions for the integration of flow proposals. (ii) Secondly, we can employ the illumination compensated brightness constancy and the illumination-compensated geometric blur constancy for creating such flow proposals from dense variational methods via de-regularization, i.e. by successively reducing the amount of smoothness. (iii) Thirdly, we can apply the illumination-aware confidence measure from the region selection step also for determining the best candidates from the previously generated flow proposals. Finally, we integrate the selected candidates into the baseline energy via the similarity term. To this end, we minimize the energy twice with an intermediate post-check that removes candidates that had a negative impact on the result.

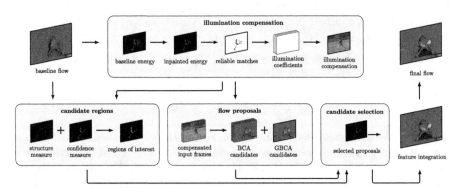

Fig. 1. Schematic overview over the pipeline at hand of the Tennis sequence [9].

3.1   Illumination Compensation

As mentioned in the previous section, all later steps of our pipeline approach rely on illumination compensated image data. Therefore, we have to estimate the illumination changes between both frames first. Since wrong matches in the initial flow $\mathbf{w}_{\text{base}}$ can significantly deteriorate this estimation, let us, however, start by discussing how potentially unreliable matches can be excluded.

Exclusion of Poor Matches. In order to detect poor matches in the baseline flow $\mathbf{w}_{\text{base}}$, we make use of the data term energy $e_{\text{base}} = \mathcal{E}_{\text{D}}(\mathbf{w}_{\text{base}})$. Here, the invariance of the gradient constancy assumption that was useful for handling potential illumination changes in the baseline flow turns into a problem. While large data term energies of such illumination-invariant constancy assumptions clearly indicate poor matches, low energies do not necessarily describe good matches, since essential information is discarded. For instance, in the case of the gradient constancy assumption, homogeneous regions of different brightness level are matched without cost. In fact, there is not even the chance that the matching cost is different form zero, since the image gradients vanishes in both frames. Fortunately, at image edges, the situation is different. Here, the image gradients are different from zero, which at least allows the resulting matching cost to be different from zero as well. Hence, we put more trust in the energy at image edges and assume that homogeneous regions are matched equally well/poorly as their enclosing contours. Using an edge indicator $\chi_{\text{inp}} = \delta \left[|\nabla f_1| > 10 \right]$, we thus inpaint the energy from edge locations into the remaining areas via

$$\mathcal{E}_{\text{inp}}(e) = \int_{\Omega} \chi_{\text{inp}} \left(e - e_{\text{base}} \right)^2 + (1 - \chi_{\text{inp}} + \varepsilon_{\text{inp}}) \, \alpha_{\text{inp}} \sum_{i=1}^{2} \Psi_{\text{S}_i} \left(|\nabla e \cdot \mathbf{r}_i|^2 \right) \, d\mathbf{x}, \quad (5)$$

where first-order anisotropic regularization [33] is used to align the energy with object boundaries. Thereby, α_{inp} is a smoothness weight, the penalizer functions Ψ_{S_i} and the directions $\mathbf{r}_i$ are defined as for the baseline and $\varepsilon_{\text{inp}} = 0.01$ is a small positive constant that guarantees a minimum amount of regularization. The final mask for excluding unreliable regions is then computed via thresholding the inpainted energy, i.e. $\chi_{\text{ill}} = \delta \left[e < 0.1 \cdot \max(e) \right]$; see first row of Fig. 2.

Illumination Estimation and Compensation. As proposed in [14] the estimation of the illumination changes is based on a parametrization in terms of local brightness transfer functions, i.e. $\Phi(\mathbf{c}, f) = \phi_0(f) + \sum_{j=1}^{n} c_j \cdot \phi_j(f)$. Here, ϕ_0 is the mean transfer function, $\phi_1, \ldots, \phi_n$ are the corresponding basis transfer functions, and $\mathbf{c}(\mathbf{x}) = (c_1(\mathbf{x}), \ldots, c_n(\mathbf{x}))^{\top}$ are spatially varying coefficients. In our case, these coefficients are estimated as minimizer of the global energy

$$\mathcal{E}_{\text{ill}}(\mathbf{c}) = \int_{\Omega} \chi_{\text{ill}} \, \Psi_{\text{D}} \left((f_2(\mathbf{x} + \mathbf{w}_{\text{base}}) - \Phi(\mathbf{c}, f_1)(\mathbf{x}))^2 \right) + \alpha_{\text{ill}} \, \|\mathcal{J}\mathbf{c}\|_F^2 \, d\mathbf{x}, \quad (6)$$

where the penalizer function Ψ_{D} is defined as in the baseline method, $\|\cdot\|_F$ denotes the Frobenius norm and α_{ill} is a smoothness weight. Moreover, the data term makes use of the previously computed mask χ_{ill} to avoid the integration of poor matches into the estimation, while the smoothness term relies on homogeneous regularization, since it aims at estimating semi-local, i.e. region-wise illumination changes. In this context, we intentionally do not seek to estimate

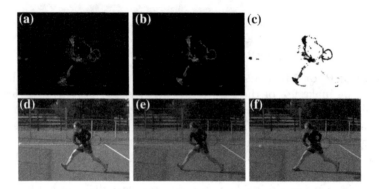

Fig. 2. First row: exclusion of poor matches. From left to right: (a) Data term energy of the baseline. (b) Energy after inpainting. (c) Mask χ_{ill} after thresholding. Second row: illumination compensation. From left to right: (d) first frame with illumination changes. (e) First frame after illumination-compensation. (f) Second frame.

very local illumination changes such as moving shadows or specular reflections as they cannot be distinguished well from brightness changes caused by mismatched large displacement objects. Furthermore, we resort to an affine parametrization, i.e. $\phi_0(f) = 0$, $\phi_1(f) = 1$, and $\phi_2(f) = f$, which is typically a good compromise between quality and complexity. After we have estimated the spatially varying coefficients $\mathbf{c}$, we are finally able to compensate the first frame via $f_{1,\mathrm{comp}} = \Phi(\mathbf{c}, f_1)$. This is illustrated in the second row of Fig. 2.

3.2 Determining Candidate Regions

As discussed before, illumination-invariant constancy assumptions like the gradient constancy discard too much information such that the resulting matching cost (data term energy) is only of conditional use as confidence measure. Now, after having compensated the image data by the computed illumination changes, however, we can directly operate on the brightness values. Hence, we are now in the position to introduce our illumination-aware confidence measure. Later on, this measure is not only used to define regions of interest for the integration of feature matches, it is also applied to select between different flow proposals.

Illumination-Aware Confidence Measure. As natural and very intuitive error measure in the context of illumination changes, we propose to use the photometrically compensated registration error

$$\rho(\mathbf{w}(\mathbf{x})) = \Psi_{\mathrm{D}}\left((f_2(\mathbf{x} + \mathbf{w}) - f_{1,\mathrm{comp}}(\mathbf{x}))^2\right). \tag{7}$$

While it aims at ignoring the illumination changes, it still keeps the idea of assessing the quality of flow fields by evaluating the matching error. Please note that, by construction, smaller values denote a higher reliability.

Regions of Interest. Let us now discuss how this confidence measure can be used to identify the regions of interest. Similarly to [26], we determine these regions relying on both the registration error and the structuredness, i.e. $\chi = \chi_{\text{reg}} \cdot \chi_{\text{struct}}$. However, instead of using the original registration error of the baseline flow as proposed in [26], we make use of our novel illumination-aware confidence measure. More precisely, to determine the registration mask χ_{reg}, we apply a double thresholding strategy on $\rho(\mathbf{w}_{\text{base}})$ based on the average confidence $\bar{\rho}(\mathbf{w}_{\text{base}})$ with thresholds $T_{\text{strict}} = 10 \cdot \bar{\rho}$ and $T_{\text{soft}} = 4 \cdot \bar{\rho}$. In order to compute the structuredness mask χ_{struct}, we follow [9] and apply single thresholding on the smaller eigenvalue λ_2 of the structure tensor [16]. In this context, we choose $T_{\text{struct}} = 0.5 \cdot \bar{\lambda}_2$, where $\bar{\lambda}_2$ denotes the average smaller eigenvalue. Finally, we eliminate isolated pixels in χ by morphological opening with a squared structuring element of size 3×3. The entire process is depicted in Fig. 3.

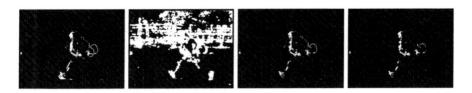

Fig. 3. From left to right: confidence mask χ_{reg}, structuredness mask χ_{struct}, intermediate mask χ before morphological opening, final mask χ after opening.

3.3 Determining Flow Proposals

Although we have computed well-localized regions of interest so far, we do not restrict the computation of flow proposals to these regions. Instead, we derive them from dense variational methods. Compared to sparse descriptor matching this allows to incorporate regularization into the estimation and thus to compute reliable matches even in homogeneous areas. This in turn improves the robustness of feature matches. Indeed, as observed in [15,26] outliers are the main source of problems when integrating such matches. Moreover, objects are more likely to prevail, if they are covered by a sufficient amount of matches.

Before we detail on the generation of our flow proposals, let us briefly summarize why common coarse-to-fine variational methods have problems with large displacements. First of all, small objects smear with their background on that coarse-to-fine-level which is appropriate to estimate their displacement. Secondly, large displacements induce large motion discontinuities, *severely* violating the smoothness assumption. The corresponding penalizer functions are typically not sufficiently robust to handle such jumps – either for convexity reasons or to avoid staircasing. Thus, even if there is enough information remaining to estimate a large displacement, it is typically cheaper to violate the data constancy assumption for a small object than to severely violate the smoothness assumption.

In the following, we address both issues by relying on a de-regularization strategy [25]. In this context, we consider two variational methods to generate the proposals – each equipped with a different constancy assumption.

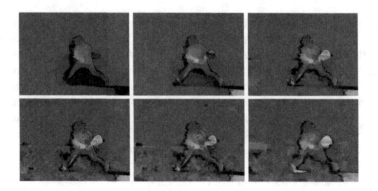

Fig. 4. Effect of de-regularization. Images 1 to 5: BCA. Image 6: GBCA.

Large Displacements via De-regularization. Let us start by explaining the general concept of de-regularization. To this end, we consider a family of energy functionals of the form

$$\mathcal{E}_{\text{cand}}(\mathbf{w}_k) = \int_{\Omega} \mathcal{E}_{\text{D,cand}}(\mathbf{w}_k) + \alpha_k \, \mathcal{E}_{\text{S,cand}}(\mathbf{w}_k) \, d\mathbf{x}. \qquad (8)$$

with successively decreasing smoothness weights α_k with $k = 1, \ldots, k_{\max}$. Evidently, decreasing the amount of regularization eases the estimation of large displacements as violating the smoothness term has less impact. Although this strategy deteriorates the average performance, since flow fields typically become very noisy, it significantly helps to improve the performance at locations with large displacements; see Fig. 4. Hence, by computing one flow field $\mathbf{w}_k$ for each of the regularization parameters α_k, we are able to generate a set of flow proposals $\mathbf{w}_1(\mathbf{x}), \ldots, \mathbf{w}_{k_{\max}}(\mathbf{x})$ per pixel from which we determine the best candidate $\mathbf{w}_{\text{P}}$ in a final selection step. While we use the same smoothness term $\mathcal{E}_{\text{S,cand}}$ as in our baseline method, we consider the following two constancy assumptions for the data term $\mathcal{E}_{\text{D,cand}}$ when generating our proposals.

Brightness Constancy Assumption (BCA). By relying on the compensated first frame, we can use pure brightness constancy in the data term:

$$\mathcal{E}_{D,1} = \Psi_{\text{D}} \left((f_2(\mathbf{x} + \mathbf{w}) - f_{1,\text{comp}}(\mathbf{x}))^2 \right). \qquad (9)$$

Apart from being robust against illumination changes, it is also rotation and scale invariant due to its locality. Realizing these properties in a feature descriptor requires much effort. Also another property of the BCA is beneficial: In contrast to gradient-like data terms, a potential violation is not only expensive at edges, but also in homogeneous parts of small fast-moving objects. This particularly complements the effect of a decreasing regularization.

Geometric Blur Constancy Assumption (GBCA). At certain locations, however, it may be more appropriate to estimate the motion using feature descriptors – in particular if the local information is not sufficient at the respective coarse-to-fine level. In this context, [24] proposed to expand the local intensity into separate intensity channels, such that each channel is resampled separately and objects of different intensities are not smeared. While this representation is more robust to resampling, it does not add any descriptiveness. In contrast, enhanced descriptiveness can be obtained by regarding neighborhood information such as in the SIFT [19], the HOG [13] and the GB descriptor [4]. Although the latter has a higher descriptiveness compared to HOG as observed in [9], so far only the HOG/SIFT descriptor has been used as data term in variational methods; see [20, 23]. According to [9], the main problem of the GB descriptor is its tendency to produce more false positives in sparse matching. This, however, is not an issue when using it in a constancy assumption of a variational method. Consequently, we propose to use GB descriptors in a feature-based data term:

$$\mathcal{E}_{D,2} = \Psi_{\mathrm{D}} \left(|GB_2(\mathbf{x} + \mathbf{w}) - GB_{1,\mathrm{comp}}(\mathbf{x})|^2 \right). \tag{10}$$

Hence, the GBCA assumes constancy on a feature (i) whose components are resampled separately following the spirit of [24], (ii) which improves descriptiveness over [23] and (iii) which overcomes the tendency of false positives stated in [9] due to the inherent regularization of the underlying variational model.

3.4   Candidate Selection

In the last section, we have generated a set of candidates that can improve the flow estimation. Let us denote them by $\mathbf{w}_{i,k}$, where $i \in \{1, 2\}$ refers to the model with data term $\mathcal{E}_{D,i}$ and k relates to the smoothness weight α_k. For each pixel within the candidate regions, it remains now to select the best candidate $\mathbf{w}_{\mathrm{P}}$ out of this set. To this end, we make once again use of our novel confidence measure based on the photometrically compensated registration error.

Essentially, we initialize our candidates with the baseline flow $\mathbf{w}_{\mathrm{base}}$ and the corresponding confidence $\rho(\mathbf{w}_{\mathrm{base}})$ and successively fuse in the candidate flows $\mathbf{w}_{i,k}$ by selecting the one with the lowest value for $\rho(\mathbf{w}_{i,k})$. Similar to [28] we thereby average the confidence within a 5×5 neighborhood. In this context, we made the following two observations regarding the effect of de-regularization: Firstly, with decreasing regularization, more image structure is necessary to obtain reliable flow vectors. Secondly, false positives become more probable. These insights inspired us to extend the initial fusion strategy only based on the confidence. To this end, we furthermore require that, if regularization is reduced, a better candidate should have more image structure and should yield a higher confidence compared to candidates with more regularization. Moreover, to account for noise, a new candidate is only used, if the confidence of the baseline flow is worse than some threshold T_{base}. Therefore, given a current pixelwise candidate state $\{\mathbf{w}_{i,k}, \alpha_{i,k}, \rho^{5 \times 5}(\mathbf{w}_{i,k})\}$, we replace it by the novel candidate state $\{\mathbf{w}_{j,l}, \alpha_{j,l}, \rho^{5 \times 5}(\mathbf{w}_{j,l})\}$ only if all of the following four conditions are met:

(i) $\rho^{5 \times 5}(\mathbf{w}_{j,l}) \leq \rho^{5 \times 5}(\mathbf{w}_{i,k})$,   (ii) $\rho^{5 \times 5}(\mathbf{w}_{j,l}) \leq \rho^{5 \times 5}(\mathbf{w}_{\mathrm{base}}) \cdot \alpha_{j,l}$,

(iii) $\rho^{5 \times 5}(\mathbf{w}_{\mathrm{base}}) \geq T_{\mathrm{base}}$,   (iv) $\lambda_2 > T_{\mathrm{struct}} \cdot (\alpha_{j,l})^{-0.5}$

Using this rule at each pixel of the candidate regions yields the proposal set $\mathbf{w}_{\mathrm{P}}$.

3.5   Final Estimation

Finally, the locally best proposal $\mathbf{w}_P$ is integrated into the extended energy functional and a final flow field $\mathbf{w}_{\text{final}}$ is estimated. The mask χ guarantees that candidate proposals are only integrated at regions of interest. In order to avoid that single bad proposals deteriorate the result, we re-compute the mask χ by excluding locations where the confidence of the final flow in a local neighborhood of size 7×7 became worse than the one of the baseline flow:

$$\chi_{\text{final}} = \delta \left[\rho^{7 \times 7}(\mathbf{w}_{\text{final}}) < \rho^{7 \times 7}(\mathbf{w}_{\text{base}}) \right]. \tag{11}$$

Using χ_{final} we then recompute the final flow field $\mathbf{w}_{\text{final}}$.

4   Evaluation

To investigate the performance of our novel strategy for adaptively integrating feature matches into variational optical flow methods, we evaluated our method on both the MPI Sintel benchmark [3] and common large displacement sequences. Additionally, we tested our method on modified versions of those sequences that additionally include illumination changes to demonstrate the robustness of this method against such changes. When running the experiments, we kept most of the parameters fixed, i.e. $\beta = 2700$, $\alpha_{\text{inp}} = 3000$, $\alpha_{\text{ill}} = 3000$, $\varepsilon_D = 0.01$, $\varepsilon_{S_1} = 0.02$, $\varepsilon_{S_2} = 0.03$, $T_{\text{base}} = 1$ fixed. In fact, using downhill simplex [27], we optimized only the parameters α, λ of the baseline and the parameters α_{cand} and k_{max} of the proposal generation. Note that we used color images within the experiments. The extension of our method straightforward [33].

4.1   MPI Sintel Training Data

In our first experiment we evaluate the performance of our integration strategy using a subset of 69 sequences of the clean dataset of the MPI Sintel training benchmark data containing images from each scene.

The outcome in Table 1 shows the superiority of our method compared to its baseline (-16%). In order to demonstrate that the benchmark contains illumination changes, we added the result for a baseline with pure brightness constancy, which is clearly inferior to the ones of the full baseline ($+7\%$).

Table 1. Overall results on the chosen MPI Sintel training subset (AEE).

Baseline		Our method
(Only BCA)	(BCA + GCA)	(Full)
7.105	6.663	**5.619**

4.2   Analysis of Components

To show the impact of the different components of our method, we made additional quantitative as well as visual experiments.

Candidate Models and De-regularization. In our second experiment, we analyze the effect of different scales of de-regularization on the overall result. To this end, we considered two candidate models (BCA, GBCA) and optimized the corresponding smoothness weight α_{cand} as well as the number of smoothness scales k_{max}. We found $\alpha_{cand} = 8$ to work well on both the MPI Sintel benchmark and other large displacement sequences. Table 2 shows the results, where the actual smoothness weight has been computed via $\alpha_k = \frac{\alpha_{cand}}{2.5 \cdot k}$.

While a certain number of smoothness scales turns out to be useful for the BCA data term, the GBCA data term seems to work best for a slight amount of de-regularization. In this context, however, one should note that the GBCA data term requires a larger smoothness weight than the BCA data term, such that the amount of de-regularization for Scale 5 of the BCA data term is actually comparable to the one of Scale 1 of the GBCA data term (also see Fig. 4). In terms of errors, the BCA allows to improve the results by almost 0.8 px (-12%), while the GCA yields improvements of almost 0.7 px (-11%).

In order to improve results further, we combined matches from both models. We made use of the GBCA model with one scale and determined the optimal value of k_{max} for the BCA model. The results can be found in Table 2. We observe, that the AEE further improves by more than 0.2 px (-4%) with both models. As already seen from our first experiment, this sums up to an improvement of more than 1 px (-16%) compared to the baseline method.

In our third experiment, we investigate the influence of the proposals from each of the candidate models on the final result at hand of the Tennis sequence. Figure 5 shows results obtained using different sets of candidate proposals. On the one hand, one can see that the Geometric Blur constancy assumption is able to capture translational and slight rotational motion, i.e. the tennis ball, the tennis racket and the arm. On the other hand, one can observe that the BCA data term is able to capture the strong rotational motion of the right foot more accurately, which complements the GB constancy. Thus, not surprisingly, combining both proposal sets also yields the best results here.

Table 2. Results for de-regularization with BCA, GBCA and the combination (AEE).

k_{max}	1	2	3	4	5	10
BCA only	6.171	5.968	5.849	**5.847**	5.893	5.909
GBCA only	**5.942**	6.094	6.607	–	–	–
$k_{max,BCA}/k_{max,GBCA}$	1/1	2/1	3/1	4/1	–	–
Both	5.783	5.716	**5.619**	5.697	–	–

Fig. 5. Influence of the candidate flows. First row, from left to right: first frame, second frame, baseline result. Second row, from left to right: our result using proposals from only the BCA data term, only the GBCA data term, both data terms.

Table 3. Impact of the illumination compensation.

	α_{ill}	AEE
Illumination compensation excluding poor matches	300	6.122
	900	5.725
	3000	**5.619**
	9000	5.768
	27000	5.914
Illumination compensation using all flow vectors	3000	7.501
No illumination compensation	–	5.948

Illumination Compensation. In our fourth experiment, we analyze the importance of the illumination compensation on the overall result. In this context, we compare a variant *without* illumination compensation, a variant where the illumination changes have been computed on *all* vectors of the baseline flow, and several variants that *excluded poor matches* from the baseline flow before computing the illumination changes.

The corresponding results are listed in Table 3. On the one hand, one can see that omitting the illumination compensation deteriorates the accuracy of the results by 0.3 px (+6%). This demonstrates that using illumination compensation is indeed useful when selecting and integrating feature matches. On the other hand, it becomes evident that simply using the entire baseline flow for estimating the illumination changes is also not a good idea. In fact, in the latter case, the results deteriorate significantly (+33%). Finally, one can observe that, when estimating the illumination changes, a moderate amount of regularization is beneficial ($\alpha_{ill} = 3000$). While a too small value for α_{ill} interprets all registration errors as local illumination changes, a too large value only allows the estimation of global illumination changes. Please note that the semi-local nature of the illumination changes is also reflected in the choice of the inpainting weight when excluding poor matches before the illumination estimation ($\alpha_{inp} = 3000$).

Post-check. In our fifth experiment, we evaluate the impact of the post-check on the final flow field. By deactivating it, an AEE of 5.657 is achieved. Hence, it has minor influence, but improves results by 0.038 px (-0.7%). This shows that we can select matches reliably using our illumination-aware confidence measure.

4.3   MPI Sintel Evaluation Data

In our sixth experiment, we compare our approach to similar methods from the literature. To this end, we submitted our results to the MPI Sintel benchmark.

As one can see from Table 4, our method shows a comparable performance as similar methods from the literature, clearly outperforming comparable approaches such as WLIF, MDP-Flow or LDOF. In this context, it is important to recall that we purely consider feature matches from dense variational methods in contrast of using sparse descriptor matches. This demonstrates that flow proposals from dense variational methods can be a serious alternative to sparse descriptor matches and matches from other large displacement methods like [12]. Moreover, it confirms the findings in [25] that use variational matches in a non-pipelined approach without considering any form of illumination compensation.

Table 4. Ranking on the MPI Sintel final evaluation data.

Method	Feature matches	AEE	AEE (s40+)
Deep+R [15]	sparse, regularized	6.769	41.687
Deep Flow [31]	sparse	7.212	44.118
Our Method	**dense, regularized**	**7.554**	**44.705**
WLIF [28]	dense, regularized	8.049	48.843
Baseline [33]	none	8.204	47.534
MDP-Flow2 [32]	sparse	8.445	50.507
LDOF [9]	sparse	9.116	57.296

4.4   Large Displacement Sequences

In our seventh experiment, we investigate the ability of our pipeline apporach to handle large displacements on three additional image sequences from the recent literature on large displacement optical flow. The results are depicted in Fig. 6(left) where they are compared to the results of the baseline. As one can see, with our strategy, all large displacements at the limbs, the racket, the tennis ball,

Fig. 6. Results for large displacement sequences. From Top to Bottom: Tennis [9], Baseball [32], Beanbags [2]. From Left to Right: First frame, second frame, baseline, our method, first frame (again), second frame with illumination changes, our method.

the bat and the beanbags have been estimated reliably. This is also reflected in a decrease of the average photometrically compensated registration errors (Tennis: 0.030 to 0.007, Baseball: 0.013 to 0.007, Beanbags: 0.050 to 0.016).

Moreover, we evaluated the ability to handle large displacements in the context of illumination changes. Hence, we changed brightness and contrast settings for the aforementioned sequences and show the results in Fig. 6(right). As one can see, results are only slightly worse than without illumination changes, but the large motion is still recovered correctly.

5   Conclusion

In this paper we have addressed the shortcomings of variational methods when estimating large displacements. In this context, we contributed in two ways. (i) On the one hand, we proposed a novel and intuitive confidence measure based on the photometrically compensated registration error. When rating the local quality of a flow field, this measure allowed us to distinguish brightness differences due to illumination changes from brightness differences due to motion. Moreover, it also enabled us to reliably select between different flow candidates before actually integrating the feature matches. (ii) Secondly, we demonstrated that flow candidates derived from dense variational methods can be a serious alternative to sparse feature matching approaches in the presence of large displacements. Combining the concept of de-regularization with suitable constancy assumptions relying on the compensated first frame allowed us to generate flow candidates that even capture the motion of small objects. Moreover, due to regularization, the flow candidates inherit the robustness of the underlying variational method. Experiments confirmed our findings. They showed clear improvements of our approach compared to the baseline method and a similar performance than comparable methods based on sparse feature matches.

Acknowledgements. We thank the German Research Foundation (DFG) for financial support within project B04 of SFB/Transregio 161.

References

1. Alvarez, L., Esclarín, J., Lefébure, M., Sánchez, J.: A PDE model for computing the optical flow. In: Proceedings of Congreso de Ecuaciones Diferenciales y
   Aplicaciones, pp. 1349–1356 (1999)
2. Baker, S., Scharstein, D., Lewis, J.P., Roth, S., Black, M.J., Szeliski, R.: A database
   and evaluation methodology for optical flow. Int. J. Comput. Vis. **92**(1), 1–31 (2010)
3. Barnes, C., Shechtman, E., Goldman, D.B., Finkelstein, A.: The generalized Patch-
   Match correspondence algorithm. In: Daniilidis, K., Maragos, P., Paragios, N.
   (eds.) ECCV 2010. LNCS, vol. 6313, pp. 29–43. Springer, Heidelberg (2010).
   https://doi.org/10.1007/978-3-642-15558-1_3
4. Berg, A., Malik, J.: Geometric blur for template matching. In: Proceedings of IEEE
   Computer Society Conference on Computer Vision and Pattern Recognition, pp.
   607–614 (2001)
5. Black, M.J., Anandan, P.: Robust dynamic motion estimation over time. In: Proceedings of IEEE Computer Society Conference on Computer Vision and Pattern
   Recognition, pp. 292–302 (1991)
6. Black, M.J., Anandan, P.: The robust estimation of multiple motions: parametric
   and piecewise smooth flow fields. Comput. Vis. Image Underst. **63**(1), 75–104 (1996)
7. Bredies, K., Kunisch, K., Pock, T.: Total generalized variation. SIAM J. Imaging
   Sci. **3**(3), 492–526 (2010)
8. Brox, T., Bruhn, A., Papenberg, N., Weickert, J.: High accuracy optical flow estimation based on a theory for warping. In: Pajdla, T., Matas, J. (eds.) ECCV 2004.
   LNCS, vol. 3024, pp. 25–36. Springer, Heidelberg (2004). https://doi.org/10.1007/
   978-3-540-24673-2_3
9. Brox, T., Malik, J.: Large displacement optical flow: descriptor matching in variational motion estimation. IEEE Trans. Pattern Anal. Mach. Intell. **33**(3), 500–513
   (2011)
10. Bruhn, A., Weickert, J.: A confidence measure for variational optic flow methods.
    In: Klette, R., Kozera, R., Noakes, L., Weickert, J. (eds.) Geometric Properties
    from Incomplete Data, Computational Imaging and Vision, vol. 31, pp. 283–297.
    Springer, Dordrecht (2006). https://doi.org/10.1007/1-4020-3858-8_15
11. Charbonnier, P., Blanc-Féraud, L., Aubert, G., Barlaud, M.: Two deterministic
    half-quadratic regularization algorithms for computed imaging. In: Proceedings of
    IEEE International Conference on Image Processing, pp. 168–172 (1994)
12. Chen, Z., Jin, H., Lin, Z., Cohen, S., Wu, Y.: Large displacement optical flow from
    nearest neighbor fields. In: Proceedings of IEEE Computer Society Conference on
    Computer Vision and Pattern Recognition, pp. 2443–2450 (2013)
13. Dalal, N., Triggs, B.: Histogram of oriented gradients for human detection. In: Pro-
    ceedings of IEEE Computer Society Conference on Computer Vision and Pattern
    Recognition, pp. 886–893 (2005)
14. Demetz, O., Stoll, M., Volz, S., Weickert, J., Bruhn, A.: Learning brightness transfer
    functions for the joint recovery of illumination changes and optical flow. In: Fleet,
    D., Pajdla, T., Schiele, B., Tuytelaars, T. (eds.) ECCV 2014. LNCS, vol. 8689, pp.
    455–471. Springer, Cham (2014). https://doi.org/10.1007/978-3-319-10590-1_30
15. Drayer, B., Brox, T.: Combinatorial regularization of descriptor matching for opti-
    cal flow estimation. In: Proceedings of British Machine Vision Conference, pp.
    42.1–42.12 (2015)
16. Förstner, W., Gülch, E.: A fast operator for detection and precise location of
    distinct points, corners and centres of circular features. In: Proceedings of ISPRS
    Intercommission Conference on Fast Processing of Photogrammetric Data, pp. 281–
    305 (1987)

17. Horn, B., Schunck, B.: Determining optical flow. Artif. Intell. **17**, 185–203 (1981)
18. Lempitsky, V., Roth, S., Rother, C.: FusionFlow: discrete-continuous optimization for optical flow estimation. In: Proceedings of IEEE Computer Society Conference on Computer Vision and Pattern Recognition, pp. 1–8 (2008)
19. Lowe, D., Bruckstein, A.M., Kimmel, R.: Distinctive image features from scale-invariant keypoints. Int. J. Comput. Vis. **60**(2), 91–110 (2004)
20. Liu, C., Yuen, J., Torralba, A., Sivic, J., Freeman, W.T.: SIFT flow: dense correspondence across different scenes. In: Forsyth, D., Torr, P., Zisserman, A. (eds.) ECCV 2008. LNCS, vol. 5304, pp. 28–42. Springer, Heidelberg (2008). https://doi.org/10.1007/978-3-540-88690-7_3
21. Nagel, H.H., Enkelmann, W.: An investigation of smoothness constraints for the estimation of displacement vector fields from image sequences. IEEE Trans. Pattern Anal. Mach. Intell. **8**, 565–593 (1986)
22. Nir, T., Bruckstein, A.M., Kimmel, R.: Over-parameterized variational optical flow. Int. J. Comput. Vis. **76**(2), 205–216 (2008)
23. Rashwan, H.A., Mohamed, M.A., García, M.A., Mertsching, B., Puig, D.: Illumination robust optical flow model based on histogram of oriented gradients. In: Weickert, J., Hein, M., Schiele, B. (eds.) GCPR 2013. LNCS, vol. 8142, pp. 354–363. Springer, Heidelberg (2013). https://doi.org/10.1007/978-3-642-40602-7_38
24. Sevilla-Lara, L., Sun, D., Learned-Miller, E.G., Black, M.J.: Optical flow estimation with channel constancy. In: Fleet, D., Pajdla, T., Schiele, B., Tuytelaars, T. (eds.) ECCV 2014. LNCS, vol. 8689, pp. 423–438. Springer, Cham (2014). https://doi.org/10.1007/978-3-319-10590-1_28
25. Stoll, M., Volz, S., Bruhn, A.: Variational large displacement optical flow without feature matches. In: Pelillo, M., Hancock, E. (eds.) EMMCVPR 2017. LNCS, vol. 10746, pp. 79–92. Springer, Cham (2017)
26. Stoll, M., Volz, S., Bruhn, A.: Adaptive integration of feature matches into variational optical flow methods. In: Lee, K.M., Matsushita, Y., Rehg, J.M., Hu, Z. (eds.) ACCV 2012. LNCS, vol. 7726, pp. 1–14. Springer, Heidelberg (2013). https://doi.org/10.1007/978-3-642-37431-9_1
27. Stoll, M., Volz, S., Maurer, D., Bruhn, A.: A time-efficient optimisation framework for parameters of optical flow methods. In: Sharma, P., Bianchi, F.M. (eds.) SCIA 2017. LNCS, vol. 10269, pp. 41–53. Springer, Cham (2017). https://doi.org/10.1007/978-3-319-59126-1_4
28. Tu, Z., Poppe, R., Veltkamp, R.C.: Weighted local intensity fusion method for variational optical flow estimation. Pattern Recogn. **50**, 223–232 (2016)
29. Volz, S., Bruhn, A., Valgaerts, L., Zimmer, H.: Modeling temporal coherence for optical flow. In: Proceedings of International Conference on Computer Vision, pp. 1116–1123 (2011)
30. Weickert, J., Schnörr, C.: A theoretical framework for convex regularizers in PDE-based computation of image motion. Int. J. Comput. Vis. **45**(3), 245–264 (2001)
31. Weinzaepfel, P., Revaud, J., Harchaoui, Z., Schmid, C.: DeepFlow: large displacement optical flow with deep matching. In: Proceedings of International Conference on Computer Vision, pp. 1385–1392 (2013)
32. Xu, L., Jia, J., Matsushita, Y.: Motion detail preserving optical flow estimation. IEEE Trans. Pattern Anal. Mach. Intell. **34**, 1744–1757 (2012)
33. Zimmer, H., Bruhn, A., Weickert, J., Valgaerts, L., Salgado, A., Rosenhahn, B., Seidel, H.-P.: Complementary optic flow. In: Cremers, D., Boykov, Y., Blake, A., Schmidt, F.R. (eds.) EMMCVPR 2009. LNCS, vol. 5681, pp. 207–220. Springer, Heidelberg (2009). https://doi.org/10.1007/978-3-642-03641-5_16

Location Uncertainty Principle: Toward the Definition of Parameter-Free Motion Estimators

Shengze Cai[1], Etienne Mémin[2], Pierre Dérian[2], and Chao Xu[1(✉)]

[1] State Key Laboratory of Industrial Control Technology and the Institute of Cyber-Systems and Control, Zhejiang University, Hangzhou 310027, Zhejiang, China
{szcai,cxu}@zju.edu.cn
[2] National Institute for Research in Computer Science and Control (Inria), Campus Universitaire de Beaulieu, 35042 Rennes, France
{etienne.memin,pierre.derian}@inria.fr

Abstract. In this paper, we propose a novel optical flow approach for estimating two-dimensional velocity fields from an image sequence, which depicts the evolution of a passive scalar transported by a fluid flow. The Eulerian fluid flow velocity field is decomposed into two components: a large-scale motion field and a small-scale uncertainty component. We define the small-scale component as a random field. Then the data term of the optical flow formulation is based on a stochastic transport equation, derived from a location uncertainty principle [17]. In addition, a specific regularization term built from the assumption of constant kinetic energy involves the same diffusion tensor as the one appearing in the data transport term. This enables us to devise an optical flow method dedicated to fluid flows in which the regularization parameter has a clear physical interpretation and can be easily estimated. Experimental evaluations are presented on both synthetic and real images. Results indicate very good performance of the proposed parameter-free formulation for turbulent flow motion estimation.

1   Introduction

Motion estimation techniques are becoming increasingly important in the study of fluid dynamics. Extracting the velocity fields from image sequences allows the researchers to get a deeper insight into the complex and unsteady fluid flows. First proposed by Horn and Schunck [13], optical flow has been intensively studied in the computer vision community, and a huge number of variations have been presented in the literature, such as [2,3,18,21]. All these methods

This work was supported by the National Natural Science Foundation of China under Grant 61473253, and the Foundation for Innovative Research Groups of the National Natural Science Foundation of China under Grant 61621002.

© Springer International Publishing AG, part of Springer Nature 2018
M. Pelillo and E. Hancock (Eds.): EMMCVPR 2017, LNCS 10746, pp. 155–171, 2018.
https://doi.org/10.1007/978-3-319-78199-0_11

rely on the fundamental assumption of a brightness conservation along a point trajectory:

$$\frac{df}{dt} = \frac{\partial f}{\partial t} + \nabla f \cdot \omega = 0. \tag{1}$$

In this transport equation, referred to as the optical flow constraint (OFC), ∇ denotes the gradient operator, f and ω the intensity of the image and the motion field, respectively. For variational optical flow approaches, the OFC equation is associated with a spatial coherency assumption [13], in order to cope with the so called aperture problem. A weighting coefficient balances these two terms in the optic-flow energy functional.

Compared to the correlation-based motion estimators, optical flow methods enable to estimate dense velocity fields and thus potentially lead to motion fields with finer details. In addition, the OFC equation (1) can be combined with various physical constraints to describe the transportation of a fluidic scalar by a motion field. Note that the classical optical flow methods are in general used for estimating rigid motions and rely on a strong smoothing constraint. This constraint is difficult to interpret physically and the weighting coefficient is hard to choose optimally.

The original Horn and Schunck (HS) formulation has been extended in various ways to cope with motion estimation from fluid images. For instance, Corpetti et al. [7] presented a fluid-flow dedicated formulation based on the integrated continuity equation (ICE) and a second-order div-curl regularizer, that allow preserving better the divergence and the vorticity of the flow. Liu and Shen [16] discussed the relation between optical flow and fluid flow, and suggested to use the projected motion equation. Among the recently-developed techniques, optical flow is also formulated in the forms of orthogonal decomposition [22], wavelet expansion with a higher-order regularization term [10,15], optimal control scheme [19] or Bayesian stochastic filtering approach [9]. A review on different fluid motion estimation techniques is presented in [12].

Despite a great deal of effort, turbulence modeling and measurement is still a challenging issue in fluid mechanics. Realistic turbulent flows contain small-scale structures that are significant for energy and mass transport. However, the sub-grid scales are not taken into account in the optical flow formulations. To overcome this problem, [5] replaced the optical flow constraint with a sub-grid transport equation by introducing an eddy-diffusivity model. The diffusion coefficient of the transport equation is selected empirically. Instead, a structural sub-grid model with an eddy viscosity for computing the small-scale diffusion factor is applied in [6]. These works show good estimation results. However, let us outline that these approaches still highly depend on a regularization parameter whose value is difficult to fix and which has no direct physical interpretation. Furthermore, an additional parameter associated with the turbulence model, which is also difficult to fix in practice, is introduced in the data term.

In this paper, we aim at proposing a novel formulation for turbulent fluid motion estimation with a different strategy. The main ideas and contributions of this work consist in reformulating the optical flow estimation problem through the introduction of turbulence modeling expressed under a location uncertainty principle. As derived in [17], the Eulerian velocity of a flow is decomposed into a large-scale component and a small-scale turbulent component. The latter one, specified as a random field and referred to as location uncertainty, gives rise to a modified transport equation [20] obtained from a stochastic expression of the Reynolds transport theorem [8,17]. The resulting stochastic optical flow constraint equation includes the effects of the unresolved (so called sub-grid) velocity component. Another constraint on the kinetic energy enables us to interpret the regularizer as a physical constraint. As we demonstrate it in this paper, all the parameters involved in this optical flow model can be optimally set or estimated. Therefore, it enables us to avoid the inescapable and cumbersome parameter tuning.

2   Methodology Description

2.1   Stochastic Transport

Let $\mathbf{x}_t = (x(t), y(t))^T$ ($\mathbf{x}_t \in \mathbb{R}^2$) denote the position of a particle in the two-dimensional (2D) domain Ω at time step t. Let us follow the basic assumption that the Eulerian velocity field of turbulent flow consists of a smooth velocity component $\omega(\mathbf{x}, t) = (u, v)^T$ and a small-scale random velocity component termed uncertainty. Accordingly, the Lagrangian stochastic displacement regarding the trajectory $\mathbf{x}_t$ reads:

$$\mathrm{d}\mathbf{x}_t = \omega(\mathbf{x}_t, t)\mathrm{d}t + \sigma(\mathbf{x}_t, t)\mathrm{d}\mathbf{B}_t. \tag{2}$$

The integral expression involves a random function, $\mathbf{B}_t$, that can be interpreted as a white noise process in space and a Brownian process in time. The spatial correlations of the velocity uncertainty are specified through a diffusion operator $\sigma(\mathbf{x}, t)$ defined through the matrix kernel $\breve{\sigma}(\cdot, \cdot, t)$ for any vectorial function $\boldsymbol{f}$ (n-dimensional) as:

$$\sigma(\mathbf{x}, t)\boldsymbol{f} \triangleq \int_\Omega \breve{\sigma}(\mathbf{x}, \mathbf{z}, t)\boldsymbol{f}(\mathbf{z}, t)\mathrm{d}\mathbf{z}. \tag{3}$$

Therefore, it can be seen that this operator is a matrix mapping from $\mathbb{R}^n$ into $\mathbb{R}^2$ at point $\mathbf{x}$. In a motion estimation context, the flow velocity field is assumed constant between two successive image frames. Then without loss of generality we can safely ignore the time variable of $\omega(\mathbf{x}, t)$ and $\sigma(\mathbf{x}, t)$. Thus we have $\mathrm{d}\mathbf{x} = \omega(\mathbf{x})\mathrm{d}t + \sigma(\mathbf{x})\mathrm{d}\mathbf{B}_t$, where $\mathrm{d}\mathbf{x} = \mathbf{x}_t - \mathbf{x}_{t-1}$ represents the displacements of particles between two successive images. The uncertainty component, $\sigma(\mathbf{x})\mathrm{d}\mathbf{B}_t$, representing the small-scale velocity, is a Gaussian random function correlated in space. The covariance tensor of the uncertainty component $\sigma\mathrm{d}\mathbf{B}_t$ (at different locations - $\mathbf{x}$ and $\mathbf{x}'$) reads:

$$Q = Cov(\mathbf{x}, \mathbf{x}')$$
$$\triangleq \mathbb{E}\left[(\sigma(\mathbf{x})\mathrm{d}\mathbf{B}_t)(\sigma(\mathbf{x}')\mathrm{d}\mathbf{B}_t)^T\right] = \int_\Omega \breve{\sigma}(\mathbf{x}, \mathbf{z})\breve{\sigma}^T(\mathbf{x}', \mathbf{z})\mathrm{d}\mathbf{z}\,\mathrm{d}t = \sigma(\mathbf{x})\sigma^T(\mathbf{x}')\,\mathrm{d}t, \tag{4}$$

where $\mathbb{E}\left[(d\mathbf{B}_t)(d\mathbf{B}_t)^T\right] = dt$. The corresponding variance tensor, $\boldsymbol{a}$, is defined by the single-point covariance of the small-scale displacement. It is a 2×2 symmetric positive definite matrix for each spatial point $\mathbf{x}$ in the 2D physical domain Ω, given by:

$$\boldsymbol{a}(\mathbf{x}) \triangleq \sigma(\mathbf{x})\sigma^T(\mathbf{x}) = \int_\Omega \breve{\sigma}(\mathbf{x},\mathbf{z})\breve{\sigma}^T(\mathbf{x},\mathbf{z})d\mathbf{z} = \frac{Cov(\mathbf{x},\mathbf{x})}{dt}. \tag{5}$$

Given the stochastic formalism, we now consider that a conserved scalar quantity f is transported by a motion field under location uncertainty. The conservation law reads:

$$f(\mathbf{x}_t + d\mathbf{x}_t, t + dt) = f(\mathbf{x}_t, t). \tag{6}$$

As f is a random function, its material derivative $D_t f \triangleq df(\mathbf{x},t) = 0$, which involves the composition of two stochastic processes, can be expanded via the generalized Ito formula (Ito-Wentzell formula). The expression is given by

$$D_t f = d_t f + \nabla f \cdot (\omega dt + \sigma d\mathbf{B}_t) + \sum_{i=1}^{2} d\langle \frac{\partial f}{\partial \mathbf{x}_i}, \mathbf{x}_i \rangle dt + \frac{1}{2}\sum_{i,j=1}^{2} a_{ij}\frac{\partial^2 f}{\partial \mathbf{x}_i \partial \mathbf{x}_j}dt, \tag{7}$$

where $d_t f$ stands for the time increment of the (non differentiable) quantity f: $d_t f = f(x, t + dt) - f(x, t)$ and the quadratic variation operator $\langle \cdot, \cdot \rangle$ is briefly presented in the Appendix A. Compared to the standard Ito formula, the expression introduces the additional co-variation terms between $\mathbf{x}$ and the gradient of the random function f. The derivation of (7) is provided in details in [17,20], here we give the conclusion: for an incompressible random velocity component (namely $\nabla \cdot \sigma d\mathbf{B}_t = 0$), the material derivative $D_t f$ has a simple form $\mathbb{D}_t$ as following

$$\mathbb{D}_t f = d_t f + \left[\nabla f \cdot \omega^* - \frac{1}{2}\nabla \cdot (\boldsymbol{a}\nabla f)\right] dt + \nabla f \cdot \sigma d\mathbf{B}_t, \tag{8}$$

which is referred to as the stochastic transport operator, where ω^* is the modified large-scale velocity that takes into account the inhomogeneity of the small-scale random velocity component:

$$\omega^* = \omega - \frac{1}{2}(\nabla \cdot \boldsymbol{a})^T. \tag{9}$$

This velocity corresponds to a correction of the large-scale velocity induced by the small scales inhomogeneity. The induced statistical velocity $(\nabla \cdot \boldsymbol{a})$ is a drift going from the variance tensor maxima to the variance tensor minima.

Compared to the deterministic material derivative, several additional terms related to the uncertainty random field are now involved in the formulation (8). A transport by the small-scale component is visible in the last right-hand side term of (8). The uncertainty term has also a mixing effect on the large-scale motion through a diffusion term along the proper directions of the variance tensor (third term of the right-hand side of (8)). Note that there are possibly a lot of degrees of freedom to define the diffusion tensor $\boldsymbol{a}$. In this study, in order to demonstrate the potential of this formalization we will only consider a simple isotropic divergence free model. This condition leads to a constant uncertainty for the whole domain between two successive samples, i.e., $\boldsymbol{a}(\mathbf{x}) = \alpha \mathbb{I}_2 = \text{const.}$, where $\mathbb{I}_2$ is the 2×2 identity matrix, hence $\nabla \cdot \boldsymbol{a}(\mathbf{x}) = 0$ and then $w^* = w$ due to (9). Therefore, in this case the stochastic transport equation (8) can be simplified as follow:

$$\mathbb{D}_t f = \mathrm{d}_t f + \left(\nabla f \cdot w - \frac{1}{2} \alpha \Delta f \right) \mathrm{d}t + \nabla f \cdot \sigma \mathrm{d}\mathbf{B}_t, \tag{10}$$

where Δ is the Laplacian operator. In the next section this expression of the material derivative enables us to derive an optical flow formulation under location uncertainty.

2.2   Data Term Based on Stochastic Model

From the previous stochastic transport equation, a novel observation term for optical flow estimation can be proposed. Hereafter, the conserved quantity f is assumed to be proportional to the image intensity. Since the Brownian random terms have zero mean, one can take the expectation of (10) to derive the mean scalar advection, namely

$$\mathbb{E}(\mathbb{D}_t f) = \mathbb{E}\left[\mathrm{d}_t f + \left(\nabla f \cdot w - \frac{1}{2} \alpha \Delta f \right) \mathrm{d}t \right]. \tag{11}$$

The data term of the motion estimation cost functional can be set as the variance of the luminance variation:

$$e_{data} = \mathbb{E}[(\mathbb{D}_t f - \mathbb{E}(\mathbb{D}_t f))^2]$$
$$\approx \int_\Omega \left[\mathrm{d}_t f + \left(\nabla f \cdot w - \frac{1}{2} \alpha \Delta f \right) \mathrm{d}t \right]^2 \mathrm{d}\mathbf{x} - \int_\Omega (\beta^2 \alpha |\nabla f|^2 \mathrm{d}t) \, \mathrm{d}\mathbf{x}. \tag{12}$$

where $| \cdot |$ denotes the Euclidean norm, i.e., $| \nabla f |^2 = (\partial_x f)^2 + (\partial_y f)^2$. The derivation in (12) is thoroughly given in Appendix A. In this expression an additional coefficient β has to be fixed or estimated. An estimate of this parameter is also provided in Appendix A. As all the quantities involved are assumed to be constant in time between two consecutive images, $\mathrm{d}t$ can be replaced by the time interval between the two frames and a dimensioned to 1. Note that if the investigated flow is fully resolved or contains no location uncertainty (i.e., the variance factor of the uncertainty component $\boldsymbol{a} = \alpha \mathbb{I}_2 = \mathbf{0}$), the simplified version of (12) boils down exactly to the classical optical flow constraint equation. The data model is the sum of two quadratic terms. The first one has the form of a modified OFC equation. This new brightness consistency model includes a diffusion of the image brightness, which represents the unresolved scales action on the transported luminance function. As for the second term, it can be observed that it corresponds to a weighting of the luminance energy dissipation. It can be thus seen as the measure of the mean transported scalar energy evolution on the time interval between two consecutive images.

2.3   Regularization Term Dedicated to Stochastic Transport

Generally, the regularization term in motion estimation cost functionals is set from a regularity condition on the solution. Such assumption is difficult to relate to kinematical or dynamical properties of the flow. In this section, we explain the spatial regularizer ensuing from an energy conservation assumption. Based on the stochastic transport presented in Sect. 2.1, a stochastic representation of the Navier-Stokes equations has been derived in [17]. By neglecting the external and conservative forces, the dynamics of the stochastic flow (namely momentum equation under location uncertainty) has the following expression:

$$\frac{\partial \omega_i}{\partial t} + (\omega \cdot \nabla)\,\omega_i - \frac{1}{2}\alpha\Delta\omega_i = 0, \tag{13}$$

where ω and $\boldsymbol{a} = \alpha\mathbb{I}_2$ are defined in the same way as (10), and i stands for the component of x-direction or y-direction, i.e., $\omega_i = u$ or v. An inner product with the velocity of Eq. (13) followed by integrations by parts provides the kinetic energy evolution:

$$\frac{\partial}{\partial t}E_{kin} = \frac{1}{2}\frac{\partial}{\partial t}\parallel \omega \parallel_2^2 = -\frac{1}{2}\int_\Omega \sum_i \alpha|\nabla\omega_i|^2 \mathrm{d}\mathbf{x}. \tag{14}$$

In order to obtain a transformation from one image to the subsequent one that tends to conserve the kinetic energy, we can expect the semi-norm appearing in the right-hand-side of (14) to be as small as possible, i.e.,

$$e_{reg} = \frac{1}{2}\int_\Omega \alpha \parallel \nabla\omega \parallel_F^2 \mathrm{d}\mathbf{x} = \frac{1}{2}\int_\Omega \alpha\left(\mid \nabla u \mid^2 + \mid \nabla v \mid^2\right)\mathrm{d}\mathbf{x}, \tag{15}$$

where $\parallel \cdot \parallel_F$ denotes the Frobenius norm for a matrix. We remark that the regularizer (15) is the same as the usual gradient penalizer except for the introduction of the diffusion factor α. Therefore, the classical gradient smoothing penalization can be interpreted as derived from a homogeneous divergence-free uncertainty random field [17]. This basic model yields a smoothing term with no preferential direction.

Remark 1. It is necessary to discuss here the unit of the covariance factor $\boldsymbol{a} = \alpha\mathbb{I}_2$. According to the principle that different terms in a physical equation should have the same unit, we now examine the units of the different terms in the momentum transport equation (13). By letting $\lceil g \rfloor$ denote the unit of function g, we have: $\lceil \frac{\partial \omega}{\partial t} \rfloor = L/T^2$, where L and T denote the basic units of length and time, respectively. That means the unit of the third term in (13) should satisfy:

$$\lceil \alpha\Delta\omega \rfloor = \lceil \alpha \rfloor \cdot \lceil \Delta\omega \rfloor = \lceil \alpha \rfloor \cdot \frac{1}{TL} = \frac{L}{T^2}.$$

Hence, we obtain the units of α, the data term and the regularization term as follows:

$$\lceil \alpha \rfloor = \frac{L^2}{T}, \quad \lceil e_{data} \rfloor = \frac{I^2}{T^2}, \quad \lceil e_{reg} \rfloor = \frac{L^2}{T^3}, \tag{16}$$

where I denotes the basic unit of the luminance function (greyscale unit), and in practice, the unit of time is adimensioned with the time difference between two samples and therefore set to 1. To balance the two terms, a weighting coefficient with unit I^2/L^2 must be introduced.

Gathering the data term and the regularization term, the final energy functional reads:

$$J = \int_\Omega \left[\left(\mathrm{d}_t f + \nabla f \cdot \omega - \frac{1}{2} \alpha \Delta f \right)^2 - \beta^2 \alpha |\nabla f|^2 \right] \mathrm{d}\mathbf{x} + \int_\Omega \frac{1}{2} \lambda \alpha \parallel \nabla \omega \parallel_F^2 \mathrm{d}\mathbf{x},$$

(17)

where λ is a positive weight coefficient. In traditional optical flow methods the weighting coefficient balancing the data term and the regularizer is a very sensible parameter that is difficult to tune since λ is not directly related to any physical quantity such as the motion amplitude. In this paper, by dimensional analysis of the objective functional (discussed in Remark 1), we find that λ should have the unit of I^2/L^2. That means λ can be related to the gradient of image intensity. Therefore, from this point of view, there are several choices can be used to formulate λ:

$$\lambda_1 = \frac{1}{\Omega} \int_\Omega \frac{(\mathrm{d}_t f)^2}{(L_{\max})^2} \mathrm{d}\mathbf{x}, \quad \lambda_2 = \frac{1}{\Omega} \int_\Omega \left(\mathrm{d}_t \mid \nabla f \mid^2 \right) \mathrm{d}\mathbf{x}, \quad \lambda_3 = \frac{1}{\Omega} \int_\Omega \left(\mid \mathrm{d}_t \nabla f \mid^2 \right) \mathrm{d}\mathbf{x},$$

(18)

where $L_{\max}$ represents a characteristic value of the length scale in the images, which can be given by the maximum magnitude of the apparent displacements. We will evaluate these candidates in the experimental section.

The resulting energy functional (17) resembles to a simple modification of the Horn & Schunk functional. However, in this new formulation, the penalization constant of the smoothing term is now interpreted as the variance of the small-scale unresolved motion. This variance parameter α now also appears as the weighting factor of two additional terms in the data adequacy terms. As explained in the next section, this property will allow us to optimally estimate the variance parameter.

3   Minimization and Implementation

In this section we detail the computation of the cost functional optimum with respect to the two unknowns: ω and α. The optimization algorithm is performed through an alternated minimization of the two variables.

3.1   Minimization with Respect to the Motion Field

Let us assume that an initial α is known (in our paper, alpha is initialized with a fixed value 0.5 for both synthetic and real-world image sequences), $\mathrm{d}_t f \triangleq$

$f_t, \nabla f = (\partial_x f, \partial_y f) \triangleq (f_x, f_y)$. By applying the calculus of variation (Euler-Lagrange equation) to (17), we have

$$
\begin{cases}
2\left(f_x u + f_y v + f_t - \dfrac{1}{2}\alpha \Delta f\right) f_x - 2 \cdot \dfrac{1}{2}\lambda \alpha \Delta u = 0, \\[2mm]
2\left(f_x u + f_y v + f_t - \dfrac{1}{2}\alpha \Delta f\right) f_y - 2 \cdot \dfrac{1}{2}\lambda \alpha \Delta v = 0.
\end{cases}
\tag{19}
$$

With the approximations of Laplacians $\Delta u \approx \kappa(\bar{u} - u)$, $\Delta v \approx \kappa(\bar{v} - v)$, where $\bar{u}, \bar{v}$ denote the local averages and κ depends on the difference scheme, (19) can be expressed as the following equations:

$$
\begin{cases}
\left(f_x^2 + \dfrac{1}{2}\lambda\alpha\right) u + f_x f_y v = \dfrac{1}{2}\lambda\alpha\bar{u} + \dfrac{1}{2}\alpha f_x \Delta f - f_x f_t, \\[2mm]
f_x f_y u + \left(f_y^2 + \dfrac{1}{2}\lambda\alpha\right) v = \dfrac{1}{2}\lambda\alpha\bar{v} + \dfrac{1}{2}\alpha f_y \Delta f - f_y f_t,
\end{cases}
\tag{20}
$$

which can be solved by the Gauss-Seidel method or the Successive Over Relation (SOR) iteration. In our algorithm, by applying elimination method to (20), the velocity vector ω can be computed by the following iterative formulation:

$$
\begin{cases}
u^{k+1} = \bar{u}^k - \dfrac{f_x \bar{u}^k + f_y \bar{v}^k + f_t - \frac{1}{2}\alpha\Delta f}{\frac{1}{2}\lambda\alpha + f_x^2 + f_y^2} f_x, \\[4mm]
v^{k+1} = \bar{v}^k - \dfrac{f_x \bar{u}^k + f_y \bar{v}^k + f_t - \frac{1}{2}\alpha\Delta f}{\frac{1}{2}\lambda\alpha + f_x^2 + f_y^2} f_y.
\end{cases}
\tag{21}
$$

3.2 Estimation of Diffusion Factor

The most important parameter to estimate the large-scale velocity field, ω, is the diffusion factor α. Since α can be regarded as an unknown in the objective functional (17), one can compute α by cancelling the energy functional gradient with respect to this variable. Therefore, we have

$$
\frac{\partial J}{\partial \alpha} = \int_{\Omega}\left[\Delta f\left(\partial_t f + \nabla f \cdot \omega - \frac{1}{2}\alpha\Delta f\right) - \beta^2 |\nabla f|^2\right] dx
$$
$$
+ \int_{\Omega} \frac{1}{2}\lambda \parallel \nabla\omega \parallel_F^2 \, dx = 0.
\tag{22}
$$

Then we readily obtain

$$
\hat{\alpha}^{k+1} = 2\frac{\int_{\Omega}\left[\Delta f\left(\nabla f \cdot \bar{\omega}^k + f_t\right) + \beta^2 \mid \nabla f \mid^2 - \frac{1}{2}\lambda \parallel \nabla\bar{\omega}^k \parallel_F^2\right] dx}{\int_{\Omega}(\Delta f)^2 dx},
\tag{23}
$$

where $\bar{\omega}^k$ is the estimated velocity vector from the previous iteration, defined by (21).

3.3   Multi-resolution Algorithm

For the basic optical flow methods, one common weakness is that the procedure can yield good results only when the magnitude of image motions is small (smaller than the shortest spatial wavelength present in the image [12]). To overcome the estimation issue due to the large displacements, we use an incremental coarse-to-fine strategy. The main idea of this strategy can be divided into several processes: (a) a multi-resolution representation through the successive Gaussian filtering and sub-sampling is applied to the images pair; (b) the optical flow is computed from the coarse-resolution level and then projected onto the next finer level of the pyramid; in the projection step, image warping is required so we only need to compute the small velocity increments at the higher resolution level; (c) this process is repeated at finer and finer spatial scales until the original image resolution is reached. The choice of the image filtering process is significant for the multi-resolution algorithm. Gaussian filters are applied to the original images to reduce the noise effect. Furthermore, median filters are applied to the estimated velocity fields after each warping step for the purpose of eliminating the outliers. For more details of the coarse-to-fine algorithm we refer to [21]. The framework of the multi-resolution strategy is shown in Algorithm 1.

4   Experimental Results

4.1   Synthetic Image Sequence

A synthetic fluid image sequence[1] is tested in this section, which is provided by [4] and generated by Direct Numerical Simulation (DNS). The phenomenon investigated is the spreading of a low diffusivity dye in a 2D homogeneous turbulent flow with Reynolds number $Re = 3000$ and Schmidt number $Sc = 0.7$. The intensity of the passive scalar images is proportional to the dye concentration. The sequence consists of 100 successive images at the resolution of 256×256 pixels. An example of the scalar image and the corresponding vorticity map are displayed in Fig. 1(a) and (b). A multi-resolution algorithm with 2 levels and 5 warping steps at each level has been implemented. As for the computation of λ, we use the first formulation in (18) by default.

To evaluate quantitatively the estimated motion fields, we follow a standard way by computing the average angular error (AAE) and root mean square error (RMSE) over N pixels of the image:

$$AAE = \frac{1}{N} \sum_{i=1}^{N} \arccos \frac{u_i^t u_i^e + v_i^t v_i^e + 1}{\sqrt{(u_i^t)^2 + (v_i^t)^2 + 1}\sqrt{(u_i^e)^2 + (v_i^e)^2 + 1}}, \tag{24}$$

$$RMSE = \sqrt{\frac{1}{N} \sum_{i=1}^{N} [(u_i^t - u_i^e)^2 + (v_i^t - v_i^e)^2]}, \tag{25}$$

where (u^t, v^t) and (u^e, v^e) denote the ground-truth velocity and the estimated velocity, respectively. The index i represents the pixel where optical flow is computed. Furthermore, vorticity maps which are computed by $(\partial_x v - \partial_y u)$ are also demonstrated.

---

[1] Available online: http://fluid.irisa.fr/.

Algorithm 1. Multi-resolution algorithm with symmetric warping for motion estimation

---

Load image pair Img_1 and Img_2;
Pre-processing;
Pyramidal generation from level 0 (original) to L (coarsest);
Compute the weighting coefficient λ;
for $l = L$ *to* 0 **do**
| **if** $l = L$ **then**
| | Set initial velocities be 0 at coarsest level $\widehat{\omega_{l+1}} = 0$;
| | Set initial estimation of α;
| **else**
| | Expand the velocities from coarser level ω_{l+1} to finer level $\widehat{\omega_{l+1}}$ by interpolation;
| | Set α the estimated α from the previous level;
| **end**
| **for** *Each warping step* **do**
| | **Symmetric warping**
| | $f_1 = warpForward(Img_1^l)$, $f_2 = warpInverse(Img_2^l)$.
| | **Optimization**
| | Compute the gradients and Laplacians f_t, ∇f, Δf;
| | Compute the estimation of β^2;
| | Estimate the motion field dω and α by iterations (21) and (23), respectively;
| | Update $\omega_l = \widehat{\omega_{l+1}} + d\omega$;
| | **Median filtering**
| **end**
end
End multi-resolution;

The estimated motion field and vorticity map at time step $t = 50$ are illustrated in Fig. 1, in comparison with the methods of [10,11,13]. We can observe from the vorticity maps that the proposed stochastic formulation performs better than the other references, especially in the area with high vorticity. The vortex structures are well recovered by the proposed optical flow formulation, whereas they are blurred by the HS method. The methods of [10,11] achieve to estimate the large-scale structures of the flow. However, they fail to provide the small-scale components in some areas. Figure 2 illustrates the result of a typical area of the scalar image with a strong vortex. A zoom in this region shows that the estimated velocity field (black vectors) is highly consistent with the ground-truth (red vectors).

Quantitative evaluations for the DNS passive scalar images are given in Fig. 3. The AAE and RMSE errors of the proposed method and the HS method are plotted in Fig. 3(a) and (b), respectively. It can be seen that the accuracy is drastically improved by more than 50% for both the AAE and the RMSE. Even

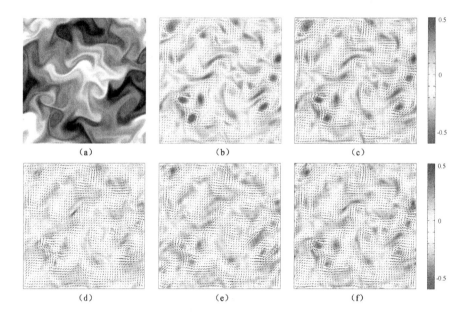

Fig. 1. Velocity fields and vorticity maps estimated from different methods on DNS passive scalar image sequence at $t = 50$: (a) scalar image; (b) ground-truth; (c) proposed method; (d) HS method; (e) method of [10]; (f) method of [11].

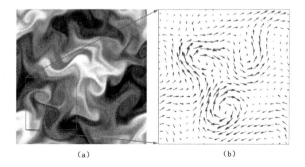

Fig. 2. Velocity vectors on DNS passive scalar image sequence at $t = 50$: (a) scalar image with zoomed area; (b) true velocity vectors (red) and the estimated vectors by the proposed method (black). (Color figure online)

compared with the state-of-the-art approaches, the proposed method shows the best performances. Error data of the other optical flow methods, including [6, 10, 11, 15, 22], are taken from [6] and displayed in Fig. 3(c) and (d). The results of our method is close to (slightly better than) the results of [6], which applies a data term based on the large eddy simulation (LES) sub-grid model and a divergence-free regularization term. Both the proposed technique and [6] outperform the other methods for the whole passive scalar image sequence, indicating that the introduction of turbulence models is significant for fluid motion estimation. The

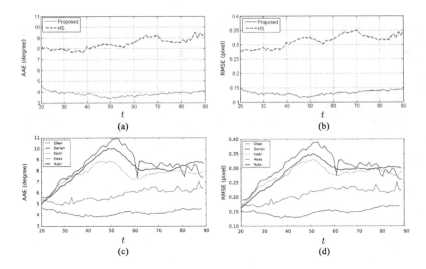

Fig. 3. AAE (left) and RMSE (right) errors of different estimators for DNS scalar image sequence. Results of the proposed method are plotted in figures (a) and (b), in comparison with the results of the HS method. Results of several state-of-the-art approaches are shown in figures (c) and (d).

method of [6] depends on several parameters: the standard deviation of a low-pass filtering applied on the sequence, the regularization coefficient, and the ratio of the Reynolds and Schmidt numbers of the flow. These constants are difficult to fix or not available with accuracy in practice, and must be adapted from one sequence to the other. On the contrary, the estimator under uncertainty proposed in this paper does not require such a tuning and can be qualified as a parameter-free approach.

Finally, Fig. 4 shows the impact of different choices for lambda, which have the same required unit. As we can see, although those candidates give different

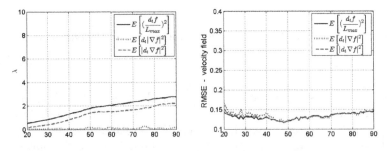

Fig. 4. The values of different choices for the weighting coefficient λ (left) and the corresponding estimated RMSE results (right) on DNS scalar images. The notation E denotes the mean value over the image domain.

values of λ, they provide similar estimation results of the velocity field. We conclude that one can choose any of them to construct a parameter-free estimator.

4.2 Real Images

An experimental image sequence of 2D turbulence is provided by [14]. The authors presented the first detailed experimental observation of the Batchelor regime [1], in which a passive scalar was dispersed by a large-scale strain. Two successive frames of this sequence are displayed in Fig. 5(a) and (b), respectively. The strong turbulent vortices can be clearly observed. We implement a multi-resolution algorithm with 5 levels and 2 warping steps at each multi-resolution level. Figure 5(c), (d) demonstrate the vorticity maps estimated from the HS method and the proposed method. It can be seen that the result of the HS method is over-smoothed, while the result of the proposed formulation shows more finer structures on the vorticity map.

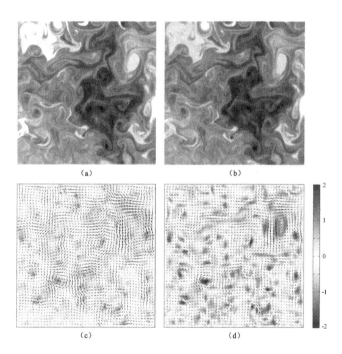

Fig. 5. Results on experimental scalar images: (a) first frame; (b) second frame; (c) estimated velocity field and vorticity map from the HS method; (d) estimated velocity field and vorticity map from the proposed method.

5   Conclusion

In this paper, we introduce a variational optical flow formulation for turbulent fluid motion estimation. This novel formulation is derived from a location uncertainty principle [17], which enables us to take into account the small-scale unresolved components of the velocity field, and also allows us to estimate explicitly the different parameters involved. The experimental results on both synthetic and real image sequences indicate the efficiency of the proposed parameter-free technique.

A   Variance of Stochastic Transport Operator

Before deriving the variance of the stochastic transport operator, we first recall briefly the notions of quadratic variation and covariation, which are important in stochastic calculus. Suppose that X_t, Y_t are stochastic processes defined on the probability space $(\Omega, \mathscr{F}, \mathbb{P})$, the quadratic covariation process denoted as $\langle X, Y \rangle_t$, is defined as the limit in probability:

$$\langle X, Y \rangle_t = \lim_{\delta t_i \to 0} \sum_{i=0}^{n-1} \left(X_{t_{i+1}} - X_{t_i} \right) \left(Y_{t_{i+1}} - Y_{t_i} \right)^T, \tag{26}$$

with $t_1 < t_2 < \cdots < t_n$ and $\delta t_i = t_{i+1} - t_i$. For the Brownian motion, the quadratic covariances can be computed by the following rules:

$$\langle B, B \rangle_t = t, \quad \langle B, h \rangle_t = \langle h, B \rangle_t = \langle h, h \rangle_t = 0, \tag{27}$$

where h is a deterministic function and B denotes a Brownian process. Now we can recall the stochastic transport of a scalar f and its expectation, i.e., Eqs. (10) and (11). Assuming a stationary distribution, then the expectation is the solution of a stationary equation and $\mathbb{E}(\mathbb{D}_t f) = 0$. Therefore, the variance of the stochastic transport operator is expressed as:

$$\begin{aligned}
Var(\mathbb{D}_t f) &= \mathbb{E}\left[(\mathbb{D}_t f - \mathbb{E}(\mathbb{D}_t f))^2 \right] = \mathbb{E}\left[(\mathbb{D}_t f)^2 \right] \\
&= \mathbb{E}\left\{ \left[d_t f + \left(\nabla f \cdot \omega - \frac{1}{2}\alpha\Delta f \right) dt \right]^2 \right\} + \mathbb{E}\left\{ (\nabla f \cdot \sigma d\mathbf{B}_t)^2 \right\} \\
&\quad + \mathbb{E}\left\{ 2 \left[d_t f + \left(\nabla f \cdot \omega - \frac{1}{2}\alpha\Delta f \right) dt \right] (\nabla f \cdot \sigma d\mathbf{B}_t) \right\},
\end{aligned} \tag{28}$$

where the second term and the third term involve a Brownian term. According to the Itô isometry, we obtain:

$$\mathbb{E}\left\{ (\nabla f \cdot \sigma d\mathbf{B}_t)^2 \right\} = \mathbb{E}\left\{ \langle \nabla f \cdot \sigma d\mathbf{B}_t, \nabla f \cdot \sigma d\mathbf{B}_t \rangle \right\} = \mathbb{E}\left\{ (\alpha |\nabla f|^2) dt \right\} \tag{29}$$

and

$$\begin{aligned}
&\mathbb{E}\left\{ 2 \left[d_t f + \left(\nabla f \cdot \omega - \frac{1}{2}\alpha\Delta f \right) dt \right] (\nabla f \cdot \sigma d\mathbf{B}_t) \right\} \\
&= 2\mathbb{E}\left\{ \left\langle d_t f + \left(\nabla f \cdot \omega - \frac{1}{2}\alpha\Delta f \right) dt, \nabla f \cdot \sigma d\mathbf{B}_t \right\rangle \right\} = 2\mathbb{E}\left\{ \langle d_t f, \nabla f \cdot \sigma d\mathbf{B}_t \rangle \right\},
\end{aligned} \tag{30}$$

where (30) represents the correlation between the martingale part of $d_t f$ and the random advection term $\nabla f \cdot \sigma d\mathbf{B}_t$. For a conserved quantity f, we have the transport equation $\mathbb{D}_t f = 0$. This implies that when separating $f = \tilde{f} + f'$ in terms of its bounded variation part and its martingale part (i.e., time scale separation in terms of dt and $d\mathbf{B}_t$, which is a unique decomposition), the transport equation can be separated into:

$$
\begin{cases}
\dfrac{\partial \tilde{f}}{\partial t} d_t + \left(\nabla f \cdot w - \dfrac{1}{2} \alpha \Delta f \right) dt = 0, \\
d_t f' + \nabla f \cdot \sigma d\mathbf{B}_t = 0.
\end{cases}
\tag{31}
$$

Thus, we have

$$
\begin{aligned}
0 &= \langle d_t f' + \nabla f \cdot \sigma d\mathbf{B}_t, \ d_t f' + \nabla f \cdot \sigma d\mathbf{B}_t \rangle, \\
&= \langle d_t f', \ d_t f' \rangle + \langle \nabla f \cdot \sigma d\mathbf{B}_t, \ \nabla f \cdot \sigma d\mathbf{B}_t \rangle + 2 \langle d_t f', \ \nabla f \cdot \sigma d\mathbf{B}_t \rangle.
\end{aligned}
\tag{32}
$$

Note that $\langle d_t f', d_t f' \rangle = \langle d_t f, d_t f \rangle$, since the quadratic variation of bounded variation functions (such as the deterministic functions) is equal to 0. Equation (31) shows that in the case of a transported quantity $d_t f' = -\nabla f \cdot \sigma d\mathbf{B}_t$. When the conservation does hold only approximately (as in the case of the brightness consistency assumption), we will assume the proportionality relation: $d_t f' = \beta \nabla f \cdot \sigma d\mathbf{B}_t$, where β has to be fixed or estimated (note that $\beta = -1$ for a strict stochastic transport). This assumption comes to assume that the highly fluctuating part of the intensity difference is explained by the transport of the luminance function by the small-scale motion up to a proportionality factor. With this assumption we have:

$$
\mathbb{E} \left\{ \langle d_t f', d_t f' \rangle \right\} = \mathbb{E} \left\{ \langle \beta \nabla f \cdot \sigma d\mathbf{B}_t, \ \beta \nabla f \cdot \sigma d\mathbf{B}_t \rangle \right\} = \beta^2 \mathbb{E} \left\{ \left(\alpha |\nabla f|^2 \right) dt \right\}.
\tag{33}
$$

An estimate of β from this equation can be readily obtained: $\beta^2 = \mathbb{E} \left[\dfrac{(d_t f')^2}{\alpha |\nabla f|^2} \right]$. In practice the fluctuation $f' = f - \tilde{f}$ is set as the difference between the luminance function and a local (spatial/temporal) mean $\tilde{f}$. For successive images, the temporal difference is thus $d_t f' = f_2' - f_1' = (f_2 - \tilde{f}_2) - (f_1 - \tilde{f}_1)$. Note that as $\mathbf{a}$ is also an unknown in the optical flow formulation. This leads to an interleaved optimization problem. Here we adopt a simpler strategy in which the proportionality coefficient is fixed from the value of $\mathbf{a}$ at the previous multi-resolution level (i.e., $\mathbf{a}^{L-1}$). Eventually, by combining (32), (33) and (29), it yields

$$
2\mathbb{E} \left\{ \langle d_t f', \ \nabla f \cdot \sigma d\mathbf{B}_t \rangle \right\} = - \mathbb{E} \left\{ \left(\alpha |\nabla f|^2 \right) dt \right\} - \beta^2 \mathbb{E} \left\{ \left(\alpha |\nabla f|^2 \right) dt \right\}.
\tag{34}
$$

Substituting these equations into (28), we finally obtain:

$$
Var(\mathbb{D}_t f) \approx \mathbb{E} \left\{ \left[d_t f + \left(\nabla f \cdot w - \dfrac{1}{2} \alpha \Delta f \right) dt \right]^2 \right\} - \mathbb{E} \left\{ \beta^2 \alpha |\nabla f|^2 dt \right\}.
\tag{35}
$$

A minimum variance estimator with a spatial averaging for the expectation or considering a homogeneous Gaussian density leads to minimize:

$$Var(\mathbb{D}_t f) \approx \int_\Omega \left[d_t f + \left(\nabla f \cdot \omega - \frac{1}{2}\alpha\Delta f \right) dt \right]^2 d\mathbf{x} - \int_\Omega \left(\beta^2\alpha|\nabla f|^2 dt \right) d\mathbf{x}. \tag{36}$$

References

1. Batchelor, G.: Small-scale variation of convected quantities like temperature in turbulent fluid. J. Fluid Mech. **5**(01), 113–133 (1959)
2. Black, M., Anandan, P.: The robust estimation of multiple motions: parametric and piecewise-smooth flow fields. Comput. Vis. Image Underst. **63**(1), 75–104 (1996)
3. Bruhn, A., Weickert, J., Schnörr, C.: Lucas/Kanade meets Horn/Schunck: combining local and global optic flow methods. Int. J. Comput. Vis. **61**(3), 211–231 (2005)
4. Carlier, J.: Second set of fluid mechanics image sequences. European Project Fluid Image Analysis and Description (FLUID) (2005). http://www.fluid.irisa.fr
5. Cassisa, C., Simoens, S., Prinet, V., Shao, L.: Subgrid scale formulation of optical flow for the study of turbulent flow. Exp. Fluids **51**(6), 1739–1754 (2011)
6. Chen, X., Zillé, P., Shao, L., Corpetti, T.: Optical flow for incompressible turbulence motion estimation. Exp. Fluids **56**(1), 1–14 (2015)
7. Corpetti, T., Mémin, E., Pérez, P.: Dense estimation of fluid flows. IEEE Trans. Pattern Anal. Mach. Intell. **24**(3), 365–380 (2002)
8. Crisan, D., Flandoli, F., Holm, D.: Solution properties of a 3D stochastic Euler fluid equation. arXiv preprint arXiv:1704.06989 (2017)
9. Cuzol, A., Mémin, E.: A stochastic filtering technique for fluid flow velocity fields tracking. IEEE Trans. Pattern Anal. Mach. Intell. **31**(7), 1278–1293 (2009)
10. Dérian, P., Héas, P., Herzet, C., Mémin, E.: Wavelets and optical flow motion estimation. Numer. Math.: Theory Methods Appl. **6**(1), 116–137 (2013)
11. Héas, P., Mémin, E., Heitz, D., Mininni, P.: Power laws and inverse motion modeling: application to turbulence measurements from satellite images. Tellus A: Dyn. Meteorol. Oceanogr. **64**, 10962 (2012)
12. Heitz, D., Mémin, E., Schnörr, C.: Variational fluid flow measurements from image sequences: synopsis and perspectives. Exp. Fluids **48**(3), 369–393 (2010)
13. Horn, B., Schunck, B.: Determining optical flow. Artif. Intell. **17**(1–3), 185–203 (1981)
14. Jullien, M., Castiglione, P., Tabeling, P.: Experimental observation of Batchelor dispersion of passive tracers. Phys. Rev. Lett. **85**(17), 3636 (2000)
15. Kadri-Harouna, S., Dérian, P., Héas, P., Mémin, E.: Divergence-free wavelets and high order regularization. Int. J. Comput. Vis. **103**(1), 80–99 (2013)
16. Liu, T., Shen, L.: Fluid flow and optical flow. J. Fluid Mech. **614**, 253–291 (2008)
17. Mémin, E.: Fluid flow dynamics under location uncertainty. Geophys. Astrophys. Fluid Dyn. **108**(2), 119–146 (2014)
18. Mémin, E., Pérez, P.: Dense estimation and object-based segmentation of the optical flow with robust techniques. IEEE Trans. Image Process. **7**(5), 703–719 (1998)
19. Papadakis, N., Mémin, E.: Variational assimilation of fluid motion from image sequence. SIAM J. Imag. Sci. **1**(4), 343–363 (2008)

20. Resseguier, V., Mémin, E., Chapron, B.: Geophysical flows under location uncertainty, part I: random transport and general models. Geophys. Astrophys. Fluid Dyn. **111**, 149–176 (2017)
21. Sun, D., Roth, S., Black, M.: Secrets of optical flow estimation and their principles. In: Proceedings of the IEEE Conference on CVPR, pp. 2432–2439. IEEE (2010)
22. Yuan, J., Schnörr, C., Mémin, E.: Discrete orthogonal decomposition and variational fluid flow estimation. J. Math. Imag. Vis. **28**(1), 67–80 (2007)

Image Processing and Segmentation

Autonomous Multi-camera Tracking Using Distributed Quadratic Optimization

Yusuf Osmanlıoğlu[1]([✉]) [iD], Bahareh Shakibajahromi[2], and Ali Shokoufandeh[2] [iD]

[1] Section of Biomedical Image Analysis,
University of Pennsylvania, Philadelphia, USA
yusuf.osmanlioglu@uphs.upenn.edu
[2] Department of Computer Science, Drexel University, Philadelphia, USA
bs643@drexel.edu, ashokouf@cs.drexel.edu

Abstract. Multi-camera object tracking is an efficient approach commonly used in security and surveillance systems. In a conventional multi-camera setup, a central computational unit processes large amounts of data in real time that is provided by distributed cameras. High network traffic, cost of storage on the central unit, scalability of the system, and vulnerability of the central unit to attacks are among the disadvantages of such systems. In this paper, we present an autonomous multi-camera tracking system to overcome these challenges. We assume cameras that are capable of limited computation for locally tracking a subset of objects in the scene, as well as peer-to-peer network connectivity among the cameras with a decent bandwidth that is sufficient for message passing to achieve coordination. We propose an efficient distributed algorithm for coordination and load-balancing among the cameras. We also provide experimental results to validate the utility of the proposed algorithm in comparison to a centralized algorithm.

Keywords: Multi-camera tracking · Distributed assignment problem
Metric labeling · Primal-dual approximation

1 Introduction

Object tracking is a fundamental problem in computer vision with various applications including security and surveillance. Tracking can be achieved using a single camera in *simple* scenes such as monitoring a highway where the traffic flows in predefined lanes. Single camera becomes insufficient for tracking in *complex* scenes that involve large number of freely roaming objects. Problems such as occlusion and inability to extract distinctive features from certain viewing angles require multiple cameras for obtaining a better coverage of the scene.

The conventional approach for multi-camera tracking is based on collecting views from various cameras at a central computing unit, which then performs the tracking with a centralized algorithm using all the available information.

© Springer International Publishing AG, part of Springer Nature 2018
M. Pelillo and E. Hancock (Eds.): EMMCVPR 2017, LNCS 10746, pp. 175–188, 2018.
https://doi.org/10.1007/978-3-319-78199-0_12

Such centralized systems has several disadvantages. First, large amount of visual data need to be collected and transferred to a central unit in real-time, which in turn induces heavy traffic over the communication network. Next, the cost of storing entire time-series on the central unit is greater than keeping and processing them in small distributed computing units. Furthermore, adding more cameras to centralized systems might require making system-wide adjustments to the extent of updating the hardware of the central unit altogether. Finally, vulnerability of a centralized system to outside attacks is high since the whole system fails by simply attacking the central unit. In this paper, we consider a multi-camera tracking system that addresses these issues while additionally achieving autonomous load-balancing across cameras.

The proposed model consists of cameras with reasonable computational capabilities, connected to each other through an ad-hoc communication network. Each camera locally tracks a subset of objects in the scene that are visually most suitable to track from its viewing advantage point while the coordination among the cameras are assessed through message passing across the network. We achieve four goals by maintaining the dynamic coordination among the cameras. First, we perform *intelligent tracking* by ensuring that each camera tracks objects that are most salient from its viewpoint, while ignoring the rest of the scene during the tracking process. Next, we *distribute* the workload as evenly as possible across the cameras without reducing the tracking quality. Thirdly, we *reduce the ambiguity* in tracking by assigning similar or closely located objects to different cameras, reducing the tracking confusion. Finally, we maintain the *coordination* among cameras dynamically, that is, the assignment of the objects to the cameras are updated as new objects enter the scene or objects leave the scene in an online fashion. This is in contrast to re-calculating the mapping from scratch as the scene changes.

The system consists of two main components. The first part involves assigning objects in the scene to the cameras using a distributed online algorithm, ensuring that each object is tracked by at least one camera. The second component involves local processing of the visual data at each camera for tracking the assigned objects. In this paper, we focus on the first component of the system by proposing a distributed algorithm for assigning objects to cameras. We will investigate the latter component in our future studies.

In our approach, we first show that the coordination among the cameras can be formulated both as a linear and quadratic optimization problem. We then utilize a modified formulation of the *metric labeling problem* for assigning objects to the cameras. This study extends our recent result of an efficient centralized primal-dual approximation algorithm [21] by proposing an equivalent distributed algorithm.

The rest of the paper is organized as follows: Sect. 2 gives an overview of the literature and lays theoretical foundations. Section 3 provides the details of the distributed approximation algorithm for the metric labeling problem. Finally, we present the experimental evaluation of our approach in Sect. 4 and conclude the document in Sect. 5.

2 Background

In this section, we first provide a brief survey of the literature on multi-camera tracking. We then present notations and definitions for assignment and related problems, and primal-dual approximation methods, which provides a basis for the algorithm that we will present in next section. Finally, we will provide a brief summary of our previous results on a centralized primal-dual approximation algorithm for the modified metric labeling problem. We refer the reader to [20, 22] for an extensive overview of the related literature on assignment problems.

2.1 Multi-camera Tracking

Robust detection and tracking of objects is a challenging problem in computer vision. Single camera trackers are not capable of accurately tracking multiple objects due to limited field of view, obstructed view, and severe occlusion that are common in crowded scenes. In contrast, multi-camera tracking methods combine information from a set of cameras overlooking the same scene. The problem is widely studied in the literature from different perspectives including the use of different camera types [1, 23], utilizing multiple plane homographies [13, 17, 18], and reformulating the problem as flow optimization [5, 26].

More recently, intelligent tracking systems are receiving increasing attention in the vision community. A common approach is to perform majority of the computations locally among distributed cameras and utilize a central computing unit to merge and finalize decisions. The system proposed in [7], for example, tracks multiple subjects using a network of smart cameras. Here, each camera estimates the position of people in their viewing range using maximum likelihood estimation, and then all estimates are merged in a central fusion algorithm to generate the final estimates. Another example is the system in [25] where the problem of finding correspondences across multiple cameras is formulated as a multi-dimensional assignment problem and a greedy randomized adaptive search procedure is used to approximate the solution to solve this NP-hard problem. Yet another approach is to use cameras for preprocessing and carrying out the demanding tasks at the central computational unit. The intelligent multi-camera video surveillance system presented in [15], for example, first extracts features from image sequences locally at cameras, and then transmits them to a central server which combines the information and reasons about the scene. Machine learning methods are also utilized for tackling the problem, such as the system in [9] where an end-to-end deep learning method for multi-view people detection has been proposed. Similarly, the system presented in [4] suggests an architecture combining convolutional neural networks and conditional random fields. The aim here is to explicitly model the ambiguities of multi-camera multi-people tracking in crowded and occluded scenes.

2.2 Graph Matching and Assignment Problems

Given two graphs G and H, finding an exact mapping among their nodes is known to be challenging. Specifically, *subgraph isomorphism* which asks whether

G contains an induced subgraph H' that is isomorphic to H, is a problem known to be NP-complete [11]. A special case of the problem, *graph isomorphism*, which is neither known to be polynomial time solvable nor NP-complete, is recently shown to be solvable in quasi-polynomial time [3]. For the past few decades, the focus of the research in the theory community has been over inexact mappings and assignments. Various approaches are devised for tackling the problem including tabu search [24], error-correcting graph matching [8], graph edit-distance based matching [6], and convex optimization formulations [2].

Another way to treat the correspondence problem is by approaching the inexact matching as classification of a set of objects into groups by minimizing a prescribed cost function, such as the *multi-way cut* [12] or the *0-extension* problems [16]. The *metric labeling problem* [19] presents a more general case for this approach where, given a set of objects and a set of labels with pairwise relationships defined among the elements of both sets, a label is assigned to each object by minimizing a cost function involving both separation and assignment costs. Separation cost penalizes assigning loosely related labels to closely related objects while assignment cost penalizes assigning an object to an unrelated label.

Taking shortest path metric as the distance measure between the nodes, problem of finding a mapping among graphs can be treated as an instance of metric labeling. Specifically, given two graphs P and L, a distance function $d :$ $L \times L \to \mathbb{R}$ denoting the shortest path between nodes of L, a weight function $w : P \times P \to \mathbb{R}$ defined as the reciprocal of shortest path distance between pairs of nodes in P, and an assignment cost function $c : P \times L \to \mathbb{R}$ defining a measure of similarity among pairs of nodes of P and L, we can restate the problem as labeling nodes of P with nodes of L.

A natural quadratic programming formulation for the metric labeling problem is as follows:

$$
\begin{aligned}
\min \quad & \sum_{\substack{p \in P \\ a \in L}} c_{pa} \cdot x_{pa} + \sum_{\substack{p,q \in P \\ a,b \in L}} w_{pq} \cdot d_{ab} \cdot x_{pa} \cdot x_{qb} \\
\text{s.t.} \quad & \textstyle\sum_{a \in L} x_{pa} = 1, \qquad \forall p \in P \\
& x_{pa} \in \{0,1\}, \qquad p \in P, a \in L
\end{aligned}
\tag{1}
$$

where x_{pa} is the indicator variable for assigning object node p to label node a and c_{pa}, w_{pq}, and d_{ab} are the cost, weight, and distance functions, respectively. The quadratic term in the separation cost renders an exact solution to the problem to be intractable. Several approximate solutions were devised through various linear programming (LP) formulations of the problem [10,19]. However, large number of constraints present in the LP formulations make the solutions unfeasible for large graphs.

2.3   A Primal-Dual Approximation of Metric Labeling

Primal-dual approximation scheme of Goemans and Williamson [14] has proven useful in finding approximate solutions to many combinatorial optimization

---

Algorithm 1. ML_{PD}: A centralized primal-dual approximation algorithm for the metric labeling problem [21].

---

procedure ML-Primal-Dual(P, L)
1: $\forall p, q \in P, a \in L : x_{pa} \leftarrow 0, P' \leftarrow P$
$\quad\quad c_{pa} \leftarrow similarity(p, a)$
$\quad\quad d_{ab} \leftarrow distance_{L}(a, b)$
$\quad\quad w_{pq} \leftarrow 1/distance_{P}(p, q)$
$\quad\quad \phi(p, a) \leftarrow c_{pa}$
2: **while** $P' \neq \emptyset$ **do**
3: $\quad$ Find $p \in P'$ that minimizes $\phi(p, a)$ for some $a \in L$
4: $\quad x_{pa} \leftarrow 1$
5: $\quad P' \leftarrow \mathcal{P}' \setminus \{p\}$
6: $\quad \forall q \in P', b \in L \setminus \{a\} : \phi(q, b) = \phi(q, b) + w_{pq} \cdot d_{ab}$
7: **end while**
8: **return** $\mathcal{X} = \{x_{pa} : \forall p \in P, a \in L\}$

---

problems that can be modeled with integer linear programs without the need for a supplementary rounding phase. The main advantages of using this approach are that the algorithm runs much faster than LP solvers and if the data changes during the course of execution, the algorithm can recover by updating only the newly violated constraints without the need for solving the problem from scratch.

Recently, we presented an efficient approximation algorithm [21] for the metric labeling by utilizing this primal-dual approximation scheme over an LP formulation of the problem. Here, we provide a short summary of our approach and refer the reader to [21] for a detailed analysis and experimental evaluation of the method.

Our approach begins by taking the dual of the LP formulation of the metric labeling due to Chekuri et al. [10]. Unlike the problems that are traditionally solved using primal-dual approximation scheme, this formulation has three interdependent types of constraints. To apply the primal-dual method, it is necessary to find an acyclic order in identifying, iterating between, and modifying violated primal constraints, their corresponding dual variables, and dependent dual variables. Strictly following the primal-dual approximation method requires assigning objects to labels in tuples since it enforces dual feasibility, which in turn may result in poor overall assignment for this setup. We overcome this problem by relaxing the dual feasibility condition for the dual constraints corresponding to the primal variables of type x_{pa} which have previously became tight.

The primal-dual approximation algorithm that we proposed for the problem, denoted ML_{PD}, is presented in Algorithm 1. The algorithm starts by initializing assignment variables x_{pa}, set of unassigned objects P', and the cost, distance, and pairwise relation functions c, d and w. It further defines an adjusted assignment cost function ϕ where the value of $\phi(p, a)$ is initially set to assignment cost of p to a (line 1). During each iteration of the loop (lines (2–7)) the algorithm makes an assignment for the object-label pair that minimizes the adjusted assignment cost function ϕ (lines 3–4). Before proceeding to the next iteration, assigned object is removed from the set P′ (line 5) and function ϕ is updated for each of the unassigned objects by the amount of separation cost with respect to the recently assigned object (line 6). The algorithm iterates until no object node remains unassigned. We have shown in [21] that running time complexity of ML_{PD} is $O(n^2m + m^2)$ where $|\mathrm{P}| = n$ and $|\mathrm{L}| = m$.

3   Distributed Assignment of Objects to Cameras

In this section, we propose algorithms for tackling the first part of the autonomous multi-camera tracking system, that is, assigning objects to cameras in a distributed fashion. In contrast to a central agent with the knowledge of the entire network layout assigning objects to cameras, our goal is to let cameras decide autonomously which objects they will track while optimizing the quality of tracking across the network. To achieve this, we parallelize the primal-dual algorithm ML_{PD} presented in the previous section.

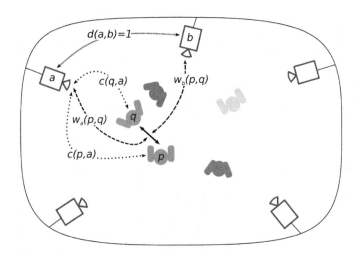

Fig. 1. Schematic overview of the autonomous multi-camera tracking system.

In our approach, P represents a set of objects and L of cameras. We assume that every camera $a \in \mathrm{L}$ has a qualitative measure $c(p, a)$ corresponding to its feasibility to track object p. Intuitively, a camera will have higher affinity for tracking an object which has better (qualitative) visibility, resulting in a low assignment cost of the object to the camera. We further assume that an ambiguity measure $w_a(\cdot, \cdot)$ is defined among objects that makes pairs of objects p and q comparable with respect to camera a. Specifically, in order to reduce ambiguity, a camera would prefer to track objects that are spatially located apart and visually dissimilar to each other. In contrast to the pairwise distance

Algorithm 2. ML_{DPD}: Distributed primal-dual algorithm for approximating metric labeling problem. The algorithm is run by each camera $a \in L$ in parallel.

---

procedure ML-Distributed-Primal-Dual(a)
1: $\forall p, q \in P : x_{pa} \leftarrow 0, P' \leftarrow P$
    $c_{pa} \leftarrow feasibility(p, a)$
    $w_{pq} \leftarrow ambiguity(p, q)$
    $\phi(p) \leftarrow c_{pa}$
2: **while** $P' \neq \emptyset$ **do**
3:     $t = argmin_{p \in P'} \phi(p)$
4:     Broadcast *assignment request (AR)* message for t
5:     $\forall b \in L \setminus a$ receive *assignment request (AR)* messages for t_b
6:     $\forall b \in L \setminus a$ send *assignment cost (AS)* messages for $\phi(t_b)$
7:     $\forall b \in L \setminus a$ receive *assignment cost (AS)* messages for $\phi_b(t)$
8:     **if** $\forall b \in L : \phi(t) < \phi_b(t)$ **then**
9:         $x_{ta} \leftarrow 1$
10:        $P' \leftarrow P' \setminus \{t\}$
11:        Broadcast *assignment confirmation (AC)* message for t
12:        $\forall q \in P' : \phi(q) = \phi(q) + w_{pq}$
13:    **else**
14:        Broadcast *no-assignment (NA)* message
15:    **end if**
16:    $\forall b \in L \setminus a$ receive *assignment confirmation (AC)* or *no assignment (NA)* messages and update P' accordingly
17: **end while**
18: **return** $\mathcal{X} = \{x_{pa} : \forall p \in P\}$

---

measure defined among the labels in metric labeling, there is no such relation among cameras in this setup, resulting in an instance of *uniform metric labeling* problem. Thus, to eliminate its contribution to the objective function in metric labeling formulation, we define the pairwise distance function as $d(a, b) = 1$ for all $a, b \in L$. A schematic overview of the system is presented in Fig. 1.

The task of assigning objects to cameras that is mainly achieved in lines 3-4 of Algorithm 1 can be parallelized. However, it is required that the cameras running the distributed algorithm keep track of assignments made network-wide to maintain an up-to-date adjusted assignment cost function ϕ, mimicking the behavior of line 6 in ML_{PD}.

We propose the round-based algorithm in Algorithm 2 (ML_{DPD}) as a distributed version of ML_{PD}. The algorithm is executed by each camera a locally and starts with initializing the assignment variables x_{pa}, set of unassigned objects $\mathcal{P}'$, and the cost, pairwise similarity, and adjusted assignment cost functions c, w and ϕ similar to ML_{PD} (line 1). During each iteration of the loop (lines 2–17), at least one assignment is guaranteed to be done network-wide. However, we note that some of the cameras might choose not to assign any object in one round. The algorithm first selects the object t which minimizes the adjusted assignment cost ϕ (line 3) and broadcasts an *assignment request* (AR) message to the network indicating its intention to track object t. It listens to the network for receiving AR messages to obtain the target objects t_b for each camera b (line 5). It then replies back to each AR message with an *assignment cost* (AS) message indicating the cost $\phi(t_b)$ of assigning the object t_b to itself (line 6). Meanwhile, it listens to the network for receiving AS messages for obtaining the cost $\phi_b(t)$ of assigning t to each camera b (line 7). After obtaining a complete list of assignment costs for object t across cameras, the algorithm checks if $\phi(t)$ is smaller than $\phi_b(t)$ for all cameras $b \in L \setminus a$ (line 8). If this is true, then the algorithm assigns the object t to the camera itself, sends *assignment confirmation* (AC) messages to the network indicating this decision, and updates the adjusted assignment cost function (lines 9–12). Otherwise, the algorithm does not make any assignment in favor of object t being tracked by another camera in a future round of the algorithm and broadcasts a *no-assignment* (NA) message (line 14). Finally, the algorithm listens to the network for receiving AC or NA messages and updates the unassigned objects list P'. The algorithm iterates until all objects are assigned.

We note that the algorithm might assign an object t to several cameras if the adjusted assignment cost of t is identical for the cameras that are considering to track it (line 8). This can be avoided by defining a simple tiebreaker function which assigns the object to the camera with the lightest load and smallest id number.

The correctness analysis of ML_{DPD} hinges upon the following observations. First, we show that the algorithm is guaranteed to terminate with a feasible solution. Next, we provide worst case analysis for its running time. Finally, we investigate the communication cost complexity of the algorithm. In the analyses to follow, we assume that $|P| = n$ and $|L| = m$.

Proposition 1. *Assuming that the cameras are not malicious, ML_{DPD} is guaranteed to terminate in $O(n)$ rounds and the resulting object assignment is a feasible solution to the metric labeling problem.*

Proof. Assume that there exist a round of the algorithm in which no object is assigned to any camera across the network. We use the labels $a_1, a_2, \ldots, a_m$ and $p_1, p_2, \ldots, p_n$, with $n >> m$, to denote cameras and objects, respectively. We also denote the most feasible object to be tracked for each camera a_i as p_i (i.e., $p_i = argmin_{p \in P'} \phi_{a_i}(p))$). Then, a camera a_i choosing not to assign object p_i to itself implies $\phi_{a_i}(p_i) > \phi_{a_j}(p_i)$ for some camera a_j. Likewise, a_j not assigning p_j to itself implies $\phi_{a_j}(p_j) > \phi_{a_k}(p_j)$ for some other camera a_k. Without loss of generality, assume $j = i + 1$, $k = j + 1$, and so on. Then, the following inequality should hold:

$$\phi_{a_1}(p_1) > \phi_{a_2}(p_1) > \phi_{a_2}(p_2) > \phi_{a_3}(p_2) > \cdots > \phi_{a_m}(p_m) > \phi_{a_1}(p_m).$$

However, $\phi_{a_1}(p_1) < \phi_{a_1}(p_m)$ must be true since the camera a_1 has chosen the object p_1 for tracking due to $p_1 = argmin_{p \in P'} \phi(p)$, which is a contradiction. Thus, at least one object is assigned to a camera across the network in each round. Since an assigned object will never get unassigned (or reassigned) in subsequent rounds, the algorithm terminates after $O(n)$ rounds in the worst case.

In order for the network-wide assignments achieved at the termination of the algorithm to be feasible, it is required that no object is left unassigned and each object is assigned to a single camera at the end. The algorithm iterates until no unassigned object is left, satisfying the first condition. A unique assignment is guaranteed for each object through the *if* condition in line 8 along with the tiebreaker function that is mentioned above. And since an object that is assigned to a camera is removed from the unassigned objects list of all cameras, it will not be reconsidered for assignment in the following rounds, guaranteeing a unique assignment for each object. Thus, the resulting solution will be feasible.   □

Proposition 2. *The running time complexity of ML_{DPD} is $O(n^2 + m)$.*

Proof. We make an aggregated running time analysis for the worst case, that is, a single object-camera assignment is made across the network at each round of the algorithm. Initialization operations in the first line takes $O(n^2 + m)$ time. Finding the object which minimizes the adjusted assignment cost in line 2 takes $O(n^2)$ time over all iterations. During each iteration of the while loop, message passing performed in lines 4–7 will each take $O(m)$ time. The **if** statement in line 8 takes $O(n)$ time to validate, $O(n + m)$ if it renders true and $O(m)$ otherwise. Finally, receiving messages at line 16 takes $O(m)$ time. Accounting for the $O(n)$ iterations of the while loop in the worst case, asymptotic running time of the algorithm sums to $O(n^2 + m)$.   □

Proposition 3. *The communication cost of ML_{DPD} is $O(nm)$ for a camera, resulting in an overall communication overhead of $O(nm^2)$ for object assignment in the entire network.*

Proof. During each round of the algorithm, each camera sends and receives $O(m)$ assignment request and assignment cost messages by executing lines 4–7. In either case of the **if** statement in line 8, a camera sends $O(m)$ assignment confirmation or no-assignment messages (lines 11 or 14). Finally, a camera receives a sum of $O(m)$ assignment confirmation or no-assignment messages at line 16. Summing the cost for all rounds, the communication cost for each camera is $O(nm)$ messages in the worst case. The overall communication overhead of the algorithm to the network is $O(nm^2)$ messages.   □

Contrasting the distributed and centralized versions of the primal-dual algorithm for metric labeling, we observe that ML_{DPD} is faster than ML_{PD} by a factor of $O(m)$ with the expense of $O(nm^2)$ messages passed across the network. We note that the overall cost of assignment obtained by ML_{DPD} is likely to be higher relative to ML_{PD}. This follows from the fact that ML_{DPD} allows assigning several objects to cameras at a round while ML_{PD} ensures that only one object is assigned to a camera at each round which would have the minimum assignment cost.

We extend ML_{DPD} to obtain distributed equivalent of ML_{PD} in Algorithm 3, denoted ML_{DPDX}. In contrast to ML_{DPD}, each camera in the extended algorithm

---

Algorithm 3. ML_{DPDX}: Extended distributed primal-dual algorithm for approximating metric labeling problem. The algorithm is run by each camera $a \in L$ in parallel.

---

    procedure ML-Extended-Distributed-Primal-Dual(a)
1: $\forall p, q \in P : x_{pa} \leftarrow 0, P' \leftarrow P$
                   $c_{pa} \leftarrow feasibility(p, a)$
                   $w_{pq} \leftarrow ambiguity(p, q)$
                   $\phi(p) \leftarrow c_{pa}$
2: **while** $P' \neq \emptyset$ **do**
3:     $t = argmin_{p \in P'} \phi(p)$
4:     Broadcast *assignment request (AR)* message for t along with $\phi(t)$
5:     $\forall b \in L \setminus a$ receive *assignment request (AR)* messages for t_b along with $\phi(t_b)$
6:     Find $b \in L$ and its corresponding t_b that minimizes $\phi_b(t_b)$
7:     **if** $b == a$ **then**
8:         $x_{ta} \leftarrow 1$
9:         $P' \leftarrow P' \setminus \{t\}$
10:        $\forall q \in P' : \phi(q) = \phi(q) + w_{pq}$
11:    **else**
12:        $P' \leftarrow P' \setminus \{t_b\}$
13:    **end if**
14: **end while**
15: **return** $\mathcal{X} = \{x_{pa} : \forall p \in P\}$

---

broadcasts the cost of assignment that it wants to make, and assigns the object to itself if its cost is minimum compared to all other assignment requests across the network. The algorithm terminates after $O(m)$ rounds in any case since it ensures that only a single assignment is made at each round. The running time of the algorithm is the same with the worst case running time of the ML_{DPD}. Although ML_{DPDX} is lighter in message passing relative to ML_{DPD}, its communication overhead is still asymptotically the same compared to the ML_{DPD}.

4 Evaluation of the Method

In this section, we provide empirical evaluation of proposed algorithms by contrasting them to the centralized primal-dual algorithm ML_{PD}. Our analysis involves the comparison of the number of rounds it takes for the algorithms to complete assignment, the cost of the assignment, and the load distribution across the cameras.

The setup for the experiment is as follows. We assume a square area $A \subset \mathbb{R}^2$ with dimensions 1000 m by 1000 m where cameras are distributed evenly among the periphery. The objects that are to be tracked are generated uniformly at random in A. Each object is randomly set a visibility score relative to each camera which is used as the assignment cost of this object to the camera. A similarity score is also randomly set among pairs of objects. The ambiguity score among object pairs is calculated as summation of the Euclidean distance and the similarity score of the pair. We normalized visibility and ambiguity scores to $[0, 1]$ and set their contribution to the objective function to be even. We repeated the experiment 1000 times with independent setups and reported the average of the scores as our final result.

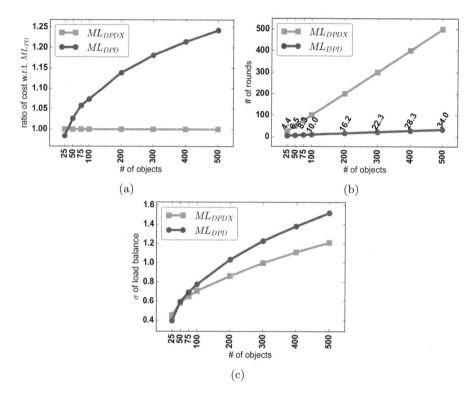

(a)

(b)

(c)

Fig. 2. Comparative performance measures of ML_{DPD} and ML_{DPDX} for (a) ratio of cost function with respect to ML_{PD}, (b) number of rounds for the algorithms to terminate, and (c) standard deviation of number of assigned objects to each camera. Number of cameras are fixed to 25 while the number of objects are in $[25, 500]$.

We present the results of the experiments in Fig. 2 for 25 cameras and number of objects varying in $[25, 500]$. Our analysis in here is threefold. First, we compare the ratio of cost of overall assignment for both distributed algorithms relative to the ML_{PD}. As shown in Fig. 2a, ML_{DPDX} is equivalent to the ML_{PD} in the cost of the object assignments. The cost of assignment is up to 25% larger for the ML_{DPD} relative to ML_{PD} for 500 objects. This is expected since the ML_{DPD} allows several assignments at each round. Next, we compare the number of rounds it take for the distributed algorithms to terminate. Since ML_{DPDX} makes a single assignment across the network at each round, the number of rounds it takes for the algorithm to terminate is tightly bounded by the number of objects, as shown in Fig. 2b. Although number of rounds for ML_{DPD} is upper bounded by the number of objects in the scene, the average number of rounds is much less than the upper bound. Finally, we compare the load balance across the cameras in terms of standard deviation of number of assigned objects per camera. Although the metric labeling formulation does not include an explicit term for adjusting load balancing, it would be ideal for each camera to have equal number of objects assigned to them. As shown in Fig. 2c, load balance for both algorithms are quite decent in that, even in the case of 500 objects, the standard deviation of the number of assigned objects is less than 2 where the average number of objects assigned per camera is 20. We also observe that ML_{DPDX} performs slightly better relative to ML_{DPD} in load balancing which might be due to improved cost performance of the latter as it was shown in Fig. 2a.

5   Conclusion and Future Work

In this paper we proposed an autonomous multi-camera object tracking system to overcome the difficulties present in conventional centralized multi-camera tracking systems such as high network traffic, cost of the storage on central unit, scalability, and vulnerability of the centralized unit to attacks. Our system assumes camera units that are capable of limited computation for locally tracking a subset of objects in the scene, as well as peer-to-peer network connectivity among the cameras for maintaining message passing to achieve coordination. Formulating the object assignment problem as an instance of the well known metric labeling problem, we proposed two efficient distributed primal-dual approximation algorithms. We also provided experimental evaluation of the proposed algorithms to demonstrate that our method achieves coordination and load-balancing among the cameras efficiently.

In this paper, we discussed the first component of our multi-camera tracking system, that is the distributed primal-dual approximation solution to the object assignment problem. We leave the second component, that is the application of the method in a real life setup along with performing object tracking, as a future work.

References

1. Alahi, A., Ramanathan, V., Fei-Fei, L.: Tracking millions of humans in crowded spaces. In: Group and Crowd Behavior for Computer Vision, pp. 115–135 (2017)
2. Almohamad, H.A., Duffuaa, S.O.: A linear programming approach for the weighted graph matching problem. IEEE Trans. Pattern Anal. Mach. Intell. **15**(5), 522–525 (1993)
3. Babai, L.: Graph isomorphism in quasipolynomial time. In: Proceedings of the Forty-Eighth Annual ACM Symposium on Theory of Computing, pp. 684–697. ACM, June 2016
4. Baqué, P., Fleuret, F., Fua, P.: Deep occlusion reasoning for multi-camera multi-target detection. arXiv preprint arXiv:1704.05775 (2017)

5. Berclaz, J., Fleuret, F., Turetken, E., Fua, P.: Multiple object tracking using k-shortest paths optimization. IEEE Trans. Pattern Anal. Mach. Intell. **33**(9), 1806–1819 (2011)
6. Berretti, S., Del Bimbo, A., Pala, P.: A graph edit distance based on node merging. In: Enser, P., Kompatsiaris, Y., O'Connor, N.E., Smeaton, A.F., Smeulders, A.W.M. (eds.) CIVR 2004. LNCS, vol. 3115, pp. 464–472. Springer, Heidelberg (2004). https://doi.org/10.1007/978-3-540-27814-6_55
7. Bo Bo, N., et al.: Robust multi-camera people tracking using maximum likelihood estimation. In: Blanc-Talon, J., Kasinski, A., Philips, W., Popescu, D., Scheunders, P. (eds.) ACIVS 2013. LNCS, vol. 8192, pp. 584–595. Springer, Cham (2013). https://doi.org/10.1007/978-3-319-02895-8_53
8. Bunke, H.: Error correcting graph matching: on the influence of the underlying cost function. IEEE Trans. Pattern Anal. Mach. Intell. **21**(9), 917–922 (1999)
9. Chavdarova, T., Fleuret, F.: Deep multi-camera people detection. arXiv preprint arXiv:1702.04593 (2017)
10. Chekuri, C., Khanna, S., Naor, J., Zosin, L.: A linear programming formulation and approximation algorithms for the metric labeling problem. SIAM J. Discrete Math. **18**(3), 608–625 (2004)
11. Cook, S.A.: The complexity of theorem-proving procedures. In: Proceedings of the Third Annual ACM Symposium on Theory of Computing, pp. 151–158. ACM (1971)
12. Dahlhaus, E., Johnson, D.S., Papadimitriou, C.H., Seymour, P.D., Yannakakis, M.: The complexity of multiway cuts (extended abstract). In: Proceedings of the 24th Annual ACM Symposium on Theory of Computing, STOC 1992, pp. 241–251. ACM, New York (1992)
13. Eshel, R., Moses, Y.: Tracking in a dense crowd using multiple cameras. Int. J. Comput. Vis. **88**(1), 129–143 (2010)
14. Goemans, M.X., Williamson, D.P.: The primal-dual method for approximation algorithms and its application to network design problems. In: Hochbaum, D.S. (ed.) Approximation Algorithms for NP-hard Problems, pp. 144–191. PWS Publishing Co., Boston (1997)
15. Hameete, P., Leysen, S., Van Der Laan, T., Lefter, I., Rothkrantz, L.: Intelligent multi-camera video surveillance. Int. J. Inf. Technol. Secur. **4**(4), 51–62 (2012)
16. Karzanov, A.V.: Minimum 0-extensions of graph metrics. Eur. J. Comb. **19**(1), 71–101 (1998)
17. Khan, S.M., Shah, M.: A multiview approach to tracking people in crowded scenes using a planar homography constraint. In: Leonardis, A., Bischof, H., Pinz, A. (eds.) ECCV 2006. LNCS, vol. 3954, pp. 133–146. Springer, Heidelberg (2006). https://doi.org/10.1007/11744085_11
18. Khan, S.M., Yan, P., Shah, M.: A homographic framework for the fusion of multi-view silhouettes. In: IEEE 11th International Conference on Computer Vision 2007, ICCV 2007, pp. 1–8. IEEE (2007)
19. Kleinberg, J., Tardos, É.: Approximation algorithms for classification problems with pairwise relationships: metric labeling and markov random fields. J. ACM **49**(5), 616–639 (2002)
20. Livi, L., Rizzi, A.: The graph matching problem. Pattern Anal. Appl. **16**(3), 253–283 (2013)
21. Osmanlıoğlu, Y., Ontañón, S., Hershberg, U., Shokoufandeh, A.: Efficient approximation of labeling problems with applications to immune repertoire analysis. In 2016 23rd International Conference on Pattern Recognition (ICPR), pp. 2410–2415. IEEE (2016)

22. Pentico, D.W.: Assignment problems: a golden anniversary survey. Eur. J. Oper. Res. **176**(2), 774–793 (2007)
23. Stillman, S., Tanawongsuwan, R., Essa, I.: Tracking multiple people with multiple cameras. In: International Conference on Audio-and Video-based Biometric Person Authentication (1999)
24. Williams, M.L., Wilson, R.C., Hancock, E.R.: Deterministic search for relational graph matching. Pattern Recogn. **32**(7), 1255–1271 (1999)
25. Wu, Z., Hristov, N.I., Hedrick, T.L., Kunz, T.H., Betke, M.: Tracking a large number of objects from multiple views. In: 2009 IEEE 12th International Conference on Computer Vision, pp. 1546–1553. IEEE (2009)
26. Zhang, L., Li, Y., Nevatia, R.: Global data association for multi-object tracking using network flows. In: IEEE Conference on Computer Vision and Pattern Recognition 2008, CVPR 2008, pp. 1–8. IEEE (2008)

Nonlinear Compressed Sensing
for Multi-emitter X-Ray Imaging

Maria Klodt and Raphael Hauser$^{(\boxtimes)}$ (iD)

Mathematical Institute, University of Oxford, Oxford OX2 6GG, UK
hauser@maths.ox.ac.uk

Abstract. Compressed sensing is a powerful mathematical modelling tool to recover sparse signals from undersampled measurements in many applications, including medical imaging. A large body of work investigates the case with linear measurements, while compressed sensing with nonlinear measurements has been considered more recently. We continue this line of investigation by considering a novel type of nonlinearity with special structure that occurs in data acquired by multi-emitter X-ray tomosynthesis systems with spatio-temporal overlap. In [15] we proposed a nonlinear optimization model to deconvolve the overlapping measurements. In this paper we propose a model that exploits the structure of the nonlinearity and a nonlinear tomosynthesis algorithm that has a practical running time of solving only two linear subproblems at the equivalent resolution. We underpin and justify the algorithm by deriving RIP bounds for the linear subproblems and conclude with numerical experiments that validate the approach.

Keywords: Nonlinear compressed sensing
Restricted isometry property · Sparse reconstruction
Image reconstruction · Medical imaging · X-ray · Tomosynthesis
Computed tomography

1 Introduction

Reconstructing a 3D image from a set of 2D projections is a typical inverse problem with applications in magnetic resonance imaging (MRI) [5,17] and X-ray computed tomography (CT) [6]. The goal of tomosynthesis is to recover the material densities in a three-dimensional domain, usually represented as a set of 2D slices. Traditionally, X-ray measurements are taken from a single moving source, yielding non-overlapping measurements (see Fig. 1(a)), which can be formulated as linear constraints [6]. A novel type of measurements arises from recent research on X-ray emitter arrays with multiple simultaneously emitting sources [12] (see Fig. 1(b)): Spatio-temporally overlapping X-rays, where rays from more than one emitter reach the same detector at the same time (see Fig. 2(b)), result in nonlinear measurements [15]. When using linear image reconstruction methods, this type of nonlinear measurements has to be avoided.

© Springer International Publishing AG, part of Springer Nature 2018
M. Pelillo and E. Hancock (Eds.): EMMCVPR 2017, LNCS 10746, pp. 189–204, 2018.
https://doi.org/10.1007/978-3-319-78199-0_13

The scanner therefore has to be designed so that spatio-temporal overlap cannot occur, and this necessitates small collimation angles and/or stand-off distances, thus imposing severe restrictions on the scanner geometry. Restrictions on the stand-off distance may be difficult to satisfy in hand-adjusted or hand-held devices for which the relative position of the emitter and detector panels cannot be controlled with high precision. Image reconstruction methodologies that can deal with spatio-temporal overlap as well as an appropriate understanding of how the reconstruction quality is affected as a function of the degree of overlap are therefore essential to improve the robustness and flexibility of the geometric design of these new X-ray tomosynthesis systems.

Since measurements are undersampled, we base our image reconstruction methodology on sparsity-regularized optimization models, or compressed sensing problems. Most of the literature studies linear compressed sensing problems, but more recently a growing body of work on nonlinear compressed sensing problems startet to appear: In 1-bit compressed sensing, the signal is constrained to lie on a hyper-sphere. [11,19] consider nonlinearity arising from noise and a generalization to quasi-linear compressed sensing, [16,21] consider compressed sensing with quadratic constraints. [3] considers more general models of nonlinearity and derives restricted isometry inequalities for nonlinear measurements as a function of a distance to a linearization. In our setting a different type of nonlinearity is produced by measurements from overlapping X-rays, where the nonlinearity arises from geometry. In [15] a method based on directly solving the nonlinear constraints in a least-squares sense is presented. In this paper we continue this investigation by exploiting the concavity of the nonlinear measurements, which makes it possible to achieve linearization via a corrective factor rather than using Taylor approximation, as proposed in other papers. We then propose an algorithm based on alternative optimization of the linear subproblems and updates of the corrective factors. The linearization renders the optimization stable and allows the use of efficient linear solvers on the linear subproblems. The practicability of the method is established in experiments both with simulated and real-world data.

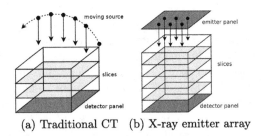

(a) Traditional CT   (b) X-ray emitter array

Fig. 1. Tomosynthesis scanners: (a) traditional CT with moving source, (b) multi-emitter panel based system without moving parts.

We consider X-rays emitted from conoidal sources arranged on a multi-emitter panel. A three-dimensional region of interest in which the object to reconstruct is located is discretized to a Cartesian grid of n voxels. A schematic 2D view is given in Fig. 2(a). A series of temporally separated measurements may be taken, actuating a different pattern of several emitters to emit photons during each aquisition interval. As long as spatio-temporal overlap does not occur, this process leads to the same measurements as if each emitter were actuated sequentially in isolation, but with the measurements taken in parallel. Each measurement j then corresponds to a pair of an emitter and a dector, the emitter emitting I_{E_j} photons in total in the direction of the detector, and the detector detecting the arrival of I_{D_j} photons emitted from the detector, the remaining photons having been absorbed by the material of the object that is to be imaged. We dicretize the paths of photons between the emitter and detector, by collapsing all these paths to a single ray j subtended by the centers of the emitter and the detector involved in measurement j. Denoting the absorption coefficient of the material occupying voxel i by $x_i \geq 0$, and writing ξ_{ji} for the length of the intersection between ray j and voxel i, the *Beer-Lambert law* implies $I_{D_j} = I_{E_j} \exp(\sum_{i=1}^{n} -\xi_{ji}x_i)$ for all j, where we assumed no ray scattering, no noise, and no spatio-temporal overlap between measurements. Taking logarithms the measurement equation can be reformulated as a linear system of equations $\log I_{D_j} - \log I_{E_j} = \sum_{i=1}^{n} -\xi_{ji}x_i$, $j = 1, \ldots, m$, which can be written as a linear system $Ax = b$ of size $m \times n$, where $a_{ji} = -\xi_{ji}$, $b \in \mathbb{R}^m$ with $b_j = \log(I_{D_j}/I_{E_j})$ and where the voxel densities were stacked into a vector $x \in \mathbb{R}^n$ with n the number of voxels in the chosen voxelization. Note that A is very sparse, because each ray intersects only a small number of voxels.

Since measurements are typically severely undersampled, prior information needs to be used to recover the original signal. Most objects to be imaged consist of material that has either density in only a small proportion of its spatial extent or has sharp interfaces between regions of high and low densities. It is known from the work of [4,8] and others that the compressed sensing problem $\min_{x \geq 0}\{\|x\|_1 : Ax = b\}$ and related models is able to reconstruct the sparsest solution x to an underdetermined linear system $Ax = b$ if matrix A satisfies the sufficient condition of a restricted isometry property (RIP), defined by the existence of $\delta_s \in (0, 1)$ for each sparsity level s such that $(1-\delta_s)\|x\|_2^2 \leq \|Ax\|_2^2 \leq (1+\delta_s)\|x\|_2^2$ holds for all s-sparse vectors x. Note that the RI constants depend on the scaling of matrix A, while the reconstructability does not. In practice, one may thus optimize the RIC over a scaling factor.

2   Nonlinear Measurements with Overlap

If multiple X-ray sources emit rays simultaneously, it may happen that two or more emitter cones can partially overlap. Figure 3 shows a schematic overview as well as example measurements at the detector for a sequential scan, and the same measurements with overlap. In the case of simultaneous emission, rays can reach the same detector at the same time (see Fig. 2(b)). When several

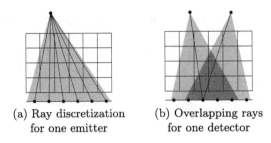

(a) Ray discretization   (b) Overlapping rays
for one emitter   for one detector

Fig. 2. Scanner set-up with emitters above the reconstruction domain, and detectors below.

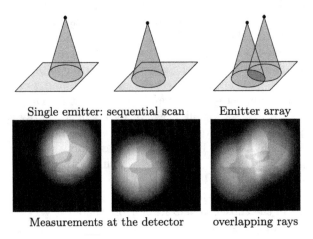

Single emitter: sequential scan   Emitter array

Measurements at the detector   overlapping rays

Fig. 3. Sequential versus parallel measurements, the latter inducing overlap of emitter cones.

rays $k \in K(j)$ reach the same detector j simultaneously, the attenuation of measurement j equals $c_j := I_{D_j} / \sum_{k \in K(j)} I_{E_k}$, where the emitter intensity of the k-th source is denoted by I_{E_k} and I_{D_j} is the total radiation measured at detector j. The *Beer-Lambert law* once again describes the relation between measurements and densities, but the constraints arising from overlapping measurements are no longer log-linear,

$$\psi_j(x) := \sum_{k \in K(j)} \lambda_{jk} \exp\left(\sum_{i=1}^{n} -\xi_{ki} x_i\right) = c_j \qquad (1)$$

where each of the summands corresponds to a different ray $k \in K(j)$. For simplicity of notation, we denote the fraction of intensity that emitter k contributes to measurement j by $\lambda_{jk} := I_{E_k} / \sum_{l \in K(j)} I_{E_l}$. The coefficients ξ_{ki} correspond to the length of intersection between voxel i and the k-th ray of measurement j. We represent them with vectors $r_{jk} \in \mathbb{R}^n$, $r_{jk} := \left(-\xi_{k1}, \ldots, -\xi_{kn}\right)^{\top}$, for all $j = 1, \ldots, m$ and $k \in K(j)$. The projection of densities along one ray is then given by $r_{jk}^\top x = \sum_{i=1}^n -\xi_{ki}x_i$. The vectors r_{jk} are sparse because each ray traverses only a small number of voxels. We remark that constraints similar to (1) are used in polychromatic reconstruction [13], where the nonlinearity arises from the polychromatic nature of the X-ray source. In our setting they arise from the geometry of rays, which yields different ξ_{ki} and x_i in each summand, in contrast to the setting of polychromatic reconstruction.

Measurements are typically severely undersampled, requiring the incorporation of prior information to recover the original signal. Assuming the signal x is sparse, minimization with ℓ_1 sparsity prior yields the reconstruction problem

$$\min_{x \geq 0} \|x\|_1 \text{ s.t. } \psi_j(x) = c_j, \ \forall j = 1, \ldots, m. \tag{2}$$

As discussed above, the ℓ_1 prior is designed to produce sparse solutions to constraints $\psi_j(x) = c_j$ $(j = 1, \ldots, m)$. Although this is justified in some applications of the X-ray scanner, such as trauma, a higher image quality is generally achieved by using a regularization term that produces sharp interfaces between regions of high and low absorption density in the object that is to be imaged. Such a term may be formulated by using the ℓ_1-norm based total variation (TV) norm, or by representing the absorption density by a D4-wavelet decompositon and imposing sparsity in the wavelet domain, thus replacing the voxelization by a wavelet discretization. We remark however that both cases can be treated in the same mathematical framework as the ℓ_1 prior discussed below, since in the case of the TV norm the term $\|x\|_1$ is replaced by $\|\Phi x\|_1$ where Φ is a matrix of size $3n \times n$ that encodes a finite-difference scheme to compute gradients, and in the case of the wavelet decomposition the coefficients ξ_{ji} of the design matrix A are replaced by the integral of wavelet i along ray j. For the sake of notational simplicity we shall henceforth only discuss the ℓ_1 regularization term, the other cases requiring minimal changes to algorithms and codes.

We will now present two different models and algorithmic approaches to solving the image reconstruction problem from overlapping rays.

2.1 A Forward-Backward Splitting Approach

The Lagrangian formulation of (2) allows for noise in the data term. In [15] we proposed to impose a quadratic penalty on the nonlinear constraints (1),

$$\min_{x \geq 0} \left\{ F_1(x) := \mu\|x\|_1 + \frac{1}{2} \sum_{j=1}^m (\psi_j(x) - c_j)^2 \right\}. \tag{3}$$

The parameter $\mu > 0$ depends on the amount of noise one is willing to tolerate in the measurements. This model interprets the error in counting the rescaled number of photons arriving at detector j as Gaussian noise, which is justified by the fact that the associated photon count is Poissonian with parameter on the order $O(10^3)$. $F_1(x)$ is a composite merit function with (3) of the form

$$\min_{x \geq 0} \{\mu f(x) + g(x)\} \tag{4}$$

with convex non-differentiable $f(x) := \|x\|_1$ and a partially convex, differentiable noise term $g(x) := \frac{1}{2}\sum_{j=1}^m (\psi_j(x) - c_j)^2$. In [15] we proved that g is convex in the region $0 \le x \le \hat{x}$, where $\hat{x}$ is a minimizer, and that ∇g is Lipschitz continuous with Lipschitz constant $L = 6mp^2h^2$, where h is the voxel size (the spatial coordinates being scaled by $1/h$ in the algorithm below).

Problem (4) can be solved via forward-backward splitting updates

$$x^{t+1} = \text{prox}_{\mu f}(x^t - \sigma \nabla g(x^t)). \tag{5}$$

The proximal operator is defined as $\text{prox}_{\mu f}(x) = \arg\min_{u \in \mathbb{R}^n} \{f(u) + \frac{1}{2\mu}\|u - x\|_2^2\}$. [9] showed that in the case $f(x) = \|x\|_1$ the prox operator is given by soft thresholding,

$$\text{prox}_{\mu f}(x) = \begin{cases} x - \mu, & x > \mu \\ 0, & |x| \le \mu \\ x + \mu, & x < -\mu \end{cases} \tag{6}$$

[7] showed that if f is lower semicontinuous convex and g is convex differentiable with Lipschitz continuous gradient, then (5) converges to a minimum of (4) at a rate $\mathcal{O}(1/t)$. [20] showed that an additional overrelaxation step yields slightly faster convergence.

For an optimization of (3) with only partially convex g, the initialization of x has to be smaller than a minimizer $\hat{x}$, for g is only guaranteed to be convex in the region $0 \le x \le \hat{x}$. Knowing that the densities x cannot be negative, we choose the lower bound $x^0 = 0$ and constrain step sizes σ so as to ensure that $x \le \hat{x}$ for all t. We enforce this by use of a backtracking line-search algorithm [1]. Although in general, $\hat{x}$ is unknown, the measurements $c_j = \psi_j(\hat{x}) + \epsilon_j$ are known, hence we constrain the line-search by the necessary (though not sufficient) condition $\psi_j(x) \ge c_j$ for all j. This leads to the following algorithm for the minimization of (3), as further discussed in [15].

Input:

$c \in \mathbb{R}^m$: vector of measurements

$r_{jk} \in \mathbb{R}^n$: sparse vectors of intersection lengths of rays and voxels

$\theta \in (0, 1)$: line search control parameter

$L = 6mp^2h^2$: Lipschitz constant for ∇g

Output:

$x \in \mathbb{R}^n$: approximate solution to (3).

Initialize $x^0 = 0$.

Iterate for $t = 0, 1, 2, \ldots$:

1. Compute search direction:

$$\nabla g(x^t) = \sum_{j=1}^m \left(\sum_{k \in K(j)} \lambda_{jk} e^{r_{jk}^\top x^t} - c_j \right) \left(\sum_{k \in K(j)} \lambda_{jk} r_{jk} e^{r_{jk}^\top x^t} \right)$$

2. Backtracking line search:

$$\sigma = 1/L$$
$$x^{new} = \text{prox}_{\mu f}(x^t - \sigma \nabla g(x^t))$$
$$\text{while } \exists j \in \{1,\ldots,m\} \text{ with } \psi_j(x^{new}) < b_j :$$
$$\sigma \leftarrow \theta \times \sigma$$
$$x^{new} = \text{prox}_{\mu f}(x^t - \sigma \nabla g(x^t))$$

3. Update x: $x^{t+1} = x^{new}$.

2.2   The Lagging Multiplier Approach

Our second approach is based on applying the logarithm to the constraints from overlapping rays (1), in analogy to what we did in the case without overlap. If there is little overlap, it is reasonable to expect that taking the logarithm transforms the problem into an optimization problem that is only slightly nonlinear. For non-zero measurements, the j-th constraint Eq. (1) thus transforms to

$$\tilde{b}_j := \log(c_j) = \log(\psi_j(x)) = \log\left(\sum_{k \in K(j)} \lambda_{jk} \exp\left(\sum_{i=1}^{n} -\xi_{ki} x_i\right)\right).$$

A regularized optimization problem with noisy measurements is now given by

$$\min_{x \geq 0} \left\{ F_2(x) := \mu\|x\|_1 + \frac{1}{2}\sum_{j=1}^{m} \left(\log(\psi_j(x)) - \tilde{c}_j\right)^2 \right\}. \tag{7}$$

The data term of this model is nonlinear, but the concavity of the logarithm implies that for any convex weights $\lambda_{jk} \in [0,1]$ with $\sum_{k \in K(j)} \lambda_{jk} = 1$ the following inequality holds,

$$\sum_{k \in K(j)}\sum_{i=1}^{n} -\lambda_{jk}\xi_{ki} x_i \leq \log\left(\sum_{k \in K(j)} \lambda_{jk} \exp\left(\sum_{i=1}^{n} -\xi_{ki} x_i\right)\right) = \log(\psi_j(x)) \leq 0.$$

Therefore, for each measurement j there exists a corrective multiplier

$$\tau_j(x) := \frac{\log\left(\sum_{k \in K(j)} \lambda_{jk} \exp\left(\sum_{i=1}^{n} -\xi_{ki} x_i\right)\right)}{\sum_{k \in K(j)}\sum_{i=1}^{n} -\lambda_{jk}\xi_{ki} x_i} \in [0,1] \tag{8}$$

such that $\tau_j \times \sum_{k \in K(j)} \sum_{i=1}^{n} -\lambda_{jk}\xi_{ki} x_i = \log(\psi_j(x))$, for $(j = 1,\ldots,m)$. The data term in (7) can thus be replaced by $\|\tau(x)\tilde{A}x - \tilde{b}\|^2$, where $\tau = \text{diag}(\tau_j)$ is the diagonal matrix with factors τ_j on the diagonal, the j-th row of matrix $\tilde{A}$ is obtained as $\tilde{a}_j = \sum_{k \in K(j)} \lambda_{jk} r_{jk}^\top$. Problem (7) can therefore be reformulated as

$$\min_{x \geq 0} \left\{ F_2(x) = \mu \|x\|_1 + \tfrac{1}{2} \|\tau(x)\tilde{A}x - \tilde{b}\|_2^2 \right\}. \tag{9}$$

Note that in contrast to (3), the data term of (9) is to be interpreted as error in the aborption due to the discretization imposed by the voxelization and to scattering. It is natural to model such noise as Gaussian.

For measurements that are only slightly nonlinear, $\tau_j(x)$ is close to 1. Instead of iteratively solving the problem (9) with its nonlinear data term, we may therefore approximate $\tau(x^{t+1})$ by $\tau(x^t)$, where x^t is the solution of the previous iteration. In other words, τ is updated with a time lag of 1 relative to the updates x^{t+1} of the densities. This leads to the following algorithm:

Input:
  $\tilde{b} \in \mathbb{R}^m$: vector of measurements
  $\tilde{A} \in \mathbb{R}^n$: sparse matrix of scaled intersection lengths of rays and voxels

Output:
  $x \in \mathbb{R}^n$: approximate solution to (9).

Initialize
  $x^0 = \tilde{A}^\top b$ (used in warm start of (10)
  $\tau^0 = \mathrm{I}_m$ (the $m \times m$ identity matrix)

Iterate for $t = 0, 1, 2, \ldots$:

  1. Update x:

$$x^{t+1} = \arg\min_{x \geq 0} \ \mu \|x\|_1 + \tfrac{1}{2} \|\tau^t \tilde{A} x - \tilde{b}\|_2^2 \tag{10}$$

  2. Update τ:

$$\tau_j^{t+1} \leftarrow \log\left(\psi_j(x^t)\right) / \tilde{a}_j x^t, \quad \forall j \in \{1, \ldots, m\} \tag{11}$$

The advantage of this method is that the update of x obtained in (10) corresponds to solving the Lasso problem and can be solved with any standard linear solver, e.g. [1]. The update of τ_j in (11) is solved explicitly in each iteration at a constant cost. The initialization $\tau_j = 1$ represents the case where all rays contribute equally to measurement j. In practice we observe that two outer iterations suffice, as the factors τ_j generally converge to near machine precision after a single update. Thus, the above described algorithm has only twice the cost of an X-ray problem without spatio-temporal overlap.

3   Restricted Isometry Bounds for Overlapping Rays

In this section we derive an analytical argument for why (9) can be expected to yield an exact sparse reconstruction as long as the average overlap per detector does not exceed a factor of 2. This will be backed up by experiments. It is conceptually important to investigate this phenomenon, because it justifies the use of the lagging multiplier method.

While related work on compressed sensing mainly investigates nonlinearity due to noise, overlapping rays yield a different type of nonlinearity. However, we note that the following derivation of restricted isometry properties for our image reconstruction model (9) was inspired by a case studied in [3], who uses first order Taylor approximation to approximate nonlinear measurements. Our case is similar, but the linearization is based on the corrective factors $\tau(x)$ rather than Taylor approximation.

To keep the argument simple, we assume that $|K(j)| = p$ for all detectors p, and that all emission intensities are the same, so that $\lambda_{jk} = 1/p$ for all $j = 1,\ldots, m$ and $k \in K(j)$. $\tilde{A} \in \mathbb{R}^{m \times n}$ is then obtained from $A \in \mathbb{R}^{mp \times n}$, the design matrix of the corresponding temporally separated measurements, by averaging the rows with indices $k \in K(j)$, as writing the p indices $k \in K(j)$ as $j1,\ldots, jp$ yields

$$
A = \begin{pmatrix} -\xi_{111} & \cdots & -\xi_{n11} \\ \vdots & & \vdots \\ -\xi_{11p} & \cdots & -\xi_{n1p} \\ -\xi_{121} & \cdots & -\xi_{n21} \\ \vdots & & \vdots \\ -\xi_{1mp} & \cdots & -\xi_{nmp} \end{pmatrix}, \quad \tilde{A} = \frac{1}{p} \begin{pmatrix} \sum_{k=1}^{p} -\xi_{11k} & \cdots & \sum_{k=1}^{p} -\xi_{n1k} \\ \sum_{k=1}^{p} -\xi_{12k} & \cdots & \sum_{k=1}^{p} -\xi_{n2k} \\ \vdots & & \vdots \\ \sum_{k=1}^{p} -\xi_{1mk} & \cdots & \sum_{k=1}^{p} -\xi_{nmk} \end{pmatrix}, \quad (12)
$$

In the following we will derive inequalities that allow us to extend RIP bounds from A to $\tau\tilde{A}$.

Lemma 1. *For all $x \in \mathbb{R}^n$ with $x_i \geq 0$, the norm of $p\tilde{A}x$ is bounded by*

$$\|Ax\|_2^2 \leq \|p\tilde{A}x\|_2^2 \leq p\|Ax\|_2^2. \quad (13)$$

Proof. Let $\alpha_{jk} \leq 0$ be defined as the negative distance weighted sum of densities along a ray between detector $j \in \{1,\ldots, m\}$ and emitter $k \in \{1,\ldots, p\}$:

$$\alpha_{jk} := r_{jk}^\top x = \sum_{i=1}^{n} -\xi_{ijk} x_i. \quad (14)$$

Then the *Multinomial Theorem* yields the lower bound

$$
\|p\tilde{A}x\|_2^2 = \sum_{j=1}^{m} \left(\sum_{k=1}^{p} \alpha_{jk} \right)^2 = \sum_{j=1}^{m} \sum_{k=1}^{p} \alpha_{jk}^2 + \sum_{j=1}^{m} \left(\sum_{k_1=1}^{p} \sum_{k_2=k_1+1}^{p} 2a_{jk_1}\alpha_{jk_2} \right)
$$

$$
= \|Ax\|_2^2 + \sum_{j=1}^{m} \left(\sum_{k_1=1}^{p} \sum_{k_2=k_1+1}^{p} 2\alpha_{jk_1}\alpha_{jk_2} \right) \quad (15)
$$

$$\geq \|Ax\|_2^2 \quad (16)$$

(16) holds because the second term in (15) contains only positve summands. The upper bound can be computed starting again from (15):

$$\|p\tilde{A}x\|_2^2 = \|Ax\|_2^2 + \sum_{j=1}^{m}\left(\sum_{k_1=1}^{p}\sum_{k_2=k_1+1}^{p} 2\alpha_{jk_1}\alpha_{jk_2}\right)$$

$$\leq \|Ax\|_2^2 + \sum_{j=1}^{m}\left(\sum_{k_1=1}^{p}\sum_{k_2=k_1+1}^{p} (\alpha_{jk_1}^2 + \alpha_{jk_2}^2)\right) \tag{17}$$

$$= \|Ax\|_2^2 + \sum_{j=1}^{m}(p-1)\sum_{k=1}^{p}\alpha_{jk}^2 \leq \|Ax\|_2^2 + (p-1)\sum_{j=1}^{m}\sum_{k=1}^{p}\alpha_{jk}^2 = p\|Ax\|_2^2$$

where (17) follows from the Binomial equation $(a-b)^2 \geq 0$ which implies $2ab \leq a^2 + b^2$ for any $a, b \in \mathbb{R}$.

Lemma 2. *Assuming that for each measurement j there exists at least one ray $k \in K(j)$ that traverses the reconstruction domain with minimal density $x_{min} > 0$, then $\tau_{min} := \min_{j=1,\dots,m} \tau_j > 0$ and $\tau_{max} := \max_{j=1,\dots,m} \tau_j \leq 1$.*

Proof. The assumptions imply that

$$\frac{1}{p}\sum_{k=1}^{p}\lambda_{jk}\exp\left(\sum_{i=1}^{n}-\xi_{ijk}x_i\right) < 1,$$

from which the result follows by strict concavity of the logarithm and the definition of τ_j in (8).

Lemma 3. *For all $x \in \mathbb{R}^n$ the following bounds apply,*

$$\tau_{min}^2\|p\tilde{A}x\|_2^2 \leq \|p\tau\tilde{A}x\|_2^2 \leq \|p\tilde{A}x\|_2^2. \tag{18}$$

Proof. Using the notation α_{jk} introduced in (14), we have

$$\|p\tau\tilde{A}x\|_2^2 = \sum_{j=1}^{m}\tau_j^2\left(\sum_{k=1}^{p}\alpha_{jk}\right)^2 \leq \tau_{max}^2\sum_{j=1}^{m}\left(\sum_{k=1}^{p}\alpha_{jk}\right)^2 = \tau_{max}^2\|p\tilde{A}x\|_2^2 \leq \|p\tilde{A}x\|_2^2$$

$$\|p\tau\tilde{A}x\|_2^2 = \sum_{j=1}^{m}\tau_j^2\left(\sum_{k=1}^{p}\alpha_{jk}\right)^2 \geq \tau_{min}^2\sum_{j=1}^{m}\left(\sum_{k=1}^{p}\alpha_{jk}\right)^2 = \tau_{min}^2\|p\tilde{A}x\|_2^2.$$

Let us now derive restricted isometry properties for the nonlinear measurements with overlap $\tau\tilde{A}$, assuming that the design matrix A of spatio-temporally separated measurements satisfies a restricted isometry inequality

$$(1 - \delta_s)\|x\|_2^2 \leq \|Ax\|_2^2 \leq (1 + \delta_s)\|x\|_2^2 \tag{19}$$

for all s-sparse vectors x, an assumption that is justfied by the work of [14].

Theorem 1 (Restricted isometry for $\tau\tilde{A}$). *Let A and $\tilde{A}$ be as defined in (12) and such that A satisfies the RIP (19). Then for all s-sparse non-negative vectors $x \in \mathbb{R}^n_+$, the rescaled matrix $p^{1/2}\tau\tilde{A}$ satisfies the RIP*

$$(1 - \alpha_s)\|x\|_2^2 \leq \left\|p^{1/2}\tau\tilde{A}x\right\| \leq (1 + \alpha_s)\|x\|_2^2$$

with $\alpha_s = 1 - \frac{\tau_{\min}^2}{p}(1 - \delta_s)$.

Proof. The lower bound follows from Lemmas 1 and 3 as follows,

$$(1 - \alpha_s)\|x\|_2^2 = \frac{\tau_{\min}^2}{p}(1 - \delta_s)\|x\|_2^2 \overset{(19)}{\leq} \tau_{\min}^2 \left\|p^{-1/2}Ax\right\|_2^2 \overset{\text{Lemma 1}}{\leq} \tau_{\min}^2 \left\|p^{1/2}\tilde{A}x\right\|_2^2$$

$$\overset{\text{Lemma 3}}{\leq} \left\|p^{1/2}\tau\tilde{A}x\right\|^2.$$

Furthermore, using $\tau_{\min}^2 \leq 1 \leq p$, we have $\delta_s \leq \alpha_s$, so that the upper bound is found as follows,

$$\left\|p^{1/2}\tau\tilde{A}x\right\|^2 \overset{\text{Lemma 3}}{\leq} \left\|p^{1/2}\tilde{A}x\right\|^2 \overset{\text{Lemma 1}}{\leq} \|Ax\|^2 \overset{(19)}{\leq} (1 + \delta_s)\|x\|_2^2 \leq (1 + \alpha_s)\|x\|_2^2,$$

4   Numerical Experiments

It follows from the results of [4,8] that sparse recovery occurs when $\alpha_{2s} < \sqrt{2}-1$. Under the realistic assumptions that $\tau_{\min} \approx 1$ and $\delta_{2s} \approx 0$, Theorem 1 then implies sparse recovery in the linear subproblems of the lagging multiplier algorithm of Sect. 2.2 generally occurs when the average overlap p per detector satisfies

$$p < \frac{\tau_{\min}^2(1 - \delta_{2s})}{2 - \sqrt{2}} \approx 1.7.$$

Our experiments indicate that this estimate is pessimistically conservative, as a little higher degree of overlap can still recover the original signal without significant deterioration of image quality.

4.1   Empirical Phase Transitions

Phase transitions have been introduced in [10] as a tool to analyse the recoverability of a k-sparse signal $x \in \mathbb{R}^n$ from measurements obtained consistent with a matrix $A \in \mathbb{R}^{m \times n}$. They present reconstruction errors from simulated signals depending on the sampling rate $\delta := m/n$ and the relative signal sparsity $\rho := k/m$. Several empirical results have shown that a relatively sharp transition can be observed between combinations of δ and ρ where the signal can be recovered and where it cannot be recovered (with high probability) [2,18]. Phase transition analysis for random Gaussian matrices has been widely investigated, e.g. [2], while more recent work has shown empirical phase transition phenomena for X-ray measurements [14].

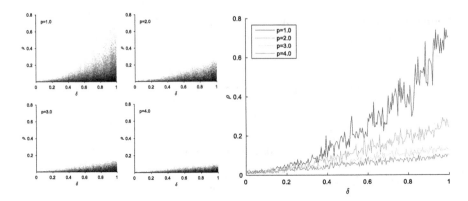

Fig. 4. Phase transitions for measurements with overlap of $p = 1$, 2, 3 and 4, depending on the sampling rate δ and signal sparsity ρ. (Color figure online)

We tested the recoverability of random signals x in a cuboidal domain which is traversed by partly overlapping X-rays. Figure 4 shows phase transitions for minimization of (2) with different degrees of overlap, defined as the number of rays simultaneously illuminating one detector. The same degree of overlap was used at all detectors. An overlap of 1 is thus equivalent to the corresponding reconstruction from sequential measurements. The four plots in the left part of the figure show phase transitions for overlap of $p = 1$, 2, 3 and 4, respectively. The grey values for each combination of δ and ρ are obtained from the following experiment: We fixed $m = 100$ and computed a reconstruction of a random signal with sparsity $k = \rho m$ and resolution of $n = m/\delta$ unknowns. Measurements were simulated by integrating the signal along m random rays through the reconstruction domain, and the simulated photon counts of overlapping rays were summed. The number of measurements m/p of each experiment was thus determined by the number of rays m and the degree p of overlap per detector. The grey value at each data point encodes the reconstruction error as the distance of the reconstruction to the original signal (while dark encodes low and bright encodes high errors). The right part of the figure shows the observed transition between recoverability and non-recoverability for given δ and ρ, obtained from thresholding the four individual plots from the left part of the figure. The phase transitions show that increasing sparsity and sampling rates yield increasing recoverability, while increasing overlap decreases the recoverability.

Our phase transition experiments show that the sparse recoverability analysis of Sect. 3 is rather conservative: If A were to be a random matrix with i.i.d. Gaussian coefficients, one of the cases for which the relation between the RIP and the sparse recoverability is well understood via quantifiable probability bounds, then $\tilde{A}$ would also be such a matrix and $\tau\tilde{A}$ nearly so (if all τ_j are close to 1), but with a sampling rate reduced by a factor $1/p$. If the sparse recoverability of system $\tau\tilde{A}x = \tilde{b}$ were thus characterized by the RIP, then thresholds on the red curve in Fig. 4 would correspond to threshold points on the blue curve at a point where the value of δ is halved and ρ is doubled. But this is not the case, and the phase transition experiment shows that the system with overlap $p = 2$ is able to recover signals a factor 2 less sparse than predicted under the RIP argument.

4.2 Comparison of Reconstruction Errors

In order to investigate how measurements with overlapping X-rays affect image reconstruction accuracies, we tested both methods discussed in this paper on simulated and real-world measurements. Simulated data allow to compute exact reconstruction errors as a distance to the ground truth, while real-world data give empirical confirmation that the methodology can be used under practical conditions that include measurement noise and scattering.

For this purpose we tested two different data sets:

1. Data set "cube" are simulated measurements from a binary test object of a cube located at the center of the reconstruction domain,
2. Data set "letters" consist of real X-ray measurements of two wooden letters stacked on top of one another. This data was kindly made available to us by Adaptix Ltd., who took the measurements in their lab in Begbroke, Oxford-shire, U.K.

In both data sets the scanner geometry was based on a panel of emitters and a parallel panel of detectors placed above and below the reconstruction domain, with parameters given in Table 1. The emitters and detectors were equally spaced in the respective panels, and all emitters had the same collimation angle (the apex angle of the emitter cone). The resolution of the reconstruction domain in z-dimension was lower than the resolution in x- and y-dimension, because its direction corresponds to the main orientation of rays, resulting in reduced recoverability along the z-axis.

Table 1. Parameters of scanner geometry and grid dimension for the two data sets.

Data set	# emitters	# detectors	# voxels	Collimation angle
"cube"	5×5	15×15	$20 \times 20 \times 20$	$20°$
"letters"	13×14	128×128	$128 \times 128 \times 20$	$\sim 12°$

Our experiments compared the following three different methods to handle overlapping X-ray measurements:

1. Discard overlapping measurements from the set of constraints and reconstruct from non-overlapping measurements only.
2. The forward-backward splitting method of Sect. 2.1.
3. The lagging multiplier method of Sect. 2.2.

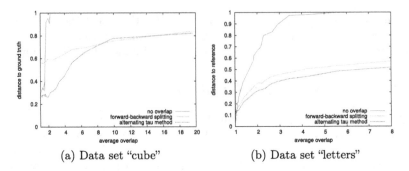

(a) Data set "cube"   (b) Data set "letters"

Fig. 5. Reconstruction error for increasing degrees of overlap, comparing reconstruction from non-overlapping measurements only (red), forward-backward splitting (green) and the alternated linearization method (blue), for the letters data set. (a) Distance to ground truth for simulated data. (b) Distance to sequential reconstruction using real X-ray measurements. (Color figure online)

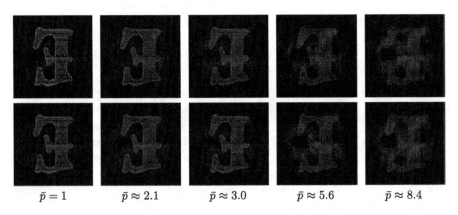

$\bar{p} = 1$ $\bar{p} \approx 2.1$ $\bar{p} \approx 3.0$ $\bar{p} \approx 5.6$ $\bar{p} \approx 8.4$

Fig. 6. Reconstructed densities (slices) of wooden letters, comparing forward-backward-splitting (first row) and the altenated optimization with τ (second row). $\bar{p}$ is the average overlap per measurement.

Figure 5 shows reconstruction errors for increasing average overlap $\bar{p}$, defined as the *average* number of rays that illuminate a detector simultaneously, as the number of overlapping rays was allowed to vary from detector to detector. Overlap was simulated by virtually firing an increasing number of randomly chosen emitters at the same time, and adding the respective photon counts at detector pixels with overlap. All other parameters of the experimental set-up were left unchanged. The relative distance $d(x)$ of a reconstruction x to the reference $\hat{x}$ was quantified as $d(x) = \|x - \hat{x}\|_2 / \|\hat{x}\|_2$. For the simulated data set "cube" the reference $\hat{x}$ is given by the ground truth. Since a ground truth is not known for real data, we compare the reconstructions of the "letters" data set to the respective reconstruction from sequential measurements, that is, sequential measurements without overlap were taken using the same rays as in the case with overlap. The real world data set was acquired using nonoverlapping sequential exposures, and overlap was imputed by adding measurements of several exposures simulated as occurring simultaneously.

The plots show that for both sets, the forward-backward splitting and the alternated optimization with the corrective factors τ yield significantly improved reconstruction qualities compared to discarding the overlapping measurements. The alternated τ-method of Sect. 2.2 yields slightly better reconstruction accuracies than the forward-backward splitting.

This observation can also be confirmed visually in Fig. 6, which shows the 5th of 20 z-slices of the reconstructed densities of the data set "letters" for increasing degrees of of average overlap $\bar{p}$.

5   Conclusions

This paper has shown practical methods to reconstruct images from measurements with overlapping X-rays and renders multi-emitter tomosynthesis robustly applicable in applications where the multi-emitter panel of a scanning system is hand held or hand-adjusted, and where an exact stand-off distance avoiding overlap cannot always be guaranteed. Both theory and practical experiments show that a moderate degree of average overlap of up to ca 2 rays per detector has no major deteriorating effect on the quality of reconstructed images.

Acknowledgments. We thank Adaptix Ltd for providing the X-ray measurements used in the experiments with real-world data. This work was supported by Adaptix Ltd and EPSRC EP/K503769/1, as well as by The Alan Turing Institute under the EPSRC grant EP/N510129/1.

References

1. Beck, A., Teboulle, M.: A fast iterative shrinkage-thresholding algorithm for linear inverse problems. SIAM J. Imaging Sci. **2**, 183–202 (2009)
2. Blanchard, J.D., Cartis, C., Tanner, J., Thompson, A.: Phase transitions for greedy sparse approximation algorithms. Appl. Comput. Harmonic Anal. **30**, 188–203 (2011)
3. Blumensath, T.: Compressed sensing with nonlinear observations and related nonlinear optimisation problems. IEEE Trans. Inf. Theory **59**, 3466–3474 (2013)
4. Candes, E.J., Romberg, J., Tao, T.: Stable signal recovery from incomplete and inaccurate measurements. Commun. Pure Appl. Math. **59**, 1207–1223 (2006)
5. Chen, C., Huang, J.: Compressive sensing MRI with wavelet tree sparsity. In: Advances in Neural Information Processing Systems, pp. 1115–1123 (2012)
6. Choi, K., Wang, J., Zhu, L., Suh, T., Boyd, S., Xing, L.: Compressed sensing based cone-beam computed tomography reconstruction with a first-order method. Med. Phys. **37**, 5113–5125 (2010)
7. Combettes, P.L., Wajs, V.R.: Signal recovery by proximal forward-backward splitting. Multiscale Model. Simul. **4**, 1168–1200 (2005)
8. Donoho, D.L.: Compressed sensing. IEEE Trans. Inform. Theory **52**, 1289–1306 (2006)

9. Donoho, D.L., Johnstone, I.M.: Minimax estimation via wavelet shrinkage. Ann. Stat. **26**, 879–921 (1998)
10. Donoho, D.L., Tanner, J.: Precise undersampling theorems. Proc. IEEE **98**, 913–924 (2010)
11. Ehler, M., Fornasier, M., Sigl, J.: Quasi-linear compressed sensing. Multiscale Model. Simul. **12**, 725–754 (2014)
12. Gonzales, B., Spronk, D., Cheng, Y., Tucker, A.W., Beckman, M., Zhou, O., Lu, J.: Rectangular fixed-gantry CT prototype: combining CNT X-ray sources and accelerated compressed sensing-based reconstruction. IEEE Access **2**, 971–981 (2014)
13. Gu, R., Dogandžić, A.: Polychromatic X-ray CT image reconstruction and mass-attenuation spectrum estimation. arXiv preprint arXiv:1509.02193 (2015)
14. Jørgensen, J.S., Sidky, E.Y.: How little data is enough? Phase-diagram analysis of sparsity-regularized X-ray computed tomography. Philos. Trans. R. Soc. Lond. Ser. A: Math. Phys. Eng. Sci. **373**, 20140387 (2015)
15. Klodt, M., Hauser, R.: 3D image reconstruction from X-ray measurements with overlap. In: Leibe, B., Matas, J., Sebe, N., Welling, M. (eds.) ECCV 2016. LNCS, vol. 9910, pp. 19–33. Springer, Cham (2016). https://doi.org/10.1007/978-3-319-46466-4_2
16. Li, X., Voroninski, V.: Sparse signal recovery from quadratic measurements via convex programming. CoRR, abs/1209.4785 (2012)
17. Lustig, M., Donoho, D.L., Santos, J.M., Pauly, J.M.: Compressed sensing MRI. IEEE Sig. Process. Mag. **25**, 72–82 (2007)
18. Monajemi, H., Jafarpour, S., Gavish, M., Donoho, D.: Deterministic matrices matching the compressed sensing phase transitions of Gaussian random matrices. Proc. Nat. Acad. Sci. **110**, 1181–1186 (2013)
19. Needell, D., Tropp, J.A.: CoSaMP: iterative signal recovery from incomplete and inaccurate samples. Commun. ACM **53**, 93–100 (2010)
20. Nesterov, Y.: Smooth minimization of non-smooth functions. Math. Program. **103**, 127–152 (2005)
21. Ohlsson, H., Yang, A., Dong, R., Sastry, S.: CPRL – an extension of compressive sensing to the phase retrieval problem. In: Advances in Neural Information Processing Systems, pp. 1367–1375 (2012)

Unified Functional Framework for Restoration of Image Sequences Degraded by Atmospheric Turbulence

Naftali Zon and Nahum Kiryati[✉]

School of Electrical Engineering, Tel Aviv University, Tel Aviv, Israel
nk@eng.tau.ac.il

Abstract. We propose a unified functional to address the restoration of turbulence-degraded images. This functional quantifies the association between a given image sequence and a candidate latent image restoration. Minimizing the functional using the alternating direction method of multipliers (ADMM) and Moreau proximity mapping leads to a general algorithmic flow. We show that various known algorithms can be derived as special cases of the general approach. Furthermore, we show that building-blocks used in turbulence recovery algorithms, such as optical flow estimation and blind deblurring, are called for by the general model. The main contribution of this work is the establishment of a unified theoretical framework for the restoration of turbulence-degraded images. It leads to novel turbulence recovery algorithms as well as to better understanding of known ones.

1 Introduction

1.1 Image Degradation by Turbulence

Taking images through atmospheric turbulence is important in long-distance surveillance, astronomy (e.g. viewing distant stars) and remote sensing. Atmospheric turbulence strongly degrades image quality. Turbulence causes spatial and temporal random-like fluctuations in the index of refraction of the atmosphere [16]. The refractive index is affected by many factors, including: temperature, humidity, air pressure and wind regime. These factors are very difficult to model. Turbulence leads to two major distortions: time and space dependent geometric deformation and blur. The distortion causes the image to perform a chaotic "dance". Certain frames are totally degraded while some regions in other frames appear quite sharp as shown in Fig. 1. Time and space dependencies rule out conventional image restoration methods.

1.2 Image Degradation Model

As in other works [5,12,15,17,24], we assume that the scene and the image sensor are both static and that observed motions are due to turbulence alone[1].

---

[1] Observation of dynamic scenes through turbulence is considered, e.g., in [4,8,23].

© Springer International Publishing AG, part of Springer Nature 2018
M. Pelillo and E. Hancock (Eds.): EMMCVPR 2017, LNCS 10746, pp. 205–219, 2018.
https://doi.org/10.1007/978-3-319-78199-0_14

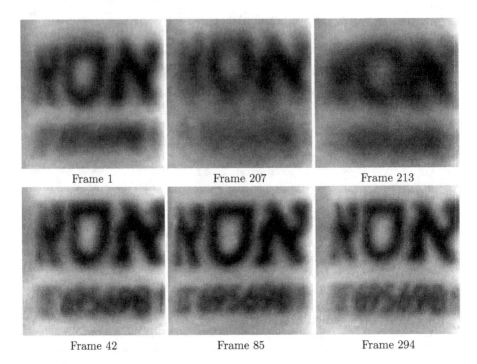

Frame 1   Frame 207   Frame 213

Frame 42   Frame 85   Frame 294

Fig. 1. An image sequence corrupted by turbulence. Top: highly corrupted crops. Bottom: moderately corrupted crops.

Let $f(\mathbf{x})$ be the latent sharp image, that is, the image that would have been captured in the absence of any blur or noise. First the latent image is degraded by the turbulence space variant kernel $h_k(\mathbf{s}; \mathbf{x})$

$$\mathcal{H}_{k,\mathbf{x}}(f)(\mathbf{x}) \equiv \int_\Omega h_k(\mathbf{s}; \mathbf{x}) f(\mathbf{s}) d\mathbf{s} \quad k = 1 \ldots K, \tag{1}$$

where $\mathcal{H}_{k,\mathbf{x}}$ is a space variant operator and the vector $\mathbf{x} = (x, y)$ stands for spatial location in frame k. Then the result is convolved with camera-optics space-invariant blur-kernel h (camera point spread function). To summarize, image acquisition through the turbulent air is modeled as a linear process

$$g_k(\mathbf{x}) = h(\mathbf{x}) * \mathcal{H}_{k,\mathbf{x}}(f)(\mathbf{x}) + n_k(\mathbf{x}), \quad k = 1 \ldots K. \tag{2}$$

The observed frame g_k is the latent image f filtered with the space variant blur kernel $h_{k,\mathbf{x}}$, then convolved with the space invariant blur kernel h plus noise n_k. Turbulence restoration aims to recover a latent image f from an observed image sequence $\{g_k\}$.

1.3 Turbulence Restoration Algorithms

In unusual cases, especially with long exposure time, the turbulence degradation can be modeled as a linear space-invariant degradation process

$$g(\mathbf{x}) = h_{\text{total}}(\mathbf{x}) * f(\mathbf{x}) + n(\mathbf{x}), \tag{3}$$

where the blur kernel h_{total} is due to both turbulence and optical blur. The resulting blind deblurring problem can be addressed by PSF estimation followed by image restoration [19]. However, in most cases the linear space-invariant model does not hold, so most turbulence reconstruction algorithms follow one of two finer approaches: *Multi-Frame Alignment* [5,12,17] and *Lucky Region Fusion* [2,3,14, 21]. Combinations of the two approaches recently appeared in [15,24].

The motivation for the multi-frame alignment method is expressing the space variant operator $\mathcal{H}_{k,\mathbf{x}}$ using a nonrigid deformation field $\mathbf{w}_k(\mathbf{x}) = (u_k(\mathbf{x}), v_k(\mathbf{x}))$:

$$\mathcal{H}_{k,\mathbf{x}}(f)(\mathbf{x}) = \int_\Omega h_k(\mathbf{s};\mathbf{x})f(\mathbf{s})d\mathbf{s} \approx f(\mathbf{x}+\mathbf{w}_k(\mathbf{x})) \quad k = 1\dots K. \quad (4)$$

Changing the order between the space variant operator $\mathcal{H}_{k,\mathbf{x}}$ and convolution with kernel h in Eq. (2) is common but formally not easy to justify. Changing the order and substituting Eq. (4) in Eq. (2) leads to

$$\begin{aligned}
g_k(\mathbf{x}) &= h(\mathbf{x}) * \mathcal{H}_{k,\mathbf{x}}(f)(\mathbf{x}) + n_k(\mathbf{x}) \\
&\approx \mathcal{H}_{k,\mathbf{x}}(h*f)(\mathbf{x}) + n_k(\mathbf{x}) \\
&= \mathcal{H}_{k,\mathbf{x}}(\bar{f})(\mathbf{x}) + n_k(\mathbf{x}) \\
&\approx \bar{f}(\mathbf{x}+\mathbf{w}_k(\mathbf{x})) + n_k(\mathbf{x}) \quad k = 1\dots K,
\end{aligned}$$

where $\bar{f} \equiv h * f$. We thus obtain the multiframe registration model

$$g_k(\mathbf{x}) = \bar{f}(\mathbf{x}+\mathbf{w}_k(\mathbf{x})) + n_k(\mathbf{x}), \quad k = 1\dots K. \quad (5)$$

Multi-frame alignment methods aim to recover the latent image f from the observed sequence $\{g_k\}$. The most straightforward approach starts with averaging the sequence: $\bar{g} = \frac{1}{K}\sum_k g_k$ [5]. Suppose that $\mathbf{w}_k$ includes *i.i.d.* random components. For $K \gg 1$, the *Law of Large Numbers* [9] states that the sample mean $\frac{1}{K}\sum_{k=1}^K \bar{f}(\mathbf{x} + \mathbf{w}_k)$ should be close to the expected value $\mathbb{E}_\mathbf{w}\bar{f}(\mathbf{x}+\mathbf{w})$.

$$\begin{aligned}
\bar{g}(\mathbf{x}) &= \frac{1}{K}\sum_{k=1}^K g_k = \frac{1}{K}\sum_{k=1}^K \bar{f}(\mathbf{x}+\mathbf{w}_k) + \bar{n} \underset{K\gg 1}{\approx} \mathbb{E}_\mathbf{w}\bar{f}(\mathbf{x}+\mathbf{w}) + \bar{n} \\
&= \int_{\Omega_\mathbf{w}} p_\mathbf{w}(\mathbf{y})\bar{f}(\mathbf{x}+\mathbf{y})d\mathbf{y} + \bar{n} = p_\mathbf{w}(-\mathbf{x}) * \bar{f}(\mathbf{x}) + \bar{n},
\end{aligned} \quad (6)$$

where $p_\mathbf{w}(\mathbf{x})$ is the probability density function of $\mathbf{w}$ and $\bar{n} = \frac{1}{K}\sum_{k=1}^K n_k$ is the average noise with reduced variance. Thus the averaging step turns the space variant distortion into space invariant blur with reduced noise. Therefore, the average image $\bar{g}$ can be expressed by convolution operations only

$$\bar{g} = p_\mathbf{w}(-\mathbf{x}) * \bar{f} = p_\mathbf{w}(-\mathbf{x}) * h * f = h_{\text{total}} * f + \bar{n} \quad (7)$$

Thus $\bar{g}$ is the latent image f convolved with an unknown blur kernel h_{total} plus noise $\bar{n}$. The subsequent blind deblurring step aims to recover the latent image f from the average image $\bar{g}$. This *Sum and Deblur* approach is summarized in Algorithm 1.

Algorithm 1. Turbulence restoration: Sum and Deblur [5]

Input: $g_k,$ $k = 1...K$
Output: $\hat{f}$
Initialization: $\bar{f}^{(0)} = \frac{1}{K} \sum_k g_k$
$\hat{f} = \texttt{blindDeblurring}(\bar{f}^{(0)})$
Return $\hat{f}$

Significant turbulence implies that $p_{\mathbf{w}}$ and the effective blur h_{total} are wide, making the blind deblurring task hard. To alleviate this problem, the averaging can be preceded by a registration step [12]. Nonrigid registration is conducted by estimating the optical flow $\hat{\mathbf{w}}_k$ between the measured frames g_k and the average frame $\bar{g}$. Then *sum and deblur* is applied on the warped frames $\{g_k(\mathbf{x} - \hat{\mathbf{w}}_k)\}_{k=1}^{K}$. This *Register Sum and Deblur* approach is summarized in Algorithm 2.

Algorithm 2. Turbulence restoration: Register Sum and Deblur [12]

Input: $g_k,$ $k = 1...K$
Output: $\hat{f}$
Initialization: $\bar{f}^{(0)} = \frac{1}{K} \sum_k g_k$
$\hat{\mathbf{w}}_k = \texttt{opticalFlow}(\bar{f}^{(0)}, g_k)$
$\bar{f}^{(1)} = \frac{1}{K} \sum_k g_k(\mathbf{x} - \hat{\mathbf{w}}_k)$
$\hat{f} = \texttt{blindDeblurring}(\bar{f}^{(1)})$
Return $\hat{f}$

Given $\{g_k\}$, Mao and Gilles [17] estimated $\bar{f}$ and $\{w_k\}$ using a variational approach. Specifically, they used *Bregman iteration* [22] and added regularization $R_{\bar{f}}(\bar{f})$, see Algorithm 3. Their approach removes only the turbulence distortion, not the optical blur h. Also note that edge preserving regularization on the blurred image $\bar{f}$ seems contradictory.

Algorithm 3. Turbulence restoration: Mao and Gilles [17]

Input: $g_k,$ $k = 1...K$
Output: $\hat{f}$
Initialization $\bar{f}^{(0)} = \frac{1}{K} \sum_k g_k,$ $f_k^{(0)} = 0,$ $k = 1...K$
for $l = 0,1,...,$ *until convergence* **do**
   $\hat{\mathbf{w}}_k^{(l+1)} = \texttt{opticalFlow}(\bar{f}^{(l)}, g_k)$
   $\bar{f}^{(l+1)} = \operatorname{argmin}_{\bar{f}} \frac{\gamma}{2} \sum_{k=1}^{K} ||\bar{f}(\mathbf{x} + \hat{\mathbf{w}}_k^{(l+1)}) - f_k^{(l)}||^2 + R_{\bar{f}}(\bar{f})$
   $f_k^{(l+1)} = f_k^{(l)} - \left(\bar{f}^{(l+1)}(\mathbf{x} + \hat{\mathbf{w}}_k^{(l+1)}) - g_k\right)$
end
Return $\hat{f} = \bar{f}^{(l)}$

Another approach, *lucky imaging*, applies image selection and fusion methods to reduce distortion caused by turbulence [2,3,10,18]. In a short exposure image sequence, the temporal integration of the turbulence distortion is small, hence "lucky" regions and frames with little blur and distortion appear in input sequence g_k. The output image is produced by combining these lucky regions and frames [10,18].

In a related method [15], image registration is first carried out to reduce geometric distortion caused by turbulence; then a lucky imaging-based weighting scheme is applied to generate a single image that is sharp everywhere. The weighting scheme is designed to balance between noise reduction and sharpness preservation. A dehazing process is finally used to enhance the visual quality.

The difficulty with the above methods is that even though turbulence blur is mostly removed through the lucky imaging process, the output still suffers from the optical blur h. Zhu and Milanfar [24] combined registration, lucky imaging and blind deconvolution: all measured frames g_k are first registered towards the average frame $\frac{1}{K}\sum_k g_k$; then weighted averaging is performed assigning higher weights to lucky (less blurred) regions; blind deconvolution is applied on the weighted averaged frame, see Algorithm 4.

Algorithm 4. Turbulence restoration: registered lucky sum and deblur [24]

Input: g_k, $k = 1...K$
Output: $\hat{f}$
Initialization: $\bar{f}^{(0)} = \frac{1}{K}\sum_k g_k$
$\hat{\mathbf{w}}_k = \texttt{opticalFlow}(\bar{f}^{(0)}, g_k)$
$\bar{f}^{(1)} = \texttt{luckyWeighting}(\{g_k(\mathbf{x} - \hat{\mathbf{w}}_k)\}_{k=1}^K)$
$\hat{f} = \texttt{blindDeblurring}(\bar{f}^{(1)})$
Return $\hat{f}$

The above algorithms serially apply image processing modules, addressing specific problems encountered in turbulence restoration pipelines. Regardless of this progress, these ad-hoc approaches, and pipelines itself, were not derived from an holistic view of the turbulence degradation model. Thus, many aspects of the problem remain unclear and the suggested solutions are hard to justify.

1.4   Contributions

We propose a novel unified variational approach to restore a sharp image from a turbulence-degraded image sequence g_k, satisfying the model (5). The suggested unified functional contributes to deeper understanding of the turbulence degradation problem. Furthermore, many modern methods are shown to be special cases of the proposed framework. Successful results on different datasets [12,13] demonstrate the applicability of the unified variational approach. The minimization of the functional is obtained using modern numerical methods (proximal mapping [6], ADMM [7]). The blind deblurring and optical flow modules found in various turbulence restoration algorithms are justified by minimization of the functional.

2   Unified Functional Framework

We propose a novel approach to restoring a single high-quality image from a sequence of frames distorted by atmospheric turbulence. We adopt the standard assumption [5,12,17,24] that the scene and the camera are static and that observed motions are due to turbulence alone. The image acquisition model is

$$g_k(\mathbf{x}) = (h * f)(\mathbf{x} + \mathbf{w}_k(\mathbf{x})) + n_k(\mathbf{x}), \quad k = 1 \ldots K, \tag{8}$$

where the latent image f is convolved with blur kernel h then deformed by the vector field $\mathbf{w}_k(\mathbf{x}) = (u_k(\mathbf{x}), v_k(\mathbf{x}))$ to obtain the measured frames g_k, $k = 1 \ldots K$. $\mathbf{w}_k$ is caused by turbulence motion and h is due to sensor optics.

2.1   Functional

Deviations from the turbulence model (8) are measured by the energy

$$E_{\text{model}} \left(\{\mathbf{w}_k\}_{k=1}^{K}, f, h \right) = \sum_k \int_{\Omega} \left(g_k(\mathbf{x} + \mathbf{w}_k) - f(\mathbf{x}) * h(\mathbf{x}) \right)^2 d\mathbf{x} \tag{9}$$

Turbulence restoration is an ill-posed problem requiring regularization. The regularization terms represent *a-priori* information about the latent functions $f, h, \{\mathbf{w}_k\}_{k=1}^{K}$. This formulation is equivalent to a maximum a-posteriori (MAP) statistical estimation problem, via the negative log-likelihood. We combine all these regularization terms into the regularization energy E_{reg}

$$E_{\text{reg}} \left(\{\mathbf{w}_k\}_{k=1}^{K}, f, h \right) = \sum_k R_{\mathbf{w}}(\mathbf{w}_k) + R_h(h) + R_f(f), \tag{10}$$

where $R_{\mathbf{w}}(\mathbf{w_k}), R_h(h), R_f(f)$ are the regularizers for functions $\mathbf{w_k}, h, f$. The total energy is the weighted sum of the data and regularization terms

$$E \left(\{\mathbf{w}_k\}_{k=1}^{K}, f, h \right) = \frac{\gamma}{2} E_{\text{model}} + E_{\text{reg}}$$

$$= \frac{\gamma}{2} \sum_k \| g_k(\mathbf{x} + \mathbf{w}_k) - f(\mathbf{x}) * h(\mathbf{x}) \|^2 + \sum_k R_{\mathbf{w}}(\mathbf{w}_k) + R_h(h) + R_f(f) \tag{11}$$

where

$$\| g_k(\mathbf{x} + \mathbf{w}_k) - f(\mathbf{x}) * h(\mathbf{x}) \|^2 = \int_{\Omega} \left(g_k(\mathbf{x} + \mathbf{w}_k) - f(\mathbf{x}) * h(\mathbf{x}) \right)^2 d\mathbf{x} \tag{12}$$

After discretizing the energy (11), the integral in norm (12) becomes the sum over all grid points.

Our main goal is finding the function f that minimizes the proposed functional. This is done by minimizing the functional with respect to the auxiliary functions $h, \{\mathbf{w}_k\}_{k=1}^K$ and the latent function f.

2.2 Minimization

We introduce the reference frame $\bar{f}$ satisfying $\bar{f} = f * h$. We rewrite energy (11) explicitly with the new auxiliary function $\bar{f}$:

$$E\left(\{\mathbf{w}_k\}_{k=1}^K, f, h, \bar{f}\right) = \frac{\gamma}{2} \sum_k \left\| g_k(\mathbf{x} + \mathbf{w}_k) - \bar{f}(\mathbf{x}) \right\|^2$$

$$+ \sum_k R_{\mathbf{w}}(\mathbf{w}_k) + R_h(h) + R_f(f), \qquad (13)$$

$$\text{s.t } \bar{f}(\mathbf{x}) = f(\mathbf{x}) * h(\mathbf{x}),$$

The resulting constrained minimization problem can be addressed with the Augmented Lagrangian method (ALM) [1]. First we write the Augmented Lagrangian:

$$\mathcal{L}\left(\{\mathbf{w}_k\}_{k=1}^K, f, h, \bar{f}, \lambda_{\bar{f}}\right) = E\left(\{\mathbf{w}_k\}_{k=1}^K, f, h, \bar{f}\right) + \frac{\alpha}{2} \left\| \bar{f}(\mathbf{x}) - h(\mathbf{x}) * f(\mathbf{x}) - \lambda_{\bar{f}}(\mathbf{x}) \right\|^2$$

$$(14)$$

where $\lambda_{\bar{f}}$ is the Lagrange multiplier function and $\alpha \geq 0$ is the penalty weight parameter. Problem (13) can be solved using ALM iterations

$$\left(\{\mathbf{w}_k^{(l+1)}\}_{k=1}^K, f^{(l+1)}, h^{(l+1)}, \bar{f}^{(l+1)}\right) = \arg \min_{\{\mathbf{w}_k\}, f, h, \bar{f}} \mathcal{L}(\{\mathbf{w}_k\}, f, h, \bar{f}, \lambda_{\bar{f}}^{(l)}) \quad (15)$$

$$\lambda_{\bar{f}}^{(l+1)} = \lambda_{\bar{f}}^{(l)} - (h^{(l+1)} * f^{(l+1)} - \bar{f}^{(l+1)}), \qquad (16)$$

where l is an iteration number. The minimization problem (15) is not trivial since it involves non-separate variables $\mathbf{w}_k, f, h, \bar{f}$. A natural way to address (15) is by applying alternate minimization with respect to $\mathbf{w}_k, f, h, \bar{f}$. ADMM is an efficient algorithm to address (15) by running just one minimization step with respect to each variable, while keeping others fixed. Thus the multivariable complex minimization (15) is transformed into the simpler minimizations

$$\mathbf{w}_k^{(l+1)} = \operatorname*{argmin}_{\mathbf{w}_k} \mathcal{L}(\mathbf{w}_k, f^{(l)}, h^{(l)}, \bar{f}^{(l)}, \lambda_{\bar{f}}^{(l)}), \quad k = 1 \ldots K,$$

$$(f^{(l+1)}, h^{(l+1)}) = \operatorname*{argmin}_{f, h} \mathcal{L}(\{\mathbf{w}_k^{(l+1)}\}_{k=1}^K, f, h, \bar{f}^{(l)}, \lambda_{\bar{f}}^{(l)}) \qquad (17)$$

$$(\bar{f}^{(l+1)}) = \operatorname*{argmin}_{\bar{f}} \mathcal{L}(\{\mathbf{w}_k^{(l+1)}\}_{k=1}^K, f^{(l+1)}, h^{(l+1)}, \bar{f}, \lambda_{\bar{f}}^{(l)})$$

The resulting minimization scheme is expressed in Algorithm 5.

Algorithm 5. Unified functional minimization

Input: $\mathcal{L}, \{g_k\}_{k=1}^K$

Output: $\hat{f}$ minimizer of (11)

1 Initialization $\bar{f}^{(0)} = \frac{1}{K}\sum_k g_k$, $\lambda_{\bar{f}}^{(0)} = 0$

2 **for** $l = 1,2,...,$ *until convergence* **do**

3    $\quad \mathbf{w}_k^{(l+1)} = \mathrm{argmin}_{\mathbf{w}_k} \frac{\gamma}{2}\|g_k(\mathbf{x}+\mathbf{w}_k) - \bar{f}^{(l)}\|^2 + R_{\mathbf{w}}(\mathbf{w}_k), \quad k = 1..K$

4    $\quad f^{(l+1)}, h^{(l+1)} = \mathrm{argmin}_{f,h} = \frac{\alpha}{2}\|\bar{f}^{(l)} - h * f - \lambda_{\bar{f}}^{(l)}\|^2 + R_h(h) + R_f(f)$

5    $\quad \bar{f}^{(l+1)} = \frac{\gamma}{\gamma+\beta}\left(\frac{1}{K}\sum_k g_k(\mathbf{x}+\mathbf{w}_k^{(l+1)})\right) + \frac{\beta}{\gamma+\beta}(h^{(l+1)} * f^{(l+1)} + \lambda_{\bar{f}}^{(l)})$

6    $\quad \lambda_{\bar{f}}^{(l+1)} = \lambda_{\bar{f}}^{(l)} - (h^{(l+1)} * f^{(l+1)} - \bar{f}^{(l+1)})$

7 **end**

8 Return $\hat{f} = f^{(l)}$

---

where $\beta = \frac{\alpha}{K}$.

Figure 2(a) shows the block diagram interpretation of Algorithm 5. First, optical flow $\{\mathbf{w}_k\}_{k=1}^K$ is calculated between the measured frames $\{g_k\}_{k=1}^K$ and the estimated reference frame $\bar{f}$, line 3 in Algorithm 5. Then the measured frames are warped toward the reference frame (using the calculated optical flow) and averaged ($\frac{1}{K}\sum_k g_k(\mathbf{x}+\mathbf{w}_k)$). The current estimate of the latent image f is obtained by applying blind deblurring on the reference image $\bar{f}$, line 4. The reference frame $\bar{f}$ is updated by applying the weighted sum of the average of the registered frames and the result of blind deblurring, line 5. These steps are repeated until convergence.

2.3   Previous Methods as Special Cases

Various turbulence restoration methods, classic and modern, can be derived as special cases of the unified functional framework.

1. The sum and deblur [5] approach, summarized in Algorithm 1, can be derived from the unified functional framework by neglecting the geometric distortions ($\mathbf{w}_k = 0$, $k = 1\ldots K$) and running only one iteration, see Fig. 2(a+b). Note that in the first iteration the weighted sum in the unified functional minimization algorithm (line 5) does nothing ($f^{(0)} = 0$, $h^{(0)} = 0$, $\lambda_{\bar{f}}^{(0)} = 0$).
2. The registered sum and deblur [12] approach, summarized in Algorithm 2, can be derived from the unified functional framework by running only one iteration, see Fig. 2(a+c).
3. The partial variational approach [17], summarized in Algorithm 3, is based on the functional

$$E\left(\{\mathbf{w}_k\}_{k=1}^K, f\right) = \frac{\gamma}{2}\sum_k \|g_k(\mathbf{x}+\mathbf{w}_k) - f(\mathbf{x})\|^2$$
$$+ \sum_k R_{\mathbf{w}}(\mathbf{w}_k) + R_f(f) \tag{18}$$

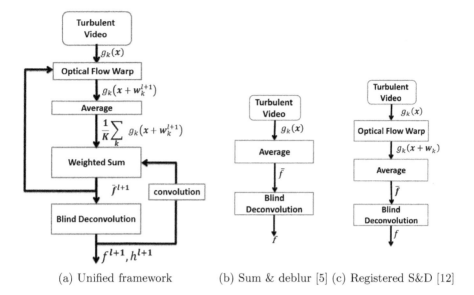

(a) Unified framework   (b) Sum & deblur [5] (c) Registered S&D [12]

Fig. 2. Previous solutions to restoration from turbulence as special cases of the proposed unified variational approach. (a) The unified functional minimization algorithm. (b) Sum and deblur [5]. (c) Registered sum and deblur [12].

Functional (18) can be obtained as a special case of the unified functional (11) by setting blur $h = \delta$, hence neglecting the blind deblurring operation in the minimization.

4. To derive the registered lucky sum and deblur [24] approach (Algorithm 4) we need to modify the proposed functional (11) and introduce lucky weighting $l_k(x, y)$. This function assigns higher weights to less blurry patches centered at (x, y). Functional

$$E\left(\{\mathbf{w}_k\}_{k=1}^{K}, f, h\right) = \frac{\gamma}{2}\sum_{k}\left\|l_k(x, y)\left(g_k(x + u_k, y + v_k) - f * h\right)\right\|^2$$
$$+ \sum_{k} R_{\mathbf{w}}(\mathbf{w}_k) + R_h(h) + R_f(f) \tag{19}$$

can be a good candidate for future work and can better exploit the lucky imaging paradigm. The weighting function $l_k(x, y)$ will induce the lucky average in the minimization of functional (19). We hypothesize that the resulting minimization algorithm will be in the form of Fig. 3 left. In that case, the lucky registered sum and deblur [24] method could be derived from functional 19 minimization, by running only one iteration, see Fig. 3.

2.4   Optical Flow and Blind Deblurring

Lines 3 and 4 of the turbulence restoration Algorithm 5 were obtained by minimization of the unified functional (11).

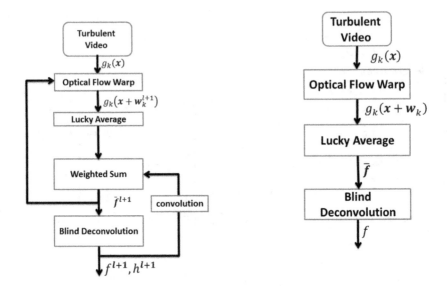

Fig. 3. Derivation of the registered lucky sum and deblur [24] method as special a case of the modified functional framework. Left: modified functional block diagram. Right: registered lucky sum and deblur block diagram.

Line 3 is readily identified as a general form of variational optical flow estimation. Substituting a specific regularization term $R_{\mathbf{w}}(\mathbf{w}_k)$ will lead to the incorporation of a particular optical flow algorithm in the unified framework for turbulence restoration. Similarly, line 4 of Algorithm 5 is a general form of variational blind deconvolution, specified by the chosen blur-kernel prior $R_h(h)$ and image prior $R_f(f)$. In practice, the optical flow estimation and blind deconvolution modules of the unified framework (Algorithm 5) can be replaced by alternative modules, not formally derived from the unified functional (11).

2.5   Results

In this paper, the main contribution is a novel unified variational framework for restoration of images viewed through atmospheric turbulence. The novelty is in the unified framework, that generalizes previous methods and leads the way to new ones. Experimental performance of the unified framework is not expected to reach the state of the art, because it does not include ad-hoc elements used by other algorithms. Nevertheless, we test the proposed unified framework to demonstrate its applicability in practical scenarios.

For image regularization we use the l_p norm on the image gradients, favoring natural images with sparse distribution of gradients. Thus,

$$R_f(f) = \sum_i ([D_x f]_i^2 + [D_y f]_i^2)^{\frac{p}{2}}, \quad 0 \le p \le 1,$$

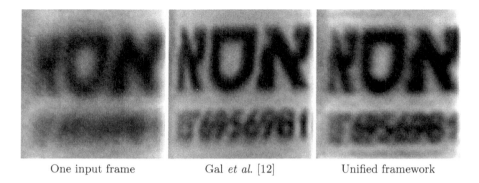

| One input frame | Gal *et al.* [12] | Unified framework |

Fig. 4. Unified framework results on *ASA* dataset [12]

where D_x and D_y are the partial derivative operators. As the blur kernel prior $R_h(h)$ we use the Laplace distribution, favoring sparse and non-negative kernel values. For the optical flow prior $R_{\mathbf{w}}(\mathbf{w})$ we used the regularization scheme proposed by Sun *et al.* [20], inducing median filtering of intermediate flow results during optimization.

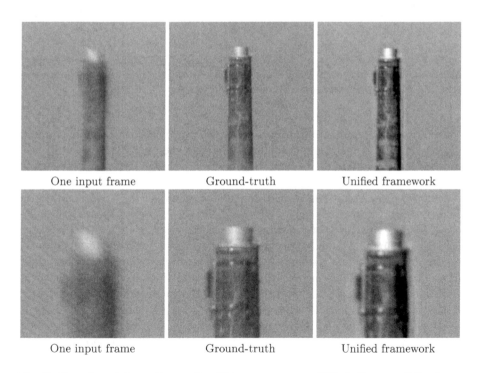

| One input frame | Ground-truth | Unified framework |

| One input frame | Ground-truth | Unified framework |

Fig. 5. Experimental results on the *Chimney* sequence [13]. Left: one of the input frames. Middle: ground-truth. Right: result obtained using the proposed unified framework. The bottom row is a zoomed version of the top row.

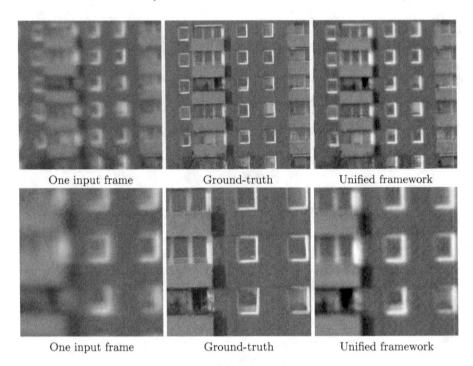

One input frame   Ground-truth   Unified framework

One input frame   Ground-truth   Unified framework

Fig. 6. Experimental results on the *Building* sequence [13]. Left: one of the input frames. Middle: ground-truth. Right: result obtained using the proposed unified framework. The bottom row is a zoomed version of the top row.

First, we show results on frames [12] captured through long range horizontal imaging through atmospheric turbulence. Figure 4 shows that the result of the proposed unified variational framework are visually close to the manually-tuned results of Gal *et al.* [12].

Next we show results on data taken under controlled but real conditions. Three image sequences, *Chimney*, *Building* and *Books* were captured through hot air exhausted by a vent, allowing comparison to clean images of the same scene [13]. Figures 5, 6 and 7 allow visual comparison between the results of proposed unified variational framework, typical input frames and the ground-truth images (captured without turbulence distortions and at ideal optical conditions). The quality of the results supports the unified variational framework and resulting algorithm. Note that slightly better results were reported on the *Chimney* and *Building* sequences in [24] using the registered lucky sum & deblur method, encouraging further research on functional (19) and the general modified functional block diagram shown in Fig. 3 (left).

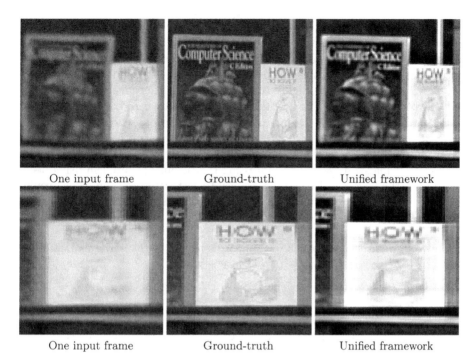

| | | |
| One input frame | Ground-truth | Unified framework |

| One input frame | Ground-truth | Unified framework |

Fig. 7. Experimental results on the *Book* sequence [13]. Left: one of the input frames. Middle: ground-truth. Right: result obtained using the proposed unified framework. The bottom row is a zoomed version of the top row.

3 Discussion

Current approaches for restoration of turbulent sequences [2, 3, 5, 12, 14, 21, 24] typically introduce a modular pipeline, and address the implementation of the modules. A few of these methods yield valuable experimental results. However, even though the processing pipelines reflect aspects of the turbulence restoration problem, they are not formally derived from a holistic, unified model of the problem. Thus, many aspects of the turbulence degradation problem remain unclear and the suggested solutions are often intuitive but lack a solid foundation.

We propose a novel variational approach to restore a sharp image from the turbulence-degraded image sequence. The suggested framework strengthens the understanding and mathematical justification of known methods, by deriving these methods as special cases of the unified variational framework. The proposed framework leads to generic blind deblurring and optical flow estimation sub-problems, where the specific solutions follow from the selection of regularization types and numerical methods. The non-smooth terms in the functional were directly addressed with modern optimization methods (ADMM, ISTA). We obtained good results on the various datasets without parameter tuning.

A natural extension of this work is incorporation of lucky imaging, see Eq. (19). The question is how to define the *lucky* weights $l_k(\mathbf{x})$. Another promising extension is replacement of the l_2 norm in the fidelity term of the unified functional by the robust l_1 norm. The l_1 norm may better handle the inevitable inaccuracy in modelling the highly complex effects of atmospheric turbulence.

State of the art optical flow estimation algorithms are based on deep learning, see e.g. Gadot and Wolf [11]. Deep learning also shows promise for blind deconvolution. It would be interesting to plug deep-learning-based optical flow and blind deconvolution modules in the unified algorithmic framework (Algorithm 5).

Finally, it makes sense to try addressing the turbulence problem using deep learning all the way, i.e., from input frames to restoration output. This would probably require the creation of a substantial database of image sequences taken through atmospheric turbulence.

Acknowledgment. This research was supported in part by the Blavatnik Interdisciplinary Cyber Research Center, Tel Aviv University.

References

1. Afonso, M.V., Bioucas-Dias, J.M., Figueiredo, M.A.T.: Fast image recovery using variable splitting and constrained optimization. IEEE Trans. Image Process. **19**, 2345–2356 (2010)
2. Aubailly, M., Vorontsov, M.A., Carhart, G.W., Valley, M.T.: Automated video enhancement from a stream of atmospherically-distorted images: the lucky-region fusion approach. In: Proceedings of the SPIE, vol. 7463 (2009)
3. Carhart, G.W., Vorontsov, M.A.: Synthetic imaging: nonadaptive anisoplanatic image correction in atmospheric turbulence. Opt. Lett. **23**, 745–747 (1998)
4. Chen, E., Haik, O., Yitzhaki, Y.: Detecting and tracking moving objects in long-distance imaging through turbulent medium. Appl. Opt. **53**, 1181–1190 (2014)
5. Cohen, B., Avrin, V., Belitsky, M., Dinstein, I.: Generation of a restored image from a video sequence recorded under turbulence effects. Opt. Eng. **36**, 3312–3317 (1997)
6. Combettes, P.L., Wajs, V.R.: Signal recovery by proximal forward-backward splitting. Multiscale Model. Simul. **4**, 1168–1200 (2005)
7. Eckstein, J., Bertsekas, D.P.: On the Douglas-Rachford splitting method and the proximal point algorithm for maximal monotone operators. Math. Program. **55**, 293–318 (1992)
8. Elkabetz, A., Yitzhaki, Y.: Background modeling for moving object detection in long-distance imaging through turbulent medium. Appl. Opt. **53**, 1132–1141 (2014)
9. Feller, W.: An Introduction to Probability Theory and Its Applications. Wiley, Hoboken (1968)
10. Fried, D.L.: Probability of getting a lucky short-exposure image through turbulence. J. Opt. Soc. Am. **68**, 1651–1658 (1978)
11. Gadot, D., Wolf, L.: Patchbatch: a batch augmented loss for optical flow. In: IEEE Conference on Computer Vision and Pattern Recognition (CVPR) (2016)
12. Gal, R., Kiryati, N., Sochen, N.A.: Progress in the restoration of image sequences degraded by atmospheric turbulence. Pattern Recogn. Lett. **48**, 8–14 (2014)

13. Hirsch, M., Sra, S., Scholkopf, B., Harmeling, S.: Efficient filter flow for space-variant multiframe blind deconvolution. In: Computer Vision and Pattern Recognition (CVPR), pp. 607–614, June 2010
14. John, S., Vorontsov, M.A.: Multiframe selective information fusion from robust error estimation theory. IEEE Trans. Image Process. **14**, 577–584 (2005)
15. Joshi, N., Cohen, M.: Seeing Mt. Rainier: lucky imaging for multi-image denoising, sharpening, and haze removal. In: Proceedings of the IEEE ICCP (2010)
16. Kopeika, N.S.: A System Engineering Approach to Imaging. SPIE Optical Engineering Press, Bellingham (1998)
17. Mao, Y., Gilles, J.: Turbulence stabilization. Proc. SPIE **8355**, 83550H–83550H-7 (2012)
18. Roggemann, M.C., Stoudt, C.A., Welsh, B.M.: Image-spectrum signal-to-noise-ratio improvements by statistical frame selection for adaptive-optics imaging through atmospheric turbulence. Opt. Eng. **33**, 3254–3264 (1994)
19. Shacham, O., Haik, O., Yitzhaky, Y.: Blind restoration of atmospherically degraded images by automatic best step-edge detection. Pattern Recogn. Lett. **28**, 2094–2103 (2007)
20. Sun, D., Roth, S., Black, M.: A quantitative analysis of current practices in optical flow estimation and the principles behind them. Int. J. Comput. Vis. **106**, 115–137 (2014)
21. Vorontsov, M.A., Carhart, G.W.: Anisoplanatic imaging through turbulent media: image recovery by local information fusion from a set of short-exposure images. J. Opt. Soc. Am. A **18**, 1312–1324 (2001)
22. Yin, W., Osher, S., Goldfarb, D., Darbon, J.: Bregman iterative algorithms for l_1-minimization with applications to compressed sensing. SIAM J. Imaging Sci. **1**, 143–168 (2008)
23. Zak, N.: Restoring an image of a moving object from a turbulence-distorted video. Master's thesis, School of Electrical Engineering, Tel Aviv University, Israel (2015)
24. Zhu, X., Milanfar, P.: Removing atmospheric turbulence via space-invariant deconvolution. IEEE Trans. Pattern Anal. Mach. Intell. **35**, 157–170 (2013)

A Convex Approach to K-Means Clustering and Image Segmentation

Laurent Condat[✉]

CNRS, GIPSA-Lab, Univ. Grenoble Alpes, 38000 Grenoble, France
laurent.condat@gipsa-lab.fr
http://www.gipsa-lab.grenoble-inp.fr/~laurent.condat/

Abstract. A new convex formulation of data clustering and image segmentation is proposed, with fixed number K of regions and possible penalization of the region perimeters. So, this problem is a spatially regularized version of the K-means problem, a.k.a. piecewise constant Mumford–Shah problem. The proposed approach relies on a discretization of the search space; that is, a finite number of candidates must be specified, from which the K centroids are determined. After reformulation as an assignment problem, a convex relaxation is proposed, which involves a kind of $l_{1,\infty}$ norm ball. A splitting of it is proposed, so as to avoid the costly projection onto this set. Some examples illustrate the efficiency of the approach.

Keywords: Image segmentation
Piecewise constant Mumford–Shah problem · K-means
Convex relaxation

1 Introduction

Data partitioning, or clustering, aims at decomposing a set of elements into groups, so as to minimize some notion of intra-group dissimilarity [1,2]. Thus, the classical *K-means* problem [3], consists in partitioning N points of $\mathbb{R}^d$ into K groups, by minimizing the sum of squared distances from every point to the nearest *centroid*, which is the center of mass of a group. For scalar data ($d = 1$), the K-means problem can be solved exactly and efficiently using dynamic programming [4,5]. By contrast, when $d \geq 2$, it is generally NP-hard [6,7]. An application is color image quantization [8,9]: one looks for the palette of K colors representing at best a given image; in that case, the points are the pixel values in $\mathbb{R}^3$, corresponding to the coordinates in some color space.

A fundamental problem in image processing and vision, which is even more difficult, is image segmentation: one wants to decompose an image of N pixels into K regions, corresponding to the objects of the scene, by favoring, in addition to intra-region similarity, some notion of spatial homogeneity [10,11]. We consider in this article the NP-hard *piecewise-constant Mumford–Shah problem* [11,12]: spatial homogeneity is obtained by penalizing the sum of the region perimeter.

© Springer International Publishing AG, part of Springer Nature 2018
M. Pelillo and E. Hancock (Eds.): EMMCVPR 2017, LNCS 10746, pp. 220–234, 2018.
https://doi.org/10.1007/978-3-319-78199-0_15

In general terms, the considered problem can be formalized as follows. The data $y = (y_n)_{n \in \Omega}$ is a 1-D signal of domain $\Omega = \{1, \ldots, N\}$ or a 2-D image of domain $\Omega = \{1, \ldots, N_1\} \times \{1, \ldots, N_2\}$ (having $N = N_1 N_2$ pixels), with values y_n in $\mathbb{R}^d$, endowed with the Euclidean norm. Given an integer $K \geq 2$, one wants to partition Ω into K regions[1] Ω_k (so $\bigcup_{k=1}^K \Omega_k = \Omega$ and $\Omega_k \cap \Omega_{k'} = \emptyset$, for all $k \neq k'$), and to find the corresponding centroids $c_k \in \mathbb{R}^d$, so as to

$$\underset{(\Omega_k)_{k=1}^K, (c_k)_{k=1}^K}{\text{minimize}} \frac{1}{2} \sum_{k=1}^K \sum_{n \in \Omega_k} \|y_n - c_k\|^2 + \frac{\lambda}{2} \sum_{k=1}^K \text{per}(\Omega_k), \qquad (1)$$

where per denotes the perimeter and $\lambda \geq 0$ is a parameter controlling the level of spatial regularization. When $\lambda = 0$, this is exactly the K-means problem; then the geometry of the domain and the indexing order do not play any role and one can think in terms of partitioning the point cloud $(y_n)_{n \in \Omega}$ in $\mathbb{R}^d$ into K groups, whose c_k are the means.

We can define the quantized or segmented signal or image $x = (x_n)_{n \in \Omega}$, with $x_n = c_k$ if $n \in \Omega_k$. So, x is a piecewise constant approximation of y, taking at most K different values. If $\lambda = 0$, we can express the problem (1) as:

$$\underset{x \in (\mathbb{R}^d)^\Omega}{\text{minimize}} \frac{1}{2} \|y - x\|_2^2 \quad \text{s.t.} \quad |\{x_n \ : \ n \in \Omega\}| \leq K, \qquad (2)$$

where $\|y - x\|_2^2 = \sum_{n \in \Omega} \|y_n - x_n\|^2$ and $|\cdot|$ denotes the cardinality of a set. Moreover, in the case of a 1-D signal, $\frac{1}{2} \sum_{k=1}^K \text{per}(\Omega_k)$ is equal to $|\{n \ : \ x_n \neq x_{n+1}\}|$, the number of jumps in the signal[2]. So, in 1-D, we can express the problem (1) as:

$$\underset{x \in (\mathbb{R}^d)^N}{\text{minimize}} \frac{1}{2} \|y - x\|_2^2 + \lambda |\{n = 1, \ldots, N-1 \ : \ x_n \neq x_{n+1}\}|$$

$$\text{s.t.} \quad |\{x_n \ : \ n = 1, \ldots, N\}| \leq K. \qquad (3)$$

The definition of the discrete perimeter in 2-D, based on finite differences, is much more problematic [13,14].

We can note that, if the regions Ω_k are fixed, the centroids c_k, solutions to the problem (1), are the means of the elements of the regions: $c_k = \frac{1}{|\Omega_k|} \sum_{n \in \Omega_k} y_n$. Conversely, if the c_k are fixed, (i) if $\lambda = 0$, we simply get the regions as the Voronoi cells $\Omega_k = \{n \in \Omega \ : \ k = \text{argmin}_{k'} \|y_n - c_{k'}\|\}$; (ii) if $\lambda > 0$, there exist efficient methods to solve the problem, by convex relaxation [13–15] or by graph cuts [16], see also [17]. Therefore, a strategy consists in alternating between updating the Ω_k at fixed c_k, and the other way around. In the case $\lambda = 0$, this yields exactly the classical K-means algorithm, due to Lloyd [18]; it must be distinguished from the K-means problem (1), for which it is a heuristic.

---

[1] The number of regions is actually at most K, and not exactly K, because some regions Ω_k could be empty. This is never the case in practical applications.

[2] We assume symmetric boundary conditions, so the boundary of the domain Ω is not counted in the perimeter.

It converges to a local minimum of the problem, but is very dependent on the initialization [3,9].

Globally convex methods, with high complexity, have been proposed to solve the problem (1), in the particular cases $K = 2$ or $K = 4$ [19–21]. The author is not aware of a generic method to approximate the global minimum of (1); existing convex relaxations of the K-means problem are discussed in Sect. 3. In this work, the problem is addressed by **discretizing** the search space of the centroids: we fix a set $\Gamma = \{a_m\}_{m=1}^M$ of M points in $\mathbb{R}^d$, called the *candidates*, and the centroids are constrained to belong to Γ, instead of to the whole space $\mathbb{R}^d$. Typically, $K \ll M$. We propose a convex formulation of the problem and an algorithm to solve it in Sect. 2. We illustrate the good performances of the approach in Sect. 4.

2   Proposed Method

2.1   Problem Reformulation by Lifting

In the segmented signal or image x, every element x_n is one of the centroids c_k, which is itself one of the candidates a_m. Thus, we can reformulate the problem (1) as an equivalent problem, whose unknown is the assignment array z, which has one more dimension than y, indexed by $m = 1, \ldots, M$; this is called *lifting* [14,22]. For every $n \in \Omega$ and $m = 1, \ldots, M$, $z_{m,n}$ is equal to 1 if $x_n = a_m$ and to 0 else. Each vector $z_{:,n} = (z_{m,n})_{m=1}^M$ belongs to the set $\mathcal{A}$ of binary assignment vectors, i.e. vectors with elements in $\{0, 1\}$ whose sum is 1.

We retrieve x from z by a simple summation:

$$x_n = \sum_{m=1}^M z_{m,n} a_m, \quad \forall n \in \Omega. \tag{4}$$

Moreover, the data fidelity term in (1) can be rewritten as

$$\frac{1}{2} \sum_{n \in \Omega} \|x_n - y_n\|^2 = \frac{1}{2} \sum_{n \in \Omega} \sum_{m=1}^M z_{m,n} w_{m,n}, \tag{5}$$

or, in short, $\frac{1}{2}\|x - y\|_2^2 = \langle z, w \rangle$, where

$$w_{m,n} = \frac{1}{2}\|y_n - a_m\|^2. \tag{6}$$

By using the coarea formula, according to which the total variation (TV) of the indicator function of a set (1 inside, 0 outside) is equal to the perimeter of that set [14], the regularization term in (1) can be rewritten as

$$\frac{\lambda}{2} \sum_{k=1}^K \mathrm{per}(\Omega_k) = \frac{\lambda}{2} \sum_{m=1}^M \mathrm{TV}(z_{m,:}), \tag{7}$$

where TV is some discrete form of the TV [13] and $z_{m,:} = (z_{m,n})_{n \in \Omega}$ is a scalar signal or image of same domain Ω as y.

It remains to reformulate the constraint that the number of regions Ω_k is at most K or, equivalently, that x takes its values in only K among the M candidates, i.e. $|\{x_n : n \in \Omega\}| \leq K$. We have the following property:

Proposition 1. *The assignment array $z \in \mathcal{A}^\Omega$ corresponds, by (4), to a signal or image $x \in \Gamma^\Omega$ taking at most K distinct values, if and only if*

$$\|z\|_{1,\infty} \leq K, \tag{8}$$

where[3] $\|z\|_{1,\infty} = \sum_{m=1}^M \max_{n \in \Omega} z_{m,n}$.

Indeed, since z takes its values in $\{0, 1\}$, a candidate a_m is assigned to at least one point x_n, and therefore is one of the centroids c_k, if and only if $z_{m,:}$ contains at least one 1; that is, if and only if $\max_{n \in \Omega} z_{m,n} = 1$.

Hence, we can rewrite the problem (1), with discrete search space Γ for the centroids, as

$$\operatorname*{minimize}_{z \in \mathcal{A}^\Omega} \langle z, w \rangle + \frac{\lambda}{2} \sum_{m=1}^M \mathrm{TV}(z_{m,:}) \text{ s.t. } \|z\|_{1,\infty} \leq K. \tag{9}$$

2.2 Convex Relaxation of the Problem

The problem (9) is not convex; more precisely, the functions and sets are all convex, except the set $\mathcal{A}$ of binary assignment vectors. So, we consider a convex relaxation, obtained by replacing $\mathcal{A}$ by its convex hull, which is the simplex Δ, i.e. the set of vectors with nonnegative elements whose sum is 1 [23]. Let us introduce the ball $\mathcal{B} = \{s \in \mathbb{R}^{M \times \Omega} : \|s\|_{1,\infty} \leq K\}$, and the convex indicator function $\imath_{\mathcal{E}}$ of a convex set $\mathcal{E}$, which takes the value 0 if its variable belongs to $\mathcal{E}$ and $+\infty$ else. The proposed convex problem is then:

$$\operatorname*{minimize}_{z \in \mathbb{R}^{M \times \Omega}} \langle z, w \rangle + \sum_{n \in \Omega} \imath_\Delta(z_{:,n}) + \frac{\lambda}{2} \sum_{m=1}^M \mathrm{TV}(z_{m,:}) + \imath_{\mathcal{B}}(z). \tag{10}$$

For conveniency, we denote by Δ^Ω the set of arrays of same size á z whose columns are on the simplex, so that $\imath_{\Delta^\Omega}(z) = \sum_{n \in \Omega} \imath_\Delta(z_{:,n})$.

We can note that another convex relaxation of the perimeter term, which is better, when z belongs to Δ^Ω, than the one in (7) we use in this paper, has been proposed in [14]. We do not consider it here because of its higher computational complexity, but there would be no difficulty in using it in our context.

The projection onto the simplex can be performed efficiently [23], see code on the author's webpage. However, the projection onto $\mathcal{B}$, which can also be performed exactly in finite time [24], is very costly. That is why we propose a (dual) splitting [25] of the maximum function:

---

[3] In this paper, we make an abuse of the terms $l_{1,\infty}$ norm and ball: the elements of z are nonnegative, so there is no need to take their absolute values.

Proposition 2. *The maximum function of a vector, or more generally of an array with N elements, s can be expressed as an infimal convolution [25]:*

$$\max_{n=1,\ldots,N} s_n = \min_{q \in \mathbb{R}} q/\sqrt{\mu N} + \imath_{\leq 0}(s - S^* q/\sqrt{\mu N}), \tag{11}$$

where $\mu > 0$ is some fixed constant, $\imath_{\leq 0}$ is the indicator function of the cone of arrays with nonpositive elements, S is the linear operator, which maps an array to the sum of its elements, and its adjoint operator S^* duplicates a real number into an array of same size as s with N identical elements.

We can note that, in Proposition 2, the norm of the linear operator $(s, q) \mapsto s - S^* q/\sqrt{\mu N}$ is $1 + 1/\mu$.

Let us introduce the constraint set $\mathcal{C} = \left\{ s \in \mathbb{R}^M \; : \; \sum_{m=1}^{M} s_m \leq K\sqrt{\mu N} \right\}$. The convex problem we propose to solve is then:

$$\operatorname*{minimize}_{z \in \mathbb{R}^{M \times \Omega}, q \in \mathbb{R}^M} \langle z, w \rangle + \imath_{\Delta^\Omega}(z) + \imath_{\mathcal{C}}(q) + \frac{\lambda}{2} \sum_{m=1}^{M} \mathrm{TV}(z_{m,:})$$

$$+ \sum_{m=1}^{M} \imath_{\leq 0}\left(z_{m,:} - S^* q_m/\sqrt{\mu N}\right). \tag{12}$$

We can remark that the use of $\|z\|_{1,\infty}$ to control the number of regions has been proposed in the same context in [26], but as a penalty and not as a constraint. Handling a $l_{1,\infty}$ constraint is a priori more difficult than regularizing with the $l_{1,\infty}$ norm, but we have seen that the splitting technique in Proposition 2 ends in the convex optimization problem (12), involving only simple terms.

2.3   Proposed Algorithm

In the following, we assume that the 'isotropic' form of the discrete TV [13] is used. There would be no difficulty in using instead the form proposed in [13], for better quality but higher computational complexity. As usual, we express the isotropic TV as the $l_{1,2}$ norm composed with a linear operator D of finite differences [13,27]. More precisely, in the case of a 2-D image with domain $\Omega = \{1, \ldots, N_1\} \times \{1, \ldots, N_2\}$, we introduce $D : \mathbb{R}^{M \times \Omega} \to \mathbb{R}^{M \times \Omega \times 2}, z \mapsto v$, with $v_{m,n,1} = z_{m,(n_1+1,n_2)} - z_{m,n}$ and $v_{m,n,2} = z_{m,(n_1,n_2+1)} - z_{m,n}$, for every $m = 1, \ldots, M$ and $n = (n_1, n_2) \in \Omega$ (using symmetric boundary conditions). Note that the operator norm of D is 8. We also introduce $\mathcal{V} = \{v \in \mathbb{R}^{M \times \Omega \times 2} \; : \; v_{m,n,1}^2 + v_{m,n,2}^2 \leq \lambda^2/4, \forall m = 1, \ldots, M, \ n \in \Omega\}$.

In the case of a 1-D signal with domain $\Omega = \{1, \ldots, N\}$, some simplifications can be made. We can set $D : \mathbb{R}^{M \times N} \to \mathbb{R}^{M \times (N-1)}, z \mapsto v$, with $v_{m,n} = z_{m,n+1} - z_{m,n}$, for every $m = 1, \ldots, M$ and $n = 1, \ldots, N - 1$, and $\mathcal{V} = \{v \in \mathbb{R}^{M \times (N-1)} \; : \; |v_{m,n}| \leq \lambda/2, \forall m = 1, \ldots, M, \ n = 1, \ldots, N - 1\}$. Note that the operator norm of D is 4 in this case, so we can set $\sigma_v := (1 - \gamma)/\tau/4$ in Algorithm 1 below.

We denote by $P_{\mathcal{E}}$ the projection onto a set $\mathcal{E}$, by $\imath_{\geq 0}$ and $P_{\geq 0}$ the indicator function of and the projection onto the cone of arrays with nonnegative elements, respectively. We extend the summation operator S to any array $u \in \mathbb{R}^{M \times \Omega}$ of same size as z, to mean summation with respect to the index $n \in \Omega$; that is, $Su = (Su_{m,:})_{m=1}^{M} \in \mathbb{R}^{M}$. Consequently, the last term in (12) can be rewritten more shortly as $\sum_{m=1}^{M} \imath_{\leq 0}(z_{m,:} - S^* q_m / \sqrt{\mu N}) = \imath_{\leq 0}(z - S^* q / \sqrt{\mu N})$.

The proposed algorithm is the over-relaxed version [28] of the Chambolle–Pock algorithm [27], applied to the problem (12), viewed as the sum of one function and two functions composed with linear operators, with the pair (z, q) as variable. With the proposed range of parameters, it is proved to converge to a solution of (12).

Algorithm 1

Input: w, K, λ, M, N, μ. Output: estimate $z^{(i+\frac{1}{2})}$ of a solution to (12).
Choose $\rho \in [1, 2)$, $\tau > 0$, $\gamma \in (0, 1)$, and the initial estimates $z^{(0)}$, $q^{(0)}$, $u^{(0)}$, $v^{(0)}$.
Set $\sigma_u := \gamma / \tau / (1 + 1/\mu)$, $\sigma_v := (1 - \gamma) / \tau / 8$.
Iterate: for $i = 0, 1, \ldots$

$$
\begin{vmatrix}
z^{(i+\frac{1}{2})} := P_{\Delta^{\Omega}}\big(z^{(i)} - \tau(u^{(i)} + w + D^* v^{(i)})\big), \\
q^{(i+\frac{1}{2})} := P_{\mathcal{C}}\big(q^{(i)} + (\tau/\sqrt{\mu N})Su^{(i)}\big), \\
u^{(i+\frac{1}{2})} := P_{\geq 0}\big(u^{(i)} + \sigma_u\big(2z^{(i+\frac{1}{2})} - z^{(i)} - S^*(2q^{(i+\frac{1}{2})} - q^{(i)})/\sqrt{\mu N}\big)\big), \\
v^{(i+\frac{1}{2})} := P_{\mathcal{V}}\big(v^{(i)} + \sigma_v D(2z^{(i+\frac{1}{2})} - z^{(i)})\big), \\
z^{(i+1)} := z^{(i)} + \rho(z^{(i+\frac{1}{2})} - z^{(i)}), \\
q^{(i+1)} := q^{(i)} + \rho(q^{(i+\frac{1}{2})} - q^{(i)}), \\
u^{(i+1)} := u^{(i)} + \rho(u^{(i+\frac{1}{2})} - u^{(i)}), \\
v^{(i+1)} := v^{(i)} + \rho(v^{(i+\frac{1}{2})} - v^{(i)}).
\end{vmatrix}
$$

The memory size for z and the dual variables u and v is $O(NM)$; it is $O(M)$ for q. The complexity of $P_{\Delta^{\Omega}}$ using the default sorting strategy is $O(NM \log M)$ [23]; it can be reduced to $O(NM)$ using a linear-time median-finding subroutine [23]. The complexity of the other operations is $O(NM)$, so the overall complexity of every iteration of the algorithm is $O(NM \log M)$. The parameters μ, τ, γ, ρ influence the convergence speed. They must be chosen on a case-by-case basis, but as a first step, one might consider $\rho = 1.9$, $\mu = 1$, $\gamma = 0.01$.

In the case of clustering or quantization, i.e. $\lambda = 0$, the algorithm can be simplified as

Algorithm 2

Input: w, K, M, N, μ. Output: estimate $z^{(i+\frac{1}{2})}$ of a solution to (12).
Choose $\rho \in [1, 2)$, $\tau > 0$, and the initial estimates $z^{(0)}$, $q^{(0)}$, $u^{(0)}$.
Set $\sigma := 1 / \tau / (1 + 1/\mu)$.
Iterate: for $i = 0, 1, \ldots$

$$
\begin{vmatrix}
z^{(i+\frac{1}{2})} := P_{\Delta^{\Omega}}\big(z^{(i)} - \tau(u^{(i)} + w)\big), \\
q^{(i+\frac{1}{2})} := P_{\mathcal{C}}\big(q^{(i)} + (\tau/\sqrt{\mu N})Su^{(i)}\big), \\
u^{(i+\frac{1}{2})} := P_{\geq 0}\big(u^{(i)} + \sigma\big(2z^{(i+\frac{1}{2})} - z^{(i)} - S^*(2q^{(i+\frac{1}{2})} - q^{(i)})/\sqrt{\mu N}\big)\big), \\
z^{(i+1)} := z^{(i)} + \rho(z^{(i+\frac{1}{2})} - z^{(i)}), \\
q^{(i+1)} := q^{(i)} + \rho(q^{(i+\frac{1}{2})} - q^{(i)}), \\
u^{(i+1)} := u^{(i)} + \rho(u^{(i+\frac{1}{2})} - u^{(i)}).
\end{vmatrix}
$$

To study the numerical convergence of Algorithm 1, it is useful to consider the dual problem [25] of the primal problem (12):

$$
\underset{u \in \mathbb{R}^{M \times \Omega}, v \in \mathbb{R}^{M \times \Omega \times 2}}{\text{minimize}} \sum_{n \in \Omega} \max_{m=1,\dots,M} (-u - D^*v - w)_{m,n} + \imath_{\mathcal{C}^*}(Su) + K(Su)_1
$$
$$
+ \imath_{\geq 0}(u) + \imath_{\mathcal{V}}(v), \tag{13}
$$

where $\mathcal{C}^* = \{s \in \mathbb{R}^M : s_1 = \cdots = s_M \geq 0\}$. At convergence, the primal and the dual cost values are opposite to each other [25]. That is, for a solution z to (12) and a solution pair (u, v) to (13),

$$
\Psi^\infty = \langle z, w \rangle + \frac{\lambda}{2} \sum_{m=1}^{M} \mathrm{TV}(z_{m,:}) = \sum_{n \in \Omega} \min_{m=1,\dots,M} (u + D^*v + w)_{m,n} - K(Su)_1. \tag{14}
$$

The iterate $z^{(i+\frac{1}{2})}$ does not belong to the $l_{1,\infty}$ ball $\mathcal{B}$, so its cost value

$$
\Psi^{\mathrm{p},i} = \langle z^{(i+\frac{1}{2})}, w \rangle + \frac{\lambda}{2} \sum_{m=1}^{M} \mathrm{TV}(z_{m,:}^{(i+\frac{1}{2})}), \tag{15}
$$

which tends to Ψ^∞ as $i \to +\infty$, is not guaranteed to be larger than Ψ^∞. Concerning the dual variables, $u^{(i+\frac{1}{2})} \geq 0$ and $v^{(i+\frac{1}{2})} \in \mathcal{V}$, but $Su^{(i+\frac{1}{2})}$ does not belong to $\mathcal{C}^*$. However, taking the maximum over m of $Su^{(i+\frac{1}{2})}$ to evaluate the cost yields a valid lower bound of Ψ^∞; that is,

$$
\Psi^{\mathrm{d},i} = \sum_{n \in \Omega} \min_{m=1,\dots,M} (u^{(i+\frac{1}{2})} + D^*v^{(i+\frac{1}{2})} + w^{(i+\frac{1}{2})})_{m,n} - K \max_{m=1,\dots,M} \sum_{n \in \Omega} u_{m,n}^{(i+\frac{1}{2})} \tag{16}
$$

is $\leq \Psi^\infty$ and tends to Ψ^∞ as $i \to +\infty$. Hence, a good way to test the convergence of Algorithm 1 is to check that

$$
\sum_{m=1}^{M} \max_{n \in \Omega} z_{m,n}^{(i+\frac{1}{2})} - K \leq \epsilon_1 \tag{17}
$$

and

$$
\Psi^{\mathrm{p},i} - \Psi^{\mathrm{d},i} \leq \epsilon_2, \tag{18}
$$

for small constants ϵ_1 and ϵ_2.

In all the examples of Sect. 4, it took several tenths of minutes to generate the results with the proposed algorithms, run several thousands of iterations.

3   Prior Work on the K-Means Problem

The convex relaxation described in the previous section turns out to be well known in operations research for the closely related K-median, a.k.a. p-median, problem, which is the same as the K-means problem, with the halved squared

Euclidean distance replaced by the Euclidean distance or any cost function with metric properties [29]. The search space for the cluster centers is discretized, as well. The classical example is a *facility location problem*: given N clients $\{y_n\}_{n=1}^N$ and a set $\Gamma = \{a_m\}_{m=1}^M$ of potential facilities, open K facilities so as to minimize the sum of the cost for each client of using its nearest open facility; typically this cost is the distance $w_{m,n} = \|y_n - a_m\|$. The 0–1 integer formulation and its convex relaxation as the linear program (12) (with $\lambda = 0$, $\mu = 1/N$, and the simplex constraint split into nonnegativity and sum-to-one constraints) can be traced back to half a century ago [30]. The variable q_m say whether facility m has to be opened or not and $z_{m,n}$ is the fraction of the demand of client n that is supplied by facility m. When the solution is not binary, several rounding strategies have been proposed [31].

In the case the candidates coincide with the data, i.e. $M = N$ and $\Gamma = y$, the K-means or K-median problem is often called the K-*medoid(s)* problem [32]; a centroid c_k is then constrained to be an exemplar from within the dataset and is called a medoid. The convex problem (12) can be simplified, because there is no need to introduce the auxiliary variables q_m: the maximum element of $z_{m,:}$ is $z_{m,m}$. Therefore, the problem can be rewritten as: minimize$_{z\in\mathbb{R}^{N\times N}}\ \langle z, w \rangle$ subject to $\mathrm{tr}(z) = K$ and $0 \le z_{m,n} \le z_{m,m}$ and $\sum_m z_{m,n} = 1$, for every m, n.

There does exist a convex relaxation of the K-means problem, which does not require discretizing the search space of the centroids. Indeed, in the solution, we have the identity $\sum_{n\in\Omega_k} \|y_n - c_k\|^2 = \sum_{(n,n')\in(\Omega_k)^2} \|y_n - y_{n'}\|^2/(2|\Omega_k|)$. Consequently, we can define the symmetric affinity matrix h of size $N \times N$, with $h_{n,n'} = \{1/|\Omega_k|$ if n and n' belong to the same cluster Ω_k in the solution, 0 else$\}$. Every column of h is on the simplex, h is positive semidefinite, its trace and rank are K. We can then reformulate the problem as minimizing $\langle h, w \rangle$ over the set of such matrices. Linear programming and semidefinite programming convex relaxations of this problem have been proposed [33–35]. They are less efficient in practice than the convex formulation (12) [35].

Another convex relaxation of the K-means problem consists in minimizing $\|x - y\|_2^2/2 + \lambda \sum_{n=1}^N \sum_{n'>n} \|x_n - x_{n'}\|_p$ over x, for some $p \in [1, +\infty]$ and some regularization parameter λ [36–40]. Then a cluster consists in all the points y_n corresponding to the same point x_n. This is a hierarchical approach: if two points are in the same cluster, they remain so when λ increases.

We can notice that the alternatives to the convex formulation (12) mentioned in this section require to store matrices of size $N \times N$ and to perform operations with complexity $O(N^2)$; this is not feasible when N is a typical image size.

4   Experimental Validation

4.1   *K*-Means Clustering

We consider the K-means problem, i.e. (1) with $\lambda = 0$, applied to the dataset A1 from https://cs.joensuu.fi/sipu/datasets/ [41], to partition the point cloud y of $N = 3000$ points in dimension $d = 2$ into $K = 20$ clusters.

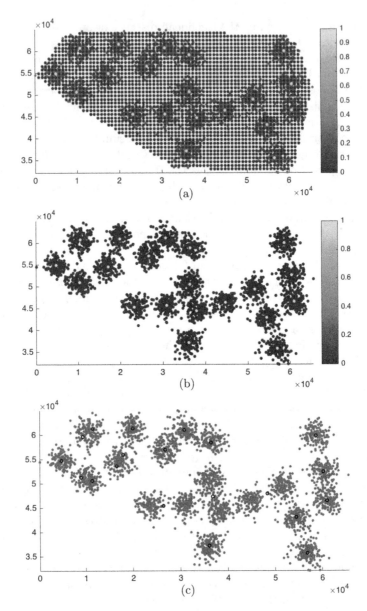

Fig. 1. Clustering of $N = 3000$ points, in red, into $K = 20$ groups. In (a), the M candidates a_m are on a uniform grid and in (b), they coincide with the data points. In both cases, they are represented with a color corresponding to the value $\max_{n \in \Omega} z_{m,n}$, where z is the solution to (12), with $\lambda = 0$. It turns out that z is binary, so it is a solution to the nonconvex problem (9). In (c), the centroids found by the kmeans algorithm of MATLAB, with default random initialization, are shown in black, whereas the centroids found by the same algorithm, but initialized with the centroids found by the proposed method, in yellow in (b), are shown in green. The latter are the global solutions of the K-means problem (1). (Color figure online)

In a first experiment, the set of candidates Γ is a uniform grid of 80×40 points, from which we kept only the $M = 2280$ points in the convex hull of y. Remarkably, Algorithm 2 converges to the exact solution of the problem (12), up to machine precision, in a finite number of iterations. It is also remarkable that the solution z, represented in Fig. 1(a), is binary, i.e. with values in $\{0, 1\}$; this can be easily verified by testing that $\min_{n \in \Omega} \max_{m=1}^{M} z_{m,n} = 1$. Therefore, the solutions to the convex problem (10) and to the nonconvex problem (9) coincide, and the proposed method yields the *global* solution of the K-means problem (for this choice of Γ).

In a second experiment, we consider the K-medoids case: $M = N$ and $\Gamma = y$. In this case too, Algorithm 2 converges in finite time to the exact solution of the problem, represented in Fig. 1(b), which is binary. So, the global solution of the K-medoids problem has been found.

We can note that it is easy to design counter-examples, where the solution of the convex problem (12) is not binary and does not yield a solution to the nonconvex K-means problem.

4.2 Color Image Quantization

We consider the color image quantization problem [9]. It is an instance of the K-means problems, which consists in partitioning the pixel values $y_n \in \mathbb{R}^3$ of an image, supposed to be coordinates in some color space. We consider here the CIELAB color space, because the Euclidean distance in it approximately matches the human perceptual distance. We first construct a palette of $M = 279$ colors, shown in Fig. 2, obtained by sampling the CIELAB space on a body centered cubic lattice. Indeed, this lattice is the one minimizing the quantization error; that is, for a given sampling density, or size of the Voronoi cell, the average squared distance between any point and the closest point in the lattice is minimized [42]. We consider three images and, for each, the three cases $K = 4$, $K = 5$, $K = 6$. The results are shown in Fig. 3. In the nine cases, Algorithm 2 converges after a finite number of iterations to the exact solution of (12), which is binary. So, the obtained images are the global solutions of the nonconvex color image quantization problem.

Fig. 2. Palette of $M = 279$ colors obtained by sampling the CIELAB color space on a body centered cubic lattice. The two last orange and red colors have been added manually. (Color figure online)

Fig. 3. In (a), original *sunflower, ladybug, parrot* images y, of size 254×168, 298×228, and 200×199, respectively. In (b)–(d), quantized images x solution to (12) ($\lambda = 0$) and (4), with $K = 6$, $K = 5$, $K = 4$, respectively, with the palette Γ of candidate colors shown in Fig. 2. (Color figure online)

4.3   Image Segmentation

We now consider the segmentation problem, i.e. (1) with $\lambda > 0$, using the same images as in Sect. 4.2 and Fig. 3(a), the palette Γ of $M = 279$ candidate colors in Fig. 2, and $\lambda = 500$. The segmented images, solution to (12) and (4), are shown in Fig. 4. z is never binary in this context and a small blur is present at edges. Indeed, each value $z_{m,n}$ can be interpreted as the proportion of the candidate a_m required to represent the pixel value x_n at location $n \in \Omega$; for a pixel at an edge between two regions, it is natural that it is soft-classified, instead of being fully assigned to one or the other region.

The method succeeds in providing images made of K colors in the palette, for the *sunflower* with $K = 4$ and $K = 6$, the *ladybug* with $K = 5$ and $K = 6$, and the *parrot* with $K = 4$ and $K = 5$; that is, $\max_{n\in\Omega} z_{m,n} = 1$ for K indexes m, and $z_{m,n} = 0$ for all the other m and all $n \in \Omega$. In these cases, up to the blur at the edges, which can be removed by rounding z to make it binary, we can consider that we have obtained the *global* solution to the segmentation problem (1). For the *sunflower* with $K = 5$, the orange at the center of the large sunflower is actually a mixture of 59% of the color $m = 80$ and 41% of the color $m = 92$. Similarly, for the *ladybug* with $K = 4$, there are two pure colors, i.e. $\max_{n\in\Omega} z_{m,n} = 1$ for $m = 155$ and $m = 107$, and mixtures of four colors, with $\max_{n\in\Omega} z_{m,n}$ equal to 0.88 for $m = 23$ and $m = 84$, and equal to 0.12 for $m = 31$ and $m = 83$. In both cases, this is not an issue, since if one really

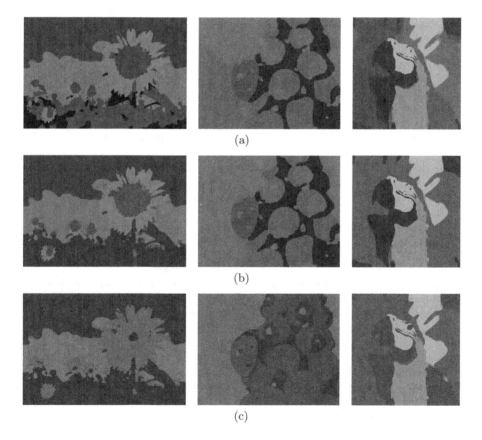

(a)

(b)

(c)

Fig. 4. Segmented images x, solution to (12) and (4), with y in Fig. 3(a), $\lambda = 500$, and the palette Γ of candidate colors shown in Fig. 2. In (a), (b), (c), $K = 6$, $K = 5$, $K = 4$, respectively. (Color figure online)

wants to identify the K colors of the palette adapted to represent the image, a post-processing step keeping the K colors with the largest value $\max_{n \in \Omega} z_{m,n}$ would be appropriate. However, in the last case of the *parrot* with $K = 6$, the method fails to provide an image with 6 dominant colors. Indeed, $\max_{n \in \Omega} z_{m,n}$ is equal to 1, 1, 1, 1, 0.66, 0.34, 0.34, 0.34, 0.33, for $m = 261$, 255, 63, 18, 155, 60, 66, 58, 68, respectively. There is no obvious rounding procedure to keep 6 out of these 9 colors. Thus, the proposed method succeeds eight times and fails once, in the nine examples considered.

We can remark that the proposed approach, which estimates the K colors and the corresponding regions with low perimeter jointly, performs better than a two-step strategy, that would first estimate the K colors using quantization and then solve the segmentation problem restricted to these $M = K$ colors. Indeed, we can see that the proposed approach yields different orange and green colors for the *sunflower* with $K = 4$, in Fig. 4(c), from the ones in Fig. 3(d). The dark gray for the *parrot* with $K = 4$, in Fig. 4(c), is also different from the dark green in Fig. 3(d). So, in these two examples, the two-step strategy would have failed to provide an image with the appropriate colors.

References.

1. Jain, A.K., Murty, M.N., Flynn, P.J.: Data clustering: a review. ACM Comput. Surv. **31**, 264–323 (1999)
2. Gan, G., Ma, C., Wu, J.: Data Clustering: Theory, Algorithms, and Applications. ASA-SIAM Series on Statistics and Applied Probability. SIAM, Philadelphia (2007)
3. Steinley, D.: K-means clustering: a half-century synthesis. Br. J. Math. Stat. Psychol. **59**(1), 1–34 (2006)
4. Wu, X.: Optimal quantization by matrix searching. J. Algorithms **12**(4), 663–673 (1991)
5. Soong, F.K., Juang, B.H.: Optimal quantization of LSP parameters. IEEE Trans. Speech Audio Process. **1**(1), 15–24 (1993)
6. Aloise, D., Deshpande, A., Hansen, P., Popat, P.: NP-hardness of Euclidean sum-of-squares clustering. Mach. Learn. **75**(2), 245–248 (2009)
7. Mahajan, M., Nimbhorkar, P., Varadarajan, K.: The planar k-means problem is NP-hard. Theor. Comput. Sci. **442**, 13–21 (2012). Special Issue on the Workshop on Algorithms and Computation (WALCOM 2009)
8. Brun, L., Trémeau, A.: Color quantization. In: Digital Color Imaging Handbook, pp. 589–638. CRC Press (2012)
9. Celebi, M.E.: Improving the performance of k-means for color quantization. Image Vis. Comput. **29**(4), 260–271 (2011)
10. Cremers, D., Rousson, M., Deriche, R.: A review of statistical approaches to level set segmentation: integrating color, texture, motion and shape. Int. J. Comput. Vis. **72**, 195–215 (2007)
11. Bar, L., Chan, T.F., Chung, G., Jung, M., Kiryati, N., Sochen, N., Vese, L.A.: Mumford and Shah model and its applications to image segmentation and image restoration. In: Scherzer, O. (ed.) Handbook of Mathematical Methods in Imaging, pp. 1095–1157. Springer, New York (2015). https://doi.org/10.1007/978-0-387-92920-0_25

12. Mumford, D., Shah, J.: Optimal approximations by piecewise smooth functions and associated variational problems. Commun. Pure Appl. Math. **42**, 577–685 (1989)
13. Condat, L.: Discrete total variation: new definition and minimization. SIAM J. Imaging Sci. **10**(3), 1258–1290 (2017)
14. Chambolle, A., Cremers, D., Pock, T.: A convex approach to minimal partitions. SIAM J. Imaging Sci. **5**(4), 1113–1158 (2012)
15. Pustelnik, N., Condat, L.: Proximity operator of a sum of functions; application to depth map estimation. IEEE Sig. Process. Lett. **24**(12), 1827–1831 (2017)
16. Yuan, J., Bae, E., Tai, X.-C., Boykov, Y.: A continuous max-flow approach to potts model. In: Daniilidis, K., Maragos, P., Paragios, N. (eds.) ECCV 2010. LNCS, vol. 6316, pp. 379–392. Springer, Heidelberg (2010). https://doi.org/10.1007/978-3-642-15567-3_28
17. Zach, C., Häne, C., Pollefeys, M.: What is optimized in convex relaxations for multilabel problems: connecting discrete and continuously inspired MAP inference. IEEE Trans. Pattern Anal. Mach. Intell. **36**(1), 157–170 (2014)
18. Lloyd, S.: Least squares quantization in PCM. IEEE Trans. Inform. Theory **28**(2), 129–136 (1982)
19. Brown, E.S., Chan, T.F., Bresson, X.: Completely convex formulation of the Chan-Vese image segmentation model. Int. J. Comput. Vis. **98**(1), 103–121 (2012)
20. Bae, E., Yuan, J., Tai, X.-C.: Simultaneous convex optimization of regions and region parameters in image segmentation models. In: Breuß, M., Bruckstein, A., Maragos, P. (eds.) Innovations for Shape Analysis: Models and Algorithms. Mathematics and Visualization, pp. 421–438. Springer, Heidelberg (2013). https://doi.org/10.1007/978-3-642-34141-0_19
21. Bae, E., Tai, X.-C.: Efficient global minimization methods for image segmentation models with four regions. J. Math. Imaging Vis. **51**(1), 71–97 (2015)
22. Pock, T., Cremers, D., Bischof, H., Chambolle, A.: Global solutions of variational models with convex regularization. SIAM J. Imaging Sci. **3**(4), 1122–1145 (2010)
23. Condat, L.: Fast projection onto the simplex and the l1 ball. Math. Program. Ser. A **158**(1), 575–585 (2016)
24. Quattoni, A., Carreras, X., Collins, M., Darrell, T.: An efficient projection for l1, ∞ regularization. In: Proceedings of ICML, Montreal, Canada, June 2009, pp. 857–864 (2009)
25. Bauschke, H.H., Combettes, P.L.: Convex Analysis and Monotone Operator Theory in Hilbert Spaces. Springer, New York (2011). https://doi.org/10.1007/978-3-319-48311-5
26. Yuan, J., Bae, E., Boykov, Y., Tai, X.-C.: A continuous max-flow approach to minimal partitions with label cost prior. In: Bruckstein, A.M., ter Haar Romeny, B.M., Bronstein, A.M., Bronstein, M.M. (eds.) SSVM 2011. LNCS, vol. 6667, pp. 279–290. Springer, Heidelberg (2012). https://doi.org/10.1007/978-3-642-24785-9_24
27. Chambolle, A., Pock, T.: A first-order primal-dual algorithm for convex problems with applications to imaging. J. Math. Imaging Vis. **40**(1), 120–145 (2011)
28. Condat, L.: A primal-dual splitting method for convex optimization involving Lipschitzian, proximable and linear composite terms. J. Optim. Theory Appl. **158**(2), 460–479 (2013)
29. Reese, J.: Solution methods for the p-median problem: an annotated bibliography. Networks **48**(3), 125–142 (2006)
30. Balinski, M.L.: On finding integer solutions to linear programs. In: Proceedings of the I.B.M. Scientific Computing Symposium on Combinatorial Problems, pp. 225–248 (1966)

31. Li, S., Svensson, O.: Approximating k-median via pseudo-approximation. In: Proceedings of the forty-Fifth Annual ACM Symposium on Theory of Computing (STOC 2013), Palo Alto, California, USA, June 2013, pp. 901–910 (2013)
32. Van der Laan, M., Pollard, K., Bryan, J.: A new partitioning around medoids algorithm. J. Stat. Comput. Simul. **73**(8), 575–584 (2003)
33. Peng, J., Xia, Y.: A new theoretical framework for K-means-type clustering. In: Chu, W., Young Lin, T. (eds.) Foundations and Advances in Data Mining. Studies in Fuzziness and Soft Computing, vol. 180, pp. 79–96. Springer, Heidelberg (2005). https://doi.org/10.1007/11362197_4
34. Peng, J., Wei, Y.: Approximating K-means-type clustering via semidefinite programming. SIAM J. Optim. **18**(1), 186–205 (2007)
35. Awasthi, P., Bandeira, A.S., Charikar, M., Krishnaswamy, R., Villar, S., Ward, R.: Relax, no need to round: integrality of clustering formulations. In: Proceedings of the 2015 Conference on Innovations in Theoretical Computer Science (ITCS), Rehovot, Israel, January 2015, pp. 191–200 (2015)
36. Pelckmans, K., De Brabanter, J., Suykens, J.A.K., De Moor, B.: Convex clustering shrinkage. In: Proceedings of Workshop on Statistics and Optimization of Clustering Workshop (PASCAL), London, UK, July 2005
37. Hocking, T., Vert, J.-P., Bach, F., Joulin, A.: Clusterpath: an algorithm for clustering using convex fusion penalties. In: Proceeding of the 28th International Conference on Machine Learning (ICML), Bellevue, USA, June 2011, pp. 745–752 (2011)
38. Lindsten, F., Ohlsson, H., Ljung, L.: Clustering using sum-of-norms regularization: with application to particle filter output computation. In: Proceedings of Statistical Signal Processing Workshop (SSP), Nice, France, June 2011, pp. 201–204 (2011)
39. Zhu, C., Xu, H., Leng, C., Yan, S.: Convex optimization procedure for clustering: theoretical revisit. In: Proceedings of NIPS, Montreal, Canada, December 2014, pp. 1619–1627 (2014)
40. Chi, E.C., Lange, K.: Splitting methods for convex clustering. J. Comput. Graph. Stat. **24**(4), 994–1013 (2015)
41. Kärkkäinen, I., Fränti, P.: Dynamic local search algorithm for the clustering problem. Technical report A-2002-6, Department of Computer Science, University of Joensuu, Joensuu, Finland (2002)
42. Barnes, E.S., Sloane, N.J.A.: The optimal lattice quantizer in three dimensions. SIAM J. Algebr. Discret. Methods **4**(1), 30–41 (1983)

Luminance-Guided Chrominance Denoising with Debiased Coupled Total Variation

Fabien Pierre[1(✉)], Jean-François Aujol[2], Charles-Alban Deledalle[2], and Nicolas Papadakis[2]

[1] Université de Lorraine, LORIA, CNRS, UMR 7503, INRIA projet Magrit, Lorraine, France
fabien.pierre@univ-lorraine.fs
[2] Univ. Bordeaux, IMB, CNRS, UMR 5251, 33400 Talence, France

Abstract. This paper focuses on the denoising of chrominance channels of color images. We propose a variational framework involving TV regularization that modifies the chrominance channel while preserving the input luminance of the image. The main issue of such a problem is to ensure that the denoised chrominance together with the original luminance belong to the RGB space after color format conversion. Standard methods of the literature simply truncate the converted RGB values, which lead to a change of hue in the denoised image. In order to tackle this issue, a "RGB compatible" chrominance range is defined on each pixel with respect to the input luminance. An algorithm to compute the orthogonal projection onto such a set is then introduced. Next, we propose to extend the CLEAR debiasing technique to avoid the loss of colourfulness produced by TV regularization. The benefits of our approach with respect to state-of-the-art methods are illustrated on several experiments.

Keywords: Colorization · Denoising · Color editing
Color assignment

1 Introduction

The representation of color images in perceptual color spaces rather than in RGB has been successfully considered in color image and video editing tasks such as colorization [17], contrast enhancement [5], color transfer [16] or JPEG compression [2]. There exists many color decompositions (HSV, YC_BC_R, YUV, Lab, $L\alpha\beta$...) that contain channels representing information such as luminance, chrominance, saturation and hue... In the following, we will consider the YUV

J.-F. Aujol is a member of Institut Universitaire de France. This study has been carried out with financial support from the French State, managed by the French National Research Agency (ANR GOTMI) (ANR-16-CE33-0010-01).

M. Pelillo and E. Hancock (Eds.): EMMCVPR 2017, LNCS 10746, pp. 235–248, 2018.
https://doi.org/10.1007/978-3-319-78199-0_16

color space, where Y denotes the luminance channel and U and V the chrominance ones that represent the color information of the image. Indeed, the YUV space being linear, the description of the chrominance range will be easier than for non-linear space, such as Lab. Color editing then consists in modifying one or several of these channels in function of the tackled application. In the image colorization problem, the original gray-scale image is considered as luminance and chrominance channel are created from examples or information given by the users. In the case of contrast enhancement, the luminance channel is modified, while the chrominance is preserved.

In this paper, we will take a point of view closer to the colorization problem, since we will only edit the chrominance, while preserving the luminance. The problem of chrominance denoising arises when restoring JPEG compressed images.

RGB range. In the colorization approaches of the literature, as well as for JPEG restoration techniques, the range of the chrominance is a problem which is rarely solved or considered. The range is defined, for a particular luminance value, as the set of the chrominance values such that the corresponding RGB colors are in the cube $[0, 255]^3$.

For instance, the iterative methods of [9] or [11] use PDE schemes on the chrominance channels, but the RGB range may be violated. Such approaches generally project the produced colors into $[0, 255]^3$, leading to a modification of the luminance. The problem of editing color while maintaining the range has been raised in [7, 10, 12]. These methods consider the problem into the 3-dimensional RGB space or use a specific parametrization of the RGB space. Editing color while maintaining the luminance channel is nevertheless a 2-dimensional problem that should be solved in the chrominance space. The two main issues for working in the chrominance space are the definition of the range for a particular luminance value and the orthogonal projection onto this set.

This range has been characterized has a convex set within the RGB color space in [12]. For colorization purposes, the authors of [12] proposed an oblique projection to maintain the RGB range while keeping the hue of the color examples. However, the use of an oblique projection into the optimization process does not ensure to compute a global optimum of the defined functional. In this paper, we work directly onto the chrominance values and the orthogonal projection on its range is designed, leading to an algorithm minimizing the proposed model.

TV regularization and loss of contrast. In order to avoid visual artefacts on the edited images, many approaches have considered the use of spatial regularization. Total Variation has therefore been extended to color images in RGB space [8] and luminance/chrominance ones for colorization [13] or JPEG image decompression or restoration [1]. In this paper, we consider the coupled TV regularization [13] designed to inpaint or to denoise chrominance channels. The results obtained with this method are efficient but they suffer from the well known bias of the Rudin, Osher and Fatemi (ROF) model [14]. In the case of gray-scale images this bias is revealed as a loss of contrast. In the case of the luminance-chrominance model of [13], the loss of contrast is visible as a loss of colorfulness of the result (see, *e.g.*, Fig. 1).

(a) Original image.   (b) Regularization of [13].   (c) Proposed model.

Fig. 1. The classical ROF model [14] produces a loss of contrast which is seen as a loss of colourfulness in the case of color images. The processed image (b) is drabber than the original one. For instance, the red part of the ladybird is less shiny. The aim of this paper is to avoid this counter-effect. (Color figure online)

An automatic debiasing strategy has been recently proposed with the CLEAR method [6] in order to deal with the loss of contrast of the original ROF model. Debiasing approaches are based on Bregman distance [3] or on the projection onto a linear space of unbiased solutions [6], but they are not designed to handle additional hard constraint such as the preservation of the luminance.

Outline and contributions. In this paper, we propose an algorithm to solve the color denoising problem with fixed luminance, while taking into account the range of admissible chrominance and debiasing the original result.

To this aim, we first characterize in Sect. 2 the acceptable convex set of chrominance values and design an algorithm to compute an orthogonal projection onto this set. We then present in Sect. 3, the chrominance denoising model related to the colorization one of [13].

Next, we propose in Sect. 4 to use the debiasing CLEAR method [6] to get rid of the loss of colourfulness issue. As the CLEAR has been designed for unconstrained problems, we extend this technique to our constrained luminance denoising problem. The benefits of this debiasing strategy are finally demonstrated in the experimental Sect. 5.

2   Range of Chrominance

The natural problem arising when editing a color while keeping its luminance or intensity constant, is the preservation of the RGB standard range of the produced image. Most of the methods of the literature work directly in the RGB space [7,10,12], since it is easier to maintain the standard range. Nevertheless, working in the RGB space needs to process 3 channels, while 2 chrominance channels are sufficient to edit a color image while maintaining the luminance.

2.1   Description of the Range

In this section, we geometrically describe the set of chrominance values which correspond to a particular luminance level, and which are contained in the RGB standard range. Let us denote by $T(y, u, v)$ the invertible linear operator mapping YUV colors onto RGB ones.

Proposition 1. *Let y be a value of luminance between 0 and 255. The set of chrominance values (u, v) that satisfy $T(y, u, v) \in [0, 255]^3$ is a convex polygon.*

Remark 1. For a given luminance, the chrominance values out of this polygon can be transformed into the RGB space, but they are out of the bounds. A truncation of the coordinates is usually done, but it generally changes both the luminance and the hue of the result.

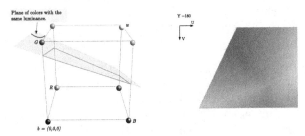

(a) Set of the RGB colors with a fixed luminance.   (b) Corresponding colors in the YUV space.

Fig. 2. The set of the RGB colors with a particular luminance is a convex polygon. The map from RGB to YUV being affine, the set of the corresponding chrominances is also a convex polygon.

Proof (of Proposition 1). The intuition of the proof is given in Fig. 2. The set of the colors in the RGB cube whose luminance is equal to a particular value y is a convex polygon (see, *e.g.*, [12]). Indeed, the set of colors with a particular luminance is an affine plane in $\mathbb{R}^3$ and the intersection of the RGB cube with it is a polygon. The transformation of the RGB values into the YUV space being affine, the set of corresponding colors is thus also a convex polygon included in the set $Y = y$.   □

2.2   Orthogonal Projection onto the Convex Range

Pixel-wise, the valid chrominances are contained in a convex polygon that has, at most, 6 edges. The numerical computation of the vertex coordinates is postponed to Appendix A. When the vertices are computed, and denoted by P1, P2, *etc.*, the orthogonal projection onto the polygon is computed as follows.

The algorithm first checks if the corresponding RGB value is between 0 and 255. If so, the point is its own orthogonal projection. If not, the orthogonal projection is onto one of the edges. So, it is computed for each edges and the closest one is retained as the solution. The algorithm is summarized in Algorithm 1 and illustrated in Fig. 3.

Algorithm 1. Algorithm computing projection $P_{\mathcal{R}}$.

Require: X: chrominance vector; Y luminance value.
1: **if** $RGB(Y, X) \notin [0, 255]^3$ **then**
2: **for** $i = 1 : n$ **do**
3:   $j \leftarrow i + 1 \bmod n$
4:   $\alpha \leftarrow \left\langle \overrightarrow{P_i P_j} | \overrightarrow{P_i X} \right\rangle / \left(\| \overrightarrow{P_i P_j} \|_2 \right)$
5:   **if** $\alpha > 1$ **then**
6:    $X_{i,j} \leftarrow P_j$
7:   **else if** $\alpha < 0$ **then**
8:    $X_{i,j} \leftarrow P_i$
9:   **else**
10:    $X_{i,j} \leftarrow P_i + \alpha \overrightarrow{P_i P_j}$
11:   **end if**
12: **end for**
13: $X \leftarrow \operatorname{argmin}_{X_{i,j}} \| X - X_{i,j} \|_2$
14: **end if**

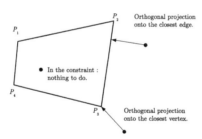

Fig. 3. To compute the orthogonal projection, different cases can appear. If the YUV color respects the constraint, the projection is the identity. Otherwise, the orthogonal projection onto the closest edge or vertex should be done.

3 Luminance-Chrominance TV-L2 Model

In the following we focus on a variational model to denoise the chrominance channels of an image while keeping the luminance unchanged. Similarly to the colorization model of [13], we want to find the minimizer $\hat{x}(y)$ of the denoising functional:

$$\hat{x}(y) = \operatorname{argmin}_{x=(U,V)} \operatorname{TV}_{\mathfrak{C}}(x) + \lambda \| x - y \|^2 + \chi_{\mathcal{R}}(x), \tag{1}$$

with

$$\mathrm{TV}_{\mathfrak{C}}(x) = \int_{\Omega} \sqrt{\gamma \|\nabla Y\|^2 + \|\nabla U\|^2 + \|\nabla V\|^2}. \qquad (2)$$

The first term is a coupled total variation which enforces the chrominance channels to have a contour at the same location as the luminance one. The fidelity data term is a classical L2 norm between chrominance channels of the unknown x and the data y. In this paper, we consider that the chrominance values live onto the convex polygon denoted by $\mathcal{R}$ and described in the previous section. This last assumption ensures that the final solution lies onto the RGB cube, avoiding final truncation that leads to modification of the luminance channel. Model (1) is convex and can be turned into a saddle-point problem of the form:

$$\min_{x \in \mathbb{R}^2} \max_{z \in \mathbb{R}^6} \frac{\lambda}{2} \|x - y\|^2 + \langle \nabla x | z_{1,\ldots,4} \rangle + \langle \gamma \nabla Y | z_{5,\ldots,6} \rangle - \chi_{B(0,1)}(z). \qquad (3)$$

The primal-dual algorithm [4] used to compute such saddle-point is recalled in Algorithm 2, where $P_{\mathcal{R}}$ is the orthogonal projection described in Algorithm 1 and $P_{\mathcal{B}}$ is defined as follows:

$$P_{\mathcal{B}}(z) = \frac{(z_{1,\ldots,4}, z_{5,6} - \sigma \nabla Y)}{\max \left(1, \|z_{1,\ldots,4}, z_{5,6} - \sigma \nabla Y\|_2\right)}. \qquad (4)$$

Algorithm 2. Minimization of (3).

1: $x^0 = y$
2: $z^0 \leftarrow \nabla x$
3: **for** $n \geq 0$ **do**
4:     $z^{n+1} \leftarrow P_{\mathcal{B}}\left(z^n + \sigma \nabla \overline{x}^n\right)$
5:     $x^{n+1} \leftarrow P_{\mathcal{R}}\left(\dfrac{x^n + \tau\left(\mathrm{div}(z^{n+1}) + \lambda y\right)}{1 + \tau \lambda}\right)$
6:     $\overline{x}^{n+1} \leftarrow 2x^{n+1} - x^n$
7: **end for**
8: set $\hat{x}(y) = x^{n+1}$ and $\hat{z} = z^{n+1}$.

The results produced by Algorithm 2 are promising but with a low data parameter λ, the results are drab, as illustrated in Fig. 1(b).

4   Constrained TV-L2 Debiasing Algorithm

In this section we present a debiased algorithm for correcting the loss of colourfulness of the solution given by the optimum of (1).

4.1 The CLEAR Method [6]

The CLEAR method [6] can be applied for debiasing estimators $\hat{x}(y)$ obtained as:

$$\hat{x}(y) \in \operatorname{argmin}_{x \in \mathbb{R}^p} F(x, y) + G(x), \tag{5}$$

where F is a convex data fidelity term with respect to data y and G is a convex regularizer. For G being the Total Variation regularization, the estimator $\hat{x}(y)$ is generally computed by an iterative algorithm, and it presents a loss of contrast with respect to the data y. In the aim of debiasing this estimator, the CLEAR method refits the data y with respect to some structural information contained in the biased estimator $\hat{x}$. This information is encoded by the Jacobian of the biased estimator with respect to the data y:

$$J_{\hat{x}}(y)d = \lim_{\varepsilon \to 0} \frac{\hat{x}(y + \varepsilon d) - \hat{x}(y)}{\varepsilon}. \tag{6}$$

For instance, when G is the anisotropic TV regularization, the Jacobian contains the information of the support of the solution $\hat{x}$, on which a projection of the data can be done to estimate.

In the general case, the CLEAR method relies on the *refitting estimator* $\mathcal{R}_{\hat{x}}(y)$ of the data y from the biased estimation $\hat{x}(y)$:

$$\mathcal{R}_{\hat{x}}(y) \in \operatorname{argmin}_{h \in \mathcal{H}} \|h(y) - y\|_2^2 \tag{7}$$

where $\mathcal{H}$ is defined as the set of maps $h : \mathbb{R}^n \to \mathbb{R}^p$ satisfying, $\forall y \in \mathbb{R}^n$:

$$h(y) = \hat{x}(y) + \rho J_{\hat{x}(y)}(y - \hat{x}(y)), \text{ with } \rho \in \mathbb{R}. \tag{8}$$

A closed formula for ρ can be given:

$$\rho = \begin{cases} \dfrac{\langle J_{\hat{x}(y)}(\delta) | \delta \rangle}{\|J_{\hat{x}(y)}(\delta)\|_2^2} & \text{if } J_{\hat{x}(y)}(\delta) \neq 0 \\ 1 & \text{otherwise.} \end{cases} \tag{9}$$

where $\delta = y - \hat{x}(y)$. In practice, the global value ρ allows to recover most of the bias in the whole image domain.

An algorithm is then proposed in [6] to compute the numerical value of $J_{\hat{x}(y)}(y - \hat{x}(y))$. The process is based on the differentiation of the algorithm providing $\hat{x}(y)$.

It is important to notice that the CLEAR method applies well for estimators obtained from the resolution of unconstrained minimization problems of the form (5). Nevertheless, it is not adapted to our denoising problem (1) that contains an additional constraint $\chi_{\mathcal{R}}(x)$ as CLEAR may violate the constraint.

4.2 Direct Extension of CLEAR to Constrained Problems

Extending the CLEAR method to the constrained Model (1) requires to take the constraint into account in the axioms of the refitting model (7). The main difference with the original model is the addition of the constraint $\chi_{\mathcal{R}}(x)$. We can first notice that the refitting axioms of [6] $h(y) = Ay + b$ for some $A \in \mathbb{R}^{p \times n}, b \in \mathbb{R}^p$ and $J_h(y) = \rho J_{\hat{x}}(y)$ for some $\rho \in \mathbb{R}$ (leading to relation (8)) are in line with the introduction of the constraint. In particular, the definition of the Jacobian $J_{\hat{x}}$ in Eq. (6) remains valid with the constraint, since $\hat{x}(y)$ and $\hat{x}(y + \varepsilon d)$ are still in the closed convex $\mathcal{R}$. The computation of the ρ parameter in Eq. (9) may nevertheless produce, from Eq. (8), an estimation out of the constraint, that has to be post-processed. This points out the main difference between the constrained and the unconstrained debiased estimator.

In [6], the value of ρ is computed from the minimization of a map from $\mathbb{R}$ to $\mathbb{R}$ defined as follows:

$$\rho \mapsto \| \left(I_d - \rho J_{\hat{x}(y)} \right) \left(\hat{x}(y) - y \right) \|_2^2. \tag{10}$$

In the case of the constrained problem, the function to be minimized is written as:

$$\rho \mapsto \| \hat{x}(y) + \rho J_{\hat{x}(y)} \left(y - \hat{x}(y) \right) - y \|_2^2 + \chi_{\mathcal{R}}(\hat{x}(y) + \rho J_{\hat{x}(y)} \left(y - \hat{x}(y) \right)). \tag{11}$$

Let us denote by ρ the value defined in Eq. (9). In the case when the constraint is fulfilled, *i.e.*, when $\hat{x}(y) + \rho J_{\hat{x}(y)} \left(y - \hat{x}(y) \right) \in \mathcal{R}$, then, the minimum of (11) is reached with ρ.

If not, let us study the function (11). This function is convex. Moreover, the value $\rho = 0$ is in the domain of the functional because $\hat{x}(y) \in \mathcal{R}$. The idea is to find the maximum value of ρ such that $\hat{x}(y) + \rho J_{\hat{x}(y)} \delta \in \mathcal{R}$. In our case, since $\mathcal{R}$ is a convex polygon, this computation can be done with a Ray-Tracing algorithm [15]. To this aim, we can parametrize the segment $[\hat{x}(y), \hat{x}(y) + \rho J_{\hat{x}(y)} \delta]$:

$$\tilde{\rho} = \max_{t \in [0,1]} t\rho \text{ such that } \hat{x}(y) + t\rho J_{\hat{x}(y)} \left(y - \hat{x}(y) \right) \in \mathcal{R}. \tag{12}$$

Equation (12) can thus be directly solved by the maximum value t such that $\hat{x}(y) + t\rho J_{\hat{x}(y)} \left(y - \hat{x}(y) \right)$ intersects the border of $\mathcal{R}$.

Direct debiasing process. Let us summarize the refitting algorithm designed for model (1). The first step consists in computing a solution of (1) with Algorithm 2. This iterative algorithm provides at convergence a first biased solution $\hat{x}(y)$ and its dual variable $\hat{z}$. Once this solution has been computed, the differentiated algorithm, presented in Algorithm 3, is applied in the direction $\delta = y - \hat{x}(y)$. This algorithm requires the definition of the operator $\Pi_{\hat{z}}(\tilde{z})$ which is the linearisation of the projection $P_{\mathcal{B}}$ around $\hat{z}$, and reads [6]:

$$\Pi_{\hat{z}}(\tilde{z}) = \begin{cases} \tilde{z} & \text{if } \|\hat{z}\| < 1 \\ \dfrac{1}{\|\hat{z}\|} \left(\tilde{z} - \dfrac{\langle \hat{z} | \tilde{z} \rangle}{\|\hat{z}\|^2} \hat{z} \right) & \text{otherwise.} \end{cases} \tag{13}$$

Finally, the Ray-Tracing is applied to obtain $\tilde{\rho}$ and get the debiased solution as $\hat{x}(y) + \tilde{\rho} J_{\hat{x}(y)}(y - \hat{x}(y))$.

Algorithm 3. Differentiation of Algorithm 2 for computing $J_{\hat{x}(y)}\delta$ from $(\hat{x}(y), \hat{z})$

1: $\tilde{x}^0 = \delta, \, \overline{\tilde{x}}^0 = \delta$

2: $\tilde{z}^0 \leftarrow \nabla\tilde{x}$

3: **for** $n \geq 0$ **do**

4: $\quad \tilde{z}^{n+1} \leftarrow \Pi_{\hat{z}}\left(\tilde{z}^n + \sigma\nabla\overline{\tilde{x}}^n\right)$

5: $\quad \tilde{x}^{n+1} \leftarrow \dfrac{\tilde{x}^n + \tau\left(\mathrm{div}(\tilde{z}^{n+1}) + \lambda\delta\right)}{1 + \tau\lambda}$

6: $\quad \overline{\tilde{x}}^{n+1} \leftarrow 2\tilde{x}^{n+1} - \tilde{x}^n$

7: **end for**

8: $J_{\hat{x}(y)}\delta = \tilde{x}^{n+1}.$

Unfortunately, this direct approach does not lead to interesting results on practical cases. Indeed, if in one particular pixel the solution $\hat{x}(y)$ is saturated, and if the debiased solution is out of $\mathcal{R}$, then $\tilde{\rho} = 0$ is the unique global ρ satisfying $\hat{x}(y) + \rho J_{\hat{x}(y)}(y - \hat{x}(y)) \in \mathcal{R}$. Thus, in practical cases, the debiased solution is equal to the biased one.

We also considered the differentiation of the projection $P_{\mathcal{R}}$, but it does not lead to better results since it does not guaranty that $\hat{x}(y) + \rho J_{\hat{x}(y)}(y - \hat{x}(y)) \in \mathcal{R}$.

Having a global ρ is thus too restrictive for our constrained problem. In the next section, we propose a model with an adaptive ρ parameter to tackle this saturated values issue.

4.3   Adaptive Debiasing Model for Constrained Problems

We first notice that for a pixel ω such that $\hat{x}(y)_\omega + \rho J_{\hat{x}(y),\omega}(y_\omega - \hat{x}(y)_\omega)$ fulfils the constraint, ρ is the best value to refit the model according to the hypothesis of model (7). Here, $J_{\hat{x}(y),\omega}$ denotes the value of $J_{\hat{x}(y)}$ in pixel ω.

On the other hand, if for a pixel ω, the value of $\hat{x}(y)_\omega$ and $J_{\hat{x}(y),\omega}(y_\omega - \hat{x}(y)_\omega)$ are such that $\hat{x}(y)_\omega + \rho J_{\hat{x}(y),\omega}(y_\omega - \hat{x}(y)_\omega) \notin \mathcal{R}$, the ρ value has to be adapted. Thus, let us define for a pixel ω the adapted $\tilde{\rho}_\omega$ as follows:

$$\tilde{\rho}_\omega = \max_{t_\omega \in [0,1]} t_\omega\rho \text{ such that } \hat{x}(y)_\omega + t_\omega\rho J_{\hat{x}(y),\omega}(y_\omega - \hat{x}(y)_\omega) \in \mathcal{R}. \tag{14}$$

The constrained refitting model is then defined pixel-wise as:

$$R_{\hat{x}}^{\mathcal{R}}(y) = \hat{x}(y)_\omega + \tilde{\rho}_\omega J_{\hat{x}(y),\omega}(y_\omega - \hat{x}(y)_\omega) \tag{15}$$

This definition ensures that the debiased estimation is in the constraint. Moreover, if the debiasing method of [6] produces an estimation in the constraint, this solution is retained. Notice however that the CLEAR hypothesis $J_h(y) = \rho J_{\hat{x}}(y)$ for some $\rho \in \mathbb{R}$ in model (7) is not fulfilled anymore. In numerical experiments, for most pixels, the values of $\tilde{\rho}_\omega$ computed with our method are the same as with Model (7).

As illustrated by Fig. 4, such local debiasing strategy realizes an oblique projection onto $\mathcal{R}$.

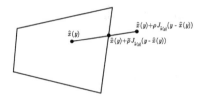

Fig. 4. An oblique projection is able to respect most of hypotheses of the Model (7) while fulfilling the constraint. The refitting of the method of [6] may be out of the constraint (in practical cases, about 5% of the refitted pixels are out of the range).

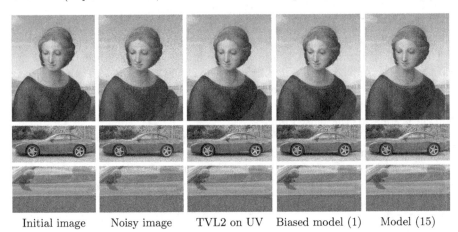

Initial image   Noisy image   TVL2 on UV   Biased model (1)   Model (15)

Fig. 5. Results of chrominance channels with a TVL2 model on chromonance, with the biased method and with the unbiased method. The debiasing algorithm produces more colorfull results. (Color figure online)

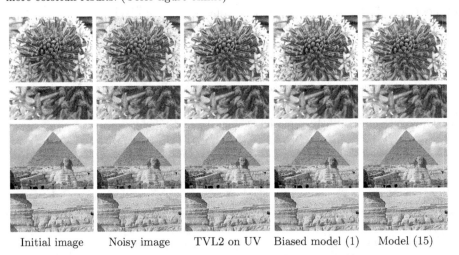

Initial image   Noisy image   TVL2 on UV   Biased model (1)   Model (15)

Fig. 6. The advantage of the coupled total variation (1) on the TVL2 model has been shown in [13]. In this work, it is refined in a better colorfulness-preserving model.

Computation of the Oblique Projection. We now describe how to compute the oblique projection when the constraint is the chrominance set for a particular value of luminance (see, *e.g.*, Sect. 2). To simplify the notation, we consider the problem for a single pixel w and set $x := \hat{x}(y)_w$, $y := J_{\hat{x}(y),w}(y_w - \hat{x}(y)_w)$ and $\rho \in \mathbb{R}$ computed by the algorithm of [6].

For $x + \rho\, y \notin \mathcal{R}$, we want to compute the maximum value of $t \in [0,1]$ such that $x + t\rho\, y \in \mathcal{R}$. We know that $x \in \mathcal{R}$, thus if $x + \rho\, y \notin \mathcal{R}$, the segment $[x, x + \rho\, y]$ cross one edge of the polygon.

Let us consider this problem by testing it into the RGB space. Indeed, the edges in the chrominance space correspond to edges in the RGB one, and the intersections between them correspond to intersections in the RGB space. In RGB, the problem of finding the intersection between an edge and the polygon is reduced to computing the intersection between the edge and the cube faces because the edges of the polygon are included in the cube by construction (see, *e.g.*, Fig. 2(a)).

Let us denote by $T_Y(x)$ the transformation of the chrominance values $x = (U, V)$ to the RGB space with the luminance value Y. From the expression of the standard transformation from RGB to YUV, we have $T_Y(x) = Y(1,1,1)^t + L(U, V)$ with L a linear function. We have the following equalities:

$$
\begin{aligned}
T_Y(x + \rho y) &= Y(1,1,1)^t + L(x + \rho y) \\
&= Y(1,1,1)^t + L(x) + \rho L(y) \\
&= T_Y(x) + \rho T_Y(y) - \rho Y(1,1,1)^t.
\end{aligned}
\tag{16}
$$

We want to compute $\tilde{\rho}$ such that $T_Y(x + \tilde{\rho}y)$ is at the boundary of the RGB cube. To this aim, we compute the 6 different values $\tilde{\rho}_c^v$ with $c \in \{R, G, B\}$ and $v \in \{0, 255\}$ corresponding to the cases where the 3 coordinates of $T_Y(x + \tilde{\rho}y)$ are equals to 0 or 255. For instance, if the first coordinate R of $T_Y(x + \tilde{\rho}y)$ is equal to 255, we have:

$$
T_Y(x + \tilde{\rho}_R^{255}y)_R = 255
\tag{17}
$$

$$
T_Y(x)_1 + \tilde{\rho}_R^{255} T_Y(y)_R - \tilde{\rho}_R^{255} Y = 255.
\tag{18}
$$

so that

$$
\tilde{\rho}_R^{255} = \frac{255 - T_Y(x)_R}{T_Y(y)_R - Y}.
\tag{19}
$$

For each of the six values $\tilde{\rho}_c^v$ computed as in Eq. (19), we can compute $t_c^v = \frac{\tilde{\rho}_c^v}{\rho}$. The values t_c^v that are between 0 and 1 correspond to an intersection of the segment $[x, x + \rho y]$ with the boundaries of $\mathcal{R}$. We finally take $t^* = \min_{t_c^v \in [0;1]} t_c^v$ and the result of Eq. (14) is given by $t^*\rho$.

5   Numerical Results

In our experimental setting, the luminance channels of the initial image is preserved, while the chrominance channels are perturbed by a Gaussian noise with a standard deviation equal to 30. Let us notice that the values of the noisy images can be out of range, but our method is able to consider such issue.

Figure 5 shows comparisons between the minimization of the total variation on UV chrominance channels, the original chrominance denoising model (1) and its debiasing version (15). In general, the naive total variation on UV channels not preserve colors. It produces halo effects as well as a loss of colourfulness. In the first line, we can see that the red cloth of the Madonna is drab with Model (1), in comparison with the original image due to the bias, whereas the debiasing approach preserves the red color. Further results are presented in Fig. 6. In the first line, we can see in the center of the image that the flower becomes drab with model (1), whereas it recovers its colourfulness with the debiasing one (see zoom on the second line). In the Sphinx image, the color of the sand recovers its colourfulness after debiasing.

The PSNR is a quantitative indicator to measure the quality of a denoised image. In Table 1, the PSNR of the different techniques proposed in Figs. 5 and 6 are presented. These values show that the debiasing technique proposed in Sect. 4.3 produces images with a better PSNR, which quantitatively confirms the visual impression of Fig. 5.

Table 1. Comparison of PSNR for techniques presented in Figs. 5 and 6. This quantitative comparison highlights the quality of the results provided by the debiasing.

Image	TV on UV	Model (1)	Model (15)
Madonna	30.45	32.37	**33.08**
Ferrari	22.50	26.05	**27.23**
Purple flowers	20.88	23.29	**24.81**
Sphinx	26.36	32.95	**33.51**

6   Conclusion

In this paper, we presented a model to denoise the chrominance channels of color images. The original model being biased, it produces a loss of colourfulness. By extending the CLEAR method of [6] to constrained denoising problems, we have proposed a new debiasing strategy. The denoising results obtained with our framework are promising, which is a relevant preliminary step for the application to JPEG restoration.

A   Computation of the Chrominance Range

In this section, we describe the numerical computation of the vertices of the polygon defined as the chrominance values of all the RGB colors which have a particular luminance values (see, *e.g.*, Fig. 2). To this aim, the RGB coordinates of the vertices are computed and then cast into the YUV space. In the following the Algorithm which computes these RGB coordinates is described.

The computation of the vertices in the RGB space consists in intersecting the edges of the RGB cube ($[0, 255]^3$) and the affine plane of the RGB colors with a particular luminance. The equation of this plane, for a luminance value equal to y is given by:

$$\mathcal{P}_Y(y) := \{(R, G, B) \text{ such that } 0.299R + 0.587G + 0.114B = y\}. \qquad (20)$$

The cube having 12 edges, the resolution of 12 systems of linear equations in 3 dimension is required. This system is composed of Eq. (20) and two additional equations describing the line in which the considered edge is included.

As an example, let us consider the intersection of the plane $\mathcal{P}_Y(150)$ of colors with a luminance equal to 150 with the edge $[(0, 0, 255), (0, 255, 255)]$. This edge is included into the line described by the equations $R = 0$ and $B = 255$. The intersection of this line with the plane $\mathcal{P}_Y(150)$ implies the resolution of:

$$\begin{cases} 150 = 0.299R + 0.587G + 0.114B \\ R = 0 \\ B = 255. \end{cases} \qquad (21)$$

The solution of this system is given by $RGB = (0, 206, 255)$. This vector being into the cube $[0, 255]^3$, the solution is a vertex of the desired polygon. When the vector is not into the cube, the solution is not a vertex of the polygon.

After computation of the vertices, the algorithm has to sort them in an order such that they represent a convex polygon. For instance, in $\mathbb{R}^2$, $(0, 0), (1, 0), (1, 1)$, $(0, 1), (0, 0)$ is a convex polygon, whereas $(0, 0), (1, 0), (0, 1), (1, 1), (0, 0)$ is not.

To tackle this issue, we remark that, given a particular vertex, the next one is a vertex with a common coordinate in RGB space because the polygon is defined as the intersection of a plane and the RGB cube. To avoid problem of equal points, we consider as the next vertex, the one with the most common coordinates. Thus, given $P_1, \ldots, P_n$ n vertices onto the edges of the RGB cube, Algorithm 4 sorts them to produce a convex polygon.

Algorithm 4. Algorithm sorting points in the order of a convex polygon.

Require: $P_1, \ldots, P_n$ n vertices onto the edges of the RGB cube.
1: **for** $i = 1 : n - 1$ **do**
2:     **for** $k = i + 1 : n$ **do**
3:         $\delta_k \leftarrow$ number of common coordinates between P_i and P_k.
4:     **end for**
5:     $k^* \leftarrow \text{argmax}_k \, \delta_k$
6:     Exchange P_{i+1} and P_{k^*}
7: **end for**

References

1. Bredies, K.: Recovering piecewise smooth multichannel images by minimization of convex functionals with total generalized variation penalty. SFB Report 6 (2012)
2. Bredies, K., Holler, M.: Artifact-free decompression and zooming of JPEG compressed images with total generalized variation. In: Csurka, G., Kraus, M., Laramee, R.S., Richard, P., Braz, J. (eds.) VISIGRAPP 2012. CCIS, vol. 359, pp. 242–258. Springer, Heidelberg (2013). https://doi.org/10.1007/978-3-642-38241-3_16
3. Brinkmann, E.M., Burger, M., Rasch, J., Sutour, C.: Bias-reduction in variational regularization. arXiv preprint arXiv:1606.05113 (2016)
4. Chambolle, A., Pock, T.: A first-order primal-dual algorithm for convex problems with applications to imaging. J. Math. Imag. Vis. **40**(1), 120–145 (2011)
5. Chien, C.L., Tseng, D.C.: Color image enhancement with exact HSI color model. Int. J. Innov. Comput. Inf. Control **7**(12), 6691–6710 (2011)
6. Deledalle, C.A., Papadakis, N., Salmon, J., Vaiter, S.: CLEAR: covariant LEAst-square re-fitting with applications to image restoration. SIAM J. Imag. Sci. **10**, 243–284 (2017)
7. Fitschen, J.H., Nikolova, M., Pierre, F., Steidl, G.: A variational model for color assignment. In: Aujol, J.-F., Nikolova, M., Papadakis, N. (eds.) SSVM 2015. LNCS, vol. 9087, pp. 437–448. Springer, Cham (2015). https://doi.org/10.1007/978-3-319-18461-6_35
8. Goldluecke, B., Cremers, D.: An approach to vectorial total variation based on geometric measure theory. In: IEEE Conference on Computer Vision and Pattern Recognition, pp. 327–333 (2010)
9. Levin, A., Lischinski, D., Weiss, Y.: Colorization using optimization. ACM Trans. Graph. **23**(3), 689–694 (2004)
10. Nikolova, M., Steidl, G.: Fast hue and range preserving histogram specification: theory and new algorithms for color image enhancement. IEEE Trans. Image Process. **23**(9), 4087–4100 (2014)
11. Peter, P., Kaufhold, L., Weickert, J.: Turning diffusion-based image colorization into efficient color compression. IEEE Trans. Image Process. **26**, 860–869 (2016)
12. Pierre, F., Aujol, J.-F., Bugeau, A., Ta, V.-T.: Luminance-hue specification in the RGB space. In: Aujol, J.-F., Nikolova, M., Papadakis, N. (eds.) SSVM 2015. LNCS, vol. 9087, pp. 413–424. Springer, Cham (2015). https://doi.org/10.1007/978-3-319-18461-6_33
13. Pierre, F., Aujol, J.F., Bugeau, A., Papadakis, N., Ta, V.T.: Luminance-chrominance model for image colorization. SIAM J. Imag. Sci. **8**(1), 536–563 (2015)
14. Rudin, L.I., Osher, S., Fatemi, E.: Nonlinear total variation based noise removal algorithms. Phys. D: Nonlinear Phenom. **60**(1), 259–268 (1992)
15. Williams, A., Barrus, S., Morley, R.K., Shirley, P.: An efficient and robust ray-box intersection algorithm. In: ACM SIGGRAPH 2005 Courses, p. 9 (2005)
16. Xiao, X., Ma, L.: Color transfer in correlated color space. In: ACM VRCIA 2006, pp. 305–309 (2006)
17. Yatziv, L., Sapiro, G.: Fast image and video colorization using chrominance blending. IEEE Trans. Image Process. **15**(5), 1120–1129 (2006)

Optimizing Wavelet Bases for Sparser Representations

Thomas Grandits$^{(\boxtimes)}$ and Thomas Pock

Institute of Computer Graphics and Vision, TU Graz, Graz, Austria
thomas.grandits@icg.tugraz.at, pock@icg.tugraz.at

Abstract. Optimization in the wavelet domain has been a very prominent research topic both for denoising, as well as compression, reflected in its use in the JPEG-2000 standard. Its performance depends to a great extent on the wavelet ψ itself, represented in the form of a filter in the case of the discrete wavelet transform. While other works solely optimize the coefficients in the wavelet domain, we will use a combined approach, optimizing the wavelet ψ and the coefficients simultaneously in order to adapt both to a given image, resulting in a better reconstruction of an image from less coefficients. We will use several orthonormal wavelet bases as a starting point, but we will also demonstrate that we can create wavelets from white Gaussian noise with our approach, which are in some cases even better in terms of performance. Experiments will be conducted on several images, demonstrating how the optimization algorithm adapts to textured, as well as more homogeneous images.

1 Introduction

The possibility of representing functions in an orthornomal basis is an idea that was already present long before the term wavelet was used. In 1909, Haar created [9] what is now often considered the oldest known wavelet. Several decades later, more complex orthogonal wavelets were discovered [13], but the discovery of an efficient scheme, called fast wavelet transform, by Mallat in [10] was arguably a milestone for the popularity of wavelets.

The wavelet transform is often seen as a frequency analysis at different scales of the signal, similar to the Fourier transform, but contrary to the Fourier transform, the wavelet transform does not lose the locality information in the process, while also begin computationally efficient compared to the windowed Fourier transform.

A lot of attention has been put into finding wavelets with desirable properties like orthogonality for easier computation and implementation, or vanishing moments, able to represent polynomials efficiently, with the main purpose of a sparser signal representation. Wavelets such as the well-known Daubechies wavelets from [8] and Coiflets from [7] are good examples of orthogonal wavelets for Multi-Resolution-Analysis (MRA), since they fulfill these former mentioned properties.

© Springer International Publishing AG, part of Springer Nature 2018
M. Pelillo and E. Hancock (Eds.): EMMCVPR 2017, LNCS 10746, pp. 249–262, 2018.
https://doi.org/10.1007/978-3-319-78199-0_17

The popularity of wavelets is reflected by the use of the bi-orthogonal Cohen-Daubechies-Feauveau wavelets [6] in the JPEG-2000 standard prior to the coding step. Wavelets are also common tools for denoising [4], dictionary learning [14], [18] and texture synthesis [16]. A lot of interest stems from the fact that wavelets are translation-invariant and the signal in question is analyzed at dyadic scales, similar to Laplace pyramids [1]. In the case of images, the wavelet also has some degree of rotation-invariance, but unfortunately, we are restricted to only few angles of analysis when applying standard wavelet analysis.

In [22], an effort was made to adapt/steer the angle of analysis of a wavelet through Riesz transform, while keeping the orthogonality in the process. They were chosen after performing a Principal Component Analysis on the pixels of an image, revealing the angles with the most intensity changes. Others put effort into estimating local geometries in [12], through segmentation of the image and subsequent estimation of the local flow of the geometry. The adapted wavelets are called Bandlets.

While these wavelets have proven to be efficient, they do not adapt to a given signal or image and rely on the fact that the analyzed signals expose the encoded structures. With the advent of statistical methods and machine learning, we want to improve the performance of a wavelet by adapting it to a given signal, allowing for even sparser representations in the wavelet domain. This will be achieved by defining the discrete constraints of a wavelet and formulating an alternating optimization problem in the wavelet coefficients and the discrete wavelet filter.

2   Related Work

The choice of wavelet ψ significantly influences the performance of the MRA, but the choice of the best wavelet basis is usually non-trivial.

Some previous works [5,20] already tried to find an optimal wavelet basis for signals, but their focus was restricted on mapping short signal sequences to a wavelet and only considered the sparse approximation implicitly, by matching the signal as best as possible. While these approach can be very efficient for periodic signals, it is unclear if it can efficiently deal with irregular signals like in some image and if it can easily be extended for the 2-dimensional case.

Swelden proposed in [19] a lifting scheme that is an efficient implementation as well as a designing tool for biorthogonal wavelets. However, the lifting scheme needs a pair of biorthogonal functions to begin with.

In [15], denoising in the wavelet domain is performed by defining a wavelet packet basis. This basis is also defined through its filters, which are optimally selected, but again these filters are assumed to be known a-priori. We will see that standard wavelets offer a very good initial performance in terms of sparsity and reconstruction quality already, but we can also show that they can be improved even more w.r.t. Peak Signal-to-Noise Ratio (PSNR) when fitting them to particular images. The benefits of a sparser representation are apparent, especially in compression, where fewer coefficients will help to decrease entropy.

In summary, our contributions are

- A combined approach for optimizing the wavelet, as well as the coefficients of the MRA
- Creation of new and sometimes better wavelets without hand crafting. They are purely generated from white noise
- Demonstration of the applicability to denoising

3 Problem Formulation and Algorithm

3.1 Preliminaries

In this paper we will solely focus on the orthogonal discrete wavelet transform, since it eases the solution of our optimization problem, as we will see later on.

Considering a wavelet-filter column vector $\mathbf{h} = (h_1, \ldots, h_K)^T$, the scaling-function ϕ and wavelet function ψ are then recursively defined as:

$$\phi(x) = \sum_{k=1}^{K} h_k \sqrt{2}\phi(2x - k) \tag{1}$$

$$\psi(x) = \sum_{k=1}^{K} g_k \sqrt{2}\phi(2x - k) \tag{2}$$

where $g_k = (-1)^k h_{K-k-1}$.

In order for ψ to be a wavelet, it must fulfill the admissibility criterion:

$$2\pi \int_0^\infty \frac{|\hat{\psi}(\omega)|^2}{\omega}\, d\omega < \infty \tag{3}$$

where $\hat{\psi}$ is the Fourier transform of ψ. From this it follows:

$$\hat{\psi}(0) = \int_{\mathbb{R}} \psi(x)\, dx = 0 \tag{4}$$

For practical reasons, we also want the transform to be normalized, therefore:

$$\int_{\mathbb{R}} \phi(x)\, dx = 1 \tag{5}$$

Since we require the transform to be orthonormal to integer translates, it follows that:

$$\int_{\mathbb{R}} \phi(x)\phi(x - n)\, dx = \delta_{0,n} \quad \forall n \in \mathbb{Z} \tag{6}$$

where $\delta_{.,.}$ is the Kronecker delta. From these continuous constraints, we can deduce our set C of feasible discrete wavelet filters $\mathbf{h}$:

$$C = \left\{ \mathbf{h} \in \mathbb{R}^K : \begin{array}{l} \sum_{k=1}^{K} g_k(\mathbf{h}) = 0, \ \sum_{k=1}^{K} h_k = \sqrt{2}, \\[2mm] \sum_{k=1}^{K} h_k\, h_{k+2n} = \delta_{0,2n} \quad \forall n \in \mathbb{Z} \end{array} \right\} \tag{7}$$

where the orthogonality constraint in the bottom row is actually a multitude of single constraints, leaving us with a high number of constraints for increasing filter lengths K. We will need these constraints later, to ensure a valid wavelet basis after each iteration of our optimization algorithm.

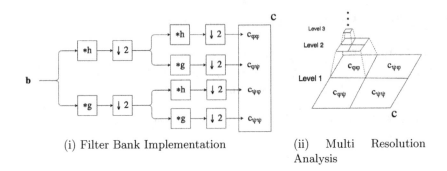

(i) Filter Bank Implementation   (ii)   Multi   Resolution Analysis

Fig. 1. Wavelet decomposition

For practical reasons, the fast wavelet transform is often depicted and implemented as a filterbank, like shown in Fig. 1, displaying the alternating sequence of filtering and downsampling steps. With all of the above conditions on $\mathbf{h}$, we can define the fast 2D-wavelet transform with a single level of decomposition as:

$$W(\mathbf{h})\mathbf{b} = \mathbf{c} \tag{8}$$

where $\mathbf{b} \in \mathbb{R}^N$ is the vectorized image and $\mathbf{c} \in \mathbb{R}^N$ are the resulting wavelet coefficients. $W(\mathbf{h}) \in \mathbb{R}^{N \times N}$ is a non-linear operator depending on the wavelet filter $\mathbf{h}$, containing all steps of a wavelet transform. For more details, we refer to [11].

The transform is unitary, hence $W^T W = I$, therefore:

$$W^T(\mathbf{h})\mathbf{c} = \mathbf{b} \tag{9}$$

3.2   Sparse Approximations

A classical model for computing sparse approximations is the least absolute shrinkage and selection operator (Lasso) model [21]:

$$\min_{\mathbf{c}} f(\mathbf{c}) = \lambda \|\mathbf{c}\|_1 + \frac{1}{2} \|A\mathbf{c} - \mathbf{b}\|_2^2 \tag{10}$$

where λ is the regularization constant, adjusting the density of the coefficients.

In case A is an inverse wavelet transform, i.e. $A = W^T(\mathbf{h})$, the solution is the particularly simple orthogonal Lasso model:

$$\mathbf{c} = \mathcal{T}_\lambda(\mathbf{c} - W(W^T\mathbf{c} - \mathbf{b})) = \mathcal{T}_\lambda(W\mathbf{b}) \tag{11}$$

where $\mathcal{T}$ is the element-wise shrinkage operator:

$$\mathcal{T}_\lambda(\mathbf{x})_i = \max(0, |x_i| - \lambda)\mathrm{sgn}(x_i) \tag{12}$$

This solution on the wavelet coefficients is the well-known wavelet shrinkage.

The main goal of this paper is to investigate the benefits of additionally optimizing for the wavelet basis $\mathbf{h}$. In this case the Lasso model we need to solve becomes:

$$\min_{\mathbf{c,h}} f(\mathbf{c,h}) = \lambda \|\mathbf{c}\|_1 + \frac{1}{2} \|W^T(\mathbf{h})\mathbf{c} - \mathbf{b}\|_2^2 \quad \text{s.t.: } \mathbf{h} \in C \tag{13}$$

This problem is considerably more complicated to minimize because the objective function is non-convex and the wavelet basis $\mathbf{h}$ is constrained to the non-convex set of wavelet filters C.

3.3   Proposed Algorithm

In order to optimize the problem (13), we use the inertial Proximal Alternating Linearized Minimization (iPALM) algorithm presented in [17], which performs alternating minimization, both in the coefficients $\mathbf{c}$ and the wavelet basis $\mathbf{h}$.

The main steps of the algorithm are outlined in Algorithm 1: We start with an arbitrary filter $\mathbf{h}_0$, that is projected onto a wavelet. This allows us to initialize a wavelet from arbitrary starting points like white Gaussian noise, as we will see in Sect. 4.

Afterwards, we iteratively compute the wavelet shrinkage $\mathbf{c}_{k+1} = \mathcal{T}_\lambda(W(\mathbf{h}_k)\mathbf{b})$, followed by an inertial step resulting in $\mathbf{y}_k$, with the author's suggested inertia of $\alpha = \frac{k-1}{k+2}$. From $\mathbf{y}_k$, we compute the gradient $\nabla_\mathbf{h} f$ and take a gradient descent step, with the step size according to the Lipschitz constant L. The resulting new point will be projected back on the wavelet space C.

We used a Lipschitz backtracking scheme like in [2] to compute the Lipschitz constant L, used as the step size for the gradient descent in Algorithm 1.

The computation of the gradient $\nabla_\mathbf{h} f$ and the projection proj_C will be outlined in more detail in Sects. 3.4 and 3.5 respectively.

Algorithm 1. iPALM

Input   : Initial filter $\mathbf{h}_0$
Output: Optimized wavelet filter $\mathbf{h}_N$
$\mathbf{h}_0 = \mathrm{proj}_C(\mathbf{h}_0)$ (Eq. (17))
while *not converged* **do**
  $\quad \mathbf{c}_{k+1} = \mathcal{T}_\lambda(W(\mathbf{h}_k)\mathbf{b})$ (Apply soft-shrinkage from Eq. (11))
  $\quad \mathbf{y}_k = \mathbf{h}_k + \alpha\,(\mathbf{h}_k - \mathbf{h}_{k-1})$
  $\quad \mathbf{h}_{k+1} \in \mathrm{proj}_C\left(\mathbf{y}_k - \frac{1}{L}\nabla_\mathbf{h} f(\mathbf{y}_k)\right)$ ($L \dots$ Lipschitz constant)
end

3.4 Computing the Gradient $\nabla_{\mathbf{h}} f$

To compute $\nabla_{\mathbf{h}} f$, we differentiate (13) w.r.t. h_i:

$$\frac{\partial}{\partial h_i} f = \left(\frac{\partial}{\partial h_i} W^T(\mathbf{h}) \mathbf{c} \right)^T \left(W^T(\mathbf{h}) \mathbf{c} - \mathbf{b} \right) \tag{14}$$

The derivation $\frac{\partial W_i}{\partial h_i}$ is simply an inverse wavelet transform with $h_j = \delta_{i,j}$, where δ is the Kronecker delta. The vector $\mathbf{h}$ is consequently the concatenation of all derived results.

To account for multiple scales, we need to redefine our problem (13) to:

$$\min_{\mathbf{c},\mathbf{h}} f_n(\mathbf{c}, \mathbf{h}) = \lambda \|\mathbf{c}\|_1 + \frac{1}{2} \left\| \left(\prod_{i=1}^{n} W_i(\mathbf{h}) \right)^T \mathbf{c} - \mathbf{b} \right\|_2^2 \quad \text{s.t.:} \mathbf{h} \in C \tag{15}$$

The optimization in $\mathbf{c}$ could be adapted to have different λ for each level, but for the sake of simplicity, we used a constant λ across all levels. Still, we need to account for multiple layers in our derivation $\nabla_{\mathbf{h}} f$ in (14):

$$\frac{\partial}{\partial h_i} f_n = \left(\frac{\partial}{\partial h_i} \left(\prod_{j=1}^{n} W_j(\mathbf{h}) \right)^T \mathbf{c} \right)^T \underbrace{\left(\left(\prod_{j=1}^{n} W_j(\mathbf{h}) \right)^T (\mathbf{h}) \mathbf{c} - \mathbf{b} \right)}_{\mathbf{e}}$$

$$= \left(\sum_{k=1}^{n} \left(\left(\prod_{j=1}^{k-1} W_j(\mathbf{h}) \right) \frac{\partial W_k(\mathbf{h})}{\partial h_i} \left(\prod_{j=k+1}^{n} W_j(\mathbf{h}) \right) \right)^T \mathbf{c} \right)^T \mathbf{e} \tag{16}$$

Our resulting gradient in Eq. (16) states that one has to derive over each layer separately and multiply the resulting sum with the approximation error vector $\mathbf{e}$. The partial derivatives need to be concatenated to receive $\nabla_{\mathbf{h}} f$.

3.5 Projection onto the Wavelet Space C

To ensure that $\mathbf{h}$ is a valid wavelet filter, we define a projection onto the set of wavelet filters C

$$\text{proj}_C(\tilde{\mathbf{h}}) = \operatorname*{argmin}_{\mathbf{h} \in C} \frac{1}{2} \left\| \tilde{\mathbf{h}} - \mathbf{h} \right\|_2^2 \tag{17}$$

To compute the projection, we rewrite Eq. (17) in the form of a Lagrangian function $\mathcal{L}$:

$$\mathcal{L}(\mathbf{h}, \Lambda) = \frac{1}{2} \left\| \tilde{\mathbf{h}} - \mathbf{h} \right\|_2^2 + \Lambda_1 \left(\mathbf{h}^T \mathbf{1} - \sqrt{2} \right) + \Lambda_2 \left(g(\mathbf{h})^T \mathbf{1} \right) + \sum_i \Lambda_i \left(\mathbf{h}^T O_i \mathbf{h} \right) \tag{18}$$

where O_i are orthogonal matrices, which describe the orthogonality constraints of the set C. Please note that we used the upper-case letter Λ for the Lagrangian multipliers, since we use λ solely for the regularization parameter in (13). $\nabla_{\mathbf{h},\Lambda}\mathcal{L}(\mathbf{h},\Lambda)$ is given as:

$$\nabla_{\mathbf{h},\Lambda}\mathcal{L}(\mathbf{h},\Lambda) = \begin{pmatrix} \dfrac{\left(\mathbf{h}-\tilde{\mathbf{h}}\right) + \Lambda_1 + \Lambda_2\nabla_h g(\mathbf{h})^T\mathbf{1} + 2\sum_i \Lambda_i O_i\mathbf{h}}{} \\ \mathbf{h}^T\mathbf{1} - \sqrt{2} \\ g(\mathbf{h})^T\mathbf{1} \\ \mathbf{h}^T O_i\mathbf{h} \end{pmatrix} \qquad (19)$$

where the vector above the line is $\nabla_{\mathbf{h}}f$ and the derivations below the line are the constraints, where the last constraint is again a vector.

If we define $\mathbf{r}(\mathbf{h},\Lambda) = \nabla_{\mathbf{h},\Lambda}\mathcal{L}(\mathbf{h},\Lambda)$, the optimality condition states that a choice of $\mathbf{h},\Lambda$, which fulfill

$$\frac{1}{2}\|\mathbf{r}(\mathbf{h},\Lambda)\|_2^2 = 0 \qquad (20)$$

is an optimal point. We solved this least square system of equations using Gauss-Newton's algorithm. The initial choice of Λ was random, but was kept for subsequent runs of the projection. If the constraints were not met, the Lagrangian multiplier were again randomly initialized.

Since the problem in (17) is a non-convex quadratic problem and relies on our initial choice of Lagrangian multipliers, we are only able to find local minima. In practice however, we are satisfied with an $\mathbf{h}$ that is close enough to our original $\tilde{\mathbf{h}}$ and which complies with all our constraints.

4   Experiments

We conducted two experiments, where the first experiment demonstrates how our algorithm improves standard wavelet shrinkage and the second is a short demonstration of the improved denoising capabilities compared to wavelet shrinkage. The advantage of the optimization over standard wavelet shrinkage comes purely from the added optimization in $\mathbf{h}$. We will see that this approach mitigates the effect of the initial choice of an wavelet $\mathbf{h}_0$ on the performance and is even able to generate wavelets from white Gaussian Noise that outperform initialization from hand-crafted wavelets.

4.1   Image Sparsity

We tested our optimization algorithm for 5 different initializations wavelet initializations $\mathbf{h}_0$: Haar, Daubechies Orthonormal Wavelets (db), Coiflets (coif), Least Asymmetric Daubechies Orthonormal Wavelets (sym) and initialization from white Gaussian noise (randn). For comparison, we use the PSNR of the reconstructed images, as well as the density of the coefficients, which we define as $\frac{\|c\|_0}{N}$, where N is the number of coefficients/pixels. We think this measure is more reliable for comparison, but we did not use it for optimization, since it is a non-convex problem. We optimized for 500 epochs for all combinations of the following configurations:

- Filter Length of 8 or 32
- Decomposition Layers of 4 or 8
- Lena, Barbara Image, or Set of McMaster Images (upsampled to 512×512) from [23]
- λ from 0.01 to 0.99 in 10 Exponential Steps
- 10 different Initialization from white Gaussian Noise

and compared the initial to the final wavelet. The images were normalized: $\mathbf{b} \in [0, 1]$.

Table 1. Optimization results on the Lena image for different initializations. Shown are the PSNR after the optimization for the corresponding Density. The values in the parentheses mark the absolute improvement over the initialization. All values in dB. Best values for given density are marked in bold.

Initialization	Densities $(\|c\|_0 / N)$		
	1/20	1/30	1/55
haar	31.64 (+2.28)	29.35 (+1.98)	27.14 (+1.74)
db	32.03 (+1.10)	29.71 (+1.06)	27.42 (+0.89)
coif	**32.11** (+0.07)	29.76 (+0.09)	27.55 (+0.13)
sym	**32.11** (+0.07)	**29.86** (+0.24)	**27.61** (+0.28)
randn	32.05 (+15.38)	29.73 (+7.45)	**27.61** (+11.35)

Table 2. The same evaluation as in Table 1, but with different Densities and on the Barbara Image

Initialization	Densities $(\|c\|_0 / N)$		
	1/9	1/14	1/24
haar	28.49 (+3.01)	26.25 (+2.22)	24.15 (+1.41)
db	29.17 (+0.81)	26.57 (+0.51)	24.24 (+0.25)
coif	29.36 (+0.42)	26.73 (+0.23)	24.34 (+0.07)
sym	**29.37** (+0.26)	**26.74** (+0.17)	24.39 (+0.10)
randn	29.27 (+11.74)	**26.74** (+9.34)	**24.41** (+8.66)

The solutions were obtained by applying Algorithm 1, which mostly leads to solutions which were superior both in terms of density and PSNR. To provide a more obvious PSNR comparison, we adapted each value of λ, for each resulting

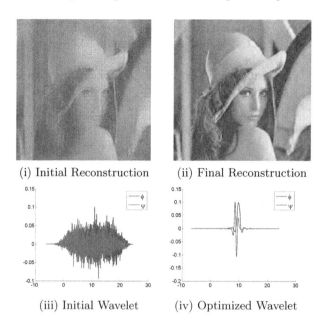

(i) Initial Reconstruction   (ii) Final Reconstruction

(iii) Initial Wavelet   (iv) Optimized Wavelet

Fig. 2. Optimization of a randomly initialized Wavelet on the Lena image for a density of $\frac{1}{55}$. The wavelet becomes much less noisy and better at describing structures of the image. The initial and final PSNR are 16.26 and 27.61 dB respectively

wavelet to offer the same density before and after optimization. This helps us to compare wavelet shrinkage to our Algorithm.

In Tables 1 and 2, one can see how much PSNR we can gain when optimizing with our approach for different initializations. We can see that we can find competitive wavelets from white noise, especially for low densities, which are even sometimes better compared to starting with a well-known standard wavelet with very good initial performance. One such evolution of a randomly initialized wavelet can be seen in Fig. 2. The initialization of the optimization is very important, as we will end up in very similar minima for similar initializations. As described, for the initialization from white noise we used 10 different random initializations for each testset and analzyed the difference between the different runs. As we can see from Fig. 3, the optimization may find very different wavelets for different initializations: The left wavelet is better at representing the textures of the image, while the right wavelet is better for edges.

The standard wavelets usually result in small varied version of themselves depending mostly on the picture and to a smaller extent on λ. In Fig. 4, we showed how such seemingly small variations can improve the reconstruction quality a lot: The Coiflet is adapted to better express the textures of the Barbara image through additional sidelobes of the wavelets. This change increases the PSNR by 0.46 dB.

Due to the space restrictions, we did not show the results for optimization on the set of images, but the values lay in the middle field between the results of the Lena and Barbara images in Tables 1 and 2. The PSNR on all images could be improved, but naturally the effect was weaker for images with heterogeneous structures.

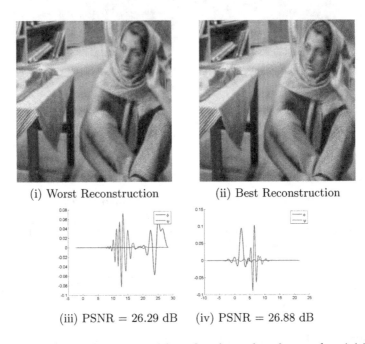

(i) Worst Reconstruction   (ii) Best Reconstruction

(iii) PSNR = 26.29 dB   (iv) PSNR = 26.88 dB

Fig. 3. Comparison of the worst and best found wavelet after random initialization wavelet for the barbara image and $\lambda = 0.128$, out of 10 runs

4.2   Image Denoising

This experiment demonstrates the generality of our approach and does not aim at offering competitive or state-of-the-art method for denoising.

As our optimization is trying to find sparser representations in the wavelet domain, we also hope to see a noise-mitigating effect. As previously discussed, wavelet shrinkage is a well known method in image denoising, since the important details of pictures are usually contained in a small subset of the coefficients with a large magnitude.

For this problem, we now define redefine our image as a sum of the original image $\tilde{\mathbf{b}}$ and white Gaussian noise ϵ:

$$\mathbf{b} = \tilde{\mathbf{b}} + \epsilon \tag{21}$$

Indeed, this part of our proposed variational method was already studied in [3] for denoising and it was shown that the soft-shrinkage on the wavelet coefficients $\mathbf{c}$

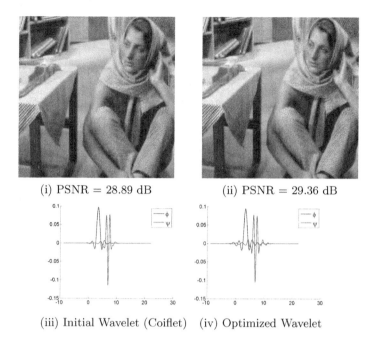

(i) PSNR = 28.89 dB   (ii) PSNR = 29.36 dB

(iii) Initial Wavelet (Coiflet) (iv) Optimized Wavelet

Fig. 4. Optimization of the Coiflet on the Barbara image for a density of $\frac{1}{9}$. The wavelet gains the ability to better express the texture of the image

(wavelet shrinkage) leads to near-optimal solutions for denoising (assuming gaussian-noise), depending on the regularization parameter λ.

We chose 8 decomposition layers and the filter length $K = 32$. Intuitively, we expect that the wavelet will adapt to the given image structures but has too little complexity to overfit on the noise. The optimization is initialized and later compared to the least asymmetric Daubechies wavelet.

The comparison of wavelet shrinkage with the least asymmetric Daubechies and our optimization for different λ can be seen in Figs. 5 and 6. Wavelet shrinkage offers good overall results already, but for the optimal choice of λ our approach can improve the result by ≈ 0.15 dB.

4.3   Runtime

The fast wavelet transform lives up to its name with a runtime complexity of $\mathcal{O}(n \log n)$ for an image with n pixels and is easily implementable on modern graphics hardware. Similarly, the wavelet shrinkage only requires $\mathcal{O}(n)$ time. The projection of an arbitrary filter on the wavelet space C however is not that easily analyzed, since we need to wait for convergence of the Gauss-Newton algorithm for a non-convex problem. Practically speaking, an iteration currently takes about 200ms on a 512×512 image with an Intel i7-5820K processor and a GTX 960 graphics card, where a more efficient projection scheme could speed up the algorithm by a great margin.

(i) Noisy Image   (ii) Wavelet Shrinkage   (iii) Our Optimization

Fig. 5. Comparison of wavelet shrinkage and our optimization for denoising

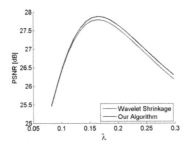

Fig. 6. Performance of denoising for different λ

5   Conclusion and Future Work

In this paper we introduced a novel approach to learn wavelets from images instead of mathematically modelling them, as well as creating sparse representation of said images in the same optimization.

The optimization proved to be an non-convex and non-linear optimization problem but still, we were able to show that our algorithm is capable of generating wavelets that show improved performance over standard wavelets. Although our optimization and experiments were centered around sparser representations mainly used in compression, denoising was also possible without any adaption. Many improvements of wavelet shrinkage have been put forward, like adaptive thresholding in [4] that could be integrated into the optimization, since our optimization in **c** is only using the standard wavelet shrinkage and not the improved versions. Additionally, we are also interested in 'steering' our wavelets to an angle that can capture local geometry in images. This approach was proposed in [22] and may help at adapting to image structures that are not aligned with the angle of analysis. An extension to bi-orthogonal wavelets could also prove very beneficial, since it decreases the constraints on our optimization and may help us find even better wavelets.

We were able to show that optimization of a wavelet to an image is an efficient approach that can improve performance of a standard wavelet, or even bring new wavelets forward from white noise, which have a non-uniform shape and are different from hand-crafted wavelets. We think with this approach, wavelets could be found that are superior in specific domains and which may be used for compression and compressed sensing. All in all, we see this work as a starting point for future research in the adaptability of wavelets to data.

References

1. Adelson, E.H., Anderson, C.H., Bergen, J.R., Burt, P.J., Ogden, J.M.: Pyramid methods in image processing. RCA Eng. **29**(6), 33–41 (1984)
2. Beck, A., Teboulle, M.: A fast iterative shrinkage-thresholding algorithm for linear inverse problems. SIAM J. Imaging Sci. **2**(1), 183–202 (2009)
3. Chambolle, A., Vore, R.A.D., Lee, N.Y., Lucier, B.J.: Nonlinear wavelet image processing: variational problems, compression, and noise removal through wavelet shrinkage. IEEE Trans. Image Process. **7**(3), 319–335 (1998)
4. Chang, S.G., Yu, B., Vetterli, M.: Adaptive wavelet thresholding for image denoising and compression. IEEE Trans. Image Process. **9**(9), 1532–1546 (2000)
5. Chapa, J., Rao, R.: Algorithms for designing wavelets to match a specified signal. Trans. Sig. Proc. **48**(12), 3395–3406 (2000)
6. Cohen, A., Daubechies, I., Feauveau, J.C.: Biorthogonal bases of compactly supported wavelets. Commun. Pure Appl. Math. **45**(5), 485–560 (1992)
7. Daubechies, I.: Ten Lectures on Wavelets. CBMS-NSF Regional Conference Series in Applied Mathematics. Society for Industrial and Applied Mathematics, Philadelphia (1992). https://doi.org/10.1137/1.9781611970104
8. Daubechies, I.: Orthonormal bases of compactly supported wavelets. Commun. Pure Appl. Math. **41**(7), 909–996 (1988)
9. Haar, A.: Zur Theorie der orthogonalen Funktionensysteme. Math. Ann. **69**(3), 331–371 (1910)
10. Mallat, S.G.: A theory for multiresolution signal decomposition: the wavelet representation. IEEE Trans. Pattern Anal. Mach. Intell. **11**(7), 674–693 (1989)
11. Mallat, S.: A Wavelet Tour of Signal Processing. The Sparse Way, 3rd edn. Academic Press, Cambridge (2008)
12. Mallat, S., Peyr, G.: A review of Bandlet methods for geometrical image representation. Num. Algorithms **44**(3), 205–234 (2007)
13. Meyer, Y.: Principe d'incertitude, bases hilbertiennes et algbres d'oprateurs. Sminaire Bourbaki, vol. 28, pp. 209–223 (1985)
14. Ophir, B., Lustig, M., Elad, M.: Multi-scale dictionary learning using wavelets. IEEE J. Sel. Topics Sig. Process. **5**(5), 1014–1024 (2011)
15. Ouarti, N., Peyr, G.: Best basis denoising with non-stationary wavelet packets. In: 2009 16th IEEE International Conference on Image Processing (ICIP), pp. 3825–3828, November 2009
16. Peyré, G.: Texture synthesis and modification with a patch-valued wavelet transform. In: Sgallari, F., Murli, A., Paragios, N. (eds.) SSVM 2007. LNCS, vol. 4485, pp. 640–651. Springer, Heidelberg (2007). https://doi.org/10.1007/978-3-540-72823-8_55
17. Pock, T., Sabach, S.: Inertial proximal alternating linearized minimization (iPALM) for nonconvex and nonsmooth problems. SIAM J. Imaging Sci. **9**(4), 1756–1787 (2016)

18. Sulam, J., Ophir, B., Zibulevsky, M., Elad, M.: Trainlets: dictionary learning in high dimensions. IEEE Trans. Sig. Process. **64**(12), 3180–3193 (2016)
19. Sweldens, W.: The lifting scheme: a custom-design construction of biorthogonal wavelets. Appl. Comput. Harmonic Anal. **3**(2), 186–200 (1996)
20. Tewfik, A.H., Sinha, D., Jorgensen, P.: On the optimal choice of a wavelet for signal representation. IEEE Trans. Inf. Theory **38**(2), 747–765 (1992)
21. Tibshirani, R.: Regression shrinkage and selection via the lasso. J. R. Stat. Soc. Ser. B Methodol. **58**, 267–288 (1996)
22. Unser, M., Chenouard, N., Van De Ville, D.: Steerable pyramids and tight wavelet frames in $L_2(R^d)$. IEEE Trans. Image Process. **20**(10), 2705–2721 (2011)
23. Zhang, L., Wu, X., Buades, A., Li, X.: Color demosaicking by local directional interpolation and nonlocal adaptive thresholding. J. Electron. Imaging **20**(2), 023016–023016-16 (2011)

Bottom-Up Top-Down Cues for Weakly-Supervised Semantic Segmentation

Qibin Hou[1], Daniela Massiceti[2(✉)], Puneet Kumar Dokania[2], Yunchao Wei[3], Ming-Ming Cheng[1], and Philip H. S. Torr[2]

[1] Nankai University, Tianjin, China
andrewhoux@gmail.com, cmm@nankai.edu.cn
[2] University of Oxford, Oxford, UK
{daniela,puneet}@robots.ox.ac.uk, phil.torr@eng.ox.ac.uk
[3] NUS, Singapore, Singapore
wychao1987@gmail.com

Abstract. We consider the task of learning a classifier for semantic segmentation using weak supervision in the form of image labels specifying objects present in the image. Our method uses deep convolutional neural networks (CNNs) and adopts an Expectation-Maximization (EM) based approach. We focus on the following three aspects of EM: (i) initialization; (ii) latent posterior estimation (E-step) and (iii) the parameter update (M-step). We show that saliency and attention maps, bottom-up and top-down cues respectively, of images with single objects (simple images) provide highly reliable cues to learn an initialization for the EM. Intuitively, given weak supervisions, we first learn to segment simple images and then move towards the complex ones. Next, for updating the parameters (M step), we propose to minimize the combination of the standard *softmax* loss and the KL divergence between the latent posterior distribution (obtained using the E-step) and the likelihood given by the CNN. This combination is more robust to wrong predictions made by the E step of the EM algorithm. Extensive experiments and discussions show that our method is very simple and intuitive, and outperforms the state-of-the-art method with a very high margin of 3.7% and 3.9% on the PASCAL VOC12 train and test sets respectively, thus setting new state-of-the-art results.

1 Introduction

Semantic segmentation performance has rapidly advanced with the use of Convolutional Neural Networks (CNNs) [5,6,24,38]. The performance of CNNs, however, is largely dependent on the availability of a large corpus of annotated training data, which is both cost- and time-intensive to acquire. The pixel-level annotation of an image takes on average 4 min [4]. This is likely a conservative estimate given that it is based on the COCO dataset [23] in which ground-truths are obtained by annotating polygon corners rather than pixels directly.

Q. Hou and D. Massiceti—These authors are contributed equally.

© Springer International Publishing AG, part of Springer Nature 2018
M. Pelillo and E. Hancock (Eds.): EMMCVPR 2017, LNCS 10746, pp. 263–277, 2018.
https://doi.org/10.1007/978-3-319-78199-0_18

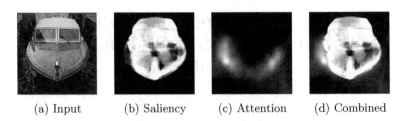

(a) Input   (b) Saliency   (c) Attention   (d) Combined

Fig. 1. Combining bottom-up (*saliency*) and top-down (*attention*) cues for simple images. Both cues complement each other by putting high probability mass on regions missed by the other.

In response, recent work has focussed on weakly-supervised semantic segmentation [4,19,27,28,30,31,36]. These works differ from the fully-supervised case in that rather than having pixel-level ground-truth segmentations, a lower degree of supervision is provided. For example, image-level labels [19,27,28,30], bounding boxes [27], and points and scribbles [4,22,35].

In our work, we address the semantic segmentation task using only image-level labels. These labels specify the object categories present in the image. Our motivation for this is two-fold. (i) The annotation of an image with the 20 PASCAL VOC object classes is estimated to take 20 s. This is at least 12 times faster than a pixel-level annotation and is also scalable. (ii) Images with their image labels or tags can easily be downloaded from the Internet, providing a rich and virtually infinite source of training data. The method we adopt, similar to [27], takes the form of Expectation-Maximization (EM) [10,25]. We focus on the three key steps of an EM-based approach: (i) initialization; (ii) latent posterior estimation (E step); and (iii) parameter update (M step). The following addresses each of these points.

We provide an informed initialization to the EM algorithm as follows. We first train a network to segment *simple* images with one object category (*ImageNet* [11] classification dataset) using an *approximate* per-pixel class distribution obtained using the combination of class-agnostic *saliency* map [15] and class-specific *attention* maps [37]. Note that obtaining saliency and attention maps does not require pixel-level semantic segmentation. We use this trained model to initialize our EM algorithm in order to learn to segment complex images. Intuitively, we first learn to segment simple images and then move towards the complex ones, similar to the work of [36]. In more detail, given a simple image, the saliency map *finds* the object (Fig. 1b) - this is a class-agnotic 'bottom-up' cue. Added to this, once provided with the class present in the image, the attention map (Fig. 1c) gives the 'top-down' class-specific regions in the image. Since both saliency and attention maps are tasked to find the same object, their combination is more powerful than if either one is used in isolation, as shown in Fig. 1d. The combined probability map is then used as the per-pixel class distribution for training an initial model for the semantic segmentation task. The trained initial model provides the initialization parameters for the follow-up

E and M steps of the EM algorithm. Notice that this initialization is in contrast to [27] where the initial model is trained for the *image classification task* on the same ImageNet dataset. To our surprise, experimentally we have found that even this initialization model, which is obtained just by training over ImageNet images in weakly supervised setting (with *no* images from PASCAL VOC12), *outperforms all the current state-of-the-art algorithms for the weakly-supervised semantic segmentation task on the PASCAL VOC12 dataset.* Note that the existing algorithms are significantly more complex and most of them rely on higher degrees of supervision such as bounding boxes, points/squiggles and superpixels. This clearly indicates the importance of learning from simple images before delving into more complex ones. With the trained initial model, we then incorporate PASCAL VOC images (with multiple objects) for the E and M Steps of our EM-based algorithm.

In the E-step, we obtain the latent posterior probability distribution by constraining (or regularizing) the CNN likelihood using image labels based prior. This reduces many false positives by redistributing the probability masses (which were initially over the 20 object categories) among only the labels present in the image and the background. In the M-step, the parameter update step, we then minimize a combination of the standard *softmax* loss (where the ground-truth is assumed to be a Dirac delta distribution) and the KL divergence [21] between the latent posterior distribution (obtained using the E-step) and the likelihood given by the CNN. In the weakly-supervised setting, this makes the approach more robust than using the *softmax* loss alone since in the case of confusing classes, the latent posterior (from the E-step) can sometimes be completely wrong. In addition to this, to obtain better CNN parameters, we add a probabilistic approximation of the Intersection-over-Union (IoU) [1,9,26] to the above loss function.

With this intuitive approach we obtain state-of-the-art results in the weakly-supervised semantic segmentation task on the PASCAL VOC 2012 [12].

2   Related Work

Work in weakly-supervised semantic segmentation has explored varying levels of supervision including combinations of image labels [19,27,28,36], annotated points [4], squiggles [22,35] and bounding boxes [27]. Papandreou et al. [27] employ an EM-based approach with supervision from image labels and bounding boxes. Their method iterates between inferring a latent segmentation (E-step) and optimizing the parameters of a segmentation network (M-step) by treating the inferred latents as the ground-truth segmentation. Similarly, [36] train an initial network using saliency maps, following which a more powerful network is trained using the output of the initial network. The MIL frameworks of [30] and [29] use fully convolutional networks to learn pixel-level semantic segmentations from only image labels. The image labels, however, provide no information about the position of the objects in an image. To address this, localization cues can be incorporated [30,31]. These can be obtained from bottom-up proposal generation methods (for example, MCG [3]) or saliency [36] and attention [37] mechanisms.

The work of [34] uses saliency maps and iteratively erases areas of the image, from most to least salient, thereby forcing the network to learn increasingly discriminative features for segmentation from image labels. Localization cues can also be obtained directly through point/squiggle annotations [4,22,35].

Our method is most similar to the EM-based approach of [27]. We use *saliency* and *attention* maps to learn a network for a simplified semantic segmentation task which provides a better initialization for the EM algorithm. This is in contrast to [27] where a network trained for a classification task is used as initialization. Also different from [27] where the latent posterior is approximated by a Dirac delta function (which we argue is too harsh of a constraint in a weakly-supervised setting), we instead propose to use the combination of the true posterior distribution and the Dirac delta function to learn the parameters.

3   The Semantic Segmentation Task

Consider an image I consisting of a set of pixels $\{y_1, \cdots, y_n\}$ where each pixel represents a random variable taking on a value from a discrete semantic label set $\mathcal{L} = \{l_0, l_1, \cdots, l_c\}$, where c is the number of classes (l_0 for the background). Under this setting, a semantic segmentation is defined as the assignment of all pixels to their corresponding semantic labels, denoted as $\mathbf{y}$.

CNNs are extensively used to model the class-conditional likelihood for this task. Specifically, assuming each random variable to be independent, a CNN models the likelihood function as $P(\mathbf{y}|I;\theta) = \prod_{m=1}^{n} p(y_m|I;\theta)$, where $p(y_m = l|I;\theta)$ is the *softmax* probability (or the marginal) of assigning label l to the m-th pixel. The *softmax* probability is obtained by applying the *softmax* [1] function to the CNN outputs $f(y_m|I;\theta)$ such that $p(y_m = l|I;\theta) \propto \exp(f(y_m = l|I;\theta))$. Given a training dataset $\mathcal{S} = \{I_i, \mathbf{y}_i\}_{i=1}^{N}$, where I_i and $\mathbf{y}_i$ represent the i-th image and its corresponding ground-truth semantic segmentation, the log-likelihood is maximized by minimizing the cross-entropy loss function using the back-propagation algorithm to obtain the optimal θ. At test time, for a given image, the learned θ is used to obtain the *softmax* probabilities for each pixel. These probabilities are either post-processed or used directly to assign semantic labels to each pixel.

4   Weakly-Supervised Semantic Segmentation

As mentioned in Sect. 3, to find the optimal θ for the semantic segmentation task, we need a dataset with ground-truth pixel-level semantic labels. Obtaining this, however, is highly time-consuming and expensive: for a given image, annotating its pixel-wise segmentation (for 20 object classes) takes nearly 239.7 s [4]. This is highly non-scalable to higher numbers of images and classes. Motivated by this, we use an EM [10,25] approach for weakly-supervised semantic segmentation using *only image-level labels*. Image-level labels tag the object classes present in

---

[1] The *softmax* function is defined as $\sigma(f_k) = \frac{e^{f_k}}{\sum_{j=0}^{c} e^{f_j}}$.

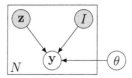

Fig. 2. The graphical model. I is the image. $\mathbf{z}$ is the set of labels present in the image. $\mathbf{y}$ is the latent variable (semantic segmentation). θ is the set of parameters.

an image and are over 10 times faster to obtain that pixel-level annotations. Let us denote $Z = \mathcal{L}\backslash l_0$, and a weak dataset as $\mathcal{D} = \{I_i, \mathbf{z}_i\}_{i=1}^{N}$ where I_i is the i-th image and $\mathbf{z}_i \subseteq Z$ is the image labels corresponding to the objects present in the i-th image. The task is to learn an optimal θ using $\mathcal{D}$.

4.1   The EM Algorithm

Similar to [27], we treat the unknown semantic segmentation $\mathbf{y}$ as the latent variable. Our probabilistic graphical model is of the following form (Fig. 2):

$$P(I, \mathbf{y}, \mathbf{z}; \theta) = P(I)P(\mathbf{y}|I, \mathbf{z}; \theta)P(\mathbf{z}), \tag{1}$$

Briefly, to learn θ while maximizing the above joint probability distribution, the three major steps of an EM algorithm are: (i) initialize the parameter θ_t; (ii) E-step: compute the expected complete-data log-likelihood $F(\theta; \theta_t)$; and (iii) M-step: update θ by maximizing $F(\theta; \theta_t)$. In what follows, we first talk about how to obtain a good initialization θ_t in order to avoid poor local maxima and then talk about optimizing parameters (E and M steps) for a given θ_t.

4.2   Initialization: Skipping Poor Local Maxima Using Bottom-Up Top-Down Cues

It is well known that if the log-likelihood has several maxima or saddle points, an EM-based approach is highly susceptible to mediocre local maxima. In such cases, a good initialization is crucial [16]. We argue that instead of initializing the algorithm with parameters learned for the classification task using the *ImageNet* dataset, as is done by most state-of-the-art methods irrespective of their nature, it is much more effective and intuitive to initialize with parameters learned for solving an easier version of the task at hand - *semantic segmentation* in our case. In the following we show how to use *ImageNet* images with image-level labels to learn parameters for the weakly-supervised semantic segmentation task. These learned parameters will be used as an initialization to the EM algorithm.

   Let us denote $\mathcal{D}(I)$ as the subset of images from the *ImageNet* dataset containing objects of the categories we are interested in (details in Sect. 5). Dataset $\mathcal{D}(I)$ contains simple images, which have mainly centered and clutter-free single objects. This is unlike the challenging PASCAL VOC 2012 dataset [12]. Given $\mathcal{D}(I)$, in order to train the initial model to obtain θ_t, we need pixel-level semantic labels which are not available in the weakly-supervised setting. To circumvent

---

Algorithm 1. Approximate ground truth distribution

---

input Image I with one object category; Image-level label z
1: $M = zeros(n)$, n is the number of pixels.
2: $\mathbf{s} \leftarrow SaliencyMap(I)$ [15]
3: $\mathbf{a} \leftarrow AttentionMap(I, z)$ [37]
4: **for** each pixel $m \in I$ **do**
5:     $M(m) = h(\mathbf{s}(m), \mathbf{a}(m))$
6: **end for**
output M

---

this, we use a class-agnostic *saliency* map [8,15] (bottom-up cue) and a class-specific *attention* map [37] (top-down cue) to *construct* the probability distribution over labels for each pixel. Intuitively, a saliency map gives the probability of each pixel belonging to *any* foreground class, and an attention map gives the probability of it belonging to a particular object class. Combining these two maps allow us to obtain an accurate probability distribution over object classes for each pixel in the image (see Fig. 1).

Precisely, as shown in Algorithm 1, for a given simple image $I \in \mathcal{D}(I)$ and its corresponding image label $z \in Z$, we combine the attention and saliency values per pixel to obtain M. $M(m) \in [0, 1]$ denotes the probability of the m-th pixel being the z-th object category. Similarly, $1 - M(m)$ denotes the probability of it being the background. The combination function $h(.,.)$ in Algorithm 1 is a user-defined function that combines the saliency and the attention maps. In this work we employ the max function which takes the union of the two maps (Fig. 1).

To construct the per-pixel distribution over classes, let us define the distribution for the m-th pixel as δ_m^I. Thus, $\delta_m^I \in [0, 1]^{|\mathcal{L}|}$, where $\delta_m^I(z) = M(m)$ at the z-th index for the object category z, $\delta_m^I(0) = 1 - M(m)$ at the 0-th index for the background, and is zero otherwise. Given δ_m^I for each pixel, we find θ_t by using a CNN and optimizing the per-pixel cross-entropy loss $\sum_{k \in \mathcal{L}} \delta_m^I(k) \log p(k|I; \theta)$ between δ_m^I and $p(y_m|I, \theta)$, where $p(y_m|I, \theta)$ is the CNN likelihood.

By using the probability value $M(m)$ directly rather than a Dirac delta distribution makes our method more robust to noisy attention and saliency maps. This can be seen as a way of mining class-specific noise-free pixels, and is motivated by the work of Bearman et al. [4] where humans annotate points and squiggles in complex images. Their work showed that the learning process can be sufficiently guided using only a few supervised pixels which are easy to obtain. We improve on [4] by completely removing the requirement for human annotators: the per-pixel label distribution can be obtained using *only* image-level labels. This makes our approach highly scalable.

4.3 Optimizing Parameters

E-Step (approximate complete-data log-likelihood). Let us now talk about how to define and optimize the expected complete-data log-likelihood $F(\theta; \theta_t)$. By definition, $F(\theta; \theta_t) = \sum_{\mathbf{y}} P(\mathbf{y}|I, \mathbf{z}; \theta_t) \log P(I, \mathbf{y}, \mathbf{z}; \theta)$, where the expectation is taken over the posterior over the latent variables at a given set of parameters θ_t. The expectation is denoted as $P(\mathbf{y}|I, \mathbf{z}; \theta_t)$. In the case of semantic segmentation, the latent space is exponentially large $|\mathcal{L}|^n$, therefore, computing $F(\theta; \theta_t)$ is infeasible. However, as will be shown, the independence assumption over the random variables, namely $P(\mathbf{y}|I; \theta) = \prod_{m=1}^{n} p(y_m|I; \theta)$, allows us to maximize $F(\theta; \theta_t)$ efficiently by decomposition. By using Eq. (1), the independence assumption, the identity $\sum_{\mathbf{y}} P(\mathbf{y}|I, \mathbf{z}; \theta_t) = 1$, and ignoring the terms independent of θ, $F(\theta; \theta_t)$ can be written in a simplified form as:

$$\bar{F}(\theta; \theta_t) = \sum_{m=1}^{n} \sum_{\mathbf{y}} P(\mathbf{y}|I, \mathbf{z}; \theta_t) \log p(y_m|I; \theta) \tag{2}$$

Without loss of generality, we can write $P(\mathbf{y}|I, \mathbf{z}; \theta_t) = P(\mathbf{y}\backslash y_m|I, \mathbf{z}, y_m; \theta_t)$ $p(y_m|I, \mathbf{z}; \theta_t)$, and using the identity $\sum_{\mathbf{y}\backslash y_m} P(\mathbf{y}\backslash y_m|I, \mathbf{z}, y_m; \theta_t) = 1$, we obtain:

$$\bar{F}(\theta; \theta_t) = \sum_{m=1}^{n} \sum_{y_m \in \mathcal{L}} p(y_m|I, \mathbf{z}; \theta_t) \log p(y_m|I; \theta) \tag{3}$$

M-Step (parameter update). The M-step parameter update, which maximises $\bar{F}(\theta; \theta_t)$ w.r.t. θ, can be written as:

$$\theta_{t+1} = \underset{\theta}{\mathrm{argmax}} \sum_{m=1}^{n} \sum_{y_m \in \mathcal{L}} p(y_m|I, \mathbf{z}; \theta_t) \log p(y_m|I; \theta) \tag{4}$$

We make the assumption that the posterior $p(y_m|I, \mathbf{z}; \theta_t)$ belongs to the exponential family distribution such that $p(y_m|I, \mathbf{z}; \theta_t) \propto \exp(f(y_m|I; \theta_t) + g(y_m, \mathbf{z}))$. Here, $f(y_m|I; \theta_t)$ is the likelihood obtained for pixel m using the CNN at a given θ_t, and $g(y_m, \mathbf{z})$ is a task-specific user-defined function which we use to *regularize* the CNN likelihood.

More specifically, we use $g(y_m, \mathbf{z})$ to explicitly impose constraints based on the image label information, namely the network should suppress the probability of objects not present in the image. For example, if we know that there are only two classes in a given training image such as 'cat' and 'person', then we would like to push the latent posterior probability $P(\mathbf{y}|I, \mathbf{z}; \theta_t)$ of absent classes to zero and increase the probability of the present classes. In order to impose the above mentioned constraints, we use $g(.,.)$ as:

$$g(y_m, \mathbf{z}) = \begin{cases} -\infty, & if \ y_m \notin \mathbf{z} \cup l_0, \\ 0, & otherwise. \end{cases} \tag{5}$$

Practically speaking, imposing the above constraint is equivalent to obtaining *softmax* probabilities for only those classes (including background l_0) present in the image and assigning a probability of zero to all other classes. In other words, the above definition of $g(.,.)$ inherently defines a uniform distribution over the object classes present in the image including the background and zero for the remaining ones. Other forms of $g(.,.)$ can also be used to impose different task-specific label-dependent constraints.

Optimizing Eq. 4 is equivalent to minimizing the cross entropy or the KL divergence between the latent posterior distribution $p(y_m|I, \mathbf{z}; \theta_t)$ and the CNN likelihood $p(y_m|I; \theta)$. [27] uses a Dirac delta approximation $\hat{p}$ of the posterior distribution, where $\hat{p}(\hat{l}_m) = 1$ at $\hat{l}_m = \arg\max_{l \in \mathcal{L}} p(y_m = l|I, \mathbf{z}; \theta_t)$ and otherwise zero. We instead propose to use the combination of the Dirac delta approximation and the actual latent posterior distribution (or the regularized likelihood) in Eq. 4 as follows:

$$J_m(I, \mathbf{z}, \theta_t; \theta) = \sum_{y_m \in \mathcal{L}} \bar{p}(y_m|I, \mathbf{z}; \theta_t) \log p(y_m|I; \theta) \tag{6}$$

where, $\bar{p}(y_m|I, \mathbf{z}; \theta_t) = (1-\epsilon)p(y_m|I, \mathbf{z}; \theta_t) + \epsilon\hat{p}(y_m)$. We argue that using a Dirac delta distribution alone imposes a *hard* constraint that is suitable only when we are very confident about the true label assignment (for example, in the fully-supervised setting). In the weakly-supervised setting where the latent posterior, which decides the label, can be noisy (mostly seen in the case of confusing classes), it is more suitable to use the true posterior distribution. Equation 6 provides the best of both worlds. We define the weighting factor ϵ as:

$$\epsilon = \begin{cases} 1, & if \ r \geq \eta, \\ r, & otherwise. \end{cases} \tag{7}$$

where $\eta \in [0, 1]$ is a hyper-parameter and $r = (p_1 - p_2)/p_1$. Values p_1 and p_2 are the highest and the second highest probabilities in the latent posterior distribution. Intuitively, $\eta = 0.05$ implies that the most probable score should be at least 5% better than the second most probable score in order to use the Dirac delta posterior alone, otherwise, the weighted combination should be used.

The IoU gain function. Along with minimizing the cross entropy losses as shown in the Eq. 6, in order to obtain a better parameter estimate, we also maximize the probabilistic approximation of the intersection-over-union (IoU) between the posterior distribution and the likelihood [1,9,26]:

$$\mathcal{J}_{IOU}(P(\mathbf{y}|I, \mathbf{z}; \theta_t), P(\mathbf{y}|I; \theta)) \approx$$
$$\frac{1}{|\mathcal{L}|} \sum_{l \in \mathcal{L}} \frac{\sum_{m=1}^n p_m^t(l) p_m^\theta(l)}{\sum_{m=1}^n \{p_m^t(l) + p_m^\theta(l) - p_m^t(l) p_m^\theta(l)\}} \tag{8}$$

where, $p_m^t(l) = p(y_m = l|I, \mathbf{z}; \theta_t)$ and $p_m^\theta(l) = p(y_m = l|I; \theta)$. Refer to [9] for further details about Eq. 8.

Overall objective function and the algorithm. Combining the cross entropy loss function (Eq. 6) and the IoU gain function (Eq. 8), the M-step parameter update problem is:

$$\theta_{t+1} = \arg\max_\theta \sum_{m=1}^n J_m(I, \mathbf{z}, \theta_t; \theta) + \mathcal{J}_{IOU} \tag{9}$$

Algorithm 2. Our final algorithm

---

input Datasets $\mathcal{D}(P)$ and $\mathcal{D}(I)$; θ_0; η; K
1: Use $\mathcal{D}(I)$ and θ_0 to obtain initialization parameter θ_t using method explained in Sec. 4.2.
2: **for** $k = 1 : K$ **do**
3:     $\theta \leftarrow \theta_t$
4:     **for** each pixel m in $\mathcal{D}(P) \cup \mathcal{D}(I)$ **do**
5:         Obtain latent posterior: $p_m(y_m|I, \mathbf{z}; \theta) \propto \exp(f_m(y_m|I; \theta) + g(y_m, \mathbf{z}))$
6:     **end for**
7:     Optimize Eq. 9 using CNN to update θ_t.
8: **end for**

We use a CNN model along with the back-propagation algorithm to optimize the above objective function. Recall that our evaluation is based on the PASCAL VOC 2012 dataset, therefore, during the M-step of the algorithm we use both the *ImageNet* $\mathcal{D}(I)$ and the PASCAL trainval $\mathcal{D}(P)$ datasets (see Sect. 5 for details). Our overall approach is summarized in Algorithm 2.

5 Experimental Results and Comparisons

We show the efficacy of our method on the challenging PASCAL VOC 2012 benchmark and outperform all existing state-of-the-art methods by a large margin. Specifically, we improve on the current state-of-the-art method [34] by 3.7% and 3.9% on the test and train sets respectively.

5.1 Setup

Dataset $\mathcal{D}(I)$ for training our initial model. To train our initial model (Sect. 4.2), we download 80,000 images from the *ImageNet* dataset. These images contain objects in the 20 foreground object categories of the PASCAL VOC 2012 segmentation task. We filter this dataset using simple heuristics. First, we discard images with width or height less than 200 or greater than 500 pixels. Using the attention model of [37], we generate a per-class *attention* map for each image and record the most probable class label with its corresponding probability. We discard images for which the most probable class label does not match the given image label and has a probability of less than 0.2. We also generate *saliency* maps using the saliency model of [15] (trained with class-agnostic saliency masks).

We then combine attention and saliency in the following way: we first generate an attention binary mask from each attention map by setting a mask pixel to 1 if its corresponding attention probability is greater than 0.5. We do the same to the saliency maps to obtain saliency binary masks. We then find the pixel-wise intersection between the saliency and the attention masks. For each object category, the images are sorted by this intersection area (i.e. the number of overlapping pixels between the two masks) with the intuition that larger intersections correspond to higher quality saliency and attention maps. The top

1500 images are then selected for each category. The only exceptions are the 'person' category where the top 2500 images are kept, and categories with fewer than 1500 images, in which case all images are kept. *This filtering process leaves us with* 24,000 *simple images* of uncluttered and mainly-centered single objects. We denote this dataset as $\mathcal{D}(I)$ and would like to *highlight that $\mathcal{D}(I)$ does not contain any additional images relative to those used by other weakly supervised-works* (see Dataset column in Table 1).

Datasets $\mathcal{D}(P)$ and $\mathcal{D}(I)$ for M-step. For the M-step, we use a filtered subset of PASCAL VOC 2012 images, denoted $\mathcal{D}(P)$, and a subset of $\mathcal{D}(I)$. To obtain $\mathcal{D}(P)$, we take complex PASCAL VOC 2012 images (10,582 in total, made up of 1,464 training images [12] and the extra images provided by [14]), and use the trained initial model (i.e. θ_t) to generate a (hard) ground-truth segmentation for each. The hard segmentations are obtained by assigning each pixel with the class label that outputs the highest probability. The ratio of the foreground area to the whole image area (where area is the sum of the number of pixels) is computed. If the ratio is below 0.05, the image is discarded. This leaves 10,000 images. We also further filter $\mathcal{D}(I)$: using the trained initial model, we generate (hard) segmentations for all simple *ImageNet* images in $\mathcal{D}(I)$. We compute the intersection area (as above) between the attention binary mask and the predicted segmentation (rather than the saliency mask as before). We then select the top 10,000 of 24,000 images based on this metric. Together $\mathcal{D}(P)$ and this subset of $\mathcal{D}(I)$ make up 20,000 images which are used for the M-step.

CNN architecture and parameter settings. Similar to [19,27,36] our initial model and our EM model are based on the largeFOV DeepLab architecture [6]. We use simple bilinear interpolation to map the downsampled feature maps to the original image size as suggested in [24]. We use the publicly available Caffe toolbox [17] for our implementation. We use weight decay (0.0005), momentum (0.9), and iteration size (10) for gradient accumulation. The learning rate is 0.001 at the beginning and is divided by 10 every 10 epochs. We use a batch size of 1 and randomly crop the input image to 321×321. Images with width or height less than 321 are padded with the mean pixel values and the corresponding places in the ground-truth are padded with ignore labels to nullify the effect of padding. We flip the images horizontally, resulting in an augmented set twice the size of the original one. We train our networks for 30K iterations by optimizing Eq. 9 as per Algorithm 2 with $\eta = 0.05$ and $K = 2$. Performance gains beyond two EM iterations were not significant compared to the computational cost.

5.2   Results, Comparisons, and Analysis

We provide Table 1 for an extensive comparison between our and current methods, their dependencies, and degrees of supervision. Regarding the dependencies of our method, our saliency network [15] is trained using salient region masks. These masks are class-agnostic, therefore, once trained the network can be used for any salient semantic object category, so there is no limit with scalability and

Table 1. Comparison table. All dependencies, datasets, and degrees of supervision used by the current methods for weakly-supervised semantic segmentation. $\mathcal{D}(I)$: *ImageNet* dataset; $\mathcal{D}(P)$: PASCAL VOC 2012 dataset (see Sect. 5); and $\mathcal{D}(F)$: 41K images from *Flickr* [36]. Note that the cross validation of CRF hyper-parameters and the training of MCG are performed using a fully-supervised pixel-level semantic segmentation dataset. Methods with equivalent supervision are underlined for fair comparison to our own method.

Method		Dataset	Dependencies	Supervision	CRF [20]	mIoU (Val)	mIoU (Test)
EM Adapt [27]		$\mathcal{D}(I), \mathcal{D}(P)$	No	Image labels	✗	–	–
					✓	38.2%	39.6%
CCNN [28]		$\mathcal{D}(I), \mathcal{D}(P)$	No	Image labels	✗	33.3%	35.6%
					✓	35.3%	–
			Class size		✗	40.5%	43.3%
					✓	42.4%	45.1%
SEC [19]		$\mathcal{D}(I), \mathcal{D}(P)$	Saliency [32] and Localization [39]	Image labels	✗	44.3%	–
					✓	50.7%	51.7%
MIL [30]		$\mathcal{D}(I)$	Superpixels [13]	Image labels	✗	36.6%	35.8%
			BBox BING [7]			37.8%	37.0%
			MCG [3]			42.0%	40.6%
WTP [4]		$\mathcal{D}(I), \mathcal{D}(P)$	Objectness [2]	Image labels	–	32.2%	–
				Image labels + 1 Point/Class		42.7%	–
				Image labels + 1 Squiggle/Class		49.1%	–
STC [36]		$\mathcal{D}(I), \mathcal{D}(P), \mathcal{D}(F)$	Saliency [18]	Image labels	✓	49.8%	51.2%
AugFeed [31]		$\mathcal{D}(I), \mathcal{D}(P)$	SS [33]	Image labels	✗	46.98%	47.8%
					✓	52.62%	52.7%
			MCG [3]		✗	50.41%	50.6%
					✓	54.34%	55.5%
AE-PSL [34]		$\mathcal{D}(P)$	Saliency [18]	Image labels	✓	55.0%	55.7%
Ours	Initial model	$\mathcal{D}(I)$	Saliency [15] and Attention [37]	Image labels	✗	53.53%	54.34%
					✓	55.19%	56.24%
	Final	$\mathcal{D}(I), \mathcal{D}(P)$			✗	56.91%	57.74%
					✓	**58.71%**	**59.58%**

no need to retrain the saliency network for new object categories. Our second dependency, the attention network [37] is trained using solely image labels.

State-of-the-art. We outperform all existing state-of-the-art methods. The most directly comparable method in terms of supervision and dependencies is *AE-PSL* [34] which uses image-level labels and a saliency network trained on bounding boxes [18] (whereas our saliency network uses class-agnostic saliency masks [15]). Our method obtains almost 4% better mIoU than *AE-PSL* on both the *val* and *test* sets. Even if we disregard 'equivalent' supervision and dependencies, our method still outperforms all existing methods.

Simplicity vs sophistication. The initial model is essential to the success of our method. We train this model in a very simple and intuitive way by learning the semantic segmentation task using a filtered subset of simple *ImageNet* images.

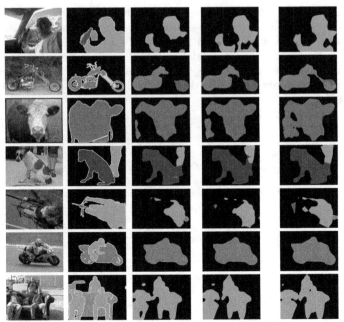

(1) Source   (2) Ground truth (3) Initial model (4) EM 1st iteration (5) EM 2nd iteration

Fig. 3. Qualitative results for weakly-supervised semantic segmentation using our proposed EM-based method. From the initial model (3rd column) to the second iteration of the EM algorithm (5th column) the segmentation quality improves incrementally. The bottom two rows show failure cases.

Importantly, this process uses only image labels and is fully automatic, requiring no human intervention. The learned θ_t provides a *very* good initialization for the EM algorithm, enabling it to avoid poor local maxima. This is shown visually in Fig. 3: the initial model (3rd column) is already a good prediction, and the 1st and 2nd EM iterations (4th and 5th columns) improve the semantic segmentation even further. We highlight that with this simple approach, surprisingly, our initial model beats *all* current state-of-the-art methods, which are more complex and often use higher degrees of supervision. By implementing this intuitive modification, we believe that many methods can easily boost their performance.

To CRF or not to CRF? Even though we employ a CRF [20] as a post-processing step, we believe that in a weakly supervised setting it should not be employed directly for the following two reasons. Firstly, CRF hyper-parameters are normally cross validated over a fully-supervised pixel-wise segmentation dataset which contradicts the "weak" supervision paradigm. This is likewise the case for MCG [3] which is trained on a pixel-level semantic segmentation dataset. Secondly, the CRF hyper-parameters are incredibly sensitive. To incorporate new object categories, therefore, would require a pixel-level annotated dataset of the new categories along with the old ones for the cross-validation of the CRF hyper-parameters. This is highly non-scalable. For completeness, however, we include our method with a CRF applied (Table 1) which boosts our accuracy by 1.8%. We note that even without a CRF, our approach exceeds the state-of-the-art [34] (which uses a CRF) by almost 2% on *train* and *test* sets.

6   Conclusions and Future Work

We have addressed weakly-supervised semantic segmentation using only image-level labels. We proposed an EM-based approach and focus on the three key components of the algorithm: (i) initialization, (ii) E-step and (iii) M-step. Using only the image labels of a filtered subset of simple *ImageNet* images, we learn a set of parameters for the semantic segmentation task which provides an informed initialization of our EM algorithm. Following this, with each EM iteration, we empirically and qualitatively verify that our method improves the segmentation accuracy on the challenging PASCAL VOC 2012 benchmark. Furthermore, we show that our method outperforms all state-of-the-art methods.

Future directions include making our method more robust to noisy labels, for example, when images downloaded from the Internet have incorrect labels, as well as better handling images with multiple classes of objects.

Acknowledgments. Qibin Hou, Yunchao Wei and Ming-Ming Cheng were sponsored by *NSFC (61620106008, 61572264)*, *CAST (YESS20150117)*, *Huawei Innovation Research Program (HIRP)*, and *IBM Global SUR award*. Daniela Massiceti, Punnet K. Dokania and Philip H.S. Torr were sponsored by *ERC grant ERC-2012-AdG 321162-HELIOS*. Ms Massiceti was also sponsored by the *Skye Foundation*. We thank all sponsors for their support.

References

1. Ahmed, F., Tarlow, D., Batra, D.: Optimizing expected intersection-over-union with candidate-constrained CRFs. In: ICCV (2015)
2. Alexe, B., Deselares, T., Ferrari, V.: Measuring the objectness of image windows. PAMI **34**(11), 2189–2202 (2012)
3. Arbelaez, P., Pont-Tuset, J., Barron, J., Marques, F., Malik, J.: Multiscale combinatorial grouping. In: CVPR (2014)
4. Bearman, A., Russakovsky, O., Ferrari, V., Fei-Fei, L.: What's the point: semantic segmentation with point supervision. In: Leibe, B., Matas, J., Sebe, N., Welling, M. (eds.) ECCV 2016. LNCS, vol. 9911, pp. 549–565. Springer, Cham (2016). https://doi.org/10.1007/978-3-319-46478-7_34
5. Chandra, S., Kokkinos, I.: Fast, exact and multi-scale inference for semantic image segmentation with deep Gaussian CRFs. In: Leibe, B., Matas, J., Sebe, N., Welling, M. (eds.) ECCV 2016. LNCS, vol. 9911, pp. 402–418. Springer, Cham (2016). https://doi.org/10.1007/978-3-319-46478-7_25
6. Chen, L.-G., Papandreou, G., Kokkinos, I., Murphy, K., Yuille, A.L.: Semantic image segmentation with deep convolutional nets and fully connected. In: ICLR (2015)

7. Cheng, M., Zhang, Z., Lin, W., Torr, P.H.S.: BING: binarized normed gradients for objectness estimation at 300fps. In: CVPR (2014)
8. Cheng, M.-M., Mitra, N.J., Huang, X., Torr, P.H.S., Hu, S.-M.: Global contrast based salient region detection. IEEE TPAMI **37**(3), 569–582 (2015)
9. Cogswell, M., Lin, X., Purushwalkam, S., Batra, D.: Combining the best of graphical models and convnets for semantic segmentation (2014). arXiv:1412.4313
10. Dempster, A.P., Laird, N.M., Rubin, D.B.: Maximum likelihood from incomplete data via the EM algorithm. J. Royal Stat. Soc. **39**(1), 1–38 (1977)
11. Deng, J., Dong, W., Socher, R., Li, L.-J., Li, K., Fei-Fei, L.: ImageNet: a large-scale hierarchical image database. In: CVPR (2009)
12. Everingham, M., Eslami, S.M.A., Gool, L.V., Williams, C., Winn, J., Zisserman, A.: The Pascal visual object classes challenge a retrospective. IJCV **111**(1), 98–136 (2015)
13. Felzenszwalb, P.F., Huttenlocher, D.P.: Efficient graph based image segmentation. IJCV **59**(2), 167–181 (2004)
14. Hariharan, B., Arbelaez, P., Bourdev, L., Maji, S., Malik, J.: Semantic contours from inverse detectors. In: ICCV (2011)
15. Hou, Q., Cheng, M.-M., Hu, X.-W., Borji, A., Tu, Z., Torr, P.: Deeply supervised salient object detection with short connections. In: IEEE CVPR (2017)
16. Jeff Wu, C.F.: On the convergence properties of the EM algorithm. Ann. Stat. **11**(1), 95–103 (1983)
17. Jia, Y., Shelhamer, E., Donahue, J., Karayev, S., Long, J., Girshick, R., Guadarrama, S., Darrell, T.: Caffe: convolutional architecture for fast feature embedding. In: ACM International Conference on Multimedia (2014)
18. Jiang, H., Wang, J., Yuan, Z., Wu, Y., Zheng, N., Li, S.: Salient object detection: a discriminative regional feature integration approach. In: CVPR (2013)
19. Kolesnikov, A., Lampert, C.H.: Seed, expand and constrain: three principles for weakly-supervised image segmentation. In: Leibe, B., Matas, J., Sebe, N., Welling, M. (eds.) ECCV 2016. LNCS, vol. 9908, pp. 695–711. Springer, Cham (2016). https://doi.org/10.1007/978-3-319-46493-0_42
20. Krahenbuhl P., Koltun, V.: Efficient inference in fully connected CRFs with Gaussian edge potentials. In: NIPS (2011)
21. Kullback, S., Leibler, R.A.: On information and sufficiency. Ann. Math. Stat. **22**(1), 79–86 (1951)
22. Lin, D., Dai, J., Jia, J., He, K., Sun, J.: ScribbleSup: scribble-supervised convolutional networks for semantic segmentation. In: CVPR (2016)
23. Lin, T.-Y., et al.: Microsoft COCO: common objects in context. In: Fleet, D., Pajdla, T., Schiele, B., Tuytelaars, T. (eds.) ECCV 2014. LNCS, vol. 8693, pp. 740–755. Springer, Cham (2014). https://doi.org/10.1007/978-3-319-10602-1_48
24. Long, J., Shelhamer, E., Darrell, T.: Fully convolutional networks for semantic segmentation. In: CVPR (2015)
25. McLachlan, G.J., Krishnan, T.: The EM Algorithm and Extensions. Wiley, Hoboken (1997)
26. Nowozin, S.: Optimal decisions from probabilistic models: the intersection-over-union case. In: CVPR (2014)
27. Papandreou, G., Chen, L.-C., Murphy, K.P., Yuille, A.L.: Weakly- and semi-supervised learning of a DCNN for semantic image segmentation. In: ICCV (2015)
28. Pathak, D., Krahenbuhl, P., Darrell, T.: Constrained convolutional neural networks for weakly supervised segmentation. In: ICCV (2015)
29. Pathak, D., Shelhamer, E., Long, J., Darrell, T.: Fully convolutional multi-class multiple instance learning. In: ICLR (2014)

30. Pinheiro, P.O., Collobert, R.: From image-level to pixel-level labeling with convolutional networks. In: CVPR (2015)
31. Qi, X., Liu, Z., Shi, J., Zhao, H., Jia, J.: Augmented feedback in semantic segmentation under image level supervision. In: Leibe, B., Matas, J., Sebe, N., Welling, M. (eds.) ECCV 2016. LNCS, vol. 9912, pp. 90–105. Springer, Cham (2016). https://doi.org/10.1007/978-3-319-46484-8_6
32. Simonyan, K., Vedaldi, A., Zisserman, A.: Deep inside convolutional networks: visualising image classification models and saliency maps. In: ICLR (2014)
33. Uijlings, J.R.R., van de Sande, K.E.A., Gevers, T., Smeulders, A.W.M.: Selective search for object recognition. IJCV **104**(2), 154–171 (2013)
34. Wei, Y., Feng, J., Liang, X., Cheng, M., Zhao, Y., Yan, S.: Object region mining with adversarial erasing: a simple classification to semantic segmentation approach. In: CVPR (2017)
35. Xu, J., Schwing, A., Urtasun, R.: Learning to segment under various forms of weak supervision. In: CVPR (2015)
36. Yunchao, W., Xiaodan, L., Yunpeng, C., Xiaohui, S., Cheng, M.-M., Yao, Z., Shuicheng, Y.: STC: a simple to complex framework for weakly-supervised semantic segmentation (2015). arXiv:1509.03150
37. Zhang, J., Lin, Z., Brandt, J., Shen, X., Sclaroff, S.: Top-down neural attention by excitation backprop. In: Leibe, B., Matas, J., Sebe, N., Welling, M. (eds.) ECCV 2016. LNCS, vol. 9908, pp. 543–559. Springer, Cham (2016). https://doi.org/10.1007/978-3-319-46493-0_33
38. Zheng, S., Jayasumana, S., Romera-Paredes, B., Vineet, V., Su, Z., Du, D., Huang, C., Torr, P.: Conditional random fields as recurrent neural networks. In: ICCV (2015)
39. Zhou, B., Khosla, A., Lapedriza, A., Oliva, A., Torralba, A.: Learning deep features for discriminative localization. In: CVPR (2016)

Vehicle X-Ray Images Registration

Abraham Marciano[1,2]([⊠]), Laurent D. Cohen[1], and Najib Gadi[2]

[1] Université Paris-Dauphine, PSL Research University, CNRS, UMR 7534,
CEREMADE, 75016 Paris, France
marciano@ceremade.dauphine.fr
[2] Smiths Detection, 94405 Vitry-sur-Seine, France

Abstract. Image registration is definitely one of the most prominent techniques at the heart of computer vision research. Applications range from medical image analysis, remote sensing or robotics to security-related tasks such as surveillance or motion tracking. In our previous work, a solution was provided to address the registration problem involving top-view radiographic images of vehicles. A unidimensional minimization scheme was formulated along with a column-wise constancy constraint on the displacement field.

In this paper, we show that the proposed method is not sufficient in case of significant vertical shifts between the cars of both moving and static images. In fact, the radiated beam is triangular, thus any translated object is projected differently according to its distance to the x-ray source. We therefore add a 1D unconstrained optimization to the previous scheme for a y-direction correction. We also demonstrate that applying the vertical correction following our 1D optimization in the x-axis yields better results than performing a simultaneous minimization on both components.

Finally, the possible apparition of artefacts in the deformed image throughout the optimization process is analyzed. Diffusion and volume-preserving schemes are considered and compared in this regard.

Keywords: Image registration · Variational approach
Energy minimization methods · Difference detection · Volume preservation

1 Introduction

Security is imposing itself as one of the most defining stakes in modern societies. Massive investments have been dedicated to find innovative solutions in order to assist the military or custom officers in their tasks. Image processing and computer vision methods are utterly ubiquitous in defense industries and provide wide-ranging applications.

Image registration is commonly referred to as the searching process of a transformation that aligns, in an optimal but "plausible" way, a moving image T with a static image R [1, 2].

Aiming at automatic threat targeting, the registration problem of top-view x-ray scans of same-model vehicles was introduced in our previous work [3]. A straightforward solution was proposed to address the nonlinear deformations issue: a unidimensional optimization scheme with a constancy constraint formulated for each column of the displacement field.

© Springer International Publishing AG, part of Springer Nature 2018
M. Pelillo and E. Hancock (Eds.): EMMCVPR 2017, LNCS 10746, pp. 278–293, 2018.
https://doi.org/10.1007/978-3-319-78199-0_19

Yet, this assumption doesn't hold in case of significant translation between both scanned cars in the vertical axis. The beam emitted from the source located at the center of the detection line is indeed triangular (see Fig. 2). Hence, any shift may be impacted differently in the projection image, according to the distance separating the source from the object.

In Fig. 1 for instance, registration artefacts at the front region prevent any threat detection capacity.

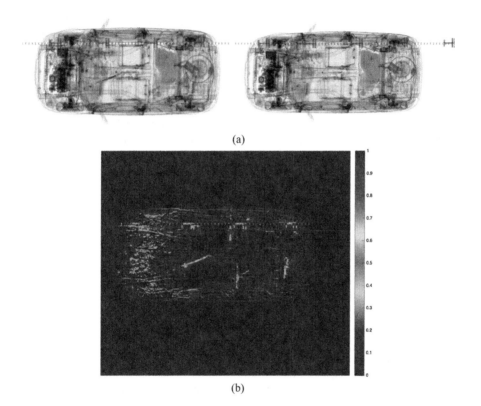

(a)

(b)

Fig. 1. (a) R containing threats and T; (b) the normalized difference map following pose estimation and unidimensional registration on x-axis. The substantial vertical shift between both acquisitions entails alignment artefacts, particularly on the front part where inspection is impossible.

Horizontal and vertical deformations being of a different nature, instead of employing "blind" non-rigid registration models, we provide a method based on the separate analysis of horizontal and vertical deformation phenomena (inherent to the scanning system features).

In this paper, we will briefly give an outline of the main energy-minimization based registration methods along with their numerical solutions.

A second part will describe the registration issue while scanning shifted cars. Then, we will show that performing a post-processing elastic registration over the vertical component of the displacement field yields satisfactory results. We will also demonstrate that the approach combining successive optimizations on horizontal and vertical components is preferable to a simultaneous minimization approach.

2 Outline of Non-linear Registration Methods

2.1 Mathematical Problem Setup

Let's consider a static image R and a moving template T; we look for a transformation applied on T such that both images align as closely as possible [1, 2]. The general formulation is given by the definition of R and T as d-dimensional images represented by the mappings $R, T : \rightarrow \mathbb{R}, \Omega \in [0, 1]^d$ in normalized coordinates. The transformation ϕ is expressed via the displacement field $u : \Omega \rightarrow \mathbb{R}^d$ with $\phi = id + u$.

Hence, given $x \in \Omega$, the transformed image is written $T(\phi(x)) = T(x + u(x))$ where u is applied on the position x of T. In other words, $T(x)$ is the intensity of T at x and $T(x + u(x))$ designates the value of T at the translated position.

Since we wish to minimize the distance between $T \circ \phi$ and R, the unified optimization framework for intensity-based registration is given by the following joint functional scheme:

$$\text{Find } u \text{ minimizing}$$
$$\mathcal{J}[u] = \mathcal{D}[T(u), R] + \alpha \mathcal{S}[u] \tag{1}$$

where $\mathcal{D}$ is referred to as the distance measure between $T \circ \phi$, the transformed image, and R. It commonly designates the data-fitting term or external force [4] to obtain an ideal alignment.

In order to tackle the ill-posedness of the optimization problem [1, 2], a regularization term $\mathcal{S}$ is introduced (internal force [4]). It is designed to keep the displacement smooth during deformation by penalizing unwanted or implausible solutions. The regularization strength is controlled by the smoothing parameter α: increasing its value emphasizes the smoothness of u and vice-versa. The choice of this parameter has been studied but remains utterly dependent on the considered application and the appreciation of the operator [5–7].

Similarly, both data-fitting and smoothing terms are selected according to the specificity of the registration task. The most popular similarity measure is the *Sum of Squared Differences (SSD)*. The *Normalized Cross Correlation (NCC)* is also widespread for matching problems of images sampled with the same modality. Information-theoretic approaches based on Shannon's entropy are employed for multimodal registration (*Mutual Information – MI*). Over the years, more sophisticated measures have also been explored such as the so-called *NGF- Normalized Gradient Fields*. See [1, 2], or [7] for further details about similarity measures.

In various registration tasks, the choice of the smoothing term is crucial [1]. It allows the embedding of physical priors or possible acquisition constraints within the optimization scheme. The next section will depict a larger overview of the most widespread regularization terms.

Note that the selection of both terms is conditioned upon the existence of a Gâteaux derivative [1, 2, 7, 8].

2.2   Registration Methods

Registration can be landmark-base, i.e. the image is transformed via the matching of sparse elements (feature points or manually annotated control landmarks) combined with TPS (Thin-Plate-Spline) interpolation to yield dense correspondences [1, 2, 6]. Although these methods have proven their efficiency in many applications e.g. in medical image analysis, they show less flexibility than the unified variational framework introduced in (1) (see [4]). In addition, despite the recent achievements on key points automatic generation (deep-learning based), unmanned landmark detection remains a challenging task.

For these reasons, we chose to focus our work on intensity-based or so-called iconic methods. These approaches generally split into two categories [1, 2, 6].

Regarding parametric image-registration techniques (*PIR*), the displacement field u is parametrized. The optimization process thereby consists in finding the parameters minimizing $\mathcal{J}$. In B-spline approaches, the displacement is defined as a linear combination of a small set of basis functions. See [2, 6] for a further description of *PIR* approaches. In general, the smoothing terms are chosen such that $\mathcal{S}[u] = 0$ with a parametrization meant to implicitly integrate regularization constraints. Though, Tychonov regularization may often be used.

In non-parametric approaches (*NPIR*), the smoothing term plays a major role while the displacement is no longer parametrized. Most common techniques employ *L2*- normed terms: diffusion, elastic, curvature or fluid regularizers [2, 4]. The demons method [9] is inspired from optical-flow techniques, where a Gaussian smoothing applied at each iteration falls within a diffusion-like regularization [4]. Topology-preserving registration is also a very active field of research. In fluid mechanics, incompressibility imposes that the determinant of the transformation Jacobian remains equal to one (see e.g. [6]). On this basis, Rohling [10] and Christensen [11] add a soft constraint to (1): $\|\det(Id + \nabla u) - 1\|^2$ or its logarithmic version to ensure volume preservation (various regularizers are also described in [10–12]). Instead, Haber and Modersitzki opt for an inequality/equality hard constraint (resp. [7, 8]). More recently, Rueckert et al. compare soft and hard-constraint techniques in [13].

Alternative approaches concentrate on the diffeomorphic framework. ϕ is then explicitly constrained to be a continuously differentiable bijection having a smooth inverse. The deformation is modeled via a time-dependent velocity vector field. See [13], the LDMM model [14] or the earlier work of [15] for detailed explanations about these schemes. See also the recent contributions of Mang et al. about mass and volume preservation constraints for a comprehensive state of the art overview [16, 17].

Other methods suggest to apply a post-processing step over the displacement field after registration. In [18], Poisson's equation is solved to maintain a so-called solenoidal displacement field u and maintain the preservation of volumes.

Similarly, rigidity constraints (hard and soft versions) have been formulated in different publications. Though, they are beyond the scope of this paper (see [19] for instance).

2.3 Numerical Solutions

In [1, 2], Modersitzki suggests a general optimization framework to solve (1). An approximation of the Gauss-Newton method is employed, usually combined with a multi-level strategy as well as a backtracking line-search method to yield an optimal step size at each iteration [2]. Matrix-free approaches are also proposed to speed-up the computations [2, 20]. The variations of the displacement between consecutive iterations are estimated by $\|\nabla u\|_2$ in order to stop the minimization process whenever it falls under a pre-set threshold ε. See [1, 2] for further details about the stopping criterion.

Alternative numerical techniques have also been presented for computationally-demanding problems. See [2] on l-BFGS or Trust-Region methods that can achieve fast quadratic convergence.

For volume-preserving constrained problems, Modersitzki et al. resort to the framework of Sequential Quadratic Methods (SQP) [7] or a variant of the log-barrier method [8].

Diffeomorphic image registration as described in [13, 14] typically uses an implicit regularization (by solving a Poisson *pde*) while gradient-descent methods are used to find an optimal solution. In their latest works, Mang et al. [15] proposed second-order numerical methods to produce diffeomorphic mappings.

In [1, 2], a discretize-then-optimize approach is used. In several applications, the choice of discretization paradigms is crucial. Staggered, nodal or cell-centered grids may be used. See [2] for more details about interpolation or discretization issues.

For an in-depth description and analysis of registration methods, please refer to [6].

3   Problem Presentation

3.1   1D Horizontal Correction Scheme Remainder

In our previous work [3], we demonstrated that rigid registration was not sufficient for top-view scans of vehicles obtained by the Smiths Detection HCVL system. Actually, despite the intrinsic rigidity of cars, non-linear deformations may occur as a result of shocks between the conveyor rollers and the back wheels. The issue was simply addressed by applying a rigid pre-processing registration (similarity pose estimation) and then running a 1D registration scheme assuming:

i. a column-wise constancy condition on the displacement field
ii. a negligible vertical component

By combining i. and ii. we had:

$$\boldsymbol{u} = (u_x(x), 0) \tag{2}$$

Eventually, using an *SSD* distance and a diffusion regularization, we formulated (1) as follows:

Find $u_x(x)$ minimizing
$$\mathcal{J}[u_x] = \tfrac{1}{2}\|\boldsymbol{T}(u_x) - \boldsymbol{R}\|_2^2 + \tfrac{1}{2}\alpha\|\nabla_x u_x\|_2^2 \tag{3}$$

These assumptions hold in most cases and yield nice registration results. See [3] for a numerical and visual appraisal of the method. Yet, in some situations, the registration accuracy achieves lower performances, as depicted in the next section.

3.2 Necessity for a Vertical Correction

In fact, the scanning system allows for an additional degree of freedom: the car translation along the trailer width (Fig. 3). Besides, the x-ray source is located at the detection line center and generates a pyramidal beam. Any significant translation w.r.t the other image therefore leads to important deformations in the y-axis. More particularly, the deformation magnitude depends on the distance/depth separating the radioactive source from the scanned object as shown in Fig. 2. Hence, hypothesis ii. is not anymore valid in this situation.

These deformations may require either small or even larger scale corrections as shown in Fig. 1.

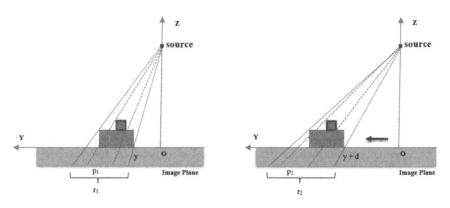

Fig. 2. Vertical shift effect description. The projected distances r_2 and p_2 after displacement are larger than r_1 and p_1. Similarly to the stereovision problem, this difference grows with the object proximity to the x-ray source: $p_2/p_1 > r_2/r_1$

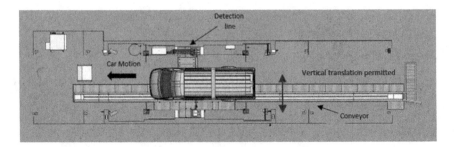

Fig. 3. Top-view of the HCVL system. Translations are permitted along the conveyor width

4 Vertical Correction Solution

4.1 Solution Description

Remaining in the variational framework, we simply intend to apply a supplementary 1D registration over the vertical component of the displacement field. Unlike our previous scheme exposed in Sect. 3.1, we do not assume any specific priors or constraints on the vertical displacement component since the required correction varies accordingly to the objects depth (Fig. 2). A row-wise constancy restriction on u_y is therefore irrelevant. Hence,

$$\boldsymbol{u} = (0, u_y(x, y)) \tag{4}$$

With a diffusion regularizer, (1) is written as follows:

$$\begin{gathered} \text{Find } u_y(\text{x,y}) \text{ minimizing} \\ \mathcal{J}[u_y] = \tfrac{1}{2} \left\| T(u_y) - R \right\|_2^2 + \tfrac{1}{2}\alpha \left\| \nabla u_y \right\|_2^2 \end{gathered} \tag{5}$$

A volume-preserving approach is also tested by adding a soft constraint to the above formula. The joint functional is then given by:

$$\begin{gathered} \text{Find } u_y(\text{x,y}) \text{ minimizing} \\ \mathcal{J}[u_y] = \tfrac{1}{2} \left\| T(u_y) - R \right\|_2^2 + \tfrac{1}{2}\alpha \left\| \nabla u_y \right\|_2^2 + \tfrac{1}{2}\beta \left\| \det(\nabla u_y + Id) - 1 \right\|_2^2 \end{gathered} \tag{6}$$

The soft constrained approach [10, 11] is preferred over hard constraints or diffeomorphisms because of its flexibility and implementation simplicity. The β parameter allows control over the balance between diffusion registration and more stringent volume preservation constraints. In alternative methods, parameter tuning turns into a tougher task.

We also wish to compare both simultaneous and successive optimization paradigms for horizontal and vertical corrections. Let's therefore formulate the above problems in two dimensions, maintaining the column-wise constancy constraint for the horizontal component of the displacement field. Thus, let's consider for $(x, y) \in \Omega$:

$$u = (u_x(x), u_y(x, y)) \qquad (7)$$

The 2D version of the diffusion registration (7) is given by:

Find u minimizing
$$\mathcal{J}[u] = \tfrac{1}{2}\left\|T(u_y) - R\right\|_2^2 + \tfrac{1}{2}\alpha\|\nabla u_x\|_2^2 + \|\nabla u_y\|_2^2 \qquad (8)$$

The two-dimensional volume-preserving optimization scheme is also written as follows:

Find u minimizing
$$\mathcal{J}[u] = \tfrac{1}{2}\|T(u) - R\|_2^2 + \tfrac{1}{2}\alpha\left(\|\nabla u_x\|_2^2\|\nabla u_y\|_2^2\right) + \tfrac{1}{2}\beta\left\|\det(\nabla u_y + Id) - 1\right\|_2^2 \qquad (9)$$

4.2   Numerical Resolution

We will describe the numerical resolution of the 2D problems (8) and (9). The resolution of (5) and (6) is then straightforward since it amounts to exclusively optimizing the vertical displacement field component after obtaining an optimal u_x^* by solving (3).

In the meantime, we resort to first-order minimization methods. Yet, in a future work, we may consider Gauss-Newton methods for speed purposes.

For the diffusion-regularization problem (10), the directional derivatives of the joint functional are given by:

$$\nabla_u \mathcal{J}[u] = \nabla T(x + u(x))(T(x + u(x)) - R(x)) - \alpha\Delta u \qquad (10)$$

with $x = (x, y) \in \Omega$ and u as defined in (7). See [1] for a comprehensive demonstration of the Gâteaux derivatives result.

Let's now look at the volume-preserving soft constraint introduced in (9). With a linearization approximation, valid for small displacements, we get:

$$\det(\nabla u + Id) - 1 \approx \partial_x u_x + \partial_y u_y = \text{div}(u) \qquad (11)$$

Consequently, the directional derivatives computation yields the following result:

$$\nabla_u \mathcal{J}[u] = \nabla T(x + u(x))(T(x + u(x)) - R(x)) - \alpha\Delta u - \beta\nabla.\text{div}(u) \qquad (12)$$

which actually corresponds to the derivative of the elastic registration functional (see [4]). In fact, by noting $\beta = \alpha + \gamma$ we retrieve the well-known Lamé coefficients α and γ. See the appendix in Albrecht et al.'s paper [21] for a detailed demonstration on how to obtain (12).

As in our previous work [3], a gradient-descent method is combined with a multi-scale approach to avoid convergence at local minima. We usually run the descent algorithm with a 4-levels Gaussian pyramid. Armijo's backtracking line-search technique is also integrated into the process to yield an optimal step size for each update of u.

Both α and β are defined for each scale by fixing desired ratios r_α and r_β between the data fitting term and the regularizers (small ratios indicate large regularization and vice versa). The parameters are estimated at the first iteration and hold for the whole level.

The optimization process is stopped as soon as the displacement variation $\|\nabla u\|_2$ falls below $\varepsilon = 10^{-4}$. Computations are performed on a 8 Go RAM - 3,1 GHz Intel Core i7 MacBook. The average calculation time on 992×1186 images with MATLAB 2016b is about 50 s for the successive minimization scheme and reaches more than two minutes for the simultaneous one (no MEX files used so far).

In the next section, we present the results of both approaches for a simultaneous and consecutive optimization fashion.

5   Results

In this chapter, we will use the example brought in Fig. 1 and focus on the front part of the vehicle for a visual and numerical assessment of the different techniques.

5.1   Horizontal-then-Vertical Optimization

In Figs. 4 and 5, we observe that the smoothing parameters/ratios must be chosen carefully. A too large regularization entails a poor registration accuracy (Figs. 4 and 5(a)) while a too soft smoothing makes objects disappear in the difference map. In this case, the threat cannot be longer detected (Fig. 4(c), more explanations are given in Subsect. 5.3). Overall, elastic registration yields better results in terms of threat visualization and MSE measures, as detailed in Table 1.

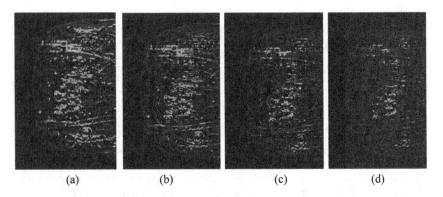

| (a) | (b) | (c) | (d) |

Fig. 4. (a) Front part difference map after 1D horizontal correction; diffusion registration with various parameters - $r_\beta = \infty$ (b) $r_\alpha = 0.1$; (c) $r_\alpha = 1$; (d) $r_\alpha = 2.5$

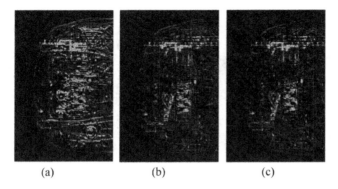

(a)   (b)   (c)

Fig. 5. Elastic registration – maps of differences with various parameters (r_α, r_β) (a) $(1, 0.1)$; (b) $(1.5, 2)$; (c) $(2.5, 1)$

In Fig. 5(c), the addition of a vertical "volume-preserving" constraint gives an accurate and reasonable difference map in contrast with the output obtained with the same r_α ratio in Fig. 4(d).

Table 1. Mean square error measures obtained with the different regularization ratios

	r_α	r_β	MSE
Diffusion	0.1	∞	3.68E−03
	1	∞	1.53E−03
	2.5	∞	7.81E−04
Elastic	1	0.1	4.21E−03
	2.5	1	1.38E−03
	1.5	2	1.54E−03

5.2   Simultaneous Optimization

As depicted in Fig. 6, we get "cleaner" difference maps with the horizontal-then-vertical registration approach where the columns are first rectified to enable a further vertical correction. In addition, by using analogous parameters, the simultaneous method yields larger MSE distances after convergence (Table 2).

5.3   Further Observations

As mentioned, regularization should be tuned carefully, a too weak smoothing may give a better registration accuracy at the expense of unreasonable deformations. This is especially true when the algorithm tends to deform T in a non-diffeomorphic fashion to "imitate" the added objects in R.

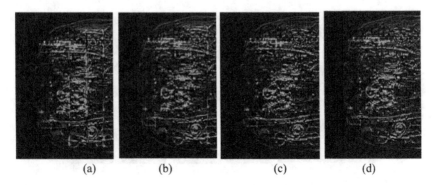

(a)   (b)   (c)   (d)

Fig. 6. Post-registration difference maps with different parameters – (r_α, r_β) (a) $(1, \infty)$; (b) $(1.5, 2)$; (c) $(2.5, 1)$; (d) $(3, 3)$

Table 2. Mean square error measures obtained with the different regularization ratios

	r_α	r_β	MSE
Diffusion	1	∞	4.59E−03
Elastic	1.5	2	4.41E−03
	2.5	1	4.41E−03
	3	3	3.80E−03

Figure 7 describes the resulting difference map with a weakly-regularized registration. The transformation is piecewise-constant (Fig. 7(b)), meaning that, in a given column, a pixel in the original image T is repeated several times in the deformed T_{final}. It actually corresponds to a non-bijective, or more generally: a non-diffeomorphic transformation.

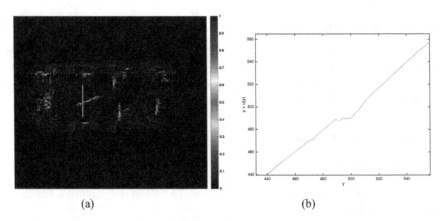

(a)   (b)

Fig. 7. Elastic registration with low smoothing $r_\alpha = 5$, $r_\beta = 5$ (a) difference map; (b) $y + u_y(x, y)$ transformation representation on localized 1D-cross-section

By using proper parameters, we prevent the transformation from being non-diffeomorphic. See the localized 1D-cross section of ϕ in Fig. 8(b): the curve is smoother and strictly increasing.

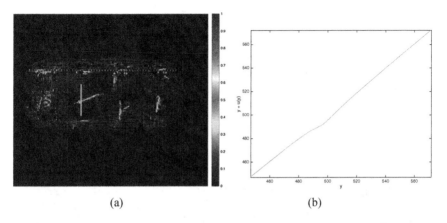

(a)   (b)

Fig. 8. Elastic registration with stronger smoothing $r_\alpha = 1.5$, $r_\beta = 2$ (a) difference map; (b) $y + u_y(x, y)$ transformation representation on localized 1D-cross-section

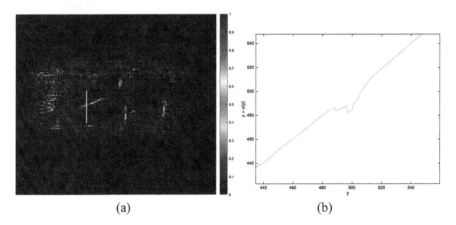

(a)   (b)

Fig. 9. Diffusion registration with $r_\alpha = 1.5$, $r_\beta = \infty$ (a) difference map; (b) $y + u_y(x, y)$ transformation representation on localized 1D-cross-section

In Fig. 9, we show the relevance of the additional volume-preserving soft constraint with respect to the diffusion registration result. On a visual aspect, the preservation of volumes is not maintained, leading to a weaker detectability of objects in the difference map. The transformation ϕ lacks regularity and is clearly non-diffeomorphic.

An additional example depicting the registration of images of a carrier is given in Fig. 10.

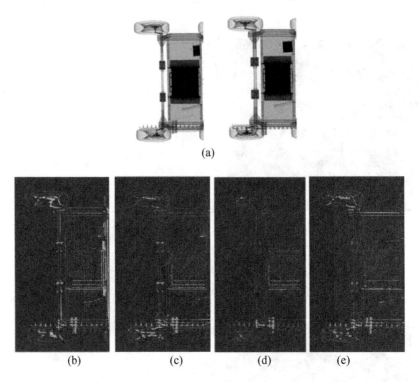

(a)

(b)   (c)   (d)   (e)

Fig. 10. (a) R and T; Difference maps following: (b) rigid registration, (c) 1D horizontal correction, (d) horizontal-then-vertical registration, (e) simultaneous registration with $r_\alpha = 1.5$, $r_\beta = 2$

On this simpler image set too, the consecutive minimization scheme generally shows a better correction than the simultaneous approach. Visually, the major differences at the wheels or the frame borders are removed. It also achieves the lowest MSE (Table 3).

Table 3. Mean square error measures obtained with the different registration methods

	Rigid Registration	X-axis Correction	X-then-Y Correction	Simultaneous
MSE	1.55E−02	4.14E−03	1.11E−03	4.90E−03

After validating our method on hundreds of scan pairs, we empirically obtain stable and accurate results with the ratio parameters $r_\alpha = 1.5$, $r_\beta = 2$ for an optimal detectability of the objects of interest targeted. In Fig. 11 and Table 4, an additional illustration of our method performances is given.

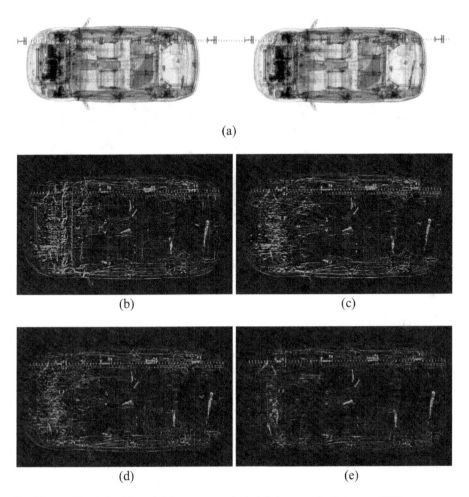

(a)

(b)   (c)

(d)   (e)

Fig. 11. (a) T and R with a visible strong vertical shift between both images; Difference maps following: (b) rigid registration, (c) 1D horizontal correction, (d) simultaneous registration, (e) horizontal-then-vertical registration with $r_\alpha = 1.5$, $r_\beta = 2$

Table 4. Mean square error measures obtained with the different registration methods

	Rigid Registration	X-axis Correction	X-then-Y Correction	Simultaneous
MSE	3.40E−03	2.80E−03	1.10E−03	2.10E−03

In Fig. 12, an extra visualization enhancement case is described.

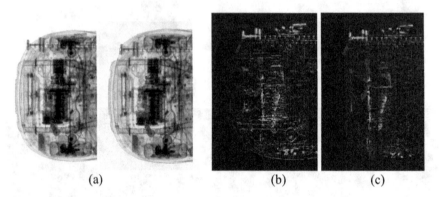

(a)   (b)   (c)

Fig. 12. (a) R and T; difference maps after: (b) 1D horizontal correction, (c) horizontal-then-vertical elastic registration with $r_\alpha = 1.5$, $r_\beta = 2$. Threat detection capacity is strongly enhanced.

6 Conclusion

In this paper, we address the issue of vertical correction for the registration of x-ray images of same-model vehicles. This correction is particularly necessary whenever a singular vertical translation occurs between the cars in R and T. Two approaches are put forward: a simple diffusion paradigm together with a linearized approximation of the smoothly-constrained volume-preserving registration (which turns into the famous elastic registration problem). A discussion is given about applying the registration simultaneously on both components of the displacement field u (with the columnwise-constancy constraint on the horizontal term). The alternative, more efficient method, consists in running the horizontal registration first, followed by the vertical correction in a separate fashion. The results are less noisy, yielding a better registration accuracy on both numerical and visual aspects, especially when elastic regularization is employed.

We also highlight the importance of an appropriate choice for the regularization parameters with the risk of smoothing out suspicious items in the difference map.

Still, resorting to second-order techniques instead of gradient descent methods should be considered for speed purposes. Similarly, matrix-free approaches should also be regarded in future works.

References

1. Modersitzki, J.: Numerical Methods for Image Registration. Oxford University Press on Demand, Oxford (2004)
2. Modersitzki, J.: FAIR: Flexible Algorithms for Image Registration, vol. 6. SIAM, Philadelphia (2009)

3. Marciano, A., Cohen, L.D., Gadi, N.: Vehicle X-ray scans registration: a one-dimensional optimization problem. In: Lauze, F., Dong, Y., Dahl, A.B. (eds.) SSVM 2017. LNCS, vol. 10302, pp. 578–589. Springer, Cham (2017). https://doi.org/10.1007/978-3-319-58771-4_46
4. Fischer, B., Modersitzki, J.: Fast image registration: a variational approach. In: Psihoyios, G. (ed.) Proceedings of the International Conference on Numerical Analysis & Computational Mathematics, pp. 69–74. Wiley (2003)
5. Haber, E., Modersitzki, J.: A multilevel method for image registration. SIAM J. Sci. Comput. **27**(5), 1594–1607 (2006)
6. Sotiras, A., Davatzikos, C., Paragios, N.: Deformable medical image registration: a survey. IEEE Trans. Med. Imaging **32**(7), 1153–1190 (2013)
7. Haber, E., Modersitzki, J.: Numerical methods for volume preserving image registration. Inverse Probl. **20**(5), 1621 (2004)
8. Haber, E., Modersitzki, J.: Image registration with guaranteed displacement regularity. Int. J. Comput. Vis. **71**(3), 361–372 (2007)
9. Thirion, J.P.: Image matching as a diffusion process: an analogy with Maxwell's demons. Med. Image Anal. **2**(3), 243–260 (1998)
10. Rohlfing, T., Maurer, C.R., Bluemke, D.A., Jacobs, M.A.: Volume-preserving nonrigid registration of MR breast images using free-form deformation with an incompressibility constraint. IEEE Trans. Med. Imaging **22**(6), 730–741 (2003)
11. Christensen, G.E., Johnson, H.J.: Consistent image registration. IEEE Trans. Med. Imaging **20**(7), 568–582 (2001)
12. Rueckert, D., Aljabar, P.: Non-rigid registration using free-form deformations. In: Paragios, N., Duncan, J., Ayache, N. (eds.) Handbook of Biomedical Imaging, pp. 277–294. Springer, Boston, MA (2015). https://doi.org/10.1007/978-0-387-09749-7_15
13. Younes, L.: Invariance, déformations et reconnaissance de formes, vol. 44. Springer Science & Business Media, Heidelberg (2003)
14. Beg, M.F., Miller, M.I., Trouvé, A., Younes, L.: Computing large deformation metric mappings via geodesic flows of diffeomorphisms. Int. J. Comput. Vis. **61**(2), 139–157 (2005)
15. Christensen, G.E., Rabbitt, R.D., Miller, M.I.: Deformable templates using large deformation kinematics. IEEE Trans. Image Process. **5**(10), 1435–1447 (1996)
16. Mang, A., Biros, G.: An inexact Newton-Krylov algorithm for constrained diffeomorphic image registration. SIAM J. Imaging Sci. **8**(2), 1030–1069 (2015)
17. Mang, A., Biros, G.: Constrained H^1-regularization schemes for diffeomorphic image registration. SIAM J. Imaging Sci. **9**(3), 1154–1194 (2016)
18. Hameeteman, R., Veenland, J.F., Niessen, W.J.: Volume preserving image registration via a post-processing stage. In: Proceedings of SPIE, vol. 6288 (2006)
19. Fischer, B., Modersitzki, J.: Ill-posed medicine—an introduction to image registration. Inverse Probl. **24**(3), 034008 (2008)
20. Rühaak, J., König, L., Tramnitzke, F., Köstler, H., Modersitzki, J.: A matrix-free approach to efficient affine-linear image registration on CPU and GPU. J. Real-Time Image Process. **13**(1), 205–225 (2017)
21. Albrecht, T., Dedner, A., Lüthi, M., Vetter, T.: Finite element surface registration incorporating curvature, volume preservation, and statistical model information. Computational and Mathematical Methods in Medicine **2013**, 14 p. (2013). http://doi.org/10.1155/2013/674273. Article no. 674273

Colour, Shading and Reflectance of Light

Superpixels Optimized by Color and Shape

Vitaliy Kurlin[1(✉)] and Donald Harvey[2]

[1] Department of Computer Science, University of Liverpool, Liverpool, UK
vitaliy.kurlin@liverpool.ac.uk
[2] Main Yard Studios, 90 Wallis Road, London E9 5LN, UK

Abstract. Image over-segmentation is formalized as the approximation problem when a large image is segmented into a small number of connected superpixels with best fitting colors. The approximation quality is measured by the energy whose main term is the sum of squared color deviations over all pixels and a regularizer encourages round shapes. The first novelty is the coarse initialization of a non-uniform superpixel mesh based on selecting most persistent edge segments. The second novelty is the scale-invariant regularizer based on the isoperimetric quotient. The third novelty is the improved coarse-to-fine optimization where local moves are organized according to their energy improvements. The algorithm beats the state-of-the-art on the objective reconstruction error and performs similarly to other superpixels on the benchmarks of BSD500.

Keywords: Superpixel · Segmentation · Approximation
Boundary Recall · Reconstruction error · Energy minimization
Coarse-to-fine optimization

1 Introduction: Motivations, Problem and Contributions

1.1 Motivations: Superpixels Speed Up Higher Level Processing

Modern cameras produce images containing millions of pixels in a rectangular grid. This pixel grid is not the most natural nor most efficient representation, because not all these pixels are needed to correctly understand an image. Moreover, processing a large image pixel by pixel is slow, and many important algorithms have the running time $O(n^2)$ in the number n of pixels. However we know of a smart vision system (called a human brain) that quickly extracts key elements of complicated scenes by skipping the vast majority of incoming light signals.

The main challenge of low-level vision is to represent a large image in a less-redundant form that can speed up the higher level processing. The central problem is the unsupervised *over-segmentation* when a pixel-based image is segmented into *superpixels* (unions of square-based pixels), which are perceptually meaningful atomic regions with consistent features such as color or texture.

Our motivations are to address the following key challenges of superpixels:

M. Pelillo and E. Hancock (Eds.): EMMCVPR 2017, LNCS 10746, pp. 297–311, 2018.
https://doi.org/10.1007/978-3-319-78199-0_20

- rigorously state the over-segmentation as an approximation problem when a large pixel-based image is approximated by a mesh of fewer superpixels;
- add constraints that superpixels are connected and have no inner holes;
- optimize superpixels in a data-driven way, e.g. by smartly choosing an original configuration and attempting steps in a good order according to their costs;
- avoid parameters whose influence on superpixels are hard to describe.

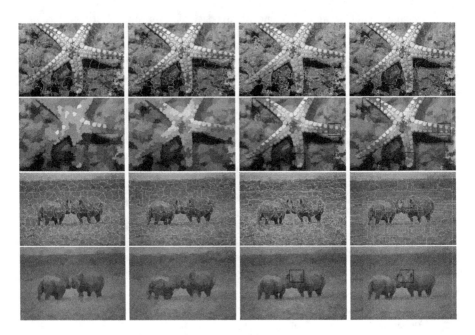

Fig. 1. Odd rows: superpixel meshes by algorithms SLIC [1], SEEDS [2], ETPS [3], ours. Even rows: reconstructed images with the average color for every superpixel. Blue rectangles show the areas where our compact superpixels better capture details. (Color figure online)

1.2   Oversegmentation by Superpixels is an Approximation Problem

The aim is to segment an image of n pixels into at most $k < n$ *superpixels* that are connected unions of pixels satisfying conditions (1.2a)–(1.2d) below.

(1.2a) The resulting superpixels with best constant colors approximate the image well, e.g. a difference between an image and its approximation is minimized.

(1.2b) By construction the superpixels are connected and have no inner holes.

(1.2c) Superpixels adhere well to object boundaries, e.g. in comparison with human-drawn contours in the Berkeley Segmentation Database BSD500 [4].

(1.2d) The only parameters are the number of superpixels and a shape coefficient for a trade-off between the accuracy of boundaries and shapes of superpixels.

Since images are often replaced by their superpixel meshes, condition (1.2a) highlights the importance to measure the quality of such an approximation. The pixelwise sum of squared differences is the standard statistical mean error and can be based on colors (as in Definition 1) or on texture information or other pixel features. Condition (1.2b) guarantees that no post-processing is needed so that a superpixel mesh can be represented by a simple graph (instead of a much larger regular grid) whose nodes are superpixels and whose links connect adjacent superpixels. Condition (1.2c) follows the tradition to evaluate superpixels on BSD benchmarks. Condition (1.2d) restricts manually chosen parameters.

1.3   Contributions to the State-of-the-Art for Superpixels

First, we introduce an *adaptive initialization* of a superpixel mesh, whose main idea of persistent edges from Definition 3 can be used in any hot spot analysis. Second, the new regularizer in Definition 2 is scale-invariant, hence the superpixels are truly optimized by shape. Third, the optimization is improved in Subsect. 4.2 and its time is justified for the first time in Theorem 12. Here are the stages of the algorithm SOCS: Superpixels Optimized by Color and Shape.

Stage 1: detecting persistent horizontal and vertical edges along object boundaries to form a non-uniform grid of rectangular blocks, see Subsect. 3.2.
Stage 2: merging blocks in a grid when a reconstruction error is minimally increased to get a non-regular initial mesh that is quickly adapted to a given image and contains a required number of superpixels, see Subsect. 3.4.
Stage 3: subdividing rectangular blocks within every superpixel into sub-blocks going from a coarse level to a finer level of optimization in Subsect. 4.1.
Stage 4: a new way to choose boundary blocks for moving to adjacent superpixels, then repeat Stage 3 until all blocks become pixels, see Subsect. 4.2.

2   A Review of Past Superpixel Algorithms

The excellent survey by Stutz et al. [5, Table 3 in Sect. 8] recommends 6 algorithms, which are reviewed below in addition to few other good methods.

A pixel-based image is represented by a graph G whose nodes are in a 1–1 correspondence with all pixels, while all edges of G represent adjacency relations between pixels, when each pixel is connected to its closest 4 or 8 neighbors.

The seminal *Normalized Cuts* algorithm by Shi and Malik [6] finds an optimal partition of G into connected components, which minimizes an energy taking into account all nodes of G. The Entropy Rate Superpixels (ERS) of Lie et al. [7] minimizes the entropy rate of a random walk on a graph. The Contour Relaxed Superpixels (CRS) by Conrad et al. [8] optimize a cost depending on texture.

The *Simple Linear Iterative Clustering* (SLIC) algorithm by Achanta et al. [1] forms superpixels by k-means clustering in a 5-dimensional space using 3 colors in CIELAB space and 2 coordinates per pixel. Because the search is restricted to a neighborhood of a given size, the complexity is $O(kmn)$, where n is the number of pixels and m is the number of iterations. This gives an average running time of about 0.2 s per image in BSD500. If a final cluster of pixels is disconnected or contains holes, post-processing is possible, but increases the runtime.

The coarse-to-fine optimisation progressively approximates a superpixel segmentation. At the initial coarse level, each superpixel consists of large rectangular blocks of pixels. At the next level, all blocks are subdivided into 4 rectangles and one rearranges the blocks to find a better approximation depending on a cost function, which continues until all blocks become pixels.

SEEDS (Superpixels Extracted via Energy-Driven Sampling) by Van de Bergh et al. [2] seems the first superpixel algorithm to use a coarse-to-fine optimization. The colors of all pixels within each fixed superpixel are put in bins, usually 5 bins for each channel. Each superpixel has the associated sum of deviations of all bins from an average bin within the superpixel. This sum is maximal for a superpixel whose pixels have colors in one bin. Then SEEDS iteratively maximizes the sum of deviations by shrinking or expanding superpixels.

The ETPS algorithm (Extended Topology Preserving Superpixels) by Yao et al. [3] minimizes a different cost function, which is the reconstruction error RE in Subsect. 3.1 plus the deviation of pixels within a superpixel from a geometric center, along with a cost proportional to the boundaries of superpixels. This regularizer encourages superpixels of small sizes, however the benchmarks on BSD500 are computed [3, Fig. 4] without the regularizer (as for SEEDS). SEEDS and ETPS satisfy topological Condition (1.2c) by construction. ETPS was highlighted as the best algorithm by Stutz et al. [5, Table 3 in Sect. 8].

Polygonal meshes of superpixels at subpixel resolution were constructed from Voronoi cells by Duan and Lafarge [9] and from constrained edges by Forsythe et al. [10]. The continuous optimization of superpixels seems harder than the traditional discrete version. However, a resulting polygonal mesh can be rendered at any higher resolution, so superpixels are no longer restricted to a pixel grid.

The related problem to find a straight-line skeleton for object boundaries [11–13] was solved with theoretical guarantees when an input is a cloud of edge pixels [14,15] based on the new methods from Topological Data Analysis [16].

3   Energy-Based Superpixels Formed by Coarse Blocks

This section explains the new adaptive initialization for coarse superpixels that are better than a uniform grid, which is used in most past superpixel algorithms. Persistent edges in a given image generate a non-uniform mesh of rectangular blocks. These blocks are iteratively merged in such a way that the energy function remains as small as possible or until we get a maximum number of superpixels.

3.1   The Energy is a Reconstruction Error of Approximation

An image I can be considered as a function from pixels to a space of colors. We consider $I(p)$ as the vector (L, a, b) of 3 colors in the CIELAB space, which is more perceptually uniform than $RGB\ space$ with red, green, blue components.

In the *CIELAB space*, L is the lightness, a represents the colors from red to green (the lowest value of a means red, the highest value of a means green). The component b similarly represents the opponent colors from yellow to blue. The OpenCV function cvtColor outputs each Lab channel in the range $[0, 255]$.

Definition 1. *Let an image I of n pixels be segmented into k superpixels. For every pixel p, denote by $S(p)$ the superpixel containing p. Then $S(p)$ has the mean* $\text{color}(S(p)) = \dfrac{1}{|S(p)|} \sum_{q \in S(p)} I(q)$. *Since I is approximated by superpixels with mean colors, the natural measure of quality is the* Reconstruction Error

$$RE = sum\ of\ squared\ color\ deviations = \sum_{p=1}^{n} \Big(I(p) - \text{color}(S(p))\Big)^2. \quad (3.1a)$$

Each of the 3 colors in the Lab space has the range $[0, 255]$. Hence the following normalized Root Mean Square of the color error in percents is shown in Fig. 6.

$$nRMS = \sqrt{\frac{RE}{3n}} \times \frac{100\%}{255} = \sqrt{\frac{1}{3n} \sum_{p=1}^{n} \Big(I(p) - \text{color}(S(p))\Big)^2} \times \frac{100\%}{255}. \quad (3.1b)$$

The Reconstruction Error RE can be written similarly to (3.1a) for other pixel properties instead of colors, e.g. for texture. The main objective advantage of $nRMS$ in (3.1b) is its independence of any subjective ground-truth.

A color term proportional to RE was used by Yao et al. [3] with the regularizer $PD = sum\ of\ squared\ pixel\ deviations = \sum_{p} \Big(p - \text{center}(S(p))\Big)^2$, where $\text{center}(S(p)) = \dfrac{1}{|S(p)|} \sum_{q \in S(p)} q$ is the geometric center of the superpixel $S(p)$. The above term PD is not invariant under scaling and penalizes large superpixels, which has motivated us to introduce the scale-invariant regularizer.

Definition 2. *The* isoperimetric quotient $IQ(S) = \dfrac{4\pi\ \text{area}(S)}{\text{perimeter}^2(S)}$ *of a superpixel S is a scale-invariant shape characteristic having the maximum value 1 for a round disk S. The IQ measure of an image over-segmentation $I = \cup S_{i=1}^{k}$ is*

$$the\ average\ IQ = \sum_{superpixels\ S} \frac{IQ(S)}{\#superpixels} = \frac{4\pi}{k} \sum_{i=1}^{k} \frac{\text{area}(S_i)}{\text{perimeter}^2(S_i)}. \quad (3.1c)$$

The SOCS algorithm will minimize the energy equal to the weighted sum

$$Energy = \frac{RE}{n} + c \times IQ,\ where\ RE\ is\ in\ (3.1a),\ c\ is\ a\ shape\ coefficient. \quad (3.1d)$$

Schick et al. [17] suggested another weighted average of isoperimetric quotients $CO = \sum_{superpixels\ S} \dfrac{\text{area}(S)}{\#superpixels} IQ(S) = \dfrac{4\pi}{k} \sum_{i=1}^{k} \dfrac{\text{area}^2(S_i)}{\text{perimeter}^2(S_i)}$, when larger superpixels are forced to have more round shapes, see graphs in Fig. 6.

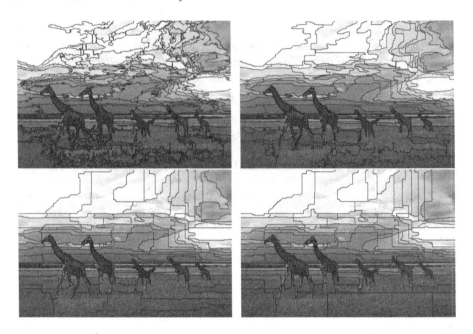

Fig. 2. The effect of c in (3.1d): **1st:** $c = 0$, **2nd:** $c = 1$, **3rd:** $c = 10$, **4th:** $c = 100$.

3.2   Stage 1: Detection of Persistent Horizontal and Vertical Edges

The SOCS algorithm starts by finding *persistent edges* along horizontal and vertical lines of a pixel grid, see Definition 3. The first step is to apply to a given image I the bilateral filter from OpenCV with the size of 5 pixels and sigma values 100 for deviations in the color and coordinate spaces. The second step is to compute the image gradients $d_x I$ and $d_y I$ using the standard 2×2 masks. For every row $j = 1, \ldots, \text{rows}(I)$ in a given image I, we have a graph of gradients $|d_y(I)|$ over $1 \leq i \leq \text{columns}(I)$. The similar graph of magnitudes $|d_x(I)|$ can be computed over every column of I. For any such graph of discrete values $f(1), \ldots, f(l)$, Definition 3 formalizes the automatic method to detect continuous intervals $a \leq t \leq b$, where the graph f has persistently high values.

Definition 3. *For any function $f : \mathbb{R} \to \mathbb{R}$ discretely sampled at $t = 1, \ldots, l$, the* strength *of a line edge $L = [a, b]$ is the sum $\sum_{t \in [a,b]} f(t)$. Figure 3 visualises the strength of an edge L as the area under the continuous graph of $f(t)$ over L.*

For any threshold v, the superlevel set *$f^{-1}[v, +\infty) = \{t \in \mathbb{R} : f(t) \geq v\}$ consists of several edges L_i. When v is decreasing, the edges L_i are growing and merge with each other until we get a single edge covering all points $t = 1, \ldots, l$.*

For any fixed v, compute the widest gap between the strengths of the edges that form the superlevel set $f^{-1}[v, +\infty)$. Find a critical level v between the median and maximum of the widest gap above is maximal, see Fig. 3. At this critical value v, the edges whose strengths are above the widest gap are called persistent.

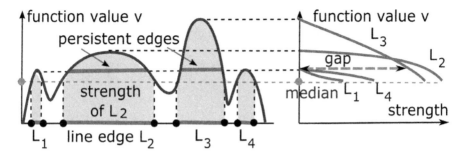

Fig. 3. Left: a superlevel set has 4 edges in green with their strengths highlighted as yellow areas at the median value of f. Right: strengths of edges are analyzed when a threshold v is decreasing, the widest gap between strengths of edges is shown in red. (Color figure online)

Proofs of claims can be replaced by more image experiments in a final version.

Lemma 4. *For any image I of a size $w \times h$ pixels, the persistent edges in all $w + h$ horizontal and vertical lines in I can be found in time $O(w \log h + h \log w)$.*

Proof. Assuming that the points $t = 1, \ldots, l$ form a connected interval graph, the segments in Definition 3 are connected components of a superlevel set $f^{-1}[v, +\infty)$. These components are maintained by a union-find structure, which requires $O(\log l)$ operations per update (creating a new segment, adding a new node to an old segment or merging 2 segments). Every update requires changes of strengths for at most 2 segments, hence $O(\log l)$ operations we if keep the ordered set of strengths in a binary tree. The time is $O(w \log h + h \log w)$ for w columns (vertical lines) of length h and h rows (horizontal lines) of length w. □

Given an expected number k of superpixels, the average area of a single superpixel is n/k. If such a superpixel is a square, its side would be $s = \sqrt{n/k}$. If an image I has a size $w \times h$, we build the a non-uniform grid of $2\lceil w/s \rceil \times 2\lceil h/s \rceil$ rectangular blocks. We select $2\lceil w/s \rceil$ columns and $2\lceil h/s \rceil$ rows that have the maximum strengths of their persistent edges from Definition 3. To avoid close edges, after selecting a current maximum along a line $x = \text{const}$ or $y = \text{const}$, we later ignore the neighboring lines at a distance less than 4 pixels. By extending persistent edges to the boundary of I, we get a non-uniform *edge grid*, see Fig. 4.

3.3   Cost of Merging Superpixels and the Superpixel Structure

The edge grid is already adapted to the image I better than the standard uniform grid used in other algorithms. However, large regions of almost constant colors such as sky can be cut by extended edges into unnecessary small blocks. Stage 2 in Subsect. 3.4 will merge rectangular blocks into a smaller number of larger superpixels without increasing the Reconstruction Error too much.

If an image is segmented into superpixels $I = \cup_{i=1}^{k} S_i$, the Reconstruction Error in formula (3.1a) decomposes as a sum of energies over all superpixels:

$$RE = \sum_{i=1}^{k} E(S_i), \text{where } E(S_i) = \sum_{p \in S_i} \left(I(p) - \text{color}(S_i) \right)^2. \qquad (3.3a)$$

The cost of merging S_i, S_j is $E(S_i, S_j) = E(S_i \cup S_j) - E(S_i) - E(S_j) \geq 0.$
$$(3.3b)$$
This cost can be 0 only if S_i, S_j have exactly the same mean $\text{color}(S_i) = \text{color}(S_j)$. Technically, two superpixels S_i, S_j may share more than one edge, e.g. a connected chain of edges. If the intersection $S_i \cap S_j$ is disconnected, e.g. one edge e and a vertex $v \notin e$, we set $E(S_i, S_j) = +\infty$, so the superpixels S_i, S_j will not merge to avoid harder cases when a superpixel may touch itself.

To prepare the coarse-to-fine optimization in Sect. 4 when superpixels are iteratively improved, we introduce the superpixel structure from our implementation with sums and pointers to 4 sub-blocks for each rectangular block.

Fig. 4. Left: red persistent edges generate the blue edge grid at stage 1. Middle: initial mesh after stage 2. Right: final mesh with 99 superpixels after stages 3–4. (Color figure online)

We split any rectangular block from the non-uniform grid into the four smaller rectangular *sub-blocks* by subdividing each side into 2 almost equal parts whose lengths differ by at most 1 pixel. We don't subdivide 1-pixel sides, so 1-pixel wide blocks are subdivided into 2 blocks. Since each block may keep pointers to its 4 sub-blocks, the superpixel structure looks like a large tree where the root points to coarsest blocks each of which points to its 4 sub-blocks and so on.

Definition 5. *The* superpixel structure *of S contains the number $|S|$ of pixels in a superpixel S, $\text{sum}(S) = \sum_{p \in S} I(p)$, $\text{sum2}(S) = \sum_{p \in S} \left(I(p) \right)^2$, the list of (x, y) indices in the block grid of blocks covered by S. Each block B has the index of its superpixel S, similar sums $|B|, \text{sum}(B), \text{sum2}(B)$ and pointers to its 4 sub-blocks.*

The color sums of a superpixel S in Definition 5 are justified by Lemma 6.

Lemma 6. *In (3.3b) the cost $S(E_i, E_j)$ of merging superpixels S_i, S_j can be computed by using the structure of S_i, S_j from Definition 5 in a constant time.*

Proof. Since $\mathrm{sum}(S) = |S|\mathrm{color}(S)$, the energy in (3.3b) becomes

$$E(S) = \sum_{p \in S} \left(\left(I(p) \right)^2 - 2\mathrm{color}(S)I(p) + \mathrm{color}(S)^2 \right) = \sum_{p \in S} \left(I(p) \right)^2$$

$$- 2\mathrm{color}(S) \sum_{p \in S} I(p) + |S| \left(\mathrm{color}(S) \right)^2 = \mathrm{sum2}(S) - \frac{\left(\mathrm{sum}(S) \right)^2}{|S|}. \tag{3.3c}$$

Since the union $S_i \cup S_j$ nicely affects the area and sums, i.e. $|S_i \cup S_j| = |S_i| + |S_j|$,

$$\mathrm{sum}(S_i \cup S_j) = \mathrm{sum}(S_i) + \mathrm{sum}(S_j), \quad \mathrm{sum2}(S_i \cup S_j) = \mathrm{sum2}(S_i) + \mathrm{sum2}(S_j),$$

(3.1c) implies that the computation of $E(S_i, S_j)$ is independent of $|S_i \cup S_j|$. □

3.4 Stage 2: Merging Adjacent Superpixels with Minimum Energy

At Stage 2 adjacent superpixels are iteratively merged starting from pairs with a minimum cost $E(S_i, S_j)$ in (3.3b). Since superpixels may share more than one edge, we associate the cost $E(S_i, S_j)$ to pairs of adjacent superpixels. Each unordered pair (S_i, S_j) has a unique $\mathrm{key}(S_i, S_j)$, e.g. formed by indices of super-pixels in the edge grid. Stage 2 finishes when the number of superpixels drops down to a given maximum m. Figure 6 shows experiments where the number of superpixels can go down to 0.25 m if $nRMS$ jumps by not more than 2%.

Lemma 7. *If an image $I = \cup_{i=1}^k S_i$ is segmented into k superpixels, there are at most $O(k)$ pairs (S_i, S_j) of adjacent superpixels. In time $O(k \log k)$ one can find and merge (S_i, S_j) with a minimal cost $E(S_i, S_j)$ updating the costs of all pairs.*

Proof. Since the common boundary of S_i, S_j grows over time, we keep the list of all common edges in the binary *edge tree* indexed by $\mathrm{key}(S_i, S_j)$, which allows a fast insertion and deletion of new pairs of adjacent superpixels. To quickly find $\mathrm{key}(S_i, S_j)$ and the corresponding pair of adjacent superpixels with a minimum cost $E(S_i, S_j)$, we put all keys into the binary *cost tree* indexed by $E(S_i, S_j)$.

All k superpixels form a planar network with f bounded faces, g edges, where each pair of adjacent superpixels is represented by one edge. Since each face f has at least $3f$ edges, the doubled number of edges $2g$ is at least $3f$, so $f \leq \frac{2}{3}g$. The Euler formula $k - g + f = 1$ gives $1 \leq k - g + \frac{2}{3}g$, hence $g \leq 3(k-1)$.

Then both binary trees above have the size $O(k)$. The first element in the cost tree has the minimum cost $E(S_i, S_j)$ and can be found and removed in a constant time. The search for the corresponding $\mathrm{key}(S_i, S_j)$ in the edge tree in time $O(\log k)$ leads to the list of common edges of the superpixels S_i, S_j.

The edge grid from is converted into a polygonal mesh using the OpenMesh library. Then each common edge is removed by the collapse and remove_edge operations from OpenMesh taking a constant time. For each of remaining $O(k)$ edges of $S_i \cup S_j$ on the boundary of another superpixel S, the cost $E(S, S_i \cup S_j)$ is computed by Lemma 6 and is added to the cost tree in time $O(\log k)$. □

4   The Coarse-to-Fine Optimization for Superpixels

This section carefully analyzes the coarse-to-fine optimization used by Yao et al. [3]. At Stage 3 each rectangular block in a current grid is subdivided as explained before Definition 5. At Stage 4 each *boundary block* that belongs to a superpixel S_i and is adjacent to another superpixel S_j is checked for a potential move from S_i to S_j. After completing this optimization for all boundary blocks, Stages 3 and 4 are repeated at the next finer level until all blocks become pixels.

4.1   Stage 3: Subdividing Rectangular Blocks into Four Sub-blocks

Lemma 8 explains how the superpixel structure from Definition 5 helps us to quickly compute color sums for all superpixels and subdivide all superpixels.

Lemma 8. *Let an image $I = \cup_{i=1}^{k} S_i$ of n pixels be segmented into k superpixels. Then all $|S_i|, \mathrm{sum}(S_i), \mathrm{sum}2(S_i)$ can be found in time $O(n)$ independent of k.*

Proof. We recursively compute all sums for each block B by adding the corresponding sums from each of 4 sub-blocks of B. Since, for each single-pixel block $B = p$, we have $|B| = 1$, $\mathrm{sum}(B) = I(p)$, $\mathrm{sum}2(B) = \left(I(p)\right)^2$, we need only $O(n) + O(n/4) + O(n/16) + \cdots = O(n)$ additions to compute the sums for all blocks. For each superpixel S_i, we find $|S_i|, \mathrm{sum}(S_i), \mathrm{sum}2(S_i)$ by adding the sums from all blocks in S_i in time $O(|S_i|)$, so the total time is $O(n)$.   □

Lemma 9. *When blocks are subdivided going from a coarse to a finer level, each superpixel S containing b blocks larger than 1×1 can be updated in time $O(b)$.*

Proof. By Definition 5 for each superpixel S, we only need to replace the list of blocks in the current grid by a longer list of blocks in the refined grid, which is done by merging the lists from the 4 sub-blocks of each block covering by S. The index of S is copied to every new sub-block, which takes $O(b)$ time.   □

4.2   Tree of Boundary Blocks and Local Connectivity of Superpixels

A block B in a superpixel S is called *boundary* if one of its 4 side neighbors belongs to a different superpixel, which can be quickly checked by comparing superpixel indices of all blocks. The ETPS algorithm puts all boundary blocks into a *priority queue* and adds any new boundary blocks to the end of this queue. We have replaced this queue by a binary tree where blocks are ordered by costs of moves so that moves are attempted according to their costs, not row by row.

Blocks in the tree are tested one by one for a potential move to an adjacent superpixel. Such a move was called *forbidden* in [3, Sect. 3] if S becomes disconnected after removing B. However, the global connectivity of $S - B$ is slow to check. A removal of a boundary block B from a superpixel S respects the *local connectivity* of S if the 8-neighborhood of B within $S - B$ is connected, see Fig. 5. The 3 pictures in [3, Fig. 3] show some (but not all) forbidden moves, so we justify below why the local connectivity can be checked in a constant time.

Fig. 5. Left: allowed moves preserves the local connectivity. Right: a forbidden move.

Lemma 10. *For any boundary block B moving from a superpixel S_i to another superpixel S_j, the local connectivity of $S_i - B$ can be checked in a constant time.*

Proof. We go around the circular 8-neighborhood $N_8(B)$, consider all blocks of $S - B$ as isolated vertices, add an edge between vertices u, v if the corresponding blocks in $N_8(B)$ share a common side. Then $S - B$ is locally connected around B if and only if the resulting graph on at most 8 vertices is connected.   □

4.3   Stage 4: Updating Superpixels in a Constant Time per Move

Let the move of $B \subset S_i$ to another superpixel S_j keep $S_i - B$ locally connected. If the Reconstruction Error in (3.1a) is decreased, we move B to S_j and will add new boundary blocks to the cost tree, otherwise we remove B from the tree.

Lemma 11. *For any block B moving from a superpixel S_i to another superpixel S_j, the structures of both superpixels S_i, S_j can be updated in a constant time.*

Proof. All sums of colors over the block B are subtracted from the corresponding sums of S_i and are added to the sums of S_j. We change the superpixel index of B from i to j. After B has moved, only its 4 neighboring blocks can change their boundary status, which is checked in a constant time by comparing superpixel indices. Any new boundary blocks are added to the binary tree of blocks.   □

The time at Stage 4 essentially depends on the number q of boundary blocks that are processed in the cost tree. Stage 4 finishes when the tree is empty or q exceeds the upper bound of n given pixels, which never happened for BSD500.

Theorem 12. *The SOCS algorithm segmenting an image of n pixels into k superpixels has the asymptotic computational complexity $O(n + k^2 \log k + q)$.*

Proof. Stage 1 has time about $O(\sqrt{n} \log n)$ by Lemma 4. At Stage 2 we merge at most $O(k)$ pairs of superpixels, each pair in time $O(k \log k)$ by Lemma 7. By Lemma 8 all superpixels are subdivided in time $O(b)$ for a grid with b blocks larger than 1×1. The number of such blocks increases to n by a factor of at least 2, hence the total time for Stages 3 is $O(n)$. By Lemmas 10 and 11 the time for Stage 4 is proportional to the number q of processed blocks in the cost tree, because each boundary block is adjacent to at most 3 other superpixels.   □

5   Comparisons with Other Algorithms on BSD500

The Berkeley Segmentation Database BSD500 [4] contains 500 natural images and human-drawn closed contours around object boundaries. Then all pixels in every image are split into disjoint *segments*, which are large unions of pixels comprising a single object. So every pixel has the index of its superpixel and some pixels are also labeled as boundary. Every image has about 5 human drawings, which vary significantly and are called the *ground-truth segmentations*.

For an image I, let $I = \cup G_j$ be a segmentation into ground-truth segments and $I = \cup_{i=1}^k S_i$ be an oversegmentation into superpixels produced by an algorithm. Each quality measure below compares the superpixel $S_1, \ldots, S_k$ with the best suitable ground-truth from the BSD500 database for every image.

Let $G(I) = \cup G_j$ be the union of ground-truth boundary pixels and $B(I)$ be the boundary pixels produced by a superpixel algorithm. For a distance ε in pixels, the Boundary Recall $BR(\varepsilon)$ is the ratio of ground-truth boundary pixels $p \in G(I)$ that are within the distance ε from the superpixel boundary $B(I)$.

$$\text{The Undersegmentation Error } USE = \frac{1}{n} \sum_j \sum_{S_i \cap G_j \neq \emptyset} |S_i - G_j| \qquad (5a)$$

was often used in the past, where $|S_i - G_j|$ is the number of pixels that are in S_i, but not in G_j. However a superpixel is fully penalized when $S_i \cap G_j$ is 1 pixel, which required ad hoc thresholds, e.g. the 5% threshold $|S_i - G_j| \geq 0.05|S_i|$ by Achanta et al. [1], or ignoring boundary pixels of S_i by Liu et al. [7].

Van den Bergh et al. [2] suggested the more accurate measure, namely

$$\text{the Corrected Undersegmentation Error } CUE = \frac{1}{n} \sum_i |S_i - G_{\max}(S_i)|, \quad (5b)$$

where, $G_{\max}(S_i)$ is the ground-truth segment having the largest overlap with S_i. Neubert and Protzel [18] introduced the Undersegmentation Symmetric Error

$$USE = \frac{1}{n} \sum_j \sum_{S_i \cap G_j \neq \emptyset} \min\{\text{in}(S_i), \text{out}(S_i)\}, \qquad (5c)$$

where $\text{in}(S_i)$ is the area of S_i inside G_j, $\text{out}(S_i)$ is the area of S_i outside G_j. To keep graphs readable, Fig. 6 compares SOCS to the 3 past algorithms ETPS, SEEDS, SLIC, coming on top of others in the evaluations by Stutz et al. [5, Table 3].

As suggested by Theorem 12, the running time of SOCS is similar to ETPS at about 1 s on average per BSD image on a laptop with 2.6 GHz and 8 GB RAM.

6   Summary and Discussion of the New SOCS Algorithm

The SOCS algorithm has a fast adaptive initialization that is based on persistent edges in an image and can substantially reduce the number of superpixels without

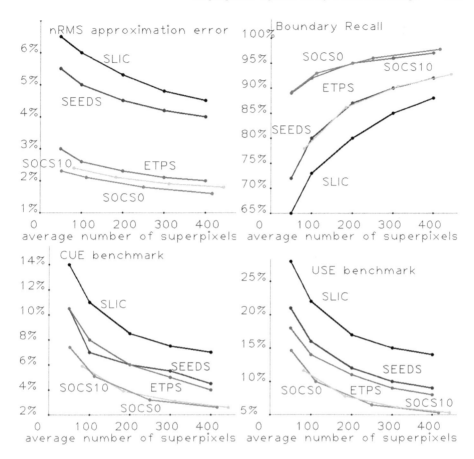

Fig. 6. Each dot is (average number of superpixels, average benchmark) on BSD500. SOCS0 in red has shape coefficient $c = 0$, SOCS10 in orange has $c = 10$, see (3.1d). (Color figure online)

compromizing the quality of approximation. The new coarse-to-fine optimization quickly converges to a minimum by moving boundary blocks of large sizes and then by subdividing them into smaller blocks many of which remain stable.

- The first theoretical contribution is the formal statement of the image over-segmentation as an approximation problem by superpixels in Subsect. 1.2.
- The adaptive initialization of superpixels consisting of large rectangular blocks can be used in many other algorithms that start from a coarse uniform grid.
- The coarse-to-fine optimization has been substantially improved by keeping boundary blocks sorted in a binary tree instead of a linear queue.
- The SOCS algorithm outperforms the state-of-the art for the approximation error (nRMS) and undersegmentation errors CUE/USE on BSD500 images.

Here are the practical advantages of the new practical SOCS algorithm.

- The output superpixels are *connected*, because the connectivity is checked in Stage 4 when boundary blocks are updated, which gives the overall speed-up.
- The SOCS algorithm can be stopped at any time after Stage 1, e.g. at any optimization step, because each update needs a constant time by Lemma 11.
- The only *essential input parameters* are the maximum number k of superpixels and the shape coefficient for a trade-off between accuracy and compactness.
- The SOCS algorithm is *modular* and allows improvements in different parts, e.g. persistent edges in Stage 1 can be found in another way, merging blocks in Stage 2 can be done by another strategy with the same Reconstruction Error.

References

1. Achanta, R., Shaji, A., Smith, K., Lucchi, A., Fua, P., Süsstrunk, S.: SLIC superpixels compared to the state-of-the-art. Trans. PAMI **34**, 2274–2282 (2012)
2. Van de Bergh, M., Boix, X., Roig, G., Van Gool, L.: SEEDS: superpixels extracted via energy-driven sampling. Int. J. Comput. Vis. **111**, 298–314 (2015)
3. Yao, J., Boben, M., Fidler, S., Urtasun, R.: Real-time coarse-to-fine topologically preserving segmentation. In: Proceedings of CVPR, pp. 216–225 (2015)
4. Arbelaez, P., Maire, M., Fowlkes, C., Malik, J.: Contour detection and hierarchical image segmenetaton. Trans. PAMI **33**, 898–916 (2011)
5. Stutz, D., Hermans, A., Leibe, B.: Superpixels: an evaluation of the state-of-the-art. Comput. Vis. Image Underst. **166**, 1–27 (2017)
6. Shi, J., Malik, J.: Normalized cuts and image segmentation. Trans. PAMI **22**, 888–905 (2000)
7. Liu, M.Y., Tuzel, O., Ramalingam, S., Chellappa, R.: Entropy rate superpixel segmentation. In: Proceedings of CVPR, pp. 2097–2104 (2011)
8. Conrad, C., Mertz, M., Mester, R.: Contour-relaxed superpixels. In: Heyden, A., Kahl, F., Olsson, C., Oskarsson, M., Tai, X.C. (eds.) Energy Minimization Methods in Computer Vision and Pattern Recognition. LNCS, vol. 8081, pp. 280–293. Springer, Heidelberg (2013). https://doi.org/10.1007/978-3-642-40395-8_21
9. Duan, L., Lafarge, F.: Image partitioning into convex polygons. In: Proceedings of CVPR, pp. 3119–3127 (2015)
10. Forsythe, J., Kurlin, V., Fitzgibbon, A.: Resolution-independent superpixels based on convex constrained meshes without small angles. In: Bebis, G., et al. (eds.) ISVC 2016. LNCS, vol. 10072, pp. 223–233. Springer, Cham (2016). https://doi.org/10.1007/978-3-319-50835-1_21
11. Chernov, A., Kurlin, V.: Reconstructing persistent graph structures from noisy images. Image-A **3**, 19–22 (2013)
12. Kurlin, V.: Auto-completion of contours in sketches, maps and sparse 2D images based on topological persistence. In: Proceedings of CTIC, pp. 594–601 (2014)
13. Kurlin, V.: A fast persistence-based segmentation of noisy 2D clouds with provable guarantees. Pattern Recogn. Lett. **83**, 3–12 (2016)
14. Kurlin, V.: A one-dimensional homologically persistent skeleton of a point cloud in any metric space. Comput. Graph. Forum **34**, 253–262 (2015)
15. Kurlin, V.: A homologically persistent skeleton is a fast and robust descriptor of interest points in 2D images. In: Azzopardi, G., Petkov, N. (eds.) CAIP 2015. LNCS, vol. 9256, pp. 606–617. Springer, Cham (2015). https://doi.org/10.1007/978-3-319-23192-1_51

16. Kurlin, V.: A fast and robust algorithm to count topologically persistent holes in noisy clouds. In: Proceedings of CVPR, pp. 1458–1463 (2014)

17. Schick, A., Fischer, M., Stifelhagen, R.: Measuring and evaluating the compactness of superpixels. In: Proceedings of ICPR, pp. 930–934 (2012)

18. Neubert, P., Protzel, P.: Compact watershed and preemptive SLIC: on improving trade-offs of superpixel segmentation algorithms. In: proceedings of the ICPR, pp. 996–1001 (2014)

Maximum Consensus Parameter Estimation by Reweighted ℓ_1 Methods

Pulak Purkait[1(✉)], Christopher Zach[1], and Anders Eriksson[2]

[1] Toshiba Research Europe, Cambridge, UK
pulak.isi@gmail.com
[2] Queensland University of Technology, Brisbane, Australia

Abstract. Robust parameter estimation in computer vision is frequently accomplished by solving the maximum consensus (MaxCon) problem. Widely used randomized methods for MaxCon, however, can only produce random approximate solutions, while global methods are too slow to exercise on realistic problem sizes. Here we analyse MaxCon as iterative reweighted algorithms on the data residuals. We propose a smooth surrogate function, the minimization of which leads to an extremely simple iteratively reweighted algorithm for MaxCon. We show that our algorithm is very efficient and in many cases, yields the global solution. This makes it an attractive alternative for randomized methods and global optimizers. The convergence analysis of our method and its fundamental differences from the other iteratively reweighted methods are also presented.

Keywords: Reweighted ℓ_1 methods · Maximum consensus
M-estimator

1 Introduction

Robust estimation of model parameters is a critical task in computer vision [1]. The literature on robust estimators is vast [2], encompassing different robust criteria and the associated algorithms. In computer vision, however, maximum consensus (MaxCon) is one of the most widely used robust criteria. Accordingly, algorithms for solving MaxCon have been researched extensively in recent years and people have developed a number of ways to solve this. In this article we seek for a fast iterative method for model estimation under MaxCon criterion.

Definition 1 MaxCon criterion. *Given a set of measurements $\mathcal{X} = \{\mathbf{x}_i\}_{i=1}^n$, find the model parameters $\boldsymbol{\theta} \in \mathbb{R}^d$ that agree with as many of the data as possible. i.e.,*

$$\max_{\boldsymbol{\theta},\, \mathcal{I} \subseteq \mathcal{X}} |\mathcal{I}| \quad subject\ to \quad r(\boldsymbol{\theta}; \mathbf{x}_i) \leq \epsilon,\ \forall \mathbf{x}_i \in \mathcal{I}, \tag{P1}$$

where $r(\boldsymbol{\theta}; \mathbf{x}_i)$ is the absolute value of the residual of $\boldsymbol{\theta}$ at the point $\mathbf{x}_i$, and ϵ is the inlier threshold. The point set $\mathcal{I}$ is called the consensus set w.r.t. $\boldsymbol{\theta}$. A data point $\mathbf{x}_i$ is called an inlier w.r.t. $\boldsymbol{\theta}$ if $\mathbf{x}_i \in \mathcal{I}$; otherwise, it is called an outlier.

© Springer International Publishing AG, part of Springer Nature 2018
M. Pelillo and E. Hancock (Eds.): EMMCVPR 2017, LNCS 10746, pp. 312–327, 2018.
https://doi.org/10.1007/978-3-319-78199-0_21

Problem (P1) can also be written by introducing slack variables, one for each data point, as follows:

$$\min_{\boldsymbol{\theta},\, \mathbf{s}} \sum_{i=1}^{n} \mathbf{1}(s_i) \quad \text{subject to} \quad r(\boldsymbol{\theta}; \mathbf{x}_i) \leq \epsilon + s_i, \quad s_i \geq 0, \qquad \text{(P2)}$$

where $\mathbf{1}(s_i)$ is an indicator function that returns 1 if s_i is *non-zero*. Effectively, a point $\mathbf{x}_i$ with a strictly positive slack s_i is regarded as an outlier. Formulation (P2) thus seeks the MaxCon solution by minimizing the number of outliers. The equivalence between the formulations (P1) and (P2) can be easily established. The optimized slack values can be interpreted as *shrinkage residuals*, to borrow a term from the area of shrinkage operators [3]. In most of the geometric problems, the residuals are linear or quasiconvex [4]. The quasiconvex functions have convex sub-level sets and the constraints $r(\boldsymbol{\theta}; \mathbf{x}_i) \leq \epsilon + s_i, s_i \geq 0$ form a convex set $\mathcal{C}$ under quasiconvex (or linear) residuals.

Now the question is whether minimizing the piecewise objective of (P2) under the convex constraints $\mathcal{C}$ is an easy problem? In the following Lemma we show that the set of stationary points of (P2) is in-fact the feasible set $\mathcal{C}$ itself. This makes the problem difficult to optimize.

Lemma 1. *Any feasible point* $(\boldsymbol{\theta}^*, \mathbf{s}^*) \in \mathcal{C}$ *is a local minimum of* (P2).

Proof. Let $\mathcal{O}^*$ be the support set of $\mathbf{s}^*$, *i.e.*, $\mathcal{O}^* = \{i : s_i^* > 0\}$. Let $s^{*+} := \min_{i \in \mathcal{O}^*} s_i^*$ and $\delta \in (0,\ s^{*+})$. Then the MaxCon objective $|\mathcal{O}| = \sum_i \mathbf{1}(s_i > 0)$ is non-decreasing in the max-norm neighbourhood

$$N_\delta := \{\mathbf{s} \in \mathbb{R}^n : s_i \geq 0, \|\mathbf{s} - \mathbf{s}^*\|_\infty \leq \delta\}.$$

The above is true because, by construction for any feasible $\mathbf{s} \in N_\delta$ has at least the same number of non-zeros as $\mathbf{s}^*$. If $\mathbf{s} \in N_\delta$ is not feasible, it leads to the infinite objective. Thus, the MaxCon objective $|\mathcal{O}|$ is not lower than the value at $(\boldsymbol{\theta}^*, \mathbf{s}^*)$ in the neighbourhood $\mathbf{s} \in N_\delta$. In summary, all feasible points $(\boldsymbol{\theta}^*, \mathbf{s}^*) \in \mathcal{C}$ are local minima of (P2) in a neighbourhood of $(\boldsymbol{\theta}^*, \mathbf{s}^*)$. $\qquad\square$

Thus MaxCon is a combinatorial optimization problem that is very challenging. It is typically approached by randomized sample-and-test methods, primarily `RANSAC` [5] and its variations [6–9]. These randomized sampling methods are limited to a "simple model", *i.e.*, would not work for Bundle Adjustment (or translation registration). Moreover, the random nature of the algorithms results in approximate solutions with no guarantees of local or global optimality; indeed sometimes the result can be far from the optimal. Presently, several globally optimal algorithms exist [10–14], however, they are usually based on branch-and-bound or brute force search, thus, they are only practical for small problem sizes n.

What is surely missing, therefore, is an *efficient* and *deterministic* algorithm for the MaxCon problem. A number of variations of `RANSAC` are available, *e.g.*, `LO-RANSAC` [15,16], nonetheless, these methods follow similar mechanism of

RANSAC. MLEsac [7,17] optimizes a (slightly) different criterion than MaxCon. Although, both are MLEs – a noise model with uniform inliers and outliers is utilized in MaxCon; in contrast, MLEsac utilizes Gaussian inliers and uniform outliers. In this work, we develop an iterative refinement scheme for MaxCon optimization (P2) that produces near optimal solutions. Thus, the proposed method lies in-between fast but very approximate solutions and superior but slow global optimal solution.

2   Iterative Reweighted ℓ_1 Methods

The convex relaxation to (P2) is the minimization of absolute sum of the shrinkage residuals (assumed bounded)

$$\min_{\boldsymbol{\theta},\,\mathbf{s}} \sum_{i=1}^{n} s_i \quad \text{subject to} \quad r(\boldsymbol{\theta};\mathbf{x}_i) \leq \epsilon + s_i, \quad s_i \geq 0, \tag{1}$$

which is also a robust estimation of the model parameters $\boldsymbol{\theta}$. Olsson et al. [18] used this formulation for outlier removal by iteratively solving (1) and removing the points with positive shrinkage residuals. Since ℓ_1 norm is linear, (1) optimizes a linear objective functional under convex constraints $\mathcal{C}$ and hence can be solved efficiently with the existing optimizers [19,20].

The difference between the objective of (P2) and (1) is in how the weighting of the magnitude of $\mathbf{s}$ affects the optimal solution. Specifically, the larger coefficients are penalized more heavily in (1) than smaller coefficients, unlike in (P2) where positive magnitudes are penalized equally.

2.1   Proposed Smooth Surrogate Function

The MaxCon (P2) cannot be solved directly due to the presence of a large number of local solutions. We utilize the regularized smooth surrogate $G_\gamma(\mathbf{s}) = \sum_{i=1}^{n} \log(s_i + \gamma)$ to reduce the number of local solutions of ℓ_0, and arrive at the following constrained concave minimization problem,

$$\min_{\boldsymbol{\theta},\,\mathbf{s}} \sum_{i=1}^{n} \log(s_i + \gamma) \quad \text{subject to} \quad r(\boldsymbol{\theta};\mathbf{x}_i) \leq \epsilon + s_i, \quad s_i \geq 0. \tag{P3}$$

γ is a parameter chosen as a small positive number to ensure the measure is bounded from below, since s_i can become vanishingly small. This damping factor γ can also be observed as a regularization of the optimization [21].

2.2   Minimization of the Smooth Surrogate Function

The general form of (P3) under the convex constraints $\mathcal{C}$

$$\min_{\mathbf{u}} f(\mathbf{u}) \quad \text{subject to} \quad \mathbf{u} \in \mathcal{C}, \tag{2}$$

where f is concave and $\mathcal{C}$ is convex. As a concave function f lies below its tangent, one can improve upon a guess $\mathbf{u}$ of the solution by minimizing a linearisation of f around $\mathbf{u}$. This yields the following iterative algorithm

$$\mathbf{u}^{(l+1)} := \arg\min_{\mathbf{u}\in\mathcal{C}} f(\mathbf{u}^{(l)}) + \left\langle \nabla f(\mathbf{u}^{(l)}), (\mathbf{u} - \mathbf{u}^{(l)}) \right\rangle := \arg\min_{\mathbf{u}\in\mathcal{C}} \left\langle \nabla f(\mathbf{u}^{(l)}), \mathbf{u} \right\rangle, \quad (3)$$

with the initialization $\mathbf{u}^0 \in \mathcal{C}$. Each iteration is now the solution to a convex problem [22]. For (P3), substituting $\nabla G_\gamma(\mathbf{s}) = [1/s_i + \gamma]$ in (3) yields

$$(\boldsymbol{\theta}^{(l+1)}, \mathbf{s}^{(l+1)}) := \underset{(\boldsymbol{\theta},\,\mathbf{s})\in\mathcal{C}}{\operatorname{argmin}} \ \sum_{i=1}^{n} s_i/(s_i^{(l)} + \gamma). \quad (4)$$

Defining $w_i^{(l)} = (s_i^{(l)} + \gamma)^{-1}$, we obtain the proposed iterative reweighted method: at each iteration it solves the following weighted problem

$$\left.\begin{aligned} (\boldsymbol{\theta}^{(l+1)}, \mathbf{s}^{(l+1)}) := \underset{\boldsymbol{\theta},\,\mathbf{s}}{\operatorname{argmin}} \sum_{i=1}^{n} w_i^{(l)} s_i \\ \text{subject to} \quad r(\boldsymbol{\theta}; \mathbf{x}_i) \le \epsilon + s_i, \ s_i \ge 0, \\ w_i^{(l)} := (s_i^{(l)} + \gamma)^{-1}. \end{aligned}\right\} \quad (\text{S1})$$

The details of the initializations are in Sect. 5. Note that computation of a step-size or a line-search is not required which can significantly speed-up the computation.

As the residuals of most of the 3D geometric problems under study are quasiconvex [23], our algorithm is guaranteed to converge (shown in Sect. 3) for any $r(\boldsymbol{\theta}; \mathbf{x}_i)$ that is quasiconvex. Thus, the proposed algorithm minimizes a linear objective under quasiconvex residuals [4]. This motivates us to call proposed algorithm IR-LP to distinguish it from traditional IRL1. Note that under linear residuals, IR-LP solves only a linear program (LP) in each iteration.

Other properties of (P3) Let $(\boldsymbol{\theta}^*, \mathbf{s}^*)$ be a minimizer of (P3). Then, the Lagrangian is given by

$$\mathcal{L}(\mathbf{s}, \boldsymbol{\theta}; \lambda, \mu) = \sum_i \left(\log(s_i + \gamma) + \lambda_i \big(r(\boldsymbol{\theta}; \mathbf{x}_i) - \epsilon - s_i\big) - \mu_i s_i \right) \quad (5)$$

where $\lambda_i \ge 0$ and $\mu_i \ge 0$ are Lagrange multipliers. The KKT conditions are as follows

$$\frac{1}{s_i^* + \gamma} - \lambda_i - \mu_i = 0, \ \sum_{i=1}^{n} \lambda_i \nabla_{\boldsymbol{\theta}} r(\boldsymbol{\theta}^*; \mathbf{x}_i) = 0$$
$$\lambda_i[r(\boldsymbol{\theta}^*; \mathbf{x}_i) - \epsilon - s_i^*] = 0, \ \mu_i s_i^* = 0 \quad (6)$$
$$r(\boldsymbol{\theta}^*; \mathbf{x}_i) \le \epsilon + s_i^*, \ s_i^* \ge 0, \ \lambda_i \ge 0, \ \mu_i \ge 0.$$

From the first condition, we know $\lambda_i + \mu_i = 1/(s_i^* + \gamma) > 0$ which implies both of μ_i and λ_i can not be zero simultaneously. Hence, for each i, one of the constraints

$s_i^* \geq 0$ or $r(\boldsymbol{\theta}^*; \mathbf{x}_i) \leq \varepsilon + s_i^*$ is always active. A local minimum $(\boldsymbol{\theta}^*, \mathbf{s}^*)$ is, thus, characterized by

$$s_i^* = \begin{cases} 0 & \text{if } i \in \mathcal{I} \\ r(\boldsymbol{\theta}^*; \mathbf{x}_i) - \varepsilon & \text{if } i \in \mathcal{O}, \end{cases}$$

where $\mathcal{O}$ is the support set of $\mathbf{s}^*$. $\mathcal{O}$ can also be considered as an outlier set as $s_i^* > 0$ corresponds to an outlier point. Thus, $\mathcal{I}$ can be considered as an inlier set. Note that for $i \in \mathcal{O}$, $\lambda_i = 1/(s_i^* + \gamma)$ and for $i \in \mathcal{I}$, $\lambda_i = 0$. Thus by (6),

$$\sum_{i: r(\boldsymbol{\theta}^*; \mathbf{x}_i) > \epsilon} \frac{\nabla_\theta r(\boldsymbol{\theta}^*; \mathbf{x}_i)}{r(\boldsymbol{\theta}^*; \mathbf{x}_i) - \epsilon + \gamma} = 0, \tag{7}$$

which says that the weighted sum of the gradients corresponding to the outliers at a minimum $(\boldsymbol{\theta}^*, \mathbf{s}^*)$ vanishes. However a direct relationship with the optimal choice of γ and the number of outliers can not be derived which would have given a potential choice of γ. The choices of γ are further discussed in Sect. 5.

Compared to (P2), where all feasible points are local minima due to Lemma 1, (P3) reduces the number of local minima by increasing γ. In Fig. 1, we display the objective of (P3) on a synthetic 2D line fitting problem, under different values of γ. As γ increases, the topographic surface of the objective function is flatten and fewer local minima are observed. This is an empirical evidence that G_γ smoothens the objective of (P2) in a sensible way. The choice of γ is discussed further in Sect. 5.

The Connection with Basis Pursuit
In the basis pursuit problem, one aims to recover the sparsest signal $\boldsymbol{\theta} \in \mathbb{R}^d$ from the measurements $\mathbf{y} \in \mathbb{R}^n$, with respect to a dictionary $\phi \in \mathbb{R}^{n \times d}$:

$$\min_{\boldsymbol{\theta}} \sum_{k=1}^{d} \mathbb{1}(\theta_k) \text{ subject to } \mathbf{y} = \phi\boldsymbol{\theta}. \tag{8}$$

Candes $et\ al.$ [24] also proposed a smooth surrogate $\sum_{k=1}^{d} \log(|\theta_k| + \gamma)$ of the objective above that results an iteratively reweighted ℓ_1-norm minimization (IRL1) algorithm for (8). Specifically, at the l-th iteration, the following weighted ℓ_1 problem is solved

$$\boldsymbol{\theta}^{(l+1)} := \operatorname*{argmin}_{\boldsymbol{\theta}} \sum_{k=1}^{d} w_k^{(l)} |\theta_k| \text{ subject to } \mathbf{y} = \phi\boldsymbol{\theta},$$
$$w_k^{(l)} := (|\theta_k^{(l)}| + \gamma)^{-1}. \tag{9}$$

Though related, (P2) and (8) are quite different problems.

- The former seeks sparsity on the shrinkage residuals $\mathbf{s}$ (parameters $\boldsymbol{\theta}$ allowed to be dense), while the latter seeks sparsity in $\boldsymbol{\theta}$.
- Further, the constraints in (P2) are usually over-determined ($n > d$), while for (8) the constraints are under-determined ($d > n$).

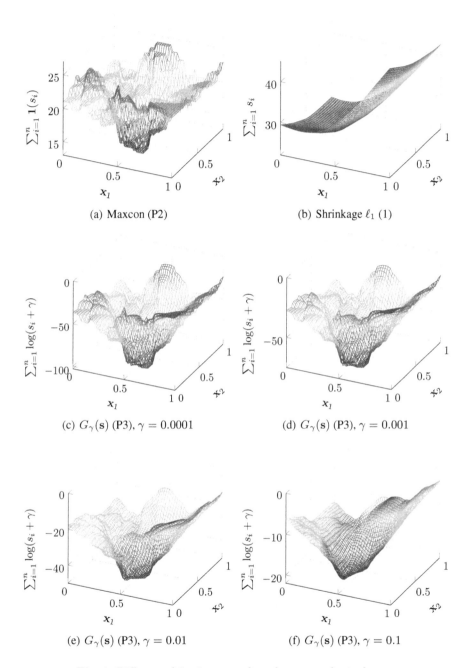

(a) Maxcon (P2)

(b) Shrinkage ℓ_1 (1)

(c) $G_\gamma(\mathbf{s})$ (P3), $\gamma = 0.0001$

(d) $G_\gamma(\mathbf{s})$ (P3), $\gamma = 0.001$

(e) $G_\gamma(\mathbf{s})$ (P3), $\gamma = 0.01$

(f) $G_\gamma(\mathbf{s})$ (P3), $\gamma = 0.1$

Fig. 1. Different objectives are plotted on a synthetic data.

- Moreover, the proposed method (S1) can also be treated as maximization of residual diversity [25, 26]. Interested readers are referred to the extended version.

Although, the proposed reweighted algorithm is inspired by Candes *et al.* [24], above set our work apart from [24] that has different theoretical underpinnings. Thus, the methods for basis pursuit problems cannot be directly adapted here.

3   Convergence Analysis

In this section, we analyse the convergence of the proposed algorithm (S1). Let $\mathcal{A} : U \to \mathcal{P}(U)$ be an algorithm defined on a set U where $\mathcal{P}(U)$ is the power set of U. Given $\mathcal{A}$, Zangwill's global convergence theorem [27] is stated as

Theorem 1. *Let $\mathcal{A} : U \to \mathcal{P}(U)$ generate a sequence $\{\mathbf{u}^{(l)}\}_{l=0}^{\infty}$ through the iteration $\mathbf{u}^{(l+1)} \in \mathcal{A}(\mathbf{u}^{(l)})$, given an initialization $\mathbf{u}^{(0)} \in U$. Let $\Gamma \subset U$ be a set called solution set. Further, let $\mathcal{A}$ satisfy the following constraints*

C1. The points in $\{\mathbf{u}^{(l)}\}$ are contained in a compact subset.
C2. If Γ is the solution space of $\mathcal{A}$, then, there is a continuous function $\mathcal{L}(\mathbf{u})$: $U \to \mathbb{R}$ satisfying

$$\begin{cases} \mathcal{L}(\mathbf{u}^{(l+1)}) < \mathcal{L}(\mathbf{u}^{(l)}) & \text{if } \mathbf{u}^{(l)} \notin \Gamma \\ \mathcal{L}(\mathbf{u}^{(l+1)}) \leq \mathcal{L}(\mathbf{u}^{(l)}) & \text{if } \mathbf{u}^{(l)} \in \Gamma \end{cases} \tag{10}$$

C3. The algorithm $\mathcal{A}$ is closed at points outside Γ.

Then, every convergent subsequence of $\{\mathbf{u}^{(l)}\}_{l=0}^{\infty}$ converges to a solution of $\mathcal{A}$.

Lemma 2. *Let us define the solution space Γ as the set of stationary points of (P3). Then sequence $\{\mathbf{s}^{(l)}\}_{l=0}^{\infty}$ generated by the proposed algorithm $\mathcal{A}$ (S1) satisfies the global convergence theorem.*

Proof. We show that the conditions for the Theorem 1 hold for the sequence $\{\mathbf{s}^{(l)}\}_{l=0}^{\infty}$ generated by the algorithm $\mathcal{A}$.
C1. Every closed and bounded set is compact. An equivalent condition is that the points in the sequence and its accumulation points are bounded. We can certainly find an upper bound of sequence $\{\mathbf{s}^{(l)}\}_{l=0}^{\infty}$ generated by $\mathcal{A}$. Such bounds exist as for a finite solution with finite points residuals cannot be arbitrary large. Moreover, the accumulation points are no greater than the bounding values. Therefore, such a compact subset S can be constructed from the bounds.
C2. Given a real number $\gamma > 0$, define $\mathcal{L}(\mathbf{s}) : \mathcal{C} \to \mathbb{R}$

$$\mathcal{L}(\mathbf{s}) = \sum_{i=1}^{n} \log(s_i + \gamma) \tag{11}$$

where $\mathcal{C}$ is the feasible region defined by the constraints in (P2). For the points $\mathbf{s}^{(l)} \notin \Gamma$

$$\frac{1}{n}\left(\mathcal{L}(\mathbf{s}^{(l+1)}) - \mathcal{L}(\mathbf{s}^{(l)})\right) = \sum_{i=1}^{n}\left(\frac{1}{n}\log(s_i^{(l+1)} + \gamma) - \frac{1}{n}\log(s_i^{(l)} + \gamma)\right)$$

$$= \sum_{i=1}^{n}\frac{1}{n}\log\frac{s_i^{(l+1)} + \gamma}{s_i^{(l)} + \gamma} < \log\left(\frac{1}{n}\sum_{i=1}^{n}\frac{s_i^{(l+1)} + \gamma}{s_i^{(l)} + \gamma}\right)$$

$$\leq \log\left(\frac{1}{n}\sum_{i=1}^{n}\frac{s_i^{(l)} + \gamma}{s_i^{(l)} + \gamma}\right) = 0 \Rightarrow \mathcal{L}(\mathbf{s}^{(l+1)}) < \mathcal{L}(\mathbf{s}^{(l)}).$$

Here the first inequality follows from the strict concavity property of the $\log(.)$ function. Note that the equality happens only when $\mathbf{s}^{(l+1)} = \mathbf{s}^{(l)}$ which implies $\langle \nabla f(\mathbf{u}^{(l)}), \mathbf{u} \rangle = 0$ (by Eq. (3)). Thus the inequality is strict for $\mathbf{s}^{(l)} \notin \Gamma$. The second inequality follows from the fact that $\mathbf{s}^{(l)}$ is obtained by minimizing $\sum_{i=1}^{n} s_i/(s_i^{(l)} + \gamma)$, $\mathbf{s} \in \mathcal{C}$ and $\log(.)$ is monotonic increasing. Moreover, for $\mathbf{s}^{(l)} \in \Gamma$

$$\begin{aligned} \mathbf{s}^{(l+1)} = \mathbf{s}^{(l)} &\implies \mathcal{L}(\mathbf{s}^{(l+1)}) = \mathcal{L}(\mathbf{s}^{(l)}) \\ \text{and} \quad \mathbf{s}^{(l+1)} \neq \mathbf{s}^{(l)} &\implies \mathcal{L}(\mathbf{s}^{(l+1)}) < \mathcal{L}(\mathbf{s}^{(l)}). \end{aligned} \quad (12)$$

Thus $\mathbf{s}^{(l)} \in \Gamma$ implies $\mathcal{L}(\mathbf{s}^{(l+1)}) \leq \mathcal{L}(\mathbf{s}^{(l)})$.

C3. A continuous mapping from a compact set to a set of real numbers is a closed map [28]. The map $\mathcal{A}$ is continuous and the set S, containing the elements of $\{\mathbf{s}^{(l)}\}_{l=0}^{\infty}$, the range of the mapping $\mathcal{A}$ in our algorithm, has already been proven as compact. $\qquad \square$

Theorem 2. *For any starting point $\{\boldsymbol{\theta}^{(0)}, \mathbf{s}^{(0)}\} \in \mathcal{C}$, there exist a subsequence of the sequence generated by (S1) converges asymptotically to a stationary point of (P3).*

Proof. The sequence $\{\mathbf{s}^{(l)}\}_{l=0}^{\infty}$ is compact. Therefore, there must exist a convergent subsequence $\{\mathbf{s}^{(p_l)}\}_{l=0}^{\infty}$ of $\{\mathbf{s}^{(l)}\}_{l=0}^{\infty}$. By Lemma 2, the convergent subsequence $\{\mathbf{s}^{(p_l)}\}_{l=0}^{\infty}$ converge to a stationary point of (P3). $\qquad \square$

The above theorem shows that the objective of (P3) generated by the sequence $\{\boldsymbol{\theta}^{(l)}, \mathbf{s}^{(l)}\}_{l=0}^{\infty}$ strictly decreases and converges to a local minimum or a saddle point of (P3). Further, by Lemma 1, any feasible solution of (P3) is also a local minimum of (P2). Thus, the proposed algorithm (S1) is guaranteed to find a local minimum of (P2).

4   Runtime Complexity

The complexity of the proposed methods IR-LP depends on the complexity of the each iteration as maximum number of iterations L is fixed. The global methods [12] and [10] that require $\mathcal{O}(k^{d+1})$ and $\mathcal{O}((d+1)^k)$ number of iterations respectively, where d is the dimension of the problem and k is the number of outliers. Note that the above numbers are enormous compared to L (choices of L are discussed in results Section of the extended version). Further, in each iteration, those global methods solve a similar linear program or convex program. Furthermore, like [12], except the initial iteration, we initialize by the solution of the previous iteration.

Linear Residuals IR-LP solves a LP in each iteration which is remarkably efficient in practice. Moreover, as the coefficient matrix is extremely sparse, it becomes an effective solver [22]. Although, there are worst-case polynomial time algorithms for solving a LP, *e.g.*Karmakar's projective algorithm $\mathcal{O}(n^{3.5})$, we utilize an approximate solution[1], which is solved in linear time [29].

Quasiconvex Residuals IR-LP minimize linear objective under convex constraints that can be solved by an interior point algorithm [30] in polynomial time.

5   Parameter Settings

Initialization

The initialization of the shrinkage residuals $\mathbf{s}^{(0)}$ can be aided using any fast approximate method. However, the initialization should not be too far from the optimal solution. In all of our experiments, unless stated otherwise, we initialize $\mathbf{s}^{(0)} = \mathbf{1}$ and then iterate the first iteration to find a suboptimal solution $\boldsymbol{\theta}^{(1)}$. Again, $\boldsymbol{\theta}^{(1)}$ is utilized to update the shrinkage residuals $\mathbf{s}^{(1)}$. A better initialization (RANSAC solution or iterative ℓ_∞ [31]) leads to a better solution in some cases, however, our chosen trivial initialization works well in most of the applications. The results under different initializations are discussed in the extended version.

Selecting γ

In the proposed algorithm, the constant γ serves to bound the smooth objective from below, and also regularizes the optimization to avoid the stiffness to the solution where $s_i^{(l)} = 0$; intuitively, note that there will be points (*i.e.*, the inliers) where the slack values are zero. In general, the algorithm works reasonably well with a small independent choice of γ. In this work, however, we chose $\gamma = 0.01$ for all the experiment reported and got satisfactory results.

In the literature of reweighted methods, some works [24,32] exhibit better performance on some datasets by adapting γ. Specifically, [24] chose $\gamma^{(l+1)} = \max\{\mathbf{s}^{(l)^+}, 0.01\}$ where $\mathbf{s}^+$ are the positive slack variables, [32] utilized an annealing schedule and forced $\gamma^{(l+1)} \to 0$. However, note that for adaptively chosen $\gamma^{(l)}$, one can no longer guarantee the convergence of the algorithm.

---

[1] Since $\mathbf{s}^{(l)}$ is only used to compute the weights $\mathbf{w}^{(l+1)}$ in the next iteration, an approximate solution, which still minimizes the objective, is sufficient to initialize $\mathbf{s}^{(l+1)}$.

Stopping Criterion

Proposed iterative reweighted method IR-LP is executed till the objective function in two consecutive iteration is less than ζ or maximum number of iterations L is exhausted. Now, if $\mathbf{s}^{(l)}$ and $\mathbf{s}^{(l+1)}$ are the shrinkage residuals of (S1) in consecutive iterations, $\sum_{i=1}^{N} w_i^{(l)} s_i^{(l)} - \sum_{i=1}^{N} w_i^{(l)} s_i^{(l+1)} \geq 0$. We terminate the iteration once the difference is less than ζ, *i.e.*,

$$0 \leq \sum_{i=1}^{N} s_i^{(l)}/(s_i^{(l)} + \gamma) - \sum_{i=1}^{N} s_i^{(l+1)}(s_i^{(l)} + \gamma) \leq \zeta$$

$$\Rightarrow 0 \leq \sum_{s_i^{(l)} > 0} (s_i^{(l)} - s_i^{(l+1)})/(s_i^{(l)}/\gamma + 1) - \sum_{s_i^{(l)} = 0} s_i^{(l+1)} \leq \gamma\zeta$$

Thus for a smaller value of γ, the above constraint enforces a small variability of $\mathbf{s}^{(l+1)}$. Notice that γ is not involved for the inlier residuals in the above expression. Thus, a small number of iteration L is required for a small choice of γ. However, in practice with the above choice of γ, the proposed method works quite well with $L = 25$ and $\zeta = 10^{-4}$.

6   Results

To evaluate the proposed method IR-LP, a number of experiments have been performed on synthetic and real datasets. We compared IR-LP against state-of-the-art approximate methods for MaxCon, namely

- IR-QP: a reweighted least square scheme obtained by replacing each iteration of (S1) by a quadratic program (QP) under linear or quasiconvex residuals (described in the extended version). Note that there is no closed form solution of each iteration and one needs to solve a convex quadratic program.
- Olsson *et al.*'s ℓ_1 method [18]; see (1).
- Sim and Hartley's ℓ_∞ method [31], where the ℓ_∞ is recursively solved and the data with the largest residuals are removed from the subsequent iterations.
- As a baseline, we ran vanilla RANSAC with confidence $\rho = 0.99$ [5].
- MLEsac method [7], that adopts similar sampling strategy as RANSAC to instantiate models, but chooses the one that maximizes the likelihood.
- We also run locally optimize LO-RANSAC [16] as a baseline. We only run our own implementation where the parameters were carefully chosen from Table 1 of [16]. The stopping criterion was considered same as vanilla RANSAC.
- For the experiments with real data, we also consider L-RANSAC – allowing vanilla RANSAC to run same amount of time as the proposed method IR-LP.
- We also execute a global method ASTAR [12][2] with maximum allowable runtime 300 seconds. Note that as the global method is terminated early, it cannot guarantee optimality.

All the methods were implemented in Matlab and executed on a *i7 CPU*.

---

[2] http://pulakpurkait.com/Data/astar_cvpr15_code.zip.

Note that when $r(\boldsymbol{\theta}; \mathbf{x}_i)$ is linear, the subproblems (each iterations) of ℓ_1 and proposed IR-LP are LPs, while for IR-QP the subproblems are QPs. The optimization toolboxes ℓ_1-magic[3] [33] and cvx[4] [34] are employed to solve the LPs and QPs. When $r(\boldsymbol{\theta}; \mathbf{x}_i)$ is quasiconvex, the subproblems of all the methods are convex programs [22]; we solved each convex program instances again with cvx.

6.1   Hyperplane Fitting

We generated $N = 250$ points around an 8-dimensional hyperplane under Gaussian noise with $\sigma_{in} = 0.1$. A number of the points (5%–80%) were then corrupted by a uniform noise (interval $[-10, 10]$) to simulate outliers. The inlier threshold was chosen as $\epsilon = 0.3$. For a chosen outlier percentage, we generated 100 instances of the data and ran the different methods. Figures 2(a) and (b) show the average consensus size and run time over the synthesized data.

While ℓ_1, ℓ_∞ and RANSAC were very fast, they usually produced lower quality results, in terms of the discrepancy with the global solution. While the solution quality of IR-LQ was better to ℓ_1 and RANSAC, it was much slower, owing to the fact that a QP needs to be solved in each iteration. MLEsac is slower than other randomized method as it has an additional inner loop to estimate the

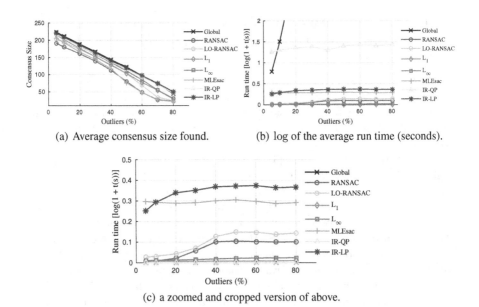

(a) Average consensus size found.   (b) log of the average run time (seconds).

(c) a zoomed and cropped version of above.

Fig. 2. Hyperplane fitting results. Proposed IR-LP clearly dominates the other methods. Please see text for details.

---

[3] http://statweb.stanford.edu/~candes/l1magic/.
[4] http://cvxr.com/cvx/.

mixing parameter. Further, unlike RANSAC, no probabilistic bounds for number of iterations has been incorporated for MLEsac and executed for 500 iterations. However, as MLEsac has different criterion (ML) for model estimation, it produces no better solution than other suboptimal methods. LO-RANSAC performs quite well for low outlier ratio. It is clear from the figures that proposed IR-LP was able to produce near optimal solutions in all the cases; in fact, we observed that IR-LP produced optimal solutions in almost 30% of the runs. Furthermore, the proposed IR-LP is most effective for the cases with (50%–70%) outlier ratio which are the most common scenarios for the real datasets.

6.2   Homography Fitting

In this experiment, we used images from the Oxford Visual Geometry Group[5], namely, Valbonne Church (image index 4 and 7), University Library (image index 1 and 2), and Keble College (image index 2 and 3). These images have been used extensively in previous works on geometric estimation. On each image pair, SIFT key-points were detected and matched using the VLFeat toolbox[6], where the second nearest neighbour test was invoked to prune wrong matches. We used the default parameters in VLFeat. Both linearised residuals and geometric (quasiconvex) residuals are considered for homography estimation, which involves estimation of an 8D parameter vector $\boldsymbol{\theta}$.

Linearised residuals. The reader is referred to [35, Sect. 4.1.2] on linearising the residuals for homography estimation. Each point-sets were normalized separately by translating to mean $= 0$ and scaling to $std = \sqrt{2}$. The inlier threshold ϵ was chosen as $\epsilon = 0.1$. Table 1 presents the results of all methods. For RANSAC and other randomized methods, the results were averaged over 100 runs. While ℓ_1 was very fast, its solution quality was very poor—this was most likely because the distribution of outliers in real data is not balanced, unlike in synthetic data where the outliers were considered to be uniformly distributed. It can also be seen that IR-QP is much slower than the other methods. We executed an efficient implementation of LO-RANSAC, but we believe, it has similar runtime complexity as RANSAC. In contrast, IR-LP always produces larger size consensus set, and while its runtime was longer than RANSAC, LO-RANSAC and ℓ_1, it was much faster than IR-QP. This proves overall better performance for IR-LP.

Quasiconvex residuals. Model estimation under quasiconvex residuals is more geometrically meaningful, and inlier thresholds can be quoted in geometric units (pixels). The reader is referred to [23] for the precise formulation of quasiconvex residuals for homography estimation.

Results under the inlier threshold $\epsilon = 1$ pixels are shown in Table 1. On average proposed IR-LP managed to return the approximate solution that is better than the other methods. Both IR-QP and IR-LP were able to significantly improved upon the other methods, and the final solution quality of

---

[5] http://www.robots.ox.ac.uk/~vgg/data/.
[6] http://www.vlfeat.org.

Table 1. First two blocks: results for homography estimation with linear and quasiconvex residuals. The last block: results for linearised fundamental matrix estimation. n: number of point correspondences, $|\mathcal{I}^*|$: consensus set size (average for the randomized methods), σ: std of the consensus set size, $|\mathcal{I}|$: consensus set size, t(s): runtime in seconds. The columns corresponding to the runtime are marked by gray, and the best values (the maximum size consensus set and the runtime) are marked with bold fonts.

	Methods	ASTAR [12]		RANSAC		ℓ_1 [18]		L_∞ [31]		MLEsac [7]		LO-RANSAC [16]		IR-QP		IR-LP		L-RANSAC	
	Datasets / n	$|\mathcal{I}^*|$	t(s)	$|\mathcal{I}|\pm\sigma$	t(s)	$|\mathcal{I}|$	t(s)	$|\mathcal{I}|$	t(s)	$|\mathcal{I}|\pm\sigma$	t(s)	$|\mathcal{I}|\pm\sigma$	t(s)	$|\mathcal{I}|$	t(s)	$|\mathcal{I}|$	t(s)	$|\mathcal{I}|\pm\sigma$	t(s)
linear	Valbonne Ch. 108	67	300	64.5±3.1	**0.02**	26	0.01	37	0.02	62.1±2.4	0.61	66.8±0.7	0.06	61	4.71	67	0.07	64.8±0.3	0.07
	University Lib. 665	552	300	542.3±12.1	**0.03**	435	0.03	251	1.30	523.4±18.7	1.34	546.8±3.8	0.27	550	5.60	553	0.21	546.2±1.8	0.21
	Keble College 399	311	300	305.7±6.0	**0.03**	102	0.02	145	0.10	224.3±26.8	1.48	308.0±0.5	0.13	310	2.37	311	0.07	307.5±0.9	0.07
	Road Sign 31	29	300	27.6±2.1	**0.01**	14	0.01	22	0.01	27.3±0.6	0.53	29.0±0.0	**0.01**	29	0.71	29	**0.01**	28.9±0.0	**0.01**
	House 492	355	300	351.0±7.8	0.03	261	**0.04**	194	0.21	344.1±9.5	1.60	349.1±2.0	0.12	351	3.14	352	0.24	353.2±2.5	0.24
	Cathedral 544	481	300	464.5±13.2	**0.02**	445	0.03	289	0.27	465.7±12.5	1.84	473.4±5.1	0.22	479	7.03	479	0.36	470.5±0.8	0.36
quasiconvex	Valbonne Ch. 108	83	21.4	75.6±6.1	0.02	26	0.19	60	1.22	71.1±4.7	1.11	82.3±2.4	0.13	**83**	6.61	83	6.23	83.9±0.1	6.23
	University Lib. 665	598	300	590.5±19.9	0.03	464	2.74	338	8.42	529.9±16.7	1.93	601.9±2.3	0.31	608	106.73	613	25.72	606.9±0.1	25.72
	Keble College 399	309	300	306.1±5.1	0.02	92	0.64	177	2.13	301.9±2.0	2.07	307.8±1.3	0.14	303	32.87	308	7.06	**309.8±0.4**	7.06
	Road Sign 31	29	300	28.5±4.1	0.01	2	0.558	23	0.242	28.7±0.4	1.39	28.3±0.8	0.03	30	3.92	**30**	1.63	30.0±0.0	1.63
	House 492	349	300	353.0±8.2	0.06	273	1.922	277	1.278	352.8±0.7	3.37	353.0±0.0	0.12	292	36.73	355	24.82	354.0±0.0	24.82
	Cathedral 544	473	300	463.0±14.7	0.02	461	1.23	438	1.28	468.0±4.9	1.72	473.7±4.7	0.19	471	28.50	**481**	9.03	479.8±0.1	9.03

	Methods	ASTAR [12]		RANSAC		ℓ_1 [18]		L_∞ [31]		MLEsac [7]		LO-RANSAC [16]		IR-QP		IR-LP		L-RANSAC	
	Datasets / n	$|\mathcal{I}^*|$	t(s)	$|\mathcal{I}|\pm\sigma$	t(s)	$|\mathcal{I}|$	t(s)	$|\mathcal{I}|$	t(s)	$|\mathcal{I}|\pm\sigma$	t(s)	$|\mathcal{I}|\pm\sigma$	t(s)	$|\mathcal{I}|$	t(s)	$|\mathcal{I}|$	t(s)	$|\mathcal{I}|\pm\sigma$	t(s)
linear	Valbonne Ch. 108	88	300	77.1±2.8	**0.06**	17	0.01	65	0.02	73.9±4.1	0.78	80.6±2.5	0.14	78	1.98	85	0.18	80.1±1.7	0.18
	Wadham Cl. 1051	365	300	287.8±18.1	**0.18**	129	0.02	213	0.09	242.5±29.5	1.15	317.0±21.3	0.22	312	13.19	344	0.38	307.8±12.9	0.38
	M. College I 577	234	300	212.6±3.6	**0.09**	79	0.01	58	0.03	197.2±4.9	0.91	212.8±1.2	0.16	211	10.83	207	0.13	216.1±18.1	0.18
	Merton Cl. III 313	214	300	176.8±6.8	**0.06**	44	0.01	174	0.01	155.4±7.2	0.59	184.9±7.0	0.08	197	7.23	210	0.24	189.3±2.6	0.24
	Corridor 124	72	300	55.3±2.5	**0.11**	13	0.01	55	0.01	41.9±3.9	0.32	59.4±2.8	0.12	56	5.39	67	0.16	57.6±0.8	0.16
	Dinosaur 156	94	300	68.6±5.3	0.08	85	0.11	23	0.02	33.8±3.7	0.04	70.0±3.9	0.01	92	0.04	78	**0.07**	82.3±2.7	0.09

IR-QP/IR-LP were much higher than iterative ℓ_1 and ℓ_∞. Under quasiconvex residuals, IR-LP is equally expensive as IR-QP due to the requirement of solving convex programs.

6.3   Fundamental Matrix Estimation

We repeat the previous experiment, for linearised fundamental matrix estimation, on the same set of image pairs. Refer to [35, Sect. 9.2.3] for the precise procedure in linearising the residual for fundamental matrix estimation. The normalizations of the individual point-sets were also performed here. $\boldsymbol{\theta}$ is also 8-dimensional and inlier threshold ϵ was chosen to be 0.1. To test the optimum performance of all methods, we did not enforce the rank-2 constraint on the resulting fundamental matrices in all the methods.

We observe that a simple choice of the initialization $\mathbf{s}^{(0)} = \mathbf{1}$ does not lead to a satisfactory local solution for this experiment. Here we initialize $\boldsymbol{\theta}$ by the solution of the iterative ℓ_∞ algorithm [31] $\boldsymbol{\theta}_\infty$. The shrinkage residuals $\mathbf{s}^{(0)}$ for all the points are then computed by evaluating residuals at $\boldsymbol{\theta}_\infty$. The RANSAC solution could also be another choice for initialization. However, iterative ℓ_∞ was chosen purely on computational basis. The results of different methods are shown in Table 1. The runtime for the iterative ℓ_∞ is added with the runtime of IR-LP and IR-QP. As the iterative ℓ_∞ method is very fast, its local refinement by proposed method is an attractive choice for fundamental matrix estimation.

7   Conclusions

In this work, we formulated the maximization of the size of a consensus set as the iterative minimization of the re-weighted ℓ_1 norm of the shrinkage residuals. Then, we illustrated different smooth surrogates of MaxCon. Followed by the minimization of a smooth surrogate that led to an iterative reweighted algorithm IR-LP. A convergent analysis and the runtime complexity of IR-LP are also discussed. Furthermore, a number of reweighted methods is derived for this task and compared with the proposed method. Experimental results show the efficiency of the proposed method compared to the existing approximate methods. Finally, we would like to draw an attention to the fact that, in the linear residual case, each iteration of our algorithm simply requires solving a single LP, and thus the method can be implemented very easily using the existing optimization tools. Thus, our method can surely be used as a replacement of the randomized methods.

References

1. Meer, P.: Robust techniques for computer vision. In: Medioni, G., Kang, S.B. (eds.) Emerging Topics in Computer Vision, pp. 107–190. Prentice Hall (2004)
2. Huber, P.J.: Robust Statistics. Springer, Heidelberg (2011). https://doi.org/10.1007/978-3-642-04898-2

3. Beck, A., Teboulle, M.: A fast iterative shrinkage-thresholding algorithm for linear inverse problems. SIAM J. Imaging Sci. **2**, 183–202 (2009)
4. Olsson, C., Kahl, F.: Generalized convexity in multiple view geometry. J. Math. Imaging Vis. **38**, 35–51 (2010)
5. Fischler, M.A., Bolles, R.C.: Random sample consensus: a paradigm for model fitting with applications to image analysis and automated cartography. Commun. ACM **24**, 381–395 (1981)
6. Choi, S., Kim, T., Yu, W.: Performance evaluation of RANSAC family. JCV **24**, 271–300 (1997)
7. Torr, P.H., Zisserman, A.: MLESAC: a new robust estimator with application to estimating image geometry. CVIU **78**, 138–156 (2000)
8. Raguram, R., Frahm, J.-M., Pollefeys, M.: A comparative analysis of RANSAC techniques leading to adaptive real-time random sample consensus. In: Forsyth, D., Torr, P., Zisserman, A. (eds.) ECCV 2008. LNCS, vol. 5303, pp. 500–513. Springer, Heidelberg (2008). https://doi.org/10.1007/978-3-540-88688-4_37
9. Raguram, R., Chum, O., Pollefeys, M., Matas, J., Frahm, J.M.: USAC: a universal framework for random sample consensus. IEEE TPAMI **35**, 2022–2038 (2013)
10. Olsson, C., Enqvist, O., Kahl, F.: A polynomial-time bound for matching and registration with outliers. In: CVPR (2008)
11. Enqvist, O., Ask, E., Kahl, F., Åström, K.: Robust fitting for multiple view geometry. In: Fitzgibbon, A., Lazebnik, S., Perona, P., Sato, Y., Schmid, C. (eds.) ECCV 2012. LNCS, vol. 7572, pp. 738–751. Springer, Heidelberg (2012). https://doi.org/10.1007/978-3-642-33718-5_53
12. Chin, T.J., Purkait, P., Eriksson, A., Suter, D.: Efficient globally optimal consensus maximisation with tree search. In: CVPR, pp. 2413–2421 (2015)
13. Li, H.: Consensus set maximization with guaranteed global optimality for robust geometry estimation. In: ICCV, pp. 1074–1080. IEEE (2009)
14. Zheng, Y., Sugimoto, S., Okutomi, M.: Deterministically maximizing feasible subsystems for robust model fitting with unit norm constraints. In: CVPR (2011)
15. Chum, O., Matas, J., Kittler, J.: Locally optimized RANSAC. In: Michaelis, B., Krell, G. (eds.) DAGM 2003. LNCS, vol. 2781, pp. 236–243. Springer, Heidelberg (2003). https://doi.org/10.1007/978-3-540-45243-0_31
16. Lebeda, K., Matas, J., Chum, O.: Fixing the locally optimized RANSAC-full experimental evaluation. In: BMVC12. Citeseer (2012)
17. Tordoff, B.J., Murray, D.W.: Guided-MLESAC: faster image transform estimation by using matching priors. IEEE TPAMI **27**, 1523–1535 (2005)
18. Olsson, C., Eriksson, A., Hartley, R.: Outlier removal using duality. In: CVPR (2010)
19. Vanderbei, R.J.: LOQO User's Manual-version 4.05. Princeton University, Princeton (2006)
20. Wächter, A., Biegler, L.T.: On the implementation of an interior-point filter line-search algorithm for large-scale nonlinear programming. Math. Prog. **106**, 25–57 (2006)
21. Chartrand, R., Yin, W.: Iteratively reweighted algorithms for compressive sensing. In: ICASSP, pp. 3869–3872. IEEE (2008)
22. Boyd, S., Vandenberghe, L.: Convex Optimization. Cambridge University Press, Cambridge (2004)
23. Kahl, F., Hartley, R.I.: Multiple-view geometry under the l_∞-norm. IEEE TPAMI **30**, 1603–1617 (2008)
24. Candes, E.J., Wakin, M.B., Boyd, S.P.: Enhancing sparsity by reweighted l_1 minimization. J. Fourier Anal. Appl. **14**, 877–905 (2008)

25. Gorodnitsky, I.F., Rao, B.D.: Sparse signal reconstruction from limited data using focuss: a re-weighted minimum norm algorithm. TSP **45**, 600–616 (1997)
26. Chartrand, R.: Exact reconstruction of sparse signals via nonconvex minimization. IEEE SPL **14**, 707–710 (2007)
27. Sriperumbudur, B.K., Lanckriet, G.R.: A proof of convergence of the concave-convex procedure using Zangwill's theory. Neural Comput. **24**, 1391–1407 (2012)
28. Tu, L.W.: An Introduction to Manifolds. Springer, New York (2010). https://doi.org/10.1007/978-1-4419-7400-6
29. Megiddo, N.: Linear programming in linear time when the dimension is fixed. JACM **31**, 114–127 (1984)
30. Ye, Y., Tse, E.: An extension of Karmarkar's projective algorithm for convex quadratic programming. Math. Prog. **44**, 157–179 (1989)
31. Sim, K., Hartley, R.: Removing outliers using the l_∞ norm. In: CVPR (2006)
32. Wipf, D., Nagarajan, S.: Iterative reweighted and methods for finding sparse solutions. JSTSP **4**, 317–329 (2010)
33. Candes, E.J., Tao, T.: Decoding by linear programming. IEEE TIT **51**, 4203–4215 (2005)
34. Grant, M., Boyd, S.: CVX: matlab software for disciplined convex programming (2008)
35. Hartley, R., Zisserman, A.: Multiple View Geometry in Computer Vision. Cambridge University Press, Cambridge (2003)

A Graph Theoretic Approach for Shape from Shading

Robert Scheffler[✉], Ashkan Mansouri Yarahmadi, Michael Breuß,
and Ekkehard Köhler

Brandenburgische Technische Universität Cottbus-Senftenberg,
Postfach 10 13 44, 03013 Cottbus, Germany
{robert.scheffler,ashkan.mansouriyarahmadi,michael.breuss,
ekkehard.koehler}@b-tu.de

Abstract. Resolving ambiguities is a fundamental problem in shape from shading (SFS). The classic SFS approach allows to reconstruct the surface locally around singular points up to an ambiguity of convex, concave or saddle point type.

In this paper we follow a recent approach that seeks to resolve the local ambiguities in a global graph-based setting so that the complete surface reconstruction is consistent. To this end, we introduce a novel graph theoretic formulation for the underlying problem that allows to prove for the first time in the literature that the underlying surface orientation problem is $\mathcal{NP}$-complete. Moreover, we show that our novel framework allows to define an algorithmic framework that solves the disambiguation problem. It makes use of cycle bases for dealing with the graph construction and enables an easy embedding into an optimization method that amounts here to a linear program.

Keywords: Shape from shading · Ambiguity · Configuration graph
Cycle basis

1 Introduction

Shape from shading (SFS) is a classic problem in computer vision which was introduced in a systematic way by Horn, cf. [10,11]. Given a single grey value input image, the task of SFS is to infer the 3D depth of the depicted objects in the photographed scene. As SFS is a photometric method it makes use of the brightness variation in the input image together with information on the lighting to achieve this goal.

The main model assumptions of the classic SFS approach are that of an orthographic camera and Lambertian surface reflectance. The light is thereby assumed to fall in a parallel way from infinity. In this setting, the solution of the SFS problem is known to be unique under some circumstances except for the convex-concave ambiguity [4,5] which is inherent to the problem, let us also refer to [8,13] for related discussions.

© Springer International Publishing AG, part of Springer Nature 2018
M. Pelillo and E. Hancock (Eds.): EMMCVPR 2017, LNCS 10746, pp. 328–341, 2018.
https://doi.org/10.1007/978-3-319-78199-0_22

The classic SFS model is formulated in terms of a partial differential equation (PDE) and gives an account of reflected light between surface normals and lighting direction. In doing this, the surface normal is determined uniquely only at the points where the surface is frontal to the light falling onto it. Under the standard assumptions, these points are identical to the brightest points in the input image and are called *singular points*. Locally around singular points, the shape is determined up to three kinds of ambiguity, namely of convex, concave or saddle type [15]. At hand of all the investigations, one may come to the conclusion that once the shape type could be assigned around a singular point in terms of convex, concave or saddle nature, the complete surface shape could be computed without any ambiguity left.

In this work we follow an idea proposed by Zhu and Shi [19]. There, a global disambiguation problem is formulated to resolve the ambiguities. It relies on a graph construction called configuration graph in the mentioned work. The underlying principle is to flip the local geometry of surface patches around singular points, so that the resulting complete surface has no kinks. In order to compute the solution on the configuration graph, Zhu and Shi propose two possible problem formulations in terms of Max-Cut problems that are solved via semidefinite programming relaxation. A related approach has been submitted earlier than [19] by Chang et al. [6] that amounts to minimize the energy model formulated for the same disambiguation goal by graph cut optimization. In the works of Abada and Aouat [1,2] the approach of Zhu and Shi is employed and algorithmically refined. However, a thorough theoretical analysis of the underlying problem has not been given in any of those papers.

Our contribution. In this paper we introduce a novel graph theoretical framework for dealing with the disambiguation problem at hand of the configuration graph. Our problem formulation is based on cycle bases of the triangulation that makes up the configuration graph. By employing our construction we prove, to our best knowledge, for the first time in the literature that the constituting problem is $\mathcal{NP}$-complete. Moreover, at hand of the cyclic graph construction we propose a new algorithm tailored to our framework. Together with a linear program we disambiguate computed height differences between singular points and achieve the global disambiguation of the SFS solution.

Let us emphasize that not only we provide a novel mathematically validated approach to the global SFS disambiguation problem. To our best knowledge, our theoretical results are also new in the graph theoretical context. They represent a contribution that could be useful independently of the presented SFS context, e.g., it is easy to see that they may be generalized to other graph classes that may arise in computer vision tasks.

Regarding the techniques dealing with the arising local SFS problems, we consider similar techniques as in the abovementioned works. We employ the well-known fast marching (FM) method as in [18] for resolving locally the 3D surface depth. The method has been used in several works for solving PDEs in the SFS context, see e.g. [16]. For obtaining the complete surface depth we employ a standard stitching technique. We do not consider this part of the complete algorithm in detail in this paper, since our aim is to focus on our novel graph-based framework here.

2   Graph Theoretical Basics

A *graph* $G = (V, E)$ consists of the set V of *vertices* and the set E of *edges* that is a subset of two-elemented subsets of V. We define $n = |V|$ as the number of vertices of G and $m = |E|$ as the number of edges. We call $u, v \in V$ *adjacent*, if the edge $\{u, v\}$ is an element of E. In that case we call u a neighbor of v and vice versa. The set of all neighbors of a vertex v is called neighborhood of v or shortly $N(v)$. The degree $deg(v)$ of a vertex v is the cardinality of its neighborhood.

A graph $H = (V', E')$ is called *subgraph* of $G = (V, E)$, if $V' \subseteq V$ and $E' \subseteq E$. H is called *induced*, if each edge between vertices of V' in E is also contained in E'. A sequence of vertices $P = (v_1, \dots, v_k)$, such that each vertex v_i is adjacent to v_{i+1} for all $1 \leq i < k$ is called *path*, if each vertex occurs only once in P. If we add the edge $\{v_k, v_1\}$ to P we get a *cycle*. A graph is *connected*, if for each pair of vertices $v_i, v_j \in V$ there exists a path between v_i and v_j. A subgraph H of a graph G is inclusion maximal connected, if there is no greater connected subgraph of G, which contains H completely. We call such a subgraph H a connected component.

If the connected graph G is not connected without the vertex v, we call v an *articulation point* of G. If a connected graph does not contain an articulation point, it is *biconnected*. The inclusion maximal biconnected subgraphs are equivalently defined. We call them *blocks*. For a better understanding see Fig. 1.

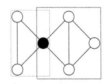

Fig. 1. A graph that is not biconnected. The articulation point is black colored. The blocks of the graph are bounded by the red and the blue rectangles. (Color figure online)

The *cycle space* $\mathscr{C}$ of a graph $G = (V, E)$ is a subspace of $\mathbb{F}_2^{|E|}$, where $\mathbb{F}_2$ is the finite field with two elements. Each element of $\mathscr{C}$ represents a Eulerian subgraph of G, i.e., a subgraph where each vertex has an even degree. Hence $\mathscr{C}$ contains each cycle of G. A basis of $\mathscr{C}$ is called *cycle basis*. For more details about this algebraic representation of cycles we refer to [3].

The graphs we considered so far are undirected, i.e., the vertices of an edge have no ordering. Now, we consider the directed case. Such graphs are called *directed graphs* or simply *digraphs*. Their set of edges is a subset of $V \times V$. We forbid edges (v, v), which are called loops. In difference to the undirected case, we now have two different neighborhoods of a vertex. The in-neighborhood $\delta_{in}(v)$ of v contains each vertex w, for that $(w, v) \in E$. The out-neighborhood $\delta_{out}(v)$ contains each vertex w, for that $(v, w) \in E$. Equivalent to the undirected case, the in- and out-degrees $deg_{in}(v)$ and $deg_{out}(v)$ are the cardinality of the in- and out-neighborhood respectively. The definitions of paths and cycles can also be adapted to the directed graphs. Here the edge (v_i, v_{i+1}) has to be contained in G. A digraph that contains no directed cycle, is called *acyclic*. To create a directed graph G^d out of a undirected graph G, we have to choose an orientation for each edge $e \in E$. Such a graph G^d is called *orientation* of G.

A graph is called *planar*, if there is an embedding into the $\mathbb{R}^2$, such that edges only intersects in their endpoints. Such an embedding is called *plane*. A planar graph, that has an embedding, where each inner face is a triangle is called *triangulation*.

3   The Distance Orientation Problem

Given a graph $G = (V, E)$ with a distance function $d : E \to \mathbb{R}_{\geq 0}$. We call a height function $h : V \to \mathbb{R}$, that assigns a height value to each vertex, a proper height function, if $\forall e = \{v, w\} \in E : |h(v) - h(w)| = d(e)$. That means, that the absolute height difference between adjacent vertices is equal to the distance of d between them. The *Distance Orientation Problem (DOP)* refers to the question, whether a given graph with its distance function has a proper height function.

Instead of searching for a height function, we can also search for an orientation of the edges. For this we define the corresponding *proper orientation* $D(h)$ of a proper height function h as follows: $(u, v) \in D(h) \Leftrightarrow h(v) - h(u) = d(u, v)$. This means in the orientation edges always point to the vertex with the greater height. So vertices with $deg_{out} = 0$ correspond to convex singular points and vertices with $deg_{in} = 0$ correspond to concave singular points. Each other vertex correspond to a saddle point, since it has neighbors that have a greater height and other neighbors with lower height.

If we have the proper height function h, it is easy to create $D(h)$. Clearly per definition there is always a unique $D(h)$. If we have a proper orientation, we uniquely define a corresponding height function by choosing one vertex and the corresponding height value for every connected component of the graph.

Let h_1 be a proper height function. Let $\alpha \in \mathbb{R}$ and $\beta \in \{0, 1\}$. Than h_2, which is defined as $h_2(v) = (-1)^\beta(h_2(v) + \alpha)$ is also a proper height function. That means, we can shift the heights on the z-axis and mirror it on the x-y-plane. Shifting does not change the corresponding orientation, mirroring inverts the orientation. We call height functions or orientations *equivalent*, which can be transformed to each other using these operations.

To characterize proper orientations without using height functions, we define cycle sums. Let $C = (v_1, \ldots, v_k)$ be a cycle of G and D an orientation of G. The cycle sum $\zeta_D(C)$ is defined as follows:

$$\zeta_D(C) = \left| \sum_{(v_i, v_{i+1}) \in D} d(v_i, v_{i+1}) - \sum_{(v_{i+1}, v_i) \in D} d(v_i, v_{i+1}) \right| \tag{1}$$

where $k + 1$ is defined as 1. It is easy to see, that in a proper orientation the cycle sum of each cycle has to be zero. To see that, we start at vertex v_1 and go through the cycle in any direction. Every time, the edge is orientated in our direction, we add the distance to our current height value, otherwise we subtract it. Since we have to reach vertex v_1 at the same height, at which we have left it at the beginning, the cycle sum has to be zero. It is easy to show, that this condition is also sufficient, i.e., if for an orientation D every cycle has a cycle sum of zero, D is a proper orientation of the graph.

Cycles are always part of a distinct block of the graph. Therefore it is sufficient to solve DOP on the blocks on a graph. Then we only have to combine the found orientations. Since we have the decision between at least two proper orientations per block, a graph with k blocks and a proper orientation has at least 2^k proper orientations.

This cycle sum condition is the same like in *Kirchhoff's voltage law*. This law states, that in an electrical circuit the potential differences or voltages have to sum up to zero in any cycle. It can be shown, that it is sufficient, to fulfill this condition on a cycle basis of G [3]. Since the inner faces of an embedding of a planar graph build a cycle basis [7], we can formulate the following lemma.

Lemma 1. *Let $G = (V, E)$ be a planar graph and $d : E \to \mathbb{R}^+$ its distance function. An orientation D of G is proper for d, if and only if there is an embedding $\mathcal{P}$ such that each inner face of $\mathcal{P}$ has a cycle sum of zero.*

3.1   Complexity

We will now prove, that the DOP for general graphs is hard to solve.

Theorem 1. *The distance orientation problem is $\mathcal{NP}$-complete.*

Proof. It is clear, that the DOP is an element of $\mathcal{NP}$ since we can prove the properness of a height function or orientation in polynomial time. To show the hardness we consider the problem PARTITION, which is $\mathcal{NP}$-complete [12]. Let $A = \{a_1, \ldots, a_k\}$ be a set of natural numbers. A is an element of PARTITION, if there is a subset $B \subseteq A$, such that $\sum_{a \in B} a = \sum_{a \notin B} a$.

Now we reduce PARTITION to our DOP. Let $A = \{a_1, \ldots, a_k\}$ be an arbitrary set of natural numbers. We construct the cycle $G = (v_1, \ldots, v_k)$. Now we define the distance function d of G as follows: $d(v_i, v_{i+1}) = a_i$ where $k + 1$ is defined as 1. Let D be a proper orientation of G. Then we define the subset $B \subseteq A$ as follows: $B := \{a_i | (v_i, v_{i+1}) \in D\}$. Since D must have a cycle sum of zero, B is a feasible solution for PARTITION. Reversely follows that each solution of the partition problem gives us a proper orientation of G. So the DOP is also $\mathcal{NP}$-complete. □

Since the reduction of the proof only uses cycles, DOP is $\mathcal{NP}$-complete not only for general graphs, but also for many interesting graph classes. For example planar graphs, bipartite graphs, and comparability graphs include cycles. Therefore, DOP is $\mathcal{NP}$-complete on these classes.

3.2   Solving the DOP for Triangulations

In the last section we have seen that the DOP is hard to solve on many graph classes. However, we will now introduce an algorithm, that solves the DOP on triangulations such as those arising in the disambiguation problem in polynomial time.

For that, we assume that each distance value is greater zero. This is not a real limitation, which one can see as follows. Assume, we have a triangulation $G = (V, E)$ and an edge $e = \{u, v\} \in E$ with $d(u, v) = 0$. Then we contract e, i.e., we merge u and v to one vertex x and get a new graph G^*. Now it is possible, that there are multiple edges between x and another vertex y. If so, we check whether both have the same distance value. If not, we know that there cannot be a proper orientation of this graph. Otherwise, we can delete the duplicate edge. Every proper orientation of G^* can be extended to a proper orientation of G by choosing an arbitrary orientation of e. On the other hand, a proper orientation of G restricted to G^* is always a proper orientation of G^*. The new graph G^* is also a triangulation, since faces that contained e in G are no faces anymore and each other face does not change its structure.

We already mentioned above, that the blocks of a graph can be oriented separately. So we only consider biconnected triangulations here. If the graph is not biconnected, we first determine its blocks. This can be done in linear time [9].

The algorithm itself uses the uniqueness of the orientation of triangles. If we look at a triangle with the edge distances a, b and c, then we know that one of the three distance values has to be the sum of the other two. So assume, $a + b = c$. If then $a + c = b$ would also hold, then a must be zero. If $b + c = a$ also holds, then b must be zero. Thus, we see that for distances >0 there is only one possibility for combining one distance as the sum of the two others. So we know that the orientation of a triangle is determined by choosing the orientation of one edge.

The first step of our algorithm is to compute a map M, which maps each edge $e \in E$ to the list of the triangles of the graph, which contain e. Then we choose an arbitrary edge e of G and give it an arbitrary orientation. According to the observation we have made above, we now know the orientation of each triangle which contains e. Hence, we can fix this orientation. Then we put every edge that got an orientation in the queue Q. From this queue, we successively take the next edge and orient each triangle that contains it. To prevent that we visit the triangles of an edge more than once, we mark the edge as visited. The described algorithm is found as pseudo code in Algorithm 1.

In order to prove that this algorithm works correctly, we introduce the novel notion of biconnected 3-covers.

Definition 1. *Let $G = (V, E)$ be a graph. A set of triangles $\mathcal{C} = \{C_1, \ldots, C_k\}$ from G is called* biconnected, *if for all $1 \leq i < j \leq k$ exists a sequence $(C_{i=i_0}, C_{i_1}, \ldots, C_{i_{l-1}}, C_{i_l=j})$ of triangles from $\mathcal{C}$, such for $1 \leq a < l$, C_{i_a} and $C_{i_{a+1}}$ have one edge in common.*

A biconnected 3-cover *of G is a biconnected triangle set $C = \{C_1, \ldots, C_k\}$ with $\bigcup_{i=1}^{k} C_i = V$. A biconnected 3-cover C is called* complete, *if every edge $e \in E$ is element of a triangle of C.*

Algorithm 1.

Input: biconnected triangulation $G = (V, E)$ with distance function $d : E \to \mathbb{R}^+$
Output: proper orientation of each edge of G or *"Not orientable"*

1 $M :=$ map from each edge $e \in E$ to the set of triangles of G, which contain e;
2 visited$(f) :=$ false $\forall f \in E$;
3 let $e \in E$ be an arbitrary edge;
4 Queue $Q := (e)$;
5 visited$(e) :=$ true;
6 choose an arbitrary orientation of e;
7 **while** $Q \neq \varnothing$ **do**
8      $\{v_i, v_j\} :=$ Q.pop;
9   **foreach** *triangle* $\{v_i, v_j, v_k\}$ **do**
10   orient triangle appropriate to the orientation of $\{v_i, v_j\}$;
11   **if** *triangle has no proper orientation* **then**
12   **return** *"Not orientable"*
13   **if** *visited(v_i, v_k) = false* **then**
14   Q.push$(\{v_i, v_k\})$;
15   visited$(v_i, v_k) :=$ true;
16   **if** *visited(v_j, v_k) = false* **then**
17   Q.push$(\{v_j, v_k\})$;
18   visited$(v_j, v_k) :=$ true;

It is easy to show that the inner faces of a triangulation build a complete biconnected 3-cover. We refrain to give a proof here due to space restrictions. So we just state here the following assertion.

Lemma 2. *A biconnected triangulation has a complete biconnected 3-cover.*

Now, we can show that Algorithm 1 solves the DOP in triangulations in polynomial time.

Theorem 2. *Algorithm 1 solves the distance orientation problem on complete 3-cover graphs in $\mathcal{O}(\min\{m^{\frac{3}{2}} + n, \ \Delta m\})$, where Δ is the maximum degree of the graph. If it founds a proper orientation, the orientation is unique, i.e., each other proper orientation is equivalent.*

Proof. At first, we have to show that at the end of the algorithm each edge has an orientation unless the algorithm returned *"Not orientable"*. We know that the graph has a complete 3-cover C. Assume $f \in E$ has not got an orientation. Let C_1 be a triangle which contains the start edge e and C_k a triangle which contains f.

Then we know, there is a biconnected sequence of triangles $(C_1, \ldots, C_k)$ in $\mathcal{C}$ that starts in C_1 and ends in C_k. Let C_i be the first triangle of that sequence, that contains an edge without orientation. We know, that it is not C_1, since we visit each triangle that contains e. Therefore, we know that each edge of C_{i-1} gets an orientation. Since C_{i-1} and C_i have one edge in common and each edge that gets an orientation is added to Q, C_i also has to be considered. So it is not possible, that it contains an unoriented edge. This is a contradiction to our assumption.

Now, we have to show that the orientation is proper, when the algorithm returns it. Because of Lemma 1, we only have to show that there is an embedding, whose inner faces all got a proper orientation. Since the graph is a triangulation, we know there is an embedding, where each inner face is a triangle. Furthermore we know that each edge and so each triangle was considered in the **foreach** loop in lines 9–18. Hence each face got a correct orientation and the whole orientation is proper.

Since each orientation of an edge, that is determined, must be chosen like the algorithm does it, the orientation of the whole graph is unique. Furthermore there is no proper orientation, when the algorithm returns *"Not orientable"*.

For the running time we firstly state that the list of all triangles is computable in $\mathcal{O}(\min\{m^{\frac{3}{2}} + n, \ \Delta m\})$ [14]. Thus we can create the map M in the same time. Additionally we have to bound the number of loop passes of the **foreach** loop in lines 9–18. Since each edge is visited only one time, each triangle is visited three times. Furthermore the number of triangles is bounded by $\mathcal{O}(\min\{m^{\frac{3}{2}}, \Delta m\})$ [14]. Therefore the **foreach** loop is passed $\mathcal{O}(\min\{m^{\frac{3}{2}}, \Delta m\})$ times.   □

Since we only need the complete biconnected 3-cover to prove the theorem, the algorithm also works for other graph classes that always contain such a cover. An example of such a graph class is given by the chordal graphs, i.e., the graphs that do not contain a cycle of length ≥ 4 as induced subgraph.

Let us sum up the results of this chapter. We have seen that the DOP is a hard problem in general. Nevertheless, the problem is easy to solve on triangulations and we gave a polynomial algorithm for that case. Furthermore we show, that for biconnected triangulations the proper orientations are always unique. So if we would be able to compute the exact height differences between singular points within a triangulation, we would be able to reconstruct the correct height relationships of that points (besides shifting and mirroring).

4   Formulation of the Optimization Problem for Describing the DOP

Any numerical method like the FM algorithm is not able to calculate exact height differences between singular points, because of discretization and rounding errors. Because of this, for our application we cannot use the decision variant of the DOP. Instead we have to consider an optimization version of this problem.

For that we have to define how we measure the quality of a height function. Let $h : V \to \mathbb{R}$ be such a height function, then we define the error of h as $\sum_{\{x,y\} \in E} ||h(x) - h(y)| - d(x,y)|$.

For a given orientation, we can find the best height function suitable to this orientation in polynomial time. For this purpose, we use the linear program which is given in Program 1. Since distances can be over- or underestimated, we have to use two error variables per edge. With ε_{ij}^+ we measure the amount of height that we have to add to the height value of vertex v_i, to fulfill the distance condition. Instead ε_{ij}^- measures, how much we have to decrease the height of v_i for that. Since we minimize the sum of all ε^+ and ε^-, it is clear that in the optimal solution for each edge $\{v_i, v_j\}$ at most one of the variables ε_{ij}^+ and ε_{ij}^- gets a value greater than zero.

Program 1

$$\min \quad \sum_{\{v_i, v_j\} \in E} \varepsilon_{ij}^+ + \varepsilon_{ij}^- \tag{2}$$

$$s.t. \quad h_i + d(v_i, v_j) = h_j + \varepsilon_{ij}^+ - \varepsilon_{ij}^- \qquad \forall (v_i, v_j) \in D \tag{3}$$

$$h_i \in \mathbb{R} \qquad \forall v_i \in V \tag{4}$$

$$\varepsilon_{ij}^+, \varepsilon_{ij}^- \geq 0 \qquad \forall \{v_i, v_j\} \in E \tag{5}$$

To find a good orientation, we can easily adapt Algorithm 1. Instead of choosing the only proper orientation of the considered triangle in line 10, we choose the orientation that fulfills the already fixed orientations and has the lowest cycle sum. Hence the algorithm always produces an orientation. If we make an assumption on the quality of the measured distances, we can show that this adapted algorithm produces the correct orientation. For that we define the *triangle condition*. Let $G = (V, E)$ be a graph and C be a triangle of G. Furthermore let d be the distance function of the real height differences and d' be another distance function. C fulfills the triangle condition for d', if the correct orientation of C in respect to d is equivalent to the orientation with the least cycle sum using the distances of d'. An example is given in Fig. 2.

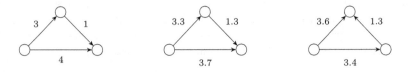

Fig. 2. An example for the triangle condition. On the left, we see the correct height differences together with one of the two correct orientations. In the middle, we see distances that fulfill the triangle condition. The given orientation has a cycle sum of 0.9. Each other nonequivalent orientation has a cycle sum of at least 1.7. So the given orientation is optimal. The distances of the triangle on the right hand side do not fulfill the condition. The given orientation has a cycle sum of 1.1. Every other orientation has a cycle sum of at least 1.5. Since the best orientation is not equivalent to the correct orientation, the cycle condition is not fulfilled.

When d' fulfills the triangle condition, the optimization variant of Algorithm 1 choose the same orientations on d' like the decision variant of Algorithm 1 does on the correct height differences d. So we can conclude the following proposition, which summarizes our developments of this section.

Proposition 1. *Let $G = (V, E)$ be a triangulation. Let $d : E \to \mathbb{R}^+$ be the real distance function and $d' : E \to \mathbb{R}^+$ be a calculated distance function. If all triangles of G fulfill the triangle condition for d', then the modified Algorithm 1 calculates on d' the proper orientation of G in respect to d.*

5   Experiments

In this section, we show how to apply the developed framework. We chose to focus on two dedicated test examples. First we consider a synthetic test on a smooth analytic function that bears all instances of possible geometries around singular points. In the second experiment we discuss the SFS solution for a real world SFS input image where many of the effects that are of interest in the context of the disambiguation problem can be found.

5.1   Synthetic Test Surface

For testing our approach on an synthetic example, we use the surface created by MATLAB shown in Fig. 3a. As indicated, it contains concave, convex and saddle points. The horizontal plane located below the displayed surface shows the two dimensional positions of corresponding singular points. To create the irradiance image in Fig. 3a, we use the Lambertian reflectance model with a light source that is located far away and on the positive side of the z-axis. As albedo ρ we use value 1. Each convex, concave or saddle point of our surface corresponds to a singular point inside the irradiance image.

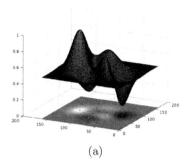
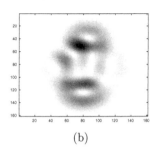

(a)   (b)

Fig. 3. (a) Our synthetic $3D$ surface containing convex, concave and saddle points whose $2D$ projections are shown on a horizontally located plane below the surface. (b) The irradiance image corresponding to the surface shown in (a).

For realizing our approach, we choose those pixels inside the irradiance image having values above a certain threshold as the singular points. In total we find eleven of such singular points, which correspond to eleven vertices of the graph.

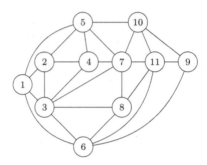

Fig. 4. The graph established by triangulating the singular points extracted from the irradiance image (Fig. 3b) corresponds to the surface shown in Fig. 3a.

In the next step, we find a triangulation of that vertices, which is shown in Fig. 4. Now we calculate the height differences by using the FM approach. In this example, the calculated distances between the singular points are quite close to the true height differences between the surface points. As we demonstrate in Fig. 5a, this is not enough information to reconstruct the surface correctly by direct integration, since we do not know which of the points is concave, convex or saddle point. Hence, we use the proposed optimization variant of Algorithm 1 to calculate an orientation of the triangulated graph with respect to the FM-based distances. The resulting orientation gives us the information which of the singular points are concave. Furthermore we know the (relative) height values of this points.

We take the estimated height of a concave point provided by our orientation and use it as the initial height of our boundary point for the FM solver. This process is repeated for all concave points. The result is given in Fig. 5b, where we show it from a slightly different point of view as the original surface. Since the calculated height differences of the singular points are quite near to the exact values, the found heights produce a set of marched surfaces from each concave point that meet each other seamlessly. So the reconstructed surface fulfills the kink-free property. The *Root mean square error* among the reconstructed surface and the original one is 0.0257.

5.2   Real Surface

To investigate the applicability of our proposed method with respect to SFS for real surfaces, we studied the 3D Beethoven face. The experiment can be found for instance in [17], and comprises a set of real images of the Beethoven bust often used in photometric stereo experiments. We chose this experiment

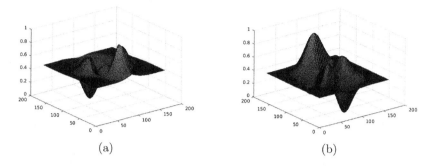

Fig. 5. (a) The reconstructed surface without using the knowledge about the geometry arrangement (especially concave points) as provided by our graph theoretic framework of our work. (b) The reconstructed seamless surface at hand of our graph theoretic framework. The reconstruction is performed by using each concave point along with its found real height as a boundary point for initializing the FM solver.

since an account of the ground truth is available. Furthermore the surface of the photographed Beethoven bust is nearly Lambertian which fits the assumptions of the classic SFS method.

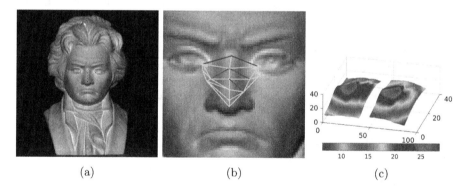

Fig. 6. (a) The beethoven bust with seven singular points located on nose and the inner corners of eyes. (b) A triangulation based on the shown singular points in (a) along with seven additional surface points located on the left and right cheeks. Note that these extra points are not necessarily singular ones. They only help us to create our triangulation. (c) The original nose (left) and the reconstructed nose (right) shown side by side. (Color figure online)

We triangulated a subset of the singular points located on and around the nose as shown in Fig. 6a. More specifically four singular points are located on the nose on a straight line connecting the tip of the nose to the center of the eye brows, and two more are located at the inner eye corners. The found singular points are accompanied with a few other surface points located on both cheeks and the result is triangulated as shown in Fig. 6b. Each edge of the graph shown in Fig. 6b is colored in green, blue or red. Those edges whose differences in real and approximated heights are less than 0.1 are green. The blue and red edges are used to visualize height differences in the range of $[0.1, 0.2)$ and $[0.2, 0.4]$, respectively.

The optimization variant of Algorithm 1 decides that the tip of the nose is the only concave point of the shown graph in Fig. 6b. The reconstruction shown in Fig. 6c is consequently done by applying the FM solver on the only found concave point. Figure 6c contains two profile views of the original nose (left) and the reconstructed one (right). It is evident that the lips area is not reconstructed in a reasonable way. The reason is that we did not look in the lips area for a singular point participating in the graph-based framework. However, the modeled nose representing the area where we defined a triangulation almost represents the geometry of the original nose, up to numerical inaccuracy.

6   Summary and Conclusion

We presented a novel graph theoretical framework built upon cycle bases. Based on this, we proposed a linear program for realizing the approach for inexact height differences. With our proceeding, we could give the first theoretical analysis of the graph-based disambiguation problem. Moreover, the results are, to our best knowledge, new in graph theory.

The triangulation of the singular points can have a great influence on the quality of the reconstructed surface. We found that in practice, automatically detected singular points not always lead to triangulations that are useful without some modification for dealing with the disambiguation problem. Due to space restrictions we have to omit this examples here. However, we think that this can be an interesting issue for future studies.

We also conjecture that our graph-based framework can be useful for achieving rigorous results for similar problems that may arise in computer vision, e.g., in other Shape-from-X-problems.

References

1. Abada, L., Aouat, S.: Tabu search to solve the shape from shading ambiguity. Int. J. Artif. Intell. Tools **24**(5) (2015)
2. Abada, L., Aouat, S.: Improved shape from shading without initial information. Front. Comput. Sci. **11**(2), 320–331 (2017)
3. Bollobás, B.: Modern Graph Theory. Graduate Texts in Mathematics, vol. 184. Springer, Heidelberg (1998). https://doi.org/10.1007/978-1-4612-0619-4
4. Bruss, A.R.: The eikonal equation: some results applicable to computer vision. J. Math. Phys. **23**(5), 890–896 (1982)
5. Bruss, A.R.: Is what you see what you get? In: Proceedings of the International Joint Conference on Artificial Intelligence, pp. 1053–1056, August 1983
6. Chang, J.Y., Lee, K.M., Lee, S.U.: Shape from shading using graph cuts. Pattern Recogn. **41**(12), 3749–3757 (2008)

7. Diestel, R.: Graph Theory. Graduate Texts in Mathematics, vol. 173, 4th edn. Springer, Heidelberg (2012). https://doi.org/10.1007/978-3-662-53622-3
8. Durou, J.D., Piau, D.: Ambiguous shape from shading with critical points. J. Math. Imaging Vis. **12**(2), 99–108 (2000)
9. Hopcroft, J., Tarjan, R.: Algorithm 447: efficient algorithms for graph manipulation. Commun. ACM **16**(6), 372–378 (1973)
10. Horn, B.K.P.: Shape from shading: a method for obtaining the shape of a smooth opaque object from one view. Ph.D. thesis, Massachusetts Institute of Technology (1970)
11. Horn, B.K.P., Brooks, M.J. (eds.): Shape from Shading. MIT Press, Cambridge (1989)
12. Karp, R.M.: Reducibility among combinatorial problems. In: Miller, R.E., Thatcher, J.W. (eds.) Complexity of Computer Computations, pp. 85–103. Plenum Press, New York (1972)
13. Kimmel, R., Bruckstein, A.M.: Global shape from shading. Comput. Vis. Image Underst. **62**(3), 360–369 (1995)
14. Köhler, E.: Recognizing graphs without asteroidal triples. J. Discret. Algorithms **2**(4), 439–452 (2004)
15. Oliensis, J.: Uniqueness in shape from shading. Int. J. Comput. Vis. **6**(2), 75–104 (1991)
16. Prados, E., Soatto, S.: Fast marching method for generic shape from shading. In: Paragios, N., Faugeras, O., Chan, T., Schnörr, C. (eds.) VLSM 2005. LNCS, vol. 3752, pp. 320–331. Springer, Heidelberg (2005). https://doi.org/10.1007/11567646_27
17. Quéau, Y., Durou, J.-D.: Edge-preserving integration of a normal field: weighted least-squares, TV and L^1 approaches. In: Aujol, J.-F., Nikolova, M., Papadakis, N. (eds.) SSVM 2015. LNCS, vol. 9087, pp. 576–588. Springer, Cham (2015). https://doi.org/10.1007/978-3-319-18461-6_46
18. Sethian, J.: Level Set Methods and Fast Marching Methods: Evolving Interfaces in Computational Geometry, Fluid Mechanics, Computer Vision, and Materials Science. Cambridge Monographs on Applied and Computational Mathematics. Cambridge University Press, Cambridge (1999)
19. Zhu, Q., Shi, J.: Shape from shading: recognizing the mountains through a global view. In: 2006 IEEE Computer Society Conference on Computer Vision and Pattern Recognition, pp. 1839–1846. IEEE, New York (2006)

A Variational Approach to Shape-from-Shading Under Natural Illumination

Yvain Quéau[1]([✉]), Jean Mélou[2,3], Fabien Castan[3], Daniel Cremers[1], and Jean-Denis Durou[2]

[1] Department of Informatics, Technical University of Munich, Munich, Germany
yvain.queau@tum.de
[2] IRIT, UMR CNRS 5505, Université de Toulouse, Toulouse, France
[3] Mikros Image, Levallois-Perret, France

Abstract. A numerical solution to shape-from-shading under natural illumination is presented. It builds upon an augmented Lagrangian approach for solving a generic PDE-based shape-from-shading model which handles directional or spherical harmonic lighting, orthographic or perspective projection, and greylevel or multi-channel images. Real-world applications to shading-aware depth map denoising, refinement and completion are presented.

1 Introduction

Standard 3D-reconstruction pipelines are based on sparse 3D-reconstruction by structure-from-motion (SFM), densified by multi-view stereo (MVS). Both these techniques require unambiguous correspondences based on local color variations. Assumptions behind this requirement are that the surface of interest is Lambertian and well textured. This has proved to be suitable for sparse reconstruction, but problematic for dense reconstruction: dense matching is impossible in textureless areas. In contrast, shape-from-shading (SFS) techniques explicitly model the reflectance of the object surface. The brightness variations observed in a single image provide dense geometric clues, even in textureless areas. SFS may thus eventually push back the limits of MVS.

However, most shape-from-shading methods require a highly controlled illumination and thus may fail when deployed outside the lab. **Numerical methods for SFS under natural illumination are still lacking.** Besides, SFS remains a classic ill-posed problem with well-known ambiguities such as the concave/convex ambiguity. Solving such ambiguities for real-world applications requires **handling priors on the surface.** There exist two main numerical strategies for solving shape-from-shading [1]. Variational methods [2] ensure smoothness through regularization. Handling priors is easy, but tuning the regularization may be tedious. Alternatively, methods based on the exact resolution of a nonlinear PDE [3], which implicitly enforce differentiability (almost everywhere), do not require any tuning, but they lack robustness and they require

M. Pelillo and E. Hancock (Eds.): EMMCVPR 2017, LNCS 10746, pp. 342–357, 2018.
https://doi.org/10.1007/978-3-319-78199-0_23

a boundary condition. To combine the advantages of each approach, **a variational solution based on PDEs would be worthwile for SFS under natural illumination**.

Contributions. This work proposes a generic numerical framework for SFS under natural illumination, which can be employed to achieve either pure shape-from-shading or shading-aware depth refinement (see Fig. 1). After reviewing existing solutions in Sect. 2, we introduce in Sect. 3 a new PDE-based model for SFS, which handles various illumination and camera models. A variational approach for solving the arising PDE is proposed in Sect. 4, which includes optional regularization terms for incorporating a shape prior or enforcing smoothness. Numerical solving is carried out using an ADMM algorithm. Experiments on synthetic datasets are presented in Sect. 5, as well as real-world applications to depth refinement and completion for RGB-D cameras or stereovision systems. Our achievements are eventually summarized in Sect. 6.

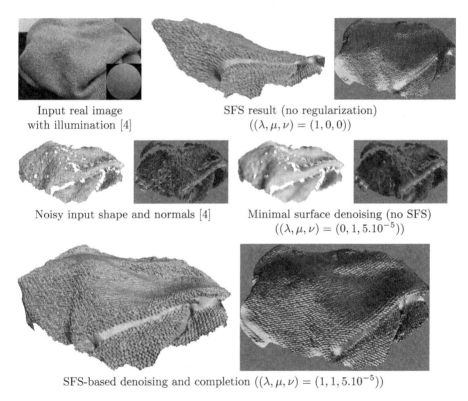

Input real image | SFS result (no regularization)
with illumination [4] | $((\lambda, \mu, \nu) = (1, 0, 0))$

Noisy input shape and normals [4] | Minimal surface denoising (no SFS)
 | $((\lambda, \mu, \nu) = (0, 1, 5.10^{-5}))$

SFS-based denoising and completion $((\lambda, \mu, \nu) = (1, 1, 5.10^{-5}))$

Fig. 1. We propose the generic variational framework (16) for shape-from-shading (SFS) under natural illumination (top row). It is able to estimate a smooth surface (out of infinitely many), which almost exactly solves the generic SFS model (2). To disambiguate SFS and improve robustness, prior surface knowledge (middle row, left) and minimal surface regularization (middle row, right) can be further included in the variational framework. These building blocks can be put together for shading-aware joint depth denoising, refinement and completion (bottom row).

2   Image Formation Model and Related Works

In the following, a 3D-frame $(Oxyz)$ is attached to the camera, O being the optical center and the axis Oz coinciding with the optical axis, such that z is oriented towards the scene. We denote $I : \Omega \subset \mathbb{R}^2 \to \mathbb{R}^C$, $(x, y) \mapsto I(x, y) = \left[I^1(x, y), \ldots, I^C(x, y)\right]^\top$ a greylevel $(C = 1)$ or multi-channel $(C > 1)$ image of a surface, where Ω represents a "mask" of the object being pictured. We assume that the surface is Lambertian, so its reflectance is completely characterized by the albedo ρ. We further consider a second-order spherical harmonic model for the lighting vector l. To deal with the spectral dependencies of reflectance and lighting, we assume a general model where both ρ and l are channel-dependent. The albedo is thus a function $\rho : \Omega \to \mathbb{R}^C$, $(x, y) \mapsto \rho(x, y) = \left[\rho^1(x, y), \ldots, \rho^C(x, y)\right]^\top$, and the lighting in each channel $c \in \{1, \ldots, C\}$ is represented as a vector $\mathbf{l}^c = [l_1^c, l_2^c, l_3^c, l_4^c, l_5^c, l_6^c, l_7^c, l_8^c, l_9^c]^\top \in \mathbb{R}^9$. Eventually, let $\mathbf{n} : \Omega \to \mathbb{S}^2 \subset \mathbb{R}^3$, $(x, y) \mapsto \mathbf{n}(x, y) = [n_1(x, y), n_2(x, y), n_3(x, y)]^\top$ be the field of unit-length outward normals to the surface. The image formation model is then written as the following extension of a well-known model [5]:

$$I^c(x, y) = \rho^c(x, y)\, \mathbf{l}^c \cdot \begin{bmatrix} \mathbf{n}(x, y) \\ 1 \\ n_1(x, y)n_2(x, y) \\ n_1(x, y)n_3(x, y) \\ n_2(x, y)n_3(x, y) \\ n_1(x, y)^2 - n_2(x, y)^2 \\ 3n_3(x, y)^2 - 1 \end{bmatrix}, \quad (x, y) \in \Omega, \ c \in \{1, \ldots, C\}. \quad (1)$$

However, let us remark that the writing (1), where both the reflectance and the lighting are channel-dependent, is abusive. Since the camera response function is also channel-dependent, this model is indeed justified only for white surfaces ($\rho^c = \rho$, $\forall c \in \{1, \ldots, C\}$) under colored lighting, or for colored surfaces under white lighting ($\mathbf{l}^c = \mathbf{l}$, $\forall c \in \{1, \ldots, C\}$). See, for instance, [6] for some discussion. In the following, we still consider the general model (1), with a view in designing a generic SFS solver handling both situations. However, in the experiments we will only consider white surfaces.

In SFS, both the reflectance values $\{\rho^c\}_{c \in \{1, \ldots, C\}}$ and the lighting vectors $\{\mathbf{l}^c\}_{c \in \{1, \ldots, C\}}$ are assumed to be known. The goal is to recover the object shape, represented in (1) by the normal field $\mathbf{n}$. Each unit-length normal vector $\mathbf{n}(x, y)$ has two degrees of freedom, thus each Eq. (1), $(x, y) \in \Omega$, $c \in \{1, \ldots, C\}$, is a nonlinear equation with two unknowns. If $C = 1$, it is impossible to solve such an equation locally: all these equations must be solved together, by coupling the surface normals in order to ensure, for instance, surface smoothness. When $C > 1$ and the lighting vectors are non-coplanar, ambiguities theoretically disappear [7]. However, under natural illumination these vectors are close to being collinear, and thus locally solving (1) is numerically challenging (bad conditioning). Again, a global solution should be preferred but this time, for robustness reasons.

There is a large amount of literature on numerical SFS, in the specific case where $C = 1$ and lighting is directional ($l_4^c = \cdots = l_9^c = 0$), see for instance [1]. However, few SFS methods deal with more general spherical harmonic lighting. First-order harmonics have been considered in [8,9], but they only capture up to 90% of natural illumination, while this rate is over 99% using second-order harmonics [10]. The latter have been used in [11], where the challenging problem of shape, illumination and reflectance from shading (SIRFS) is tackled (this method is also applicable to SFS if albedo and lighting are fixed). However, all these works heavily rely on multi-scale or regularization mechanisms, and not only for disambiguation or for handling noise. For instance, SIRFS "fails badly" [11] without a multi-scale strategy, and the method of [9] becomes unstable without depth regularization (see Fig. 2). Although regularization mechanisms somewhat circumvent such numerical instabilities in practice, an ideal numerical solver would rely on regularization only for disambiguation and for handling noise, not for enforcing numerical stability. In order to design such a solver, a variational approach based on PDEs may be worthwile. In the next section, we thus rewrite (1) as a nonlinear PDE.

Input synthetic image and illumination   Fixed point [9] without regularization

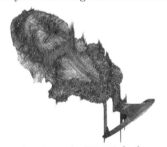

Single-scale SIRFS [11]   Proposed (without regularization)

Fig. 2. Greylevel shape-from-shading using first-order spherical harmonics. Linearization strategies such as the fixed point one used in [9] fail if regularization is not employed. Similar issues arise in SIRFS [11] when the multi-scale approach is not used. Our SFS method can use regularization for disambiguation and for improving robustness, but it remains stable even without. In these three experiments, the same initial shape was used (the "Realistic initialization" of Fig. 3).

3   A Generic PDE-Based Model for Shape-from-Shading

We assume hereafter that lighting and albedo are known (in our experiments, the albedo is assumed uniformly white and colored lighting is estimated from a gross surface approximation). These assumptions are usual in the SFS literature. They could be relaxed by simultaneously estimating shape, illumination and reflectance [11], but we leave this as future work and focus only on shape estimation. This is the most challenging part anyways, since (1) is linear in the lighting and the albedo, but is generally nonlinear in the normal.

In order to comply with the discussion above, Eq. (1) should be solved *globally* over the entire domain Ω. To this end, we do not estimate the normals but rather the underlying depth map, through a PDE-based approach [3]. This has the advantage of implicitly enforcing smoothness (almost everywhere) without requiring any regularization term (regularization will be introduced in Sect. 4, but only for the sake of disambiguation and robustness against noise). We show in this section the following result:

Proposition 1. *Under both orthographic and perspective projections, the image formation model (1) can be rewritten as the following nonlinear PDE in z:*

$$\mathbf{a}^c_{(\nabla z)} \cdot \nabla z + b^c_{(\nabla z)} = I^c \quad over \ \Omega, \ c \in \{1, \ldots, C\} \tag{2}$$

with $z : \Omega \to \mathbb{R}$ a map characterizing the shape, $\nabla z : \Omega \to \mathbb{R}^2$ its gradient, and where $\mathbf{a}^c_{(\nabla z)} : \Omega \to \mathbb{R}^2$ and $b^c_{(\nabla z)} : \Omega \to \mathbb{R}$ are a vector field and a scalar field, respectively, which depend in a nonlinear way on ∇z.

Proof. The 3D-shape can be represented as a patch over the image domain, which associates each pixel $(x, y) \in \Omega$ to its conjugate 3D-point $\mathbf{x}(x, y)$ on the surface:

$$
\begin{aligned}
\mathbf{x} : \quad & \Omega \ \to \mathbb{R}^3 \\
& (x, y) \mapsto \begin{cases} [x, y, \tilde{z}(x, y)]^\top & \text{under orthographic projection,} \\ \tilde{z}(x, y) \left[\frac{x - x_0}{\tilde{f}}, \frac{y - y_0}{\tilde{f}}, 1 \right]^\top & \text{under perspective projection,} \end{cases}
\end{aligned} \tag{3}
$$

with $\tilde{z}$ the *depth map*, $\tilde{f} > 0$ the *focal length*, and $(x_0, y_0) \in \Omega$ the coordinates of the *principal point* in the image plane.

Using this parameterization, the normal to the surface in a surface point $\mathbf{x}(x, y)$ is the unit-length, outgoing vector proportional to the cross product $\mathbf{x}_x(x, y) \times \mathbf{x}_y(x, y)$, where $\mathbf{x}_x$ (resp. $\mathbf{x}_y$) is the partial derivative of $\mathbf{x}$ along the x (resp. y)-direction. After a bit of algebra, the following formula is obtained, which relates the normal field to the depth map:

$$
\begin{aligned}
\mathbf{n} : \quad & \Omega \ \to \mathbb{S}^2 \subset \mathbb{R}^3 \\
& (x, y) \mapsto \frac{1}{d_{(\nabla z)}(x, y)} \begin{bmatrix} f \, \nabla z(x, y) \\ -1 - [\tilde{x}, \tilde{y}]^\top \cdot \nabla z(x, y) \end{bmatrix},
\end{aligned} \tag{4}
$$

where

$$(z, f, \tilde{x}, \tilde{y}) = \begin{cases} (\tilde{z}, 1, 0, 0) & \text{under orthographic projection,} \\ (\log \tilde{z}, \tilde{f}, x - x_0, y - y_0) & \text{under perspective projection,} \end{cases} \tag{5}$$

and where the map $d_{(\nabla z)}$ ensures the unit-length constraint on $\mathbf{n}$:

$$d_{(\nabla z)} : \quad \Omega \quad \to \mathbb{R}$$
$$(x, y) \mapsto \sqrt{f^2 \|\nabla z(x, y)\|^2 + \left(1 + [\tilde{x}, \tilde{y}]^\top \cdot \nabla z(x, y)\right)^2}. \tag{6}$$

Note that $\|d_{(\nabla z)}\|_{\ell^1(\Omega)}$ is the total area of the surface, which will be used in Sect. 4 for designing a regularization term.

By plugging (4) into (1), we obtain the nonlinear PDE (2), if we denote:

$$\mathbf{a}^c_{(\nabla z)} : \quad \Omega \quad \to \mathbb{R}^2$$
$$(x, y) \mapsto \frac{\rho^c(x, y)}{d_{(\nabla z)}(x, y)} \begin{bmatrix} f\, l^c_1 - \tilde{x}\, l^c_3 \\ f\, l^c_2 - \tilde{y}\, l^c_3 \end{bmatrix}, \tag{7}$$

$$b^c_{(\nabla z)} : \quad \Omega \quad \to \mathbb{R}$$

$$(x, y) \mapsto \rho^c \begin{bmatrix} l^c_3 \\ l^c_4 \\ l^c_5 \\ l^c_6 \\ l^c_7 \\ l^c_8 \\ l^c_9 \end{bmatrix} \cdot \begin{bmatrix} \dfrac{-1}{d_{(\nabla z)}(x,y)} \\ 1 \\ \dfrac{f^2 z_x(x,y) z_y(x,y)}{\left(d_{(\nabla z)}(x,y)\right)^2} \\ \dfrac{f z_x(x,y)(-1 - (\tilde{x}, \tilde{y}) \cdot \nabla z(x,y))}{\left(d_{(\nabla z)}(x,y)\right)^2} \\ \dfrac{f z_y(x,y)(-1 - (\tilde{x}, \tilde{y}) \cdot \nabla z(x,y))}{\left(d_{(\nabla z)}(x,y)\right)^2} \\ \dfrac{f^2 \left(z_x(x,y)^2 - z_y(x,y)^2\right)}{\left(d_{(\nabla z)}(x,y)\right)^2} \\ \dfrac{3(-1 - (\tilde{x}, \tilde{y}) \cdot \nabla z(x,y))^2}{\left(d_{(\nabla z)}(x,y)\right)^2} - 1 \end{bmatrix}. \tag{8}$$

$\square$

When $C = 1$, the camera is orthographic and the lighting is directional and frontal (*i.e.*, $l_3 < 0$ is the only non-zero lighting component), then (2) becomes the *eikonal equation* $\frac{\rho|l_3|}{\sqrt{1 + \|\nabla z\|^2}} = I$. Efficient numerical methods for solving this nonlinear PDE have been suggested, using for instance semi-Lagrangian schemes [12]. Such techniques can also handle perspective camera projection and/or nearby point light source illumination [13]. Still, existing PDE-based methods require a boundary condition, or at least a state constraint, which is rarely available in practice. In addition, the more general PDE-based model (2), which handles both orthographic or perspective camera, directional or second-order spherical harmonic lighting, and greylevel or multi-channel images, has not been tackled so far. A variational solution to this generic SFS problem, which is inspired by the classical method of Horn and Brooks [2], is presented in the next section.

4   Variational Formulation and Optimization

The C PDEs in (2) are in general incompatible due to noise. Thus, an approximate solution must be sought. If we assume that the image formation model (1) is satisfied up to an additive, zero-mean and homoskedastic, Gaussian noise, then the maximum likelihood solution is attained by estimating the depth map z which minimizes the following least-squares cost function:

$$\mathcal{E}(\nabla z; I) = \sum_{c=1}^{C} \left\| \mathbf{a}^c_{(\nabla z)} \cdot \nabla z + b^c_{(\nabla z)} - I^c \right\|^2_{\ell^2(\Omega)}. \tag{9}$$

In recent works on shading-based refinement [9], it is suggested to minimize a cost function similar to (9) iteratively, by freezing the nonlinear fields $\mathbf{a}^c$ and b^c at each iteration. This strategy must be avoided. For instance, it cannot handle the simplest case of orthographic projection and directional, frontal lighting: this yields $\mathbf{a}^c \equiv \mathbf{0}$ according to (7), and thus (9) does not even depend on the unknown depth z if b^c is frozen. Even in less trivial cases, Fig. 2 shows that this strategy is unstable, which explains why regularization is employed in [9]. We will also resort to regularization later on, but only for the sake of disambiguating SFS and handling noise: our proposal yields a stable solution even in the absence of regularization (see Fig. 2). Let us first sketch the proposed solver in the regularization-free case, and discuss its connection with Horn and Brooks' variational approach to SFS.

4.1   Horn and Brooks' Method Revisited

In [2], Horn and Brooks introduce a variational approach for solving the eikonal SFS model, which is a special case of (2). They promote a two-stages shape recovery method, which first estimates the gradient and then integrates it into a depth map. That is to say, the energy (9) is first minimized in terms of the gradient $\theta := \nabla z$, and then θ is integrated into a depth map z.

Since local gradient estimation is ambiguous, they put forward the introduction of the so-called integrability constraint for the first stage. Indeed, θ is the gradient of a function z: it must be a conservative field. This implies that it should be irrotational (zero-curl condition). Introducing the divergence operator $\nabla\cdot$, the latter condition reads

$$\underbrace{\left\| \nabla \cdot \begin{bmatrix} 0 & 1 \\ -1 & 0 \end{bmatrix} \theta \right\|^2_{\ell^2(\Omega)}}_{:=\mathcal{I}(\theta)} = 0. \tag{10}$$

In practice, they convert the hard-constraint (10) into a regularization term, introducing two hyper-parameters $(\lambda, \mu) > (0, 0)$ to balance the adequation to the images and the integrability of the estimated field:

$$\widehat{\theta} = \underset{\theta:\, \Omega \to \mathbb{R}^2}{\mathrm{argmin}}\ \lambda \mathcal{E}(\theta; I) + \mu \mathcal{I}(\theta). \tag{11}$$

After solving (11), $\widehat{\theta}$ is integrated into a depth map z, by solving $\nabla z = \widehat{\theta}$. However, since integrability is not strictly enforced but only used as regularization, there is no guarantee for $\widehat{\theta}$ to be integrable. Therefore, Horn and Brooks recast the integration task as another variational problem (see [14] for an overview of this problem):

$$\min_{z:\,\Omega\to\mathbb{R}} \left\| \nabla z - \widehat{\theta} \right\|_{\ell^2(\Omega)}^2. \tag{12}$$

This two-stages approach consisting in solving (11), and then (12), is however prone to bias propagation: any error during gradient estimation may have dramatic consequences for the integration stage. We argue that such a sequential approach is not necessary. Indeed, θ is conservative *by construction*, and hence integrability should not even have to be invoked. We put forward an integrated approach, which infers shape clues from the image using local gradient estimation as in Horn and Brooks' method, but which explicitly constrains the gradient to be conservative. That is to say, we simultaneously estimate the depth map and its gradient, by turning the minimization of (9) into a constrained variational problem:

$$\min_{\substack{\theta:\,\Omega\to\mathbb{R}^2 \\ z:\,\Omega\to\mathbb{R}}} \mathcal{E}(\theta; I) \tag{13}$$
$$\text{s.t. } \nabla z = \theta$$

The variational problem (13) can be solved using an augmented Lagrangian approach. In comparison with Horn and Brooks' method, this avoids tuning the hyper-parameter μ in (11), as well as bias propagation due to the two-stages approach. Besides, this approach is easily extended in order to handle regularization terms, if needed. In the next paragraph, we consider two types of regularization: one which represents prior knowledge of the surface, and another one which ensures its smoothness.

4.2   Regularized Variational Model

In some applications such as RGB-D sensing, or MVS, a depth map z^0 is available, which is usually noisy and incomplete but may represent a useful "guide" for SFS. We may thus consider the following prior term:

$$\mathcal{P}(z; z^0) = \left\| z - z^0 \right\|_{\ell^2(\Omega^0)}^2, \tag{14}$$

where $\Omega^0 \subseteq \Omega \subset \mathbb{R}^2$ is the image region for which prior information is available.

In order not to interpret noise in the image as geometric artifacts, one may also want to improve robustness by explicitly including a smoothness term. However, standard total variation regularization, which is often considered in image processing, may tend to favor piecewise fronto-parallel surfaces and thus induce staircasing. We rather penalize the total area of the surface, which has recently been shown in [15] to be better suited for depth map regularization. To this end, let us remark that in differential geometry terms, the map $d_{(\nabla z)}$ defined in (6) is the square root of the determinant of the first fundamental form of function z (metric tensor). Its integral over Ω is exactly the area of the surface, and thus the following smoothness term may be considered:

$$\mathcal{S}(\nabla z) = \left\| d_{(\nabla z)} \right\|_{\ell^1(\Omega)}. \tag{15}$$

Putting altogether the pieces (9), (14) and (15), and using the same change of variable $\theta := \nabla z$ as in (13), we obtain the following constrained variational approach to shape-from-shading:

$$\begin{aligned} &\min_{\substack{\theta:\, \Omega \to \mathbb{R}^2 \\ z:\, \Omega \to \mathbb{R}}} \lambda\, \mathcal{E}(\theta; I) + \mu\, \mathcal{P}(z; z^0) + \nu\, \mathcal{S}(\theta) \\ &\text{s.t. } \nabla z = \theta \end{aligned} \tag{16}$$

which is a regularized version of the pure shape-from-shading model (13) where $(\lambda, \mu, \nu) \geq (0, 0, 0)$ are user-defined parameters controlling the respective influence of each term.

Let us remark that our variational model (16) yields a pure SFS model if $\mu = \nu = 0$, a depth denoising model similar to that in [15] if $\lambda = 0$ and $\Omega^0 = \Omega$, and a shading-aware joint depth refinement and completion if $\lambda > 0$, $\mu > 0$ and $\Omega^0 \subsetneq \Omega$.

4.3   Numerical Solution

The change of variable $\theta := \nabla z$ in (16) has a major advantage when it comes to numerical solving: it separates the difficulty induced by the nonlinearity (shape-from-shading model and minimal surface prior) from that induced by the global nature of the problem (dependency upon the depth gradient).

Optimization can then be carried out by alternating nonlinear, yet local gradient estimation and global, yet linear depth estimation. To this end, we make use of the ADMM procedure, a standard approach to constrained optimization which dates back to the 70s [16]. We refer the reader to [17] for a recent overview of this method.

The augmented Lagrangian functional associated to (16) is defined as follows:

$$\mathcal{L}_\beta(\theta, z, \Psi) = \lambda\, \mathcal{E}(\theta; I) + \mu\, \mathcal{P}(z; z^0) + \nu\, \mathcal{S}(\theta) + \langle \Psi, \nabla z - \theta \rangle + \frac{\beta}{2} \left\| \nabla z - \theta \right\|_2^2, \tag{17}$$

with $\Psi : \Omega \to \mathbb{R}^2$ the field of Lagrange multipliers, $\langle \cdot \rangle$ the scalar product induced by $\|\cdot\|_2$ over Ω, and $\beta > 0$.

ADMM iterations are then written:

$$\theta^{(k+1)} = \underset{\theta}{\operatorname{argmin}}\ \mathcal{L}_{\beta^{(k)}}\left(\theta, z^{(k)}, \Psi^{(k)}\right), \tag{18}$$

$$z^{(k+1)} = \underset{z}{\operatorname{argmin}}\ \mathcal{L}_{\beta^{(k)}}\left(\theta^{(k+1)}, z, \Psi^{(k)}\right), \tag{19}$$

$$\Psi^{(k+1)} = \Psi^{(k)} + \beta^{(k)}\left(\nabla z^{(k+1)} - \theta^{(k+1)}\right). \tag{20}$$

where $\beta^{(k)}$ can be determined automatically [17].

Problem (18) is a pixelwise non-trivial optimization problem which is solved using a Newton method with an L-BFGS stepsize. As for (19), it is discretized by first-order, forward finite differences with Neumann boundary condition. This yields a linear least-squares problem whose normal equations provide a symmetric, positive definite (semi-definite if $\mu = 0$) linear system. It is sparse, but too large to be solved directly: conjugate gradient iterations should be preferred. In our experiments, the algorithm stops when the relative variation of the energy in (16) falls below 10^{-3}.

This ADMM algorithm can be interpreted as follows. During the θ-update (18), local estimation of the gradient (*i.e.*, of the surface normals) is carried out based on SFS, while ensuring that the gradient map is smooth and close to the gradient of the current depth map. Unlike in the fixed point approach [9], local surface orientation is inferred from the whole model (2), and not only from its linear part. In practice, we observed that this yields a much more stable algorithm (see Fig. 2). In the z-update (19), these surface normals are integrated into a new depth map, which should stay close to the prior.

Given the non-convexity of the shading term $\mathcal{E}$ and of the smoothness term $\mathcal{S}$, convergence of the ADMM algorithm is not guaranteed. However, in practice we did not observe any particular convergence-related issue, so we conjecture that a convergence proof could eventually be provided, taking inspiration from the recent studies on non-convex ADMM [18,19]. However, we leave this as future work and focus in this proof of concept work on sketching the approach and providing preliminary empirical results.

The next section shows quantitatively the effectiveness of the proposed ADMM algorithm for solving SFS under natural illumination, and introduces qualitative results on real-world datasets.

5 Experiments

5.1 Quantitative Evaluation of the Proposed SFS Framework

We first validate in Fig. 3 the ability of the proposed variational framework to solve SFS under natural illumination *i.e.*, to solve (1). Our approach is compared against SIRFS [11], which is the only method for SFS under natural illumination whose code is freely available. For fair comparison, albedo and lighting estimations are disabled in SIRFS, and its multi-scale strategy is used, in order to avoid the artifacts shown in Fig. 2.

Since we only want to compare here the ability of both methods to explain a shaded image, our regularization terms are disabled ($\mu = \nu = 0$), as well as those from SIRFS. To quantify this ability, we measure the RMSE between the input images and the reprojected ones, as advised in [1].

To create these datasets, we use the public domain "Joyful Yell" 3D-shape, considering orthographic projection for fair comparison (SIRFS cannot handle perspective projection). Noise-free images are simulated under three lighting scenarios. We first consider greylevel images, with a single-order and then a

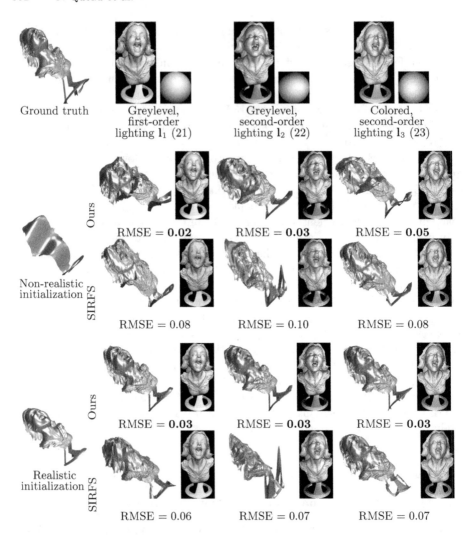

Fig. 3. Evaluation of our SFS approach against the multi-scale one from SIRFS [11], in three different lighting situations and using two different initial 3D-shapes (the first one is Matlab's "peaks" function, while the second one is a smoothed version of the ground truth). For each experiment, we show the estimated depth map and the reprojected image, and provide the root mean square error (RMSE) between the input synthetic image and the reprojection (the input images are scaled between 0 and 1). Our variational framework solves SFS under natural illumination more accurately than state-of-the-art.

second-order lighting vectors. Eventually, we consider a colored, second-order lighting vector. These lighting vectors are defined, respectively, by:

$$\mathbf{l}_1 = [0.1, -0.25, -0.7, 0.2, 0, 0, 0, 0, 0]^\top, \tag{21}$$

$$\mathbf{l}_2 = [0.2, 0.3, -0.7, 0.5, -0.2, -0.2, 0.3, 0.3, 0.2]^\top, \tag{22}$$

$$\mathbf{l}_3 = \begin{bmatrix} -0.2 & -0.2 & -1 & 0.4 & 0.1 & -0.1 & -0.1 & -0.1 & 0.05 \\ 0 & 0.2 & -1 & 0.3 & 0 & 0.2 & 0.1 & 0 & 0.1 \\ 0.2 & -0.2 & -1 & 0.2 & -0.1 & 0 & 0 & 0.1 & 0 \end{bmatrix}^\top. \tag{23}$$

To illustrate the underlying ambiguities, we consider two different initial 3D-shapes: one very different from the ground truth (Matlab's "peaks" function), and one close to it (obtained by applying a Gaussian filter to the ground truth).

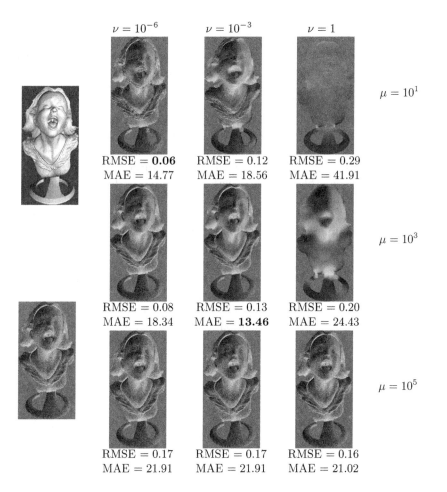

Fig. 4. Left: input noisy image ($\sigma_I = 2\%$ of the maximum greylevel) and noisy prior shape ($\sigma_z = 0.2\%$ of the maximum depth), represented by a normal map to emphasize the details. Right: estimated shape with $\lambda = 1$ and various values of μ and ν. The RMSE between the image and the reprojection is minimal when μ and ν are minimal, but the mean angular error (MAE, in degrees) between the estimated shape and the ground truth one is not.

Interestingly, although $\mu = 0$ for the tests in Fig. 3, our method does not drift too much from the latter: the shape is qualitatively satisfactory as soon as a good initialization is available.

In all the experiments, the images are better explained using our framework, which shows that the proposed numerical strategy solves the challenging, highly nonlinear SFS model (1) in a more accurate manner than state-of-the-art. Besides, the runtimes of both methods are comparable: a few minutes in all cases (on a standard laptop using Matlab codes), for images having around 150.000 pixels inside Ω. Unsurprisingly, initialization matters a lot, because of the inherent ambiguities of SFS.

In Fig. 4, we illustrate the influence of the hyper-parameters μ and ν which control, respectively, the shape prior and the smoothness term. We consider the same dataset as in the second experiment of Fig. 3, but with additive, zero-mean, homoskedastic Gaussian noise on the image and on the depth forming the shape prior (we use the "Realistic initialization" as prior). If $\lambda = 1$ and $(\mu, \nu) = (0, 0)$, then pure SFS is carried out: high-frequency details are perfectly recovered, but the surface might drift from the initial 3D-shape and interpret image noise as unwanted geometric artifacts. If $\mu \to +\infty$, the initial estimate (which exhibits reasonable low-frequency components, but no geometric detail) is not modified. If $\nu \to +\infty$, then only the minimal surface term matters, hence the result is over-smoothed. In this experiment, we also evaluate the accuracy of the 3D-reconstructions through the mean angular error (MAE) on the normals: it is minimal when the parameters are tuned appropriately, not when the image error (RMSE) is minimal since minimizing the latter comes down to estimating geometric details explaining the image noise.

The appropriate tuning of μ and ν depends on how trustworthy the image and the shape prior are. Typically, in RGB-D sensing, the depth may be noisier than in this synthetic experiment: in this case a low value of μ should be used. On the other hand, natural illumination is generally colored, so the three image channels provide redundant information: regularization is less important and a low value of the smoothness parameter ν can be used. We found that $(\lambda, \mu, \nu) = (1, 1, 5.10^{-5})$ provides qualitatively nice results in all our real-world experiments.

5.2   Qualitative Evaluation on Real-World Datasets

The importance of initialization is further confirmed in the top rows of Figs. 1 and 5. In these experiments, our SFS method ($\mu = \nu = 0$) is evaluated, under perspective projection, on real-world datasets obtained using an RGB-D sensor [4], considering a fronto-parallel surface as initialization. Although fine details are revealed, the results present an obvious low-frequency bias, and artifacts due to the image noise occur. This illustrates both the inherent ambiguities of SFS, and the need for depth regularization.

In order to illustrate the practical disambiguation of SFS using a shape prior, we next consider as initialization $z^{(0)}$ and prior z^0 the depth provided by the RGB-D sensor. It is both noisy and incomplete, but with our framework it can be denoised, refined and completed in a shading-aware manner, by tuning the

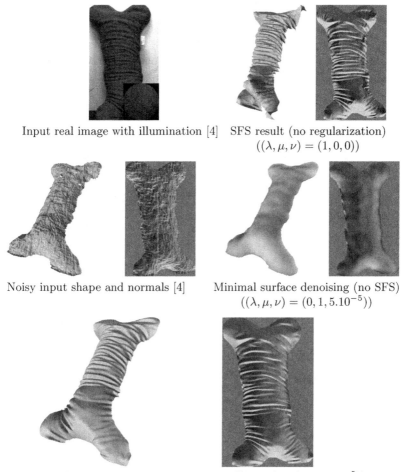

Input real image with illumination [4]   SFS result (no regularization)
$$((\lambda, \mu, \nu) = (1, 0, 0))$$

Noisy input shape and normals [4]   Minimal surface denoising (no SFS)
$$((\lambda, \mu, \nu) = (0, 1, 5.10^{-5}))$$

SFS-based denoising and completion $((\lambda, \mu, \nu) = (1, 1, 5.10^{-5}))$

Fig. 5. Results on three computer vision problems: SFS, "blind" (not shading-based) depth refinement, and shading-based depth refinement. The shape estimated by SFS is distorted (due to the ambiguities of SFS), and artifacts occur (due to noise), but it contains the fine-scale details. Although the depth map provided by the RGB-D sensor is denoised without considering shading, thin structures are missed. With the proposed method, noise is removed and fine details are revealed. (Color figure online)

parameters μ (prior) and ν (smoothness). Second and third rows of Figs. 1 and 5 illustrate the interest of SFS for depth refinement, in comparison with "blind" methods based solely on depth regularization [15].

Eventually, Fig. 6 demonstrates an application to stereovision, using a real-world dataset from [20]. This time, the initial depth map is obtained by a multi-view stereo (MVS) algorithm [21]. We estimated lighting from this initial depth map, assuming uniform albedo. Then, we let our algorithm recover the thin geometric structures, which are missed by MVS. The initial depth map contains a lot of missing data and discontinuities, which is challenging for our algorithm: ambiguities arise inside the large holes, and our model favors smooth surfaces. Indeed, the concavities are not well recovered, and the discontinuities are partly smoothed. Still, nice details are recovered, and the overall surface seems reasonable.

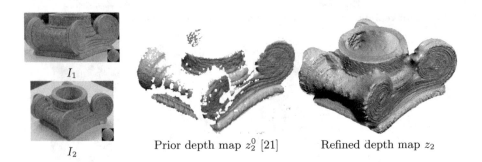

Prior depth map z_2^0 [21]   Refined depth map z_2

Fig. 6. Left: two (out of 30) images I_1 and I_2 of the "Figure" object [20]. Middle: depth map z_2^0 obtained by the CMPMVS method [21] (before meshing). Right: refined and completed depth map z_2.

6   Conclusion and Perspectives

We have introduced a generic variational framework for SFS under natural illumination, which can be applied in a broad range of scenarios. It relies on a tailored PDE-based SFS formulation which handles a variety of models for the camera and the lighting. To solve the resulting system of PDEs, we introduce an ADMM algorithm which separates the difficulty due to nonlinearity from that due to the dependency upon the gradient. Shape prior and nonlinear smoothing terms are easily included in this variational framework, allowing disambiguation of SFS as well as practical applications to depth map refinement and completion for RGB-D sensors or stereovision systems.

As future work, we plan to investigate the convergence of the proposed ADMM algorithm for our non-convex problem, and to include reflectance and lighting estimation. With these extensions, we have good hope that the proposed variational framework may be useful in other computer vision applications, such as shading-aware dense multi-view stereo.

References

1. Durou, J.D., Falcone, M., Sagona, M.: Numerical methods for shape-from-shading: a new survey with benchmarks. Comput. Vis. Image Underst. **109**, 22–43 (2008)
2. Horn, B.K.P., Brooks, M.J.: The variational approach to shape from shading. Comput. Vis. Graph. Image Process. **33**, 174–208 (1986)

3. Lions, P.L., Rouy, E., Tourin, A.: Shape-from-shading, viscosity solutions and edges. Numer. Math. **64**, 323–353 (1993)
4. Han, Y., Lee, J.Y., Kweon, I.S.: High quality shape from a single RGB-D image under uncalibrated natural illumination. In: Proceedings of the ICCV (2013)
5. Basri, R., Jacobs, D.P.: Lambertian reflectances and linear subspaces. IEEE Trans. Pattern Anal. Mach. Intell. **25**, 218–233 (2003)
6. Quéau, Y., Mecca, R., Durou, J.D.: Unbiased photometric stereo for colored surfaces: a variational approach. In: Proceedings of the CVPR (2016)
7. Johnson, M.K., Adelson, E.H.: Shape estimation in natural illumination. In: Proceedings of the CVPR (2011)
8. Huang, R., Smith, W.A.P.: Shape-from-shading under complex natural illumination. In: Proceedings of the ICIP (2011)
9. Or-El, R., Rosman, G., Wetzler, A., Kimmel, R., Bruckstein, A.: RGBD-fusion: real-time high precision depth recovery. In: Proceedings of the CVPR (2015)
10. Frolova, D., Simakov, D., Basri, R.: Accuracy of spherical harmonic approximations for images of Lambertian objects under far and near lighting. In: Pajdla, T., Matas, J. (eds.) ECCV 2004. LNCS, vol. 3021, pp. 574–587. Springer, Heidelberg (2004). https://doi.org/10.1007/978-3-540-24670-1_44
11. Barron, J.T., Malik, J.: Shape, illumination, and reflectance from shading. IEEE Trans. Pattern Anal. Mach. Intell. **37**, 1670–1687 (2015)
12. Cristiani, E., Falcone, M.: Fast semi-Lagrangian schemes for the eikonal equation and applications. SIAM J. Numer. Anal. **45**, 1979–2011 (2007)
13. Breuß, M., Cristiani, E., Durou, J.D., Falcone, M., Vogel, O.: Perspective shape from shading: ambiguity analysis and numerical approximations. SIAM J. Imaging Sci. **5**, 311–342 (2012)
14. Quéau, Y., Durou, J.D., Aujol, J.F.: Normal integration: a survey. J. Math. Imaging Vis. (2017)
15. Graber, G., Balzer, J., Soatto, S., Pock, T.: Efficient minimal-surface regularization of perspective depth maps in variational stereo. In: Proceedings of the CVPR (2015)
16. Glowinski, R., Marroco, A.: Sur l'approximation, par éléments finis d'ordre un, et la résolution, par pénalisation-dualité d'une classe de problèmes de Dirichlet non linéaires. Revue française d'automatique, informatique, recherche opérationnelle. Analyse numérique **9**, 41–76 (1975)
17. Boyd, S., Parikh, N., Chu, E., Peleato, B., Eckstein, J.: Distributed optimization and statistical learning via the alternating direction method of multipliers. Found. Trends Mach. Learn. **3**, 1–122 (2011)
18. Li, G., Pong, T.K.: Global convergence of splitting methods for nonconvex composite optimization. SIAM J. Optim. **25**, 2434–2460 (2015)
19. Hong, M., Luo, Z.Q., Razaviyayn, M.: Convergence analysis of alternating direction method of multipliers for a family of nonconvex problems. SIAM J. Optim. **26**, 337–364 (2016)
20. Zollhöfer, M., Dai, A., Innman, M., Wu, C., Stamminger, M., Theobalt, C., Nießner, M.: Shading-based refinement on volumetric signed distance functions. ACM Trans. Graph. **34**, 96:1–96:14 (2015)
21. Jancosek, M., Pajdla, T.: Multi-view reconstruction preserving weakly-supported surfaces. In: Proceedings of the CVPR (2011)

Sharpening Hyperspectral Images Using Spatial and Spectral Priors in a Plug-and-Play Algorithm

Afonso M. Teodoro, José M. Bioucas-Dias, and Mário A. T. Figueiredo

Instituto de Telecomunicações, Instituto Superior Técnico,
Universidade de Lisboa, Lisbon, Portugal
`mario.figueiredo@tecnico.ulisboa.pt`

Abstract. This paper proposes using both spatial and spectral regularizers/priors for hyperspectral image sharpening. Leveraging the recent *plug-and-play* framework, we *plug* two Gaussian-mixture-based denoisers into the iterations of an *alternating direction method of multipliers* (ADMM): a spatial regularizer learned from the observed multispectral image, and a spectral regularizer trained using the hyperspectral data. The proposed approach achieves very competitive results, improving the performance over using a single regularizer. Furthermore, the spectral regularizer can be used to classify the image pixels, opening the door to class-adapted models.

Keywords: Data fusion · Hyperspectral imaging
Spatial-spectral regularization · Plug-and-play · Gaussian mixture

1 Introduction

Hyperspectral images (HSI) are very rich sources of information, with a diverse range of applications, such as surveillance, agriculture, and remote sensing [13]. What characterizes HSI is their spectral resolution, that is, instead of having the three standard RGB channels, they have hundreds of bands, each covering a narrow range of the electromagnetic spectrum, giving HSI the potential to identify or discriminate, for example, different land types or materials [5]. The development of classification and segmentation techniques that exploit the rich information carried by HSI has been a very active research topic over the past couple of decades [12].

Due to the limited amount of energy reaching the sensors, there is a trade-off between the spectral and spatial resolutions: at a given signal-to-noise ratio, increasing the spectral resolution implies reducing the spatial resolution, and vice versa. In practice, it may be possible to detect the presence of a given material, but impossible to accurately pinpoint its location. Hyperspectral image sharpening, or hyperspectral fusion, is an approach that has been proposed to overcome this limitation: the idea is to take an HSI (which has high spectral

© Springer International Publishing AG, part of Springer Nature 2018
M. Pelillo and E. Hancock (Eds.): EMMCVPR 2017, LNCS 10746, pp. 358–371, 2018.
https://doi.org/10.1007/978-3-319-78199-0_24

resolution, but low spatial resolution) and a *panchromatic* (PAN) or *multispectral image* (MSI) of the same scene (which has low spectral resolution, but high spatial resolution) and combine/fuse them into an image with high spectral **and** spatial resolutions. Hyperspectral image sharpening, or hyperspectral fusion, has been an active research area [13].

The current state-of-the-art in hyperspectral image sharpening uses Bayesian or variational approaches [17,25], relying on variable splitting algorithms, namely the *alternating direction method of multipliers* (ADMM) [6], to solve the resulting optimization problem. Recently, ADMM has also been used in a *plug-and-play* (PnP) framework, together with state-of-the-art denoisers for grayscale images, and showed very competitive results [22,23]. Still, the denoiser has been used only to impose prior knowledge on the spatial features of the image being estimated.

Here, we build upon our previous work [22,23], and show that the *plugged-in* denoiser can also be applied to the spectral signatures of each pixel, imposing prior knowledge on the spectra. Furthermore, it is straighforward to apply spatial and spectral regularization together, using the ADMM algorithm and PnP priors. We claim that, in order to obtain better performance, we should leverage not only the spatial properties of the data, but also the spectral characteristics, which provide such useful information in the case of HSI.

Moreover, for simplification purposes, we use the same type of denoiser (or prior) on the spatial and spectral characteristics, and illustrate, with a toy example, that the proposed method is able to perform classification of the pixel spectral signatures and improve the resolution of HSI, thus combining two important tasks (sharpening and classification) into one framework.

This paper is organized as follows. Section 2 formulates the hyperspectral fusion problem using both spatial and spectral regularizers. Section 3 provides a detailed description of the proposed method, while Sect. 4 reports experimental results. Section 5 concludes the paper.

2   Problem Formulation

Following the description of HSI and MSI, the underlying assumptions are that an HSI is a spatially blurred and downsampled version of a *target image* we wish to infer, whereas the corresponding MSI is a spectrally degraded version of the same target image. To formally express this observation model, let $\mathbf{Y}_h \in \mathbb{R}^{L_h \times n_h}$ denote the matrix collecting the spectra at all the n_h pixels of the HSI, each of which has L_h bands; similarly, $\mathbf{Y}_m \in \mathbb{R}^{L_m \times n_m}$ is the matrix with the spectra of all the n_m pixels of the MSI, each of which has L_m bands. (Notice that, in both cases, columns index pixels and rows index bands.) As explained above, we typically have $L_h \gg L_m$ (in the extreme case of PAN images, $L_m = 1$); conversely, $n_h \ll n_m$. With this notation, the observation model can be written compactly as

$$\mathbf{Y}_h = \mathbf{ZBM} + \mathbf{N}_h, \tag{1}$$
$$\mathbf{Y}_m = \mathbf{RZ} + \mathbf{N}_m, \tag{2}$$

where $\mathbf{Z} \in \mathbb{R}^{L_h \times n_m}$ is the target image, $\mathbf{B}$ is a spatial low-pass filter, $\mathbf{M}$ is a spatial down-sampling acting on each row (*i.e.*, each band) of $\mathbf{Z}$, and $\mathbf{R}$ represents the spectral responses of the multispectral sensor; finally, $\mathbf{N}_h$ and $\mathbf{N}_n$ are additive noises, herein assumed to be zero-mean Gaussian with known variances σ_h^2 and σ_m^2, respectively (other types of noise can also be considered [13,25]). Matrices $\mathbf{B}$ and $\mathbf{R}$ are assumed known, *i.e.*, we consider a non-blind scenario.

One of the difficulties in the hyperspectral fusion problem is its high dimensionality. Exploiting the fact that the bands tend to be highly correlated, this difficulty may be mitigated by assuming that the spectra in $\mathbf{Z}$ live in a low-dimensional subspace, say L_s-dimensional, with $L_s \ll L_h$. (The low-dimensionality of the spectra may be seen as resulting from the number of different materials in a scene often being much smaller than the number of bands [5].) This assumption allows shifting the problem from estimating L_h-dimensional spectra to estimating L_s-dimensional representations of these spectra. Several methods have been proposed to identify a suitable subspace, such as *principal component analysis* (PCA) [11], *singular value decomposition* (SVD), *maximum noise fraction* (MNF) [10], and *Hysime* [4]. In this paper, we adopt *Hysime* [4], which estimates both the dimension L_s and a basis of the subspace, and is completely parameter-free. Letting $\mathbf{E} = [\mathbf{e}_1, \ldots, \mathbf{e}_{L_s}] \in \mathbb{R}^{L_h \times L_s}$ contain a basis of this subspace, we may write $\mathbf{Z} = \mathbf{EX}$, where each column of $\mathbf{X} \in \mathbb{R}^{L_s \times n_m}$ is a representation (vector of coefficients) of the corresponding column of $\mathbf{Z}$. Instead of estimating $\mathbf{Z}$ directly, the problem is reformulated as that of obtaining an estimate $\widehat{\mathbf{X}}$, from which an estimate $\widehat{\mathbf{Z}} = \mathbf{E}\widehat{\mathbf{X}}$ can be obtained. Replacing $\mathbf{Z}$ with $\mathbf{EX}$ in (1) and (2) yields

$$\mathbf{Y}_h = \mathbf{EXBM} + \mathbf{N}_h, \tag{3}$$
$$\mathbf{Y}_m = \mathbf{REX} + \mathbf{N}_m. \tag{4}$$

Often, even after dimensionality reduction, the number of unknowns, $L_s\, n_m$, is larger than the number of observations, $L_h\, n_h + L_m\, n_m$, making the problem ill-posed. In this paper, we adopt the Bayesian *maximum a posteriori* (MAP) criterion (equivalently, a regularization criterion) and estimate $\mathbf{X}$ via

$$\widehat{\mathbf{X}} = \operatorname*{argmin}_{\mathbf{X}} \frac{1}{2\sigma_h^2} \|\mathbf{EXBM} - \mathbf{Y}_h\|_F^2 + \frac{1}{2\sigma_m^2}\|\mathbf{REX} - \mathbf{Y}_m\|_F^2 + \phi(\mathbf{X}), \tag{5}$$

where $\|\cdot\|_F^2$ denotes squared Frobenius norm, the noise matrices $\mathbf{N}_h$ and $\mathbf{N}_m$ are assumed independent (conditioned on $\mathbf{X}$), and $\phi(\mathbf{X})$ is the regularizer; equivalently, $p(\mathbf{X}) \propto \exp(-\phi(\mathbf{X}))$ is the prior on $\mathbf{X}$. Dividing the objective function by σ_h^2 and denoting $\lambda_m = 1/(\sigma_h^2 \sigma_m^2)$ and $\lambda_m = 1/\sigma_h^2$ yields

$$\widehat{\mathbf{X}} \in \operatorname*{argmin}_{\mathbf{X}} \frac{1}{2}\|\mathbf{EXBM} - \mathbf{Y}_h\|_F^2 + \frac{\lambda_m}{2}\|\mathbf{REX} - \mathbf{Y}_m\|_F^2 + \lambda_\phi \phi(\mathbf{X}), \tag{6}$$

where the first two terms are data fidelity terms and parameters λ_m and λ_ϕ control the relative weight of each term.

2.1   Choice of Prior

The prior, or regularizer, is used to promote particular characteristics on $\widehat{\mathbf{X}}$. Typical choices include sparsity-inducing norms on some representation (such as a wavelet frame or a learned dictionary), total-variation [16], low-rank constraints, probabilistic models, and others. In fact, all of the above have shown good results on grayscale image denoising, where they were first proposed, and afterwards applied to more involved imaging inverse problems. The recent *plug-and-play* (PnP) framework [24] provides a very flexible way to leverage state-of-the-art denoising methods to tackle deblurring [20,21], compressive imaging [20], super-resolution [8,9], and other problems [9,18,24]. As the name suggests, in a PnP scheme, the denoiser is *plugged* into an iterative algorithm, which decomposes the original problem (6) into a sequence of simpler problems, one of which is a denoising problem. Other frameworks, such as RED (*regularization by denoising*, [15]), exploit similar ideas.

Recently, we have used a *Gaussian mixture model* (GMM) as a prior for hyperspectral fusion, in a PnP scheme, and shown very competitive results [22, 23]. However, in our previous work, we applied the GMM denoiser to each row of $\mathbf{X}$, that is, each image of coefficients, in an independent way. In this paper, we claim that we may be able to improve the results if, instead of imposing only a low-dimensionality constraint by representing the target image on subspace, we regularize the spectra, or even the low dimensional versions thereof.

In (6), the form of the regularizer or prior is yet to be defined. In order to include both spatial and spectral priors, we split function $\phi(\mathbf{X})$ into two terms, namely $\phi_{\mathrm{spatial}}(\mathbf{X})$ and $\phi_{\mathrm{spectral}}(\mathbf{X}^T)$, respectively. The transpose in the spectral prior is used to stress that the regularization is applied to the columns of $\mathbf{X}$, that is, to each image pixel, in contrast with the spatial prior applied to each row, *i.e.*, to (patches of) each band. The final formulation becomes

$$\widehat{\mathbf{X}} \in \operatorname*{argmin}_{\mathbf{X}} \frac{1}{2}\|\mathbf{EXBM} - \mathbf{Y}_h\|_F^2 + \frac{\lambda_m}{2}\|\mathbf{REX} - \mathbf{Y}_m\|_F^2$$
$$+ \lambda_{\mathrm{spatial}}\,\phi_{\mathrm{spatial}}(\mathbf{X}) + \lambda_{\mathrm{spectral}}\,\phi_{\mathrm{spectral}}(\mathbf{X}^T), \tag{7}$$

where we introduce parameters $\lambda_{\mathrm{spatial}}$ and $\lambda_{\mathrm{spectral}}$ to control the strength of each of the two regularizers.

3   Method

As mentioned above, in order to solve (7), we resort to a PnP scheme built into an ADMM algorithm, which is supported on a so-called *variable splitting* procedure, yielding a constrained optimization problem equivalent to (6):

$$\widehat{\mathbf{X}}, \widehat{\mathbf{V}}_1, \widehat{\mathbf{V}}_2, \widehat{\mathbf{V}}_3, \widehat{\mathbf{V}}_4 \in \operatorname*{argmin}_{\mathbf{X},\mathbf{V}_1,\mathbf{V}_2,\mathbf{V}_3,\mathbf{V}_4} \frac{1}{2}\|\mathbf{EV}_1\mathbf{M} - \mathbf{Y}_h\|_F^2 + \frac{\lambda_m}{2}\|\mathbf{REV}_2 - \mathbf{Y}_m\|_F^2$$
$$+ \lambda_{\mathrm{spatial}}\,\phi_{\mathrm{spatial}}(\mathbf{V}_3) + \lambda_{\mathrm{spectral}}\,\phi_{\mathrm{spectral}}(\mathbf{V}_4^T)$$

$$\text{subject to } \mathbf{V}_1 = \mathbf{XB}, \ \mathbf{V}_2 = \mathbf{X}, \ \mathbf{V}_3 = \mathbf{X}, \ \mathbf{V}_4 = \mathbf{X}. \tag{8}$$

The corresponding *augmented Lagrangian* is

$$\mathcal{L}(\mathbf{X}, \mathbf{V}, \mathbf{D}) = \frac{1}{2}\|\mathbf{EV}_1\mathbf{M} - \mathbf{Y}_h\|_F^2 + \frac{\lambda_m}{2}\|\mathbf{REV}_2 - \mathbf{Y}_m\|_F^2$$

$$+\lambda_{\text{spatial}}\,\phi_{\text{spatial}}(\mathbf{V}_3) + \lambda_{\text{spectral}}\,\phi_{\text{spectral}}(\mathbf{V}_4^T) + \frac{\mu}{2}\|\mathbf{XB} - \mathbf{V}_1 - \mathbf{D}_1\|_F^2$$

$$+\frac{\mu}{2}\Big(\|\mathbf{X} - \mathbf{V}_2 - \mathbf{D}_2\|_F^2 + \|\mathbf{X} - \mathbf{V}_3 - \mathbf{D}_3\|_F^2 + \|\mathbf{X} - \mathbf{V}_4 - \mathbf{D}_4\|_F^2\Big),$$

where $\mathbf{D} = (\mathbf{D}_1, \mathbf{D}_2, \mathbf{D}_3, \mathbf{D}_4)$ are the scaled dual variables (or Lagrange multipliers, see [6]), and $\mathbf{V} = (\mathbf{V}_1, \mathbf{V}_2, \mathbf{V}_3, \mathbf{V}_4)$. For simplicity, we use the same penalty parameter μ for all the constraints. ADMM works by minimizing $\mathcal{L}$ alternatingly w.r.t. to each of the primal variables $\mathbf{X}, \mathbf{V}_1, \mathbf{V}_2, \mathbf{V}_3$, and $\mathbf{V}_4$ (while keeping the others fixed), and then updating the dual variables [6]. For completeness, we present here the sequence of subproblems to be solved at each step of the ADMM algorithm, and emphasize the main aspects concerning each of them. In summary, each iteration of the PnP-ADMM for solving (7) takes the form

$$\mathbf{X}^{k+1} = \underset{\mathbf{X}}{\arg\min}\ \|\mathbf{XB} - \mathbf{V}_1^k - \mathbf{D}_1^k\|_F^2 + \|\mathbf{X} - \mathbf{V}_2^k - \mathbf{D}_2^k\|_F^2$$

$$+ \|\mathbf{X} - \mathbf{V}_3^k - \mathbf{D}_3^k\|_F^2 + \|\mathbf{X} - \mathbf{V}_4^k - \mathbf{D}_4^k\|_F^2, \tag{9}$$

$$\mathbf{V}_1^{k+1} = \underset{\mathbf{V}_1}{\arg\min}\ \frac{1}{2}\|\mathbf{EV}_1\mathbf{M} - \mathbf{Y}_h\|_F^2 + \frac{\mu}{2}\|\mathbf{X}^{k+1}\mathbf{B} - \mathbf{V}_1 - \mathbf{D}_1^k\|_F^2, \tag{10}$$

$$\mathbf{V}_2^{k+1} = \underset{\mathbf{V}_2}{\arg\min}\ \frac{\lambda_m}{2}\|\mathbf{REV}_2 - \mathbf{Y}_m\|_F^2 + \frac{\mu}{2}\|\mathbf{X}^{k+1} - \mathbf{V}_2 - \mathbf{D}_2^k\|_F^2, \tag{11}$$

$$\mathbf{V}_3^{k+1} = \underset{\mathbf{V}_3}{\arg\min}\ \lambda_{\text{spatial}}\,\phi_{\text{spatial}}(\mathbf{V}_3) + \frac{\mu}{2}\|\mathbf{X}^{k+1} - \mathbf{V}_3 - \mathbf{D}_3^k\|_F^2, \tag{12}$$

$$\mathbf{V}_4^{k+1} = \underset{\mathbf{V}_4}{\arg\min}\ \lambda_{\text{spectral}}\,\phi_{\text{spectral}}(\mathbf{V}_4^T) + \frac{\mu}{2}\|\mathbf{X}^{k+1} - \mathbf{V}_4 - \mathbf{D}_4^k\|_F^2, \tag{13}$$

followed by an update of the dual variables

$$\mathbf{D}_1^{k+1} = \mathbf{D}_1^k + \mathbf{V}_1^{k+1} - \mathbf{X}^{k+1}\mathbf{B}, \tag{14}$$

$$\mathbf{D}_2^{k+1} = \mathbf{D}_2^k + \mathbf{V}_2^{k+1} - \mathbf{X}^{k+1}, \tag{15}$$

$$\mathbf{D}_3^{k+1} = \mathbf{D}_3^k + \mathbf{V}_3^{k+1} - \mathbf{X}^{k+1}, \tag{16}$$

$$\mathbf{D}_4^{k+1} = \mathbf{D}_4^k + \mathbf{V}_4^{k+1} - \mathbf{X}^{k+1}. \tag{17}$$

Subproblems (9)–(11) are quadratic and thus have closed-form solutions, which are given by

$$\mathbf{X}^{k+1} = \big[(\mathbf{V}_1^k + \mathbf{D}_1^k)\,\mathbf{B}^T + (\mathbf{V}_2^k + \mathbf{D}_2^k) + (\mathbf{V}_3^k + \mathbf{D}_3^k) + (\mathbf{V}_4^k + \mathbf{D}_4^k)\big]\,\big[\mathbf{BB}^T + 3\mathbf{I}\big]^{-1}, \tag{18}$$

$$\mathbf{V}_1^{k+1} = \big[\mathbf{E}^T\mathbf{E} + \mu\mathbf{I}\big]^{-1}\big[\mathbf{E}^T\mathbf{Y}_h + \mu\,(\mathbf{X}^{k+1}\mathbf{B} - \mathbf{D}_1^k)\big] \odot \mathbf{M}$$

$$+ (\mathbf{X}^{k+1}\mathbf{B} - \mathbf{D}_1^k) \odot (1 - \mathbf{M}), \tag{19}$$

$$\mathbf{V}_2^{k+1} = \big[\lambda_m\mathbf{E}^T\mathbf{R}^T\mathbf{RE} + \mu\mathbf{I}\big]^{-1}\big[\lambda_m\mathbf{E}^T\mathbf{R}^T\mathbf{Y}_m + \mu\,(\mathbf{X}^{k+1} - \mathbf{D}_2^k)\big], \tag{20}$$

where $\odot$ denotes entry-wise (Hadamard) product. A few comments are in order; for more details about the derivation of these expressions, we refer the reader to [17,22,23]. Assuming periodic boundary conditions, computing (18) can be done efficiently in the Fourier domain, using the *fast Fourier transform* (FFT). Moreover, the matrix inversions in (19) and (20) involve matrices of dimension $L_s \times L_s$, where L_s is the dimension of the subspace. Since we assumed that the subspace is smaller than the number of bands in the HSI, the cost is low. Also, if parameter μ is kept fixed, the inverses can be pre-computed, stored, and re-used at each iteration.

For convex regularizers, subproblems (12) and (13) are the *Moreau proximity operators* (MPO) [3] of (scaled versions of) functions ϕ_{spatial} and ϕ_{spectral}, computed at $(\mathbf{X}^{k+1} - \mathbf{D}_3^k)$ and $(\mathbf{X}^{k+1} - \mathbf{D}_4^k)$, respectively. In more general terms, (12) and (13) can be interpreted as denoising problems, where $(\mathbf{X}^{k+1} - \mathbf{D}^k)$ is the noisy observed data, λ/μ is the Gaussian noise variance, and ϕ is the regularizer, or negative log-prior. Under a PnP perspective [24], we should leverage current state-of-the-art denoisers to solve (12) and (13). As we build upon our previous work in [22,23], our choice of denoiser to solve both problems is the GMM-based minimum mean squared error (MMSE) denoiser. Naturally, different denoisers could be used to solve either or both of these sub-problems.

Having a probabilistic model to represent the spectral signatures provides a straightforward way to classify each image pixel. Assuming there are several classes in an scene, for example, several different materials, we can train a GMM for each of the classes. A similar approach was proposed in [21], yet the training of the models relied on external datasets of images. In this paper, we estimate a rough initial image segmentation with C classes, based on *simple linear iterative clustering* (SLIC) [2], followed by clustering the mean values of the segments. Then, we leverage the segmentation to learn a K-component GMM for each of the classes, and denoise each pixel spectral signature, $\mathbf{x}_i$, using the model from the class that maximizes the likelihood, that is,

$$\hat{c}_i = \arg \max_{c \in \{1,\ldots C\}} p(\mathbf{x}_i|c), \tag{21}$$

where

$$p(\mathbf{x}_i|c_i) = \sum_{m=1}^{K^{(c_i)}} \alpha_m^{(c_i)} \, \mathcal{N}(\mathbf{x}_i; \boldsymbol{\mu}_m^{(c_i)}, \mathbf{C}_m^{(c_i)}). \tag{22}$$

By using a model targeted to a specific class, we claim we can achieve a better reconstruction than if we used a so-called generic prior that does not take this additional information, *i.e.* the class, into account.

3.1   Training the GMM

Our choice of using two regularizers stems from the fact that the HSIs have distinct features from the MSIs, which we aim to fuse. Hence, each GMM should be trained to capture the characteristics of each type of data. Furthermore, training a model jointly for the spatial and spectral characteristics would be impractical due to the size of the data involved. In particular, the size of the patches would increase by a factor of L_h (or L_s, considering the lower dimensional representation of the data), and the covariance matrices by L_h^2 (respectively, L_s^2).

On the one hand, concerning the spatial GMM, we follow the methodology proposed in [23]. We use the high spatial resolution MSIs to train a patch-based scene-adapted GMM, and use the GMM as a prior to denoise each image of coefficients, that is, each row of $\mathbf{X}$. For more details on the patch-based GMM denoiser, see [19].

On the other hand, for the spectral GMM, we use the high spectral resolution HSIs. Note that, in this case, we are not dealing with patches, but with individual spectral signatures. Moreover, as we are dealing with the subspace representation of the HSIs, the GMM is low dimensional as well, that is, for each mixture component, the mean vector is of size L_s and the covariance matrix $L_s \times L_s$, thus reducing the number of samples needed for training, as well as the computational complexity.

3.2   Convergence   ·

In [23], we showed that the ADMM algorithm with a plugged-in scene-adapted GMM denoiser as a spatial regularizer is guaranteed to converge to a global minimum of a cost function, implicitly defined by the choice of prior. In a nutshell, the proof consists in showing that the denoiser is a non-expansive operator and the sub-gradient of a convex function, shown in [18] to be sufficient conditions. In fact, such conditions are necessary and sufficient to ensure that the denoising operator is actually a proximity operator, and thus maximal monotone [3]. However, for these conditions to be met, we needed to remove the nonlinearity in the weights, β (see [23]) that represent the posterior probability of a patch being generated by each component of the mixture. We accomplished this by keeping the weights β fixed, in light of the scene-adaptiveness proposed therein.

Using the same arguments, we can guarantee convergence of ADMM with two denoisers, provided that both of them satisfy the same conditions. As mentioned before, the spatial regularizer is the same as in [23], and thus we need only to remove the nonlinearity in the spectral regularizer. Two possible ways to do so are: (i) after a number of iterations, keep the weights β fixed; (ii) instead of computing the *minimum mean square error* (MMSE) estimates of the spectral signatures, as proposed in [19], compute the approximate MAP by selecting a single mixture component (with corresponding weight equal to 1) [27,28]. In our experiments (Sect. 4), we keep the non-linearity of the weights, as we always observe convergence in practice, anyway.

4   Experimental Results

In this section, we present and discuss the experimental results. The experiments were designed to provide answers to two main questions: (i) is there any advantage in using two regularizers, one spatial and the other one spectral, instead of just one of them? (ii) can we leverage the classification/segmentation of the image to improve the quality of the fusion results?

In order to answer these questions, we assess the fusion accuracy according to three standard quality metrics, namely average *peak signal-to-noise ratio* (PSNR), *spectral angle mapper* (SAM), and *erreur relative globale adimensionnelle de synthèse* (ERGAS) (see [13,26], for details).

4.1   Image Fusion

In this section, we address the first question, that is, we illustrate the accuracy of the proposed hyperspectral and multispectral image fusion algorithm, on different datasets, and compare it against using only one of the priors. We also include the results obtained with the *HySure* algorithm [17], to provide a baseline of the current state-of-the-art. Other algorithms may be able to achieve slightly better results but according to a recent comparative review [26], HySure is the most robust. Moreover, the proposed method shares with HySure the same core algorithm (an instance of ADMM known as *SALSA* [1]), differing only in the choice of regularizer, allowing us to draw conclusions on the usefulness of GMM priors.

Tables 1 and 2 summarize the obtained results. A description of the datasets, including sensor specifications, wavelength range, image size, water absorption bands, can be found in [26], along with the simulation procedure. Here, we replicated the same setting, and repeated the results for the HySure algorithm to account for possible differences, namely in parameter values. Briefly, the simulation procedure starts by generating the observed HSI and MSI from a clean high spatial and spectral resolution reference image, using known blur kernels, downsampling factors, sensor responses, and adding Gaussian noise so that the signal-to-noise ratio on all bands is 35 dB. Afterwards, we estimate the target high resolution image and compare it to the reference image, with regard to the three quality metrics mentioned above.

In Table 1, we let the Hysime algorithm [4] determine the dimension of the subspace, and the basis vectors, *i.e.*, the columns of $\mathbf{E}$, as mentioned before. The parameters of the GMM were learned from either the observed MSI or the subspace representation of the HSI, using the standard *expectation-maximization* (EM) algorithm, but thresholding the singular values of the covariance matrices to be positive, thus ensuring positive-definite covariances. Both spatial and spectral GMM have 20 components, and for the spatial GMM we considered non-overlapping, 8×8 patches. Using non-overlapping patches significantly decreases the amount of computation, but the performance is not compromised. This happens because the denoising step is only one of the steps of ADMM, hence the blocking artifacts that are typical in non-overlapping patch-based denoising methods get diluted with the other global operations. The remaining parameters that need to be set are μ, λ_m, λ_{spatial}, and $\lambda_{\text{spectral}}$ (see (7)). We selected the combination of values that provides the best results on each dataset using grid search in the following set of possible values: $\{10^{-3}, 10^{-2}, 10^{-1}, 1\} \times \{10^{-2}, 10^{-1}, 1, 10\} \times \{10^{-6}, 10^{-5}, 10^{-4}\} \times \{10^{-6}, 10^{-5}, 10^{-4}\}$, respectively. For the single GMM prior and HySure, we simply removed one of the regularizers and the corresponding parameter.

Table 1. Results obtained for several datasets (for PSNR, higher values are better, whereas for SAM and ERGAS, lower values are better), with **E** obtained by the *Hysime* algorithm [4]. Boldface denotes the best result of each measure, for each dataset, while Italic denotes the second best result.

Method	HySure			Spatial			Spectral			Spatial-Spectral		
Dataset	PSNR	SAM	ERGAS	PSNR	SAM	ERGAS	PSNR	SAM	ERGAS	PSNR	SAM	ERGAS
AVIRIS Indian Pines	39.46	0.83	0.50	*43.10*	**0.60**	**0.31**	**43.22**	*0.61*	**0.31**	42.77	*0.61*	*0.32*
AVIRIS Cuprite	39.74	0.93	0.38	*43.97*	*0.50*	0.23	42.66	0.59	0.28	**44.09**	**0.48**	0.23
AVIRIS Moffett Field	33.76	2.17	5.56	*34.67*	1.95	*5.23*	34.38	*1.85*	5.28	**36.28**	**1.50**	**4.49**
HYDICE W. DC Mall	30.75	4.40	5.75	*34.59*	3.44	*4.66*	34.19	*1.75*	6.20	**37.16**	**1.53**	**3.94**
HyperSpec Chikusei	35.05	2.57	2.37	*44.70*	*1.49*	*1.53*	41.34	1.70	1.98	**44.76**	**1.39**	**1.48**
ROSIS-3 Univ. Pavia	31.11	5.78	2.07	*39.35*	4.49	*1.29*	37.99	*3.20*	1.34	**40.84**	**3.01**	**0.97**
CASI Univ. Houston	38.49	4.27	3.58	**44.86**	2.17	*1.65*	41.32	*2.01*	2.16	*44.65*	**1.89**	**1.59**
Average	35.48	2.99	2.89	*40.75*	2.09	*2.13*	39.30	*1.67*	2.51	**41.51**	**1.49**	**1.86**

Comparing the GMM-based methods, we see that ADMM with the spatial GMM-denoiser performs better than the one using only the spectral regularizer. A possible cause is the low-dimensionality that is imposed on the spectra, which effectively also provides spectral regularization. Moreover, we conclude that, in most cases, using both spatial and spectral regularization improves the results over using only a single prior.

The results obtained with the HySure algorithm [17] are considerably worse because SVD-like bases, such as those obtained with Hysime, are orthonormal and concentrate most of the signal energy on a single (or very few) direction(s). Thus, the final results are highly dependent on the performance of the method for this particular coefficient. Furthermore, allowing the basis vectors to have positive or negative entries causes each image of coefficients to have different amplitudes, which may be difficult to handle with a TV regularizer.

Table 2 reports the second set of experiments, this time using the VCA algorithm [14] to determine the basis for the subspace. VCA is an unmixing strategy, thus the basis vectors have nonnegative entries only, and the images of

Table 2. Results obtained for several datasets, with **E** provided by the VCA algorithm [14] (for PSNR, higher values are better, whereas for SAM and ERGAS, lower values are better). Boldface denotes the best result of each measure, for each dataset, while Italic denotes the second best result.

Method	HySure			Spatial			Spectral			Spatial-Spectral		
Dataset	PSNR	SAM	ERGAS	PSNR	SAM	ERGAS	PSNR	SAM	ERGAS	PSNR	SAM	ERGAS
AVIRIS Indian Pines	*42.39*	0.62	*0.33*	**42.52**	0.59	0.32	41.70	0.65	0.35	42.34	*0.60*	*0.33*
AVIRIS Cuprite	43.81	*0.50*	0.22	*43.91*	**0.48**	0.22	42.83	0.55	*0.25*	**44.38**	**0.48**	0.22
AVIRIS Moffett Field	33.66	**1.59**	4.44	*36.14*	1.62	4.66	35.92	1.89	4.70	**36.28**	1.62	*4.65*
HYDICE W. DC Mall	**38.21**	**1.73**	**3.54**	37.91	1.83	3.60	37.82	1.87	3.62	*37.95*	*1.80*	*3.65*
HyperSpec Chikusei	41.48	**1.20**	1.52	**46.06**	*1.22*	1.56	40.89	1.37	1.63	*46.05*	1.22	1.57
ROSIS-3 Univ. Pavia	38.63	**2.69**	*0.83*	*41.75*	2.70	**0.80**	38.47	2.73	*0.83*	**42.71**	2.71	**0.80**
CASI Univ. Houston	43.82	1.39	*1.13*	**47.29**	1.47	*1.13*	43.49	1.47	1.14	*47.28*	*1.45*	**1.12**
Average	40.29	**1.39**	**1.71**	*42.23*	1.42	*1.76*	40.16	1.50	1.79	**42.43**	*1.41*	*1.76*

coefficients have similar ranges, boosting the performance of HySure. Still, the proposed method performs on the same level, revealing robustness with respect to the choice of subspace basis.

4.2   Simultaneous Fusion and Segmentation: A Toy Example

In this section, we target the second question: can we leverage the classification/segmentation of the image to improve fusion accuracy? To this end, we created an image with several well-defined segments, each of which with a particular texture and spectral signature. We started by selecting 10 end-members from the USGS 1995 library[1]. Then, for every class, we built abundance matrices using three of the selected end-members, by drawing from a Dirichlet distribution. Multiplying the abundance matrix by the end-members, we obtained our reference dataset (with high spatial and spectral resolutions). Afterwards, we repeat the simulation procedure from the previous section, that is, we generate the observed MSI and HSI, add noise, combine them, and measure the quality of the output by comparing to the reference image. In Fig. 1, we show the segmentation ground truth, a false color representation of the reference image, and the average spectral signature of each class.

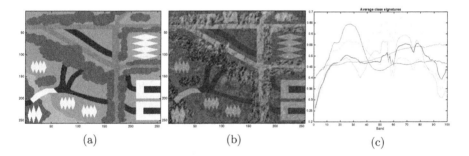

(a)   (b)   (c)

Fig. 1. Synthetic hyperspectral data for image segmentation: (a) ground truth; (b) false color; (c) average class spectral signatures. (Color figure online)

Once again, the goal of this experiment is to leverage the class knowledge, in order to train better priors, denoted class-adapted [20,21], and consequently improve the reconstruction results. To evaluate the impact of the class-adapted models, we start by running each method with the same parameters as before: 20 components in each GMM, 8×8 patches for the spatial prior, and using Hysime to determine the dimension of the subspace and the basis vectors. The results are summarized in Table 3. Afterwards, we repeat the experiments but taking into account the segmentation, that is, instead of training a single 20-component GMM for all spectral signatures, we train a model using only the pixels of each

---

[1] https://speclab.cr.usgs.gov/spectral-lib.html.

class, and use the corresponding model in the spectral denoising step; hence, the class-adapted designation. At first, we considered that the classes were known, as proof of concept. If the fusion results, did not improve in this situation, our claim would have no support but, from Table 4, column labelled *True*, we see that using class-adapted models improves the results, as expected. Only when we use a single Gaussian to model each class, we observe a drop in performance. This is because a single Gaussian is not flexible enough to capture the intraclass variability.

Then, we introduced the classification step (21), into ADMM: at each iteration and for each pixel, we use the selected model. Once again, from the results in Table 4, we see that using the locally selected priors, as in [21], also improves the performance of the proposed method. However, this improvement is not as significant as before. Note: Table 4 does not show results concerning the use of only the spatial prior because the classification feature described here is particular to the use of the spectral prior. Other classifiers, based on the spatial features could be exploited here, but we chose to leverage the spectral information as is typically done in hyperspectral unmixing and classification [5].

Table 3. Quality measures for synthetic dataset, without classification.

Method	HySure			Spatial			Spectral			Spatial-Spectral		
Dataset	PSNR	SAM	ERGAS	PSNR	SAM	ERGAS	PSNR	SAM	ERGAS	PSNR	SAM	ERGAS
Synthetic	35.22	1.25	0.72	36.65	1.02	0.57	*38.17*	*0.80*	*0.45*	**38.68**	**0.77**	**0.42**

Table 4. Quality measures for synthetic dataset, with classification.

Method	Spectral							Spatial-Spectral						
Classes	True			Estimated				True			Estimated			
Dataset	K	PSNR	SAM	ERGAS	PSNR	SAM	ERGAS	OA	PSNR	SAM	ERGAS	PSNR	SAM	ERGAS	OA
Synthetic	1	28.72	1.10	1.32	30.01	1.02	1.13	77.11	28.89	1.11	1.30	33.01	1.09	0.80	**78.62**
	5	38.53	0.76	0.42	38.24	0.79	0.44	75.06	39.00	0.74	0.39	38.83	0.75	0.41	75.51
	20	39.11	0.72	0.39	38.58	0.75	0.42	76.32	39.36	0.71	0.38	**38.99**	**0.73**	**0.40**	77.25

Finally, we evaluate the *overall accuracy* (OA) of the proposed method, with respect to the initial estimate of 74.98%. As mentioned before, the initial segmentation estimate is obtained using SLICs [2], followed by k-means clustering of the mean values of the segments into classes. Although the final segmentation is not very good, we are able to improve not only the initial estimate, but also the reconstruction, which is our main goal. Other, more sophisticated, methods could also be employed to obtain a better segmentation (Fig. 2), such as the α-expansion graph-cuts algorithm [7].

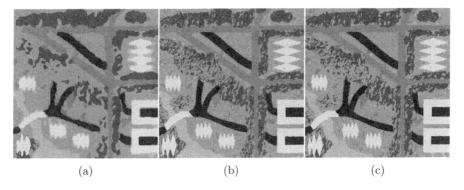

(a)   (b)   (c)

Fig. 2. Synthetic hyperspectral data for image segmentation: (a) initial estimate (OA = 74.98%); (b) final estimate using maximum likelihood (OA = 77.25%); (c) final estimate using α-expansion (OA = 79.85%).

5   Conclusion

This paper proposed a simultaneous spatial and spectral regularization approach to hyperspectral sharpening, under the so-called *plug-and-play* framework. The method leverages two GMM-based denoisers: the spatial GMM is scene-adapted [22], whereas the spectral GMM is learned from the coefficient representation of the observed HSI on a subspace. Results show that with the two regularizers we are able to slightly improve the performance, over using only a single regularizer.

Moreover, introducing the spectral GMM opens the door to simultaneous fusion and segmentation of the data, which may be useful for other applications. In this paper, we showed that we can leverage the segmentation to develop class-adapted models, thus improving further the reconstruction.

Acknowledgments. This work was partially supported by *Fundação para a Ciência e Tecnologia* (FCT), grants BD/102715/2014, UID/EEA/5008/2013, and ERANETMED/0001/2014. The authors would like to thank Prof. N. Yokoya for providing the datasets [26].

References

1. Afonso, M., Bioucas-Dias, J., Figueiredo, M.: Fast image recovery using variable splitting and constrained optimization. IEEE Trans. Image Process. **19**, 2345–2356 (2010)
2. Achanta, R., Shaji, A., Smith, K., Lucchi, A., Fua, P., Ssstrunk, S.: SLIC superpixels compared to state-of-the-art superpixel methods. IEEE Trans. Pattern Anal. Mach. Intell. **34**, 2274–2282 (2012)
3. Bauschke, H., Combettes, P.: Convex Analysis and Monotone Operator Theory in Hilbert Spaces. Springer, Heidelberg (2011). https://doi.org/10.1007/978-1-4419-9467-7

4. Bioucas-Dias, J., Nascimento, J.: Hyperspectral subspace identification. IEEE Trans. Geosci. Remote Sens. **46**, 2435–2445 (2008)

5. Bioucas-Dias, J., Plaza, A., Dobigeon, N., Parente, M., Du, Q., Gader, P., Chanussot, J.: Hyperspectral unmixing overview: geometrical, statistical, and sparse regression-based approaches. IEEE J. Sel. Topics Appl. Earth Obs. Remote Sens. **5**, 354–379 (2012)

6. Boyd, S., Parikh, N., Chu, E., Peleato, B., Eckstein, J.: Distributed optimization and statistical learning via the alternating direction method of multipliers. Found. Trends Mach. Learn. **3**, 1–122 (2011)

7. Boykov, Y., Veksler, O., Zabih, R.: Fast approximate energy minimization via graph cuts. IEEE Trans. Pattern Anal. Mach. Intell. **23**, 1222–1239 (2001)

8. Brifman, A., Romano, Y., Elad, M.: Turning a denoiser into a super-resolver using plug and play priors. In: IEEE ICIP (2016)

9. Chan, S., Wang, X., Elgendy, O.: Plug-and-play ADMM for image restoration: fixed point convergence and applications. IEEE Trans. Comput. Imaging **PP**(99), 1 (2016)

10. Green, A., Berman, M., Switzer, P., Craig, M.: A transformation for ordering multispectral data in terms of image quality with implications for noise removal. IEEE Trans. Geosci. Remote Sens. **26**, 65–74 (1988)

11. Jolliffe, I.: Principal Component Analysis. Springer, New York (1986). https://doi.org/10.1007/b98835

12. Landgrebe, D.: Signal Theory Methods in Multispectral Remote Sensing. Wiley, Hoboken (2003)

13. Loncan, L., Almeida, L., Bioucas-Dias, J., Briottet, X., Chanussot, J., Dobigeon, N., Fabre, S., Liao, W., Licciardi, G., Simões, M., Tourneret, J.-Y., Veganzones, M., Vivone, G., Wei, Q., Yokoya, N.: Hyperspectral pansharpening: a review. IEEE Geosci. Remote Sens. Mag. **3**, 27–46 (2015)

14. Nascimento, J., Bioucas-Dias, J.: Vertex component analysis: a fast algorithm to unmix hyperspectral data. IEEE Trans. Geosci. Remote Sens. **43**, 898–910 (2005)

15. Romano, Y., Elad, M., Milanfar, P.: The little engine that could: regularization by denoising (RED) arXiv:1611.02862 (2016)

16. Rudin, L., Osher, S., Fatemi, E.: Nonlinear total variation based noise removal algorithms. Phys. D: Nonlinear Phenom. **60**, 259–268 (1992)

17. Simões, M., Bioucas-Dias, J., Almeida, L., Chanussot, J.: A convex formulation for hyperspectral image superresolution via subspace-based regularization. IEEE Trans. Geosci. Remote Sens. **55**, 3373–3388 (2015)

18. Sreehari, S., Venkatakrishnan, S., Wohlberg, B., Buzzard, G., Drummy, L., Simmons, J., Bouman, C.: Plug-and-play priors for bright field electron tomography and sparse interpolation. IEEE Trans. Comput. Imaging **2**(4), 408–423 (2016)

19. Teodoro, A., Almeida, M., Figueiredo, M.: Single-frame image denoising and inpainting using Gaussian mixtures. In: ICPRAM, pp. 283–288 (2015)

20. Teodoro, A., Bioucas-Dias, J., Figueiredo, M.: Image restoration and reconstruction using variable splitting and class-adapted image priors. In: IEEE-ICIP (2016)

21. Teodoro, A., Bioucas-Dias, J., Figueiredo, M.: Image restoration with locally selected class-adapted models. In: IEEE-MLSP (2016)

22. Teodoro, A., Bioucas-Dias, J., Figueiredo, M.: Sharpening hyperspectral images using plug-and-play priors. In: Tichavský, P., Babaie-Zadeh, M., Michel, O., Thirion-Moreau, N. (eds.) LVA/ICA 2017. LNCS, vol. 10169, pp. 392–402. Springer, Cham (2017). https://doi.org/10.1007/978-3-319-53547-0_37

23. Teodoro, A., Bioucas-Dias, J., Figueiredo, M.: Hyperspectral sharpening using scene-adapted Gaussian mixture priors. Preprint arXiv:1702.02445 (2017)

24. Venkatakrishnan, S., Bouman, C., Chu, E., Wohlberg, B.: Plug-and-play priors for model based reconstruction. In: IEEE GlobalSIP, pp. 945–948 (2013)
25. Wei, Q., Bioucas-Dias, J., Dobigeon, N., Tourneret, J.-Y.: Hyperspectral and multispectral image fusion based on a sparse representation. IEEE Trans. Geosci. Remote Sens. **53**, 3658–3668 (2015)
26. Yokoya, N., Grohnfeldt, C., Chanussot, J.: Hyperspectral and multispectral data fusion: a comparative review. IEEE Geosci. Remote Sens. Mag. **5**, 29–56 (2017)
27. Yu, G., Sapiro, G., Mallat, S.: Solving inverse problems with piecewise linear estimators: from Gaussian mixture models to structured sparsity. IEEE Trans. Image Process. **21**, 2481–2499 (2012)
28. Zoran, D., Weiss, Y.: From learning models of natural image patches to whole image restoration. In: IEEE-CVPR, pp. 479–486 (2011)

Inverse Lightfield Rendering for Shape, Reflection and Natural Illumination

Antonin Sulc$^{(\boxtimes)}$, Ole Johannsen, and Bastian Goldluecke

University of Konstanz, Konstanz, Germany
`antonin.sulc@uni-konstanz.de`

Abstract. We propose an inverse rendering model for light fields to recover surface normals, depth, reflectance and natural illumination. Our setting is fully uncalibrated, with the reflectance modeled with a spatially-constant Blinn-Phong model and illumination as an environment map. While previous work makes strong assumptions in this difficult scenario, focusing solely on specific types of objects like faces or imposing very strong priors, our approach leverages only the light field structure, where a solution consistent across all subaperture views is sought. The optimization is based primarily on shading, which is sensitive to fine geometric details which are propagated to the initial coarse depth map. Despite the problem being inherently ill-posed, we achieve encouraging results on synthetic as well as real-world data.

Keywords: Lightfield · Inverse rendering · BRDF
Natural illumination

1 Introduction

The irradiance of a ray received by a sensor depends on many factors, in particular surface properties like geometry and reflectance, but also the illumination from the environment. The difficult ill-posed problem of inverse rendering aims at recovering all such variables which lead to the formation of an image of a scene. While there has been tremendous progress in the recovery of each of these variables separately [13,15] or under very restrictive conditions [1,2], the general question of reconstructing a geometry, natural illumination and non-Lambertian reflectance from an image has not been addressed yet. The main problem is that an image can be explained in numerous ways, therefore strong priors or additional information such as multiple views are required. Thus, most of previous works based on traditional images focus on recovery of an object either with priors which restrict the set of possible solutions [3], restrict the type of analyzed object [2] or require additional information either about illumination [15] or shape [13].

In contrast, we rely only on the structure of the lightfield. Lightfield imaging is an efficient way to acquire a structured array of multiple views with a small baseline which attracted lots of attention in the past decade [8,11,22,23,25].

© Springer International Publishing AG, part of Springer Nature 2018
M. Pelillo and E. Hancock (Eds.): EMMCVPR 2017, LNCS 10746, pp. 372–388, 2018.
https://doi.org/10.1007/978-3-319-78199-0_25

Despite the fact that the idea of lightfield is relatively old, the recent availability of hand-held cameras like the Lytro Illum or Raytrix gives a chance to answer many questions which for 2D imaging were ill-posed in practice using comparably simple off-the-shelf devices. Significant progress has, for example, been made in Lambertian depth estimation [12], where the appearance of a surface point does not change for different viewpoints. However, it has also been shown that due to the rigid structure of a 4D lightfield, recovery of surface normals and single-lobe BRDF is robustly possible for non-Lambertian objects in case of a calibrated single distant light source [24].

The problem of inverse rendering is severely ill-posed but the specific structure of the 4D lightfield as a regular grid of views gives us an opportunity to enforce a prior which encourages consistent solutions between views and does not suffer from the bias towards a certain class of solutions. Furthermore, although most algorithms for depth estimation fail when the Lambertian assumption is violated, a loss in accuracy is usually visible mostly on geometric details on fine scales, while the depth map is on a coarse level accurate enough. This is sufficient as an initial guess of geometry, which is necessary to initially estimate other missing variables as a starting point for further refinement.

Contributions. In this work, we present the first complete inverse lightfield rendering pipeline which can recover shape, reflectance, and natural illumination. The model is inspired by [13], where they formulated estimation of natural illumination and reflectance given surface shape. However, by leveraging the lightfield structure, we can in addition impose priors on geometry, i.e. depth and normals. We do this by employing regularization based on [21,25], which encourages the normal field to be consistent with depth and in particular respect the epipolar geometry across subaperture views. Overall, we obtain a unified energy-based formulation, where we can optimize for all variables in turn. In contrast to previous work based on individual images [3], our model does not overly restrict the space of solutions.

2   Related Work

There have been tremendous efforts in lightfield depth estimation, nevertheless most approaches allow only very small deviation from the Lambertian assumption, if any. For an in-depth review of recent Lambertian methods, we refer to [12]. A recent lightfield based method which does not assume Lambertian reflectance is [24], where they use differential motion to estimate normals under the assumptions of Blinn-Phong reflectance and a single calibrated point light source placed at infinity. They can also recover BRDF parameters, but in contrast to their work, we estimate unknown natural illumination. In [15] they proposed an approach for shape estimation under the assumption of known natural illumination. They used strong geometric priors in order to enforce smoothness while maintaining similarity from exhaustive search in a discrete set of normals. Since the normals are estimated from a look-up table, the accuracy is highly dependent on the quality of input illumination. The work of [15] is further extended to multiple views [16] and can recover full 3D geometry of the object.

Shape-from-shading based methods underwent promising development under challenging conditions like natural illumination [10]. Here, light is modeled with spherical harmonics and the objects are assumed to be Lambertian. In [3], they search for the most likely parameters for geometry, reflectance, and illumination that would explain a single input image. In contrast to our work, they employ very strong priors to prune the set of possible solutions.

Our work is partially based on [13]. They formulated the problem of estimation of natural illumination and reflectance as a maximum a-posteriori problem. In contrast to approaches where they use spherical harmonics [10] to represent illumination, they discretize distant illumination on the space of illumination directions. However, in contrast to their work, we do not assume known geometry, which makes the problem much harder. In [19] they proposed a method for BRDF estimation from a single image of a known shape in an unknown natural illumination. They leveraged a statistic about real-world illumination and estimated the reflectance that is most likely under a distribution of probable illumination environments.

3 Inverse Lightfield Rendering Model

3.1 Lightfield Parametrization and Rendering Equation

In this paper, we adopt the absolute two-plane parametrization for a lightfield. The rays are parameterized as an intersection of two parallel planes, a *focal plane* Π and *image plane* Ω, where the focal plane Ω is closer to the scene at distance f from Π. Each ray is described by a four-dimensional vector $r = [u, v, s, t] \in \mathbf{R}^4$. The intuition is that the coordinates $(s, t) \in \Pi$ select the focal point of an ideal pinhole camera with a corresponding *sub-aperture view*, while $(u, v) \in \Pi$ corresponds to image coordinates within this view, see Fig. 1 for an illustration.

The way how each image is formed is described by the *rendering equation*. The radiance R observed for a ray r is a function of several variables. It depends (i) on the surface normal $n(r)$ at the point in 3D space where the ray intersects the scene geometry, (ii) on the reflectance function (BRDF) at this location, given by a global parametric model depending on parameters Ψ, and (iii), on global illumination L. In particular, all rays which intersect in the same 3D point have the same normal associated to them, a fact which we employ later to impose consistency. We have

$$R(r, L, \Psi) = \int_{H_{n(r)}} \mathrm{BRDF}_{\Psi}(v(r), n(r), l) \, L(l) \, \mathrm{d}l, \tag{1}$$

where integration is performed over the hemisphere $H_{n(r)}$ of illumination directions l which are oriented into the direction of the surface normal $n(r)$, i.e. $l \cdot n(r) \geq 0$, and $v(r)$ is the viewing direction for the ray. Global illumination L

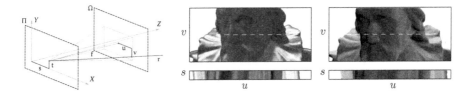

Fig. 1. Light field and epipolar plane images. Left: an incident ray r is parametrized by its intersections with the *focal plane* Π and the *image plane* Ω (red dots). The planes are parallel with distance equal to the focal length f. The intersection coordinates (s, t) are given in relation to the origin of the world coordinate system. The coordinates (u, v) are given relative to the intersection of the optical axis of a virtual camera placed at $(s, t, 0)$ in Z direction with the second plane (green dot). Each of these virtual cameras gives a subaperture view of the light field. Middle: an example center view of a non-Lambertian object. One can observe in the EPI below that not all lines exhibit constant values, which is caused by viewpoint dependency of the material. Right: illustration of the normal map n for the center view. The restriction n_{tv} for fixed (t, v)-coordinates below shows that while for the input image the appearance of some points may change with viewpoint, normals stay constant and exhibit the line pattern. (Color figure online)

is assumed to be distant and thus just a function of direction l. Note that we can easily specialize to a setting with just one point light source by setting illumination to a delta distribution over the set of directions, in which case the integral reduces to a single evaluation of the reflectance function. In the following, we discuss the specific parametric models we use for BRDF and illumination, as well as our priors on geometry imposed by the lightfield structure.

Reflectance and Illumination

Reflectance is the ratio of the radiance of light which was reflected by material and the irradiance of incident light $L(\omega_i)$. In order to keep the model simple, we model the reflectance with a spatially uniform Blinn-Phong parametric model, but model can be easily replaced with a more complex reflectance. The specular lobe is parametrized with respect to the half-angle vector $\mathbf{h}$ and the normal $\mathbf{n}$. The constant $\alpha > 0$ is the *shininess* of the material, while $i^d, i^s \in \mathbb{R}^3$ are diffuse and specular colors respectively. The resulting BRDF is

$$\text{BRDF}_{\Psi}(v, n, l) = i^d \left(n^T l\right) + i^s \left(n^T h\right)^{\alpha} \tag{2}$$

where the $h = n^T v$ is half vector between normal and viewing direction [20].

The general assumptions about illumination are that it is natural, infinitely far from an observer. We adopt a similar approach to [13,15], where the illumination is discretized on the domain of l. A linear combination of spherical harmonics [10] can be considered as an alternative for the illumination. See Sect. 4 for further details.

Depth and Normal Maps

We recover depth $z = z(r)$ for each individual ray as the distance of the focal plane to the 3D point on the scene surface intersected by the ray. Knowing depth immediately gives us 3D coordinates of the point, and we can in turn recover disparity, i.e. the ratio of shift in the (u, v) plane vs. shift in the (s, t) plane when changing viewpoint [25]. In addition, we require normal vectors $\boldsymbol{n}(\boldsymbol{r})$ for each ray to evaluate the rendering equation (1). While normals could be computed directly from the depth map z via suitable derivatives, in our approach, we decided to introduce two separate variables and link them via soft constraints. One of the important motivations is that normals are more sensitive on fine geometric details, which we are lacking on the smoothed input depth map. Therefore, the results of an optimized normal field provide a valuable clue about fine geometric details.

Our priors for geometry follow directly from the structure of the lightfield. We further impose area minimization and piecewise smoothness of the normal map to regularize the surface geometry. For this, we need to leverage and combine several insights from previous work. In [7], it was shown that if depth z is reparametrized in a new variable $\zeta := \frac{1}{2}z^2$, then the linear operator N given by

$$N(\zeta) = \begin{bmatrix} -\zeta_u/f \\ -\zeta_v/f \\ \hat{u}\zeta_u/f + \hat{v}\zeta_v/f + 2\zeta/f^2 \end{bmatrix} \tag{3}$$

maps a depth map ζ to the map of corresponding normals scaled with the local area element of the parametrized surface. Above, $(\hat{u}, \hat{v})$ are the homogenous image coordinates of the ray where the normal is computed, in particular, N varies across rays, while ζ_u and ζ_v denote partial derivatives of reparametrized depth with respect to the spatial view coordinates.

We leverage this map in several ways. First, we follow [7] and introduce a minimal surface regularizer by encouraging small total $\|N\zeta\|$, which corresponds to the surface area. Second, we follow [21], and encourage similarity of the normal map $\boldsymbol{n}$ scaled by local area element α and $N\zeta$. They enforce consistency of normals and depth as a soft constraint by penalizing $\|N\zeta - \alpha\boldsymbol{n}\|_2^2$. In particular, they propose an optimization strategy to jointly optimize for depth and α, which we make use of in our framework, see below for more details.

Consistency with the Lightfield Structure

The rigid structure of the lightfield becomes visible when considering *epipolar plane image space*, which is the restriction of ray space to 2D horizontal (s, u) or vertical (t, v) coordinates, respectively. In particular, a non-occluded point of a perfectly Lambertian surface has the same radiance regardless of viewpoint, thus all its observations in the lightfield should be the same. As it is well known [4,25], projections of a single point lie on a line in horizontal and vertical epipolar plane image space, whose orientation $\boldsymbol{q} = [d, 1]^T$ can be computed from disparity d. This becomes immediately visible when considering the horizontal and vertical epipolar plane images (EPIs) I_{su} and I_{tv} of a Lambertian lightfield I, see Fig. 1, which consists of patterns of lines. Determining their slope is the same as reconstructing disparity [25].

In [25], this constancy was encouraged by means of an anisotropic regularizer J_q, which for a vector-valued function $u : \Omega \to \mathbb{R}^n$ and a point-wise orientation field can be defined as

$$J_q(u) = \sum_{i=1}^{n} \int_{\Omega} \left\| \nabla u_i^T (qq^T) \nabla u_i \right\|_2 \, dx. \tag{4}$$

This regularizer prefers functions which are constant along q. In our case, it is not the radiance which is constant for a surface point, as we are considering non-Lambertian scenes. However, an invariant property is the surface normal. Thus, a well-defined reconstruction of normals n should be consistent with the disparity map in the sense that it is constant along the corresponding directions on epipolar plane image space. We thus define a regularization term to obtain a consistent normal field as

$$J_d(n) = \int J_{[d,1]}(n_{su}) \, d(s, u) + \int J_{[d,1]}(n_{tv}) \, d(t, v), \tag{5}$$

where we sum up the contributions of all epipolar plane images. See [25] for details.

Final Inverse Rendering Model

Putting together the terms from the previous subsections, we arrive at the energy

$$E(\zeta, n, \Psi, L) = E_{\text{data}}(n, \Psi, L) + E_{\text{depth}}(\zeta, n) + E_{\text{normals}}(n),$$

$$\text{with} \quad E_{\text{data}}(n, \Psi, L) = \int_{\mathcal{R}} \left(R(n(r), L, \Psi) - I(r) \right)^2 dr, \tag{6}$$

$$E_{\text{depth}}(\zeta, n) = \| N\zeta - \alpha n \|_2^2 + \| N\zeta \|_{2,1},$$

$$E_{\text{normals}}(n) = J_{d(\zeta)}(n),$$

to be minimized for ray-wise depth ζ, surface normal n, global BRDF parameters Ψ, and global illumination L.

Above, the data term E_{data} enforces similarity of the rendering with the input lightfield I. It is defined as an integral of the squared error over all rays. However, as we currently have only one global BRDF, we have to restrict to the rays which actually hit the object, and perform a semi-automatic pre-segmentation of the lightfield using [26]. The depth term E_{depth} links the depth map to the estimated normals, and is used for depth refinement. For depth regularization, we minimize the area of the scene surface as suggested in [7]. The scaling factor α, equal to the local area element, is recovered during optimization [21]. Finally, the normal term E_{normals} regularizes the normal map and makes it consistent with the epipolar plane image structure encoded in the disparity map, as detailed above. Note that disparity can be computed directly from ζ using the standard lightfield projection equations [25].

In the following section, we describe the strategy for optimization, which has to be very carefully performed as the energy is highly non-convex.

4 Optimization

In this section we describe how we minimize energy (6) to recover depth reparametrized as ζ, normals n, BRDF parameters $\boldsymbol{\Psi}$ and illumination L. Overall, it follows three main iterative steps. The first is optimization with respect to variables $n, L, \boldsymbol{\Psi}$ which appear in the rendering equation (1). For BRDF parameters and illumination, we compute a global solution given all of the other variables. For the normals, we compute a descent step. The second is refinement of depth ζ given the normal field, i.e. descent with respect to E_{depth} with fixed n. In the third step, we finally adapt the normals n to make them consistent with the lightfield structure. This is a descent step with respect to E_{normals}. All three steps are iterated until convergence to a local minimum. For initialization, we require an initial geometry. For this, we compute a coarse depth map under the assumption that the scene is approximately Lambertian. In practice, we use [21], and apply strong regularization. The algorithm is summarized in Fig. 3. We now describe the descent steps in more detail.

Optimization for E_{data}
In our framework, the data term E_{data} in the unknowns $n, \boldsymbol{\Psi}$ and L is optimized in a relatively straight-forward manner by computing the gradients of the rendering equation finding a maximum-likelihood solution.

Reflectance. The Blinn-Phong reflectance is defined via the shininess parameter α, the color of specular lobe $i^s = \begin{bmatrix} i_r^s, i_g^s, i_b^s \end{bmatrix}^T$ and color of diffuse albedo $i^d = \begin{bmatrix} i_r^d, i_g^d, i_b^d \end{bmatrix}^T$, which make up the seven unknowns summarized in the vector $\boldsymbol{\Psi}$. Because reflectance parameters occur only in the data-term, optimization is simple. The rendering equation is differentiable in all variables, and we can easily calculate explicit derivatives with respect to $\boldsymbol{\Psi}$. In each iteration of reflectance estimation, the parameters $\boldsymbol{\Psi}$ are optimized with trust-region optimization [5] with constraints $\alpha > 0$, $i^d \geq 0$ and $i^s \geq 0$.

Illumination. The illumination L is distant and thus a function of illumination direction l. In our implementation, it is discretized as a 2D high-dynamic-range image, where coordinates represent azimuth and height of l. Same as the reflectance, the illumination occurs only in data-term and explicit derivatives of L are easy to calculate.

It is important to highlight that if the object's surface is not a perfect mirror, material plays the role of a low-pass filter [18]. Therefore, there is a loss in high frequency components of illumination, which makes it impossible to recover fine details in incoming illumination. Each iteration of illumination estimation entails a few iterations of a limited memory BFGS-B [6].

Normals. Neither reflectance nor illumination is likely yet accurate enough to render a correct normal, as the energy landscape for normals is drastically influenced by both illumination and reflectance, see Fig. 2 for an illustration. For this reason, we do not look for a global minimum, put perform only a few steps of ray-wise gradient descent for the normals, with explicit exact derivatives calculated

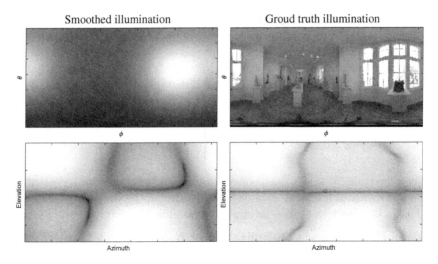

Fig. 2. Influence of illumination on energy landscape for normals. The top images show two different input illumination maps parametrized with respect to a normal with the direction $[0; 0; 1]$. Right is the ground truth, left a smoothed version of the ground truth. The figures at the bottom show how the log of the dataterm for different orientations of a normal for these two illumination maps. The correct normal is indicated by a red circle. One can see that the correct normal orientation can be coarsely identified in the case of ground truth illumination, but optimization is still quite prone to get stuck in a local minimum. For the incorrect smoothed illumination, the global minimum actually gives the wrong result. (Color figure online)

Initialization Calculate initial coarse depth map $\zeta^{(0)}$,
$$n^{(0)} = N\zeta^{(0)}/\|N\zeta^{(0)}\|, \quad L^{(0)} = 0.$$

Iterate K times:

- Illumination update $L^{(i+1)} \leftarrow \operatorname{argmin}_L E_{\text{data}}(n^{(i)}, L, \Psi^{(i)})$ using the method in [13].
- Reflectance update $\Psi^{(i+1)} \leftarrow \operatorname{argmin}_\Psi E_{\text{data}}(n^{(i)}, L^{(i+1)}, \Psi)$ using trust-region optimization [5].
- Normal update $n^{(i+1)} \leftarrow n^{(i)} - \tau \frac{\partial E_{\text{data}}}{\partial n}(n^{(i)}, L^{(i+1)}, \Psi^{(i+1)})$.
- Depth update $\zeta^{(i+1)} = \operatorname{prox}_{\tau E_{\text{depth}}(\cdot, n)}(\zeta^{(i)})$ using [21].
- Normal consistency update $n^{(i+1)} = \operatorname{prox}_{\tau E_{\text{normals}}}(n^{(i)})$ using [26,28].

Fig. 3. Overview of the optimization algorithm for energy (6).

for the rendering equation in a straight-forward way. Unit length of the normals is enforced with a Lagrange multiplier applied during these gradient descent steps. Thus, we give the regularizer a better chance to propagate information across the lightfield in the next step, and avoid local minima.

Refinement of Depth by Optimizing E_{depth}

We want to update our perspective depth map ζ to account for the currently estimated normals $\boldsymbol{n}$. The framework for this was established in [21], which makes use of the linearized relationship of depth to normals from [7]. The optimization problem we solve in this step is an implicit subgradient descent for E_{data} in ζ given the current solution $\zeta^{(i)}$, i.e.

$$\zeta^{(i+1)} = \text{prox}_{\tau E_{\text{depth}}(\cdot, \boldsymbol{n})}(\zeta^{(i)}) = \underset{\zeta}{\text{argmin}} \left\{ \frac{\|\zeta - \zeta^{(i)}\|_2^2}{2\tau} + E_{\text{depth}}(\zeta, \boldsymbol{n}) \right\}. \quad (7)$$

Once the operator N in E_{depth} has been computed, minimization is essentially a straight-forward implementation of [17]. Details can be found in [7,21]. Note that E_{depth} is convex in ζ, so a solution can be uniquely determined.

Consistent Normals via Optimization of E_{normals}

Finally, we compute an implicit subgradient descent step for the normal map regularizer J_d in E_{normals} to enforce consistent normals across the lightfield. This means we have to solve essentially a L^2-denoising problem for the current normal map $\boldsymbol{n}^{(i)}$,

$$\boldsymbol{n}^{(i+1)} = \text{prox}_{\tau E_{\text{normals}}}(\boldsymbol{n}^{(i)}) = \underset{\boldsymbol{n}}{\text{argmin}} \left\{ \frac{\|\boldsymbol{n} - \boldsymbol{n}^{(i)}\|_2^2}{2\tau} + J_d(\boldsymbol{n}) \right\}. \quad (8)$$

This is done by using the idea in [27] of successively linearizing and updating the normals around the current solution to deal with the unit length constraint, and embed it in the framework of [25] for consistent regularization with respect to the lightfield structure imposed by current disparity d. For details, we refer to the respective papers and [21]. While the energy E_{normals} is formally convex in $\boldsymbol{n}$, the constraint $\|\boldsymbol{n}\|_2 = 1$ is not, adding to the overall non-convexity and difficulty of the problem.

5 Experiments

For our experiments, we render several lightfields using a high-resolution HDR environment map for illumination[1]. In addition, we recorded a real-world scene using a Baumer LXG-200C camera designed for industrial inspection, mounted on a high-accuracy gantry. The camera was calibrated to minimize distortion and the disparity lies in $[-1.5; 1.5]$.

For the synthetic scenes first normals for all subaperture views are generated using the light field Blender Addon [9] and afterwards each view rendered with the engine provided in [13]. The calculations are performed with CUDA custom kernels on a Geforce GTX 1080 Ti GPU. For the initial estimation of geometry, we use the method in [25], which is simple but achieves an accurate enough initial estimate of depth map $\zeta^{(0)}$ after applying strong TV-L^2 denoising [17].

---

[1] Available online http://gl.ict.usc.edu/Data/HighResProbes/.

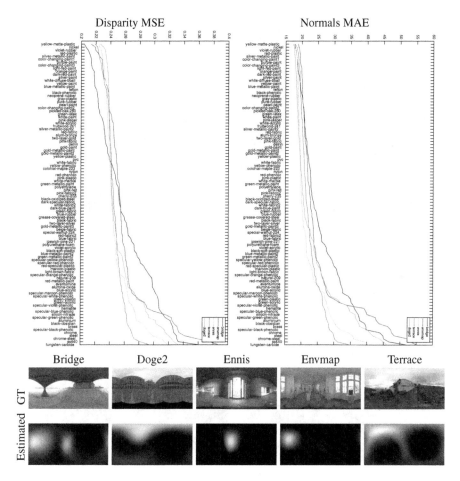

Fig. 4. Evaluation of a *ball* dataset MERL BRDF database [14]. The first row shows disparity MSE and normals MAE of our method on all MERL BRDFs with various environment maps. The second row shows ground truth environment maps used for reddening. The last row shows respective estimated environment maps averaged over all MERL materials

The inverse rendering problem is of course highly ill-posed, which is underlined by two observations. The first is that it is very complicated to separate color of illumination from the color of reflectance, and there is inherent ambiguity in how strong a material reflects vs. Brightness of illumination. While possible in theory, this works only for very accurate geometry [13], which is difficult to achieve from our rough initial guess. The second problem arises from the nature of the BRDF. The BRDF is a low-pass filter on the illumination [18] and limits possible higher details of illumination.

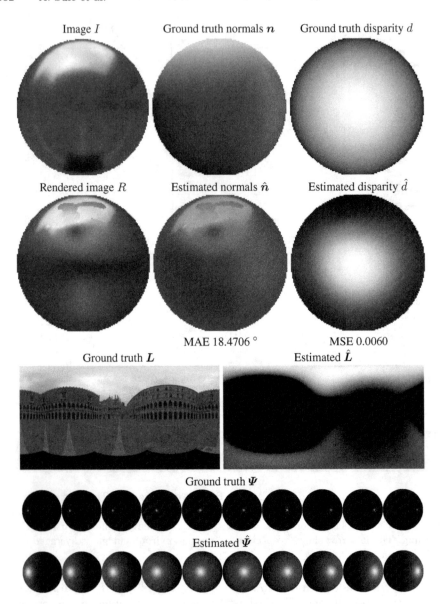

Fig. 5. Synthetic dataset *ball* with *nickel* reflectance from the MERL BRDF database [14] and Doge2 illumination. The first row shows the center views of the input image I, ground truth normals n and disparity d. The second row shows center views of re-rendered image for the estimated parameters R, estimated normals $\hat{n}$ and disparity $\hat{d}$. The third row shows ground truth illumination L and the estimated illumination $\hat{L}$. And last two rows show comparison of ground truth Ψ versus estimated reflectance $\hat{\Psi}$ where the spheres are rendered with a point light which moves from left to right.

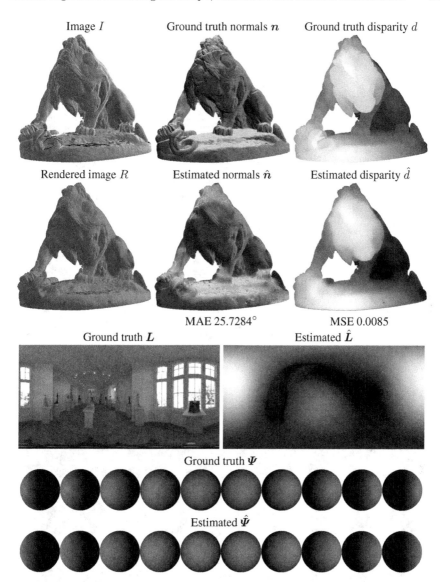

Image I Ground truth normals n Ground truth disparity d

Rendered image R Estimated normals $\hat{n}$ Estimated disparity $\hat{d}$

MAE 25.7284°   MSE 0.0085

Ground truth L Estimated $\hat{L}$

Ground truth Ψ

Estimated $\hat{\Psi}$

Fig. 6. Synthetic dataset *lion* with *light-red-paint* reflectance from the MERL BRDF database [14] and Envmap illumination. The first row shows the center views of the input image I, ground truth normals n and disparity d. The second row shows center views of re-rendered image for the estimated parameters R, estimated normals $\hat{n}$ and disparity $\hat{d}$. The third row shows ground truth illumination L and the estimated illumination $\hat{L}$. And last two rows show comparison of ground truth Ψ versus estimated reflectance $\hat{\Psi}$ where the spheres are rendered with a point light which moves from left to right. (Color figure online)

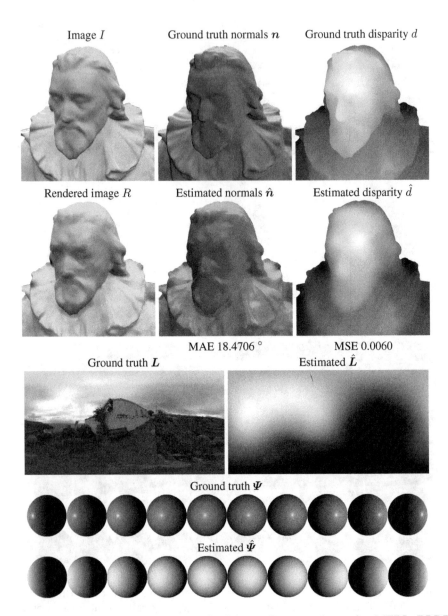

Fig. 7. Synthetic dataset *cotton1* with *delrin* reflectance from the MERL BRDF database [14] and Terrace illumination. The first row shows the center views of the input image I, ground truth normals n and disparity d. The second row shows center views of re-rendered image for the estimated parameters R, estimated normals $\hat{n}$ and disparity $\hat{d}$. The third row shows ground truth illumination L and the estimated illumination $\hat{L}$. And last two rows show comparison of ground truth Ψ versus estimated reflectance $\hat{\Psi}$ where the spheres are rendered with a point light which moves from left to right. (Color figure online)

Input image I Rendered image R Estimated normals

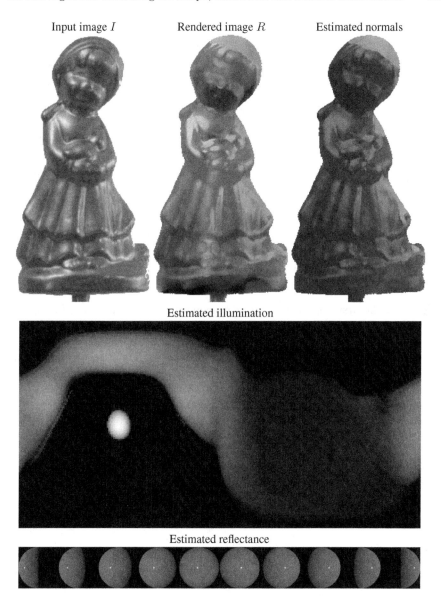

Estimated illumination

Estimated reflectance

Fig. 8. Results on a real world dataset recorded with our gantry. The left image shows the input image, the one in the middle is re-rendered image with parameters estimated using our pipeline, and the one on the right shows estimated normals. The second row shows estimated illumination and last row shows renderings of the unit sphere using estimated BRDF parameters with a point light moving from left to right.

We evaluate our method on three different synthetic models, a simple smooth sphere and two objects with complex geometry. The first lightfield shows the sphere, see Fig. 5. We evaluated mean angular error (MAE) of the estimated normal field and mean squared error (MSE) of the estimated disparity on all materials from MERL database [14] and five different environment maps, see Fig. 4. The second lightfield *lion* shows a statue with more complex geometry, rendered with the *light-red-paint* MERL BRDF and environment map *Envmap*, see Fig. 6. The third lightfield *cotton1* is rendered with the *delrin* MERL BRDF and environment map *Terrace*, see Fig. 7. Reflectance and illumination have very similar colors, therefore reflectance suffers from the above-mentioned illumination-reflectance ambiguity. The relatively small value of MAE is attributed to the good identification of strong light sources in the scene, however, some persistent inaccuracies in geometry cause the shininesses α to be too small.

Finally, Fig. 8 shows results for a real-world lightfield. The dot in the left part of illumination shows a correctly reconstructed point light source, while the smooth right part is from light reflected from a slightly reflective table.

6   Conclusion

We propose the first approach to recover geometry, reflectance and illumination from a single lightfield. Our approach is based on energy-minimization and starts with a coarse initial depth map, computed under the Lambertian assumption, to provide an initial clue to roughly initialize all variables. Then we iteratively optimize for each of the variables in turn, imposing the inverse rendering dataterm, enforcing depth to be consistent with normals, and enforcing normals to be consistent with the epipolar plane image structure. We tested our approach on synthetic as well as real-world data, and achieve encouraging results despite having a substantially ill-posed problem with a non-convex energy.

Acknowledgement. This work was supported by the ERC Starting Grant "Light Field Imaging and Analysis" (LIA 336978, FP7-2014).

References

1. Aldrian, O., Smith, W.A.P.: Inverse rendering of faces on a cloudy day. In: Fitzgibbon, A., Lazebnik, S., Perona, P., Sato, Y., Schmid, C. (eds.) ECCV 2012. LNCS, vol. 7574, pp. 201–214. Springer, Heidelberg (2012). https://doi.org/10.1007/978-3-642-33712-3_15
2. Aldrian, O., Smith, W.A.: Inverse rendering of faces with a 3D Morphable model. IEEE Trans. Pattern Anal. Mach. Intell. **35**(5), 1080–1093 (2013)
3. Barron, J.T., Malik, J.: Shape, illumination, and reflectance from shading. IEEE Trans. Pattern Anal. Mach. Intell. **37**(8), 1670–1687 (2015)
4. Bolles, R., Baker, H., Marimont, D.: Epipolar-plane image analysis: an approach to determining structure from motion. Int. J. Comput. Vis. **1**(1), 7–55 (1987)
5. Branch, M.A., Coleman, T.F., Li, Y.: A subspace, interior, and conjugate gradient method for large-scale bound-constrained minimization problems. SIAM J. Sci. Comput. **21**(1), 1–23 (1999)

6. Byrd, R.H., Lu, P., Nocedal, J., Zhu, C.: A limited memory algorithm for bound constrained optimization. SIAM J. Sci. Comput. **16**(5), 1190–1208 (1995)

7. Graber, G., Balzer, J., Soatto, S., Pock, T.: Efficient minimal-surface regularization of perspective depth maps in variational stereo. In: Proceedings of the International Conference on Computer Vision and Pattern Recognition (2015)

8. Heber, S., Pock, T.: Shape from light field meets robust PCA. In: Fleet, D., Pajdla, T., Schiele, B., Tuytelaars, T. (eds.) ECCV 2014. LNCS, vol. 8694, pp. 751–767. Springer, Cham (2014). https://doi.org/10.1007/978-3-319-10599-4_48

9. Honauer, K., Johannsen, O., Kondermann, D., Goldluecke, B.: A dataset and evaluation methodology for depth estimation on 4D light fields. In: Lai, S.-H., Lepetit, V., Nishino, K., Sato, Y. (eds.) ACCV 2016. LNCS, vol. 10113, pp. 19–34. Springer, Cham (2017). https://doi.org/10.1007/978-3-319-54187-7_2

10. Huang, R., Smith, W.A.: Shape-from-shading under complex natural illumination. In: 2011 18th IEEE International Conference on Image Processing (ICIP), pp. 13–16. IEEE (2011)

11. Jeon, H., Park, J., Choe, G., Park, J., Bok, Y., Tai, Y., Kweon, I.: Accurate depth map estimation from a lenslet light field camera. In: Proceedings of the International Conference on Computer Vision and Pattern Recognition (2015)

12. Johannsen, O., Honauer, K., Goldluecke, B., Alperovich, A., Battisti, F., Bok, Y., Brizzi, M., Carli, M., Choe, G., Diebold, M., Gutsche, M., Jeon, H.G., Kweon, I.S., Park, J., Schilling, H., Sheng, H., Si, L., Strecke, M., Sulc, A., Tai, Y.W., Wang, Q., Wang, T.C., Wanner, S., Xiong, Z., Yu, J., Zhang, S., Zhu, H.: A taxonomy and evaluation of dense light field depth estimation algorithms. In: Proceedings of the 2nd Workshop on Light Fields for Computer Vision at CVPR (2017)

13. Lombardi, S., Nishino, K.: Reflectance and natural illumination from a single image. In: Fitzgibbon, A., Lazebnik, S., Perona, P., Sato, Y., Schmid, C. (eds.) ECCV 2012. LNCS, vol. 7577, pp. 582–595. Springer, Heidelberg (2012). https://doi.org/10.1007/978-3-642-33783-3_42

14. Matusik, W., Pfister, H., Brand, M., McMillan, L.: A data-driven reflectance model. ACM Trans. Graph. **22**(3), 759–769 (2003)

15. Oxholm, G., Nishino, K.: Shape and reflectance from natural illumination. In: Fitzgibbon, A., Lazebnik, S., Perona, P., Sato, Y., Schmid, C. (eds.) ECCV 2012. LNCS, vol. 7572, pp. 528–541. Springer, Heidelberg (2012). https://doi.org/10.1007/978-3-642-33718-5_38

16. Oxholm, G., Nishino, K.: Multiview shape and reflectance from natural illumination. In: Proceedings of the IEEE Conference on Computer Vision and Pattern Recognition, pp. 2155–2162 (2014)

17. Pock, T., Chambolle, A.: Diagonal preconditioning for first order primal-dual algorithms in convex optimization. In: International Conference on Computer Vision (ICCV 2011) (2011)

18. Ramamoorthi, R., Hanrahan, P.: A signal-processing framework for inverse rendering. In: Proceedings of the 28th Annual Conference on Computer Graphics and Interactive Techniques, pp. 117–128. ACM (2001)

19. Romeiro, F., Zickler, T.: Blind reflectometry. In: Daniilidis, K., Maragos, P., Paragios, N. (eds.) ECCV 2010. LNCS, vol. 6311, pp. 45–58. Springer, Heidelberg (2010). https://doi.org/10.1007/978-3-642-15549-9_4

20. Rusinkiewicz, S.M.: A new change of variables for efficient BRDF representation. In: Drettakis, G., Max, N. (eds.) Rendering Techniques '98, pp. 11–22. Springer, Heidelberg (1998). https://doi.org/10.1007/978-3-7091-6453-2_2

21. Strecke, M., Alperovich, A., Goldluecke, B.: Accurate depth and normal maps from occlusion-aware focal stack symmetry. In: Proceedings of the International Conference on Computer Vision and Pattern Recognition (2017)
22. Tao, M., Hadap, S., Malik, J., Ramamoorthi, R.: Depth from combining defocus and correspondence using light-field cameras. In: Proceedings of the International Conference on Computer Vision (2013)
23. Wang, T., Efros, A., Ramamoorthi, R.: Occlusion-aware depth estimation using light-field cameras. In: Proceedings of the IEEE International Conference on Computer Vision, pp. 3487–3495 (2015)
24. Wang, T.C., Chandraker, M., Efros, A.A., Ramamoorthi, R.: SVBRDF-invariant shape and reflectance estimation from light-field cameras. In: Proceedings of the IEEE Conference on Computer Vision and Pattern Recognition, pp. 5451–5459 (2016)
25. Wanner, S., Goldluecke, B.: Variational light field analysis for disparity estimation and super-resolution. IEEE Trans. Pattern Anal. Mach. Intell. **36**(3), 606–619 (2014)
26. Wanner, S., Straehle, C., Goldluecke, B.: Globally consistent multi-label assignment on the ray space of 4D light fields. In: Proceedings of the International Conference on Computer Vision and Pattern Recognition (2013)
27. Zeisl, B., Zach, C., Pollefeys, M.: Variational regularization and fusion of surface normal maps. In: Proceedings of the International Conference on 3D Vision (3DV) (2014)

Shadow and Specularity Priors
for Intrinsic Light Field Decomposition

Anna Alperovich$^{(\boxtimes)}$, Ole Johannsen, Michael Strecke, and Bastian Goldluecke

University of Konstanz, Konstanz, Germany
anna.alperovich@uni-konstanz.de

Abstract. In this work, we focus on the problem of intrinsic scene decomposition in light fields. Our main contribution is a novel prior to cope with cast shadows and inter-reflections. In contrast to other approaches which model inter-reflection based only on geometry, we model indirect shading by combining geometric and color information. We compute a shadow confidence measure for the light field and use it in the regularization constraints. Another contribution is an improved specularity estimation by using color information from sub-aperture views. The new priors are embedded in a recent framework to decompose the input light field into albedo, shading, and specularity. We arrive at a variational model where we regularize albedo and the two shading components on epipolar plane images, encouraging them to be consistent across all sub-aperture views. Our method is evaluated on ground truth synthetic datasets and real world light fields. We outperform both state-of-the art approaches for RGB+D images and recent methods proposed for light fields.

1 Introduction

Intrinsic image decomposition is one of the fundamental problems in computer vision, and has been studied extensively [3,23]. For Lambertian objects, where an input image is decomposed into albedo and shading components, numerous solutions have been presented in the literature. Depending on the input data, the approaches can be divided into those dealing with a single image [14,18,36], multiple images [24,39], and image + depth methods [9,20]. However, in the real world, there are few scenes with only Lambertian objects. Instead, they have specular surfaces, which makes the decomposition problem harder due to the complicated nature of specular reflection.

For these scenes, the recent work [1] proposes an intrinsic light field model for non-Lambertian scenes, where decomposition is performed with respect to albedo, shading, and specularity. The authors exploit the structure of the light field: while albedo and shading are view-independent and therefore constant

Electronic supplementary material The online version of this chapter (https://doi.org/10.1007/978-3-319-78199-0_26) contains supplementary material, which is available to authorized users.

M. Pelillo and E. Hancock (Eds.): EMMCVPR 2017, LNCS 10746, pp. 389–406, 2018.
https://doi.org/10.1007/978-3-319-78199-0_26

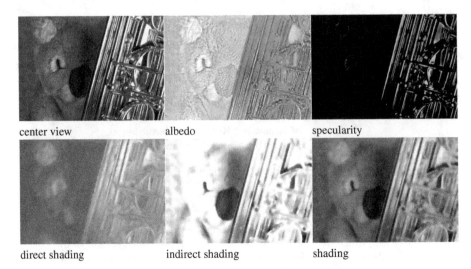

center view   albedo   specularity

direct shading   indirect shading   shading

Fig. 1. We present a variational framework to decompose light field radiance into the intrinsic components albedo, shading and specularity. By further breaking down shading into direct and indirect contributions, we achieve superior performance compared to previous work. This light field is captured with Lytro Illum plenoptic camera, 9×9 views, resolution 434×625.

along projections of the same scene point in the epipolar plane images (EPIs), specularity is view-dependent and shows a different behaviour.

While the previous model thus correctly identifies specular regions, there is still room for improvements in modeling the albedo and shading components. Shading is modeled by relying only on surface geometry, which might suffer from inaccurate disparity estimation. Furthermore, the previous model does not provide any information about inter-reflections and cast shadows in the scene.

Contributions. In this work, we propose a model for intrinsic light field decomposition which combines geometrical and color information to define a novel shading prior. We design a shadow detection model with light field data, which to our knowledge is the first time this problem has been addressed for light fields. We apply the estimated shadow score to model cast shadows and inter-reflections explicitly, which results in more consistent shading compared to previous approaches, with better identification of soft and hard shadows. Furthermore, we use pre-estimated specularity positions to make both albedo and shading priors less affected by highlights. For this, we compute a specular free representation of the input light field with [33], and use this information to model albedo and shading. A thorough evaluation of our algorithm on synthetic datasets with ground truth as well as the real world examples demonstrates that we outperform previous approaches based on RGB+D images or light fields.

2   Related Work

Intrinsic Images. Intrinsic image decomposition was first introduced by Barrow and Tenenbaum [3], who propose to decompose an input image into two components, albedo and shading. Since then, plenty of algorithms were proposed for intrinsic image decomposition. One of the first was Retinex, introduced by Land and McCann [23], where they assume that large discontinuities in image derivatives correspond to albedo or reflectance, and the remainder accounts for shading. Over the years, the basic Retinex algorithm was improved by many authors [15,18,36]. Besides the Retinex approach, there are many other interesting methods for single image decomposition [14,21,30].

$N_{3D}^{loc}(\boldsymbol{r})$	k-nearest neighbors of $\boldsymbol{r}$ in 3D space.
$N_{3D}^{glob}(\boldsymbol{r})$	random subset of k among the M-nearest neighbors of $\boldsymbol{r}$ in 3D space.
$N_{6D}^{loc}(\boldsymbol{r})$	k-nearest neighbors of $\boldsymbol{r}$ in 6D space (spatial coordinate + normal).
$N_{6D}^{glob}(\boldsymbol{r})$	random subset of k among the M-nearest neighbors of $\boldsymbol{r}$ in 6D space (spatial coordinate + normal).
$N_{obj}(\boldsymbol{r})$	subset of k among the M-nearest neighbors of $\boldsymbol{r}$ in 3D space within the same object as $\boldsymbol{r}$.

Fig. 2. Different neighborhoods used to define the smoothness priors. They differ by (i) where the nearest neighbor search is performed, either in 3D spatial domain or 6D domain of spatial coordinates and normal vectors, (ii) whether the neighborhood is local, i.e. the k-nearest neighbors are chosen directly, or global, i.e. a random subset of k elements is chosen from the M-nearest neighbors, with a large number M, and (iii) whether we try to sample rays within the same object. To identify objects, we use a simple low-level approach based on Triantafyllidis et al. [37]. Throughout all experiments, we set $k = 20$ and $M = 900$. For global neighborhood in the energy for the indirect shading (17) $k = 80$.

Intrinsic image decomposition was improved a lot by using additional information, for example image sequences [11,39], video [6,42] and RGB+D sensors. Lee [25] recover intrinsic components from a video sequence, Jeon et al. [20] and Chen and Koltun [9] use RGB+D images. Barron and Malik [2] propose the Scene-SIRFS model that outputs an improved depth map, a set of surface normals, a reflectance image, a shading image, and a spatially varying model of illumination. Garces et al. [13] extend the Retinex theory to light fields, where they perform intrinsic light field decomposition for 4D rays instead of 2D images. Alperovich and Goldluecke [1] propose intrinsic light field decomposition for non-Lambertian scenes, where they introduce an additional term to model specularity.

Shadow Detection and Removal. There are several approaches to tackle this problem. They can be divided into user-assistant methods, shape-based, and color-based techniques. Xiao et al. [40] use RGB+D images to construct cast shadow priors, which utilize geometrical information, particularly spatial locations and corresponding outer normals of scene points. Finlayson et al. [12] use an illumination invariant 1D representation of images to detect and remove shadows with Retinex. Lalonde et al. [22] use edge information to detect shadows on the ground for outdoor scenes. Panagopoulos et al. [28] use the concept of a bright channel cue to obtain shadow confidence from a single image.

Specularity Detection. The model for decomposition into specular and diffuse components or *dichromatic model* was introduced by Shafer [29]. This model was adopted for light fields by Tao et al. [35]. They propose a depth estimation and specular removal method that utilizes angular consistency of the diffuse component. Sulc et al. [32] proposed a specular removal approach for light fields based on sparse coding and specular flow. Wang et al. [38] remove specularity by clustering specular pixels into "unsaturated" and "saturated" categories with further color refinement.

3   Intrinsic Light Field Decomposition

We review the intrinsic light field model proposed by Alperovich and Goldluecke [1], and extend it to a model with an additional prior for cast shadows.

Light Field Structure. We start with briefly reviewing notation and basic definitions of the light field structure. A light field is defined on 4D ray space $\mathcal{R} = \Pi \times \Omega$, where a ray is identified by four coordinates $r = (s, t, u, v)$, which describe the intersections with two parallel planes. Here (s, t) are view point coordinates on the focal plane Π, and (u, v) are coordinates on the image plane Ω. Epipolar plane images (EPIs) can be obtained by restricting 4D ray space to 2D slices. For more information and a thorough introduction on light field geometry, we refer to [5,17,26].

Input to the Model. Our model is based on reconstructed 3D geometry of the scene, in particular we use disparity, spatial locations of the 3D points and surface normals. Since we work with non-Lambertian scenes, disparity estimation might not be accurate in the specular regions, thus the resulting 3D geometry suffers from noise and artifacts. To obtain consistent 3D geometry, we estimate depth with an occlusion-aware algorithm that exploits focal stack symmetry and offers joint regularization of depth and surface normals [31]. We assume that objects have a piecewise smooth geometry, thus surface normals point in similar directions for spatially close points [27,43].

Light Field Decomposition. According to dichromatic reflection model [10,29] the total radiance L of reflected light is the sum of two independent parts, the radiance of the reflected light at the surface body (diffuse reflection) and radiance of the reflected light at interface (specular reflection). By further breaking down diffuse component into albedo and shading, our intrinsic light field model assumes that the radiance L of a ray r can be decomposed as

$$L(r) = A(r)S(r) + H(r), \tag{1}$$

where A is an albedo component that represents the surface color independently of illumination and angular camera position, S is a shading component that represents intensity changes due to surface geometry and illumination conditions, and H is specularity or highlights. All intrinsic components are modeled as tri-chromatic vectors, corresponding to color channels R, G, B.

We consider the case that the scene is illuminated with a single white light source. In case of non-white illumination or multiple light sources the model should be multiplied with the illumination color, which can be estimated with color constancy algorithms [34, 41].

Inspired by the approach by Chen and Koltun [9], we extend the model (1) by introducing an additional term for cast shadows and inter-reflections. Thus, the new decomposition model

$$L(r) = A(r)s_d(r)S_i(r) + H(r) \tag{2}$$

has components s_d and S_i that describe direct and indirect shading, see Fig. 1. Direct shading can be understood as the shading that an object would have if it were alone in the scene, i.e. without shadows or reflected light. The second component S_i models inter-reflections and cast shadows. However, in contrast to Chen and Koltun [9], where S_i is modeled to be smooth in 3D space, we model it using a shadow confidence measure that is proportional to shadow intensities in the scene. This allows us to handle hard shadows, where the spatial smoothness assumption is violated. Also, since even under white light inter-reflections depend on the object colors, we model S_i as tri-chromatic, while s_d is mono-chromatic.

To obtain a linear decomposition equation from (2), we apply the logarithm,

$$L^{log}(r) = A^{log}(r) + 1s_d^{log}(r) + S_i^{log}(r) + H^{log}(r, A, s_d, S_i, H)$$
$$\text{with } H^{log} = 1 + \frac{H(r)}{A(r)s_d(r)S_i(r)}, \tag{3}$$

and ignore the dependence of the specular component H on albedo and shading by treating H^{log} as another independent variable.

4   Optimization

The decomposition (3) is ambiguous, and we thus need strong prior assumptions on all of the unknown intrinsic components. In addition, we need to leverage the light field structure, and enforce a consistent decomposition across all subaperture views. We thus reformulate the decomposition problem as minimization of the energy

$$\underset{(A^{log}, s_d^{log}, s_c^{log}, H^{log})}{\arg\min} \left\{ \|D(r)\|_2^2 + E_a(A^{log}) + E_d(s_d^{log}) + E_i(S_i^{log}) + \dots \right.$$
$$\left. + E_s(H^{log}) + J(A^{log}, s_d^{log}, S_i^{log}) \right\}. \tag{4}$$

The first data term enforces consistency of the decomposition, as the residual $D = L^{log} - A^{log} - 1s_d^{log} - S_i^{log} - H^{log}$ should be small. To reduce complexity of the problem (4), we substitute the specularity variable for the difference $L^{log} - A^{log} - 1s_d^{log} - S_i^{log}$ between the input light field and other intrinsic components. The energies E denote the aforementioned priors, to be explained later, while the term J enforces consistency with the light field structure. The idea is to encourage smoothness of albedo and the two shading components along the projections of scene points in the epipolar plane images, and spatial smoothness on each subaperture view by means of a total generalized variation regularizer [7]. We refer to [16] for details on definition and implementation.

The problem (4) can be rewritten as

$$\underset{(A^{log}, s_d^{log}, s_c^{log}, H^{log})}{\arg\min} \left\{ F(A^{log}, s_d^{log}, S_i^{log}, H^{log}) + J(A^{log}, s_d^{log}, S_i^{log}) \right\}. \quad (5)$$

where $F = \|D(r)\|_2^2 + E_a(A^{log}) + E_d(s_d^{log}) + E_i(S_i^{log}) + E_s(H^{log})$ is a convex and differentiable functional on the ray space $\mathcal{R}$. The regularizer

$$J = \mu J_{xs} + \mu J_{yt} + \lambda J_{st} \quad (6)$$

is convex and closed as the sum of convex and closed functionals, but not differentiable, while λ and μ are constants determining smoothness.

The problem (5) is now solved in three steps:

1. Pre-compute all data necessary for albedo, shading, and specularity terms: spatial neighborhoods, chromaticity, the specular and shadow masks that will be discussed in the following sections.
2. Initialize abledo A^{log} and direct shading s_d^{log} variables using values obtained from Color Retinex [18], initialize indirect shading S_i^{log} and specularity H^{log} with zeros.
3. Iteratively solve the problem within the framework proposed by Goldluecke and Wanner [16] for inverse problems on light fields:
   - Optimize F using gradient descent with respect to all intrinsic variables A^{log}, s_d^{log}, S_i^{log} and H^{log} for every $(s, t) \in \Pi$.
   - Enforce angular consistency of A^{log}, s_d^{log} and S_i^{log} by optimizing J_{yt} and J_{xs} with subgradient descent.
   - Regularize A^{log} with total variation (TV), and s_d^{log}, S_i^{log} with total generalized variation (TGV) for every subaperture view $(s, t) \in \Pi$.

To speed up convergence, we recompute the step size of the gradient descent in each iteration such that the cost is minimized as fast as possible, then we perform implicit subgradient descent for the regularizers with the same step size.

What is left to describe is how we model the prior energies for albedo, shading, and specularity. For this, we first start with specular confidence estimation and computation of the diffuse light field in Sect. 5. This is followed up with estimation of a shadow confidence map in Sect. 6. Finally, in Sect. 7, we can assemble the energy terms for all intrinsic components.

5   Specularity Detection

The aim of this section is to first compute a specular mask h, which takes values in $[0, 1]$ and can be understood as the probability of a ray to be specular. Afterwards, we use this mask to obtain pseudo-diffuse radiance L_d for the complete light field.

Confidence Mask h. Our main modeling assumptions are that specularity either is view-dependent or "non-saturated", and thus has high variation in pixel intensities over different sub-aperture views, or it is "saturated", and thus appears bright through all sub-aperture views. We mask out occlusion boundaries while computing the specular mask, thus we assume that there is no specularity on object edges. Dealing with specularity on occlusion boundaries is left for future work.

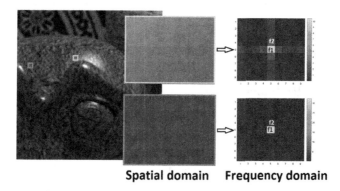

Spatial domain   Frequency domain

Fig. 3. Specularity detection mechanism. Left: part of the center view of the light field with a diffuse point in the red rectangle and a specular point in the yellow rectangle. Middle: angular patches for the diffuse and specular points. Right: Fourier transform of one color channel with non-zero frequencies for the specular point and almost zero frequencies for the diffuse point. (Color figure online)

Under the assumption of known disparity l, we construct angular patches $S(x, l(x))$ for every scene point x. The angular patch represents the corresponding 3D point's color over all sub-aperture views, see Fig. 3. Contrary to [1,38], we not only divide points within $S(x, l(x))$ into specular and specular free clusters, which might be inaccurate due to the small baseline of light field cameras, which results in low separability between clusters. Instead, we apply a Fourier transform to every angular patch $S(x, l(x))$, and search for significant low-frequency oscillations in every color channel of the patch. If such oscillations exist, then we conclude that the angular patch represents a specular point. We assume that high frequency oscillations are caused by noise, and are not caused by specularity. See Fig. 3 for an overview of the proposed technique. Our algorithm proceeds in four steps:

1. Apply Fourier transform to logarithm of $S(x, l(x))$ for each color channel independently.
2. Compute the ratio $r(x) = \frac{f_2(x)}{f_1(x)}$ between amplitude of the lowest frequency $f_2(x)$ and amplitude of the frequency $f_1(x)$ that corresponds to the mean value of the angular patch.
3. Choose the largest ratio $r(x)$ over the color channels. If $r(x)$ is close to 1, then there is a significant low-frequency component which we interpret as specularity. Usually, specularity exhibits high pixel intensity in at least one sub-aperture view. Thus, we multiply $r(x)$ with the maximum brightness of the patch $S(x, l(x))$ to filter out other low frequency color variations.
4. Compute a specular mask $h = \begin{cases} 1, & \text{if } r \geq \tau_f \\ 0, & \text{otherwise} \end{cases}$ by thresholding low values of $r(x)$.

To identify saturated points, we search for the brightest regions in every sub-aperture view that are connected to specular mask h via an eight-neighborhood, and mark them as specular. We define the p-th brightness percentile as the threshold for a point to be bright. Finally, we apply a Gaussian filter with standard deviation σ_h, see Fig. 4(a) for the result.

Pseudo-diffuse Radiance L_d. To model intrinsic components, we do not need actual diffuse color of the scene point. Instead, we need to know whenever chromaticities of the neighboring points are similar. For this purpose, we compute the pseudo-diffuse color $L_d(r)$ with the algorithm proposed in [33], where the authors suggest to compute a specular-free component that differs from the true diffuse component only in surface color. The authors assume that the scene contains only dielectric materials, and is illuminated under a white light source.

Fig. 4. Left: specular confidence mask with brighter regions corresponding to the higher probability of surface to be specular. Middle: point-wise shadow confidence β_{pw}. Right: resulting final shadow score β after regularization and incorporating the brightness channel cue.

The specular-free representation L_d is computed according to

$$L_\mathrm{d} = L - \frac{1}{3}\left(\sum_{i=R,G,B} L_i - \frac{\tilde{L}(3\tilde{c}-1)}{\tilde{c}(3\tilde{\Lambda}-1)} \right), \quad \tilde{L} = \max_{c=\{R,G,B\}} L_c, \tag{7}$$

with diffuse maximum chromaticity $\tilde{\Lambda} = 0.5$ and $\tilde{c} = \max\limits_{c=\{R,G,B\}} \frac{L_c}{(L_R+L_G+L_B)}$.
For saturated rays, we instead compute the average color over the k-nearest non-specular ($h = 0$) spatial neighbors $N_{3D}^{loc}(r)$ of L_{d}. See the supplementary material for an example of the estimated pseudo-diffuse chromaticity.

Note that throughout the paper, we employ different neigbourhoods of rays for different prior terms. These are summarized in Fig. 2 for convenience.

6   Cast Shadow Detection

In order to identify regions with cast shadows and remove them from the direct shading component, we compute a shadow score $\beta \in [0,1]$. For this, we first estimate the point-wise confidence β_{pw} of each ray to be shadowed, and then encourage shadowing to be consistent across all sub-aperture views. From these, we will later compute a shadow boundary score to set up the smoothness priors for the intrinsic shading components.

Point-wise Shadow Confidence. Inspired by the work of Xiao et al. [40], we use the spatial locations d, surface normals n, and chromaticity χ of the scene to decide which rays are likely to be shadowed. Since chromaticity images still may contain shadows, we follow previous work [28] and use several illumination-free representations based on RGB, HSV, and $c1c2c3$ color spaces to compute the chromaticity difference. In addition we compute RGB chromaticity from pseudo-diffuse representation L_d that was described in Sect. 5. The chromaticity is computed by dividing each channel of L_d by the sum of all channels $R+G+B$. For the different representations, we obtain different ray-wise chromaticity value candidates $\chi(r)$, which all enter the final score below.

Our main assumption is that if two points are spatially close to each other, share the same orientation, and have similar values in chromaticity space, but different intensities, the point with lower intensity is likely to be shadowed. To put this into formulas, we compute a weight $\theta_{r,q}$ that measures likely similarity in shading between rays r and q, in the sense that $\theta_{r,q}$ is close to one if the two points should have similar shading, We set

$$\theta_{r,q} = w_{r,q}^\chi \, w_{r,q}^n \, w_{r,q}^d, \text{where} \quad w_{r,q}^\chi = \exp\left(-\frac{\min_\chi \|\chi(r) - \chi(q)\|^2}{\sigma_\chi^2}\right),$$

$$w_{r,q}^n = \exp\left(\frac{\cos(n_r, n_q) - 1}{\sigma_n^2}\right) \quad \text{and} \quad w_{r,q}^d = 1 - \frac{\|d_r - d_q\|^2}{\max\limits_{\tilde{q}\in N_{obj}(r)} \|d_r - d_{\tilde{q}}\|^2}. \tag{8}$$

The minimum for w^χ is computed over the different ways to compute chromaticity values introduced above. Based on the similarity scores, we can then estimate color

$$m(r) = \left(\sum_{q\in N_{obj}(r)} \theta_{r,q} L_{\mathrm{d}}(q)\right) \Big/ \left(\sum_{q\in N_{obj}(r)} \theta_{r,q}\right) \tag{9}$$

for each ray by computing a weighted average over its neighbors. See Fig. 2 for the definition of $N_{obj}(\boldsymbol{r})$. This neighborhood is designed to restrict comparison to suitable rays within the same object.

Finally, the point-wise shadow confidence score β_{pw} is based on the observation that the estimated and actual intensities $v(\cdot)$ should be the same for an unshadowed ray. Here by intensity we mean average over all color channels R, G and B. Thus, we set ray-wise

$$\beta_{pw} = 1 - \exp\left(-\frac{1}{\sigma_v^2}\max\left(v(m) - v(L_d), 0\right)\right), \qquad (10)$$

with $\sigma_v^2 = 0.25$. See Fig. 4(b) for an example of a point-wise shadow confidence map.

Consistency Across the Light Field. While the locations of shadows are correct in most of the cases, point-wise estimation suffers from noise, and some shadow-free surfaces are classified as shadowed. To correct the initial shadow estimation, we minimize an energy $E(\beta)$, where we impose smoothness and consistency by embedding it in the optimization framework [16] similar to the main energy minimization problem (4).

In addition to light-field consistency, shadowed regions β are required to coincide with those regions where not only point-wise shadow confidence is high, but also intensity is low. For this, we combine the point-wise confidence with a shadow score derived from the bright channel concept for shadow detection introduced by Panagopoulos et al. [28]. We want our shadow score to be equal to the inverse bright channel cue

$$\overline{br}(\boldsymbol{r}) = 1 - \max_{c \in \{R,G,B\}} L_c(\boldsymbol{r}), \qquad (11)$$

in regions where point-wise shadow confidence β_{pw} is high. We also include a sparsity prior to remove regions with low confidence altogether.

Thus, the final energy we minimize is

$$E(\beta) = \lambda_d \int_{\mathcal{R}} \beta_{pw}(\boldsymbol{r}) \|\beta(\boldsymbol{r}) - \overline{br}(\boldsymbol{r})\|^2 \, d\boldsymbol{r} + \lambda_\beta^s \|\beta(\boldsymbol{r})\|_1 + J(\beta), \qquad (12)$$

defined over all sub-aperture views. In the regularizer J, we weight the spatial regularization with $1 - w_{r,q}^\chi$, since shadow should be the same if chromaticity is changing, while discontinuities that are related to reflectance changes should be smoothened. Figure 4(c) shows final shadow confidence after optimization.

7   Prior Energies for Intrinsic Components

In this section, we can finally define the prior energies E for albedo, shading and specularity for our model (4). In particular, we show how to employ the previously defined shadow and specularity confidence measures to improve upon the regularization terms.

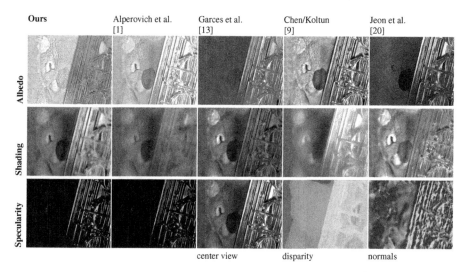

Fig. 5. Qualitative comparisons on te real world light field from Fig. 1. This scene has highly specular saxophone and almost Lambertian koala toy. Background and koala have similar pallets, which creates difficulties in albedo estimation. Our method successfully removes cast shadows, especially on the wall behind the koala and from saxophone. Our specular component contains all highlights that appear on saxophone. (Color figure online)

Albedo. We model albedo energy based on the state-of-the art approaches also used in [9, 20] and many other works. We assume that points with close chromaticity values are likely to have similar albedo. Thus, we define the albedo term as

$$E_a(A^{log}) = \lambda_{ch} \int_{\mathcal{R}} \sum_{q \in N_{3D}^{glob}(r)} \alpha_{r,q} \| A^{log}(r) - A^{log}(q) \|^2 \, dr. \qquad (13)$$

However, in contrast to previous work, to compute weights $\alpha_{r,q}$, we use flexible thresholds for chromaticity differences, based on our novel specular confidence h. To remove artifacts caused by specularity from the albedo component, we give more weight to points that have high specularity confidence. Thus, we model final regularization weights as

$$\alpha_{r,q} = \left(1 + \gamma_a \max\left(h(r), h(q)\right)\right) w_{r,q}^{\chi}. \qquad (14)$$

The weights for the pairs of rays above are defined in (8).

Direct Shading. To model direct shading, we assume that spatially close surface points that share the same orientation should have similar shading. We thus formulate the corresponding direct shading term

$$E_d(s_d^{log}) = \lambda_d^l \int_{\mathcal{R}} \sum_{q \in N_{6D}^{loc}(r)} \|s_d^{log}(r) - s_d^{log}(q)\|^2 \, dr + \dots$$

$$+ \lambda_d^g \int_{\mathcal{R}} \sum_{q \in N_{6D}^{glob}(r)} w_{r,q}^{glob} \|s_d^{log}(r) - s_d^{log}(q)\|^2 \, dr \tag{15}$$

as the combination of local and non-local direct shading priors. See Fig. 2 for a definition of the neighborhoods, which here live in the six-dimensional space of spatial locations d and normals n corresponding to the individual rays. For the global direct shading consistency, we also weight the similarity between two rays r and q with the angular difference of the outer normals and spatial locations,

$$w_{r,q}^{glob} = (1 + \gamma_d \max(\beta(r), \beta(q))) \, w_{r,q}^n w_{r,q}^d. \tag{16}$$

Since direct shading should be free of cast shadows, we add additional weight for neighbors with high cast shadow score β.

Indirect Shading. We model indirect shading by means of the shadow confidence measure β, whose detailed computation is described in Sect. 6. The main modeling assumptions are:

A1. Shading is spatially smooth except on shadow boundaries, i.e. near discontinuities of β.
A2. Assume two points are spatially close to each other, none of them specular, they share the same orientation and their chromaticity is similar. Then $\theta_{r,q} = 1$ in (8), and their shadow free representation should be the same.
A3. The distribution of indirect shading is sparse except inside the areas within cast shadows.

These assumptions lead to the indirect shading term

$$E_i(S_i^{log}) = \lambda_i^l \int_{\mathcal{R}} \sum_{q \in N_{3D}^{loc}(r)} \delta_{r,q} \|S_i^{log}(r) - S_i^{log}(q)\|^2 \, dr + \lambda_i^g \int_{\mathcal{R}} \sum_{q \in N_{3D}^{glob}(r)} \theta_{r,q} \beta(r) \cdot \dots$$

$$\cdot \|(L^{log}(r) - S_i^{log}(r)) - (L^{log}(q) - S_i^{log}(q))\|^2 \, dr + \lambda_i^s \int_{\mathcal{R}} (1 - \beta(r))\|S_i^{log}(r)\|^2 \, dr. \tag{17}$$

We define δ_β to be the norm of the Gaussian filtered gradient of β, and thus the weight $\delta_{r,q} := (1 - \max(\delta_\beta(r), \delta_\beta(q)))$ is small if and only if any of the two rays is close to a shadow boundary.

Table 1. Main parameters for intrinsic light field decomposition.

Specularity detection	Shadow detection	$\sigma_n^2 = 0.1$	Weights	$\lambda_d^g = 0.1$	Iterations
$\tau_f = 0.1$	$\lambda_d = 1$	$Priors$	$\lambda_s = 10$	$\lambda_i^l = 1$	global iter = 50
$\sigma_h = 2$	$\lambda_\beta \in [0.1, 0.5]$	$\gamma_a = 10$	$\lambda_{ch} = 0.1$	$\lambda_i^g = 5$	local iter = 10
$p = 85$	$\sigma_{Xi}^2 = 0.01$	$\gamma_d = 10$	$\lambda_d^l = 1$	$\lambda_i^s = 10$	$\lambda_h = 5$

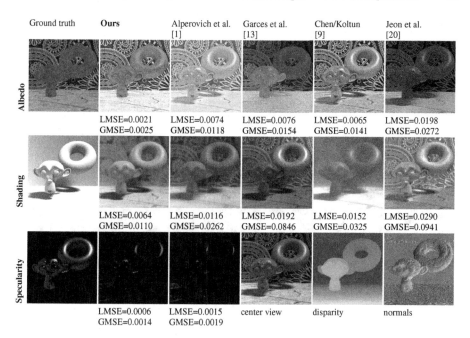

Fig. 6. Quantitative comparison on synthetic data rendered with *Blender*. The scene illustrates how the algorithms can cope with hard shadows, where chromaticity might vary in the shaded area. The disparity range is from −1.5 to 1.5, light field size is $9 \times 9 \times 512 \times 512$. We find that our proposed method outperforms the other approaches in all measures. In the visual comparison, we see that our approach successfully removed shadow from albedo component.

Specularity. The specularity prior we define as

$$E_s(H^{log}) \;=\; \lambda_s \int_{\mathcal{R}} (1 - h(\boldsymbol{r})) \|H^{log}(\boldsymbol{r})\|^2 \, d\boldsymbol{r} + \lambda_h \|H^{log}(\boldsymbol{r})\|_1. \qquad (18)$$

Specularity is encouraged to have non-zero values only inside the specular mask h.

8   Results

We validate our decomposition method on ground truth synthetic data sets as well as real world light fields. Our main goal is to achieve superior results for the center view of the light field, thus we present decomposition and evaluations only for the center view. Overall, our approach makes use of a light field with 9×9 views. Since estimated disparity is less accurate for the edge views we solve the final optimization (4) for a cross-hair shaped subset of 13 views from a light field with 7×7 views, where edge views are excluded. In the main paper we present a single synthetic data set (Fig. 6) and two real world light fields (Figs. 1, 5 and 8). For more results, we refer to the supplementary material.

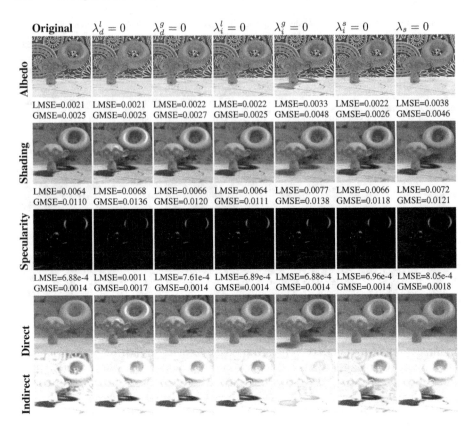

Fig. 7. The leftmost column shows the original result with experimentally determined parameter settings as specified in Table 1. The other columns show the consequences of excluding certain priors from the model. We can see that local direct and indirect shading priors λ_d^l and λ_i^l, respectively, influence the smoothness of the shading component, while the global indirect shading prior λ_i^g controls the amount of cast shadows left in the albedo component. The specularity prior λ_s is required to prevent over-smoothing of albedo and shading.

Note that since there is currently no ground truth available for intrinsic light field decomposition, and benchmark data sets for intrinsic image decomposition presented in [4,8,18] are not applicable to our setting, we created our own ground truth data. Synthetic datasets were generated with Blender using the Cycles rendering engine and a light field plugin [19]. Internally, Blender combines direct shading D, indirect shading I and object color C for diffuse and glossy reflection separately by evaluating $(D_{\text{diffuse}} + I_{\text{diffuse}})C_{\text{diffuse}} + (D_{\text{glossy}} + I_{\text{glossy}})C_{\text{glossy}}$. The separate components can be stored individually to use as ground truth for evaluation. Albedo corresponds to the diffuse color, shading is the sum $D_{\text{diffuse}} + I_{\text{diffuse}}$ of direct and indirect illumination components, and specularity is the difference $H = L - AS$ between input light field and product of albedo and shading.

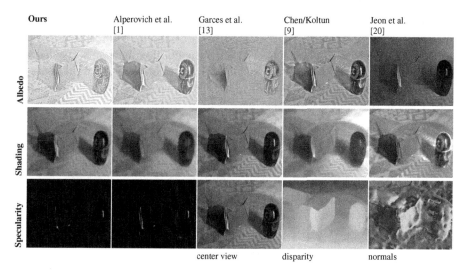

Fig. 8. Qualitative comparisons on te real world data set captured with Lytro Illum plenoptic camera, 9×9 views, resolution 434×625. Proposed method outperforms the previous approaches, especially albedo component has much less cast shadows, and shading is more smooth and contains less texture.

Our definition of direct and indirect shading differs from one that used in Blender, thus for quantitative evaluations we compute resulting shading $S = s_d S_i$ and then compare to the ground truth shading $S_{ground\ truth} = D_{\text{diffuse}} + I_{\text{diffuse}}$.

We compare our method to the two RGB+D approaches proposed by Chen and Koltun [9] and Jeon et al. [20], as well as recent methods for light fields proposed by Alperovich and Goldluecke [1] and Garces et al. [13]. For quantitative evaluations, we selected two error measures. The first is local mean-squared error (LMSE) [18] computed patch-wise. The idea is to reduce scaling ambiguity by adjusting the brightness of the image patch such that it corresponds to that of the ground truth, and then to compute MSE. In our experiments, we use rectangular overlapping patches with a size of 20% of the total image size.

Although the LMSE is more reasonable measure than pure MSE, sometimes it can produce non-sensible results in shaded areas. Thus, we also compute global mean-squared error (GMSE) that adjusts the test image brightness value to the ground truth for the whole image. With GMSE, if shaded regions are present in the albedo image, the influence of their error will be more reasonably reflected in the error measure - MSE would otherwise be very small for dark regions. Since aggregated quantitative evaluation can only give a partial picture about the performance, we also visualize all decomposition results for qualitative evaluations. Resulting direct and indirect shading computed with our method for the above datasets can be observed in Figs. 1 and 7. In order to asses individual contributions of proposed priors we present results where some prior are excluded by setting corresponding weights to zero Fig. 7.

9   Conclusion

In this work, we propose a new model for intrinsic non-Lambertian scene decomposition with light field data, where we fully use information available in the light field to model priors for albedo, shading, and specularity. One of our main contributions is a novel shading term that describes cast shadows and inter-reflections. By means of this term, we recover consistent shading components, both in case of hard as well as soft shadows. Moreover, we improve albedo and specularity estimation by embedding specularity information in the albedo prior, which makes albedo less affected by highlights. We introduce ground truth evaluations on synthetic light fields, where we show qualitatively and quantitatively that our method significantly outperforms existing algorithms. In addition, we perform qualitative evaluations for the real world examples captured with Lytro Illum plenoptic camera.

Acknowledgments. This work was supported by the ERC Starting Grant "Light Field Imaging and Analysis" (LIA 336978, FP7-2014).

References

1. Alperovich, A., Goldluecke, B.: A variational model for intrinsic light field decomposition. In: Lai, S.-H., Lepetit, V., Nishino, K., Sato, Y. (eds.) ACCV 2016. LNCS, vol. 10113, pp. 66–82. Springer, Cham (2017). https://doi.org/10.1007/978-3-319-54187-7_5
2. Barron, J.T., Malik, J.: Intrinsic scene properties from a single RGB-D image. IEEE TPAMI **38**(4), 690–703 (2015)
3. Barrow, H.G., Tenenbaum, J.M.: Recovering intrinsic scene characteristics from images. Comput. Vis. Syst. **23**(1), 3–26 (1978)
4. Bell, S., Bala, K., Snavely, N.: Intrinsic images in the wild. ACM Trans. Graph. (SIGGRAPH) **33**(4), 159:1–159:12 (2014)
5. Bolles, R., Baker, H., Marimont, D.: Epipolar-plane image analysis: an approach to determining structure from motion. Int. J. Comput. Vis. **1**(1), 7–55 (1987)
6. Bonneel, N., Sunkavalli, K., Tompkin, J., Sun, D., Paris, S., Pfister, H.: Interactive intrinsic video editing. ACM Trans. Graph. **33**(6), 197 (2014)
7. Bredies, K., Kunisch, K., Pock, T.: Total generalized variation. SIAM J. Imaging Sci. **3**(3), 492–526 (2010)
8. Butler, D.J., Wulff, J., Stanley, G.B., Black, M.J.: A naturalistic open source movie for optical flow evaluation. In: Fitzgibbon, A., Lazebnik, S., Perona, P., Sato, Y., Schmid, C. (eds.) ECCV 2012. LNCS, vol. 7577, pp. 611–625. Springer, Heidelberg (2012). https://doi.org/10.1007/978-3-642-33783-3_44
9. Chen, Q., Koltun, V.: A simple model for intrinsic image decomposition with depth cues. In: Proceedings of the ICCV (2013)
10. Finlayson, G., Schaefer, G.: Solving for colour constancy using a constrained dichromatic reflection model. Int. J. Comput. Vis. **42**(3), 127–144 (2001)
11. Finlayson, G.D., Drew, M.S., Lu, C.: Intrinsic images by entropy minimization. In: Pajdla, T., Matas, J. (eds.) ECCV 2004. LNCS, vol. 3023, pp. 582–595. Springer, Heidelberg (2004). https://doi.org/10.1007/978-3-540-24672-5_46

12. Finlayson, G.D., Hordley, S.D., Lu, C., Drew, M.S.: On the removal of shadows from images. IEEE Trans. Pattern Anal. Mach. Intell. **28**(1), 59–68 (2006)
13. Garces, E., Echevarria, J.I., Zhang, W., Wu, H., Zhou, K., Gutierrez, D.: Intrinsic light field images. Comput. Graph. Forum **36**, 589–599 (2017)
14. Garces, E., Munoz, A., Lopez-Moreno, J., Gutierrez, D.: Intrinsic images by clustering. Comput. Graph. Forum (Proc. EGSR 2012) **31**(4), 1415–1424 (2012)
15. Gehler, P.V., Rother, C., Kiefel, M., Zhang, L., Schölkopf, B.: Recovering intrinsic images with a global sparsity prior on reflectance. In: Proceedings of the NIPS (2011)
16. Goldluecke, B., Wanner, S.: The variational structure of disparity and regularization of 4D light fields. In: Proceedings of the CVPR (2013)
17. Gortler, S., Grzeszczuk, R., Szeliski, R., Cohen, M.: The Lumigraph. In: Proceedings of the SIGGRAPH, pp. 43–54 (1996)
18. Grosse, R., Johnson, M.K., Adelson, E.H., Freeman, W.T.: Ground truth dataset and baseline evaluations for intrinsic image algorithm. In: Proceedings of the ICCV (2009)
19. Honauer, K., Johannsen, O., Kondermann, D., Goldluecke, B.: A dataset and evaluation methodology for depth estimation on 4D light fields. In: Lai, S.-H., Lepetit, V., Nishino, K., Sato, Y. (eds.) ACCV 2016. LNCS, vol. 10113, pp. 19–34. Springer, Cham (2017). https://doi.org/10.1007/978-3-319-54187-7_2
20. Jeon, J., Cho, S., Tong, X., Lee, S.: Intrinsic image decomposition using structure-texture separation and surface normals. In: Fleet, D., Pajdla, T., Schiele, B., Tuytelaars, T. (eds.) ECCV 2014. LNCS, vol. 8695, pp. 218–233. Springer, Cham (2014). https://doi.org/10.1007/978-3-319-10584-0_15
21. Kim, S., Park, K., Sohn, K., Lin, S.: Unified depth prediction and intrinsic image decomposition from a single image via joint convolutional neural fields. In: Leibe, B., Matas, J., Sebe, N., Welling, M. (eds.) ECCV 2016. LNCS, vol. 9912, pp. 143–159. Springer, Cham (2016). https://doi.org/10.1007/978-3-319-46484-8_9
22. Lalonde, J.-F., Efros, A.A., Narasimhan, S.G.: Detecting ground shadows in outdoor consumer photographs. In: Daniilidis, K., Maragos, P., Paragios, N. (eds.) ECCV 2010. LNCS, vol. 6312, pp. 322–335. Springer, Heidelberg (2010). https://doi.org/10.1007/978-3-642-15552-9_24
23. Land, E.H., McCann, J.J.: Lightness and retinex theory. J. Opt. Soc. Am. **61**(1), 1–11 (1978)
24. Lee, K.J., Zhao, Q., Tong, X., Gong, M., Izadi, S., Lee, S.U., Tan, P., Lin, S.: Estimation of intrinsic image sequences from image+depth video. In: Fitzgibbon, A., Lazebnik, S., Perona, P., Sato, Y., Schmid, C. (eds.) ECCV 2012. LNCS, vol. 7577, pp. 327–340. Springer, Heidelberg (2012). https://doi.org/10.1007/978-3-642-33783-3_24
25. Lee, T.S.: Top-down influence in early visual processing: a Bayesian perspective. Physiol. Behav. **77**, 645–650 (2002)
26. Levoy, M.: Light fields and computational imaging. Computer **39**(8), 46–55 (2006)
27. Olsson, C., Boykov, Y.: Curvature-based regularization for surface approximation. In: Proceedings of the CVPR (2012)
28. Panagopoulos, A., Wang, C., Samaras, D., Paragios, N.: Estimating shadows with the bright channel cue. In: Kutulakos, K.N. (ed.) ECCV 2010. LNCS, vol. 6554, pp. 1–12. Springer, Heidelberg (2012). https://doi.org/10.1007/978-3-642-35740-4_1
29. Shafer, S.: Using color to separate reflection components. Color Res. Appl. **10**(4), 210–218 (1985)

30. Shen, L., Tan, P., Lin, S.: Intrinsic image decomposition with non-local texture cues. In: Proceedings of the CVPR (2008)
31. Strecke, M., Alperovich, A., Goldluecke, B.: Accurate depth and normal maps from occlusion-aware focal stack symmetry. In: Proceedings of the CVPR (2017)
32. Sulc, A., Alperovich, A., Marniok, N., Goldluecke, B.: Reflection separation in light fields based on sparse coding and specular flow. In: Proceedings of the VMV (2016)
33. Tan, R.T., Ikeuchi, K.: Separating reflection components of textured surfaces using a single image. In: Proceedings of the ICCV (2003)
34. Tan, R.T., Nishino, K., Ikeuchi, K.: Color constancy through inverse-intensity chromaticity space. J. Opt. Soc. Am. A: **21**(3), 321–334 (2004)
35. Tao, M., Su, J.C., Wang, T.C., Malik, J., Ramamoorthi, R.: Depth estimation and specular removal for glossy surfaces using point and line consistency with light-field cameras. IEEE TPAMI **38**(6), 1155–1169 (2015)
36. Tappen, M.F., Freeman, W.T., Adelson, E.H.: Recovering intrinsic images from a single image. IEEE TPAMI **27**(9), 1459–1472 (2005)
37. Triantafyllidis, G., Dimitriou, M., Kounalakis, T., Vidakis, N.: Detection and classification of multiple objects using an RGB-D sensor and linear spatial pyramid matching. ELCVIA Electron. Lett. Comput. Vis. Image Anal. **12**(2), 78–87 (2013)
38. Wang, H., Xu, C., Wang, X., Zhang, Y., Peng, B.: Light field imaging based accurate image specular highlight removal. PLoS ONE **11**(6), e0156173 (2016)
39. Weiss, Y.: Deriving intrinsic images from image sequences. In: Proceedings of the ICCV (2001)
40. Xiao, Y., Tsougenis, E., Tang, C.K.: Shadow removal from single RGB-D images. In: Proceedings of the CVPR (2014)
41. Yang, K., Gao, S., Li, Y.: Efficient illuminant estimation for color constancy using grey pixels. In: Proceedings of the CVPR (2015)
42. Ye, G., Garces, E., Liu, Y., Dai, Q., Gutierrez, D.: Intrinsic video and applications. ACM Trans. Graph. (SIGGRAPH 2014) **33**(4), 80 (2014)
43. Zeisl, B., Zach, C., Pollefeys, M.: Variational regularization and fusion of surface normal maps. In: 3DV (2014)

Propagation and Time-Evolution

Modelling Stable Backward Diffusion and Repulsive Swarms with Convex Energies and Range Constraints

Leif Bergerhoff[1]([✉]) [iD], Marcelo Cardénas[1], Joachim Weickert[1],
and Martin Welk[2] [iD]

[1] Mathematical Image Analysis Group, Saarland University,
Campus E1.7, 66041 Saarbrücken, Germany
{bergerhoff,cardenas,weickert}@mia.uni-saarland.de
[2] Institute of Biomedical Image Analysis, Private University for Health Sciences,
Medical Informatics and Technology, Eduard-Wallnöfer-Zentrum 1,
6060 Hall/Tyrol, Austria
martin.welk@umit.at

Abstract. Backward diffusion and purely repulsive swarm dynamics are generally feared as ill-posed, highly unstable processes. On the other hand, it is well-known that minimising strictly convex energy functionals by gradient descent creates well-posed, stable evolutions. We prove a result that appears counterintuitive at first glance: We derive a class of one-dimensional backward evolutions from the minimisation of strictly convex energies. Moreover, we stabilise these inverse evolutions by imposing range constraints. This allows us to establish a comprehensive theory for the time-continuous evolution, and to prove a stability condition for an explicit time discretisation. Prototypical experiments confirm this stability and demonstrate that our model is useful for global contrast enhancement in digital greyscale images and for modelling purely repulsive swarm dynamics.

Keywords: Convex optimisation · Inverse processes
Dynamical systems · Diffusion filtering · Swarm dynamics

1 Introduction

Backward parabolic partial differential equations such as inverse diffusion are classical representatives of ill-posed processes. For nonsmooth initial data, they may have no solution at all. Even if a solution exists, it is highly sensitive and intrinsically unstable: Already the smallest perturbations of the initial data can cause huge deviations during the evolution.

Nevertheless, since forward parabolic equations can blur or smooth images, there have been a number of attempts to invert these evolutions for deblurring or sharpening degraded imagery. This, however, requires additional stabilisation. The most widely used strategy is to impose constraints at extrema that aim at

© Springer International Publishing AG, part of Springer Nature 2018
M. Pelillo and E. Hancock (Eds.): EMMCVPR 2017, LNCS 10746, pp. 409–423, 2018.
https://doi.org/10.1007/978-3-319-78199-0_27

enforcing a maximum–minimum principle. One example is the inverse diffusion filter of Osher and Rudin [4], which implements backward diffusion everywhere except at extrema, where the evolution is set to zero. Another example is the so-called forward-and-backward (FAB) diffusion of Gilboa et al. [3]. It differs from the closely related Perona–Malik filter [5] by the fact that it uses negative diffusivities for a specific range of gradient magnitudes. However, at extrema where the gradient vanishes, it always avoids explosions by imposing forward diffusion. So far any attempt to adequately implement inverse diffusions with forward or zero diffusion at extrema requires sophisticated numerical schemes [4,8].

A second, less widely-used class of stabilisation attempts adds a fidelity term that prevents the backward evolution to move too far away from the original image [7] or from the average grey value of the desired range [6]. In this case the range of the filtered image obviously depends on the weights of the fidelity and the backward diffusion term.

In conclusion, we see that handling backward diffusion in practice is problematic and requires specific care to keep some sort of stability.

In order to gain new insights how to end up with stable backward evolutions, let us for a moment turn our attention to forward diffusion processes. For simplicity we consider a simple 1-D evolution for signal smoothing. It regards the original signal $f : [a,b] \rightarrow \mathbb{R}$ as initial state of the diffusion equation

$$\partial_t u = \partial_x \big(g(u_x^2)\,u_x\big) \tag{1}$$

where $u = u(x,t)$ is a filtered version of the original signal $u(x,0) = f(x)$, $u_x = \partial_x u$, and reflecting boundary conditions at $x = a$ and $x = b$ are imposed. Larger diffusion times t create simpler representations. The diffusivity function g is nonnegative. In order to smooth less at signal edges than in more homogeneous regions, Perona and Malik [5] propose to choose g as a decreasing function of the contrast u_x^2. If the flux function $\Phi(u_x) := g(u_x^2)\,u_x$ is strictly increasing in u_x we have a forward diffusion process that cannot sharpen edges. Then the diffusion process can be seen as a gradient descent evolution for minimising the energy

$$E[u] = \int_a^b \Psi(u_x^2)\,dx \tag{2}$$

with a potential function $\tilde{\Psi}(u_x) = \Psi(u_x^2)$ that is strictly convex in u_x, increasing in u_x^2, and satisfies $\Psi'(u_x^2) = g(u_x^2)$. Since the energy functional is strictly convex, it has a unique minimiser. This minimiser is given by the (flat) steady state $(t \rightarrow \infty)$ of the gradient descent method, and the gradient descent/diffusion evolution is well-posed. Due to this classical appearance of well-posed forward diffusion as a consequence of strictly convex energies, one might be tempted to believe that backward diffusion is necessarily connected to nonconvex energies. Interestingly, this is not correct! Understanding this connection better opens new ways to design stable backward processes.

Our Contribution. We consider a space-discrete model where we admit *globally negative* diffusivities, corresponding to decreasing penalisers Ψ. However, we require $\Psi(u_x^2)$ to be strictly convex in u_x. We do not rely on stabilisations through zero or forward diffusivities at extrema, and we do not incorporate fidelity terms explicitly. Stabilisation will be achieved in our model on a *global* level by bounding the range of u. This is achieved by imposing reflecting boundary conditions in the co-domain. We show that this is sufficient to stabilise the inverse diffusion in the space-discrete and time-continuous setting. We also prove that a straightforward explicit time discretisation inherits this stability, if it satisfies a suitable time step size restriction. Our model opens up interesting applications in signal and image filtering, where it can be used for global contrast enhancement, as well as in swarm-like particle systems with purely repulsive interactions.

Structure of the Paper. In Sect. 2 we introduce our novel one-dimensional model and present a comprehensive analysis of its theoretical properties. The third section establishes stability bounds for an explicit time discretisation. Section 4 deals with the application of our model to image enhancement and the modelling of swarm behaviour. Finally, Sect. 5 gives conclusions and an outlook on future challenges.

2   Model and Theory

2.1   Discrete Variational Model

We start by introducing a dynamical system that is motivated from a spatial discretisation of the energy functional (2) with a decreasing penaliser function $\Psi : \mathbb{R}_0^+ \to \mathbb{R}$ and a global range constraint on u. The corresponding flux function Φ is given by $\Phi(s) := \Psi'(s^2)s$.

We consider vectors $\boldsymbol{v} = (v_1, \ldots, v_N)^{\mathrm{T}} \in (0,1)^N$, where v_i for $i = 1, \ldots, N$ are assumed to be distinct. We extend such $\boldsymbol{v}$ with the additional coordinates $v_{N+1}, \ldots, v_{2N}$ defined as $v_{2N+1-i} = 2 - v_i \in (1,2)$. For this extended $\boldsymbol{v} \in [0,2]^{2N}$, we consider the energy function

$$E(\boldsymbol{v}) = \frac{1}{4} \cdot \sum_{i=1}^{2N} \sum_{j=1}^{2N} \Psi\big((v_j - v_i)^2\big). \tag{3}$$

which models the repulsion potential between all positions v_i and v_j. A typical scenario for Ψ is illustrated in Fig. 1. First, the function $\Psi(s^2)$ is defined as a continuously differentiable, decreasing, and strictly convex function for $s \in [0,1]$ with $\Psi(0) = 0$ and $\Phi_-(1) = 0$ (left-sided derivative). It is then extended to $[-1,1]$ by symmetry and to $\mathbb{R}$ by periodicity $\Psi\big((2+s)^2\big) = \Psi(s^2)$. As a result, $\Psi(s^2)$ is continuously differentiable everywhere except at even integers, where it is still continuous. Note that $\Psi(s^2)$ is increasing on $[-1,0]$ and $[1,2]$. The flux Φ is continuous and increasing in $(0,2)$ with jump discontinuities at 0 and 2 (see Fig. 1). Furthermore, we have that $\Phi(s) = -\Phi(-s)$ and $\Phi(2+s) = \Phi(s)$. A gradient descent for (3) is given by

$$\partial_t v_i = -\partial_{v_i} E(\boldsymbol{v}) = \sum_{\substack{j=1 \\ j \neq i}}^{2N} \Phi(v_j - v_i), \quad i = 1, \ldots, 2N, \tag{4}$$

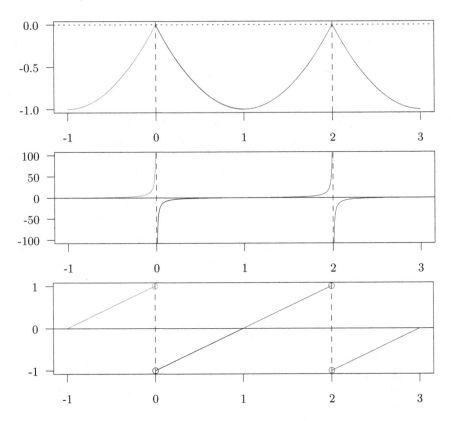

Fig. 1. Top: Exemplary penaliser $\tilde{\Psi}(s) := \Psi(s^2)$ with $\tilde{\Psi}(s) = (s-1)^2 - 1$ for $s \in [0,1]$, extended to the interval $[-1,3]$ by imposing symmetry and $\Psi((2+s)^2) = \Psi(s^2)$. Middle: Corresponding diffusivity $\tilde{g}(s) := g(s^2) = \Psi'(s^2)$ where $\tilde{g}(s) := 1 - 1/s$ for $s \in (0,1]$. Bottom: Corresponding flux $\Phi(s) = \Psi'(s^2)s$ with $\Phi(s) = s - 1$ for $s \in (0,1]$. (Colour figure online)

where v_i now are functions of the time t. Note that for $1 \le i,j \le N$, thus $|v_j - v_i| < 1$, the flux $\Phi(v_j - v_i)$ is negative for $v_j > v_i$ and positive otherwise, thus driving v_i always away from v_j. This implies that we have negative diffusivities Ψ' for all $|v_j - v_i| < 1$. Due to the convexity of $\Psi(s^2)$, the absolute values of the repulsive forces Φ are decreasing with the distance between v_i and v_j. We remark that the jumps of Φ at 0 and 2 are not problematic here as the v_i are required to be distinct.

Let us discuss shortly how the interval constraint for the v_i, $i = 1,\dots,N$, is enforced in (3) and (4). First, notice that v_{2N+1-i} for $i = 1,\dots,N$ is the reflection of v_i on the right interval boundary 1. For v_i and v_{2N+1-j} with $1 \le i,j \le N$ and $v_{2N+1-j} - v_i < 1$ there is a repulsive force due to $\Phi(v_{2N+1-j} - v_i) < 0$ that drives v_i and v_{2N+1-j} away from the *right* interval boundary. The closer v_i and v_{2N+1-j} come to this boundary, the stronger is the repulsion. For $v_{2N+1-j} - v_i > 1$, we have $\Phi(v_{2N+1-j} - v_i) > 0$. By $\Phi(v_{2N+1-j} - v_i) = \Phi((2 - v_j) - v_i) = \Phi((-v_j) - v_i)$, this can equally be interpreted as a repulsion between v_i and $-v_j$ where $-v_j$ is the reflection of v_j at the left interval boundary 0. In this case the interaction between v_i and v_{2N+1-j} drives v_i and $-v_j$ away from the *left* interval boundary. Recapitulating both possible cases, it becomes clear that every v_i is either repelled from the reflection of v_j at the left or at the right interval boundary, but never from both at the same time.

As $\partial_t v_{2N+1-i} = -\partial_t v_i$ holds in (4), the symmetry of $\boldsymbol{v}$ is preserved. Dropping the redundant entries $v_{N+1}, \ldots, v_{2N}$, Eq. (4) can be rewritten as

$$\partial_t v_i = \sum_{\substack{j=1 \\ j \neq i}}^{N} \Phi(v_j - v_i) - \sum_{j=1}^{N} \Phi(v_i + v_j), \quad i = 1, \ldots, N, \tag{5}$$

where the second sum represents the repulsions between original and reflected coordinates in a more symmetric way.

Given an initial vector $\boldsymbol{f} \in (0,1)^N$ with distinct entries f_i, and initialising $v_i(0) = f_i$, $v_{2N+1-i}(0) = 2 - f_i$ for $i = 1, \ldots, N$, the gradient descent (4) or (5) evolves $\boldsymbol{v}$ towards a minimiser of E.

A more detailed analysis below shows that in the course of the evolution, no v_i can reach the interval boundaries 0 or 1, and no v_i, v_j with $i \neq j$ can ever become equal. Thus the initial rank-order of v_i is preserved throughout the evolution. Each of the $N!$ possible rank-orders constitutes a connected component of the configuration space for $\boldsymbol{v}$. There is a unique minimiser of E in the interior of each connected component due to the strict convexity of $\Psi(s^2)$.

Theorem 1 (Avoidance of Boundaries). *The N initially distinct positions $v_i \in (0,1)$ evolving according to (5) never reach the domain boundaries 0 and 1.*

Proof. The definition of Ψ implies that

$$\lim_{h \to 0^+} \frac{\Psi\left((0+h)^2\right) - \Psi(0)}{h} < 0 \quad \text{and} \quad \lim_{h \to 0^-} \frac{\Psi(2^2) - \Psi\left((2+h)^2\right)}{h} > 0, \tag{6}$$

from which it follows that

$$\lim_{v_i \to 0^+} \left(-\Phi(2v_i)\right) > 0 \quad \text{and} \quad \lim_{v_i \to 1^-} \left(-\Phi(2v_i)\right) < 0. \tag{7}$$

Equation (5) can be written as

$$\partial_t v_i = \sum_{\substack{j=1 \\ j \neq i}}^{N} \left(\Phi(v_j - v_i) - \Phi(v_i + v_j)\right) - \Phi(2v_i). \tag{8}$$

Since for $j = 1, \ldots, N$ and $j \neq i$ one has

$$\lim_{v_i \to 0^+} \Phi(v_j - v_i) - \Phi(v_i + v_j) = 0 \quad \text{and} \quad \lim_{v_i \to 1^-} \Phi(v_j - v_i) - \Phi(v_i + v_j) = 0, \tag{9}$$

it follows that

$$\lim_{v_i \to 0^+} \partial_t v_i > 0 \quad \text{and} \quad \lim_{v_i \to 1^-} \partial_t v_i < 0. \tag{10}$$

Consequently, v_i can never reach the left interval boundary 0 because it will move to the right when getting closer to it. The same holds for the right domain boundary 1 where v_i will move to the left before reaching it.   □

Theorem 2 (Nonequality of v_i and v_j). *Among N initially distinct positions $v_i \in (0,1)$ evolving according to (5), no two ever become equal.*

Proof. Using (5) it is possible to derive the difference

$$\partial_t (v_j - v_i) = 2 \cdot \Phi(v_i - v_j) + \sum_{\substack{k=1 \\ k \neq i,j}}^{N} \left(\Phi(v_k - v_j) - \Phi(v_k - v_i) \right)$$
$$- \sum_{k=1}^{N} \left(\Phi(v_j + v_k) - \Phi(v_i + v_k) \right) \qquad (11)$$

where $1 \leq i, j \leq N$. Assume w.l.o.g. that $v_j > v_i$ and consider (11) in the limit $v_j - v_i \to 0$. Then we have

$$\lim_{v_j - v_i \to 0} \partial_t(v_j - v_i) = \lim_{v_j - v_i \to 0} 2 \cdot \Phi(v_i - v_j) > 0. \qquad (12)$$

The latter inequality follows from the fact that $\Phi(s) > 0$ for $s \in (-1, 0)$. This means that v_j will always start moving away from v_i (and vice versa) when the difference between both gets sufficiently small. Since the initial positions are distinct, it follows that $v_i \neq v_j$ for $i \neq j$ for all times t.   □

Theorem 3 (Explicit Steady-State Solution). *Under the assumption that (v_i) is in increasing order and that $\Psi(s^2)$ is twice continuously differentiable in $(0, 2)$ the unique minimiser of (3) is given by $\boldsymbol{v}^* = (v_1^*, \ldots, v_{2N}^*)^\mathrm{T}$, $v_i^* = (i - 1/2)/N$, $i = 1, \ldots, 2N$.*

Proof. Equation (3) can be rewritten without the redundant entries of $\boldsymbol{v}$ as

$$E(\boldsymbol{v}) = 2 \cdot \sum_{i=1}^{N-1} \sum_{j=i+1}^{N} \Psi\left((v_j - v_i)^2\right) + \sum_{i=1}^{N} \Psi(4v_i^2) + 2 \cdot \sum_{i=1}^{N-1} \sum_{j=i+1}^{N} \Psi\left((v_i + v_j)^2\right) \quad (13)$$

from which one can verify by straightforward, albeit lengthy calculations that $\nabla E(\boldsymbol{v}^*) = 0$, and the Hessian of E at $\boldsymbol{v}^*$ is

$$\mathrm{D}^2 E(\boldsymbol{v}^*) = \sum_{k=1}^{N} \boldsymbol{A}_k \Phi'\left(\frac{k}{N}\right) \qquad (14)$$

with sparse symmetric $N \times N$-matrices

$$\boldsymbol{A}_k = 4\boldsymbol{I} - 2\boldsymbol{T}_k - 2\boldsymbol{T}_{-k} + 2\boldsymbol{H}_{k+1} + 2\boldsymbol{H}_{2N-k+1}, \quad k = 1, \ldots, N-1, \quad (15)$$
$$\boldsymbol{A}_N = 2\boldsymbol{I} + 2\boldsymbol{H}_{N+1} \qquad (16)$$

where the unit matrix I, single-diagonal Toeplitz matrices T_k and single-anti-diagonal Hankel matrices H_k are defined as

$$I = \left(\delta_{i,j}\right)_{i,j=1}^{N}, \qquad T_k = \left(\delta_{j-i,k}\right)_{i,j=1}^{N}, \qquad H_k = \left(\delta_{i+j,k}\right)_{i,j=1}^{N}. \qquad (17)$$

Here, $\delta_{i,j}$ denotes the Kronecker symbol, $\delta_{i,j} = 1$ if $i = j$, and $\delta_{i,j} = 0$ otherwise. All A_k, $k = 1,\dots,N$ are weakly diagonally dominant with positive diagonal, thus positive semidefinite by Gershgorin's Theorem. Moreover, the tridiagonal matrix A_1 is of full rank, thus even positive definite. By strict convexity of $\Psi(s^2)$, all $\Phi'(k/N)$ are positive, thus $D^2 E(v^*)$ is positive definite.

As a consequence, the steady state of the gradient descent (5) for any initial data f (with arbitrary rank-order) can be computed directly by sorting the f_i: Let σ be the permutation of $\{1,\dots,N\}$ for which $(f_{\sigma^{-1}(i)})_{i=1,\dots,N}$ is increasing (this is what a sorting algorithm computes), the steady state is given by $v_i^* = (\sigma(i) - 1/2)/N$ for $i = 1,\dots,N$. □

Theorem 4 (Convergence). *For $t \to \infty$ any initial configuration $v \in (0,1)^N$ with distinct entries converges to a unique steady state v^* which is the global minimiser of the energy given in (13).*

Proof. As a sum of convex functions, (13) is convex. Therefore the function $V(v) := E(v) - E(v^*)$ (where v^* is the equilibrium point) is a Lyapunov function with $V(v^*) = 0$ and $V(v) > 0$ for all $v \neq v^*$. Furthermore, we have

$$\partial_t V(v) = -\sum_{i=1}^{N} \left(\partial_{v_i} E(v)\right)^2 \leq 0. \qquad (18)$$

Note that due to the positive definiteness of (14) we know that $E(v)$ has a strict (global) minimum which implies that the inequality in (18) becomes strict except in case of $v = v^*$. This guarantees asymptotic Lyapunov stability of v^* and thus convergence to v^* for $t \to \infty$. □

2.2 Generalisation with Weights

Let us now consider a generalised version of our model that allows for localisation and different treatment of distinct v_i. For this purpose we make use of vectors $w = (w_1,\dots,w_N)^T \in (0,\infty)^N$ and $x = (x_1,\dots,x_N)^T \in (\mathbb{R}^n)^N$ which we extend – similar to v – with the coordinates $w_{N+1},\dots,w_{2N}$ and $x_{N+1},\dots,x_{2N}$. Both are defined as $w_{2N+1-i} = w_i$ and $x_{2N+1-i} = x_i$. Each w_i denotes the weight, or importance, of the corresponding v_i, whereas the x_i provide additional n-dimensional position information which will become relevant in our future research. Neither w_i nor x_i change over time. Additionally, we introduce a weighting function $\gamma(|x|)$ which is 1 for $|x| \leq \varrho$ and 0 else for $\varrho > 0$. Now regard the adapted variant of (3) given by

$$E(p,x,w) = \frac{1}{4}\sum_{i=1}^{2N}\sum_{j=1}^{2N} w_i \cdot w_j \cdot \gamma(|x_j - x_i|) \cdot \Psi\left(\left(\frac{p_j}{\sqrt{w_j}} - \frac{p_i}{\sqrt{w_i}}\right)^2\right), \qquad (19)$$

where we make use of the coordinate transform $p_i := \sqrt{w_i} \cdot v_i$, $i = 1, \ldots, 2N$. Referring to (4), a gradient descent can be formulated as

$$\partial_t p_i = \sqrt{w_i} \cdot \sum_{\substack{j=1 \\ j \neq i}}^{2N} w_j \cdot \gamma(|\boldsymbol{x}_j - \boldsymbol{x}_i|) \cdot \Phi\left(\frac{p_j}{\sqrt{w_j}} - \frac{p_i}{\sqrt{w_i}}\right) \qquad (20)$$

for $i = 1, \ldots, 2N$. Since $\partial_t p_{2N+1-i} = -\partial_t p_i$, we can drop the redundant entries and rewrite (20) with $\partial_t p_i = \sqrt{w_i} \cdot \partial_t v_i$ for $i = 1, \ldots, N$ as

$$\partial_t v_i = \sum_{\substack{j=1 \\ j \neq i}}^{N} w_j \cdot \gamma(|\boldsymbol{x}_j - \boldsymbol{x}_i|) \cdot \Phi(v_j - v_i) - \sum_{j=1}^{N} w_j \cdot \gamma(|\boldsymbol{x}_j - \boldsymbol{x}_i|) \cdot \Phi(v_i + v_j). \quad (21)$$

Properties of the Generalised Model. Proceeding similar as in Sect. 2.1 it can be shown that Theorem 1 also holds for the generalised model. Theorem 2 applies for all pairs v_i, v_j with $|\boldsymbol{x}_i - \boldsymbol{x}_j| \leq \varrho$. A minimiser $\boldsymbol{p}^*$ for $E(\boldsymbol{p}, \boldsymbol{x}, \boldsymbol{w})$ depends in general on the definition of Ψ. As evident from Theorem 3 this dependency vanishes in the special case $\gamma = 1$, $w_i = 1$, for $i = 1, \ldots, N$. For nontrivial w_i we assume for the moment that Φ belongs to the class of linear functions, i.e. $\Phi(s) = a \cdot (s - 1)$, $a > 0$ (cf. Fig. 1). For adequate ϱ – implying $\gamma = 1$ for all pairs $(\boldsymbol{x}_i, \boldsymbol{x}_j)$ – our model acts globally and we get

$$p_i^* = \sqrt{w_i} \cdot v_i^* = \sqrt{w_i} \cdot \frac{\sum\limits_{j=1}^{i} w_j - \frac{1}{2} w_i}{\sum\limits_{j=1}^{N} w_j}, \quad i = 1, \ldots, N \qquad (22)$$

as a sufficient condition for the elements of a global minimiser $\boldsymbol{p}^*$ (and $\boldsymbol{v}^*$). A restriction to linear functions Φ also allows to prove convergence in accordance with Theorem 4.

2.3   Relation to Variational Signal and Image Filtering

We will interpret $v_1, \ldots, v_N$ as samples of a smooth 1D signal $u : \Omega \to [0, 1]$ over an interval Ω of the real axis, taken at sampling positions $x_i = x_0 + i h$ with grid mesh size $h > 0$. We consider the model (19) with all w_i fixed to 1.

Theorem 5 (Space-Continuous Energy). *Equation* (19) *with $w_i = 1$ for all i can be considered as a discretisation of*

$$E[u] = \frac{1}{2} \int_{\Omega} \left(W(u_x^2) + B(u)\right) dx \qquad (23)$$

with penaliser $W(u_x^2) \approx C\,\Psi(u_x^2)$ and barrier function $B(u) \approx D\,\Psi(4u^2)$, where C and D are positive constants.

Remark 1. The function W represents a decreasing penaliser convex in u_x, whereas B denotes a convex barrier function that enforces the interval constraint on u by favouring values u away from the interval boundaries. The discrete penaliser Ψ generates both the penaliser W for derivatives and the barrier function B.

Remark 2. Note that by construction of W the diffusivity $g(u_x^2) := W'(u_x^2) \sim \Psi'(u_x^2)$ has a singularity at 0 with $-\infty$ as limit.

Remark 3. The cut-off of γ at radius ϱ implies the locality of the functional (23) that can thereby be linked to a diffusion equation of type (1). Without a cut-off, a nonlocal diffusion equation would arise instead.

Proof (of Theorem 5). We notice first that $v_j - v_i$ and $v_i + v_j$ for $1 \le i, j \le N$ are first-order approximations of $(j - i) h\, u_x(x_i)$ and $2u(x_i)$, respectively.

Derivation of the Penaliser W. Assume first for simplicity that $\Psi(s^2) = -\kappa s$, $\kappa > 0$ is linear in s on $[0, 1]$ (thus not strictly convex). Then we have for a part of the inner sums of (19) corresponding to a fixed i

$$\frac{1}{2}\left(\sum_{j=1}^{N}\gamma(|x_j - x_i|)\Psi\big((v_j - v_i)^2\big) + \sum_{j=N+1}^{2N}\gamma(|x_j - x_{2N+1-i}|)\Psi\big((v_j - v_{2N+1-i})^2\big)\right)$$

$$= \sum_{j=1}^{N}\gamma(|x_j - x_i|)\cdot\Psi\big((|v_j - v_i|)^2\big) \approx -\kappa\, h\, u_x(x_i)\sum_{j=1}^{N}\gamma(|j - i|\, h)\cdot|j - i|$$

$$= h\,\Psi\big(u_x(x_i)^2\big)\sum_{k=1-i}^{N-i}|k|\,\gamma(|k|\, h) \approx h\,\Psi\big(u_x(x_i)^2\big)\lfloor\varrho\rfloor\,(\lfloor\varrho\rfloor + 1) \tag{24}$$

where in the last step the sum over $k = 1 - i, \ldots, N - i$ has been replaced with a sum over $k = -\lfloor\varrho\rfloor, \ldots, \lfloor\varrho\rfloor$, thus introducing a cutoff error for those locations x_i that are within the distance ϱ from the interval ends. Summation over $i = 1, \ldots, N$ approximates $\int_{\Omega}\lfloor\varrho\rfloor(\lfloor\varrho\rfloor + 1)\,\Psi(u_x^2)\,dx$ from which we can read off $W(u_x^2) \approx \lfloor\varrho\rfloor(\lfloor\varrho\rfloor + 1)\Psi(u_x^2)$.

For $\Psi(s^2)$ that are non-linear in s, $\Psi(u_x(x_i)^2)$ in (24) is changed into a weighted sum of $\Psi\big((ku_x(x_i))^2\big)$ for $k = 1, \ldots, N - 1$, which still amounts to a decreasing function $W(u_x^2)$ that is convex in u_x. Qualitatively, W' then behaves the same way as before.

Derivation of the Barrier Function B. Collecting the summands of (19) that were not used in (24), we have, again for fixed i,

$$\frac{1}{2}\left(\sum_{j=N+1}^{2N}\gamma(|x_j - x_i|)\Psi\big((v_j - v_i)^2\big) + \sum_{j=1}^{N}\gamma(|x_j - x_{2N+1-i}|)\Psi\big((v_j - v_{2N+1-i})^2\big)\right)$$

$$= \sum_{j=1}^{N}\gamma(|x_j - x_i|)\Psi\big((v_i + v_j)^2\big) \approx (2\lfloor\varrho\rfloor + 1)\,\Psi\big(4u(x_i)^2\big), \tag{25}$$

and thus after summation over i analogous to the previous step $\int_\Omega B(u)\,\mathrm{d}x$ with $B(u) \approx h^{-1}(2\lfloor\varrho\rfloor + 1)\,\Psi(4u^2)$.   □

Similar derivations can be made for patches of $2D$ images. A point worth noticing is that the barrier function B is bounded. This differs from usual continuous models where such barrier functions tend to infinity at the interval boundaries. However, for each given sampling grid and patch size the barrier function is just strong enough to prevent W from pushing the values out of the interval.

3   Explicit Time Discretisation

Using forward differences to approximate the time derivative in (21) and using $\gamma_{i\ell} := \gamma(|\boldsymbol{x}_\ell - \boldsymbol{x}_i|)$ the explicit scheme of the generalised model reads

$$v_i^{k+1} = v_i^k + \tau\cdot\sum_{\substack{\ell=1\\ \ell\neq i}}^{N} w_\ell\cdot\gamma_{i\ell}\cdot\Phi(v_\ell^k - v_i^k) - \tau\cdot\sum_{\ell=1}^{N} w_\ell\cdot\gamma_{i\ell}\cdot\Phi(v_i^k + v_\ell^k), \quad i = 1,\ldots,N, \quad (26)$$

where τ denotes the time step size and an upper index k refers to the time $k\tau$.

Theorem 6 (Stability Guarantees for the Explicit Scheme). *Let L_Φ be the Lipschitz constant of Φ restricted to the interval $(0,2)$. Moreover, assume that the time step size used in the explicit scheme (26) satisfies $\tau < \left(2L_\Phi\sum_{i=1}^{N} w_i\right)^{-1}$. Then the following stability properties hold:*

(i) If $0 < v_i^k < 1$, then $0 < v_i^{k+1} < 1$, for every $1 \leq i \leq N$.
(ii) If $\gamma = 1$ and $0 < v_i^k < v_j^k < 1$, then $v_i^{k+1} < v_j^{k+1}$.

Proof. (*i*). Let $0 < v_i^k, v_j^k < 1$. We have the following three cases:
If $v_i^k < v_j^k$ then $v_j^k - v_i^k, v_i^k + v_j^k \in (0,2)$. Thus,

$$\left|\Phi(v_i^k + v_j^k) - \Phi(v_j^k - v_i^k)\right| < L_\Phi\cdot 2v_i. \quad (27)$$

If $v_j^k < v_i^k \leq \frac{1}{2}$, then $v_j^k - v_i^k \in (-1,0)$ and $v_j^k + v_i^k \in (0,1)$. Thus,

$$0 \leq \Phi(v_j^k - v_i^k) - \Phi(v_i^k + v_j^k) \quad \text{and} \quad 0 \leq -\Phi(2v_i). \quad (28)$$

Finally, if $v_j^k < v_i^k$ and $\frac{1}{2} < v_i^k$, using $\Phi(1) = 0$ and the periodicity of Φ we get

$$\left|\Phi(v_j^k - v_i^k) - \Phi(v_i^k + v_j^k)\right| = \left|\Phi(v_i^k + v_j^k) - \Phi(2 + v_j^k - v_i^k)\right| < 2v_iL_\Phi \quad \text{and}$$
$$\left|\Phi(2v_i)\right| = \left|\Phi(2v_i) - \Phi(1)\right| \leq |2v_i - 1|\,L_\Phi \leq 2v_iL_\Phi. \quad (29)$$

Combining (27), (28) and (29), we obtain that

$$-\tau\cdot\sum_{\substack{\ell=1\\ \ell\neq i}}^{N} w_\ell\cdot\gamma_{i\ell}\cdot\left(\Phi(v_i^k + v_\ell^k) - \Phi(v_\ell^k - v_i^k)\right) - \tau\cdot w_i\cdot\Phi(2v_i^k)$$

$$< \tau\cdot L_\Phi\cdot 2v_i^k\cdot\sum_{\ell=1}^{N} w_\ell < v_i^k. \quad (30)$$

This, together with (26) shows that $v_i^{k+1} > 0$, as claimed.
The proof of $v_i^{k+1} < 1$ proceeds in a similar way.

(*ii*). Considering the explicit discretisation of (26) for $\partial_t v_i$ and $\partial_t v_j$ and $\gamma = 1$, we obtain for $i, j = 1, \ldots, N$

$$v_j^{k+1} - v_i^{k+1} = v_j^k - v_i^k + \tau \cdot (w_i + w_j) \cdot \Phi(v_i^k - v_j^k) +$$

$$\tau \cdot \sum_{\substack{\ell=1 \\ \ell \neq i,j}}^{N} w_\ell \cdot \left(\Phi(v_\ell^k - v_j^k) - \Phi(v_\ell^k - v_i^k) \right) -$$

$$\tau \cdot \sum_{\ell=1}^{N} w_\ell \cdot \left(\Phi(v_\ell^k + v_j^k) - \Phi(v_\ell^k + v_i^k) \right). \tag{31}$$

Using the fact that Φ is Lipschitz in the interval $(0, 2)$, we also know that

$$\tau \cdot \sum_{\substack{\ell=1 \\ \ell \neq i,j}}^{N} w_\ell \cdot \left| \Phi(v_\ell^k - v_j^k) - \Phi(v_\ell^k - v_i^k) \right| + \tau \cdot \sum_{\ell=1}^{N} w_\ell \cdot \left| \Phi(v_\ell^k + v_j^k) - \Phi(v_\ell^k + v_i^k) \right|$$

$$< \tau \cdot L_\Phi \cdot 2 \left| v_j^k - v_i^k \right| \cdot \sum_{\ell=1}^{N} w_\ell < v_j^k - v_i^k. \tag{32}$$

Finally, since $0 < \Phi(v_i^k - v_j^k)$, (31) and (32) imply that $0 < v_j^{k+1} - v_i^{k+1}$, as claimed. □

4   Applications

Image Enhancement. Similar to the approach proposed in [1] we apply our model to enhance the *global* contrast of digital grey value images $f : \{1, \ldots, n_x\} \times \{1, \ldots, n_y\} \to [0, 1]$. As illustrated in Fig. 2, this can be achieved in two ways. The first option uses the explicit scheme (26) to describe the evolution of grey values v_i up to some time t (see Fig. 2(b) and (c)) where the weights w_i reflect the multiplicity of each grey value v_i. Note that all grey values are mapped to the interval $(0, 1)$ beforehand to ensure the validity of our model and γ is fixed to 1. The amount of contrast enhancement grows with increasing values of t. In our experiments an image size of 683×384 pixels and the application of a flux function Φ with $L_\Phi = 1$ (see Fig. 1) imply an upper bound of $1/(2 \cdot 683 \cdot 384)$ for τ and allow us to achieve the presented results after just one iteration.

If one is only interested in an enhanced version of the original image with maximum global contrast there is an alternative, namely the derived steady state solution for linear flux functions (22). The result is shown in Fig. 2 (d). Figure 2 also confirms that the solution of the explicit scheme (26) converges to the steady state solution (22) for $t \to \infty$. From (22) it is clear that the steady state is equivalent to histogram equalisation. It is therefore interesting to compare our evolution with the histogram modification flow introduced in [6] which can have

(a) Original image. (b) Result for $t = 5 \cdot 10^{-7}$.

(c) Result for $t = 10^{-6}$. (d) Steady state following (22).

Fig. 2. Processing a photography of "Flatowturm (Potsdam)" taken by the authors.

the same steady state. Indeed, the flow from [6] can also be translated into a combination of repulsion among grey-values and a barrier function. However, as in [6] the repulsive force is constant, and the barrier function quadratic, they cannot be derived from the same kind of interaction between the v_i and their reflected counterparts.

Repulsive Swarm Dynamics. Recent swarm models with individual particles often employ a pairwise potential $U : \mathbb{R}^n \to \mathbb{R}$ to model the attraction and repulsion behaviour among swarm mates (see e.g. [2] and the references therein). Physically simplified models describe the particle velocity directly (first order models):

$$\partial_t v_i = -\sum_{\substack{j=1 \\ j \neq i}}^{N} \nabla U\big(|v_i - v_j|\big), \quad i = 1, \ldots, N. \tag{33}$$

These models are often inspired by biology and describe long-term attractive and short-term repulsive behaviour between particles. The interplay of attractive and repulsive forces leads to flocking and allows to gain stability for the swarm. In the following we show that our model can be understood in terms of a purely repulsive first order swarm model. To the best of our knowledge there exists no model so far which restricts itself to pure repulsion among particles.

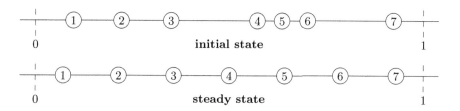

Fig. 3. Application of the model to a system of 7 particles with weights $w_i = 1$.

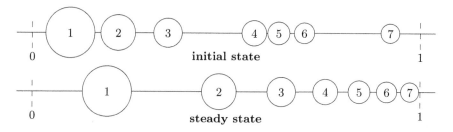

Fig. 4. Application of the model to a system of 7 particles with weights $w_i = 1/i$.

Our Purely Repulsive Swarm Model. Let $\boldsymbol{v} \in (0,2)^{2N}$ denote the extended vector of particle positions and $\boldsymbol{w} \in (0,\infty)^{2N}$ a constant vector containing the corresponding particle weights (extension of both vectors as described in Sects. 2.1 and 2.2). Then the evolution of particles is given by

$$\partial_t v_i = \sum_{\substack{j=1 \\ j \neq i}}^{2N} w_j \cdot \Phi(v_j - v_i) = \sum_{\substack{j=1 \\ j \neq i}}^{2N} w_j \cdot k(v_j - v_i) \cdot (v_j - v_i) , \quad i = 1,\ldots,2N, \quad (34)$$

where $k(s) = \Phi(s)/s$ for $s \neq 0$. The latter kernel function k describes the amount of repulsion between two particles and can also be interpreted in terms of a diffusivity $\Psi'(s^2)$. Comparing (33) and (34) it becomes clear that our model represents a first order swarm model which defines purely repulsive forces among N particles and their N reflections at the domain boundaries. A specific characteristic of our model compared to others is that it describes the repulsive movement of N particles in a closed system.

Results. Our experiments on purely repulsive swarm behaviour presented in Figs. 3 and 4 illustrate the basic properties of our model. We start with a random initial particle distribution for $N = 7$ assuming that $v_i \in (0,1)$. In our first experiment (Fig. 3) the initial weights w_i are set to 1 which results – as described in Sect. 2.1 – in a uniform distribution of the particles in the steady state. In the second experiment (Fig. 4) we assume that the initial weights are given by $w_i = 1/i$, for $i = 1,\ldots,N$. These weights are illustrated by the area of the particles in Fig. 4. We apply the linear flux function Φ from Fig. 1. We notice that in the steady state given by (22), all particles are still in the same order as before. However, the distance between neighbouring particles varies and depends on the particle weights: the larger the weight, the larger the distance.

5   Summary and Conclusions

In our paper we have shown an unexpected result: Pure backward diffusion and fully repulsive swarm behaviour can be modelled as gradient descent of strictly convex energies. Moreover, we have demonstrated that it is neither necessary to impose forward or zero diffusion at extrema nor to add classical fidelity terms: Stability can already be guaranteed by reflecting boundary conditions in the diffusion co-domain or the domain of the positions of the swarm members. This stability carries over directly to a straightforward explicit scheme. No sophisticated numerics is required. A multi-dimensional extension of our model was left out for reasons of clarity and simplicity and is part of our current research.

We expect that our two key ingredients – convex energies combined with range constraints – are not only beneficial for modelling backward diffusion and repulsive swarm dynamics: They may pave the road to a number of new backward models for visual computing applications. This is part of our ongoing research as well.

Acknowledgement. Our research activities have been supported financially by the *Deutsche Forschungsgemeinschaft (DFG)* through a *Gottfried Wilhelm Leibniz Prize* for Joachim Weickert. This is gratefully acknowledged.

References

1. Bergerhoff, L., Weickert, J.: Modelling image processing with discrete first-order swarms. In: Pillay, N., Engelbrecht, A.P., Abraham, A., du Plessis, M.C., Snášel, V., Muda, A.K. (eds.) Advances in Nature and Biologically Inspired Computing. AISC, vol. 419, pp. 261–270. Springer, Cham (2016). https://doi.org/10.1007/978-3-319-27400-3_23
2. Carrillo, J.A., Fornasier, M., Toscani, G., Vecil, F.: Particle, kinetic, and hydrodynamic models of swarming. In: Naldi, G., Pareschi, L., Toscani, G. (eds.) Mathematical Modeling of Collective Behavior in Socio-Economic and Life Sciences, pp. 297–336. Birkhäuser, Boston (2010)
3. Gilboa, G., Sochen, N.A., Zeevi, Y.Y.: Forward-and-backward diffusion processes for adaptive image enhancement and denoising. IEEE Trans. Image Process. **11**(7), 689–703 (2002)
4. Osher, S., Rudin, L.: Shocks and other nonlinear filtering applied to image processing. In: Tescher, A.G. (ed.) Proceedings of SPIE Applications of Digital Image Processing XIV, vol. 1567, pp. 414–431. SPIE Press, Bellingham (1991)
5. Perona, P., Malik, J.: Scale space and edge detection using anisotropic diffusion. IEEE Trans. Pattern Anal. Mach. Intell. **12**, 629–639 (1990)
6. Sapiro, G., Caselles, V.: Histogram modification via differential equations. J. Differ. Eqn. **135**, 238–268 (1997)

7. Sochen, N.A., Zeevi, Y.Y.: Resolution enhancement of colored images by inverse diffusion processes. In: Proceeding of 1998 IEEE International Conference on Acoustics, Speech and Signal Processing, pp. 2853–2856, Seattle, WA (May 1998)
8. Welk, M., Weickert, J.: An efficient and stable two-pixel scheme for 2D forward-and-backward diffusion. In: Lauze, F., Dong, Y., Dahl, A.B. (eds.) SSVM 2017. LNCS, vol. 10302, pp. 94–106. Springer, Cham (2017). https://doi.org/10.1007/978-3-319-58771-4_8

Euler-Lagrange Network Dynamics

Jianjia Wang$^{(\boxtimes)}$, Richard C. Wilson, and Edwin R. Hancock

Department of Computer Science,
University of York, York YO10 5DD, UK
jw1157@york.ac.uk

Abstract. In this paper, we investigate network evolution dynamics using the Euler-Lagrange equation. We use the Euler-Lagrange equation to develop a variational principle based on the von Neumann entropy for time-varying network structure. Commencing from recent work to approximate the von Neumann entropy using simple degree statistics, the changes in entropy between different time epochs are determined by correlations in the degree difference in the edge connections. Our Euler-Lagrange equation minimises the change in entropy and allows to develop a dynamic model to predict the changes of node degree with time. We first explore the effect of network dynamics on three widely studied complex network models, namely (a) Erdős-Rényi random graphs, (b) Watts-Strogatz small-world networks, and (c) Barabási-Albert scale-free networks. Our model effectively captures the structural transitions in the dynamic network models. We also apply our model to a time sequence of networks representing the evolution of stock prices on the New York Stock Exchange (NYSE). Here we use the model to differentiate between periods of stable and unstable stock price trading and to detect periods of anomalous network evolution. Our experiments show that the presented model not only provides an accurate simulation of the degree statistics in time-varying networks but also captures the topological variations taking place when the structure of a network changes violently.

Keywords: Dynamic networks · Approximate von Neumann entropy
Euler-Lagrange equation

1 Introduction

The study of network evolution plays an increasingly crucial role in modelling and predicting the structural variance of the complex networks [1]. Previous studies have addressed this problem from the perspectives of both the local and the global characterization of network structure. At the local level, the aim is to model how the detailed connectivity structure changes with time [2,5]. Specifically, networks grow and evolve with the addition of new components and connections, or the rewiring of connections from one component to another [3,9]. On the other hand, at the global level, the aim is to model the evolution of characteristics which capture the structure of a network and allow different types

© Springer International Publishing AG, part of Springer Nature 2018
M. Pelillo and E. Hancock (Eds.): EMMCVPR 2017, LNCS 10746, pp. 424–438, 2018.
https://doi.org/10.1007/978-3-319-78199-0_28

of network to be distinguished from one to another. Thermodynamic analysis of network structure allows the macroscopic properties of network structure to be described in terms of variables such as temperature, associated with the internal structural [2]. There are also models developed to learn the patterns of network evolution. Examples here include generative and autoregressive models which allow the detailed evolution of edge connectivity structure to be estimated from noisy or uncertain input data [4].

However, both the global and the local methods require to us to develop models that can be fitted to the available data by estimating their parameters, which describe how vertices interact through edges and how this interaction evolves with time. There are few simple and effective methods to predict the network structure. Motivated by the need to fill this gap in the literature and to augment the methods available for understanding the evolution of time-varying networks, we extend the generative model by capturing probabilistically which can be described by various forms of regressive or autoregressive models [4,6]. However, these essentially local models are parameter intensive and a simpler approach is to coach the model in terms of how different node degrees co-occur on the edges connecting them [2,14].

In recent work we have addressed the problem by detailing a generative model of graph-structure [4] and have shown how it can be applied to network time series using an autoregressive model [6,7]. One of the key elements of this model is a means of approximating the von Neumann entropy of both directed and undirected graphs [15]. Von Neumann entropy is the extension of the Shannon entropy defined over the re-scaled eigenvalues of the normalised Laplacian matrix. A quadratic approximation of the von Neumann entropy gives a simple expression for the entropy associated with the degree combinations of nodes forming edges. In accordance with intuition, those edges that connect high degree vertices have the lowest entropy, while those connecting low degree vertices have the highest entropy [2,14]. Making connections between low degree vertices is entropically unfavourable. Moreover, the fitting of the generative model to dynamic network structure involves description length criterion which describes both the likelihood of the goodness of fit to the available network data together with the approximate von Neumann entropy of the fitted network which regulates the complexity of the fitted structure [1,6]. The change in entropy of an edge between different epochs depends on the product of the degree of one vertex and the degree change of the second vertex. In other words, the change in entropy depends on the structure of the degree change correlations.

The aim of this paper is to explore whether our model of network entropy can be extended to model the way in which the node degree distribution evolves with time, taking into account the effect of degree correlations caused by the degree structure of edges. We exploit this property by modelling the evolution of network structure using the Euler-Lagrange equations. Our variational principle is to minimise the changes in entropy during the evolution. Using our approximation of the von Neumann entropy, this leads to update equations for the node degree which include the effects of correlations induced by the edges of the network. It is effectively a type of diffusion process that models how the degree distribution propagates across the network. In fact, it has elements similar to preferential attachment [9], since it favours edges that connect high degree nodes [13,14].

This model extends our earlier work reported in [11]. Here we developed a Euler-Lagrange equation to based on an analysis of small changes in node degree. Specifically, we considered just the first order the correlation $d_u \Delta_v + d_v \Delta_v$, where d_u is the degree of node u at a given time epoch and Δ_u is the change in degree of node u at a later time epoch. In this paper, we provide a complete analysis which includes second order terms and leads to a more complex model which has better predictive capabilities.

The remainder of the paper is organized as follows. In Sect. 2, we provide a detailed analysis of entropic quantity in dynamic networks and develop the degree statistical model by minimising the von Neumann entropy using the Euler-Lagrange equation. In Sect. 3, we conduct the numerical experiments to the synthetic and real-world time-varying networks and apply the resulting characterization of network evolution. Finally, we conclude the paper and make suggestions for future work.

2 Variational Principle on Graphs

2.1 Preliminaries

Let $G(V, E)$ be an undirected graph with node set V and edge set $E \subseteq V \times V$, and let $|V|$ represent the total number of nodes on graph $G(V, E)$. The adjacency matrix A of a graph is defined as

$$A = \begin{cases} 1 & \text{if } (u, v) \in E \\ 0 & \text{otherwise.} \end{cases} \tag{1}$$

Then the degree of node u is $d_u = \sum_{v \in V} A_{uv}$.

The normalized Laplacian matrix $\tilde{L}$ of the graph G is defined as $\tilde{L} = D^{-\frac{1}{2}} L D^{\frac{1}{2}}$, where $L = D - A$ is the Laplacian matrix and D denotes the degree diagonal matrix whose elements are given by $D(u, u) = d_u$ and zeros elsewhere. The element-wise expression of $\tilde{L}$ is

$$\tilde{L}_{uv} = \begin{cases} 1 & \text{if } u = v \text{ and } d_u \neq 0 \\ -\frac{1}{\sqrt{d_u d_v}} & \text{if } u \neq v \text{ and } (u, v) \in E \\ 0 & \text{otherwise.} \end{cases} \tag{2}$$

2.2 Network Entropy

Passerini and Severini [16] exploit the concept of density matrix ρ from quantum mechanics in the network domain. He obtains the density matrix for a network by scaling the combinatorial Laplacian matrix by the reciprocal of the number of nodes in the graph, i.e. $\rho = \frac{\tilde{L}}{|V|}$. Then the von Neumann entropy of the network is defined as the Shannon entropy of the scaled Laplacian eigenvalues $\lambda_1, \ldots, \lambda_V$ and is given by

$$S = -Tr[\rho \log \rho] = -\text{Tr}(\rho \log \rho) = -\sum_{i=1}^{|V|} \frac{\hat{\lambda}_i}{|V|} \log \frac{\hat{\lambda}_i}{|V|} \tag{3}$$

Because of the overheads involved in computing the Laplacian eigensystem, the complexity of the above entropy is cubic in the number of nodes. To overcome this problem Han et al. [15] reduce the cubic complexity by making a second order approximation to the Shannon entropy, and re-expressing it in terms of the traces of the normalised Laplacian and its square. The resulting approximate von Neumann entropy depends on the degree of pairs of nodes forming edges, and is given by

$$S = 1 - \frac{1}{|V|} - \frac{1}{|V|^2} \sum_{(u,v) \in E} \frac{1}{d_u d_v} \tag{4}$$

The approximation of von Neumann entropy has a quadratic complexity which allows it to be used to efficiently compute the entropy of networks. It has been shown to be an effective tool for characterizing structural properties of networks, with extremal values for the cycle and fully connected graphs [4]. Since Eq. (4) contains the term $1\backslash d_u d_v$, it illustrates that the high degree vertices have the lowest entropy, while those connecting low degree vertices have the highest entropy.

Suppose that two undirected graphs $G_t = (V_t, E_t)$ and $G_{t+1} = (V_{t+1}, E_{t+1})$ represent the structure of a time-varying complex network at two consecutive epochs t and $t + 1$ respectively. Then the change of von Neumann entropy between two undirected graphs can be written

$$\Delta S = S(G_{t+1}) - S(G_t) = \frac{1}{|V|^2} \sum_{(u,v) \in E, E'} \frac{d_u \Delta_v + d_v \Delta_u + \Delta_u \Delta_v}{d_u (d_u + \Delta_u) d_v (d_v + \Delta_v)} \tag{5}$$

where G_t and G_{t+1} are two undirected graphs at time points t and $t + 1$. Δ_u is the change of degree for node u, i.e., $\Delta_u = d_u^{t+1} - d_u^t$; Δ_v is similarly defined as the change of degree for node v, i.e., $\Delta_v = d_v^{t+1} - d_v^t$. The entropy change is sensitive to degree correlations for pairs of nodes connected by an edge.

2.3   Euler-Lagrange Equation

We would like to understand the dynamics of a network which evolves so as to minimise this entropy change between different sequential epochs. To do this we cast the evolution process into a variational setting of the Euler-Lagrange equation, and consider the system which optimises the functional

$$\mathcal{E}(q) = \int_{t_1}^{t_2} \mathcal{G}\left[t, q(t), \dot{q}(t)\right] dt \tag{6}$$

where t is time, $q(t)$ is the variable of the system as a function of time, and $\dot{q}(t)$ is the time derivative of $q(t)$. Then, the Euler-Lagrange equation is given by

$$\frac{\partial \mathcal{G}}{\partial q}[t, q(t), \dot{q}(t)] - \frac{d}{dt}\frac{\partial \mathcal{G}}{\partial \dot{q}}[t, q(t), \dot{q}(t)] = 0 \tag{7}$$

Here we consider an evolution which changes just the edge connectivity structure of the vertices and does not change the number of vertices in the graph. As a result, the factors $1 - \frac{1}{|V|}$ and $\frac{1}{|V|^2}$ are constants and do not affect the solution of the Euler-Lagrange equation.

We aim to study evolutions that minimise the entropy associated with the structure of the degree change correlations, i.e. minimise the entropy change between time intervals. Previously, to simplify the calculation, we ignored the product term, i.e. $d_u d_v$, in the denominator [11]. In order to represent the change entropy more accurately, here, we approximate the denominator in Eq. (5) to the quadratic term and apply the Euler-Lagrange equation $\mathcal{G} = \Delta S$ with the entropy change to obtain

$$\mathcal{G}[t, u(t), \Delta_u(t), v(t), \Delta_v(t)] = \frac{d_u \Delta_v + d_v \Delta_u + \Delta_u \Delta_v}{d_u^2 d_v^2} \tag{8}$$

For the vertex indexed u with degree d_u the Euler-Lagrange equation in Eq. (7) gives,

$$\frac{\partial \mathcal{G}}{\partial d_u} - \frac{d}{dt}\frac{\partial \mathcal{G}}{\partial \Delta_u} = 0 \tag{9}$$

First, solving for the partial derivative of the degree d_u, we find

$$\frac{\partial \mathcal{G}}{\partial d_u} = -\frac{d_u \Delta v + 2d_v \Delta u + 2\Delta u \Delta v}{d_u^3 d_v^2} \tag{10}$$

Our earlier model was limited to a first order solution in terms of the degree change in the numerator [11]. The more detailed analysis above not only involves the terms to first order in the node degree change but also those of second order, i.e. degree difference correlations of the form $\Delta u \Delta v$.

Then computing the partial time derivative to the first order degree difference Δ_u, we obtain

$$\frac{\partial \mathcal{G}}{\partial \Delta_u} = \frac{d_v + \Delta v}{d_u^2 d_v^2} \tag{11}$$

Substituting Eqs. (10) and (11) into Eq. (9),

$$\frac{\partial \mathcal{G}}{\partial d_u} - \frac{d}{dt}\frac{\partial \mathcal{G}}{\partial \Delta_u} = \frac{2\Delta^2 u - d_u \dot{\Delta} u}{d_u^3 d_v^2} = 0 \tag{12}$$

The solution for Euler-Lagrange equation in terms of node degree difference is

$$\Delta_u = \left(\frac{d_u}{d_v}\right)^2 \Delta_v + C \tag{13}$$

where C is the constant term coming from the integral of differential equation. This leads to a more detailed degree update equation compared to the previous solution in [11]. It involves a square term of d_u/d_v and plus a constant C. The new solution proves to be more accurate in predicting the degree distribution since it considers the effects of second order terms in the change of von Neumann entropy.

As a result, solving the Euler-Lagrange equation which minimises the change in entropy over time gives a relationship between the degree changes of nodes connected by an edge. Since we are concerned with understanding how network structure changes with time, the solution of the Euler-Lagrange equation provides a way of modelling the effects of these structural changes on the degree distribution across nodes in the network. The update equation for the node degree is at time epochs t and $t + 1$ is

$$d_u^{t+1} = d_u^t + \sum_{v \sim u} \dot{\Delta}_v \Delta_t = d_u^t + \sum_{v \sim u} \left(\frac{\Delta_u}{\Delta_t} \right)_v \Delta_t \qquad (14)$$

In other words by summing over all edges connected to node u, we increment the degree at node u due to changes associated with the degree correlations on the set of connecting edges. We then leverage the solution of the Lagrange equation to simplify the degree update equation, to give the result

$$d_u^{t+1} = d_u^t + \sum_{v \sim u} \left(\frac{d_u}{d_v} \right)^2 \Delta_v + C \qquad (15)$$

which is compared to the previous first-order degree update equation in [11], i.e.

$$d_u^{t+1} = d_u^t + \sum_{v \sim u} \left(\frac{d_u}{2d_v} \right) \Delta_v \qquad (16)$$

This can be viewed as a more accurate type of diffusion process, which updates edge degree so as to satisfy constraints on degree change correlation so as to minimise the entropy change between time epochs. Specifically, the update of degree reflects the effects of correlated degree changes between nodes connected by an edge.

3   Experimental Evaluation

3.1   Data Sets

Synthetic Networks: We generate three kinds of complex network models, namely, (a) Erdős-Rényi random graph model, (b) Watts-Strogatz small-world model [12], and (c) Barabási-Albert scale-free model [9,10]. These are created with the fixed number of vertices with changing the parameters with the network structure evolution. For the Erdős-Rényi random graph, the connection probability is monotonically increasing at the constant rate of 0.005. Similarly, the link rewiring probability in the small-world model [12] increases constantly between 0 to 1 as the network evolution. For the scale-free model [10], one vertex is added to the connection at each time step.

Real-world Networks: We test our method on data provided by the New York Stock Exchange. This dataset consists of the daily prices of 3,799 stocks traded continuously on the New York Stock Exchange over 6000 trading days. The stock prices were obtained from the Yahoo! financial database [17]. A total of 347 stock were selected from this set, for which historical stock prices from January 1986 to February 2011 are available. In our network representation, the nodes correspond to stock and the edges indicate that there is a statistical similarity between the time series associated with the stock closing prices [8]. To establish the edge-structure of the network we use a time window of 20 days is to compute the cross-correlation coefficients between the time-series for each pair of stock.

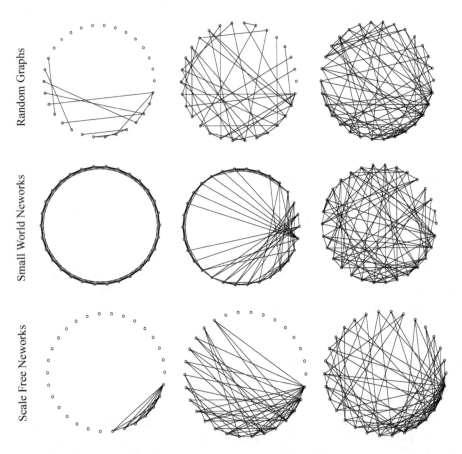

Fig. 1. Visualization of dynamic network structures in time evolution for three network models (Erdős-Rényi random graphs, Watts-Strogatz small-world networks, Barabási-Albert scale-free networks)

Connections are created between a pair of stock if the cross-correlation exceeds an empirically determined threshold. In our experiments, we set the correlation coefficient threshold to the value to $\xi = 0.85$. This yields a time-varying stock market network with a fixed number of 347 nodes and varying edge structure for each of 6,000 trading days. The edges of the network, therefore, represent how the closing prices of the stock follow each other.

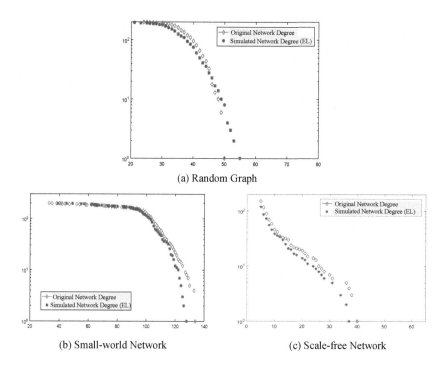

(a) Random Graph

(b) Small-world Network

(c) Scale-free Network

Fig. 2. Degree distribution of original networks and simulated networks for three network models. The red line is for the original observed networks and the blue line is for the results simulated with the second order Euler-Lagrant analysis. (Erdős-Rényi random graphs, Watts-Strogatz small-world networks, Barabási-Albert scale-free networks). We set the same initial state for three network models and update 100 time steps. We compare the degree distribution of final states and the simulations. (Color figure online)

3.2   Synthetic Networks

Using the degree update equation derived from the principle of minimum entropy change and the Euler-Lagrange equation in Eq. 15, we turn our attention to synthetic network data to characterize the structural variance in network models. Figure 1 shows the visualization of the time evolution for three complex networks.

During the evolution process, we fix the number of vertices to be a constant. For random graphs, the networks evolve from an initially sparse set of edges with a low value of the connection probability. As the connection probability increases, the structure of the random graph exhibits a phase transition to a state with a high density of connection and a giant connected component. A phase transition can also be observed for the Watts-Strogatz small-world model, as the rewiring probability evolves with time. Commencing from a regular ring lattice, the network structure evolves to a small-world network with high rewiring probability, and then to an Erdős-Rényi random graph structure with unit rewiring probability. For the scale-free network, the evolution takes place via preferential attachment. The nodes with the highest degree have the largest probability to receive new links. This process produces several high degree nodes or hubs in the network structure.

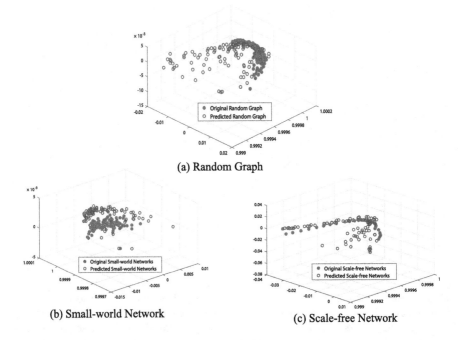

(a) Random Graph

(b) Small-world Network

(c) Scale-free Network

Fig. 3. Visualization of degree distribution in network evolution with principle component analysis (Erdős-Rényi random graphs, Watts-Strogatz small-world networks, Barabási-Albert scale-free networks). (Color figure online)

Now we explore whether the Euler-Lagrange equation can capture structural properties in the time evolution. We use our model to predict the network structure at subsequent time steps and simulate the degree distribution. We then to compare the predicted degree distribution with that from the original time series. Figure 2 shows the simulation results and degree distribution comparisons. The predicted degree distribution resulting from Euler-Lagrange dynamics for the simulated networks fit quite well to the observed distributions. This provides empirical evidence that the Euler-Lagrange equation accurately predicts the short-term evolution of the different network models.

To visualise how the different networks evolve over extended time intervals, we apply the principal component analysis of the degree distribution to project the degree distribution sequences for the networks into a low dimensional space. To commence, we normalize the degree distributions so that the bin contents sum to unity, and then we construct a long vector from the normalised bin contents. We then construct the covariance matrix for the set of long vectors representing the observed degree distributions for the sample of networks. Finally, we apply principal component analysis to the sample covariance matrix for the sample of observed vectorised network degree distributions. We project both the observed and predicted distributions into the principal component space spanned by the leading three eigenvectors of the covariance matrix. In this way, we visualise the evolution of the observed and predicted degree distributions in the principal component space. The results are shown in Fig. 3. The red points are the original network distributions and the blue ones are the predicted ones. Figure 3 clearly shows that for all three network models the predicted network degree distribution evolves in a similar manner to the observed network degree distribution.

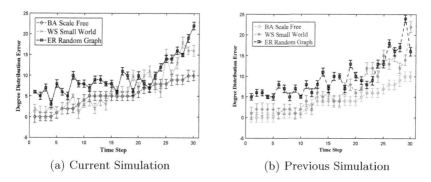

(a) Current Simulation   (b) Previous Simulation

Fig. 4. The degree distribution error with different value of time steps for three network models (Erdős-Rényi random graphs, Watts-Strogatz small-world networks, Barabási-Albert scale-free networks). The degree prediction error in with previous first order simulation (right) increases quickly after time step $\Delta t = 20$ compared to current second order method (left).

Finally, we explore the effect of length of time step on the performance of the degree distribution prediction accuracy. Figure 4(a) shows the degree distribution error with a different value of the time step for the three different network models. The prediction error shown is the standard error over the normalised bin contents (the standard deviation of the difference in observed and predicted bin contents, divided by the square root of the number of bins). The longer the time intervals Δt, the higher the prediction error in the degree distribution. For the random graph, the errors sharply increase around the step $\Delta t = 20$. This is because, during the evolution, the random graph undergoes a phase transition from being sparsely connected to containing a giant connected component. At large time intervals, the predictions fail because of the presence of the giant component.

A similar behaviour can be observed in the sample of small-world networks. As the time step interval increases, there are two instants in time separating three different evolution models. The first event occurs around $\Delta t = 15$ and the second at $\Delta t = 25$. The reason is that, during the evolution, the structure of network changes from a regular lattice at the beginning to a small-world network, and then finally takes on a similar structure to a random graph. These three epochs and the associated of structural transition impact on the performance of degree distribution prediction. Finally, the degree prediction error for the scale-free network grows slowly and smoothly with the time step, since there are no significant structure transitions during the evolution and as a result, the topology of the scale-free network remains stable.

A similar tendency of degree prediction error is observed in Fig. 4(b) which are the result coming from our previous simulation in [11]. The difference between the update equation presented in this paper and our previous one is that as the time step increases, the degree error in the previous model increases rapidly compared to the method described in this paper. Overall, increasing the value of the time interval results in a reduction of the prediction accuracy. Our new model is capable of capturing the local trends arising from the structural changes during the evolution.

3.3 Real-World Networks

Next, we simulate the behaviour of the financial market networks. Here we focus on how the degree distribution evolves with time. We compare the simulated structure and the observed network properties and provide a way to identify the consequence of structural variations in time-evolving networks.

Our procedure is as follows. We select a network at a particular epoch from the time series and simulate its evolution using the degree update equation in Eq. 15. Then we compare the degree distributions for the real network sampled at a subsequent time and the simulation of the degree distribution after an identical elapsed time. One of the most salient events in the NYSE is Black Monday. This event occurred on October 19, 1987, during which the world stock markets crashed, dropping in value in a very short time.

We compare the prediction of consecutive time steps at different epochs, before/after and during the Black Monday crisis. The results are shown in Fig. 5. The most obvious feature is that the degree distribution for the networks before and after Black Monday is quite different to that during the crisis period. During the Black Monday crisis, a large number of vertices in the network is disconnected. This results in a power-law degree distribution. However, for time epochs before and after Black Monday, the disconnected nodes recover their interactions to one another. This increases the number of connections among vertices and causes departures from the power law distribution. This phenomenon is also

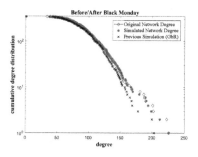
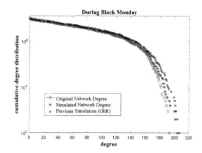

Fig. 5. Degree distribution of original observed networks and simulated networks before/after Black Monday (left) and during Black Monday (right). Around the Black Monday, the network is highly connected with large number of nodes having high degree. During Black Monday, the network is becomes disconnected and most vertices are disjoint, which results in the degree distribution following the power-law.

observed in the networks simulated networks using our degree update equation. This is an important result that shows empirically that the simulated networks reflect the structural properties of the original networks from which they are generated. Moreover, our dynamic model can reproduce the topological changes that occur during the financial crisis.

In Fig. 6, we show network visualizations corresponding to three different instants of time around the Black Monday crisis. In order to compare the simulated network structures resulting from the new second-order model presented in this paper and our earlier first order model, we show the connected components (community structures) at three-time epochs. As the network approaches the crisis, the network structure changes violently, and the community structure substantially vanishes. Only a single highly connected cluster at the centre of the network persists. These features can be observed in both the second and first order simulations of the time evolution of the networks. At the crisis epoch, most stocks are disconnected, meaning that the prices evolve independently without strong correlations to the remaining stock. During the crisis, the persistent connected component exhibits a more homogeneous structure as shown in Fig. 2. Compared to the first order model, our new second order network prediction gives structures that more closely resemble the original network structure. We quantify the prediction accuracy using degree sequence. For the previous model, the degree prediction accuracies are 90.3%, 81.2%, 85.6%, respectively. For the current new model, at three critical time points, these accuracies increase to 92.7%, 85.8% and 86.0%. After the crisis, the network preserves most of its existing community structure and begins to reconnect again. This result also agrees with findings in other literature concerning the structural organization of financial market networks [8].

Finally, we explore the anomaly detection in dynamic networks. We validate our framework by analysing the entropy differences between simulated networks and actual stock market networks on the New York Stock Exchange (NYSE).

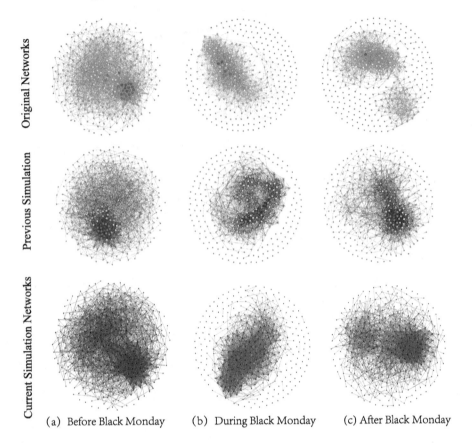

Original Networks

Previous Simulation

Current Simulation Networks

(a) Before Black Monday   (b) During Black Monday   (c) After Black Monday

Fig. 6. The visualization of network structure at three specific days in Black Monday financial crisis. The red line corresponds to the original networks; the blue line represents the first order model simulated networks [11]; and the gray line is the second order Euler-Lagrange model. (Color figure online)

In order to quantitatively investigate the relationship between a financial crisis and network entropy changes, we analyse a set of well-documented crisis periods. These periods are marked alongside the curve of the first order entropy difference in Fig. 7, for all business days in our dataset.

Overall, the most striking observation is that the largest peaks of entropy difference can be used to identify the corresponding financial crisis both in the original and simulated networks. This shows that the entropy difference is sensitive to significant structural changes in networks. The financial crises are characterized by significant entropy changes, whereas outside these critical periods the entropy difference remains stable.

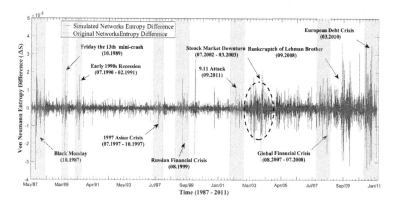

Fig. 7. The von Neumann entropy difference in NYSE (1987–2011) for original financial networks and simulated networks. Critical financial events, i.e., Black Monday, Friday the 13th mini-crash, Early 1990s Recession, 1997 Asian Crisis, 9.11 Attacks, Downturn of 2002–2003, 2007 Financial Crisis, the Bankruptcy of Lehman Brothers and the European Debt Crisis, are associated with large entropy differences.

4   Conclusion

In conclusion, we apply the Euler-Lagrange equation to minimise the change of von Neumann entropy in the network structures. This treatment leads to predict the degree statistics varying with time and captures the effects of degree change correlations introduced by the edge-structure of the network. In other words, because of these correlations, the variety of one degree determines the translation in connected nodes. We conduct numerical experiments for the synthetic and real-world network data in time evolution. Our model is capable of simulating the degree distribution and detecting significant variations in the network structure.

References

1. Wolstenholme, R.J., Walden, A.T.: An efficient approach to graphical modeling of time series. IEEE Trans. Signal Process. **63**, 3266–3276 (2015). ISSN 1053-587X
2. Ye, C., Torsello, A., Wilson, R.C., Hancock, E.R.: Thermodynamics of time evolving networks. In: Liu, C.-L., Luo, B., Kropatsch, W.G., Cheng, J. (eds.) GbRPR 2015. LNCS, vol. 9069, pp. 315–324. Springer, Cham (2015). https://doi.org/10.1007/978-3-319-18224-7_31
3. Ernesto, E., Naomichi, H.: Communicability in complex networks. Phys. Rev. E **77**, 036111 (2008)
4. Han, L., Wilson, R.C., Hancock, E.R.: Generative graph prototypes from information theory. IEEE Trans. Pattern Anal. Mach. Intell. **37**(10), 2013–2027 (2015)
5. Lacasa, L., Luque, B., Ballesteros, F., Luque, J., Nuno, J.C.: From time series to complex networks: the visibility graph. Proc. Nat. Acad. Sci. **105**(13), 4972–4975 (2008)
6. Loukas, A., Simonetto, A., Leus, G.: Distributed autoregressive moving average graph filters. IEEE Signal Process. Lett. **22**(11), 1931–1935 (2015)

7. Ye, C., Wilson, R.C., Hancock, E.R.: Correlation network evolution using mean reversion autoregression. In: Robles-Kelly, A., Loog, M., Biggio, B., Escolano, F., Wilson, R. (eds.) S+SSPR 2016. LNCS, vol. 10029, pp. 163–173. Springer, Cham (2016). https://doi.org/10.1007/978-3-319-49055-7_15

8. Silva, F.N., Comin, C.H., Peron, T.K.D., Rodrigues, F.A., Ye, C., Wilson, R.C., Hancock, E., Costa, L.F.: Modular dynamics of financial market networks, pp. 1–13 (2015)

9. Barabási, A.L., Albert, R.: Emergence of scaling in random networks. Science **286**(5439), 509–512 (1999)

10. Barabási, A.L., Albert, R., Jeong, H.: Mean-field theory for scale free random networks. Phys. A **272**, 173–187 (1999)

11. Wang, J., Wilson, R.C., Hancock, E.R.: Minimising entropy changes in dynamic network evolution. In: Foggia, P., Liu, C.-L., Vento, M. (eds.) GbRPR 2017. LNCS, vol. 10310, pp. 255–265. Springer, Cham (2017). https://doi.org/10.1007/978-3-319-58961-9_23

12. Watts, D., Strogatz, S.: Collective dynamics of 'small world' networks. Nature **393**, 440–442 (1998)

13. Wang, J., Wilson, R., Hancock, E.R.: Network entropy analysis using the Maxwell-Boltzmann partition function. In: The 23rd International Conference on Pattern Recognition (ICPR), pp. 1–6 (2016)

14. Wang, J., Wilson, R.C., Hancock, E.R.: Spin statistics, partition functions and network entropy. J. Complex Netw. **5**, 1–25 (2017). https://doi.org/10.1093/comnet/cnx017

15. Han, L., Escolano, F., Hancock, E.R., Wilson, R.C.: Graph characterizations from von Neumann entropy. Pattern Recognit. Lett. **33**, 1958–1967 (2012)

16. Passerini, F., Severini, S.: The von Neumann entropy of networks. Int. J. Agent Technol. Syst. **1**, 58–67 (2008)

17. Yahoo! Finance. http://finance.yahoo.com

Slack and Margin Rescaling as Convex Extensions of Supermodular Functions

Matthew B. Blaschko$^{(\boxtimes)}$ (iD)

Center for Processing Speech and Images, Departement Elektrotechniek,
KU Leuven, Leuven, Belgium
matthew.blaschko@esat.kuleuven.be

Abstract. Slack and margin rescaling are variants of the structured output SVM, which is frequently applied to problems in computer vision such as image segmentation, object localization, and learning parts based object models. They define convex surrogates to task specific loss functions, which, when specialized to non-additive loss functions for multi-label problems, yield extensions to increasing set functions. We demonstrate in this paper that we may use these concepts to define polynomial time convex extensions of arbitrary supermodular functions, providing an analysis framework for the tightness of these surrogates. This analysis framework shows that, while neither margin nor slack rescaling dominate the other, known bounds on supermodular functions can be used to derive extensions that dominate both of these, indicating possible directions for defining novel structured output prediction surrogates. In addition to the analysis of structured prediction loss functions, these results imply an approach to supermodular minimization in which margin rescaling is combined with non-polynomial time convex extensions to compute a sequence of LP relaxations reminiscent of a cutting plane method. This approach is applied to the problem of selecting representative exemplars from a set of images, validating our theoretical contributions.

1   Introduction

Structured output support vector machines [17] are commonly applied to a range of structured prediction problems in computer vision. A key open question is the tightness of the loss surrogate, with negative results from statistical learning theory indicating that neither slack rescaling nor margin rescaling variants lead to statistical consistency [14]. Nevertheless, due to their good empirical performance, they are frequently applied in practice, indicating the importance of the analysis of these surrogates.

In this work, we explore strategies analogous to margin and slack rescaling in structured output support vector machines [17] for set function minimization in general, and substantially advance the theory of supermodular set function minimization as well as provide insights into possible improvements to structured output support vector machines. In particular, we advance the theory of convex extensions of supermodular extensions by showing an explicit form for

© Springer International Publishing AG, part of Springer Nature 2018
M. Pelillo and E. Hancock (Eds.): EMMCVPR 2017, LNCS 10746, pp. 439–454, 2018.
https://doi.org/10.1007/978-3-319-78199-0_29

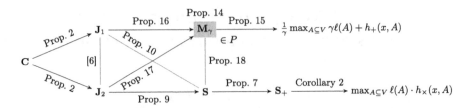

Fig. 1. A summary of the main theoretical contributions in this paper. **C** denotes the convex closure, $\mathbf{J}_1$ and $\mathbf{J}_2$ are extensions based on known modular upper bounds to supermodular functions [6,9], and **M** and **S** indicate variants of extensions derived from margin rescaling and slack rescaling, directed edges indicate that the parent dominates the child, and red undirected edges indicate that there is no ordering between a pair of nodes. $\mathbf{M}_\gamma$ is the only convex extension whose computation always corresponds to a supermodular maximization and is therefore known to be polynomial time computable. (Color figure online)

the closest convex extension to the convex closure within a specific multiplicative family (Proposition 6), and an additive family (Proposition 15). We prove that extension operators derived from modular upper bounds [6,9] on submodular functions dominate both slack and margin rescaling, but computation of these tighter convex extensions do not correspond with supermodular maximization problems. This in general suggests a trade-off between the tightness of the extension and the tractability of its computation [7], with only margin rescaling demonstrated to correspond to submodular minimization. We develop a minimization strategy where the computation of margin rescaling is used to optimize other surrogates for which optimization is non-submodular. We summarize key theoretical contributions in Fig. 1 through a partial ordering over convex extensions of set functions considered in this paper using the notion of restricted operator inequality (Definition 5). If a fixed finite set of convex extensions does not have a total ordering, the pointwise maximum of these extensions is a convex extension that dominates them all.

Supermodular minimization (submodular maximization) has been widely studied both in the context of relaxations as well as for fully combinatorial approximation algorithms [11]. Notable contributions include a 1/2-approximation algorithm for unconstrained non-negative submodular maximization [1] and the multi-linear extension [18,19], which, although non-convex, has been used in an approximate minimization framework. A very general treatment of the development of concave extensions of submodular functions viewed through superdifferentials of submodular functions has been developed in [7]. A large amount of work has been done on relaxations approximate energy minimization in the context of random field models with low maximal clique size [12], including roof duality [10], or by making additional assumptions on the structure of the graph such as balancedness [20]. The connection between the structured output SVM [17] and convex extensions of set functions seems to be due to [21]. The comparative difficulty of optimizing slack rescaling vs. margin rescaling is known, and an interesting approach to the optimization of slack rescaling via multiple margin rescaling operations can be found in [2]. In the sequel, we first develop basic definitions and results for supermodular set functions and LP relaxations (Sect. 2), we then provide the main theoretical analysis in Sect. 3 before demonstrating empirical results consistent with the theory (Sect. 4).

2 Mathematical Preliminaries

Definition 1 (Set function). *A set function ℓ is a mapping from the power set of a base set V to the reals:*

$$\ell : \mathcal{P}(V) \to \mathbb{R}. \tag{1}$$

We will assume wlog that $\ell(\emptyset) = 0$ for all set functions in the sequel.

Definition 2 (Extension). *An extension of a set function is an operator that yields a continuous function over the p-dimensional unit cube (where $p = |V|$) such that the function value on each vertex x of the unit cube is equal to $\ell(A)$ for A such that $x_i = 1 \iff i \in A$.*

Definition 3 (Submodular set function). *A set function f is said to be submodular if for all $A \subseteq B \subset V$ and $x \in V \setminus B$,*

$$f(A \cup \{x\}) - f(A) \geq f(B \cup \{x\}) - f(B). \tag{2}$$

A set function is said to be supermodular if its negative is submodular, and a function is said to be modular if it is both submodular and supermodular. We denote the set of all submodular functions $\mathcal{S}$, the set of all supermodular functions $\mathcal{G} := \{g | -g \in \mathcal{S}\}$, and $\mathcal{G}_+ := \{g | g(S) \geq 0, \forall S \subseteq V\} \cap \mathcal{G}$ denotes the set of all non-negative supermodular functions. Additional properties of submodular functions can be found in [5].

Lemma 1. *Any $g \in \mathcal{G}$ such that $g(\{x\}) \geq 0, \forall x \in V$ is an increasing function, i.e. $g(A \cup \{x\}) \geq g(A), \forall A \subset V, x \in V \setminus A$.*

Proof. We have that $g(\emptyset) = 0$ and $g(\{x\}) \geq 0, \forall x \in V$ by assumption. From the supermodularity of g, $\forall A \subset V, x \in V \setminus A$,

$$g(A) + \overbrace{g(\{x\})}^{\geq 0} \leq g(A \cup \{x\}) + \overbrace{g(A \cap \{x\})}^{=0} \implies g(A) \leq g(A \cup \{x\}) \tag{3}$$

$\square$

Corollary 1. *Any $g \in \mathcal{G}_+$ is an increasing function.*

Definition 4 (Convex closure). *The convex closure of a set function $\mathbf{C}\ell$ is defined for all $x \in [0,1]^{|V|}$ as*

$$\mathbf{C}\ell(x) := \min_{\alpha \in \mathbb{R}^{2^{|V|}}} \sum_{A \subseteq V} \alpha_A \ell(A), \text{ s.t. } \sum_{A \subseteq V} \alpha_A \mathbf{1}_A = x, \sum_{A \subseteq V} \alpha_A = 1, \alpha_A \geq 0 \; \forall A \subseteq V, \tag{4}$$

where $\mathbf{1}_A$ denotes the binary vector of length $|V|$ whose entries indexed by the elements in A are 1.

In general, the convex closure is not polynomial time solvable, with the notable exception of the Lovász extension for submodular functions [5, 13].

Definition 5 (Restricted operator inequality). *For two convex extensions* **A** *and* **B***, we write* **A** < **B** *(analogously* **A** ≤ **B***) if* $\mathbf{A}\ell(x) < \mathbf{B}\ell(x) \, \forall \ell, x \in [0,1]^p$. *We may subscript the inequality sign with a class of set functions (e.g.* **B** $\leq_S$ **L***) if the inequality holds for all set functions in that class.*

Proposition 1 (Transitivity of restricted operator inequality). *For two function sets* $\mathcal{A}$ *and* $\mathcal{B}$ *and three operators* **F***,* **G***, and* **H***,*

$$\mathbf{F} \leq_\mathcal{A} \mathbf{G} \leq_\mathcal{B} \mathbf{H} \implies \mathbf{F} \leq_{\mathcal{A} \cap \mathcal{B}} \mathbf{H}. \tag{5}$$

Proposition 2. B ≤ **C** *for all convex extensions* **B***.*

Proof. Assume that there exists some convex extension $\mathbf{B}\ell$ such that $\mathbf{B}\ell(x) > \mathbf{C}\ell(x)$ for some x in the p-dimensional unit cube. If we map all $A \subseteq V$ to vertices of the p-dimensional unit cube, and identify ℓ with 2^p distinct points in $\{0,1\}^p \times \mathbb{R}$, then the convex closure is the lower hull of these points taken with respect to the dimension corresponding to the function value. As each of the 2^p points corresponding to the function are vertices of this lower hull, any $\mathbf{B}\ell$ that has a point greater than a corresponding point of $\mathbf{C}\ell$ is not a lower hull and cannot contain all of these vertices, which contradicts the definition of a set function extension. □

Definition 6 ([6,9]). *The following two modular functions upper bound any submodular function,* ℓ*:*

$$m_{V,A}\ell(B) = \ell(A) - \sum_{j \in A \setminus B} (\ell(V) - \ell(V \setminus \{j\})) + \sum_{j \in B \setminus A} (\ell(A \cup \{j\}) - \ell(A)) \tag{6}$$

$$m_{\emptyset,A}\ell(B) = \ell(A) - \sum_{j \in A \setminus B} (\ell(A) - \ell(A \setminus \{j\})) + \sum_{j \in B \setminus A} (\ell(\{j\}) - \underbrace{\ell(\emptyset)}_{=0}). \tag{7}$$

It is straightforward to show that these modular functions become upper bounds if ℓ is supermodular by the fact that $m_{\emptyset,A}\ell = -m_{\emptyset,A}(-\ell)$, and $m_{V,A}\ell = -m_{V,A}(-\ell)$. It is also straightforward to check that $m_{\emptyset,A}\ell(A) = m_{V,A}\ell(A) = \ell(A)$ and that the supremum with respect to A of these lower bounds therefore recovers the original function.

We now use Definition 6 to construct convex extensions of supermodular functions by (i) constructing the convex closures of the modular upper bounds $\mathbf{C}m_{\emptyset,A}\ell(\cdot)$ and $\mathbf{C}m_{V,A}\ell(\cdot)$, which are linear functions, and (ii) taking the pointwise maximum over A. The resulting extensions are given in the following definition:

Definition 7. *The following two operators yield convex extensions for $g \in \mathcal{G}$, but the arg max does not correspond to supermodular maximization in general:*

$$\mathbf{J}_1 g(x) = \arg\max_{A \subseteq V} g(A) + \sum_{i \in V \setminus A} x_i \left(g(A \cup \{i\}) - g(A)\right) \tag{8}$$
$$- \sum_{i \in A}(1 - x_i)\left(g(V) - g(V \setminus \{i\})\right),$$

$$\mathbf{J}_2 g(x) = \arg\max_{A \subseteq V} g(A) + \sum_{i \in V \setminus A} x_i g(\{i\}) - \sum_{i \in A}(1 - x_i)\left(g(A) - g(A \setminus \{i\})\right). \tag{9}$$

2.1 LP Relaxations

One of the key applications of convex set function extensions is their application to (approximate) minimization of set functions through LP relaxations. If a relaxation is a convex extension, we have the useful result that an integral minimizer of the relaxation is guaranteed to be an optimal solution to the orignal discrete optimization problem. In Proposition 3 we formalize the unsurprising but important result that dominating convex extensions result in integral solutions to LP relaxations with higher probability. For some set $S \subseteq \mathbb{R}^d$, we denote $\text{int}(S) := [S \cap \{0,1\}^d \neq \emptyset]$, where we have used Iverson bracket notation.

Proposition 3. *For two operators that yield convex extensions for all set functions $\ell \in \mathcal{A}$, $\mathbf{F} \leq_{\mathcal{A}} \mathbf{G} \leq_{\mathcal{A}} \mathbf{C}$, and for all distributions $p : \mathcal{A} \to \mathbb{R}_+$,*

$$\mathbb{E}_{\ell \sim p}\left\{\text{int}\left(\arg\min_{x \in [0,1]^{|V|}} \mathbf{F}\ell(x)\right)\right\} \leq \mathbb{E}_{\ell \sim p}\left\{\text{int}\left(\arg\min_{x \in [0,1]^{|V|}} \mathbf{G}\ell(x)\right)\right\}, \tag{10}$$

where arg min is defined as a map to the set of minimizers of an expression.

Proof. It is sufficient to demonstrate that

$$\text{int}\left(\arg\min_{x \in [0,1]^{|V|}} \mathbf{F}\ell(x)\right) \Longleftarrow \text{int}\left(\arg\min_{x \in [0,1]^{|V|}} \mathbf{G}\ell(x)\right). \tag{11}$$

Denote x^* an integral optimum of $\arg\min_{x \in [0,1]^{|V|}} \mathbf{C}\ell(x)$, As $\mathbf{C}\ell(x)$ is total dual integral [4, Sect. 4.6], we have that $\mathbf{C}\ell(x^*) = \min_{A \subseteq V} \ell(A)$. If $\ell(x^*) = \mathbf{G}\ell(x^*) \neq \min_{x \in [0,1]^{|V|}} \mathbf{G}\ell(x)$, there must be a negative directional gradient $\nabla_v \mathbf{G}\ell(x^*)$ for some $v \in \mathbb{R}^{|V|}$ pointing into the unit cube from x^*. From the definition of set function extensions and the fact that $\mathbf{F} \leq_{\mathcal{A}} \mathbf{G}$, we have that $0 > \nabla_v \mathbf{G}\ell(x^*) \geq \nabla_v \mathbf{F}\ell(x^*)$ and therefore $x^* \notin \arg\min_{x \in [0,1]^{|V|}} \mathbf{F}\ell(x)$. $\square$

3 Slack and Margin Rescaling

In these sections we develop two convex extensions of increasing set functions that are based on the surrogate loss functions of two variants of the structured output support vector machine [17]. Their relationship to convex extensions of set functions is due to [21], which developed results analogous to Propositions 5, and 11.

3.1   Slack Rescaling

Definition 8 (Slack rescaling for increasing set functions [17,21,22]). *For an increasing set function ℓ,*

$$\mathbf{S}_+\ell(x) = \max_{A \subseteq V} \ell(A) \left(1 - |A| + \sum_{i \in A} x_i \right). \tag{12}$$

Proposition 4. *Computation of $\mathbf{S}_+\ell(x)$ corresponds to a non-supermodular maximization problem for $x \in [0,1]^{|V|}$ and supermodular increasing ℓ.*

Proof. To show that the optimization is not in general a supermodular maximization, we may consider the following counterexample: $V = \{a, b\}$, $\ell(\emptyset) = \ell(\{a\}) = \ell(\{b\}) = 0$, $\ell(\{a, b\}) = \varepsilon$. For $\varepsilon > 0$ this is strictly supermodular. Denote

$$\tilde{\ell}(A) := \ell(A) \left(1 - |A| + \sum_{i \in A} x_i \right) \tag{13}$$

For $x = \mathbf{0}$, we have that $\tilde{\ell}(\emptyset) = \tilde{\ell}(\{a\}) = \tilde{\ell}(\{b\}) = 0$ and $\tilde{\ell}(\{a, b\}) = \varepsilon(1 - 2 + 0) = -\varepsilon$, but this indicates that $\ell = -\tilde{\ell}$ and as ℓ is not modular they cannot both be supermodular.   □

Proposition 5 (Proposition 1 of [21]). *Definition 8 yields a convex extension for all increasing ℓ.*

Proof. Equation (12) is convex: The r.h.s. is a maximum over linear functions of x, and is therefore convex in x.

Equation (12) is an extension: We will use the notation set : $\{0,1\}^{|V|} \to \mathcal{P}(V)$ to denote the set associated with an indicator vector. We must have that $\mathbf{S}_+\ell(x) = \ell(A)$ whenever $x \in \{0,1\}^{|V|}$ and $\mathrm{set}(x) = A$. For $\mathrm{set}(x) = A$

$$\ell(A)\Big(1 - |A| + \overbrace{\sum_{i \in A} x_i}^{=|A|} \Big) = \ell(A). \tag{14}$$

We now must show that when $\mathrm{set}(x) = A$

$$\ell(A) \geq \ell(B) \left(1 - |B| + \sum_{i \in B} x_i \right) = \ell(B)\,(1 - |B| + |A \cap B|). \tag{15}$$

For any $B \not\subseteq A$: $\overbrace{\ell(B)}^{\geq 0}\,\overbrace{(1 - |B| + |A \cap B|)}^{\leq 0} \leq 0 \leq \ell(A)$, and for any $B \subseteq A$: $\overbrace{\ell(B)}^{\leq \ell(A)}\,\overbrace{(1 - |B| + |A \cap B|)}^{=1} \leq \ell(A)$.   □

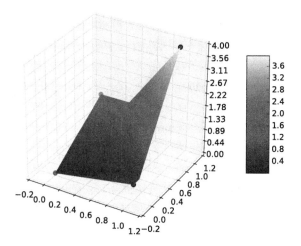

Fig. 2. An example of the convex extension achieved with Definition 8 applied to the set function $\ell(\emptyset) = 0$, $\ell(\{a\}) = 0.5$, $\ell(\{b\}) = 1.5$, $\ell(\{a,b\}) = 4$.

An example of the extension is shown in Fig. 2.

We can see that Definition 8 is an instance of a general strategy for multiplicative extensions of increasing functions:

$$\mathbf{B}\ell(x) = \max_{A \subseteq V} \ell(A) \cdot h_\times(x, A) \tag{16}$$

where $h_\times : [0,1]^{|V|} \times \mathcal{P}(V) \to \mathbb{R}$ is a function convex in x. We require $h_\times(x, A) \le \frac{\ell(\mathrm{set}(x))}{\ell(A)}$ for integral x and all A in order to ensure that $\ell(A)h_\times(x, A) \le \ell(\mathrm{set}(x))$, which is the general form of the inequality in Eq. (15). This strategy is quite general if $h_\times$ can have a dependency on ℓ as the convex closure is recovered simply by setting $h_\times(x, A) = \frac{\mathbf{C}\ell(x)}{\ell(A)}$; In Proposition 6 we will limit ourselves to $h_\times$ that have no explicit dependency on ℓ beyond that it is an increasing function. For increasing ℓ, $\frac{\ell(\mathrm{set}(x))}{\ell(A)}$ is bounded below by 1 for all $A \subseteq \mathrm{set}(x)$ and is bounded below by 0 for all $A \not\subseteq \mathrm{set}(x)$. If we additionally assume that $\ell \in \mathcal{G}_+$, these bounds remain unchanged. If we take these bounds with respect to all increasing ℓ, or indeed all $g \in \mathcal{G}_+$, the bounds are sharp.

Proposition 6. $h_\times(x, A) = \left(1 - |A| + \sum_{i \in A} x_i\right)_+$ *is the largest possible convex function over the unit cube satisfying the constraints that for integral x, $h_\times(x, A) \le 1 \; \forall A \subseteq \mathrm{set}(x)$, $h_\times(x, A) \le 0 \; \forall A \not\subseteq \mathrm{set}(x)$, and $h_\times(x, \mathrm{set}(x)) = 1$, where $(\cdot)_+ = \max(0, \cdot)$.*

Proof. Each constraint is an equality constraint or an upper bound, and the values of the function are constrained only at the vertices of the unit cube, so the maximum convex function satisfying the bounds must be the lower hull of the constraint points, that is $h_\times(x, A) = \mathbf{C}[A \subseteq \cdot](x)$, where we have employed

Iverson bracket notation in our definition of the function to which we apply $\mathbf{C}$:

$$\mathbf{C}[A \subseteq \cdot](x) = \min_{\alpha \in \mathbb{R}^{2^{|V|}}} \sum_{B \subseteq V} \alpha_B [A \subseteq B] = \min_{\alpha \in \mathbb{R}^{2^{|V|}}} \sum_{B \supseteq A} \alpha_B \tag{17}$$

$$\text{s.t.} \sum_{B \subseteq V} \alpha_B \mathbf{1}_B = x, \sum_{B \subseteq V} \alpha_B = 1, \alpha_B \geq 0 \; \forall B. \tag{18}$$

For all $|V| \in \mathbb{Z}_+$ and $A \subseteq V$, we have that

$$\mathbf{C}[A \subseteq \cdot](x) = 0, \; \forall x \in \text{conv}\left(\{y \in \{0,1\}^{|V|} | \, \text{set}(y) \not\supseteq A\}\right). \tag{19}$$

For $x \in \text{conv}\left(\{y \in \{0,1\}^{|V|} | \, \text{set}(y) \supseteq A \vee \exists i \in V, \text{set}(y) \cup \{i\} \supseteq A\}\right)$, we must have a linear function that interpolates between 0 at all integral points x such that $\text{set}(x) \not\supseteq A$ and 1 at points such that $\text{set}(x) \supseteq A$. This linear function is exactly $1 - |A| + \sum_{i \in A} x_i$. $\qquad\square$

In practice, we do not need to threshold $h_\times$ at zero (i.e. we do not need to apply the $(\cdot)_+$ operation above) as this is redundant with the maximization in Eq. (16) being achieved by $A = \emptyset$.

Remark 1. The constraints in Proposition 6 are necessary and sufficient conditions on h for this multiplicative family to yield a convex extension for all increasing ℓ. They are also necessary and sufficient to yield a convex extension for all $g \in \mathcal{G}_+$.

Corollary 2 ($\mathbf{S}_+$ dominates the multiplicative family of extensions). *For all $\mathbf{B}$ that can be written as in Eq. (16) and h having no oracle access to ℓ,*

$$\mathbf{B} \leq_+ \mathbf{S}_+. \tag{20}$$

Definition 9 ($\mathrm{m}\,g$). *For a set function g, we may construct a modular function m such that*

$$m(\{x\}) = g(\{x\}), \; \forall x \in V, \tag{21}$$

and denote this function $\mathrm{m}\,g$.

For non-increasing supermodular g, we may extend the definition of slack rescaling as follows:

Definition 10 (Slack rescaling for all supermodular g). *Assume $g \in \mathcal{G}$. For a modular function m, we will denote $\text{vec}(m) \in \mathbb{R}^{|V|}$ the vectorization of m, and we define*

$$\mathbf{S}g(x) = \langle \text{vec}(\mathrm{m}\,g), x \rangle + \max_{A \subseteq V}(g(A) - \mathrm{m}\,g(A))\left(1 - |A| + \sum_{i \in A} x_i\right). \tag{22}$$

Examples of the convex extension achieved by Definition 10 are given in Fig. 3. We now show that not only does $\mathbf{S}$ yield an extension for non-increasing supermodular functions, it yields a tighter extension for increasing supermodular functions:

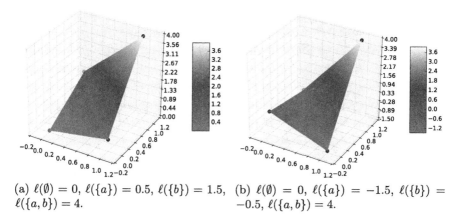

(a) $\ell(\emptyset) = 0$, $\ell(\{a\}) = 0.5$, $\ell(\{b\}) = 1.5$, (b) $\ell(\emptyset) = 0$, $\ell(\{a\}) = -1.5$, $\ell(\{b\}) = $
$\ell(\{a, b\}) = 4$. -0.5, $\ell(\{a, b\}) = 4$.

Fig. 3. An example of the convex extension achieved with Definition 10 applied to positive and non-positive supermodular functions. In this special case where $|V| = 2$, the extension achieves the convex closure.

Proposition 7. $\mathbf{S}_+ \leq_{\mathcal{G}_+} \mathbf{S}$.

Proof. For $g \in \mathcal{G}_+$,

$$\mathbf{S}g(x) - \mathbf{S}_+g(x) = \max_{A \subseteq V} \left(\langle \text{vec}(\text{m} \, g), x \rangle + (g(A) - \text{m} \, g(A)) \left(1 - |A| + \sum_{i \in A} x_i \right) \right) \tag{23}$$

$$- \max_{A \subseteq V} g(A) \left(1 - |A| + \sum_{i \in A} x_i \right).$$

To show that this difference is greater than zero, we will demonstrate that the difference of each element indexed by $A \subseteq V$ is greater than zero.

$$\langle \text{vec}(\text{m} \, g), x \rangle + (g(A) - \text{m} \, g(A)) \left(1 - |A| + \sum_{i \in A} x_i \right) - g(A) \left(1 - |A| + \sum_{i \in A} x_i \right)$$

$$= \sum_{i \in V} g(\{i\}) x_i - \sum_{i \in A} g(\{i\}) + \sum_{i \in A} (1 - x_i) g(\{i\}) \tag{24}$$

$$+ \overbrace{\sum_{i \in A} \sum_{j \in A} g(\{i\})(1 - x_j) - \sum_{i \in A} (1 - x_i) g(\{i\})}^{= \xi \geq 0} = \sum_{i \in V \setminus A} g(\{i\}) x_i + \xi \geq 0 \tag{25}$$

$\square$

Proposition 8. *Slack rescaling is invariant to scaling of g:*

$$\mathbf{S}g = \gamma^{-1}\mathbf{S}(\gamma g), \quad \forall \gamma > 0. \tag{26}$$

Proof

$$\gamma^{-1}\mathbf{S}(\gamma g)(x) = \frac{1}{\gamma}\left(\langle \mathrm{vec}(\mathrm{m}\,\gamma g), x\rangle + \max_{A \subseteq V}(\gamma g(A) - \mathrm{m}\,\gamma g(A))\left(1 - \sum_{i \in A}(1 - x_i)\right)\right) \tag{27}$$

$$= \mathbf{S}g(x) \tag{28}$$

$\square$

Proposition 9. $\mathbf{J}_2 \geq_{\mathcal{G}} \mathbf{S}$.

Proof. We first add a modular function to $g \in \mathcal{G}$ such that it is increasing. The difference between each of the corresponding arg max elements in the definition of $\mathbf{J}_2$ and $\mathbf{S}$, respectively, is $\sum_{i \in V \setminus A} x g(\{i\}) + \sum_{i \in A}(1 - x_i)g(A \setminus \{i\})$. Each of these terms is non-negative by assumption. $\square$

Proposition 10. $\mathbf{J}_1 \not\geq_{\mathcal{G}} \mathbf{S}$, $\mathbf{S} \not\geq_{\mathcal{G}} \mathbf{J}_1$.

Proof. For g increasing supermodular with $g(\{i\}) = 0 \;\forall i \in V$, the difference between each of the corresponding arg max elements in the definition of $\mathbf{S}$ and $\mathbf{J}_1$, respectively, is $\sum_{i \in A}(1 - x_i)(g(V) - g(V \setminus \{i\}) - g(A)) - \sum_{j \in V \setminus A} x_j(g(A \cup \{j\}) - g(A))$.

$\mathbf{S} \not\geq_{\mathcal{G}} \mathbf{J}_1$: Each of the terms in the second sum is non-negative, meaning that the second term is non-positive. Setting $x_i = 1$ for all $i \in A$, $x_j > 0$ for all $j \in V \setminus A$ gives the result.

$\mathbf{J}_1 \not\geq_{\mathcal{G}} \mathbf{S}$: Set $x_j = 0$ for all $j \in V \setminus A$. For the difference to be positive, we need to find an increasing $g \in \mathcal{G}$ such that $g(V) > g(A) + g(V \setminus \{i\})$ for some $A \subset V$ and $i \in V$. Consider the function $g(A) = 0, \;\forall A \subset V, g(V) = 1$. $\square$

3.2   Margin Rescaling

We may similarly define the following convex extension based on margin rescaling:

Definition 11 (Margin rescaling for supermodular g [17,21]).

$$\mathbf{M}g(x) := \langle \mathrm{vec}(m), x\rangle + \max_{A \subseteq V} g(A) - m(A) - |A| + \sum_{i \in A} x_i. \tag{29}$$

It has been demonstrated that up to a strictly positive scale factor, margin rescaling yields an extension in the convex surrogate loss setting [21, Proposition 2]. We prove this next and in doing so show necessary and sufficient conditions on the scaling of the loss function for $\mathbf{M}$ to yield an extension.

Proposition 11 (Proposition 2 of [21]). *For every increasing function ℓ, there exists a scalar $\gamma > 0$ such that $\gamma^{-1}\mathbf{M}(\gamma\ell)$ is an extension of ℓ. We will denote $\mathbf{M}_\gamma := \gamma^{-1}\mathbf{M}(\gamma \cdot)$ for short.*

Proof. Wlog, we may assume that $\ell(\{i\}) = 0$, $\forall i \in V$. For $\mathbf{M}_\gamma$ to yield an extension, we need that for all $A, B \subseteq V$, $\text{set}(x) = A$,

$$\gamma g(A) \geq \gamma g(B) - |B| + \sum_{i \in B} x_i = \gamma g(B) - |B| + |A \cap B|. \tag{30}$$

$$\gamma(g(A) - g(B)) + |B| - |A \cap B| \geq 0 \tag{31}$$

$$\Longleftrightarrow \gamma \leq \frac{|B| - |A \cap B|}{g(B) - g(A)} \ \forall (A, B) \text{ s.t. } g(A) < g(B) \tag{32}$$

where the restriction to $g(A) < g(B)$ is due to the fact that $g(B) < g(A)$ reverses the direction of the inequality as we multiply both sides by $\frac{1}{g(B)-g(A)}$. As g is increasing, we cannot simultaneously have that $|A \cap B| = |B|$ and $g(A) < g(B)$ so the numerator is strictly positive and the ratio is therefore strictly positive. $\square$

Proposition 12. *For $0 < \gamma_1 < \gamma_2$ both satisfying the conditions in Eq. (32), $\mathbf{M}_{\gamma_1} \leq_+ \mathbf{M}_{\gamma_2}$.*

Proof. Wlog, we will assume that g is increasing and ignore the modular transformation in Definition 11.

$$0 \leq \frac{1}{\gamma_2} \max_{A \subseteq V} \left(\gamma_2 g(A) - |A| + \sum_{i \in A} x_i \right) - \frac{1}{\gamma_1} \max_{A \subseteq V} \left(\gamma_1 g(A) - |A| + \sum_{i \in A} x_i \right)$$

$$\tag{33}$$

$$\Longleftarrow 0 \leq \frac{1}{\gamma_2} \left(\gamma_2 g(A) - |A| + \sum_{i \in A} x_i \right) - \frac{1}{\gamma_1} \left(\gamma_1 g(A) - |A| + \sum_{i \in A} x_i \right) \tag{34}$$

$$= \underbrace{\left(\frac{1}{\gamma_2} - \frac{1}{\gamma_1} \right)}_{<0} \underbrace{\left(-|A| + \sum_{i \in A} x_i \right)}_{\leq 0} \tag{35}$$

$\square$

Proposition 13. *For $g \in \mathcal{G}_+$, the optimal γ satisfying Eq. (32) is*

$$\min_{A, B \subseteq V, g(B) > g(A)} \frac{|B| - |A \cap B|}{g(B) - g(A)} = \left(g(V) - \min_{i \in V} g(V \setminus \{i\}) \right)^{-1}. \tag{36}$$

Proof. For a given B in Eq. (36), we may show by induction that A must be $B \setminus \{i\}$ for i that maximizes $g(B) - g(A)$. The inductive step can be shown from the property that by supermodularity $g(B \setminus \{i\}) - g(B \setminus \{i, j\}) \leq g(B) - g(B \setminus \{j\}) \leq g(B) - g(B \setminus \{i\}) \implies g(B) - g(B \setminus \{i\}) \geq g(B \setminus \{i\}) - g(B \setminus \{i, j\})$. This implies $g(B) - g(B \setminus \{i, j\}) \leq 2(g(B) - g(B \setminus \{i\}))$, so any A smaller than $|B| - 1$ will increase $|B| - |A \cap B|$ more than the increase in the denominator. From supermodularity, the maximum of $g(B) - g(B \setminus \{i\})$ is achieved for $B = V$. □

Remark 2. We note that Eq. (36) bears a strong similarity to the notion of submodular curvature [3, 8, 19], and connections to curvature are interesting directions for future research.

Proposition 14. $\mathbf{M}_\gamma g(x)$ *is polynomial time computable for $g \in \mathcal{G}$ polynomial time computable and $x \in [0, 1]^{|V|}$.*

Proof. Computation of mg and γ require a linear number of calls to g. The arg max in the definition of $\mathbf{M}_\gamma$ is a maximization of the sum of supermodular and modular functions. □

As in Eq. (16), we may view margin rescaling as a special case of an additive strategy for computing convex extensions in which

$$\mathbf{B}\ell(x) = \frac{1}{\gamma} \max_{A \subseteq V} \left(\gamma \ell(A) + h_+(x, A) \right) \tag{37}$$

where as in Eq. (16), h_+ is some convex function. We have seen from Proposition 11 that γ is necessary to guarantee that the operator yields an extension for members of this family. We again restrict ourselves to h_+ that do not have access to ℓ except the assumption that it is an increasing function satisfying scaling constraints.

Proposition 15. *Margin rescaling, $\mathbf{M}_\gamma$, dominates the family of additive extensions following Eq. (37), where $h_+(x, A)$ has no oracle access to ℓ.*

Proof. For Eq. (37) to yield an extension, we require that $\gamma \ell(A) + h_+(x, A) \geq \gamma \ell(\text{set}(x))$ for integral $x \in [0, 1]^{|V|}$, which is satisfied for $h_+(x, A) \geq \gamma(\ell(\text{set}(x)) - \ell(A))$. For increasing ℓ, $\gamma(\ell(\text{set}(x)) - \ell(A))$ is bounded below by 0 for all $A \subseteq \text{set}(x)$, and for $A \nsubseteq \text{set}(x)$ we have from Eq. (30) that $\gamma(\ell(\text{set}(x)) - \ell(A)) \geq -|A| + |\text{set}(x) \cap A|$. These yield a set of constraints similar to those in Proposition 6 but shifted downwards by one, and without thresholding to zero. In Proposition 6 the maximal convex function satisfying the constraints had the form $\left(1 - |A| + \sum_{i \in A} x_i\right)_+$, and we may conclude that the optimal function for the additive family of extensions is achieved by $h_+(x, A) = -|A| + \sum_{i \in A} x_i$. This indicates that $\mathbf{M}_\gamma \ell$ with maximal γ satisfying Eq. (32) (Proposition 12) is the additive convex extension closest to the convex closure of ℓ. □

Proposition 16. $\mathbf{J}_1 \geq_{\mathcal{G}} \mathbf{M}_\gamma$.

Proof. For g increasing supermodular with $g(\{i\}) = 0 \,\forall i \in V$ and scaling satisfying Proposition 13, the difference between each of the corresponding arg max elements in the definition of $\mathbf{M}$ and $\mathbf{J}_1$, respectively, is $- \sum_{i \in V \setminus A} x_i \left(g(A \cup \{i\}) - g(A)\right) + \sum_{i \in A}(1 - x_i)\left(g(V) - g(V \setminus \{i\}) - 1\right)$. The first summation is non-positive as $x_i \geq 0$ and g is increasing. The second term is non-positive as $g(V) - g(V \setminus \{i\}) \leq 1$ from Proposition 13. $\qquad\square$

Proposition 17. $\mathbf{J}_2 \geq_{\mathcal{G}} \mathbf{M}_\gamma$.

Proof. For g increasing supermodular and appropriately scaled, the difference between each of the corresponding arg max elements in the definition of $\mathbf{M}$ and $\mathbf{J}_2$, respectively, is $\sum_{i \in A}(1 - x_i)(g(A) - g(A \setminus \{i\}) - 1) - \sum_{j \in V \setminus A} x_j g(\{i\})$. The second term is clearly non-positive, and the first term is non-positive if $g(A) - g(A \setminus \{i\}) \leq 1$, which follows from Proposition 13 and the supermodularity of g. $\qquad\square$

We next demonstrate that neither margin rescaling nor slack rescaling dominate the other, and neither is guaranteed to be closer to the convex closure everywhere in the unit cube.

Proposition 18. *Neither slack rescaling nor margin rescaling dominates the other:* $\mathbf{M}_\gamma \not\leq_{\mathcal{G}} \mathbf{S}, \ \mathbf{S} \not\leq_{\mathcal{G}} \mathbf{M}_\gamma$.

Proof. Define g to be a symmetric set function with $V =_{\cdot} \{a, b, c, d\}$ and $g(\emptyset) = 0$, $g(\{a\}) = -0.5$, $g(\{a, b\}) = -\frac{2}{3}$, $g(\{a, b, c\}) = -0.5$, $g(\{a, b, c, d\}) = 0$. It is straightforward to show numerically that $\mathbf{S}g(x) > \mathbf{M}_\gamma g(x)$ in the neighborhoods around where $|\operatorname{set}(x)| = 2$, and that $\mathbf{M}_\gamma g(x) > \mathbf{S}g(x)$ in the neighborhood around where $|\operatorname{set}(x)| = 4$. $\qquad\square$

4   Experimental Validation

In this section, we validate the theoretical results by optimizing an objective for exemplar selection from a set of images. The objective to be optimized is

$$\arg\min_{A \subseteq V} - \sum_{i \in V} \left[\min_{j \in A} \|y_i - y_j\| \leq \varepsilon \right], \text{ s.t. } |A| \leq C, \tag{38}$$

which is supermodular as it is simply the negative of a coverage function. We have optimized this objective for a random subsample of images from [16], where ε was set to the 10th percentile of all pairwise distances. For the LP relaxations, the discrete constraint $|A| \leq C$ was relaxed to $\|x\|_1 \leq C$. Discretization of LP relaxations was achieved by setting A to contain the C largest indices of x. LP relaxations began with the tightest budget constraint, and each increased budget was initialized with the cutting planes of the previous iteration. Results are shown in Fig. 4. The margin rescaling LP bound is very loose, at a budget of

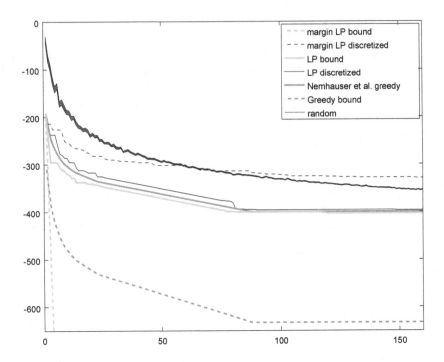

Fig. 4. Empirical results optimizing Eq. (38) on a sample of 400 images, where the horizontal axis represents the budget C.

$C = 160$ the bound is -8195.7 although the minimum of the coverage objective is -400. By contrast, the LP that incorporates the constraints from Definition 7 gives very tight optimality bounds close to the empirical performance of the Nemhauser et al. greedy algorithm, and a huge improvement over the known $(1 - 1/e)$ multiplicative offline bound [15]. Minimization for a single budget took a few seconds on a MacBook Pro for the LP incorporating constraints from $\mathbf{J}_1$ and $\mathbf{J}_2$, while a single budget could take tens of minutes for minimization of $\mathbf{M}_\gamma$.

5   Discussion and Conclusions

In this work, we have formally analyzed slack and margin rescaling as convex extensions of supermodular set functions. Although they were originally developed for empirical risk minimization of (increasing) functions, we show that they can be adapted to non-increasing supermodular functions as well. Supermodularity can be exploited to first transform a set function into an increasing function with a linear number of accesses to the set function. We have shown that neither slack rescaling nor margin rescaling dominates the other in the sense that it is strictly closer to the convex closure. However, computation of slack rescaling for a supermodular function corresponds to a non-supermodular maximization problem. Margin rescaling, by contrast, remains tractable. We have further shown that margin and slack rescaling correspond to optimal additive and multiplicative extensions, respectively, given a computational budget of one oracle access to the loss function. Still, slack and margin rescaling are dominated by extensions derived from known modular bounds on submodular functions (Definition 7). Empirically, incorporating cutting planes from these extensions gives nearly optimal bounds on an objective that measures coverage on a budget, indicating that this approach may be of wider interest for obtaining very tight empirical optimality guarantees for submodular maximization. At the same time, this indicates that novel structured prediction algorithms with improved performance over structured output SVMs may be developed based on the extensions in Definition 7. Complete source code for the experiments is available from https://github.com/blaschko/supermodularLP.

Acknowledgments. This work is funded by Internal Funds KU Leuven, FP7-MC-CIG 334380, and the Research Foundation - Flanders (FWO) through project number G0A2716N.

References

1. Buchbinder, N., Feldman, M., Naor, J.S., Schwartz, R.: A tight linear time (1/2)-approximation for unconstrained submodular maximization. In: FOCS (2012)
2. Choi, H., Meshi, O., Srebro, N.: Fast and scalable structural SVM with slack rescaling. In: AISTATS, pp. 667–675 (2016)
3. Conforti, M., Cornuéjols, G.: Submodular set functions, matroids and the greedy algorithm: tight worst-case bounds and some generalizations of the Rado-Edmonds theorem. Discret. Appl. Math. **7**(3), 251–274 (1984)
4. Conforti, M., Cornuéjols, G., Zambelli, G.: Integer Programming. Springer, Cham (2014). https://doi.org/10.1007/978-3-319-11008-0
5. Fujishige, S.: Submodular Functions and Optimization. Elsevier, Amsterdam (2005)
6. Iyer, R., Bilmes, J.: Algorithms for approximate minimization of the difference between submodular functions, with applications. In: UAI, pp. 407–417 (2012)
7. Iyer, R.K., Bilmes, J.A.: Polyhedral aspects of submodularity, convexity and concavity. CoRR, abs/1506.07329 (2015)
8. Iyer, R.K., Jegelka, S., Bilmes, J.A.: Curvature and optimal algorithms for learning and minimizing submodular functions. In: NIPS, pp. 2742–2750 (2013)
9. Jegelka, S., Bilmes, J.A.: Submodularity beyond submodular energies: coupling edges in graph cuts. In: CVPR, pp. 1897–1904 (2011)
10. Kahl, F., Strandmark, P.: Generalized roof duality. Discret. Appl. Math. **160**(16–17), 2419–2434 (2012)
11. Krause, A., Golovin, D.: Submodular function maximization. In: Bordeaux, L., Hamadi, Y., Kohli, P. (eds.) Tractability: Practical Approaches to Hard Problems. Cambridge University Press, Cambridge (2014)
12. Kumar, M.P., Kolmogorov, V., Torr, P.H.S.: An analysis of convex relaxations for MAP estimation of discrete MRFs. JMLR **10**, 71–106 (2009)
13. Lovász, L.: Submodular functions and convexity. In: Bachem, A., Korte, B., Grötschel, M. (eds.) Mathematical Programming The State of the Art, pp. 235–257. Springer, Berlin (1983). https://doi.org/10.1007/978-3-642-68874-4_10

14. McAllester, D.: Generalization bounds and consistency for structured labeling. In: Bakır, G., Hofmann, T., Schölkopf, B., Smola, A., Taskar, B., Vishwanathan, S. (eds.) Predicting Structured Data. MIT Press, Cambridge (2007)
15. Nemhauser, G.L., Wolsey, L.A., Fisher, M.L.: An analysis of approximations for maximizing submodular set functions-I. Math. Prog. **14**(1), 265–294 (1978)
16. Torralba, A., Fergus, R., Freeman, W.T.: 80 million tiny images: a large data set for nonparametric object and scene recognition. PAMI **30**(11), 1958–1970 (2008)
17. Tsochantaridis, I., Joachims, T., Hofmann, T., Altun, Y.: Large margin methods for structured and interdependent output variables. JMLR **6**, 1453–1484 (2005)
18. Vondrák, J.: Optimal approximation for the submodular welfare problem in the value oracle model. In: STOC, pp. 67–74 (2008)
19. Vondrák, J.: Submodularity and curvature: the optimal algorithm. RIMS Kôkyûroku Bessatsu **23**, 253–266 (2010)
20. Weller, A., Sontag, D., Rowland, M.: Tightness of LP relaxations for almost balanced models. In: AISTATS, pp. 47–55 (2016)
21. Yu, J., Blaschko, M.B.: Learning submodular losses with the Lovász hinge. In: ICML, pp. 1623–1631 (2015)
22. Yu, J., Blaschko, M.B.: A convex surrogate operator for general non-modular loss functions. In: AISTATS, pp. 1032–1041 (2016)

Multi-object Convexity Shape Prior
for Segmentation

Lena Gorelick[✉] and Olga Veksler

University of Western Ontario, London, Canada
{ygorelic,olga}@csd.uwo.ca

Abstract. Convexity is known as an important cue in human vision and has been recently proposed as a shape prior for segmenting a single foreground object. We propose a mutli-object convexity shape prior for multilabel image segmentation. We formulate a novel multilabel discrete energy function. To optimize our energy, we extend the trust region optimization framework recently proposed in the context of binary optimization. To that end we develop a novel graph construction. In addition to convexity constraints, our model includes L^1 color separation term between the background and the foreground objects. It can also incorporate any other multilabel submodular energy term. Our formulation can be used to segment multiple convex objects sharing the same appearance model, or objects consisting of multiple convex parts. Our experiments demonstrate general usefulness of the proposed convexity constraint on real image segmentation examples.

Keywords: Image segmentation · Convexity shape prior
High-order functionals · Trust region · Graph cuts

1 Introduction

Convexity is known to be an important cue in human vision [10,11]. Many natural images and medical images have convex or nearly convex objects. Convexity shape prior for binary segmentation was recently introduced in [7,8,15].

In [15], an object is modeled as an n-sided convex polygon. In this approach there is one foreground and n background labels. The drawback of this approach is that an accurate segmentation of an arbitrary convex object, e.g. a circle, requires a finer discretization with more background parts. This affects the runtime. Moreover, [15] is formulated within continuous optimization framework and therefore requires GPU for efficiency.

To the best of our knowledge, the method in [13] is the only related work modeling segmentation of multiple convex objects. They extend min-cost multicut problem by a set of constraints and use branch-and-cut of ILP solver to optimize the resulting integer linear problem. Such approach is feasible for small images, but an efficient solver of multilabel convexity shape prior is still elusive.

© Springer International Publishing AG, part of Springer Nature 2018
M. Pelillo and E. Hancock (Eds.): EMMCVPR 2017, LNCS 10746, pp. 455–468, 2018.
https://doi.org/10.1007/978-3-319-78199-0_30

Fig. 1. Multi-object convexity shape prior example. Left to right: input image, user scribbles, initial solution, final result.

In this paper we extend the single object convexity shape prior in [7,8] proposed in the context of binary energies to handle segmentation of multiple convex objects. We develop an energy-based formulation for multi-object convexity prior in discrete optimization framework, see Fig. 1. Our energy can combine multi-part convexity prior with hard-constraints, unary appearance [2], boundary length [1], color separation [16], and any other multilabel submodular term.

Given that an efficient solver for the binary convexity shape prior exists [7,8], it would be natural to use α-expansion [4] for our multilabel energy. However, as we explain in Sect. 2 enforcing convexity constraints for multiple objects is not possible in a standard binary expansion move. It can only guarantee convexity for the label that is being expanded, but can violate convexity of other objects.

Instead, following [7,8] we develop an efficient optimization algorithm based on trust region optimization framework [5,6,17]. Trust region framework is an iterative optimization algorithm. In each iteration the original energy is approximated around current solution by a tractable model. The approximation is only trusted in a certain region around current solution, called trust region. The approximate model is then optimized within the trust region, leading to a new solution. This step is called the trust region sub-problem. The size of the trust region is subsequently adjusted based on the quality of the approximation.

Unlike [7,8], where trust region sub-problem is a binary energy, in our case the trust region sub-problem involves optimization of a multilabel energy. To that end we design a novel multi-layer graph construction that allows solving the trust region sub-problem in a single graph-cut. The main advantage of our graph construction is that we can enforce convexity constraints for all objects simultaneously in contrast to the expansion.

Our formulation can be used to segment multiple convex objects. Alternatively, it can be applied to segmentation of a single object that consists of multiple convex parts. In both cases, our model penalizes discontinuities between the background and any foreground label. There is no penalty for discontinuities between any two object labels.

Our model trivially allows incorporation of any standard unary appearance terms and hard constraints. However, in this paper we resort to another popular approach that does not require modeling appearance. In [16] they use L^1 color separation term between the foreground and the background. We show how to incorporate a similar color separation term in our multilabel graph construction. For multi object segmentation, we seek the best color separation between the background and all foreground objects (or object parts).

The paper is organized as follows: Sect. 2 defines the multilabel segmentation energy, Sect. 3 reviews trust region framework, provides details on our optimization algorithm, including energy approximation and the novel graph construction required to solve each trust region sub-problem. Finally, we show experimental results in Sect. 4 and conclude in Sect. 5.

2 Energy

We now formulate multi-object segmentation with convex shape prior as an energy minimization problem. Denote by $\Omega \subset \mathcal{R}^2$ the set of all image pixels and let $\mathcal{M} = \{0\} \cup \{1, 2, ..., M\}$ be the set of labels representing the background and the foreground objects respectively. Let $f_p \in \mathcal{M}$ be the variable representing the label of pixel $p \in \Omega$ and $\mathbf{f}$ be the vector of all integer variables. Denote by $S \subset \Omega$ the set of all pixels that belong to the foreground, i.e. $S = \{p \in \Omega | f_p \neq 0\}$ and by $S_m \subset S$ the object m, $S_m = \{p \in S | f_p = m, m \in \mathcal{M}, m \neq 0\}$. Let $\mathbf{x}^m$ be the vector of indicator binary variables for object m, that is $x_p^m \in \{0, 1\}, x_p = 1 \Leftrightarrow f_p = m \Leftrightarrow p \in S_m$. We collect all $\mathbf{x}^m$ into one set $\mathbf{x}$. Note that there is a one-to-one correspondence between the set $\mathbf{x}$ and multilabel variables $\mathbf{f}$.

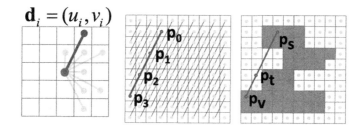

Fig. 2. Left: example of discretized orientations given by a 5×5 stencil. One orientation d_i is highlighted. Middle: set L_i of all discrete lines on image grid that are parallel to d_i. Right: example of a triple clique (p_s, p_t, p_v) that violates convexity constraint.

In continuous case, object S_m is convex if and only if for any $p, r \in S_m$ there is no $q \in \Omega$ on a line between them s.t. $q \notin S_m$. In discrete case, following [8], we approximate convexity constraints using triple clique potentials defined over all possible triplets of ordered pixels along any straight line in any orientation.

First, as in [8] we define the set of all discrete lines. Let $i \in \{1, ..., T\}$ enumerate discrete orientations, see Fig. 2 (left). Let $p \in \Omega$ be a pixel, and let l_p^i be a discrete line passing through p in orientation $d_i = (u_i, v_i) \in \mathcal{R}^2$. That is,

$$l_p^i = \{p_t | \, p_t = p + t \cdot d_i, \, t \in \mathcal{Z}, \, p_t \in \Omega\}. \tag{1}$$

We define $L_i = \{l_p^i \subset \Omega | p \in \Omega\}$ as the set of all *discrete lines* l_p^i with orientation d_i. Figure 2 (middle) illustrates set L_i for one orientation d_i and highlights one

l_p^i. To avoid double indexing we use f_t and x_t instead of f_{p_t} and x_{p_t} to denote the multilabel and binary variables for pixel p_t on a line[1].

One way to represent discrete convexity constraint is based on potential $\phi :$ $\{0, 1, \cdots M\}^3 \to \mathcal{R}$ defined for all triplets of *ordered* pixels (p_s, p_t, p_v), $s < t < v$ along any discrete line $l \in \bigcup L_i$

$$\phi(f_s, f_t, f_v) = \begin{cases} \infty & \text{if } (f_s = f_v) \wedge (f_t \neq f_s) \wedge (f_s \neq 0) \\ 0 & \text{otherwise.} \end{cases}$$

We redefine potential ϕ separately for each part m in terms of its indicator variables $\mathbf{x}^m$

$$\phi(x_s^m, x_t^m, x_v^m) = \begin{cases} \infty & \text{if } (x_s^m, x_t^m, x_v^m) = (1, 0, 1) \\ 0 & \text{otherwise.} \end{cases}$$

In practice we use a finite penalty ω and rewrite it algebraically as

$$\phi(x_s^m, x_t^m, x_v^m) = \omega \cdot x_s^m (1 - x_t^m) x_v^m. \tag{2}$$

The convexity energy $E_{convexity}(\mathbf{f})$ integrates this triple clique potential over all parts, all orientations, all lines and all triplets:

$$E_{convexity}(\mathbf{f}) = \sum_{m \in \mathcal{M} \setminus \{0\}} \sum_{l \in \bigcup L_i} \sum_{\substack{(p_s, p_t, p_v) \in l \\ s < t < v}} \phi(x_s^m, x_t^m, x_v^m) = E_{convexity}(\mathbf{x}). \tag{3}$$

As discussed in [8], even in the case of binary segmentation ($|\mathcal{M}| = 2$), convexity energy (3) is hard to optimize for two reasons: it is non-submodular and it has a prohibitively large number of cliques. In our case, the energy (3) is multilabel, which makes it even harder.

Our overall segmentation energy $E(\mathbf{f})$ combines the convexity term $E_{convexity}(\mathbf{f})$ in (3) with any multilabel submodular term $E_{sub}(\mathbf{f})$, see definition in [14].

$$E(\mathbf{f}) = E_{convexity}(\mathbf{f}) + E_{sub}(\mathbf{f}). \tag{4}$$

In the reminder of the paper we use a specific term

$$E_{sub}(\mathbf{f}) = E_{length}(\mathbf{f}) + E_{color}(\mathbf{f}). \tag{5}$$

Here, the first term regularizes the boundary length between the background and the foreground objects

$$E_{length}(\mathbf{f}) = \sum_{\{p,q\} \in \mathcal{N}} w_{pq} \mathbf{1}_{[\mathbf{f_p} = 0 \wedge \mathbf{f_q} \neq 0]} \tag{6}$$

where $\mathcal{N}$ is the set of pairs of neighboring pixels in Ω and w_{pq} is a contrast sensitive coefficient. Note, that general multilabel Potts model between all labels is

---

[1] Note, pixel $p \in \Omega$ has unique index t_l on each line l passing through it.

not multilabel submodular. Therefore, we do not penalize discontinuity between different objects. This corresponds to an assumption that if different foreground labels touch, they are different parts of the same object. This naturally allows us to use the same energy to segment objects consisting of multiple convex parts.

The second term in (5) is a version of the L_1 color separation term [16] which we extend to handle multiple foreground objects. In [16] the appearance of foreground and background is modeled using discrete color histograms for foreground and for background. The color separation term penalizes any color bin that is split between the foreground and background object. In our multilabel case we combine histograms of all foreground objects into one histogram. Let $\theta^{\mathbf{f}=0}$ be the histogram of the background and $\theta^{\mathbf{f}\neq0}$ be the histogram of the foreground objects induced by labeling $\mathbf{f}$. Let K denote the number of bins and $n_k^{\mathbf{f}=0}$ and $n_k^{\mathbf{f}\neq0}$, $1 \le k \le K$ denote the corresponding pixel counts for bin k. Then, the color separation term

$$E_{color}(\mathbf{f}) = E_{color}(\theta^{\mathbf{f}=0}, \theta^{\mathbf{f}\neq0}) = -\|\theta^{\mathbf{f}=0} - \theta^{\mathbf{f}\neq0}\|_{L_1} \tag{7}$$

measures the L_1 distance between the two histograms. It can be shown [16] that minimizing (7) is equivalent to minimizing

$$E_{color}(\theta^{\mathbf{f}=0}, \theta^{\mathbf{f}\neq0}) = \sum_{k=1}^{K} \min(n_k^{\mathbf{f}=0}, n_k^{\mathbf{f}\neq0}) - \frac{|\Omega|}{2}. \tag{8}$$

The term above encourages to keep all pixels that belong to the same color bin together, either in the background or in the foreground objects. Note that there is no penalty when the bin is split between foreground objects (parts).

As shown in [8], our convexity shape prior $E_{convexity}(\mathbf{f})$ is non-submodular even for the case of binary segmentation [8] and therefore is hard to optimize. We use iterative trust region (TR) framework [17], which has been shown promising for various non-submodular energies [5,6,12] including convexity shape prior for binary segmentation [8]. In the next section we describe our optimization algorithm for the energy (4).

3   Energy Minimization

It is tempting to apply α-expansion for optimization of multilabel energy in (3), especially since there exist a solver for binary segmentation with convexity shape prior [8]. However, it is easy to show that standard α-expansion would not work.

First, let us briefly review α-expansion algorithm. In a single α-expansion move, each pixel can either switch to new label α or keep the old label. This move can be formulated as optimization of binary energy over new binary variables $\mathbf{y}$, where the old label is represented with 0 and the new label α is represented with 1. The problem is that there is no distinction in the expansion energy between the background and the objects that are not being expanded on. These foreground objects have to be convex, while the background can assume any shape, but both have label 0.

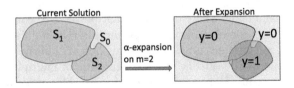

Fig. 3. Intuitive explanation why α-expansion cannot optimize $E(\mathbf{f})$ in (4). Expanding on label $m = 2$ using the construction in [8] enforces convexity for part S_2 but introduces new violations of convexity for part S_1. This is because α-expansion does not distinguish between background, which does not have to be convex and the remaining objects, which have to be convex. Both have label 0 during the binary expansion step.

Consider Fig. 3. When expanding on label $\alpha = 2$ (object S_2), the construction in [8] could enforce convexity prior for that part. This part would have label 1 in the binary energy representing the expansion move. However, new violations of convexity might be introduced to object S_1. This object has label 0 in the binary expansion move and, thus, is indistinguishable from the background.

In this paper we develop a multilabel optimization method based on trust region framework. The advantage of our method over the expansion algorithm is that each object maintains a distinct label, allowing incorporating convexity constraints for all foreground objects throughout.

3.1 Multilabel Trust Region Framework

We now review and generalize the method in [8] to handle multi-part shape convexity. In each iteration of trust region, we construct an approximate tractable $\tilde{E}^k$ of the energy E in (4) near current solution $\mathbf{f}^k$. Unlike the binary case in [7,8], our energy is multilabel and, therefore, the approximation $\tilde{E}^k$ must be multilabel submodular in order to be tractable. The approximate model $\tilde{E}^k$ is assumed to be accurate within a small region around $\mathbf{f}^k$ called "trust region". It is then optimized within the trust region to obtain a candidate solution. This optimization step is called *trust region (TR) sub-problem*. The size of the trust region is adjusted in each iteration based on the quality of the current approximation. See [17] for a review of the trust region technique.

Algorithm 1 summarizes our approach. Line 4 computes unary approximate energy E_{apprx}^k for the non-submodular $E_{convexity}$ around $\mathbf{f}^k$. See Sect. 3.2 for details of the approximation. Line 5 combines E_{apprx}^k with the multilabel submodular E_{sub}. The resulting $\tilde{E}^k$ is multilabel submodular and coincides with the exact energy E on $\mathbf{f}^k$. The TR sub-problem requires minimization of $\tilde{E}^k$ within a small region $||\mathbf{f} - \mathbf{f}^k|| \leq d_k$ around $\mathbf{f}^k$. Unfortunately, minimizing $\tilde{E}^k$ under distance constraints is NP-hard [5].

Instead, we use a simpler formulation of the TR sub-problem proposed in [5,6,8] based on unconstrained optimization of multilabel submodular Lagrangian

$$L^k(\mathbf{f}) = \tilde{E}^k(\mathbf{f}) + \lambda_k||\mathbf{f} - \mathbf{f}^k||. \tag{9}$$

Algorithm 1. TRUST REGION CONVEXITY

---

1 $\mathbf{f}^0 \longleftarrow \mathbf{f}_{\text{init}}, \lambda_0 \longleftarrow \lambda_{\text{init}}, \text{convergedFlag} \longleftarrow 0$
2 **Repeat Until** convergedFlag
3   //**Approximate** $E_{convexity}(\mathbf{f})$ around $\mathbf{f}^k$
4   Compute $E_{apprx}^k(\mathbf{f})$ (see Sect. 3.2)
5      $\tilde{E}^k(\mathbf{f}) = E_{apprx}^k(\mathbf{f}) + E_{sub}(\mathbf{f})$ // keep the multilabel submodular part
6   //**Trust region sub-problem**
7      $\mathbf{f}^* \longleftarrow \text{argmin}_\mathbf{f} L^k(\mathbf{f})$ (9) // (see Sect. 3.3)
8   Evaluate $E(\mathbf{f}^k), E(\mathbf{f}^*)$ (see [8])
9   Evaluate $E_{apprx}^k(\mathbf{f}), E_{apprx}^k(\mathbf{f}^*)$ (see [8])
10     $R = E(\mathbf{f}^k) - E(\mathbf{f}^*)$ // actual reduction in energy
11     $P = \tilde{E}_k(\mathbf{f}^k) - \tilde{E}^k(\mathbf{f}^*)$ // predicted reduction in energy
12   **If** $P = 0$ // meaning $\mathbf{x}^* = \mathbf{x}^k$ and $\lambda > \lambda_{\max}$
13         $\lambda_k \longleftarrow \lambda_{\max}$
14   //**Trust region sub-problem:**
15         $\mathbf{f}^* \longleftarrow \text{argmin}_\mathbf{f} L^k(\mathbf{f})$ (9) // (see Sect. 3.3)
16   Evaluate $E(\mathbf{f}^k), E(\mathbf{f}^*)$ (see [8])
17   Evaluate $E_{apprx}^k(\mathbf{f}^k), E_{apprx}^k(\mathbf{f}^*)$ (see [8])
18         $R = E(\mathbf{f}^k) - E(\mathbf{f}^*)$ // actual reduction in energy
19         $P = \tilde{E}^k(\mathbf{f}^k) - \tilde{E}^k(\mathbf{f}^*)$ // predicted reduction in energy
20   //**Update current solution**
21         $\mathbf{f}^{k+1} \longleftarrow \begin{cases} \mathbf{f}^* \text{ if } R > 0 \\ \mathbf{f}^k \text{ otherwise} \end{cases}$
22   convergedFlag $\longleftarrow (R \le 0)$
23   **Else**   // meaning $\mathbf{f}^* \ne \mathbf{f}^k$ and $\lambda \le \lambda_{\max}$
24   //**Update current solution**
25         $\mathbf{f}^{k+1} \longleftarrow \begin{cases} \mathbf{f}^* \text{ if } R > 0 \\ \mathbf{f}^k \text{ otherwise} \end{cases}$
26   //**Adjust the trust region**
27         $\lambda_{k+1} \longleftarrow \begin{cases} \lambda_k/\alpha \text{ if } R/P > \tau_2 \\ \lambda_k \cdot \alpha \text{ otherwise} \end{cases}$

---

28 we use $\alpha = 10$, $\tau_2 = 0.25$

Here parameter λ_k controls the trust region size indirectly instead of distance d_k. We use L_2 distance expressed with unary terms, see [5,6]. Therefore $L_k(\mathbf{f})$ is multilabel submodular. Line 7 solves (9) for some fixed λ_k. To that end, we propose a novel multi-layer graph construction in Sect. 3.3 that allows us to solve this step using one graph cut.

The candidate solution $\mathbf{f}^*$ is accepted whenever the original energy decreases (line 25). To avoid prohibitively expensive computation for evaluation of all triple clique potentials in (3), we use dynamic programming, see details in [7,8]. The Lagrange multiplier λ_k is adaptively changed (line 27) based on the quality of the current approximation as motivated by empirical inverse proportionality relation between λ_k and d_k (see [6]). In each iteration of the trust region, either the energy decreases or the trust region size is reduced. When the trust region is so small that it does not contain a single discrete solution, namely $\mathbf{f}^* = \mathbf{f}^k$ (Line 12), one more attempt is made using $\lambda_{\max}$, where $\lambda_{\max} = \sup\{\lambda | \mathbf{f}^* \neq \mathbf{f}^k\}$ (see [6]). If there is no reduction in energy with smallest discrete step $\lambda_{\max}$ (Line 21), we are at a local minimum [3] and we stop (Line 22).

3.2   Linear Approximation of $E_{convexity}$

In this section we derive linear approximation $E_{apprx}^k(\mathbf{f})$ for $E_{convexity}(\mathbf{f})$ in (3) around current solution $\mathbf{f}^k$. That is, we are interested in linear function

$$E_{apprx}^k(\mathbf{f}) = \sum_{p \in \Omega} D_p^k(f_p) \tag{10}$$

where $D_p^k(f_p)$ is the unary term for pixel p having label f_p. Recall that for each labeling $\mathbf{f}$ there is a corresponding set of indicator variables $\mathbf{x}$ such that:

$$E_{convexity}(\mathbf{f}) = E_{convexity}(\mathbf{x}) = \sum_{m \in \mathcal{M} \setminus \{0\}} \sum_{l \in \bigcup L_i} \sum_{\substack{(p_s, p_t, p_v) \in l \\ s < t < v}} \phi(x_s^m, x_t^m, x_v^m).$$

First, we derive linear approximation $\phi^k(x_s^m, x_t^m, x_v^m)$ for each term $\phi(x_s^m, x_t^m, x_v^m)$ around $\mathbf{x}^k$ and then we rewrite $E_{apprx}^k(\mathbf{f})$ as:

$$E_{apprx}^k(\mathbf{x}) = \sum_{m \in \mathcal{M} \setminus \{0\}} \sum_{l \in \bigcup L_i} \sum_{\substack{(p_s, p_t, p_v) \in l \\ s < t < v}} \phi^k(x_s^m, x_t^m, x_v^m). \tag{11}$$

Let us focus on one particular object m. Assume current configuration of the corresponding indicator variables is $\mathbf{x}^{m,k}$. For convenience and to avoid triple indexes, let us rename $\mathbf{x}^m$ as $\mathbf{y}$ and $\mathbf{x}^{m,k}$ as $\mathbf{y}^k$. We can use the linear approximation for the binary convexity energy in [8]. Omitting constants, Taylor expansion $\phi^k(y_s, y_t, y_v)$ of a potential $\phi(y_s, y_t, y_v)$ around current $\mathbf{y}^k$ is

$$\phi^k(y_s, y_t, y_v) = (1 - y_t^k) \cdot y_v^k \cdot y_s - y_v^k \cdot y_s^k \cdot y_t + (1 - y_t^k) \cdot y_s^k \cdot y_v. \tag{12}$$

The components in (12) have an intuitive interpretation. Consider the first component $(1 - y_t^k) \cdot y_v^k \cdot y_s$. Recall that pixels p_s, p_t, p_v are on a line and p_t is between p_s and p_v. If the current configuration $\mathbf{y}^k$ is such that $y_t^k = 0$, and $y_v^k = 1$, then assigning label 1 to y_s violates convexity of object m, assuming y_t and y_v keep their labels unchanged from $\mathbf{y}^k$. The unary term $(1 - y_t^k) \cdot y_v^k \cdot y_s$ penalizes this violation: assignment $y_s = 1$ carries a penalty, whereas $y_s = 0$ is not penalized. The other two components in (12) have similar intuitive interpretations.

Approximation in (12) gives three unary terms for each triple clique. Consider pixel $p_s \in l$. It can be either the leftmost, middle, or rightmost member of a clique on that line. We need to sum the terms from all triple cliques on line l involving pixel p_s. First with p_s being on the left, then in the middle and finally on the

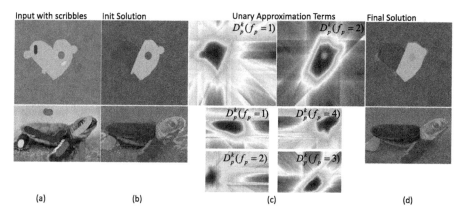

Fig. 4. Unary approximation of $E_{convexity}$ for synthetic (row one) and natural (row two) images: (a) Input image with user scribbles - red for the background ($f_p = 0$) and blue, green, purple, cyan for different object parts ($f_p = 1, 2, 3, 4$), (b) Initial solution, (c) Unary approximation terms for different object parts as in (10) during the first iteration of the trust region. For each pixel p and object part m, blue-cyan colors encode preference to have label $f_p = m$ and orange-red colors encode preference to belong to other object parts or background, (d) Final segmentation superimposed on image where the object is decomposed into its constituent convex parts. (Color figure online)

right of the clique. All these terms contribute to the unary potential $u_s^l(y_s)$ for a single pixel p_s.

$$u_{p_s}^l(y_s) = \sum_{\substack{(p_t,p_v)\in l \\ s<t<v}}(1 - y_t^k) \cdot y_v^k \cdot y_s - \sum_{\substack{(p_t,p_v)\in l \\ t<s<v}} y_t^k \cdot y_v^k \cdot y_s + \sum_{\substack{(p_t,p_v)\in l \\ t<v<s}} (1 - y_v^k) \cdot y_t^k \cdot y_s. \quad (13)$$

For object m, the full Taylor based unary term for pixel $p \in \Omega$ sums the above expression over all lines passing through p giving a unary term:

$$u_p(y_p) = \sum_{l \in \cup L_i} u_p^l(y_p).$$

Due to one-to-one correspondence between the indicator binary variables $\mathbf{y}^k$, $\mathbf{x}^{m,k}$ and the multi-variable labeling $\mathbf{f}^k$ this gives the unary term $D_p^k(f_p = m)$. Similarly, we can compute linear terms for other objects.

Therefore, the overall approximation $E_{apprx}^k(\mathbf{x})$ in (11) is a linear function of $\mathbf{x}$ that coincides with the original convexity term $E_{convexity}(\mathbf{x})$ on each of current configurations $\mathbf{x}^{m,k}$ for each object.

To avoid prohibitively expensive computation for approximation of all triple clique potentials in (10), we use dynamic programming, see details in [7,8].

Figure 4 illustrates the resulting unary terms arising from such approximation. They encourage any holes or concavities in each part to be filled in, and any protrusions/disconnected components to be erased.

3.3   Graph Construction for Multi-label Submodular $L_k(\mathbf{f})$

Below we show how to construct a directed weighted graph $G = (V, A, c)$ such that minimum cut on this graph corresponds to the optimal solution of the Lagrangian $L^k(\mathbf{f})$ in (9). In particular we explain how to incorporate the unary terms in (9) as well as multilabel submodular terms E_{length} and E_{color} in (5). Our construction extends the ideas in [9,14,16]. For illustration see Fig. 5.

Let $z^s, z^t \in V$ denote the special source and sink nodes. An s-t cut $C(S, T)$ partitions the nodes of the graph into two disjoint sets $S, T \subset V, S \cap T \neq \emptyset$ such that $z^s \in S, z^t \in T$. The cost of a cut is defined as sum of arc costs between the sets S and T. That is $\text{cost}(S, T) = \sum_{(z_1, z_2) \in S \times T \cap A} c(z_1, z_2)$. Informally we say arc $(z_1, z_2) \in A$ is in the cut $C(S, T)$ if $z_1 \in S$ and $z_2 \in T$.

First we construct the set of nodes V. In addition to the source and sink nodes, for each pixel $p \in \Omega$ we construct M nodes $z_p^m, 1 \leq m \leq M$, (see blue nodes in Fig. 5(a)). Therefore, our graph consists of M layers $V = \{V_1 \cup \ldots \cup V_M\}$ where V_m is the set of all nodes on layer m of the graph, $V_m = \{z_p^m | p \in \Omega\}$. It is convenient to denote $z_p^0 = z^s$ and $z_p^M = z^t$.

The set of all arcs A consists of three disjoint subsets of edges $A = A_1 \cup A_2 \cup A_3$. The arcs in A_1 connect vertices corresponding to the same pixel and model the unary terms in $L^k(\mathbf{f})$. The arcs in A_2 connect vertices corresponding to neighboring pixels in the image domain and model E_{length}. Finally, the arcs in A_3 model the color separation term E_{color}.

First we describe A_1. For $0 \leq r \leq M$ we connect z_p^r to z_p^{r+1} with unary cost $D_p(r)$ as in (9) (black vertical arrows in (a)) and z_p^{r+1} to z_p^r with infinite cost (red vertical arrows). This ensures that for a finite cost cut and for each pixel p there is exactly one arc of type (z_p^r, z_p^{r+1}) in the cut. As a result there is a natural correspondence between the set of all finite cuts and the set of all multi-label labelings:

$$(z_p^r, z_p^{r+1}) \text{ is in } C(S, T) \Leftrightarrow f_p = r. \tag{14}$$

Notice that the sum of all arc costs in A_1 that are cut by $C(S, T)$ is equal to the sum of the unary terms D_p of the labeling $\mathbf{f}$ induced by $C(S, T)$, yielding the linear term of $L^k(\mathbf{f})$ in (10).

We now describe the second set of edges A_2. For each pair of neighboring pixels in the image domain $\{p, q\} \in \mathcal{N}$, there is a corresponding pair of nodes in V_1. We construct two arcs $(z_p^1, z_q^1) \in A_2$ and $(z_q^1, z_p^1) \in A_2$ with cost w_{pq} as in (5) (see black horizontal arrows in the top layer of the graph in Fig. 5(a)). Let $C(S, T)$ be a cut with a finite cost. Suppose $z_p^1 \in T$ and $z_q^1 \in S$. Using (14), that implies that corresponding $f_p = 0$, meaning pixel p is in the background and $f_q \neq 0$ meaning q is in the foreground. In this case arc (z_q^1, z_p^1) must be in $C(S, T)$. Similarly, if $q \in S$ and $p \in T$, then arc (z_p^1, z_q^1) must be in $C(S, T)$. If both $p, q \in S$ or $p, q \in T$, no arcs between them are cut. Therefore, the sum of all arc costs in A_2 that are cut by $C(S, T)$ is equal to the sum of the pairwise terms w_{pq} of the labeling $\mathbf{f}$ induced by $C(S, T)$, yielding $E_{length}(\mathbf{f})$ in (5).

In addition to boundary length regularization between the background and the set of foreground objects, E_{sub} in (5) incorporates the color separation term

$E_{color}(\mathbf{f})$, see Fig. 5(b) for the corresponding graph construction. For each color bin i we create an auxiliary node b_i. For each image pixel p in bin i, we connect its node $z_p^1 \in V_1$ to the auxiliary node b_i. We refer to the set of all such edges as A_3. It is convenient to denote by $V_1^i \subset V_1$ the set of nodes representing pixels in bin i in layer V_1 of the graph.

Let $C(S,T)$ be a finite cost cut. Assume that the cut splits the nodes in V_1^i such that n_i^{bg} nodes are in S and n_i^{fg} nodes are in T. Using (14), that implies that there are n_i^{bg} pixels in the background and n_i^{fg} pixels are in the foreground. This implies that either n_i^{bg} or n_i^{fg} edges in A_3 must be cut. The minimum cost cut will cut $\min(n_i^{bg}, n_i^{fg})$ corresponding to E_{color} in (5). Such construction has a bias to keeping pixels of the same color together, either in the foreground or in the background. See [16] for more details.

We have shown that the total cost of any finite cut $C(S,T)$ is exactly the energy value of the corresponding labeling $\mathbf{f}$. Thus, solving the min-cut problem on graph G gives an optimal labeling $\mathbf{f}^*$ with respect to energy $L^k(\mathbf{f})$.

Hard constraints can also be easily incorporated. For a specific pixel p that must be assigned label $f_p = m$, we set the arc $c(z_p^r, z_p^{r+1}) = 0$ for $r = m$, and $c(z_p^r, z_p^{r+1}) = \infty$ for $r \neq m$.

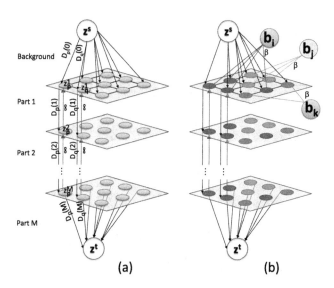

Fig. 5. Graph construction for minimizing energy $L^k(\mathbf{f})$ (9): (a) illustrates the nodes V and arcs $A_1 \cup A_2$ of the graph. (b) illustrates implementation of the color separation term E_{color} in (8) between the background and all foreground labels using edges in A_3. See text for details. (Color figure online)

4   Experimental Evaluation

Below we apply the proposed method to multilabel segmentation. We experiment both with images containing multiple convex objects (see Fig. 6, column 1) and with images containing objects consisting of multiple object parts (see Fig. 7, column 1.) In all the experiments below we use user scribbles as hard constraints (see color-coded scribbles in column 2). For the color separation term we use sixteen bins per color channel, discarding unoccupied bins. We used coefficients $\lambda_{color} = 1$, $\lambda_{length} = 15$ and $\lambda_{convexity} = 1$ for the corresponding energy terms and annealed the latter two terms as in [8]. The total computation time per image is under half a minute.

All experiments are initialized the same way. First we perform binary segmentation by minimizing $E_{color} + E_{length}$ subject to the hard constraints, as in [16]. The resulting binary solution is transformed into multilabel initial solution as follows. We keep the background pixels unchanged, but split the foreground pixels between the foreground objects/parts based on the location of the scribbles. We assign each foreground pixel the label of an object/part whose scribble is the closest. See initial solutions in column 3 of Figs. 6 and 7. The final results containing convex objects or parts are shown in the last column of the figures.

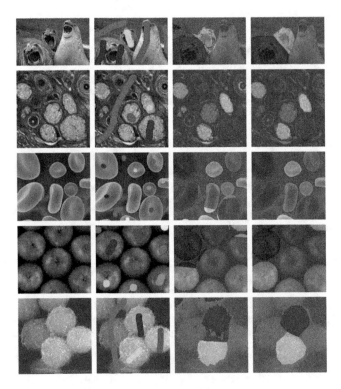

Fig. 6. Experimental results for segmentation of multiple convex object: Left to right: input image, color coded user scribbles, initial solution, final result.

Fig. 7. Experimental results for segmentation of objects with multiple convex parts: Left to right: input image, color coded user scribbles, initial solution, final result.

5   Conclusions

We proposed a mutli-object convexity shape prior for multi-label image segmentation. We formulated a novel multi-label discrete energy function and extended the trust region optimization framework to handle the multilabel case. For that, we developed a novel graph construction that allows solving a trust region subproblem step in a single graph-cut. In addition to convexity constraints, our model can incorporate any multi-label submodular energy term. In particular we showed how to incorporate L^1 color separation term between the background and the foreground objects. Our formulation can be used to segment multiple convex objects sharing the same appearance model, or objects consisting of multiple convex parts.

Our method has several limitations. As with any local optimization, trust region framework can be sensitive to initialization. Furthermore, our method does not model occlusion and therefore cannot segment convex objects that overlap in the image domain. Such occlusion might create concavities in the shape of the occluded objects that should be accounted for. Another limitation is that our construction does not allow boundary length penalty between different object labels. This can be an advantage in case of multi-part objects. Also, this is not an issue when objects do not touch, since there a penalty for boundary length between the background and the foreground objects. However, when multiple objects do touch, this may be a drawback.

References

1. Boykov, Y., Kolmogorov, V.: Computing geodesics and minimal surfaces via graph cuts. In: International Conference on Computer Vision, pp. 26–33 (2003)
2. Boykov, Y., Jolly, M.P.: Interactive graph cuts for optimal boundary and region segmentation of objects in N-D images. In: IEEE International Conference on Computer Vision (ICCV) (2001)
3. Boykov, Y., Kolmogorov, V., Cremers, D., Delong, A.: An integral solution to surface evolution pdes via geo-cuts. In: Leonardis, A., Bischof, H., Pinz, A. (eds.) ECCV 2006. LNCS, vol. 3953, pp. 409–422. Springer, Heidelberg (2006). https://doi.org/10.1007/11744078_32
4. Boykov, Y., Veksler, O., Zabih, R.: Fast approximate energy minimization via graph cuts. IEEE Trans. Pattern Anal. Mach. Intell. **23**(11), 1222–1239 (2001)
5. Gorelick, L., Boykov, Y., Veksler, O., Ben Ayed, I., Delong, A.: Submodularization for binary pairwise energies. In: IEEE Conference on Computer Vision and Pattern Recognition (CVPR), pp. 1154–1161, June 2014
6. Gorelick, L., Schmidt, F.R., Boykov, Y.: Fast trust region for segmentation. In: IEEE Conference on Computer Vision and Pattern Recognition (CVPR), Portland, Oregon, pp. 1714–1721, June 2013
7. Gorelick, L., Veksler, O., Boykov, Y., Nieuwenhuis, C.: Convexity shape prior for segmentation. In: Fleet, D., Pajdla, T., Schiele, B., Tuytelaars, T. (eds.) ECCV 2014. LNCS, vol. 8693, pp. 675–690. Springer, Cham (2014). https://doi.org/10.1007/978-3-319-10602-1_44
8. Gorelick, L., Veksler, O., Boykov, Y., Nieuwenhuis, C.: Convexity shape prior for binary segmentation. Trans. Pattern Anal. Mach. Intell. **39**(2), 258–271 (2017)
9. Ishikawa, H.: Exact optimization for markov random fields with convex priors. IEEE Trans. Pattern Anal. Mach. Intell. **25**(10), 1333–1336 (2003)
10. Liu, Z., Jacobs, D., Basri, R.: The role of convexity in perceptual completion: beyond good continuation. Vis. Res. **39**, 4244–4257 (1999)
11. Mamassian, P., Landy, M.: Observer biases in the 3D interpretation of line drawings. Vis. Res. **38**, 2817–2832 (1998)
12. Nieuwenhuis, C., Töppe, E., Gorelick, L., Veksler, O., Boykov, Y.: Efficient regularization of squared curvature. In: IEEE Conference on Computer Vision and Pattern Recognition (CVPR), pp. 4098–4105, June 2014
13. Royer, L.A., Richmond, D.L., Rother, C., Andres, B., Kainmueller, D.: Convexity shape constraints for image segmentation. In: 2016 IEEE Conference on Computer Vision and Pattern Recognition, CVPR 2016, Las Vegas, NV, USA, 27–30 June 2016, pp. 402–410 (2016)
14. Schlesinger, D.: Exact solution of permuted submodular minsum problems. In: Yuille, A.L., Zhu, S.-C., Cremers, D., Wang, Y. (eds.) EMMCVPR 2007. LNCS, vol. 4679, pp. 28–38. Springer, Heidelberg (2007). https://doi.org/10.1007/978-3-540-74198-5_3
15. Strekalovskiy, E., Cremers, D.: Generalized ordering constraints for multilabel optimization. In: International Conference on Computer Vision (ICCV) (2011)
16. Tang, M., Gorelick, L., Veksler, O., Boykov, Y.: Grabcut in one cut. In: International Conference on Computer Vision (2013)
17. Yuan, Y.: A review of trust region algorithms for optimization. In: The Fourth International Congress on Industrial and Applied Mathematics (ICIAM) (1999)

Fast Asymmetric Fronts Propagation for Voronoi Region Partitioning and Image Segmentation

Da Chen$^{(\boxtimes)}$ and Laurent D. Cohen

CEREMADE, CNRS, University Paris Dauphine, PSL Research University,
UMR 7534, 75016 Paris, France
chenda@ceremade.dauphine.fr

Abstract. In this paper, we introduce a generalized asymmetric fronts propagation model based on the geodesic distance maps and the Eikonal partial differential equations. One of the key ingredients for the computation of the geodesic distance map is the geodesic metric, which can govern the action of the geodesic distance level set propagation. We consider a Finsler metric with the Randers form, through which the asymmetry and anisotropy enhancements can be taken into account to prevent the fronts leaking problem during the fronts propagation. These enhancements can be derived from the image edge-dependent vector field such as the gradient vector flow. The numerical implementations are carried out by the Finsler variant of the fast marching method, leading to very efficient interactive segmentation schemes.

1 Introduction

Fronts propagation models have been considerably developed since the original level set framework proposed by Osher and Sethian [1]. Guaranteed by the solid mathematical background, the fronts propagation models lead to strong abilities in a wide variety of computer vision tasks such as image segmentation [2–5]. In their basic formulation, the boundaries of an object are modelled as closed contours, each of which can be obtained by evolving an initial closed curve till the stopping criteria are reached. The use of curve evolution scheme for image segmentation can be back-track to the original active contour model [6].

Let $\Omega \subset \mathbb{R}^2$ be an open bounded domain. Based on the level set framework [1], a closed contour γ can be retrieved by identifying the zero level set line of a function $\phi : \Omega \to \mathbb{R}$ such that $\gamma := \{\mathbf{x} \in \Omega; \phi(\mathbf{x}) = 0\}$. By this curve representation, the curve evolution is carried out by evolving the function ϕ

$$\partial \phi / \partial t = \xi \, \|\nabla \phi\|, \tag{1}$$

where $\xi : \Omega \to \mathbb{R}$ is a speed function and t denotes the time. At any time t, the curve γ can be recovered by identifying the zero-level set lines of ϕ. Using the level set evolutional Eq. (1), the contours splitting and merging can be adaptively handled. The main drawback of the level set-based front propagation

© Springer International Publishing AG, part of Springer Nature 2018
M. Pelillo and E. Hancock (Eds.): EMMCVPR 2017, LNCS 10746, pp. 469–484, 2018.
https://doi.org/10.1007/978-3-319-78199-0_31

method is its expensive computational burden. In order to alleviate this problem, Adalsteinsson and Sethian [7] suggested to restrict the computation for the update of the level set function ϕ within a narrow band such that only the values of ϕ at the points within this narrowband are updated according to Eq. (1). Moreover, the distance-preserving level set method [8] is able to avoid level set reinitialization by enforcing the level set function ϕ as a signed Euclidean distance function from the current curves during the evolution.

Despite the efforts devoted to the reduction of the computation burden, the classical level set-based fronts propagation scheme (1) is still impractical especially for realtime applications. In order to solve this issue, Malladi and Sethian [9] proposed a new geodesic distance-based fronts propagation model for real time image segmentation. It relies on a geodesic distance map $\mathcal{U}_\mathfrak{s} : \Omega \to \mathbb{R}^+ \cup \{0\}$ associated to a collection $\mathfrak{s} \subset \Omega$ of source points. The value of $\mathcal{U}_\mathfrak{s}(\mathbf{x})$ in essence equals to the minimal geodesic length between $\mathbf{x}$ and a source point $\mathbf{s} \in \mathfrak{s}$ associated to an isotropic Riemannian metric. The numerical computation of $\mathcal{U}_\mathfrak{s}$ can be carried out by the fast marching method [10,11]. The efficiency of the fast marching methods provide the possibility of real time segmentation application. Based on the geodesic distance map $\mathcal{U}_\mathfrak{s}$, a curve can be denoted by the T-level set of $\mathcal{U}_\mathfrak{s}$, where $T > 0$ is a geodesic distance thresholding value. In other words, a curve γ can be characterized at the distance value T such that

$$\gamma := \{\mathbf{x} \in \Omega; \mathcal{U}_\mathfrak{s}(\mathbf{x}) = T\}. \tag{2}$$

One difficulty suffered by the geodesic distance-based fronts propagation scheme is that the fronts may leak outside the targeted regions before all the points of these regions have been visited by the fronts. The leaking problem sometimes occurs near the boundaries close to the source positions or in weak boundaries, especially when handling long and thin structures. The main reason for this leaking problem is the positivity constraint required by the Eikonal equation. Chen and Cohen [12] considered an anisotropic Riemannian metric for fronts propagation, where the path orientations are taken into account to mitigate the leaking problem. Arbeláez and Cohen [13] and Bai and Sapiro [14] made use of the concept of Voronoi index map which is constructed by the geodesic distance associated to the orientation-dependent pseudo path metric, the values of which are allowed to be zero. The image segmentation can be characterized by the Voronoi regions, each of which involves all the points with the same voronoi index. In this case, the contours indicating the tagged object edges are common boundaries of the adjacent voronoi regions. Li and Yezzi [15] proposed a dual fronts propagation model for active contours evolution, where the geodesic metric comprises both edge and region statistical information. The basic idea of [15] is to propagate the fronts simultaneously from the exterior and interior boundaries of the narrowband. The optimal contours can be recovered from the positions where the two fast marching fronts meet. These meeting interfaces also correspond to the boundaries of the adjacent voronoi regions.

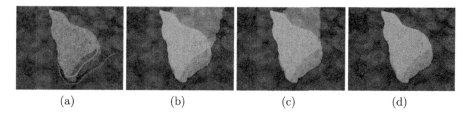

(a)	(b)	(c)	(d)

Fig. 1. Image segmentation through different metrics. (a) The original image and seeds. (b–d) Segmentation results via isotropic Riemannian metric, anisotropic Riemannian metric and the proposed Finsler metric.

In this paper we extend the geodesic distance-based fronts propagation framework to the Finsler case, where the edge anisotropy and asymmetry are taken into account simultaneously. Moreover, we present a way to construct the Finsler metric with respect to foreground and background segmentation. The existing fronts propagation methods invoking either Riemannian metric [12,16] or pseudo path metric [13,14], do not take into account the edge asymmetry information. This may lead to leaking problem when the seeds are close to the targeted boundaries. We show an example of such problems in Fig. 1, where the seeds are shown in Fig. 1a with green and red brushes indicating background and foreground. It can be seen that the segmentation contours from the Riemannian metrics shown in Figs. 1b and c cross the boundaries before the whole object has been covered by the fronts. In contrast, the segmentation results from the proposed Finsler metric case can catch the desired boundary (see Fig. 1d).

1.1 Paper Outline

The remaining of this paper is organized as follows: In Sect. 2, we introduce the geodesic distance map associated to a general Finsler metric, the Voronoi regions and the relevant numerical tool. Section 3 presents construction principle for the asymmetric Finsler metric. The numerical considerations for the Finsler metric-based fronts propagation are introduced in Sect. 4. The experimental results and the conclusion are respectively presented in Sects. 5 and 6.

2   Background on Geodesic Distance Map

A Finsler geodesic metric $\mathcal{F} : \Omega \times \mathbb{R}^2 \to [0, +\infty]$ is a continuous function over the domain $\Omega \times \mathbb{R}^2$. For each fixed point $\mathbf{x} \in \Omega$, the geodesic metric $\mathcal{F}(\mathbf{x}, \mathbf{v})$ can be characterized by an asymmetric norm of $\mathbf{v} \in \mathbb{R}^2$. In other words, $\mathcal{F}$ is convex and 1-homogeneous on its second argument. It is also potentially asymmetric such that $\exists \mathbf{x} \in \Omega$ and $\exists \mathbf{v} \in \mathbb{R}^2$, the inequality $\mathcal{F}(\mathbf{x}, \mathbf{v}) \neq \mathcal{F}(\mathbf{x}, -\mathbf{v})$ is held.

The curve length associated to the metric $\mathcal{F}$ along a Lipschitz continuous curve $\mathcal{C}$ can be expressed by $\ell_{\mathcal{F}}(\mathcal{C}) := \int_{\mathcal{C}} \mathcal{F}(\mathcal{C}(s), \mathcal{C}'(s)) \, ds$ with s the arc-length parameter of $\mathcal{C}$. It is possible for the geodesic length $\ell_{\mathcal{F}}$ to take into account both the path directions and image data [17]. Letting $\mathfrak{s} \subset \Omega$ be the set of the source points, the minimal curve length from $\mathbf{y}$ to $\mathbf{x}$ associated to the metric $\mathcal{F}$ is defined by

$$\mathcal{D}_{\mathcal{F}}(\mathbf{y}, \mathbf{x}) = \inf_{\mathcal{C} \in \mathcal{A}_{\mathbf{y}, \mathbf{x}}} \ell_{\mathcal{F}}(\mathcal{C}), \tag{3}$$

where $\mathcal{A}_{\mathbf{y}, \mathbf{x}}$ is the set of Lipschitz continuous curves linking $\mathbf{y}$ to $\mathbf{x} \in \Omega$. The geodesic distance map $\mathcal{U}_{\mathfrak{s}}$ associated to the metric $\mathcal{F}$ can be defined by

$$\mathcal{U}_{\mathfrak{s}}(\mathbf{x}) := \inf_{\mathbf{s} \in \mathfrak{s}} \mathcal{D}_{\mathcal{F}}(\mathbf{s}, \mathbf{x}), \tag{4}$$

It is the unique viscosity solution to the Eikonal equation such that $\mathcal{U}_{\mathfrak{s}}(\mathbf{x}) = 0$, $\forall \mathbf{x} \in \mathfrak{s}$, and

$$\max_{\|\mathbf{v}\| \neq 0} \frac{\langle \nabla \mathcal{U}_{\mathfrak{s}}(\mathbf{x}), \mathbf{v} \rangle}{\mathcal{F}(\mathbf{x}, \mathbf{v})} = 1, \forall \mathbf{x} \in \Omega \backslash \mathfrak{s}, \tag{5}$$

where $\nabla \mathcal{U}_{\mathfrak{s}}(\cdot)$ denote the Euclidean gradient vector and $\langle \cdot, \cdot \rangle$ is the standard Euclidean scalar product in $\mathbb{R}^2$. The Eikonal equation (5) can be interpreted by the Bellman's optimality principle such that

$$\mathcal{U}_{\mathfrak{s}}(\mathbf{x}) = \min_{\mathbf{y} \in \partial \Lambda(\mathbf{x})} \{ \mathcal{D}_{\mathcal{F}}(\mathbf{y}, \mathbf{x}) + \mathcal{U}_{\mathfrak{s}}(\mathbf{y}) \}, \tag{6}$$

where $\Lambda(\mathbf{x})$ is a neighbourhood of point $\mathbf{x}$ and $\partial \Lambda(\mathbf{x})$ is the boundary of $\Lambda(\mathbf{x})$.

2.1   Voronoi Index Map

We consider a more general case for which a family of source point sets, denoted by $\mathfrak{s}_k$, are provided. These sets are indexed by $k \in \{1, 2, \cdots, n\}$ with n the total number of source point sets. For the sake of simplicity, we note $\mathfrak{s} = \cup_{k=1}^{n} \mathfrak{s}_k$. A Voronoi index map can be defined as a labelling function $\mathcal{L} : \Omega \to \{1, 2, 3, \ldots, n\}$, which satisfies that $\mathcal{L}(\mathbf{x}) = k$, $\forall \mathbf{x} \in \mathfrak{s}_k$. In the sense of the geodesic distance $\mathcal{U}_k$, the Voronoi index map $\mathcal{L}$ assigns a label identical to the index of its closest source point set such that

$$\mathcal{L}(\mathbf{x}) = \arg\min_{1 \leq k \leq n} \mathcal{U}_k(\mathbf{x}). \tag{7}$$

By the map $\mathcal{L}$, one can partition the domain Ω into n Voronoi regions $\mathcal{V}_k \subset \Omega$

$$\mathcal{V}_k := \{ \mathbf{x} \in \Omega; \mathcal{L}(\mathbf{x}) = k \}. \tag{8}$$

The common boundary $\Gamma_{i,j} := \partial \mathcal{V}_i \cap \partial \mathcal{V}_j$ of two adjacent voronoi regions $\mathcal{V}_i$ and $\mathcal{V}_j$ is comprised of a set of equidistant points to the collections $\mathfrak{s}_i$ and $\mathfrak{s}_j$. The geodesic distance map $\mathcal{U}_{\mathfrak{s}}$ associated to $\mathfrak{s} = \cup_k \mathfrak{s}_k$ can be computed by

$$\mathcal{U}_{\mathfrak{s}}(\mathbf{x}) = \min_{1 \leq k \leq n} \mathcal{U}_k(\mathbf{x}). \tag{9}$$

2.2   Fast Marching Method

The fast marching method is a very efficient way to estimate the geodesic distance map. One key point of the fast marching method is the stencil map Λ, where $\Lambda(\mathbf{x})$ defines the neighbourhood of a grid point $\mathbf{x}$. The original fast marching methods [10, 11] are established on the regular 4-connectivity neighbourhood system, which may suffer some difficulties for the general Finsler metric. Alternatively, the Finsler variant of the fast marching method [18] make use of a complicated neighbourhood system depending on the metric.

The Finsler invariant of the fast marching method [18] estimates the distance values on a discretization grid $\mathbb{Z}^2$ of the domain Ω. It makes use of the Hopf-Lax operator to approximate the Eikonal equation (6) such that

$$\mathcal{U}_\mathfrak{s}(\mathbf{x}) = \min_{\mathbf{y} \in \partial \Lambda(\mathbf{x})} \{\mathcal{F}(\mathbf{x}, \mathbf{x} - \mathbf{y}) + \mathbb{I}_{\Lambda(\mathbf{x})} \mathcal{U}_\mathfrak{s}(\mathbf{y})\}, \tag{10}$$

where $\Lambda(\mathbf{x})$ denotes the stencil of $\mathbf{x}$ involving a set of vertices in $\mathbb{Z}^2$ and $\mathbb{I}_{\Lambda(\mathbf{x})}$ is a piecewise linear interpolation operator in the neighbourhood $\Lambda(\mathbf{x})$. The minimal curve length $\mathcal{D}_\mathcal{F}$ of a short geodesic from $\mathbf{y}$ to $\mathbf{x}$ is approximated by the value of $\mathcal{F}(\mathbf{x}, \mathbf{x} - \mathbf{y})$. The distance value $\mathcal{U}_\mathfrak{s}(\mathbf{y})$ in Eq. (6) is estimated by the piecewise linear interpolation operator $\mathbb{I}_{\Lambda(\mathbf{x})}$ at $\mathbf{y}$ located at the stencil boundary $\partial \Lambda(\mathbf{x})$. It is comprised of a set $\mathcal{T}_\mathbf{x}$ of 1-dimensional simplexes or line segments. Each simplex $\mathbb{T}_i \in \mathcal{T}_\mathbf{x}$ connects two adjacent vertices which are involved in the stencil $\Lambda(\mathbf{x})$. The solution $\mathcal{U}_\mathfrak{s}$ to the Hopf-Lax operator (10) can be attained by

$$\mathcal{U}_\mathfrak{s}(\mathbf{x}) = \min_{\mathbb{T}_i \in \mathcal{T}_\mathbf{x}} U_i(\mathbf{x}), \tag{11}$$

where U_i is the solution to the minimization problem

$$U_i(\mathbf{x}) = \min_{\mathbf{y} \in \mathbb{T}_i} \{\mathcal{F}(\mathbf{x}, \mathbf{x} - \mathbf{y}) + \mathbb{I}_{\Lambda(\mathbf{x})} \mathcal{U}_\mathfrak{s}(\mathbf{y})\}. \tag{12}$$

For each simplex $\mathbb{T}_i \in \mathcal{T}_\mathbf{x}$ which joins two vertices $\mathbf{z}_1$ and $\mathbf{z}_2$, the minimization problem (12) can be approximated by Tsitsiklis' theorem [11] such that

$$U_i(\mathbf{x}) = \min_\lambda \mathcal{F}\left(\mathbf{x}, \mathbf{x} - \sum_{i=1}^2 \lambda_i \mathbf{z}_i\right) + \sum_{i=1}^2 \lambda_i \mathcal{U}_\mathfrak{s}(\mathbf{z}_i), \tag{13}$$

where $\boldsymbol{\lambda} = (\lambda_1, \lambda_2)$ subject to $\lambda_1, \lambda_2 \geq 0$ and $\sum_i^2 \lambda_i = 1$.

Fast Marching Update Scheme. The fast marching method estimates the geodesic distance map $\mathcal{U}_\mathfrak{s}$ in a wave front propagation manner. The fast marching fronts propagation is coupled with a procedure of label assignment operation, during which all the grid points are classified into three categories: *Accepted* points (for which the values of $\mathcal{U}_\mathfrak{s}$ have been estimated and frozen), *Far* points (for which the values of $\mathcal{U}_\mathfrak{s}$ are unknown), and *Trial* points (the remaining grid points in $\mathbb{Z}^2$ which form the fast marching *fronts*). A *Trial* point will be assigned a label of *Accepted* if it has the minimal geodesic distance value among all the *Trial* points. In the course of the geodesic distance estimation, each grid point $\mathbf{x} \in \mathbb{Z}^2\backslash\mathfrak{s}$ will be visited by the monotonically advancing fronts which expand from the source points involved in $\mathfrak{s}$. The values of $\mathcal{U}_\mathfrak{s}$ for all the *Trial* points are stored in a priority queue in order to quickly find the point with minimal $\mathcal{U}_\mathfrak{s}$. The label assignment procedure[1] can be carried out by a binary map $b : \mathbb{Z}^2 \to \{Accepted, Far, Trial\}$.

Suppose that $\mathfrak{s} = \cup_k \mathfrak{s}_k$ with $\mathfrak{s}_k$ a source point set. The geodesic distance map $\mathcal{U}_\mathfrak{s}$ and the Voronoi index map $\mathcal{L}$ can be simultaneously computed [19,20], where the computation scheme in each iteration can be divided into two steps.

Voronoi index update. In each geodesic distance update iteration, among all the *Trial* points, a point $\mathbf{x}_{\min}$ that globally minimizes the geodesic distance map $\mathcal{U}_\mathfrak{s}$ is chosen and tagged as *Accepted*. We set $\mathcal{L}(\mathbf{x}_{\min}) = k$ if $\mathbf{x}_{\min} \in \mathfrak{s}_k$. Otherwise, the geodesic distance value $\mathcal{U}_\mathfrak{s}(\mathbf{x}_{\min})$ can be estimated in the simplex $\mathbb{T}^* \in \mathcal{T}_{\mathbf{x}_{\min}}$ (see Eq. (11)), where the vertices relevant to $\mathbb{T}^*$ are respective $\mathbf{z}_1$ and $\mathbf{z}_2$. This is done by finding the solution to (13) with respect to the simplex $\mathbb{T}^*$, where the minimizer is $\boldsymbol{\lambda}^* = (\lambda_1^*, \lambda_2^*)$. Then the Voronoi index map $\mathcal{L}$ can be computed by

$$\mathcal{L}(\mathbf{x}_{\min}) = \begin{cases} \mathcal{L}(\mathbf{z}_1), & \text{if } \lambda_1^* \geq \lambda_2^*, \\ \mathcal{L}(\mathbf{z}_2), & \text{otherwise.} \end{cases} \qquad (14)$$

Local geodesic distance update. For a grid point $\mathbf{x}$, we denote by $\Lambda_\star(\mathbf{x}) := \{\mathbf{z} \in \mathbb{Z}^2; \mathbf{x} \in \Lambda(\mathbf{z})\}$ the reverse stencil. The remaining step in this iteration is to update $\mathcal{U}_\mathfrak{s}(\mathbf{z})$ for each grid point $\mathbf{z}$ such that $\mathbf{z} \in \Lambda_\star(\mathbf{x}_{\min})$ and $b(\mathbf{z}) \neq Accepted$ through the solution $\hat{\mathcal{U}}_\mathfrak{s}(\mathbf{z})$ to the Hopf-Lax operator (10). This is done by assigning to $\mathcal{U}_\mathfrak{s}(\mathbf{z})$ the smaller value between the solution $\hat{\mathcal{U}}_\mathfrak{s}(\mathbf{z})$ and the current geodesic distance value of $\mathcal{U}_\mathfrak{s}(\mathbf{z})$. Note that the solution $\hat{\mathcal{U}}_\mathfrak{s}(\mathbf{z})$ to (10) is attained using the stencil $\Lambda(\mathbf{z})$ [18]. The algorithm for the fast marching method is described in Algorithm 1. In this algorithm, the computation of a map $\mathfrak{C}_{\mathrm{dyn}}$ in Line 12 of Algorithm 1 is not necessary for the general fast marching fronts propagation scheme, but required by our method as discussed in Sect. 4.2.

3   Finsler Metrics Construction

Definition 1. *Let S_2^+ be the collection of all the positive definite symmetric matrices with size 2×2. For any matrix $M \in S_2^+$, we define a norm $\|\mathbf{u}\|_M = \sqrt{\langle \mathbf{u}, M\mathbf{u}\rangle}, \forall \mathbf{u} \in \mathbb{R}^2$.*

In this section, we present the construction method of the Finsler metric which is suitable for fronts propagation and image segmentation. Suppose that a vector field $\mathfrak{g} : \Omega \to \mathbb{R}^2$ has been provided such that $\mathfrak{g}(\mathbf{x})$ points to the object edges at least when $\mathbf{x}$ is nearby them. In this case, the orthogonal vector field $\mathfrak{g}^\perp$ indicates the tangents of the edges.

---

[1] Initially, each source point $\mathbf{x} \in \mathfrak{s}$ is tagged as *Trial* and the remaining grid points are tagged as *Far*.

Algorithm 1. Fast Marching Fronts Propagation

Input: Source points set $\mathfrak{s} = \cup_k \mathfrak{s}_k$.
Output: Geodesic distance map $\mathcal{U}_\mathfrak{s}$ and Voronoi index map $\mathcal{L}$.
1: $\forall \mathbf{x} \in \Omega \backslash \mathfrak{s}$, set $\mathcal{U}_\mathfrak{s}(\mathbf{x}) \leftarrow \infty$ and $b(\mathbf{x}) \leftarrow Far$.
2: $\forall \mathbf{x} \in \mathfrak{s}$, set $\mathcal{U}_\mathfrak{s}(\mathbf{x}) \leftarrow 0$ and $b(\mathbf{x}) \leftarrow Trial$.
3: $\forall \mathbf{x} \in \mathfrak{s}_k$, set $\mathcal{L}(\mathbf{x}) = k$.
4: **while** there remains at least one *Trial* point **do**
5:     Find a *Trial* point $\mathbf{x}_{\min}$ globally minimizing $\mathcal{U}_\mathfrak{s}$.
6:     Set $b(\mathbf{x}_{\min}) \leftarrow Accepted$.
7:     **if** $\mathbf{x}_{\min} \notin \mathfrak{s}$ **then**
8:         Update the Voronoi index $\mathcal{L}(\mathbf{x}_{\min})$ by Eq. (14).
9:     **end if**
10:     **for** all $\mathbf{z} \in \mathbb{Z}^2$ such that $\mathbf{x}_{\min} \in \Lambda(\mathbf{z})$ **do**
11:         **if** $b(\mathbf{z}) \neq Accepted$ and $\mathbf{z} \notin \mathfrak{s}$ **then**
12:             /* Update some map $\mathfrak{C}_{\text{dyn}}(\mathbf{z})$ if necessary. */
13:             Find $\hat{\mathcal{U}}(\mathbf{z})$ by evaluating the Hopf-Lax formula (11).
14:             Set $\mathcal{U}_\mathfrak{s}(\mathbf{z}) \leftarrow \min\{\mathcal{U}_\mathfrak{s}(\mathbf{z}), \hat{\mathcal{U}}(\mathbf{z})\}$ and $b(\mathbf{z}) \leftarrow Trial$.
15:         **end if**
16:     **end for**
17: **end while**

Basically, the Eikonal equation-based fronts propagation models [9] perform the segmentation scheme through a geodesic distance map. In order to find a good solution for image segmentation, the used geodesic metric should be able to reduce the risk of front leaking problem. For this purpose, we search for a direction-dependent metric $\mathcal{F}_\mathfrak{g}$ satisfying the following inequality such that for an edge point $\mathbf{x}$

$$\mathcal{F}_\mathfrak{g}(\mathbf{x}, \mathfrak{g}^\perp(\mathbf{x})) < \mathcal{F}_\mathfrak{g}(\mathbf{x}, \mathfrak{g}(\mathbf{x})) < \mathcal{F}_\mathfrak{g}(\mathbf{x}, -\mathfrak{g}(\mathbf{x})). \tag{15}$$

Recall that for an edge point $\mathbf{x}$, both the feature vectors $\mathfrak{g}^\perp(\mathbf{x})$ or $-\mathfrak{g}^\perp(\mathbf{x})$ are propositional to the tangent of the edge at $\mathbf{x}$. When the fast marching front arrives at the vicinity of image edges, it prefers to travel along the edge feature vectors $\mathfrak{g}^\perp(\mathbf{x})$ and $-\mathfrak{g}^\perp(\mathbf{x})$, instead of passing through the edges, i.e., prefers to travel along the direction $-\mathfrak{g}(\mathbf{x})$.

The inequality (15) requires the geodesic metric $\mathcal{F}_\mathfrak{g}$ to be anisotropic and asymmetric with respect to its second argument. Thus, we consider a Finsler metric with a Randers form [21] involving a symmetric quadratic term and a linear asymmetric term for any $\mathbf{x} \in \mathbb{R}^2$ and any vector $\boldsymbol{u} \in \mathbb{R}^2$

$$\mathcal{F}(\mathbf{x}, \boldsymbol{u}) := \mathfrak{C}(\mathbf{x}) \left(\|\boldsymbol{u}\|_{\mathcal{M}_\mathfrak{g}(\mathbf{x})} - \langle \boldsymbol{\omega}_\mathfrak{g}(\mathbf{x}), \boldsymbol{u} \rangle \right), \tag{16}$$

where $\mathcal{M}_\mathfrak{g} : \Omega \to S_2^+$ is a positive symmetric definite tensor field and $\boldsymbol{\omega}_\mathfrak{g} : \Omega \to \mathbb{R}^2$ is a vector field that is sufficiently small. The function $\mathfrak{C} : \Omega \to \mathbb{R}^+$ is a positive scalar-valued potential which gets small values in the homogeneous regions and large values around the image edges. It can be derived from the image data such as the coherence measurements of the image features, which will be discussed in detail in Sect. 4.2. The tensor field $\mathcal{M}_{\mathfrak{g}}$ and the vector field $\boldsymbol{\omega}_{\mathfrak{g}}$ should satisfy the constraint

$$\|\boldsymbol{\omega}_{\mathfrak{g}}(\mathbf{x})\|_{\mathcal{M}_{\mathfrak{g}}^{-1}(\mathbf{x})} < 1, \quad \forall \mathbf{x} \in \Omega, \tag{17}$$

in order to guarantee the positiveness [18] of the Randers metric $\mathcal{F}$.

We reformulate the Randers metric $\mathcal{F}_{\mathfrak{g}}$ in Eq. (16) as

$$\mathcal{F}_{\mathfrak{g}}(\mathbf{x}, \mathbf{u}) = \mathfrak{C}(\mathbf{x}) \, \mathcal{G}_{\mathfrak{g}}(\mathbf{x}, \mathbf{u}), \tag{18}$$

where $\mathcal{G}_{\mathfrak{g}} : \Omega \times \mathbb{R}^2 \to [0, \infty]$ is still a Randers metric formulated by

$$\mathcal{G}_{\mathfrak{g}}(\mathbf{x}, \mathbf{u}) = \|\mathbf{u}\|_{\mathcal{M}_{\mathfrak{g}}(\mathbf{x})} - \langle \boldsymbol{\omega}_{\mathfrak{g}}(\mathbf{x}), \mathbf{u} \rangle. \tag{19}$$

The remaining part of this section will be devoted to the construction of the Randers metric $\mathcal{G}_{\mathfrak{g}}$ in terms of the vector field $\mathfrak{g}$ which is able to characterize the directions orthogonal to the image edges.

Let us define a new vector field $\bar{\mathfrak{g}} : \Omega \to \mathbb{R}^2$ by $\bar{\mathfrak{g}}(\mathbf{x}) := \mathfrak{g}(\mathbf{x})/\|\mathfrak{g}(\mathbf{x})\|^2$. The tensor field $\mathcal{M}_{\mathfrak{g}}$ used in Eq. (16) can be constructed dependently on two scalar-valued coefficient functions η_1 and η_2 such that

$$\mathcal{M}_{\mathfrak{g}}(\mathbf{x}) = \eta_1^2(\mathbf{x})\bar{\mathfrak{g}}(\mathbf{x}) \otimes \bar{\mathfrak{g}}(\mathbf{x}) + \eta_2(\mathbf{x})\bar{\mathfrak{g}}^{\perp}(\mathbf{x}) \otimes \bar{\mathfrak{g}}^{\perp}(\mathbf{x}), \tag{20}$$

where $\bar{\mathfrak{g}}^{\perp}(\mathbf{x})$ is the orthogonal vector of $\bar{\mathfrak{g}}(\mathbf{x})$ and $\otimes$ denotes the tensor product, i.e., $\mathbf{u} \otimes \mathbf{u} = \mathbf{u}\mathbf{u}^T$. Note that the eigenvalues of $\mathcal{M}_{\mathfrak{g}}(\mathbf{x})$ are $\eta_1^2(\mathbf{x})/\|\mathfrak{g}(\mathbf{x})\|^2$ and $\eta_2(\mathbf{x})/\|\mathfrak{g}(\mathbf{x})\|^2$, respectively corresponding to the eigenvectors $\mathfrak{g}(\mathbf{x})/\|\mathfrak{g}(\mathbf{x})\|$ and $\mathfrak{g}^{\perp}(\mathbf{x})/\|\mathfrak{g}(\mathbf{x})\|$. The vector $\boldsymbol{\omega}_{\mathfrak{g}}(\mathbf{x})$ is positively collinear to field $\mathfrak{g}(\mathbf{x})$ for all $\mathbf{x} \in \Omega$

$$\boldsymbol{\omega}_{\mathfrak{g}}(\mathbf{x}) = \tau(\mathbf{x}) \, \bar{\mathfrak{g}}(\mathbf{x}), \tag{21}$$

where $\tau : \Omega \to \mathbb{R}$ is a scalar-valued coefficient function.

We estimate the coefficient functions η_1, η_2 and τ through two cost functions $\psi_{\mathrm{f}}, \psi_{\mathrm{b}} : \Omega \to (1, +\infty)$, which assign the cost values $\psi_{\mathrm{f}}(\mathbf{x})$, $\psi_{\mathrm{b}}(\mathbf{x})$ and 1 to the Randers metric $\mathcal{G}_{\mathfrak{g}}$ respectively along the directions $\mathfrak{g}(\mathbf{x})$, $-\mathfrak{g}(\mathbf{x})$ and $\mathfrak{g}^{\perp}(\mathbf{x})$ for any point $\mathbf{x} \in \Omega$ such that

$$\mathcal{G}_{\mathfrak{g}}(\mathbf{x}, \mathfrak{g}(\mathbf{x})) = \psi_{\mathrm{f}}(\mathbf{x}), \quad \mathcal{G}_{\mathfrak{g}}(\mathbf{x}, -\mathfrak{g}(\mathbf{x})) = \psi_{\mathrm{b}}(\mathbf{x}), \quad \mathcal{G}_{\mathfrak{g}}(\mathbf{x}, \mathfrak{g}^{\perp}(\mathbf{x})) = 1. \tag{22}$$

Combining Eqs. (20) and (22) yields that

$$\eta_1(\mathbf{x}) - \tau(\mathbf{x}) = \psi_{\mathrm{f}}(\mathbf{x}), \quad \eta_1(\mathbf{x}) + \tau(\mathbf{x}) = \psi_{\mathrm{b}}(\mathbf{x}), \quad \eta_2(\mathbf{x}) = 1, \quad \forall \mathbf{x} \in \Omega. \tag{23}$$

The positive symmetric definite tensor field $\mathcal{M}_{\mathfrak{g}}$ and the vector field $\boldsymbol{\omega}_{\mathfrak{g}}$ thus can be respectively expressed in terms of the cost functions ψ_{f} and ψ_{b} by

$$\mathcal{M}_{\mathfrak{g}}(\mathbf{x}) = \frac{1}{4}(\psi_{\mathrm{f}}(\mathbf{x}) + \psi_{\mathrm{b}}(\mathbf{x}))^2 \, \bar{\mathfrak{g}}(\mathbf{x}) \otimes \bar{\mathfrak{g}}(\mathbf{x}) + \bar{\mathfrak{g}}^{\perp}(\mathbf{x}) \otimes \bar{\mathfrak{g}}^{\perp}(\mathbf{x}), \tag{24}$$

$$\boldsymbol{\omega}_{\mathfrak{g}}(\mathbf{x}) = \frac{1}{2}(\psi_{\mathrm{b}}(\mathbf{x}) - \psi_{\mathrm{f}}(\mathbf{x})) \, \bar{\mathfrak{g}}(\mathbf{x}). \tag{25}$$

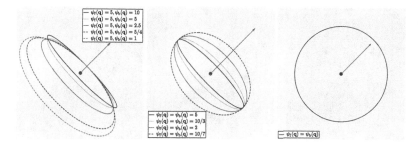

Fig. 2. Control sets for different metrics corresponding to different values of $\psi_f(\mathbf{x}_0)$ and $\psi_b(\mathbf{x}_0)$. The blue dots and the contours denote the origins and the boundaries of these balls, respectively. (Color figure online)

Based on the tensor field $\mathcal{M}_\mathfrak{g}$ and the vector field $\boldsymbol{\omega}_\mathfrak{g}$ respectively formulated in Eqs. (24) and (25), the positiveness constraint (17) is satisfied due to the assumption that $\psi_f(\mathbf{x}) > 1$ and $\psi_b(\mathbf{x}) > 1$, $\forall \mathbf{x} \in \Omega$. The cost functions ψ_f and ψ_b can be derived from the image edge information such as the image gradients, which will be discussed in Sect. 4.

Note that if we set $\psi_f \equiv \psi_b$, the vector field $\boldsymbol{\omega}_\mathfrak{g}$ will vanish, i.e., $\boldsymbol{\omega}_\mathfrak{g} \equiv \mathbf{0}$ (see Eq. (25)). In this case, one has $\langle \boldsymbol{\omega}_\mathfrak{g}(\mathbf{x}), \mathbf{u} \rangle = 0$ for any point $\mathbf{x} \in \Omega$ and any vector $\mathbf{u} \in \mathbb{R}^2$, leading to a special form of the Randers metric $\mathcal{G}_\mathfrak{g}$. This special form is a symmetric (potentially anisotropic) Riemannian metric $\mathcal{R}(\mathbf{x}, \mathbf{u}) = \|\mathbf{u}\|_{\mathcal{M}_\mathfrak{g}(\mathbf{x})}$ which depends only on the tensor field $\mathcal{M}_\mathfrak{g}$.

Tissot's indicatrix. A basic tool for studying and visualizing the geometry distortion induced from a geodesic metric is the Tissot's indicatrix defined as the collection of control sets in the tangent space [22]. For an arbitrary geodesic metric $\mathcal{F} : \Omega \times \mathbb{R}^2 \to [0, \infty]$, the control set $\mathcal{B}(\mathbf{x})$ for any point $\mathbf{x} \in \Omega$ is defined as the unit ball centered at $\mathbf{x}$ such that $\mathcal{B}(\mathbf{x}) := \{\mathbf{u} \in \mathbb{R}^2; \mathcal{F}(\mathbf{x}, \mathbf{u}) \leq 1\}$. We demonstrate the control sets $\mathcal{B}(\mathbf{q})$ in Fig. 2 for the Randers metric $\mathcal{G}_\mathfrak{g}(\mathbf{q}, \cdot)$ with different values of $\psi_f(\mathbf{q})$ and $\psi_b(\mathbf{q})$ at a point $\mathbf{q} \in \Omega$. The vector $\mathfrak{g}(\mathbf{q})$ is set as $\mathfrak{g}(\mathbf{q}) = \left(\cos\left(\frac{\pi}{4}\right), \sin\left(\frac{\pi}{4}\right)\right)^T$. In Fig. 2a, we show the control sets for the Randers metric $\mathcal{G}_\mathfrak{g}$ with respect to different values of $\psi_b(\mathbf{q})$ and a fixed value $\psi_f(\mathbf{q}) = 5$. One can point out that the common origin of these control sets have shifted from the original center of the ellipses[2] due to the asymmetric property. In Fig. 2b, the control sets for the Randers metric $\mathcal{G}_\mathfrak{g}$ associated to $\psi_f(\mathbf{q}) = \psi_b(\mathbf{q}) > 1$ are demonstrated, where $\mathcal{G}_\mathfrak{g}(\mathbf{q}, \cdot)$ gets to be anisotropic and symmetric on its second argument. When $\psi_f(\mathbf{q}) = \psi_b(\mathbf{q}) = 1$, the values of the Randers metric $\mathcal{G}_\mathfrak{g}(\mathbf{q}, \boldsymbol{u})$ turn to be invariant with respect to $\boldsymbol{u}$ as shown in Fig. 2c. In this case, the tensor $\mathcal{M}_\mathfrak{g}(\mathbf{q})$ is propositional to the identity matrix. In Fig. 3, we show the geodesic distance maps associated to $\mathcal{G}_\mathfrak{g}$ with different values of the cost functions ψ_f and ψ_b. The vector field $\mathfrak{g}$ is set to $\mathfrak{g} \equiv \left(\cos\left(\frac{\pi}{4}\right), \sin\left(\frac{\pi}{4}\right)\right)^T$.

---

[2] These ellipses are the boundaries of the control sets.

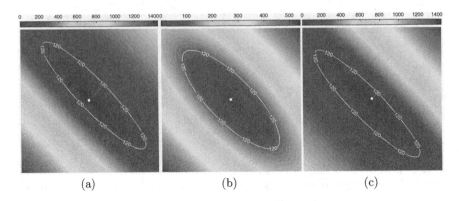

Fig. 3. Geodesic distance maps associated to the Randers metric $\mathcal{G}_{\mathfrak{a}}$ with different values of ψ_f and ψ_b. The red arrow indicate the vector $(\cos(\pi/4), \sin(\pi/4))^T$. The white dots are the source points. Each white curve indicates a level set line of the respective geodesic distance map. (a) shows the geodesic distance map associated to $\psi_f \equiv 3$ and $\psi_b \equiv 8$. (b) shows the geodesic distance map associated to $\psi_f \equiv 3$ and $\psi_b \equiv 3$. (c) shows the geodesic distance map associated to $\psi_f \equiv 8$ and $\psi_b \equiv 3$. (Color figure online)

In Figs. 3a and c, we can see that the geodesic distance maps have strongly asymmetric and anisotropic appearance. In Fig. 3b, we observe that the geodesic distance map appears to be symmetric and strongly anisotropic. This is because the respective propagation speed of the fast marching fronts along the directions $(\cos(\pi/4), \sin(\pi/4))^T$ and $-(\cos(\pi/4), \sin(\pi/4))^T$ are identical to each other.

4   Numerical Considerations

Let $\mathbf{I} = (I_1, I_2, I_3) : \Omega \to \mathbb{R}^3$ be a vector-valued image in the chosen color space and let G_σ be a Gaussian kernel with variance σ (we set $\sigma = 1$ through all the experiments of this paper). The gradient of the image $\mathbf{I}$ at each point $\mathbf{x} = (x, y)$ is a 2×3 Jacobian matrix $\nabla \mathbf{I}_\sigma(\mathbf{x}) = \nabla G_\sigma * \mathbf{I}(\mathbf{x})$ involving the Gaussian-smoothed first-order derivatives along the axis directions x and y

$$\nabla \mathbf{I}_\sigma(\mathbf{x}) = \begin{pmatrix} \partial_x G_\sigma * I_1 & \partial_x G_\sigma * I_2 & \partial_x G_\sigma * I_3 \\ \partial_y G_\sigma * I_1 & \partial_y G_\sigma * I_2 & \partial_y G_\sigma * I_3 \end{pmatrix}(\mathbf{x}). \tag{26}$$

Let $\rho : \Omega \to \mathbb{R}$ be an edge saliency map. It has high values in the vicinity of image edges and low values inside the flatten regions. For each domain point $\mathbf{x} \in \Omega$, the value of $\rho(\mathbf{x})$ can be computed by the Frobenius norm of the Jacobian matrix $\nabla \mathbf{I}_\sigma(\mathbf{x})$

$$\rho(\mathbf{x}) = \sqrt{\sum_{i=1}^{3} \left(|\partial_x G_\sigma * I_i(\mathbf{x})|^2 + |\partial_y G_\sigma * I_i(\mathbf{x})|^2 \right)}. \tag{27}$$

For a gray level image $I : \Omega \to \mathbb{R}$, the edge saliency map ρ can be simply computed by the norm of the Euclidean gradient of the image I such that

$$\rho(\mathbf{x}) = \sqrt{|\partial_x G_\sigma * I(\mathbf{x})|^2 + |\partial_y G_\sigma * I(\mathbf{x})|^2}. \tag{28}$$

4.1   Construction of the Vector Field $\mathfrak{g}$

We use the gradient vector flow method [23] to compute the vector field $\mathfrak{g}$ for the construction of the Randers metric $\mathcal{F}_\mathfrak{g}$. This can be done by minimizing the following functional $\mathcal{E}_{gvf}$ with respect to a vector field $\boldsymbol{h} = (h_1, h_2)^T : \Omega \to \mathbb{R}^2$, where $\mathcal{E}_{gvf}$ can be expressed as

$$\mathcal{E}_{gvf}(\boldsymbol{h}) = \epsilon \, \mathcal{E}_{reg}(\boldsymbol{h}) + \mathcal{E}_{data}(\boldsymbol{h}), \tag{29}$$

where $\epsilon \in \mathbb{R}^+$ is a constant and

$$\mathcal{E}_{reg}(\boldsymbol{h}) = \int_\Omega \left(\|\nabla h_1(\mathbf{x})\|^2 + \|\nabla h_2(\mathbf{x})\|^2 \right) d\mathbf{x}, \tag{30}$$

$$\mathcal{E}_{data}(\boldsymbol{h}) = \int_\Omega \|\nabla\rho(\mathbf{x})\|^2 \|\boldsymbol{h}(\mathbf{x}) - \nabla\rho(\mathbf{x})\|^2 \, d\mathbf{x}. \tag{31}$$

The parameter ϵ controls the balance between the regularization term $\mathcal{E}_{reg}$ and the data fidelity term $\mathcal{E}_{data}$. As discussed in [23], the values of ϵ should depend on the image noise levels such that a large value of ϵ is able to suppress the effects from image noise. We set $\epsilon = 0.1$ through all the numerical experiments of this paper. The minimization of the functional $\mathcal{E}_{gvf}$ can be carried out by solving the Euler-Lagrange equations of the functional $\mathcal{E}_{gvf}$ with respect to the components h_1 and h_2. The gradient vector flow $\boldsymbol{h}$ is more dense and smooth than the original gradient filed $\nabla\rho$. The vector field $\mathfrak{g}$ for the construction of the Randers metric $\mathcal{G}_\mathfrak{g}$ can be obtained by normalizing the vector field $\boldsymbol{h}$

$$\mathfrak{g}(\mathbf{x}) = \boldsymbol{h}(\mathbf{x})/\|\boldsymbol{h}(\mathbf{x})\|, \quad \forall \mathbf{x} \in \Omega. \tag{32}$$

The cost functions ψ_f and ψ_b used in Eq. (22) for the foreground and background segmentation application can be expressed for any $\mathbf{x} \in \Omega$ by

$$\psi_f(\mathbf{x}) = \exp\left(\alpha_f \, \rho(\mathbf{x})/\|\rho\|_\infty \right), \quad \psi_b(\mathbf{x}) = \exp\left(\alpha_b \, \rho(\mathbf{x})/\|\rho\|_\infty \right) \psi_f(\mathbf{x}), \tag{33}$$

where α_f and α_b are non-negative constants dominating how anisotropic and asymmetric the Randers metric $\mathcal{G}_\mathfrak{g}$ is. Once the cost functions ψ_f and ψ_b have been computed by Eq. (33), we can construct the tensor filed $\mathcal{M}_\mathfrak{g}$ and the vector field $\omega_\mathfrak{g}$ respectively via Eqs. (24) and (25). Indeed, one has $\psi_f(\mathbf{x}) \approx \psi_b(\mathbf{x}) \approx 1$ for the points $\mathbf{x}$ located in the homogeneous region of the image $\mathbf{I}$ where $\rho(\mathbf{x}) \approx 0$. In this case, the data-driven Randers metric $\mathcal{G}_\mathfrak{g}(\mathbf{x}, \cdot)$ in Eq. (19) approximates to be an isotropic Riemannian metric. For each point $\mathbf{x}$ around the image edges where the value of $\rho(\mathbf{x})$ is large, the Randers metric $\mathcal{G}_\mathfrak{g}(\mathbf{x}, \cdot)$ will appear to be strongly anisotropic and asymmetric.

4.2   Computing the Potential $\mathfrak{C}$

We present the computation methods for the potential function $\mathfrak{C}$ used by the data-driven Randers metric in Eq. (18). Basically, the function $\mathfrak{C}$ should have small values in the flatten regions and large values in the vicinity of image edges. The potential function $\mathfrak{C}$ can be expressed by

$$\mathfrak{C}(\mathbf{x}) = \exp\left(\beta_s\,\rho(\mathbf{x})/\|\rho\|_\infty\right)\mathfrak{C}_{\mathrm{dyn}}(\mathbf{x}), \qquad (34)$$

where β_s is a positive constant and ρ is the edge saliency map defined in Eqs. (27) or (28). The term $\exp(\beta_s\,\rho(\mathbf{x}))$ which depends only on the edge saliency map ρ will keep invariant during the fast marching fronts propagation. The dynamic potential function $\mathfrak{C}_{\mathrm{dyn}}$ relies on the positions of the fronts. It will be updated in the course of the geodesic distances computation in terms of some consistency measure of image features [14]. Basically, the values of the dynamic potential $\mathfrak{C}_{\mathrm{dyn}}$ should be small in the homogeneous regions. We use a feature map $\mathfrak{F}$: $\Omega \to \mathbb{R}^n$ with n the dimensions of the feature vector to establish the dynamic potential $\mathfrak{C}_{\mathrm{dyn}}$. The feature map $\mathfrak{F}$ can be the image color vector $(n = 3)$, the image gray level $(n = 1)$, or the scalar probability map $(n = 1)$ as used in [14].

Recall that in each fast marching distance update iteration, the latest *Accepted* point $\mathbf{x}_{\min}$ is chosen by searching for a *Trial* point with minimal distance value $\mathcal{U}_\mathfrak{s}$ ($\mathfrak{s}$ is the set of the source points), i.e.,

$$\mathbf{x}_{\min} := \underset{\mathbf{x}:b(\mathbf{x})=Trial}{\arg\min}\ \mathcal{U}_\mathfrak{s}(\mathbf{x}). \qquad (35)$$

Then the value of $\mathfrak{C}_{\mathrm{dyn}}(\mathbf{z})$ for each point $\mathbf{z} \in \mathbb{Z}^2\backslash\mathfrak{s}$ such that $\mathbf{x}_{\min} \in \Lambda(\mathbf{z})$ and $b(\mathbf{z}) \neq Accepted$ can be updated by evaluating the Euclidean distance between $\mathfrak{F}(\mathbf{z})$ and $\mathfrak{F}(\mathbf{x}_{\min})$ (see Line 12 of Algorithm 1). In other words, one can compute the dynamic potential $\mathfrak{C}_{\mathrm{dyn}}$ in each fast marching update iteration by

$$\mathfrak{C}_{\mathrm{dyn}}(\mathbf{z}) = \exp(\beta_d\,\|\mathfrak{F}(\mathbf{z}) - \mathfrak{F}(\mathbf{x}_{\min})\|) \qquad (36)$$

for all grid points $\mathbf{z} \in \mathbb{Z}^2\backslash\mathfrak{s}$ such that $\mathbf{x}_{\min} \in \Lambda(\mathbf{z})$ and $b(\mathbf{z}) \neq Accepted$, where β_d is a positive constant. Note that we initialize the dynamic potential $\mathfrak{C}_{\mathrm{dyn}}$ by $\mathfrak{C}_{\mathrm{dyn}}(\mathbf{x}) = 1,\quad \forall \mathbf{x} \in \mathfrak{s}$.

5   Experimental Results

The anisotropy and asymmetry of the Randers metric $\mathcal{G}_\mathfrak{g}$ is determined by the parameters α_f and α_b (see Eq. (33)). We denote by $\mathcal{G}_\mathfrak{g}^\alpha$ the Randers metric $\mathcal{G}_\mathfrak{g}$ with a pair of parameters $\boldsymbol{\alpha} = (\alpha_f, \alpha_b)$. In this case, the corresponding Randers metric $\mathcal{F}_\mathfrak{g}$ in Eq. (18) can be noted by $\mathcal{F}_\mathfrak{g}^\alpha$. The potential function $\mathfrak{C}$ relies on two parameters β_s and β_d. We fix $\beta_d = 10$ through all the experiments, except in Fig. 5 for which we set $\beta_d = 5$. The values of β_s are individually set for each experiment. Note that the parameter $\boldsymbol{\alpha} = (0,0)$ means that the metric $\mathcal{G}_\mathfrak{g}^{(0,0)}$

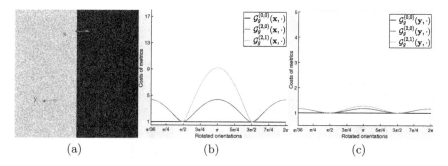

Fig. 4. (a) shows a synthetic image. The blue dots indicate two sampled points. The arrows indicate the directions $\mathfrak{g}(\mathbf{x})$ and $\mathfrak{g}(\mathbf{y})$. (b) and (c) respectively plot the cost values of $\mathcal{G}_{\mathfrak{g}}^{(0,0)}$, $\mathcal{G}_{\mathfrak{g}}^{(2,0)}$ and $\mathcal{G}_{\mathfrak{g}}^{(2,1)}$ at points $\mathbf{x}$ and $\mathbf{y}$ along different directions. (Color figure online)

is isotropic with respect to its second argument. Furthermore, when $\alpha = (a, 0)$ with $a \in \mathbb{R}^+$, the metric $\mathcal{G}_{\mathfrak{g}}^{(a,0)}$ gets to be the anisotropic Riemannian cases[3].

The interactive foreground and background segmentation task can be converted to the problem of identifying the Voronoi index map or Voronoi regions in terms of geodesic distance [13, 14]. Let $\mathfrak{s}_1$ and $\mathfrak{s}_2$ be the sets of source points which are respectively located at the foreground and background regions. The Voronoi regions $\mathcal{V}_1$ and $\mathcal{V}_2$, indicating foreground and background regions respectively, can be determined by the Voronoi index map $\mathcal{L}$ through Eq. (8) such that $\mathcal{V}_i := \{\mathbf{x} \in \Omega; \mathcal{L}(\mathbf{x}) = i\}, \quad i = 1, 2$.

Let us consider a synthetic image as shown in Fig. 4a with two sampled points $\mathbf{x}$ and $\mathbf{y}$ indicated by blue dots. The arrows respectively indicate the directions of $\mathfrak{g}(\mathbf{x})$ and $\mathfrak{g}(\mathbf{y})$, where $\mathbf{x}$ is near the edges and $\mathbf{y}$ is located inside the homogeneous region. In Fig. 4b, we plot the cost values of the metrics $\mathcal{G}_{\mathfrak{g}}^{(0,0)}(\mathbf{x}, \mathbf{u}_j)$, $\mathcal{G}_{\mathfrak{g}}^{(2,0)}(\mathbf{x}, \mathbf{u}_j)$ and $\mathcal{G}_{\mathfrak{g}}^{(2,1)}(\mathbf{x}, \mathbf{u}_j)$, along different directions $\mathbf{u}_j \in \mathbb{R}^2$. The directions $\mathbf{u}_j$ are obtained by rotation such that $\mathbf{u}_j = M(j\,\theta_s)\,\mathfrak{g}(\mathbf{x})$, $j = 1, 2, \ldots, 72$, where $\theta_s = \pi/36$ is the angle resolution and $M(j\,\theta_s)$ is a rotation matrix with angle $j\,\theta_s$ in a count-clockwise order. In Fig. 4c, we plot the cost values for the metrics $\mathcal{G}_{\mathfrak{g}}^{(0,0)}(\mathbf{y}, \mathbf{v}_j)$, $\mathcal{G}_{\mathfrak{g}}^{(2,0)}(\mathbf{y}, \mathbf{v}_j)$ and $\mathcal{G}_{\mathfrak{g}}^{(2,1)}(\mathbf{y}, \mathbf{v}_j)$ with $\mathbf{v}_j = M(j\,\theta_s)\mathfrak{g}(\mathbf{y})$. In Fig. 4b, we can see that all of the three metrics get low values around the directions $M(\pi/2)\,\mathfrak{g}(\mathbf{x})$ and $M(3\pi/2)\,\mathfrak{g}(\mathbf{x})$, which are orthogonal to the direction $\mathfrak{g}(\mathbf{x})$. However, around the direction $-\mathfrak{g}(\mathbf{x})$, the Randers metric $\mathcal{G}_{\mathfrak{g}}^{(2,1)}$ attains much larger values than the Riemannian cases $\mathcal{G}_{\mathfrak{g}}^{(0,0)}$ and $\mathcal{G}_{\mathfrak{g}}^{(2,0)}$. Such an asymmetric property is able to reduce the risk of front leakages.

In Fig. 5, we show the fronts propagation results on a synthetic image. In the first column of Fig. 5, we initialize the sets of the source points in different locations, which are indicated by green and blue brushes. The columns 2 to 4 of Fig. 5 are the segmentation results from the isotropic Riemannian metric

---

[3] Note that metric $\mathcal{G}_{\mathfrak{g}}^\alpha$ has the identical anisotropy and asymmetry properties to $\mathcal{F}_{\mathfrak{g}}^\alpha$.

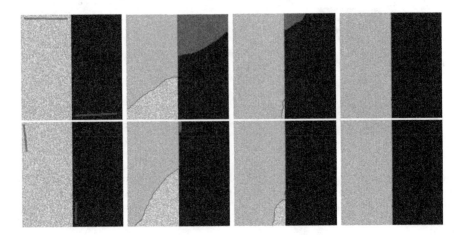

Fig. 5. Image segmentation via different geodesic metrics on a synthetic image. Column 1 shows the initializations, where the green and blue brushes indicating the seeds in different regions. Columns 2–4 show the segmentation results by the fronts propagation associated to the isotropic Riemannian metric $\mathcal{F}_{\mathfrak{g}}^{(0,0)}$, the anisotropic Riemannian metric $\mathcal{F}_{\mathfrak{g}}^{(2,0)}$ and the Randers metric $\mathcal{F}_{\mathfrak{g}}^{(2,3)}$, respectively. (Color figure online)

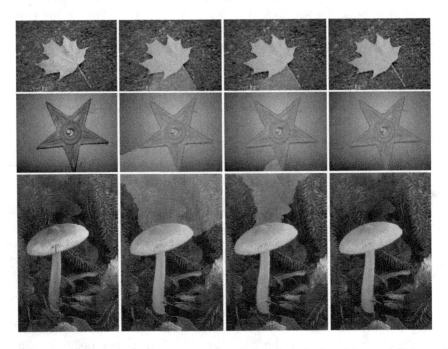

Fig. 6. Image segmentation via different geodesic metrics on real images. Column 1 shows the initializations, where the green and blue brushes are the seeds for background and foreground regions. Columns 2–4 show the segmentation results by the metrics $\mathcal{F}_{\mathfrak{g}}^{(0,0)}$, $\mathcal{F}_{\mathfrak{g}}^{(2,0)}$ and $\mathcal{F}_{\mathfrak{g}}^{(2,3)}$, respectively. (Color figure online)

$\mathcal{F}_{\mathfrak{g}}^{(0,0)}$, the anisotropic Riemannian metric $\mathcal{F}_{\mathfrak{g}}^{(2,0)}$ and the Randers metric $\mathcal{F}_{\mathfrak{g}}^{(2,3)}$, respectively. For the purpose of fair comparisons, the three metrics used in this experiment share the same potential function $\mathfrak{C}$ defined in Eq. (34). One can point out that the results from the metrics $\mathcal{F}_{\mathfrak{g}}^{(0,0)}$ and $\mathcal{F}_{\mathfrak{g}}^{(2,0)}$ suffer from the leaking problem, while the final boundaries (red curves) associated the proposed Randers metric $\mathcal{F}_{\mathfrak{g}}^{(2,3)}$ are able to catch the expected edges thanks to the asymmetric enhancement. In this experiment, we choose $\beta_{\mathrm{d}} = 5$.

In Fig. 6, we compare the interactive image segmentation results via different geodesic metrics on real images [24,25]. The final segmentation results are derived from the boundaries of the corresponding Voronoi index maps. In column 1, we show the initial images with seeds indicating by green and blue brushes respectively inside the foreground and background regions. In columns 2 to 4 of Fig. 6, we demonstrate the segmentation results obtained via the isotropic Riemannian metric $\mathcal{F}_{\mathfrak{g}}^{(0,0)}$, the anisotropic Riemannian metric $\mathcal{F}_{\mathfrak{g}}^{(2,0)}$ and the Randers Metric $\mathcal{F}_{\mathfrak{g}}^{(2,3)}$. For the results from the isotropic and anisotropic Riemannian metrics (shown in columns 2 and 3), the final contours leak into the background regions. In contrast, the segmentation contours associated to the Randers metric $\mathcal{F}^{(2,3)}$ are able to follow the desired object boundaries.

6   Conclusion

In this paper, we extend the fronts propagation framework from the Riemannian case to a general Finsler case with applications to image segmentation. The Finsler metric with a Randers form allows us to take into account the asymmetric and anisotropic image features in order to reduce the risk of the leaking problem during the fronts propagation. We presented a method for the construction of the Finsler metric with a Randers form using a vector field derived from the image edges. This metric can also integrate with a feature coherence penalization term updated in the course of the fast marching fronts propagation. Experimental results show that the proposed model indeed produces promising image segmentation results.

Acknowledgment. The authors would like to thank all the anonymous reviewers for their detailed remarks that helped us improve the presentation of this paper. The authors thank Dr. Jean-Marie Mirebeau from Université Paris-Sud for his fruitful discussion and creative suggestions. The first author also thanks Dr. Gabriel Peyré from ENS Paris for his financial support. This work was partially supported by the European Research Council (ERC project SIGMA-Vision).

References

1. Osher, S., Sethian, J.A.: Fronts propagating with curvature-dependent speed: algorithms based on Hamilton-Jacobi formulations. JCP **79**(1), 12–49 (1988)
2. Caselles, V., Catté, F., Coll, T., Dibos, F.: A geometric model for active contours in image processing. Numer. Math. **66**(1), 1–31 (1993)

3. Malladi, R., Sethian, J., Vemuri, B.C.: Shape modeling with front propagation: a level set approach. TPAMI **17**(2), 158–175 (1995)
4. Caselles, V., Kimmel, R., Sapiro, G.: Geodesic active contours. IJCV **22**(1), 61–79 (1997)
5. Yezzi, A., Kichenassamy, S., Kumar, A., Olver, P., Tannenbaum, A.: A geometric snake model for segmentation of medical imagery. TMI **16**(2), 199–209 (1997)
6. Kass, M., Witkin, A., Terzopoulos, D.: Snakes: active contour models. IJCV **1**(4), 321–331 (1988)
7. Adalsteinsson, D., Sethian, J.A.: A fast level set method for propagating interfaces. JCP **118**(2), 269–277 (1995)
8. Li, C., Xu, C., Gui, C., Fox, M.D.: Distance regularized level set evolution and its application to image segmentation. TIP **19**(12), 3243–3254 (2010)
9. Malladi, R., Sethian, J.A.: A real-time algorithm for medical shape recovery. In: Proceeding of ICCV, pp. 304–310 (1998)
10. Sethian, J.A.: Fast marching methods. SIAM Rev. **41**(2), 199–235 (1999)
11. Tsitsiklis, J.N.: Efficient algorithms for globally optimal trajectories. TAC **40**(9), 1528–1538 (1995)
12. Chen, D., Cohen, L.D.: Vessel tree segmentation via front propagation and dynamic anisotropic Riemannian metric. In: Proceedings of ISBI, pp. 1131–1134 (2016)
13. Arbeláez, P.A., Cohen, L.D.: Energy partitions and image segmentation. JMIV **20**(1), 43–57 (2004)
14. Bai, X., Sapiro, G.: Geodesic matting: a framework for fast interactive image and video segmentation and matting. IJCV **82**(2), 113–132 (2009)
15. Li, H., Yezzi, A.: Local or global minima: flexible dual-front active contours. TPAMI **29**(1), 1–14 (2007)
16. Cohen, L.D., Deschamps, T.: Segmentation of 3D tubular objects with adaptive front propagation and minimal tree extraction for 3D medical imaging. CMBBE **10**(4), 289–305 (2007)
17. Melonakos, J., Pichon, E., Angenent, S., Tannenbaum, A.: Finsler active contours. TPAMI **30**(3), 412–423 (2008)
18. Mirebeau, J.M.: Efficient fast marching with Finsler metrics. Numer. Math. **126**(3), 515–557 (2014)
19. Benmansour, F., Cohen, L.D.: Fast object segmentation by growing minimal paths from a single point on 2D or 3D images. JMIV **33**(2), 209–221 (2009)
20. Cohen, L.: Multiple contour finding and perceptual grouping using minimal paths. JMIV **14**(3), 225–236 (2001)
21. Randers, G.: On an asymmetrical metric in the four-space of general relativity. Phys. Rev. **59**(2), 195 (1941)
22. Chen, D., Mirebeau, J.M., Cohen, L.D.: Global minimum for a Finsler elastica minimal path approach. IJCV **122**(3), 458–483 (2017)
23. Xu, C., Prince, J.L.: Snakes, shapes, and gradient vector flow. TIP **7**(3), 359–369 (1998)
24. Alpert, S., Galun, M., Brandt, A., Basri, R.: Image segmentation by probabilistic bottom-up aggregation and cue integration. TPAMI **34**(2), 315–327 (2012)
25. Rother, C., Kolmogorov, V., Blake, A.: Grabcut: interactive foreground extraction using iterated graph cuts. ToG **23**(3), 309–314 (2004)

PointFlow: A Model for Automatically Tracing Object Boundaries and Inferring Illusory Contours

Fang Yang[1]([✉]), Alfred M. Bruckstein[2], and Laurent D. Cohen[1]

[1] CEREMADE, CNRS, UMR 7534, Université Paris Dauphine,
PSL Research University, 75016 Paris, France
`fang.yang@dauphine.eu,fangyang.whu@gmail.com`
[2] Computer Science Department, Technion – IIT,
718 Taub Building, 32000 Haifa, Israel

Abstract. In this paper, we propose a novel method for tracing object boundaries automatically based on a method called "PointFlow" in image induced vector fields. The PointFlow method comprises two steps: edge detection and edge integration. Basically, it uses an ordinary differential equation for describing the movement of points under the action of an image-induced vector field and generates induced trajectories. The trajectories of the flows allow to find and integrate edges and determine object boundaries. We also extend the original PointFlow method to make it adaptable to images with complicated scenes. In addition, the PointFlow method can be applied to infer certain illusory contours.

We test our method on real image dataset. Compared with the other classical edge detection and integration models, our PointFlow method is better at providing precise and continuous curves. The experimental results clearly exhibit the robustness and effectiveness of the proposed method.

Keywords: Automatically tracing object boundaries
Induced vector field · PointFlow · Differential equations
Illusory contours

1 Introduction

Edge detection focuses on finding the sharp discontinuities and aims at capturing places where important changes occur in images. It plays an important role in image processing and computer vision. Since the early years of image processing, numerous methods were proposed to detect edges, such as the Sobel operator, Prewitt operator, Canny and the Haralick edge detector and so on [12]. The early edge detectors focus on the change of image brightness or image graylevel. However, many visually salient edges or contours do not simply correspond to gradients, but also to the texture edges or illusory contours. Therefore, researchers try to apply multiple features such as brightness, color, texture *etc.* to their edge detectors [1,9,10].

© Springer International Publishing AG, part of Springer Nature 2018
M. Pelillo and E. Hancock (Eds.): EMMCVPR 2017, LNCS 10746, pp. 485–498, 2018.
https://doi.org/10.1007/978-3-319-78199-0_32

In [4], a semi-automatic method to detect boundaries of objects is proposed by using simulation of particle motion in an image induced vector field. But users of the method were required to provide the location of starting points and the number of time steps to be carried out. In addition, users needed to adjust the parameters to achieve a good result. In [7], a similar model is used to track and detect the most important edges, in order to produce artistic one-liner renderings of objects appearing in images. In [6] a c-evolute model is presented for the particle motion in [4] to approximate the edge curves. More recently, Kimmel and Bruckstein [5] proposed to incorporate the Haralick/Canny edge detector into a variational edge integration process.

In this paper, we are interested in providing a process of tracking object boundaries automatically. Imagine that a magnetic vector field is induced by the image. As the input, a number of points which are arranged randomly in the image and are considered as small magnetized iron pieces. When the field is applied, the iron pieces start moving following the direction of the magnetic field. We record the trajectories of these points and use them to obtain the edges on the images. The movement of the points can be described by an ordinary differential equation.

The construction of the vector field is a crucial step for PointFlow method. To trace the image boundaries, the vector field must be designed edge-oriented. After obtaining the vector field, we initiate the flow from a number of random points in the image plane. These points are then attracted towards and along the significant edges in the image. Note that the flow process will not end until a stopping criterion is met. An integration process allows to refine the trajectories and make the result more robust.

Additionally, the PointFlow method can be extended to infer the illusory contours under the assumption that the illusory contours are a missing part of a regular shape and it can be approximated by an arc of certain radian.

The contribution of this paper is that we propose a PointFlow method to detect and integrate edges. Commendably, the edges that are extracted are continuous and precise. Moreover, the PointFlow method can be applied to the inference of illusory contours.

The paper is organised as follows: Sect. 2 introduces the core algorithm of the PointFlow method. Section 3 details how to construct the vector field. Section 4 shows some experiments results on real images. Section 5 describes how the PointFlow method could be used to infer the illusory contours. Section 6 displays some inference results by using the PointFlow method. Section 7 provides some concluding remarks and possible directions for the future work.

2   PointFlow Method

The PointFlow method first uses an ordinary differential equation (ODE) to describe the motion of a moving point under a vector field $\mathbf{V}$ within a period of time. In a $2D$ domain, the ODE is defined as follows:

$$\frac{d(P(t))}{dt} = \mathbf{V}(P(t)) \tag{1}$$

where $P(t) = (x(t), y(t))$ is a point function which describes the location of a moving point at time t and starting from a given point p_0 at time $t = 0$. Within time ΔT, the trajectory of P under the effect of the vector field $\mathbf{V}$ will be recorded, where $\mathbf{V}$ is a vector field that controls the speed and direction of the points. The flow will not stop until some stopping criterion is met. Figure 1 displays the flowchart of applying the PointFlow method to trace image boundaries.

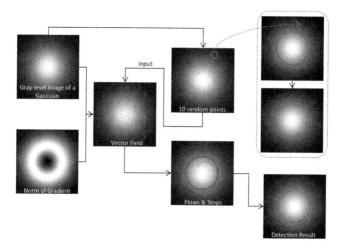

Fig. 1. The flowchart of the PointFlow. The left column presents a 2D Gaussian image and the magnitude of its gradient. A linear combination of their gradients are used to form the vector field $\mathbf{V}$. 10 random points are used as the starting points, moving under the effect of the vector field. The top-right dashed-box shows the movement of a single point, the blue curve is the trajectory and the red point is the end point. The flow terminates when it hit its own trajectory. Then, we re-initiate the flow from the end (red) point to obtain a complete and precise contour of the Gaussian image. The other points move in the same way. Right below shows the result. (Color figure online)

3   Construction of Vector Field

3.1   Vector Field by Using Graylevel

To design an edge-oriented vector field, two kinds of forces are necessary: one towards the edges and the other one along the edges. This can be realized simply by combining the tangent vector of the points on the image (along the edges) $\mathbf{V}_1 = \nabla I^\top$ and the gradient of the magnitude of the gradient of the image $\mathbf{V}_2 = \nabla(\|\nabla I\|)$ (towards the edges). Then the vector field is written as: $\mathbf{V} = \zeta\mathbf{V}_1 + \xi\mathbf{V}_2$ or $\mathbf{V}' = \zeta\mathbf{V}_1 - \xi\mathbf{V}_2$, empirically, $\zeta = \xi = 0.5$.

Three stopping and reflowing criteria are defined for the flow:

1. When the trajectory of flow hits itself. The kind of end points are the first type of end points, labelled as "*E1*". For points labeled as "*E1*", we re-initiate the movement from these points in the same vector field **V**. The trajectories by re-initiating will result in closed curves, which are boundaries or contours of objects in images.
2. When the flow hits the boundary of the image. These kinds of end points are the second type of end points, labelled as "*E2*". For points labeled as "*E2*", we re-initiate the movement from these points in the vector field **V′**. Thus a complete boundary from the boundary of the image will be detected.
3. When the flow hits a pixel where the gradient is zero. If it is at the source point where the gradient is very close to zero, we will remove this source point.

For the other cases, we will not consider the flow unless it merges with other flows. This is a case for junction points. Generally speaking, if a junction point is not revisited by any flow, it will start moving on both direction **V** and **V′** until it meets a stopping criterion.

Figure 2 displays some tests on real natural gray images. From (a) and (b), we can see that nearly all the curves and boundaries are perfectly extracted. But in (c), the contour of the seastar is not extracted while a lot of white spots on its back are detected. This is because in (a) and (b) the scenes are monotonous, and the subject can be easily distinguished by the graylevel. While in (c), the most salient changes in graylevel are the differences between the white spots and the background.

(a)   (b)   (c)

Fig. 2. Experiment on some gray images.

3.2 Vector Field by Using Multiple Features

As described above, in natural images, it is not always sufficient by only using the gradient of the graylevel to form the vector field. It is necessary to combine some other features to form the vector field. In this section, the BSDS500 dataset [1] is used for extracting the high level feature such as the texture and spectral features.

In [9], the authors proposed a *Pb* (probability of boundary) edge detector where they introduced several features, i.e. color, texture to detect the contours.

In [1], the authors not only use the features in [9] but also introduce a spectral feature and combine them into a global feature to detect the contours, the detector is known as the gPb (global probability of boundary) contour detector. To compute the gradient on each pixel of each feature channel, it is implemented by setting discs on each pixel (x, y) of each feature channel and comparing the histogram difference of each half-disc g and h using different orientations.

$$\chi^2 = \frac{1}{2} \sum_i \frac{(g(i) - h(i))^2}{g(i) + h(i)} \tag{2}$$

Normally, eight orientations are used to describe the directions of the gradient, where the orientation $\theta = [0, \pi/8, \pi/4, 3\pi/8, \pi/2, 5\pi/8, 3\pi/4, 7\pi/8]$.

To improve the performance of PointFlow in complicated scenes, we use the same feature in [1]. The color channels contain three channels which correspond to the CIE Lab colorspace including the brightness, color a and color b channels. The fourth channel is the texture channel, to obtain the texture feature, the primary thing is to compute the texton map. Firstly, each input training image from the BSDS dataset is convolved with a set of 17 Gaussian derivative, shown in Fig. 3.

Fig. 3. 17 filters for computing the texture [1].

Then each pixel is associated with a 17-dimensional vector of responses. A K-means cluster is then used to define the clustering center, with $K = 64$. For the test image, we convolve them with the set of 17 Gaussian derivatives, too. The pixel which is nearest to one of the 64 cluster center will be given the corresponding texton label.

Let us denote the gradient signal $G_i(x, y, \theta)$, where i represents the feature channel and θ is the orientation. We also use the multi-scale and spectral features that were described in [1].

For the multi-scale features, each channel gradient G_i is computed at three scales $s = [\frac{\sigma}{2}, \sigma, 2\sigma]$, where s controls the radius of the disc for computing the histogram difference in Eq. (2). $\sigma = 5$ pixels for brightness channel and $\sigma = 10$ pixels for color a, color b and texture channel. The multi-scale gradient signals at each orientation can be combined linearly as follows:

$$mG(x, y, \theta) = \sum_s \sum_i \alpha_{i,s} G_{i,s}(x, y, \theta) \tag{3}$$

where $\alpha_{i,s}$ are the weights at different scale s and different feature channel i and it is learned by gradient ascent on the F-measure on the BSDS500 dataset.

The spectral features are used to obtain the most salient curves and it is obtained by a spectral clustering technique. A sparse symmetric matrix W which describes the affinity of two pixels of which the distance is within a fixed radius r is provided according to mG. The spectral signals sG then are obtained by solving a Laplacian system of W, details can be found in [1].

By combining the multi-scale and spectral features, we can get the global signal at each orientation $gG(x, y, \theta)$.

Figure 4 shows the feature map gG of eight orientations.

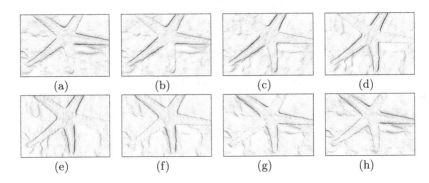

Fig. 4. Gradient magnitude of eight orientations, from (a) to (h), the orientation $\theta = [0, \pi/8, \pi/4, 3\pi/8, \pi/2, 5\pi/8, 3\pi/4, 7\pi/8]$

After computing the multiple features at each orientation, we can construct the vector field based on these features. For each single pixel, we will retain the largest value among the eight values as the gradient magnitude, and the corresponding orientation denotes the direction of the gradient:

$$\mathbf{v}(x, y) = gG_{\max}(x, y, \theta) * (\cos(\theta), \sin(\theta)) \tag{4}$$

so that: $\mathbf{V}_2 = \mathbf{v}^{\perp}$. To compute the second-order derivative, we use the gradient of $gG_{\max}(x, y)$, that is: $\mathbf{V}_1 = \nabla(gG_{\max})$.

Figure 5 shows an example of using multiple features to construct the vector field and its corresponding contour detection result. Compared with Fig. 2(c), the result in Fig. 5(c) is more clean and complete.

Fig. 5. Experiment on a real color image. From left to right are the original color image, the trajectories generated from 2000 random points and the result of detection of contours. (Color figure online)

4   Experiments on Edge Detection

4.1   Data Settings

We test our PointFlow method on the images from the Berkeley Segmentation Dataset [8]. The number of random source points is set to be $N = 2000$ for each image. For vector fields using graylevel, we use a Gaussian filter to smooth the images, the standard deviation of the filter is $\sigma = 2$, and the size is 4×4.

In addition, there are numerous ways to approximate numerically the solution to the first-order ODE Eq. (1), such as the Euler method, backward Euler method, first-order exponential integrator method and so on [2]. To make the solutions smoother and with less oscillations, we use the Runge-Kutta 4 algorithm to solve Eq. (1).

4.2   Experiments Results and Discussions

Figure 6 are some results on the vector field generated by using only graylevel. From the results, we can see that for simple scenes, such as (a) and (d), by using graylevel to construct the vector is enough to obtain a complete and clean detection result. While in the complicated scenes, though that a lot of details are extracted, the result are not quite satisfying, this is because the object boundaries are affected by a lot of factors, e.g. colors, shadows, textures and so on.

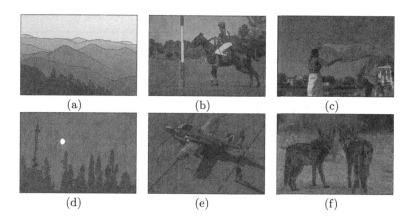

(a)   (b)   (c)

(d)   (e)   (f)

Fig. 6. Some results on the BSD dataset.

In addition, Fig. 7 shows some results obtained by using different vector fields. The top row displays the results by only using graylevel feature to construct the vector field and the bottom row shows the results by using multiple features to construct the vector field. From Fig. 7, we can see that by using the multiple features (obtained from the gPb detector), we can get more complete and clean results, while by using the single feature (graylevel or color), we are able to obtain more details. Note that not by using multiple features is better than by using graylevel, there is no conflict of these two features. For example, in Fig. 7(f), around the wheel the global feature is too weak to detect, it is not easy to distinguish the wheel from the ground by the multiple features while we can obtain more details by using graylevel (c).

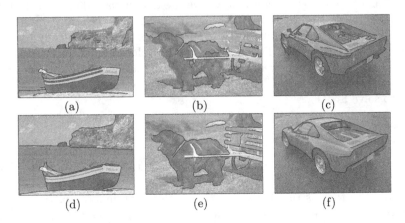

Fig. 7. Some comparisons between using single feature (graylevel) and multiple features. Top row displays the results by only using color feature to construct the vector field; bottom row shows the results by using multiple features to construct the vector field.

5   Inference of Illusory Contours

The illusory contours (or subjective contours) are visual illusions that evoke the perception of an edge without brightness or color changes on that edge. Speaking of the illusory contours, the first image comes into our mind would be the Kanizsa's Triangle, shown in Fig. 12(a). We can perceive from the spatially separated fragments in Fig. 12(a) that there are hidden triangles and circles, but in the view of images, there are no complete triangles or circles. Then we cannot help to ask that why we can perceive these unreal outline, can we make the computers sense these phenomena. According to [11], it is believed that for us human beings, the early visual cortical regions such as the primary visual cortex (V1) and secondary visual cortex (V2) are responsible for forming illusory contours.

For the computers, how can they perceive the illusory contours like humans remains our problem. By observing the illusory contours shown in Figs. 11(a) and 12(a), (d), we can find that the illusory contours are usually an unseen curve that connects two corner points. Here comes the question: is it possible for a point to flow on the illusory contour rather than on the boundary of an object when it meets a corner point? The answer is yes.

5.1 Inertia-Driven PointFlow and Circle Hough Transform

In our opinion, the illusory contours are always small missing parts of regular and predictable shapes. For these kind of illusory contours, it can be completed by an arc with a specific radian. Even for a straight line, it can be described as an arc of a circle of which the radius is infinite. So the main task of inferring the illusory contours is to find an appropriate arc that connects two specific terminals.

In this paper, we assume that the terminals of an illusory contour are always caused by abrupt changes in curvature of a closed curve. These terminals are also called the corner points.

After obtaining the corner points, we make the point flow at the same velocity (angular velocity) $\mathbf{v}$ as it did before it hits the corner point, instead of flowing on the image induced vector field.

Due to the fact that the point flow on a field without any forces after hitting the corner point, the behavior of the point is like inertia-driven, so we call this process "Flow with inertia".

The idea that a missing curve in illusory contour images can be completed by an arc reminds us of the Circle Hough Transform (CHT) [3]. The CHT is a technique for detecting circular objects in images. In a 2D space, a circle of which the center locates at (a, b) with a radius r can be described by:

$$(x - a)^2 + (y - b)^2 = r^2 \tag{5}$$

The parameter space is 3D, (a, b, r). And the corresponding circle parameters can be identified by the intersection of a number of conic surfaces which are caused by the circle on the image.

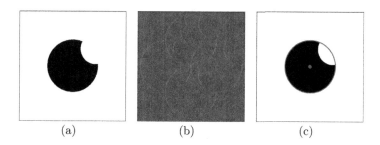

(a)   (b)   (c)

Fig. 8. Detection of circles by CHT, from left to right, (a) the original image; (b) projection on (a, b, r) space ($r = 95$); (c) circle detection result, shown in red. (Color figure online)

From the example in Fig. 8, we can see that the CHT has the ability to infer obvious circles. In the experiment section, we will compare the circle detection results by our inertia-driven flow and the circle hough transform.

5.2   Corner Points via Flowing

The trajectories formed by the motion of these points are continuous, thus we can detect the corner points by taking the change of curvature into consideration, where the curvature at a point p of a curve l is defined as:

$$k(p) = \left\| \frac{\mathrm{d}A}{\mathrm{d}T} \right\| \tag{6}$$

where A is the angle and $\mathrm{d}A$ stands for the change of angle, T represents time and $\mathrm{d}T$ is the change of time. In this paper, each point driven by the vector field moves in a unit time every step.

A potential corner point may exist at places where there is abrupt change of curvature of a trajectory. It can be determined that whether the potential corner point is a true corner point or not by setting a threshold for the curvature. Figure 9 demonstrates the process of finding corners by taking the curvature into consideration.

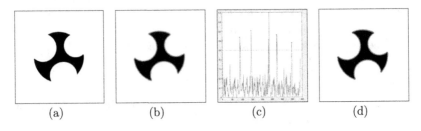

(a)   (b)   (c)   (d)

Fig. 9. An example for detecting the corner points, (a) the original image; (b) the cyan curve is the detection result by using PointFlow method; (c) the plot of curvature of the detected curve; (d) the corresponding corner points. (Color figure online)

5.3   Flow with Inertia

After obtaining the corner points, we can infer the illusory contours by deciding that whether two corner points can be connected by an arc. Each movement of a single point p takes a unit time, then after ΔT, there should be $\Delta T + 1$ points on the trajectory. In addition, we are introducing a kind of "inertia" here, which means that when a point flows to a corner point, it will no longer move on the vector field, but continues to flow according to the trajectory it just passed through.

The following three steps describe how to infer the illusory contours by Point-Flow in detail.

1. Compute the angular velocity. For a trajectory L_j which passes through two corner points p_{cor_1} and p_{cor_2} successively, in order to guarantee the accuracy and robustness, N points $\{p_n\}, n = [1, ..., N]$ are chosen on L_j to compute

the angular velocity. First, the angular displacement from these N points to corner point p_{cor_2} is computed.

$$\phi(l_n) = \text{acos}\left(\frac{\langle \vec{a}, \vec{b} \rangle}{\|\vec{a}\| \|\vec{b}\|}\right).$$

where l_n is a piece of L_j from p_n to p_{cor_2}, $\vec{a}$ stands for the tangent vector of the curve at the corner point p_{cor_2} and $\vec{b}$ represents the tangent vector at nth point. A toy model can be found in Fig. 10(a). So the angular velocity for l_n is:

$$\omega(l_n) = \frac{\phi(l_n)}{\text{length}(p_{cor_2}, p_n)} \tag{7}$$

2. Flow with inertia. Based on the assumption that the missing part of the illusory contours can be simulated by an arc, we propose a kind of "inertia" which contains an angular velocity $\omega(l_n)$ and the initial speed $\mathbf{v}_0 = \mathbf{V}(p_{cor_2})$. Now p_{cor_2} is considered as the starting point of the trajectory generated by the inertia. With the inertia, the velocity $\mathbf{v}$ of the point can be obtained by a rotation matrix:

$$\mathbf{v}_t = \mathbf{v}_0 \begin{bmatrix} \cos(\omega(l_n)) & \sin(\omega(l_n)) \\ -\sin(\omega(l_n)) & \cos(\omega(l_n)) \end{bmatrix}^t \tag{8}$$

where t is the time of flowing, and it also represents the number of flowing steps. When $\omega(l_n) = 0$, the inertia will generate a straight line, when $\omega(l_n) \neq 0$, the inertia will generate an arc with corresponding radian.

3. Let us denote the trajectory generated by inertia l_{ine}, and the starting point of l_{ine} is the corner point p_{cor_2}. After time ΔT_1, which is a time limit provided by the users, if l_{ine} is still not connected to any corner point, l_{ine} will be ignored. If it is connected to some corner point according to the three conditions below, it is an illusory contour. Decisions will be made at each move of l_{ine}, and the end point of l_{ine} at each step is labeled as p_{cur}. To decide whether l_{ine} can be an illusory contour or not, there are three conditions:
   - The length of l_{ine} is larger than a threshold that we set:

$$\text{len}(l_{ine}) > \beta$$

   - The distance between p_{cur} and a corner point p_{cor} should be small enough:

$$\text{len}(p_{cur}, p_{cor}) < \epsilon$$

   - l_{ine} can be connected to p_{cor} in a smooth enough way. Let us take Fig. 10(b) as an example. Let $\vec{c}$ be the tangent vector of l_{ine} at p_{cur} (the blue arrow), $\vec{d}$ the tangent vector at p_{cor_2} (the red arrow). The angle φ between $\vec{c}$ and $\vec{d}$ can be represented by the cosine value: $\varphi = (\frac{\langle \vec{c}, \vec{d} \rangle}{\|\vec{c}\| \|\vec{d}\|})$. If $\|\varphi\| > \delta$, $\delta = 0.9$ empirically, then the direction of the flow l_{ine} is in accordance with L_j.

If the above three conditions are satisfied, l_{ine} can be considered as the arc which fill in the missing part of the illusory contours.

In addition, if the trajectory generated by the inertia could not form an illusory contour, we will not take it into consideration anymore.

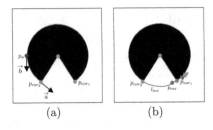

(a)   (b)

Fig. 10. An example for the flow driven by inertia, (a) describes how the angular displacement is obtained in Step 1; (b) shows how l_{ine} can be an illusory contour. (Color figure online)

6   Experiment on the Inference of Illusory Contours

6.1   Data Settings

We test our method for inferring the illusory contours on three illusory images.
The first image is shown in Fig. 11(a), the size is 400×400. We can perceive that there are 4 circles in this image: a big black one and three small white ones. For inferring the circles, we use twice the Circle Hough Transform (of which the radius ranges from 30 to 100 pixels and 50 to 120 pixels) and our inference method respectively.
The other two images are shown in Fig. 12(a) and (d), the size of both images is 400×400. It can be perceived by users that in Fig. 12(a) there exist a hidden triangle and three small circles and in Fig. 12(d) there are four circles as well as a square.

6.2   Experiment Result and Analysis

Figure 11(b) and (c) display the circle detection result by using the CHT and PointFlow respectively. From the result (b), we can see that the CHT detect only the small circles. Due to that the big black circle has a lot of missing parts, no matter what radius we use, it cannot be detected. While in (c), our method has inferred and completed all the four circles. By using the CHT to detect circles, users have to provide a range of radii and a very large radius may result in bad results. While inference by PointFlow is automatical and effective.
From Fig. 12(c) and (f), it can be seen that all the illusory parts (either the straight lines of the triangle and the square or the curved arcs of the circles) are inferred by using our inertia-driven PointFlow method.

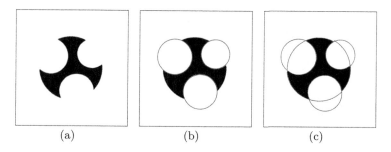

Fig. 11. Experiment on a synthetic image, from left to right, (a) the original image (b) the circle detection result by using CHT (c) the inference result by using our method.

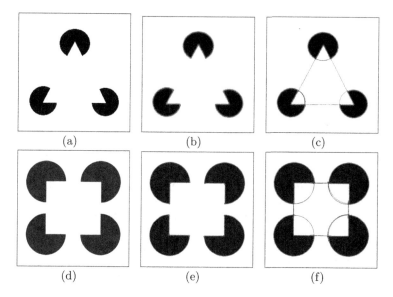

Fig. 12. Experiment on synthetic images, from left to right: (a) and (d) are the original image, (b) and (e) are the corresponding detected corner points, (c) and (f) the inferred illusory contours.

The results in Figs. 11 and 12 demonstrate that our inference algorithm is efficient at inferring the straight and smooth arcs. The disadvantage of this method lies in that it will fail when there is an obstacle on an illusory contours, or the illusory contours could no longer be represented by an arc.

7   Conclusion and Future Work

This paper proposes a novel method called PointFlow for automatically modeling the process of tracking object boundaries in images and inferring illusory contours. The PointFlow method is realized by using an ordinary differential equation, where an image-induced vector field is the most crucial factor for boundary detection and integration. In this paper, the vector field can be constructed simply based on the gradient of the graylevel of the image, or via combining multiple features. Compared with the classical edge detectors, the PointFlow achieves a very satisfying result. For the inference of the illusory contours, we have introduced an "inertia-driven" flow, it has an ability to find the missing part of the illusory contours. In future, we would like to adapt the PointFlow method to infer more complicated illusory contours by using higher-level features.

References

1. Arbelaez, P., Maire, M., Fowlkes, C., Malik, J.: Contour detection and hierarchical image segmentation. IEEE Trans. Pattern Anal. Mach. Intell. **33**(5), 898–916 (2011)
2. Butcher, J.C.: Numerical methods for ordinary differential equations in the 20th century. J. Comput. Appl. Math. **125**(1), 1–29 (2000)
3. Duda, R.O., Hart, P.E.: Use of the hough transformation to detect lines and curves in pictures. Commun. ACM **15**(1), 11–15 (1972)
4. Eua-Anant, N., Udpa, L.: Boundary detection using simulation of particle motion in a vector image field. IEEE Trans. Image Process. **8**(11), 1560–1571 (1999)
5. Kimmel, R., Bruckstein, A.M.: Regularized laplacian zero crossings as optimal edge integrators. Int. J. Comput. Vis. **53**(3), 225–243 (2003)
6. Lu, C., Chi, Z., Chen, G., Feng, D.: Geometric analysis of particle motion in a vector image field. J. Math. Imaging Vis. **26**(3), 301–307 (2006)
7. Makhervaks, V., Barequet, G., Bruckstein, A.: Image flows and one-liner graphical image representation. Ann. N.Y. Acad. Sci. **972**(1), 10–18 (2002)
8. Martin, D., Fowlkes, C., Tal, D., Malik, J.: A database of human segmented natural images and its application to evaluating segmentation algorithms and measuring ecological statistics. In: Proceeding on 8th International Conference of Computer Vision, vol. 2, pp. 416–423, July 2001
9. Martin, D.R., Fowlkes, C.C., Malik, J.: Learning to detect natural image boundaries using local brightness, color, and texture cues. IEEE Trans. Pattern Anal. Mach. Intell. **26**(5), 530–549 (2004)
10. Rubner, Y., Tomasi, C.: Coalescing texture descriptors. In: ARPA Image Understanding, Workshop, pp. 927–935 (1996)
11. Von Der Heyclt, R., Peterhans, E., Baurngartner, G.: Illusory contours and cortical neuron responses. Science **224**, 1260–1262 (1984)
12. Ziou, D., Tabbone, S.: Edge detection techniques - an overview. Int. J. Pattern Recogn. Image Anal. **8**, 537–559 (1998)

An Isotropic Minimal Path Based Framework for Segmentation and Quantification of Vascular Networks

Emmanuel Cohen[1,2]([✉]), Laurent D. Cohen[1], Thomas Deffieux[2],
and Mickael Tanter[2]

[1] CEREMADE, PSL Research University, Université Paris Dauphine, CNRS,
UMR 7534, 75016 Paris, France
`cohen.emm@gmail.com`
[2] Institut Langevin, PSL Research University, ESPCI ParisTech, CNRS,
UMR 7587, INSERM U979, 75005 Paris, France

Abstract. Minimal path approaches for image analysis aim to extract curves minimizing an energy functional. The energy of a path corresponds to its weighted curve length according to a relevant metric function. In this study, we design a binary isotropic metric model with the use of a Hessian-based vascular enhancement filter in order to extract geometrical features from vascular networks. We introduce a constrained keypoint search method able to extract subpixel vessel centrelines, diameters and bifurcations. Experiments on retinal images demonstrated that the proposed framework achieves similar even better segmentation performances as compared with methods using more sophisticated metric designs.

1 Introduction

Vascular networks in the human body own anatomical characteristics that are crucial to analyse for diverse purposes, e.g. in biology for a better understanding of the vascular architecture, or in medicine for the diagnosis of many diseases such as vessel tortuosity based analyses [5,15]. In Cohen et al. [9], we also showed that the vascular anatomy represent a useful positioning landmark for neuron-avigation and image registration. Therefore, in this paper, we aim to develop numerical methods for accurately extracting geometrical features from the vascular data, such as vessel centrelines, diameters, bifurcations, etc.

Geodesic methods for minimal path extraction in images have widely demonstrated their worth [10,21], in particular to extract path from tubular structures for medical applications [4,13]. A minimal path or *geodesic* joining two points in an image is a curve that globally minimizes a well chosen energy among all curves joining these two points. The energy of a path corresponds to its curve length weighted by a potential or metric function. The metric assigns to every pixel a scalar weight (isotropic case) by privileging pixels of interest like those

© Springer International Publishing AG, part of Springer Nature 2018
M. Pelillo and E. Hancock (Eds.): EMMCVPR 2017, LNCS 10746, pp. 499–513, 2018.
https://doi.org/10.1007/978-3-319-78199-0_33

from tubular-vessel structures by assigning them low weights. Thus, minimal paths preferably follows vessel structures of the image.

The choice of the metric represents a major issue. Several authors have already proposed different types of metrics according to the type of processed images as summarized in [21]. Isotropic intensity-based metrics favour salient pixels according to their intensity and position only. As a consequence, minimal paths can deviate and miss some vessels or take a shortcut generally due to image noise. To overcome this problem, the metric must take into account the geometry of vessels in order to better discriminate them. Thus, Benmansour and Cohen [4] introduced an anisotropic metric design based on the optimally oriented flux (OOF) filter [18] to incorporate the local orientation of vessels.

Another important issue is the way to extract the entire vascular tree. Benmansour and Cohen [3] proposed a minimal path method with keypoints detection (MPWKD) all along the curves of interest given a single source (start) point. Instead of collecting many pairs of source and endpoints, a unique path grows iteratively through the entire vascular network. However, the method suffers from detection of outlier keypoints or paths due to the isotropy of the metric and the simplicity of the stopping criterion. An improvement of the method is proposed in [7] with the use of an OOF based anisotropic metric and a more specific stopping criterion.

Beyond the problem of extracting relevant minimal paths, methods for quantification of vascular networks require also the paths to be centred enabling for instance evaluation of vessel tortuosity, and furthermore a local estimation of vessel diameters. In Li et al. [19], they solved the minimal path problem by adding an extra dimension to the classical Fast Marching scheme [24]. They characterized a point by its euclidean coordinates added by the radius at this point. This allows to extract both the centred vessel paths and the corresponding diameters. Yet, despite the accuracy of the method, this one dimension higher scheme increases computing times, a serious drawback regarding real time applications.

In this paper, we propose a minimal path based method to extract entire vascular trees with accurate centrelines, diameters, and bifurcations, that competes with the above state of the art while keeping an isotropic metric and a classical Fast Marching scheme. This prevents from dealing with the problem of high anisotropy ratios [20], and the numerical scheme preserves its classical dimension.

Instead of incorporating the vessel orientation inside an anisotropic metric, we propose to use an isotropic binary metric classifying pixels as vessel or non-vessel. As a consequence, the extracted minimal paths are necessary constrained to follow a priori segmented vascular structures. This binary metric can be simply obtained by thresholding a *vesselness* map i.e. the response of a vascular enhancement filter. We could use any successful filter such as OOF or Hessian-based methods [14]. We chose the most recent and efficient Hessian-based filter introduced by Jerman et al. [16]. To centre the extracted paths and deduce the corresponding diameters, we constrained the MPWKD method to detect only centred keypoints and paths by using a pre-computing distance-to-boundary map. At the end, we get a complete graph representation of vascular networks with subpixel precise centred vessel paths, local diameters, bifurcations, and taking into account cycles or closed curves in the graph.

The paper is organized as follows. In Sect. 2, we recall some backgrounds on the minimal path extraction method. Section 3 describes the proposed framework. In Sect. 4, we show some experimental results on 2D retinal images from the DRIVE database [26]. Conclusion and future works are given in Sect. 5.

2   Background

This study deals with 2D images for simplicity of visualization and validation of the results but one can straightforwardly extend the method to 3D e.g. on the basis of [13, 16]. Let Ω be a closed subset of $\mathbb{N}^2$ and $I : \Omega \to [0, 1]$ a 2D image.

2.1   Minimal Path Extraction

Given an isotropic (scalar) potential or metric function $\mathcal{P} : \Omega \to \mathbb{R}$, the minimal path γ^* joining a source point $s \in \Omega$ to an endpoint $e \in \Omega$ is the global minimizer of the following energy

$$\mathcal{E}(\gamma) = \int_0^1 \mathcal{P}(\gamma(t), \gamma'(t)) dt \tag{1}$$

among all smooth curves $\gamma : [0, 1] \to \Omega$ joining $\gamma(0) = s$ to $\gamma(1) = e$. To solve this minimization problem, let define the geodesic distance map $\mathcal{U}_S : \Omega \to \mathbb{R}$ for any set S of source points by

$$\mathcal{U}_S(x) = \min_{\substack{\gamma(0) \in S \\ \gamma(1) = x}} \mathcal{E}(\gamma). \tag{2}$$

Inspired from works on the viscosity solutions of Hamilton-Jacobi equations [11, 22], Cohen and Kimmel [10] noticed that $\mathcal{U}_S$ satisfies the Eikonal equation and used the Fast Marching numerical scheme introduced by Sethian [23] to solve it. The Fast Marching is a front propagation approach computing iteratively the values of $\mathcal{U}_S$ in increasing order from the source points verifying

$$\mathcal{U}_S(s) = 0 \,, \, \forall s \in S. \tag{3}$$

Thus, γ^* can be easily extracted by performing a gradient descent on $\mathcal{U}_S$ from e to s.

To get a more accurate numerical solution for $\mathcal{U}_S$, we prefer to use the numerical scheme introduced by Tsitsiklis [27] also detailed in [17].

2.2   Automatic Keypoint Detection

In the previous section, the user needs to provide source and endpoints. When dealing with complex tubular structures with many bifurcations like vascular networks, we need an automatic procedure. The Benmansour and Cohen [3] method

(MPWKD) has the benefit to extract automatically from one single source point several successive minimal paths following the entire tree structure. This is achieved by iteratively detecting keypoints along the tree and at each iteration tracing back to the previous keypoint by gradient descent on $\mathcal{U}_S$ (see also [19]).

The criterion to detect a keypoint is based on the computation of the euclidean distance map $\mathcal{L}_S$ in parallel with the geodesic distance map. Starting from the source points $S = \{s_0, \ldots, s_{n_S}\}$ with $\mathcal{U}_S(s) = 0$ and $\mathcal{L}_S(s) = 0$ for all $s \in S$, the front propagates according to the Fast Marching algorithm with metrics $\mathcal{P}$ and $\tilde{\mathcal{P}} = 1$ respectively for $\mathcal{U}_S$ and $\mathcal{L}_S$, until a point p_0 satisfying $\mathcal{L}_S(p_0) \geq \lambda$ is designated as a new keypoint. The crucial part of the algorithm is now to update $\mathcal{U}_S(p_0) = 0$ and $\mathcal{L}_S(p_0) = 0$ so that p_0 becomes also a new source point and $S \leftarrow S \cup \{p_0\}$. This process is iterated to obtain a set of keypoints $\{s_0, \ldots, s_{n_S}, p_0, \ldots, p_{n_K}\}$ until the total euclidean distance $\mathcal{L}_S^T$ reaches another given parameter λ_T. The computation of $\mathcal{L}_S^T$ follows the Fast Marching procedure with a unit potential as for $\mathcal{L}_S$ but without updating its values at each iteration.

The $\mathcal{U}_S, \mathcal{L}_S$ update step modifies the natural propagation of the front. To correct and update the modified values of the front, a Voronoi index map $\mathcal{V} : \Omega \rightarrow \mathbb{N}$, computed in parallel of $\mathcal{U}_S$, is defined for a current set of source points S as

$$\forall x \in \mathcal{R}_j \, , \, \mathcal{V}(x) = j \tag{4}$$

where

$$\mathcal{R}_j = \{x \in \Omega \, , \, \mathcal{U}_{s_j}(x) \leq \mathcal{U}_{s_i}(x) \, , \, \forall i \in \{1, \ldots, n_S\} \, , \, i \neq j\} \tag{5}$$

Every detected keypoint j has its own Voronoi region $\mathcal{R}_j$ containing its closest points according to geodesic distance. When a keypoint is detected, the algorithm continues to update the $\mathcal{U}_S, \mathcal{L}_S$ values in $\mathcal{R}_j$ as detailed in [3].

2.3   Hessian-Based Vascular Enhancement Filtering

A well-known local approximation of a vessel consists in a tube elongated in the direction of the vessel and with a Gaussian profile in its orthogonal plane. Let denote by λ_1, λ_2 with $|\lambda_1| \leq |\lambda_2|$ and v_1, v_2 respectively the two eigenvalues and two eigenvectors of the Hessian matrix at the centre of such 2D tubular structure. One can notice that v_1 is aligned with the direction of the vessel and v_2 with its normal direction, and the eigenvalues verify $|\lambda_1| \ll |\lambda_2|$. The design of an enhancement function or vesselness map $\mathcal{F} : \Omega \rightarrow \mathbb{R}$ based on the last property allows to characterize vascular structures in an image. As well summarized in [16], many authors have proposed different variants for $\mathcal{F}$ e.g. Frangi et al. [14]. Very recently, Jerman et al. [16] improved the state of the art methods by designing a more robust enhancement filter. They proposed to regularize λ_2 as follows

$$\lambda_\rho(\sigma) = \begin{cases} \lambda_2 & \text{if } \lambda_2 > \tau \max_{x \in \Omega} \lambda_2(x, \sigma) \\ \tau \max_{x \in \Omega} \lambda_2(x, \sigma) & \text{if } 0 < \lambda_2 \leq \tau \max_{x \in \Omega} \lambda_2(x, \sigma) \\ 0 & \text{otherwise} \end{cases} \tag{6}$$

where $\tau \in [0, 1]$ is a threshold parameter and σ is the standard deviation of the Gaussian used to compute the Hessian image. In fact, the Hessian of the image is computed on a Gaussian scale space by convolving I with second derivatives of Gaussian. Then, they defined the scale dependent version of $\mathcal{F}$ by

$$
\mathcal{F}_\sigma = \begin{cases} 0 & \text{if } \lambda_2 \leq 0 \text{ or } \lambda_\rho \leq 0 \\ 1 & \text{if } \lambda_2 \geq \lambda_\rho/2 > 0 \\ \lambda_2^2 (\lambda_\rho - \lambda_2) \left(\frac{3}{\lambda_2 + \lambda_\rho} \right)^3 & \text{otherwise} \end{cases} \tag{7}
$$

Finally, $\mathcal{F}$ is obtained for all $x \in \omega$ by

$$
\mathcal{F}(x) = \sup \{ \mathcal{F}_\sigma(x), \sigma_{\min} \leq \sigma \leq \sigma_{\max} \}. \tag{8}
$$

3 The Proposed Minimal Path Based Framework

3.1 Binary Metric Design

In vessel extraction applications, an isotropic metric is a function of the image value that must assign low values to vessels. On real images like retinal ones, the keypoint detection method (Sect. 2.2) with classical isotropic metric models (as described e.g. in [13,21]) generally produces many outlier keypoints. Indeed, those simple models only consider pixel intensities without any a priori knowledge on vascular structures leading to false positive detection. Therefore, before solving the minimal path problem, we pre-process the image with a vascular enhancement filter and threshold the obtained vesselness map $\mathcal{F}$ with threshold parameter δ to get a binary mask $\mathcal{F}_\delta$ depicting the segmented vessels. This mask is then used to design a binary metric $\mathcal{P}_b$ as follows: the background is put to infinity in order to stop the front propagation at vessel boundaries, whereas the vascular shape is equipped with a unit metric. This constrains the minimal paths to lie only on vascular patterns without any use of sophisticated stopping criteria and anisotropic models.

The vascular enhancement filter used is the filter introduced by Jerman et al. [16] as described in Sect. 2.3. Figure 1(b) and (c) respectively illustrate the

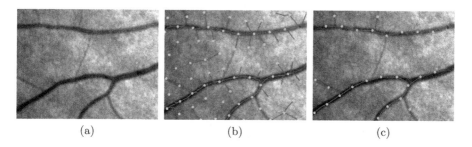

| (a) | (b) | (c) |

Fig. 1. (a) Original cropped retinal image. (b) Outlier keypoints with an intensity-based isotropic metric. (c) The proposed solution using an Hessian-based binary metric.

problem of false positive keypoints when using an intensity-based metric of the form $\mathcal{P} = \mathcal{P}_0 + |I - c|$ with c an approximate value of the vessel pixels, and the improvement result with the binary metric $\mathcal{P}_b$.

3.2 Centred Keypoint Detection

If we apply the keypoint detection algorithm directly with $\mathcal{P}_b$, some problems may persist. First, the paths are not necessarily centred as observed on Fig. 1(c). Secondly, outlier keypoints may appear inside the mask for small values of the parameter λ (see Sect. 2.2) as shown Fig. 2 on a synthetic example. Thus, we propose to constrain the MPWKD at the centre of vessels. The algorithm now designates a point p as a keypoint if p satisfies two conditions:

$$\begin{cases} \mathcal{L}_S(p) \geq \lambda \text{ as in 2.2} \\ p \text{ lies on a vessel centreline} \end{cases}$$

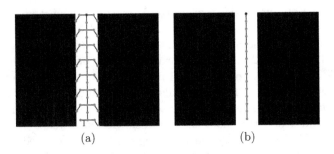

(a)   (b)

Fig. 2. Outlier keypoints problem with the MPWKD on a binary metric (left). The proposed solution is on the right.

To check that p lies on a centreline, we compute the skeleton of $\mathcal{P}_b$ and check if p is inside. The skeleton can be approximated for instance by thresholding the gradient magnitude of the distance-to-boundary map $\mathcal{D} : \Omega \to \mathbb{R}_+$

$$\text{Skeleton} = \{x \in \Omega \, , \, \|\nabla \mathcal{D}\| \leq \tau_s\}. \tag{9}$$

Indeed, maxima of $\mathcal{D}$ are reached at centrelines with very small values of $\|\nabla \mathcal{D}\|$. Decreasing τ_s makes the skeleton thinner but too small values may ignore some branches. As explained in [13], to compute $\mathcal{D}$, two different front propagations using the $\mathcal{P}_b$ metric are needed. The first one starts from one source point inside the vascular shape, and because of the infinite background, the front is automatically frozen at vessel boundaries. Then, those boundary points are used as source points in a second front propagation whose resulting geodesic distance map $\mathcal{U}_S$ is exactly the desired distance-to-boundary map $\mathcal{D}$.

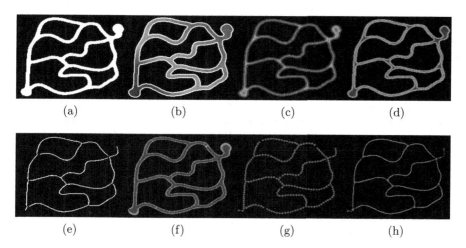

Fig. 3. Centred keypoint detection. (a) Original synthetic tubular binary image. (b) Boundaries detection (in white). (c) Distance to boundary map $\mathcal{D}$ (high values in red). (d) $\mathcal{P}_c$ metric. (e) Skeleton. (f) Geodesic distance map $\mathcal{U}_S$ computed by centred keypoint detection. (g) Centred keypoint detection result. (h) Diameters and graph representation (bifurcations and leaves resp. in green and blue) (Color figure online)

Once the keypoint detection has been constrained on the skeleton, no more outlier keypoints should appear. However, the presence of noise on the boundaries of $\mathcal{P}_b$ can induce some errors in the skeleton such as new outlier branches as described in [1]. Fortunately, keypoints are extracted along the skeleton with a quasi constant spacing equal to λ. This avoids to select outlier keypoints close to the boundary by choosing λ greater than typical radius values inside the vascular shape.

Regarding the extracted minimal paths, they still remain not centred. Therefore, similarly to [13], instead of using $\mathcal{P}_b$ we propagate the MPWKD with a new metric function based on $\mathcal{D}$ defined by

$$\mathcal{P}_c = \begin{cases} +\infty & \text{in the background } (\mathcal{P}_b = +\infty) \\ \mathcal{P}_0 + e^{-\alpha \mathcal{D}} & \text{elsewhere } (\mathcal{P}_b = 1) \end{cases} \quad (10)$$

where $\alpha \in \mathbb{R}_+$ controls the exponential. Thus, the lowest metric weights are attributed to centrelines (high values of $\mathcal{D}$) and the extracted minimal paths are constrained to lie on centrelines as well. Figures 3(a)–(g) illustrates the proposed centred keypoint detection method on a synthetic example.

Note that $\mathcal{P}_b$ and $\mathcal{P}_c$ may contain different sets of connected pixels describing the entire vascular structure. In this case, we apply the method with multiple source points arbitrary selected inside each set of pixels.

3.3 Cycles and Graph Representation

An important assumption of the Fast Marching is that $\mathcal{U}_S$ can only increase with the front propagation. As a consequence, if a keypoint p has been already detected and its $\mathcal{U}_S$ value frozen to zero, all the next detected keypoints will be different from p. Thus, the centred MPWKD described in the last section is still unable to extract closed curves. This can be observed on Figs. 3(g) and 4(a) where five pairs of keypoints remain unconnected, while the synthetic tubular shape is constituted of five cycles (or loops) needing to be closed.

The problem of detecting cycles in vascular networks is particularly important to characterize specific vascular anomalies encountered for instance in tumours where the excessive formation of new vessels can lead to the apparition of cycles [12]. Therefore, cycles detection must be incorporated in the proposed framework especially for 3D applications, whereas in 2D images some cycles may be caused by the superposition of vessels inducing wrong bifurcations. The problem of 2D wrong bifurcations is not addressed in this study but solutions may be found for instance in [2].

The notion of cycles leads us to represent vascular networks as graphs with nodes and edges. The keypoint detection method is by construction a tree structure approach where keypoints are the nodes and the extracted minimal paths are edges. When a keypoint p_1 is originated from the keypoint p_0 i.e. the minimal path extracted by the detection of p_1 has reached after gradient descent on $\mathcal{U}_S$ the keypoint p_0, we say that p_0 is the father of p_1, and p_1 the child of p_0. Let notice that a keypoint cannot have more than one father. We also define a leaf by a keypoint that has no child, and a bifurcation by a keypoint with at least two children. A keypoint with only one child is not considered as a node but as a sample point of a minimal path joining two nodes. Regarding source points, they must be carefully treated. With one, two or more than three children, a source point is considered respectively as a leaf, a sample point, or a bifurcation.

On Fig. 4(a), we notice that keypoints with missing connections are all leaves. We begin by identifying them. Among all leaves, we keep only pairs of keypoints $\{(s_1, e_1), \ldots, (s_n, e_n)\}$ (here $n = 5$) whose Voronoi regions have at least one pixel in common, as shown on Fig. 4(c). Then, we propagate the Fast Marching with the metric $\mathcal{P}_c$ and the s_i as source points. As soon as a point e_i is reached by the front, the minimal path joining e_i to s_i is extracted by gradient descent on $\mathcal{U}_S$. The front stops to propagate when all the n paths have been extracted. Finally, the n paths are added to their corresponding edges to form a complete graph structure. The obtained result is shown on Fig. 3(h) where leaves and bifurcations are respectively coloured in blue and green.

In some cases, two leaves may have their Voronoi Regions side by side but there is no need to connect them. This is for example the case of some close leaves on the synthetic tree Fig. 6. To avoid such problems of outlier cycles, we use the criterion introduced by Kaul et al. [17] to treat general closed curves. Let $f(s_i)$ be the father of s_i. We impose to join e_i to s_i only if

$$\left| \|e_i - f(s_i)\| - \|e_i - s_i\| - \lambda \right| \leq \epsilon \tag{11}$$

where $\epsilon/\lambda \sim 0.2$.

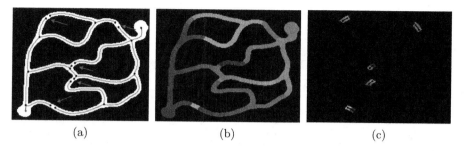

Fig. 4. Detection and closure of cycles on a synthetic tubular binary image. (a) Centred keypoint detection result. (b) Voronoi index map $\mathcal{V}$ (low indexes in blue, high ones in red). (c) The five pairs of neighbour leaves with their computed connections and Voronoi regions. (Color figure online)

3.4 Subpixel Vessel Extraction

Minimal paths γ^* are extracted by gradient descent on $\mathcal{U}_S$ according to the following equation

$$\frac{d\gamma^*(t)}{dt} = -\frac{\nabla \mathcal{U}_S(\gamma^*(t))}{\|\nabla \mathcal{U}_S(\gamma^*(t))\|}. \tag{12}$$

Therefore, the sampling points describing γ^* have subpixel coordinates. This makes the paths more smooth and regular. Nevertheless, in the keypoint detection approach, the computation of $\nabla \mathcal{U}_S$ may induce some errors in the path extraction. Indeed, the paths extraction is done while the map $\mathcal{U}_S$ is not completely computed yet. Thus, some $\mathcal{U}_S$ values have still an infinite value causing troubles in the gradient computation, such as stagnancy of the path before reaching the source illustrated on Fig. 5.

To overcome this kind of problems, we modify the classical computation of the gradient to take into account the infinite values of $\mathcal{U}_S$ at boundaries. Let $x \in \mathbb{R}^2$ be a sampling point of γ^*. The gradient $\nabla \mathcal{U}_S(x)$ at x is computed by bilinear interpolation from the 4 pixel neighbours of x. Let $p_{i,j} \in \Omega$ (line i, column j) be a pixel neighbour of x, and $p_{\min}$ the neighbour of $p_{i,j}$ with the smallest $\mathcal{U}_S$ value. If $\mathcal{U}_S(p_{i,j}) = +\infty$, we impose the coordinate of $\nabla \mathcal{U}_S(p_{i,j})$ in the direction of $p_{\min}$ to be 1 and the other coordinate to be 0. Otherwise, we look at the 4 neighbours of $p_{i,j}$. For instance in the vertical direction, if $\mathcal{U}_S(p_{i-1,j}) = +\infty$ or $\mathcal{U}_S(p_{i+1,j}) = +\infty$, then $\partial_x \mathcal{U}_S(p_{i,j})$ is computed respectively by a forward or a backward finite difference; if both $\mathcal{U}_S(p_{i-1,j}) = +\infty$ and $\mathcal{U}_S(p_{i+1,j}) = +\infty$, then $\partial_x \mathcal{U}_S(p_{i,j}) = 0$; otherwise, we use a central finite difference. We do the same in the horizontal direction. At the end, if $\partial_x \mathcal{U}_S(p_{i,j})$ and $\partial_y \mathcal{U}_S(p_{i,j})$ are both very close to zero, we impose to move in the direction of $p_{\min}$ as for the case $\mathcal{U}_S(p_{i,j}) = +\infty$. We repeat this procedure for each neighbour of x and then we can interpolate.

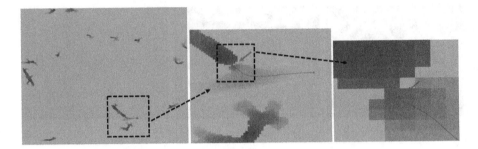

Fig. 5. Stagnation of the gradient descent before reaching the source point on a retinal image example. The path is superimposed on the geodesic distance map $\mathcal{U}_S$ with low and high values respectively in blue and yellow. (Color figure online)

3.5   Vessel Diameter Estimation

From the distance-to-boundary map $\mathcal{D}$, we can estimate vessel diameters. At a sample point x belonging to a centred minimal path, the local diameter can be defined by $2\mathcal{D}(x)$. Since minimal paths are subpixel curves whereas $\mathcal{D}$ is a pixel mapping, we interpolate $\mathcal{D}(x)$ by a bilinear interpolation on the 4 pixel neighbours of x. This local diameter estimation is realized for every edge of the extracted graph. Figure 6 illustrates the performance of the proposed framework respectively on a synthetic tree and a cropped retinal image.

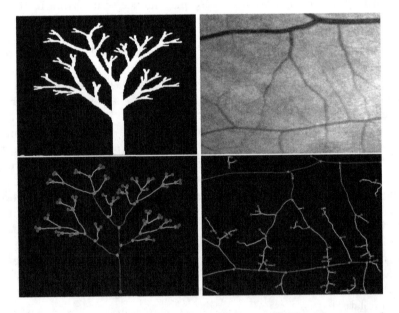

Fig. 6. Illustration of the method on a synthetic tree (left) and a cropped retinal image (right). Top line: original images. Bottom line: our vessel extraction method; high diameters are in yellow/red, low ones in blue/green; bifurcations and leaves are respectively in green and blue. (Color figure online)

4   Experiments on Retinal Images

The proposed framework was tested on the DRIVE database [26] composed of 40 retinal images. The database is divided in two sets of 20 images for training and test. For each image, a manual segmentation serves as groundtruth and a mask delineating the FOV (field of view) is also available. For the test set, two manual segmentations are provided.

To evaluate the results, we first measure the accuracy of the segmentation results, i.e. the sum of the true positive and true negative pixels divided by the number of pixels in the FOV. The training set is first used to estimate the optimal values of parameters τ in Eq. (6) and δ in Sect. 3.1. We found $(\tau, \delta) = (0.25, 0.9)$ by maximizing the average accuracy of the training images. The other main parameters are fixed to $\sigma = \{0.1, 0.2, ..., 2.9, 3\}$, $\lambda = 5$, $\mathcal{P}_0 = 0.1$, $\alpha = 0.5$, $\tau_s = 0.5$, $\epsilon = 0.2\lambda$.

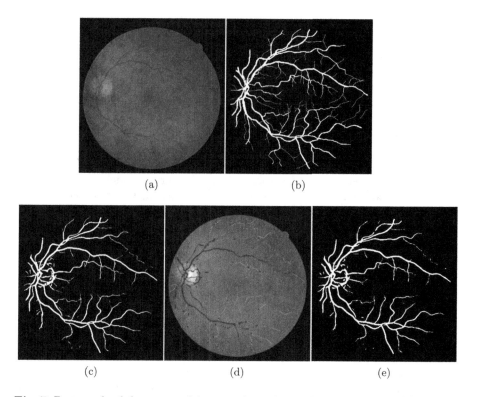

(a)   (b)

(c)   (d)   (e)

Fig. 7. Best result of the proposed framework on the DRIVE test dataset. (a) Original test image. (b) Groundtruth. (c) Thresholded vesselness map $\mathcal{F}_\delta$ with the Hessian-based filter [16]. (d) Centrelines and diameters with the proposed method. (e) Segmented image reconstructed from (d).

Table 1. Accuracy measures on the 20 retinal images of the DRIVE test dataset.

Methods	Maximum	Minimum	Average	Standard deviation
Benmansour and Cohen [4]	0.947	0.927	0.9372	0.0054
Chen and Cohen [6]	0.949	0.930	0.9397	0.0052
Jerman et al. [16]	0.954	0.924	0.9410	0.0096
Our reconstruction from centrelines	0.951	0.923	0.9382	0.0089

For each test image, two segmented images are evaluated. The first one is the thresholded vesselness map $\mathcal{F}_\delta$ introduced by Jerman et al. [16], and the second one is a segmented image reconstructed from vessel centrelines and diameters extracted with the proposed framework. With $\mathcal{F}_\delta$, we evaluate the performance of the vascular enhancement filter used [16]. With the reconstructed image, we evaluate the accuracy of the vessel centrelines and diameters extracted by our method. Figure 7 shows the different segmented images from the test image giving the maximal accuracy.

Results on test images are presented in Table 1. We compare our result with two other methods, Benmansour and Cohen [4] and Chen and Cohen [6], where an anisotropic metric and a one dimension higher numerical scheme for estimation of the diameters are used as detailed in Sect. 1. Their performance measures were taken from [6]. We used the same experimental conditions: only the second manual segmentations are used as groundtruth and the FOV mask is eroded by 11 pixels to remove the effect of the FOV boundary on $\mathcal{F}_\delta$. With higher maximal and average accuracies than [4] and higher maximal accuracy than [6], we see the pretty good performance of our reconstruction from centrelines demonstrating the quality of our vessel extraction method, even if we used an isotropic metric model and a classical numerical scheme. Also, the advantage of pre-segmenting the images with the Jerman et al. [16] filter is clearly pointed out by its outperformance.

Because of the large number of negatives in the background of segmented vessel images, as explained in [25], we usually prefer sensitivity (recall) and precision to measure vessel detection performance. Therefore, we plot on Fig. 8-left the sensitivity of our results as a function of (1-precision) for all test images. We see how the points corresponding to Jerman and to our reconstruction lie closely to the first manual segmentations of the DRIVE dataset, except few images with lower sensitivity. We also observe that our centrelines reconstruction algorithm lacks of sensitivity comparing to Jerman. This is due to some unreconstructed small vessel branches. However, its performance remains enough similar to Jerman to ensure that centrelines are pretty well extracted. On Fig. 8-right, a ROC curve whose points correspond to several values of parameters (τ, δ) applied on the first test image illustrates a performing profile.

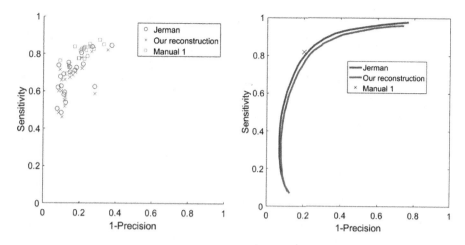

Fig. 8. Sensitivity vs (1-Precision) graph comparing Jerman and our reconstruction to the first manual segmentations of the DRIVE dataset. Left: each point corresponds to one image of the test set. Right: ROC curve of the first test image.

5 Conclusion

In this study, we showed how to design a binary isotropic metric with vascular a priori knowledge and introduced a new centred keypoint detection method to accurately extract centrelines, diameters and bifurcations of a vascular network. This framework has the advantage to use the classical minimal path based methods and in the same time to produce performance being similar even better to the recent methods as shown on the retinal images DRIVE database. Besides, it can easily be extended to 3D. As future works, the robustness of the proposed graph description of vascular networks to outlier bifurcations and outlier cycles should be studied. Also, the proposed framework will be applied on 3D ultrasensitive Doppler images to improve our recent results presented in [8].

Acknowledgement. We would like to particularly thank Dr. Da Chen for his precious help and advice.

References

1. Attali, D., Boissonnat, J.D., Edelsbrunner, H.: Stability and computation of medial axes-a state-of-the-art report. In: Möller, T., Hamann, B., Russell, R.D. (eds.) Mathematical Foundations of Scientific Visualization, Computer Graphics, and Massive Data Exploration. MATHVISUAL, pp. 109–125. Springer, Heidelberg (2009). https://doi.org/10.1007/b106657_6
2. Bekkers, E.J., Duits, R., Mashtakov, A., Sanguinetti, G.R.: Data-driven subriemannian geodesics in SE(2). In: Aujol, J.-F., Nikolova, M., Papadakis, N. (eds.) SSVM 2015. LNCS, vol. 9087, pp. 613–625. Springer, Cham (2015). https://doi.org/10.1007/978-3-319-18461-6_49

3. Benmansour, F., Cohen, L.D.: Fast object segmentation by growing minimal paths from a single point on 2D or 3D images. J. Math. Imaging Vis. **33**(2), 209–221 (2009)
4. Benmansour, F., Cohen, L.D.: Tubular structure segmentation based on minimal path method and anisotropic enhancement. Int. J. Comput. Vis. **92**(2), 192–210 (2011)
5. Bullitt, E., Gerig, G., Pizer, S.M., Lin, W., Aylward, S.R.: Measuring tortuosity of the intracerebral vasculature from mra images. IEEE Trans. Med. Imaging **22**(9), 1163–1171 (2003)
6. Chen, D., Cohen, L.D.: Piecewise geodesics for vessel centerline extraction and boundary delineation with application to retina segmentation. In: Aujol, J.-F., Nikolova, M., Papadakis, N. (eds.) SSVM 2015. LNCS, vol. 9087, pp. 270–281. Springer, Cham (2015). https://doi.org/10.1007/978-3-319-18461-6_22
7. Chen, D., Mirebeau, J.M., Cohen, L.D.: Vessel tree extraction using radius-lifted keypoints searching scheme and anisotropic fast marching method. J. Algorithms Comput. Technol. **10**(4), 224–234 (2016)
8. Cohen, E., Deffieux, T., Demené, C., Cohen, L.D., Tanter, M.: 3D vessel extraction in the rat brain from ultrasensitive doppler images. In: Gefen, A., Weihs, D. (eds.) Computer Methods in Biomechanics and Biomedical Engineering. LNB, pp. 81–91. Springer, Cham (2018). https://doi.org/10.1007/978-3-319-59764-5_10
9. Cohen, E., Deffieux, T., Tiran, E., Demene, C., Cohen, L., Tanter, M.: Ultrasensitive doppler based neuronavigation system for preclinical brain imaging applications. In: 2016 IEEE International Ultrasonics Symposium (IUS), pp. 1–4. IEEE (2016)
10. Cohen, L.D., Kimmel, R.: Global minimum for active contour models: a minimal path approach. Int. J. Comput. Vision **24**(1), 57–78 (1997)
11. Crandall, M.G., Lions, P.L.: Viscosity solutions of hamilton-jacobi equations. Trans. Am. Math. Soc. **277**(1), 1–42 (1983)
12. Demené, C.: Cartographie vasculaire et fonctionnelle du cerveau par échographie Doppler ultrarapide chez le petit animal et le nouveau-né. Ph.D. thesis, Paris 7 (2015)
13. Deschamps, T., Cohen, L.D.: Fast extraction of minimal paths in 3D images and applications to virtual endoscopy. Med. Image Anal. **5**(4), 281–299 (2001)
14. Frangi, A.F., Niessen, W.J., Vincken, K.L., Viergever, M.A.: Multiscale vessel enhancement filtering. In: Wells, W.M., Colchester, A., Delp, S. (eds.) MICCAI 1998. LNCS, vol. 1496, pp. 130–137. Springer, Heidelberg (1998). https://doi.org/10.1007/BFb0056195
15. Hart, W.E., Goldbaum, M., Côté, B., Kube, P., Nelson, M.R.: Measurement and classification of retinal vascular tortuosity. Int. J. Med. Inform. **53**(2), 239–252 (1999)
16. Jerman, T., Pernus, F., Likar, B., Spiclin, Z.: Enhancement of vascular structures in 3D and 2D angiographic images. IEEE Trans. Med. Imaging **35**(9), 2107 (2016)
17. Kaul, V., Yezzi, A., Tsai, Y.: Detecting curves with unknown endpoints and arbitrary topology using minimal paths. IEEE Trans. Pattern Anal. Mach. Intell. **34**(10), 1952–1965 (2012)
18. Law, M.W.K., Chung, A.C.S.: Three dimensional curvilinear structure detection using optimally oriented flux. In: Forsyth, D., Torr, P., Zisserman, A. (eds.) ECCV 2008. LNCS, vol. 5305, pp. 368–382. Springer, Heidelberg (2008). https://doi.org/10.1007/978-3-540-88693-8_27

19. Li, H., Yezzi, A., Cohen, L.: 3D multi-branch tubular surface and centerline extraction with 4D iterative key points. In: Yang, G.-Z., Hawkes, D., Rueckert, D., Noble, A., Taylor, C. (eds.) MICCAI 2009. LNCS, vol. 5762, pp. 1042–1050. Springer, Heidelberg (2009). https://doi.org/10.1007/978-3-642-04271-3_126
20. Mirebeau, J.M.: Anisotropic fast-marching on cartesian grids using lattice basis reduction. SIAM J. Numer. Anal. **52**(4), 1573–1599 (2014)
21. Peyré, G., Péchaud, M., Keriven, R., Cohen, L.D.: Geodesic methods in computer vision and graphics. Found. Trends® Comput. Graph. Vis. **5**(3–4), 197–397 (2010)
22. Rouy, E., Tourin, A.: A viscosity solutions approach to shape-from-shading. SIAM J. Numer. Anal. **29**(3), 867–884 (1992)
23. Sethian, J.A.: A fast marching level set method for monotonically advancing fronts. Proc. Nat. Acad. Sci. **93**(4), 1591–1595 (1996)
24. Sethian, J.A.: Level Set Methods and Fast Marching Methods: Evolving Interfaces in Computational Geometry, Fluid Mechanics, Computer Vision, and Materials Science, vol. 3. Cambridge University Press, Cambridge (1999)
25. Sofka, M., Stewart, C.V.: Retinal vessel centerline extraction using multiscale matched filters, confidence and edge measures. IEEE Trans. Med. Imaging **25**(12), 1531–1546 (2006)
26. Staal, J., Abramoff, M., Niemeijer, M., Viergever, M., van Ginneken, B.: Ridge based vessel segmentation in color images of the retina. IEEE Trans. Med. Imaging **23**(4), 501–509 (2004)
27. Tsitsiklis, J.N.: Efficient algorithms for globally optimal trajectories. IEEE Trans. Autom. Control **40**(9), 1528–1538 (1995)

Inference, Labeling and Relaxation

Limited-Memory Belief Propagation
via Nested Optimization

Christopher Zach[⊠]

Toshiba Research Europe, Cambridge, UK
christopher.m.zach@gmail.com

Abstract. In this work we express resource-efficient MAP inference as joint optimization problem w.r.t. (i) messages (i.e. reparametrizations) and (ii) surrogate potentials that are upper bounds for the problem of interest and allow efficient inference. We show that resulting nested optimization task can be solved on trees by a convergent and efficient algorithm, and that its loopy extension also returns convincing MAP solutions in practice. We demonstrate the utility of the method on dense correspondence and image completion problems.

Keywords: MAP inference · Belief propagation
Markov random fields

1 Introduction

Maximum a-posteriori (MAP) inference in graphical models, and especially in random fields defined over image domains, is one of the most useful tools in computer vision and related fields. Sometimes non-linear optimization is the method of choice for best efficiency, if all potentials defining the objective are of parametric (or better convex) shape. On the other hand, if the potentials are of non-parametric shape (or highly non-convex), then well established methods such as loopy belief propagation (BP) or its convex variants are the method of choice. BP and related algorithms are highly successful and relatively efficient if the underlying state space is rather small, but suffer from two limitations in case of large state spaces: first, the intrinsic message passing step has at least linear but generally superlinear run-time complexity in terms of the state space size. Thus, the runtime of these methods does not scale well with the number of states. Further, even computing all terms (potentials) needed to define the problem instance can already be prohibitive. Second, the memory consumption grows linearly with the state space size, since belief propagation requires to maintain *messages* for each state. These shortcomings of BP are well known, and several approaches proposed in the literature (which will be reviewed in Sect. 1.1) attempt to overcome these limitations.

In this work we derive an algorithm similar in spirit to particle belief propagation (PBP, and its more recent flavors) by considering the following nested

© Springer International Publishing AG, part of Springer Nature 2018
M. Pelillo and E. Hancock (Eds.): EMMCVPR 2017, LNCS 10746, pp. 517–532, 2018.
https://doi.org/10.1007/978-3-319-78199-0_34

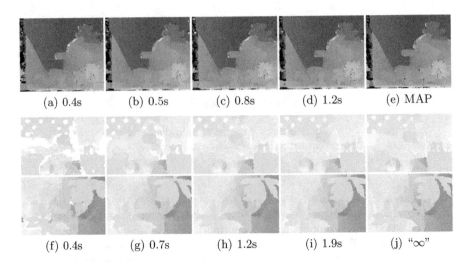

(a) 0.4s   (b) 0.5s   (c) 0.8s   (d) 1.2s   (e) MAP

(f) 0.4s   (g) 0.7s   (h) 1.2s   (i) 1.9s   (j) "∞"

Fig. 1. (a)–(d) Extracted primal solutions for a stereo correspondence problem after 2, 4, 8 and 16 passes over the image. (e) Primal solution obtained by full scale MAP inference. (f)–(i) Extracted primal solutions for optical flow problems after 2, 4, 8 and 16 passes. (j) Result after 256 iterations (22 s) as approximation to intractable full scale MAP inference. [Best viewed on screen.]

optimization problem: for a given inference problem determine the surrogate problem (represented by its potentials) among the set of "easy" problems (for which MAP inference is efficient) that is most similar to the given one. This is a nested problem since both the surrogate problem and the MAP solution need to be determined jointly. One contribution of this work is, that the joint search over potentials and MAP solutions can be carried out efficiently, and the resulting algorithm has a striking resemblence with PBP by alternating state selection and message passing steps. In contrast to PBP, which is rooted in approximating intractable distributions, our method aims directly to find an accurate MAP solution.

We also show that our proposed method is always convergent if applied on problems defined on tree graphs, and experimentally we found convergent behavior of its loopy extension to lattices. To our knowledge no such guarantee is available for any version of particle-based BP.

We illustrate the performance of the method in Fig. 1 on dense correspondence problems, where the evolution of label assignments returned by our method after the given numbers of passes over the image are shown. It can be seen that after 16 passes (<2 s CPU runtime) the label assignment has essentially converged. More important than the run-time is memory consumption: in the optical flow scenario our method requires about 32 MB (which compares to 16 GB—or 694 MB if the data terms are computed on the fly—required by a message passing method exploiting the specific product label structure [20]). This makes our approach also appealing for embedded and mobile implementations.

1.1   Related Work

There is a huge amount of literature related to MAP inference algorithms and efficient implementations, and we focus on methods specifically addressing large state spaces. A number of approaches approximate full-scale messages required in belief propagation by sets of particles to cope with large or even continuous state spaces, e.g. [3,12,14,16,17,24]. These methods augment MAP inference with a stochastic particle selection step to approximate otherwise intractable distributions (or message functions) arising in the propagation step. They differ largely in how new candidates states (particles) are sampled and filtered, and some approaches are specifically designed for dense correspondence estimation. While these methods show good performance in practice, only few (and often weak) quantitative results are known for message passing in a stochastic context (e.g. [9,15,16]).

"Active MAP inference" [18] explicitly addresses the case when the computation of unary potentials is the main run-time bottleneck: unary potentials can be queried during MAP inference, and states selected for instantiating the unary potentials are ranked based on aggregated belief. In order to rank "unseen" states repeated MAP inference is necessary. In contrast to that particle max-product methods (and our approach) interleave MAP inference and candidate state scoring and is therefore very efficient for large state spaces. A different nested approach for MAP inference with large or continuous state spaces is considered in [17,27], but in contrast to this work the surrogate problems are constructed to be lower bounds for the target inference problem. [21] mentions, but does not explore further a primal "majorize-minimize" framework, which shares the use of majorizing potentials (but in a different context) with our approach.[1]

1.2   Background

This section presents some material on MAP inference and the underlying linear programming relaxation. A labeling or MAP inference problem is determining the optimal label $x_s \in \mathcal{L}$ assigned at each node $s \in \mathcal{V}$, where the objective is over unary terms and pairwise terms[2],

$$\mathbf{x}^* \stackrel{\text{def}}{=} \arg\min_{\mathbf{x}} \sum_{s \in \mathcal{V}} \theta_s(x_s) + \sum_{(s,t) \in \mathcal{E}} \theta_{st}(x_s, x_t) = \arg\min_{\mathbf{x}} \sum_{\alpha} \theta_\alpha(\mathbf{x}_\alpha), \qquad (1)$$

where $\mathbf{x} = (x_s)_{s \in \mathcal{V}} \in \mathcal{L}^{|\mathcal{V}|}$, and $\mathcal{E}$ is a problem-specific set of undirected edges. θ are the *potentials* or costs for assigning particular states to nodes and edges. We will use $\alpha \in \mathcal{V} \cup \mathcal{E}$ to index unary and pairwise cliques in a unified way. The above label assignment problem is generally intractable to solve, and one

---

[1] Stricly speaking we do not use a *majorizer*, but only (less constrained) upper bounds.
[2] We focus on problems with at most pairwise cliques, which are most relevant in practice, but everything can be generalized to higher order cliques straightforwardly.

highly successful approach to approximately solve this problem is to employ the corresponding linear programming (LP) relaxation (see e.g. [26])[3],

$$E_{\mathrm{MAP}}(\tau; \theta) \stackrel{\mathrm{def}}{=} \sum_{\alpha, \mathbf{x}_\alpha} \theta_\alpha(\mathbf{x}_\alpha) \tau_\alpha(\mathbf{x}_\alpha) \qquad \text{s.t. } \tau_s(x_s) = \sum_{\mathbf{x}_\alpha \setminus x_s} \tau_\alpha(\mathbf{x}_\alpha), \qquad (2)$$

$\sum_{x_s} \tau_s(x_s) = 1$ and $\tau_\alpha(\mathbf{x}_\alpha) \geq 0$. The unknowns $\{\tau_s\}_{s \in \mathcal{V}}$ and $\{\tau_{st}\}_{(s,t) \in \mathcal{E}}$ are "one-hot" encodings of the assigned labels, e.g. if τ^* is the optimal solution of E_{MAP} and the relaxation is tight, then $\tau_s(x_s)$ is ideally 1 iff state x_s is the optimal label at node s, and 0 otherwise. The constraints linking τ_s and τ_α are usually called *marginalization constraints*, and the unit sum constraint is typically referred as *normalization constraint*. The linear program in Eq. 2 is not unique, since redundant non-negativity and normalization constraints can be added to E_{MAP} without affecting the optimal solution or value. Consequently, different duals are considered in the literature. A particular dual program is the following,

$$E_{\mathrm{MAP}}^*(\lambda; \theta) = \sum_\alpha \min_{\mathbf{x}_\alpha} \left\{ \theta_\alpha(\mathbf{x}_\alpha) - \sum_{s \in \alpha} \lambda_{\alpha \to s}(x_s) \right\} \qquad (3)$$

subject to $\sum_{\alpha \ni s} \lambda_{\alpha \to s}(x_s) = 0$ for all s and x_s. Recall that cliques α range over nodes and edges. The dual unknowns $\{\lambda_{\alpha \to s}\}_{s \in \mathcal{V}, \alpha \ni s}$ are the Lagrange multipliers for the marginalization constraints, which are often termed "messages" [7,22,23,26].

2   A Nested Approach to MAP Inference

This section describes our framework designed for MAP inference in large state spaces. First, we present a generic, high-level view, which will be made more concrete to obtain efficient implementable algorithms in subsequent sections.

For given potentials θ and ϕ we will write $\theta \preceq \phi$ if $\theta_\alpha(\mathbf{x}_\alpha) \leq \phi_\alpha(\mathbf{x}_\alpha)$ for all cliques α and clique states $\mathbf{x}_\alpha$. Hence, $\phi \succeq \theta$ is an element-wise upper bound of θ. Let $\Phi \subseteq \{\phi : \phi \succeq \theta\}$ be a set of potentials upper bounding θ. It will be described later in more detail, but in general the potentials $\phi \in \Phi$ should also allow very efficient minimization of the MAP objective $E_{\mathrm{MAP}}(\cdot; \phi)$ (compared to θ). Let

$$v^*(\theta) := \min_\tau E_{\mathrm{MAP}}(\tau; \theta) \qquad (4)$$

be the optimal value of interest (i.e. the optimal value for the MAP problem instance we want to solve). We introduce

$$\bar{v}(\Phi) := \min_{\phi \in \Phi} \min_\tau E_{\mathrm{MAP}}(\tau; \phi), \qquad (5)$$

---

[3] The expression $\mathbf{x}_\alpha \setminus x_s$ is shorthand for $\{\mathbf{x}_\alpha' : x_s' = x_s\}$. We will also write compactly $\alpha \ni s$ instead of $\{\alpha : s \in \alpha\}$.

Algorithm 1. Abstract nested MAP inference

Require: Φ such that $\phi \succeq \theta$ for all $\phi \in \Phi$
 1: Initialize $\phi \in \Phi$
 2: **while** not a stopping criterion reached **do**
 3:     Compute $\hat{v} \leftarrow \min_\tau E_{\mathrm{MAP}}(\tau; \phi)$
 4:     Choose $\phi' \in \Phi$ and compute $\check{v} \leftarrow \min_\tau E_{\mathrm{MAP}}(\tau; \phi')$
 5:     **if** $\check{v} < \hat{v}$ **then** $\phi \leftarrow \phi'$ **end if**
 6: **end while**

---

i.e. the smallest MAP objective obtained by any potential $\phi \in \Phi$. Since by assumption $\phi \succeq \theta$, we have $v^*(\theta) \leq \overline{v}(\Phi)$. Thus, the goal is to find a $\phi^* \in \Phi$ such that the optimal value $\min_\tau E_{\mathrm{MAP}}(\tau; \phi^*)$ is closest to $v^*(\theta)$.

Let us rephrase the meaning of $\overline{v}(\Phi)$: if Φ contains potentials that permit efficient inference, the task is to determine $\phi^* \in \Phi$ (and the corresponding primal solution $\mathbf{x}^*$) with the optimal value nearest to the one of the original problem, $v^*(\theta)$. In that sense ϕ^* is the most faithful surrogate for θ among the elements in Φ. Since $\overline{v}(\Phi)$ is defined as optimal value of a nested optimization problem, the requirement is that the additional cost of optimizing over $\phi \in \Phi$ is more than compensated by the efficient computation of $E_{\mathrm{MAP}}(\cdot; \phi)$.

A generic descent algorithm that is guaranteed not to increase the maintained estimate $\hat{v}$ of $v^*(\Phi)$ is given in Algorithm 1. The algorithm iteratively probes new potentials $\phi' \in \Phi$ and replaces the current one ϕ, if the optimal value $\check{v} = \min_\tau E_{\mathrm{MAP}}(\tau; \phi')$ is smaller than $\hat{v} = \min_\tau E_{\mathrm{MAP}}(\tau; \phi)$. Note that computing $\hat{v}$ and $\check{v}$ will be usually performed indirectly by a message passing or belief propagation algorithm, which iteratively or in closed form determine a set of messages certifying optimality of the returned solution. The generic algorithm Algorithm 1 leaves many design choices open, which will influence its practical performance (and even tractability):

- How should Φ be designed? Elements $\phi \in \Phi$ need to allow very efficient MAP inference, and they need to be bounded from below by the given potentials θ. Further, it must be easy to select elements from Φ. Below we describe a natural choice for Φ in detail.
- Given the current value of ϕ, how should the next $\phi' \in \Phi$ be selected? In general the choice of ϕ' given ϕ is application specific. In our setting ϕ' will differ from ϕ only in potentials associated with a currently active node s, i.e. $\phi'_\alpha(\mathbf{x}_\alpha) = \phi_\alpha(\mathbf{x}_\alpha)$ for all cliques α not containing a node s. Thus, the minimization w.r.t. $\phi \in \Phi$ is therefore made compositional by successively optimizing only over $(\phi_\alpha)_{\alpha \ni s}$ for a node s currently in focus. As we will see later in Sect. 3, the BP messages needed to compute $v^*(\phi)$ also provide useful information on how to select ϕ'. Further details on the transition $\phi \rightsquigarrow \phi'$ are given in Sects. 3 and 5.
- How can we efficiently compute $\check{v}$? In Sect. 3 we will show that $\check{v}$ can be computed very efficiently if ϕ and ϕ' differ only locally. In this setting most BP messages used to determine $\hat{v}$ in the min-sum BP algorithm can be reused to compute $\check{v}$. In Sect. 3 it is shown that Algorithm 1 (together with our choice

of Φ and the transition $\phi \rightsquigarrow \phi'$) condenses to alternating between locally modifying ϕ and local message passing steps.

The following sections specify more implementation details to convert the abstract meta-algorithm in Algorithm 1 into a practical implementation.

Design of Φ: The requirement is that elements $\phi \in \Phi$ need to allow efficient computation of $v^*(\phi) = \min_\tau E_{\mathrm{MAP}}(\tau; \phi)$ in order for the nested algorithm (Algorithm 1) to be competitive in run-time (compared to directly inferring $\min_\tau E_{\mathrm{MAP}}(\tau; \theta)$). Intuitively, elements in Φ shall be "uninformative" for most (clique) states in order to allow significant runtime and memory savings for computing and representing messages. One choice of Φ that allows very efficient MAP inference is given as follows: the set Φ_K consists of elements which are indexed by the cartesian product of "resident sets"[4] $\mathbf{R} = (R_s)_{s \in V}$, where $R_s \subseteq \mathcal{L}$ has exactly K elements (with $K \in \{1, \ldots, |\mathcal{L}|\}$ being a user-specified parameter):

$$\Phi_K \overset{\mathrm{def}}{=} \{\theta[\mathbf{R}] : \forall s : R_s \subseteq \mathcal{L} : |R_s| = K\} \qquad (6)$$

with

$$\theta_\alpha[R_\alpha](\mathbf{x}_\alpha) \overset{\mathrm{def}}{=} \begin{cases} \theta_\alpha(\mathbf{x}_\alpha) & \text{if } \forall s \in \alpha : x_s \in R_s \\ \overline{\theta}_\alpha & \text{otherwise} \end{cases} \qquad (7)$$

and $\theta[\mathbf{R}] \overset{\mathrm{def}}{=} (\theta_\alpha[R_\alpha])_\alpha$. Here $\overline{\theta}_\alpha$ is a value large enough to guarantee $\phi \succeq \theta$ for all $\phi \in \Phi_K$, e.g. $\overline{\theta}_\alpha = \max_{\mathbf{x}_\alpha} \theta_\alpha(\mathbf{x}_\alpha)$. These values are typically known from the (application-specific) design of θ.

Since for a $\phi \in \Phi_K$ non-resident states have all equal potentials $\overline{\theta}_\alpha$, they are indistinguishable. This also implies that we can enforce equality of BP messages for non-resident states, e.g. $\mu_{\alpha \to s}(x_s) = \mu_{\alpha \to s}(x'_s)$ for $x_s, x'_s \notin R_s$, without affecting the optimal solution. This property can be shown to hold also for messages λ of the dual objective E^*_{MAP}. [27] Overall, the $|\mathcal{L}|$-dimensional message vector (e.g. $\mu_{\alpha \to s}(\cdot)$) can be losslessly represented to a $K+1$-dimensional messages $(\mu_{\alpha \to s}(x_s))_{x_s \in R_s \cup \{*\}}$, where $*$ represents all (indistinguishable) non-resident states $x_s \notin R_s$. Further, message computation is reduced from $O(|\mathcal{L}|^{|\alpha|})$ in general to $O(K^{|\alpha|})$. Both memory and runtime savings can therefore be significant.

Φ_K has another property that will be useful in practical implementations: given one element $\phi \in \Phi_K$, we can create another $\phi' \in \Phi_k$ by modifying one of the resident sets R_s, and every element in Φ_K can be generated by a sequence of local alterations of resident sets. This property allows us to construct ϕ' from the current potentials ϕ in Algorithm 1 solely by local changes.

3   Nested MAP Inference on Trees

When the underlying graph is a tree, exact MAP inference can be performed efficiently by min-sum belief propagation on a tree, which is essentially an instance

---

[4] We prefer the term "resident set" over "particles", since—in contrast to particle message passing—our method maintains messages also for non-resident states.

of dynamic programming. In this section we demonstrate how the message pass-ing steps of min-sum BP can be interleaved with updating the currently active potentials ϕ in order to reduce the target objective $\overline{v}(\Phi)$. The generic method (Algorithm 1) is specialized in Algorithm 2 for MAP inference problems on trees.

Algorithm 2. Nested MAP inference on a tree.

1: Choose initial $\phi \in \Phi$ and initial root s
2: Compute all upward messages $\mu_{u \to t}$ towards the root s:

$$\mu_{u \to t}(x_t) \leftarrow \min_{x_u} \left\{ \phi_{ut}(x_u, x_t) + \sum_{v \in ch(u)} \mu_{v \to u}(x_u) \right\}$$

3: **while** not a stopping criterion reached **do**
4: $\quad \hat{v} \leftarrow \min_{x_s} \left\{ \sum_{t \in ch(s)} \mu_{t \to s}(x_s) \right\}$ $\qquad\qquad\qquad\qquad \triangleright \hat{v} = v^*(\phi)$
5: $\quad$ Select K^+ states C_s with $C_s \cap R_s = \emptyset$; enrich $\phi'_{st} = \theta_{st}[R_s \cup C_s]$
6: $\quad$ Compute min-marginals $\forall x_s \in R_s \cup C_s$

$$\mu'_{t \to s}(x_s) \leftarrow \min_{x_t} \left\{ \phi'_{st}(x_s, x_t) + \sum_{v \neq s} \mu_{v \to t}(x_t) \right\}$$

$$\nu'_s(x_s) \leftarrow \sum_{t \in N(s)} \mu'_{t \to s}(x_s)$$

$$\check{v} \leftarrow \min_{x_s \in R_s \cup C_s} \nu'_s(x_s) \qquad\qquad\qquad\qquad \triangleright \check{v} = v^*(\phi') \leq \hat{v}$$

7: $\quad$ Rank $x_s \in R_s$ w.r.t. $\nu'_s(x_s): x_s^{(1)}, x_s^{(2)}, \ldots$
8: $\quad R_s \leftarrow \{x_s^{(1)}, \ldots, x_s^{(K-K^+)}\} \cup C_s$ and $\phi \leftarrow \theta[\mathbf{R}]$
9: $\quad$ Choose a child t of s as the new root: $\hat{s} \leftarrow s$
10: $\quad$ Reorient the tree edges towards t
11: $\quad$ Update $\mu_{\hat{s} \to s}$ from $\hat{s}$ to the current root s,

$$\mu_{\hat{s} \to s}(x_s) \leftarrow \min_{x_{\hat{s}}} \left\{ \phi_{\hat{s}s}(x_{\hat{s}}, x_s) + \sum_{u \in ch(\hat{s})} \mu_{u \to \hat{s}}(x_{\hat{s}}) \right\}$$

12: **end while**
13: Update downward messages away from the current root s
14: **return** $\mathbf{x}^*$ with $x_u^* \in \arg\min_{x_u} \sum_{v \in N(u)} \mu_{v \to u}(x_u)$ $\quad \triangleright \hat{v} = E_{\text{MAP}}(\mathbf{x}^*; \phi) = v^*(\phi)$

The current estimates $\hat{v}$ (line 4) and $\check{v}$ (line 7) for $\overline{v}(\Phi)$ are efficiently com-puted and updated via min-sum BP. Note that these values are not actually used in the algorithm (other than for tracking and reporting progress). In order to obtain $\hat{v}$ (and $\check{v}$) it is only necessary to compute upward messages towards the root. Further, if ϕ and ϕ' differ only in their potentials at the root node s (i.e. $\phi_\alpha = \phi'_\alpha$ if $s \notin \alpha$), then all upward messages other then the ones incoming at s can be reused when computing $\check{v}$.

Moreover, when a new root (e.g. $\hat{s}$) is selected, then only messages along the (directed) path from s to $\hat{s}$ will need recomputation in order to make all upward messages valid again. Hence, if $\hat{s}$ is a direct neighbor of s (i.e. a child of s), then only the messages on a single edge $\hat{s} \to s$ are invalidated and need recomputation.

It is sensible to traverse neighboring nodes, since updated potentials at node s will have the strongest impact on adjacent nodes. Overall we can state the following: *message passing steps updating μ and minimization steps w.r.t. ϕ can be interleaved yielding an efficient implementation of the generic descent method in Algorithm* 1. The most important invariant of the method is, that upward messages are always consistent with the current potentials ϕ. Since $\hat{v}$ computed in line 4 is non-increasing by construction and lower bounded by $v^*(\theta)$, we obtain the following

Proposition 1. *The sequence of values $\hat{v}$ generated in line 4 of Algorithm 2 converges.*

The algorithm is presented for the particular choice of $\Phi = \Phi_K$ with K user-specified. A further design choice is that ϕ and ϕ' differ only in their resident sets R_s and R'_s by $K^+ < K$ elements. By exploiting the particular structure of Φ, we can actually test $\binom{K}{K-K^+}$ elements from Φ in each step simultaneously. For notational brevity we present the algorithm with vanishing unary potentials, i.e. $\theta_s = \phi_s \equiv 0$. This can be always achieved w.l.o.g. by reparametrizing the original potential via $\theta_{st}(x_s, x_t) \leftarrow \theta_{st}(x_s, x_t) + \theta_s(x_s)/\deg(s) + \theta_t(x_t)/\deg(t)$ and $\theta_s(x_s) \leftarrow 0$ for all x_s, x_t.

Candidate set C_s: In line 5 the selection of K^+ candidate states C_s is left unspecified. Given that the objective is to lower $\hat{v}$ as much as possible (and therefore make $\check{v}$ as small is possible), C_s should contain states with small min-marginals,

$$\sum_{t \in ch(s)} \min_{x_t} \left\{ \theta_{st}(x_s, x_t) + \sum_{r \in ch(t)} \mu_{r \to t}(x_t) \right\}. \tag{8}$$

This criterion assumes knowledge of the true unary and pairwise potentials associated with (a small number of) candidate states. In our implementations, given an enlarged candidate set $\tilde{C}_s$ of states C_s is obtained by taking the K^+ states from $\tilde{C}_s$ with the smallest min-marginals. There are many options to generate $\tilde{C}_s$. In our experiments we determine $\tilde{C}_s$ by generating one sample for each state in one neighboring resident set, i.e. $\tilde{C}_s = \{H(x_t) : x_t \in \bigcup_{t \in N(s)} R_t\}$. H is a stochastic function that aims to model pairwise interactions, that is $-\log P(H(x_t), x_t) \approx \theta_{st}(H(x_t), x_t)$. Further details of the implementations are given in Sect. 5.

Limit points: While Algorithm 2 is guaranteed to be non-increasing in its estimate $\hat{v}$, this does not imply that $\hat{v} \to \bar{v}(\Phi)$. In will be shown below that e.g. iterated conditional modes is an instance of our framework, hence a poor limit point could be returned by Algorithm 2. In practice we found that in our experiments setting $K \geq 3$ is sufficient to provide results often indistinguishable from the true MAP solution.

Connection to move-making methods: At this point we are able to discuss move-making algorithms for MAP inference such as ICM (iterated conditional modes [2]), graph-cuts [4] and fusion moves [13] (or any other primal move-making framework) are instances of the generic method in Algorithm 1. Since these algorithms iteratively update a single primal label assignment, the set of potential vectors is given by a variation of Φ_1,

$$\Phi_1^\infty := \left\{ \phi : \phi_\alpha(\mathbf{x}_\alpha) = \begin{cases} \theta_\alpha(\mathbf{x}_\alpha) & \text{if } \forall s \in \alpha : x_s \in R_s \\ \infty & \text{otherwise} \end{cases}, \ R_s \subseteq \mathcal{L} : |R_s| = 1 \right\}, \quad (9)$$

where there is exactly one resident state per node (the best solution found so far). The upper bound $\overline{\theta}_\alpha$ is always set to ∞. The transition from ϕ to ϕ' in all methods is done by first augmenting ϕ to $\hat{\phi}$ by allowing a larger resident set (adding all potentials for a single node in ICM, adding a constant state at each node in graph-cuts, and expanding the resident set by a global proposal in fusion moves), followed by a "projection" of $\hat{\phi}$ to an element $\phi' \in \Phi_1^\infty$ that decreases the MAP objective. In graph-cuts and fusion moves this projection to find the optimal $\phi' \in \Phi_1^\infty$ is the computationally expensive step. Since $K = 1$ and $\overline{\theta} = \infty$, the messages are uninformative, and the determination of ϕ' is therefore agnostic to any non-resident state.

In light of these observations, one can interpret Algorithm 2 as variant of ICM with the following important differences: the resident sets contain more than one element, and by maintaining meaningful messages for non-resident states, sensible, finite estimates for the values (i.e. min-marginals) of even non-resident states are available. In contrast to other richer moving making algorithms the proposed method requires only local proposals.

4   Nested MAP Inference on Loopy Graphs

Most MAP instances are defined on graphs contain cycles, most prominently 4- or 8-neighborhood lattices used in computer vision applications. Algorithm 2 is not directly applicable to loopy graphs, since it relies crucially on the properties of belief propagation on a tree. There are two natural ways to extend the algorithm to general graphs, which are described in the following.

4.1   Limited Memory SGM

One straightforward option is to use a fixed tree decomposition and run Algorithm 2 independently on trees, and to extract the desired label assignment using the obtained min-marginals. This approach can be seen as limited-memory version of a semi-global method [8], which itself can be interpreted as one iteration of TRW message passing [5]. Let $\mathcal{T}$ be a tree decomposition such that each $T \in \mathcal{T}$ is a tree over $\mathcal{E}$, and $\{\theta^T\}$ are potentials such that $\sum_{T \in \mathcal{T}} \theta^T = \theta$. For each $T \in \mathcal{T}$ let $\hat{v}^T$ be the sequence of estimates for $v^*(\theta^T)$ maintained by Algorithm 2, then in light of Proposition 1 we conclude

Proposition 2. *The quantity $\sum_{T \in \mathcal{T}} \hat{v}^T$ is convergent.*

In contrast to standard tree decomposition for MAP inference the value of $\sum_{T \in \mathcal{T}} \hat{v}^T$ is not necessarily a lower bound for $v^*(\theta)$, since the attained surrogate potentials ϕ^T may be strictly larger than θ^T. In the supplementary material [28] it is shown how to obtain a (rather loose) upper bound for $v^*(\theta)$.

Figure 2 illustrates results for dense stereo and optical flow using horizontal, vertical and diagonal chains as suggested in [8]. We refer to Sect. 5.1 for a description of the employed unary and pairwise potentials. SGM relies on informative data terms to reach agreement of tree solutions in one step, which explains why it performed poorly for image completion tasks exhibiting weak data terms (discussed in Sect. 5.2) in our experiments. For dense correspondence problems with relatively informative unary potentials the MAP solutions obtained by limited memory SGM are qualitatively similar to the results returned by the method described in the following.

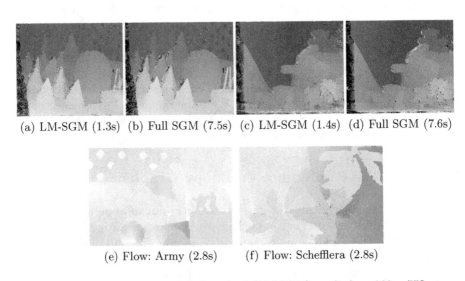

(a) LM-SGM (1.3s) (b) Full SGM (7.5s) (c) LM-SGM (1.4s) (d) Full SGM (7.6s)

(e) Flow: Army (2.8s)   (f) Flow: Schefflera (2.8s)

Fig. 2. Limited memory semi-global method (LM-SGM) applied on 900×750 stereo images (a, c) and on optical flow (e, f). We compare to SGM over the full label space (b, d). Best viewed on screen.

4.2   Loopy Message Passing

Often a static tree decomposition does not yield satisfactory results for MAP problems, and proper inference on graphs with cycles is required. In such scenarios computing $\hat{v} = \min_\tau E_{\mathrm{MAP}}(\tau; \phi) = v^*(\phi)$ and $\check{v} = \min_\tau E_{\mathrm{MAP}}(\tau; \phi') = v^*(\phi')$ *exactly* in general requires an iterative method that converges to the desired value, even for "efficient" potentials coming from Φ_K. Many methods to (approximately) solve $\min_\tau E_{\mathrm{MAP}}(\tau; \phi)$ strongly resemble min-sum belief propagation (e.g. [6,7,10,25], and therefore it is natural to consider respective adaptations of Algorithm 2: in short the messages μ in Algorithm 2 are replaced by reparametrizations λ appearing in the dual program $E^*_{\mathrm{MAP}}(\lambda;\theta)$ (or one of its variants), and the message passing steps (e.g. in line 11) are replaced by the respective message updates derived via the dual program. The modification of Algorithm 2 to loopy graphs is given in Algorithm 3 below.

Algorithm 3. Nested MAP inference on a loopy graph.

1: Choose initial $\phi \in \Phi$
2: Initialize $\lambda \equiv 0$
3: Optional: run loopy message passing to have better estimate for $E^*_{\mathrm{MAP}}(\lambda;\phi)$
4: **for** $s \in$ sequence over $\mathcal{V}$ **do**
5:     Select K^+ states C_s with $C_s \cap R_s = \emptyset$
6:     Compute min-marginals $\forall x_s \in R_s \cup C_s$

$$\nu_s(x_s) \leftarrow \sum_{t \in N(s)} \min_{x_t}\left\{\phi_{st}(x_s, x_t) + \sum_{v \neq s}\lambda_{v \to t}(x_t)\right\}$$

$$\check{v} \leftarrow \min_{x_s \in R_s \cup C_s} \nu_s(x_s)$$

7:     Rank $x_s \in R_s \cup C_s$ w.r.t. $\nu_s(x_s) : x_s^{(1)}, x_s^{(2)}, \ldots$
8:     $R_s \leftarrow \{x_s^{(1)}, \ldots, x_s^{(K-K^+)}\} \cup C_s$ and $\phi \leftarrow \theta[\mathbf{R}]$
9:     Update $\lambda_{t \to s}(x_s)$ for all $K+1$ states x_s w.r.t. ϕ (e.g. using "star updates")
10: **end for**
11: **return** $\mathbf{x}^*$ with $x_u^* \in \arg\min_{x_u} \sum_{v \in N(u)} \lambda_{v \to u}(x_u)$

If we go back to Algorithm 1, then the main difficulty to apply it on loopy graphs is that only lower and upper bounds for $\hat{v}$ and $\check{v}$ are easily available by convex duality: the current dual objective is always a lower bound for the optimal value, and an upper bound is available by extracting a primal solution $\tilde{\mathbf{x}}(\lambda)$ for given messages λ, i.e. $E_{\mathrm{MAP}}(\tilde{\mathbf{x}}(\lambda);\phi) \geq v^*(\phi)$. Assume we have a lower bound $\hat{v}_{\mathrm{LB}}$ for $\hat{v} = v^*(\phi)$ and an upper bound for $\check{v}_{\mathrm{UB}}$ for $\check{v} = v^*(\phi')$. If $\check{v}_{\mathrm{UB}} < \hat{v}_{\mathrm{LB}}$, then from the following chain of inequalities $v^*(\phi') \leq \check{v}_{\mathrm{UB}} < \hat{v}_{\mathrm{LB}} \leq v^*(\phi)$ we conclude that $v^*(\phi') < v^*(\phi)$. Thus, we obtain the following proposition:

Proposition 3. *If $\phi \leftarrow \phi'$ is only executed if $\check{v}_{\mathrm{UB}} < \hat{v}_{\mathrm{LB}}$ in line 5 of Algorithm 1, then the sequence $\hat{v}$ generated by the algorithm is non-increasing.*

Unfortunately the condition $\check{v}_{\mathrm{UB}} < \hat{v}_{\mathrm{LB}}$ is a very strong one and requires the duality gap $E_{\mathrm{MAP}}(\tilde{\mathbf{x}}(\lambda);\phi) - E^*_{\mathrm{MAP}}(\lambda;\phi)$ to be extremely small in order for the condition to be satisfied. The theoretically justified extensions of Algorithms 1 and 2 to loopy graphs are not performing satisfactory in practice.

Hence, we propose to use the dual objectives $E^*_{\mathrm{MAP}}(\cdot;\phi)$ and $E^*_{\mathrm{MAP}}(\cdot;\phi')$ as surrogates for $v^*(\phi)$ and $v^*(\phi')$, respectively. Since we use lower bounds for both $\hat{v}$ and $\check{v}$, monotonic behavior of $\hat{v} = v^*(\phi)$ is not guaranteed. In practice, $\hat{v}$ generally decreases monotonically and shows minor oscillations (as shown in the supplementary material).[5] In some applications (most notably image completion in Sect. 5.2) we observed a rather large duality gap between the dual objective $\hat{v}$ and the value of the extracted primal solution. In this case we run a standard TRW-S like message passing algorithm to lower the primal-dual gap after Algorithm 2 completed a full traversal of nodes in the graph. In future work we aim to obtain stronger estimates for $v^*(\phi')$ via tree-block updates [23].

5   Results

In this section we show approximate MAP solutions obtained by our method applied in two very different application scenarios: the first application is dense correspondence estimation (including stereo disparity estimation and optical flow), and the second application is image completion via a random field model. In both settings the underlying graph is defined as 4-connected grid. We use the loopy extension of Algorithm 2 in our experiments. We fix the resident set size K to 5 in all application, hence the memory consumption required for the messages is $(5+1) \times 4 = 24$ floating point values per node. The algorithm is implemented in straightforward C++ with OpenMP enabled, and the runtimes are reported for a dual Xeon E5-2690 system with 64 GB of main memory. GPU acceleration is not employed.

5.1   Dense Correspondence

We demonstrate results on dense correspondence problems from the Middlebury benchmark datasets [1,19]. Note that the goal is not to propose a competing new model for dense correspondence estimation, but to demonstrate the effectiveness of our method for inferring MAP estimates in these problems.

Dense disparity estimation: For the chosen dense disparity instances the state space has 120 or 240 elements (corresponding to integral disparities). The data term (unary potential) attains values in $[0, 1]$ and is given by $\min\{0, \frac{1-ZNCC}{2}\}$, where $ZNCC$ is the zero-mean NCC of 5×5 gray-scale images patches. We use the 3-way P_1-P_2 pairwise potential from [8] (with $P_1 = 1/2$ and $P_2 = 1$). The sampling function H to determine $\tilde{C}_s$ is given as follows: from ground truth disparity maps we estimate the relative frequencies of events $x_s = x_t$ ($\approx 94\%$), $x_s = x_t \pm 1$ ($\approx 5.8\%$), and $|x_s - x_t| \geq 2$ ($\approx 0.2\%$) for neighboring pixels s, t, hence for a random variable $u \sim U[0, 1]$ we have

$$H(x; u) = \begin{cases} x & \text{if } u < 0.94 \\ x \pm 1 & \text{if } u < 0.998 \\ U[x_{min}, x_{max}] & \text{otherwise.} \end{cases} \quad (10)$$

---

[5] One can make the algorithm (trivially) convergent e.g. by conditionally updating the primal solution, such that the solution with minimal primal objective so far is always reported.

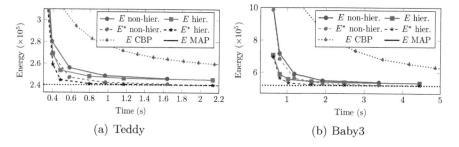

(a) Teddy   (b) Baby3

Fig. 3. Evolution of primal E and dual E^*_{MAP} energies for dense disparity estimation with respect to wall time. E MAP is the converged primal energy of full scale BP.

Figure 3 shows the evolution of the attained primal objective with respect to wall clock time for a full scale convex BP implementation and two variants of the proposed method: the first one uses a non-hierarchical ("flat") approach and initializes R_s using uniform sampling, and the second one uses a multi-scale framework (with 5 octaves), and one element in the initial R_s is fixed to the label propagated from the coarser level. Clearly, our method achieves a lower energy much faster than full-scale BP (which also requires about 20–40 times more memory). Figure 1(a)–(d) shows corresponding labeling results returned by the hierarchical approach after 2, 4, 8, and 16 passes. Further results are available in the supplementary material [28].

Optical flow: We ran similar numerical experiments for optical flow instances [1]. The state space contains 129^2 flow vectors corresponding to a 16 pixel search range at quarter-pixel accuracy. We upscale the original grayscale images to 400% and use the same ZNCC-based score as for dense stereo (but computed on 7×7 patches from the upscaled images). The pairwise smoothness term is the P_1-P_2 model applied separately in the horizontal and vertical component of the motion vector. The decrease in primal energy for the solution returned after the respective number of passes with respect to wall clock time is shown in Fig. 4. In

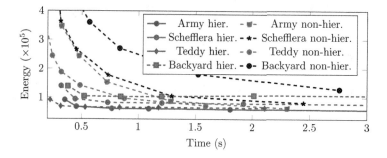

Fig. 4. Evolution of primal energies for dense optical flow problems.

this case the memory consumption is $6/129^2$ or less than 0.04% of running full inference,[6] and usable motion fields are obtained after about one second of CPU time. Visualizations of the corresponding flow fields are depicted in Fig. 1(f)–(i) and in the supplementary material [28].

5.2   Image Completion

Dense correspondence problems can usually rely on relatively strong unary potentials (image matching scores), and the role of pairwise potentials is to provide a regularizer where the unary potential leads to ambiguous estimates. Hence, we show in this section results on the very complementary problem of image completion, where unary potentials are only available near the border of the missing region, and the labels assigned in the interior entirely rely on the pairwise potentials to avoid visual artefacts. We use a similar framework and settings as proposed in [11], where the image completion problem was formulated as MAP inference problem with the state space being patches extracted from the known parts of the given image. Both the unary and pairwise potentials for this problem are non-parametric and instance dependent. For all results in Fig. 5 we used 13×13 patches to compute the unary potentials (compatibility with known

Fig. 5. Masked source images and image completion results [Best viewed on screen.]

---

[6] And 4.7% when using the weaker product label space relaxation [20] with unary potentials being computed on the fly.

pixels) and pairwise potentials (compatibility where patches assigned to adjacent nodes overlap). As in [11] the region to be filled is represented by a subsampled grid in the inference problem such that adjacent patches have approximately 50% overlap. Depending on the provided image the size of the state space varies between 10,000 and 50,000 patches that are available for image completion. The sampling function H proposes the appropriately shifted patch in the source image (60%), one of the 5 nearest neighbors in patch appearance (20%), and a random source image patch (20%). The results shown in Fig. 5 are obtained after running 100 passes of the loopy min-max inference method, which take about 10 s using a straightforward OpenMP accelerated C++ implementation. We did not investigate in further speedups such as using FFT suggested in [11]. The resulting images are generated by simply averaging pixels covered by multiple patches, and are qualitatively comparable to the ones reported in [11].

6   Conslusion and Future Work

In this work we formulate resource-efficient MAP inference as joint optimization problem over messages and over upper bounding potentials, which allow efficient inference. We show that this nested min-max problem can be efficiently minimized on trees yielding a convergent algorithm, and that its loopy extension also returns convincing MAP solutions in practice. Future work will address (i) theoretical aspects such as a better understanding of the method on loopy graphs, and (ii) practical aspects such as the impact of diverse [16] and non-local [14] label proposals.

References

1. Baker, S., Scharstein, D., Lewis, J., Roth, S., Black, M.J., Szeliski, R.: A database and evaluation methodology for optical flow. IJCV **92**(1), 1–31 (2011)
2. Besag, J.: On the statistical analysis of dirty pictures. J. R. Stat. Soc. B **48**, 259–302 (1986)
3. Besse, F., Rother, C., Fitzgibbon, A., Kautz, J.: PMBP: patchmatch belief propagation for correspondence field estimation. IJCV **110**(1), 2–13 (2014)
4. Boykov, Y., Veksler, O., Zabih, R.: Fast approximate energy minimization via graph cuts. IEEE Trans. Pattern Anal. Mach. Intell. **23**(11), 1222–1239 (2001)
5. Drory, A., Haubold, C., Avidan, S., Hamprecht, F.A.: Semi-global matching: a principled derivation in terms of message passing. In: Jiang, X., Hornegger, J., Koch, R. (eds.) GCPR 2014. LNCS, vol. 8753, pp. 43–53. Springer, Cham (2014). https://doi.org/10.1007/978-3-319-11752-2_4
6. Globerson, A., Jaakkola, T.: Fixing max-product: convergent message passing algorithms for MAP LP-relaxations. In: NIPS (2007)
7. Hazan, T., Shashua, A.: Norm-prodcut belief propagtion: primal-dual message-passing for LP-relaxation and approximate-inference. IEEE Trans. Inform. Theory **56**(12), 6294–6316 (2010)
8. Hirschmüller, H.: Accurate and efficient stereo processing by semi-global matching and mutual information. In: Proceedings of the CVPR, pp. 807–814 (2005)

9. Ihler, A., McAllester, D.: Particle belief propagation. In: AISTATS, pp. 256–263 (2009)
10. Kolmogorov, V.: Convergent tree-reweighted message passing for energy minimization. IEEE Trans. Pattern Anal. Mach. Intell. **28**(10), 1568–1583 (2006)
11. Komodakis, N., Tziritas, G.: Image completion using efficient belief propagation via priority scheduling and dynamic pruning. IEEE Trans. Image Proc. **16**(11), 2649–2661 (2007)
12. Kothapa, R., Pacheco, J., Sudderth, E.: Max-product particle belief propagation. Technical report, Master's project report, Brown University, Department of Computer Science (2011)
13. Lempitsky, V., Rother, C., Roth, S., Blake, A.: Fusion moves for Markov random field optimization. IEEE Trans. Pattern Anal. Mach. Intell. **32**(8), 1392–1405 (2010)
14. Li, Y., Min, D., Brown, M.S., Do, M.N., Lu, J.: SPM-BP: sped-up patchmatch belief propagation for continuous MRFs. In: Proceedings of the ICCV, pp. 4006–4014 (2015)
15. Noorshams, N., Wainwright, M.J.: Stochastic belief propagation: a low-complexity alternative to the sum-product algorithm. IEEE Trans. Inform. Theory **59**(4), 1981–2000 (2013)
16. Pacheco, J., Sudderth, E.: Proteins, particles, and pseudo-max-marginals: a submodular approach. In: Proceedings of the ICML, pp. 2200–2208 (2015)
17. Peng, J., Hazan, T., McAllester, D., Urtasun, R.: Convex max-product algorithms for continuous MRFs with applications to protein folding. In: Proceedings of the ICML (2011)
18. Roig, G., Boix, X., Nijs, R.D., Ramos, S., Kuhnlenz, K., Gool, L.V.: Active map inference in CRFs for efficient semantic segmentation. In: Proceedings of ICCV, pp. 2312–2319 (2013)
19. Scharstein, D., Szeliski, R.: High-accuracy stereo depth maps using structured light. In: Proceedings of the CVPR, pp. 195–202 (2003)
20. Shekhovtsov, A., Kovtun, I., Hlaváč, V.: Efficient MRF deformation model for non-rigid image matching. CVIU **112**(1), 91–99 (2008)
21. Shekhovtsov, A., Reinbacher, C., Graber, G., Pock, T.: Solving dense image matching in real-time using discrete-continuous optimization. arXiv preprint arXiv:1601.06274 (2016)
22. Sontag, D., Globerson, A., Jaakkola, T.: Introduction to dual decomposition for inference. In: Optimization for Machine Learning. MIT Press (2011)
23. Sontag, D., Jaakkola, T.: Tree block coordinate descent for MAP in graphical models. J. Mach. Learn. Res. (2009)
24. Trinh, H., McAllester, D.: Unsupervised learning of stereo vision with monocular cues. In: Proceedings of the BMVC, pp. 72–81 (2009)
25. Wainwright, M.J., Jaakkola, T.S., Willsky, A.S.: MAP estimation via agreement on trees: message-passing and linear programming. IEEE Trans. Inf. Theory **51**(11), 3697–3717 (2005)
26. Werner, T.: A linear programming approach to max-sum problem: a review. IEEE Trans. Pattern Anal. Mach. Intell. 29(7) (2007)
27. Zach, C.: A principled approach for coarse-to-fine MAP inference. In: Proceedings of the CVPR, pp. 1330–1337 (2014)
28. Zach, C.: Limited-memory belief propagation via nested optimization (2017). Supplementary material https://sites.google.com/site/christophermzach/home/pdf/lmbp_supp.pdf

Geometric Image Labeling with Global Convex Labeling Constraints

Artjom Zern[1,2]($\boxtimes$) , Karl Rohr[2] , and Christoph Schnörr[1]

[1] Image and Pattern Analysis Group, Heidelberg University,
Heidelberg, Germany
`artjom.zern@iwr.uni-heidelberg.de`
[2] Biomedical Computer Vision Group, BIOQUANT,
Heidelberg University, Heidelberg, Germany

Abstract. In [2], a smooth geometric labeling approach was introduced by following the Riemannian gradient flow of a given objective function on the so-called assignment manifold. The approach evaluates a user-defined data term and performs spatial regularization by Riemannian averaging of the assignment vectors. In this paper, we extend this approach in order to impose global convex constraints on the labeling results based on linear filter statistics in the label space. The smoothness of the approach is preserved by using logarithmic barrier functions to handle the new constraints. We discuss how suitable filters can be determined from example data of a given image class, and we demonstrate numerically the effectiveness of the constraints in several academic labeling scenarios.

Keywords: Image labeling · Assignment manifold
Statistical label constraints · Riemannian gradient flow
Information geometry

1 Introduction

The *discriminative* power of *filter statistics* for object detection and classification is well known [7,8] and has been widely explored in the literature. The *generative* power of filter statistics for representing image structure, on the other hand, has been less explored during the recent years. The present paper focuses on a mathematically sound and numerically tractable approach to impose filter statistics on *labelled* image structure.

Early seminal work on generative aspects of filter statistics includes [10,14] and many references in these papers. In the former case, heavy-tailed empirical filter statistics are imposed on the variational problem of learning the parameters of a Gibbs-Boltzmann distribution. In the latter case, several hundred filter constraints form nonlinear submanifolds (level sets) onto which a given image has to be projected. While both works impressively demonstrate the generative power of filter statistics, exploiting these statistics as prior knowledge for inference and

M. Pelillo and E. Hancock (Eds.): EMMCVPR 2017, LNCS 10746, pp. 533–547, 2018.
https://doi.org/10.1007/978-3-319-78199-0_35

reproducibility of results has remained a challenge from the viewpoint of algorithm design and numerical optimization. This assessment also applies to current mainstream research with a focus on the engineering of deep networks [13], without denying the remarkable quality of corresponding experimental results.

Contribution. The present work conforms to this research direction but deviates in the following aspects:

1. We focus on filter statistics in *label space* rather than in image space. The simplest constraint, for example, imposes lower and upper bounds on the area occupied by some label, without specifying the corresponding locations, of course. More general constraints arise from replacing the 'identity filter' by linear filters learned offline through a simple generalized eigenvalue technique, and imposing similar linear statistical constraints. While such statistical moments can be taken into account as constraints using graphical models, in principle, this would again lead like [14] to maximum-entropy distributions in Gibbs-Boltzmann form [5], that are intractable regarding both learning and inference, due to the *global* nature of these constraints. In fact, a recent assessment of approaches to inference with discrete graphical models [6] revealed the limited capability of established state-of-the-art solvers in this respect, i.e. to handle cliques of *large* size of the underlying graph.

2. We focus on a numerically *tractable and reproducible* way to incorporate such constraints into an algorithm for image labeling. To this end, we adopt the recent approach [2] to image labeling based on simple geometric averaging induced by the Fisher-Rao metric on the so-called assignment manifold, i.e. the relative interior of a product of probability simplices, whose vertices represent discrete decisions as is common with graphical models and convex variational relaxations. A key aspect of the approach [2] is that the usual two-step procedure of (i) solving the LP relaxation [12] by some iterative method, and (ii) projecting back the solution to the set of integral solutions, is combined into a *single smooth* process that converges to integral solutions (labelings). The objective of the present paper is to show that *filtered label statistics* can be taken into account in a straightforward and comprehensible way by using standard log-barrier constraints [4,9] and geometric numerical integration [11].

Organization. Section 2 sketches the works [2,11] on which the present paper is based. Section 3 details our contribution: learning filters for label statistics and taking corresponding empirical constraints into account during inference for image labeling. We do not focus on any specific application in this paper. Rather, the proof-of-concept experiments discussed in Sect. 4 are supposed to demonstrate how statistics gathered by linear filters of small support can *enhance image labeling*, represent *primitive shape information* and support *spatial pattern formation*, by extending the geometric non-convex approach [2] through corresponding *convex* constraints.

Basic Notation. Functions and binary operations are applied component-wise to vectors and matrices, i.e. for $u, v \in \mathbb{R}^n$ we have $\sqrt{u} = (\sqrt{u_1}, \ldots, \sqrt{u_n})^\top, u \cdot v =$

$(u_1 v_1, \ldots, u_n v_n)^\top$ and similar for $e^u, \log(u)$ and $\frac{u}{v}$. We set $[n] = \{1, \ldots, n\}$ and $\mathbb{1}_n = (1, \ldots, 1)^\top \in \mathbb{R}^n$ as well as $\mathbb{1}_{m \times n} = (\mathbb{1}_m, \ldots, \mathbb{1}_m) \in \mathbb{R}^{m \times n}$. By $\langle \cdot, \cdot \rangle$ we will denote the Euclidean inner product on $\mathbb{R}^n$ or the Frobenius inner product on $\mathbb{R}^{m \times n}$. For a matrix $W \in \mathbb{R}^{m \times n}$ we will denote the i-th row by $W_i \in \mathbb{R}^n$ and the j-th column by $W^j \in \mathbb{R}^m$. Elements of a sequence are indexed with an upper script index enclosed in brackets, for example, $W^{(k)} \in \mathbb{R}^{m \times n}$.

2   Image Labeling on the Assignment Manifold

We briefly summarize the smooth label assignment approach introduced in [2]. This approach will be extended in the next section in order to handle global constraints imposed on labelings.

Given an image with m pixels and a set $\mathcal{L} = \{l^{(1)}, \ldots, l^{(n)}\}$ of n predefined labels, the task is to assign each pixel $i \in [m]$ one label in $\mathcal{L}$. The labeling problem can be formulated as finding an optimal assignment matrix in

$$\overline{\mathcal{W}}^* = \{W \in \mathbb{R}^{m \times n} : W_i \in \{e^{(1)}, \ldots, e^{(n)}\} \subset \mathbb{R}^n, \forall i \in [m]\}, \qquad (2.1)$$

where each label $l^{(j)} \in \mathcal{L}$ is represented by a vertex $e^{(j)} \in \mathbb{R}^n$ of the probability simplex. In [2], a smooth geometric approach was presented which is defined on the assignment manifold

$$\mathcal{W} := \{W \in \mathbb{R}_{>0}^{m \times n} : \langle W_i, \mathbb{1}_n \rangle = 1, \forall i \in [m]\} \subset \mathbb{R}_{>0}^{m \times n}, \qquad (2.2)$$

that is the set of all row-stochastic matrices with full support. This is a smooth manifold with tangent space at $W \in \mathcal{W}$ given by

$$\mathcal{T} := T_W \mathcal{W} = \{V \in \mathbb{R}^{m \times n} : \langle V_i, \mathbb{1}_n \rangle = 0, \forall i \in [m]\}. \qquad (2.3)$$

The assignment manifold $\mathcal{W}$ is turned into a Riemannian manifold by equipping it with the Fisher-Rao metric

$$g_W^{\mathcal{W}}(U, V) = \langle \tfrac{U}{\sqrt{W}}, \tfrac{V}{\sqrt{W}} \rangle \quad \text{for} \quad U, V \in T_W \mathcal{W}, \quad W \in \mathcal{W} \qquad (2.4)$$

with *componentwise multiplication* (and subdivision) of vectors and matrices (with strictly positive support).

Input data for the assignment approach are pixel neighborhoods $\mathcal{N}(i) = \{j \in [m] : i \sim j\}, i \in [m]$ defined by the adjacency relation (edges) of an underlying graph, and a distance matrix $D \in \mathbb{R}^{m \times n}$ whose components D_{ij} store the application-specific distance between the image data observed at pixel $i \in [m]$ and label $l^{(j)} \in \mathcal{L}$. The goal is to find an assignment $W \in \overline{\mathcal{W}}$ which is spatially consistent with respect to neighborhood assignments, on the one hand, and reflects the data represented by the distance matrix D as closely as possible, on the other hand. This is accomplished by computing a curve $W(t) \in \mathcal{W}, t \geq 0$ on the assignment manifold that converges to an *integral* solution and locally minimizes a functional $J(W)$ which accounts for the given data and regularization.

The ingredients for defining a corresponding sequence

$$W^{(k)} = W(t_k), \qquad W^{(0)} = W(0) = C := \frac{1}{n}\mathbb{1}_{m\times n} \qquad (2.5)$$

are the *barycenter* C of $\mathcal{W}$, an approximation of the exponential mapping of $\mathcal{W}$ given by

$$\exp_W : \mathcal{T} \to \mathcal{W}, \qquad \exp_W(V)_i = \frac{W_i \cdot e^{V_i}}{\langle W_i, e^{V_i}\rangle}, \qquad \forall i \in [m], \qquad (2.6)$$

and the orthogonal projection onto the tangent space (2.3)

$$\Pi_{\mathcal{T}} : \mathbb{R}^{m\times n} \to \mathcal{T}, \qquad \Pi_{\mathcal{T}}(D) = D - \frac{1}{n}D\mathbb{1}_{n\times n}. \qquad (2.7)$$

The data D is taken into account by the likelihood matrix

$$L(W) = \exp_W\bigl(\Pi_{\mathcal{T}}(-D)\bigr) \in \mathcal{W}, \qquad (2.8)$$

whereas regularization is performed by computing approximate Riemannian means of the assignment vectors $\{L(W)_j\}_{j\in\mathcal{N}(i)}$ over spatial neighborhoods $\mathcal{N}(i)$, for each pixel $i \in [m]$. We refer to [2] for details.

We adopt the general numerical scheme suggested by [11],

$$\dot{V}(t) = -\Pi_{\mathcal{T}}\bigl[\nabla J\bigl(W(t)\bigr)\bigr], \quad W(t) = \exp_C\bigl(V(t)\bigr), \quad V(0) = 0, \qquad (2.9)$$

which enables to apply standard algorithms for integrating the flow $V(t)$ on the tangent space $\mathcal{T}$ so as to determine a minimizing path $W(t)$ on the manifold $\mathcal{W}$. For example, combining the simplest integration method, i.e. explicit Euler steps, with smooth rounding to an integral solution leads to a sequence (2.5) given by

$$W_i^{(k+\frac{1}{2})} = \frac{W_i^{(k)} \cdot e^{-h\nabla_{W_i} J(W^{(k)})}}{\langle W_i^{(k)}, e^{-h\nabla_{W_i} J(W^{(k)})}\rangle}, \qquad (2.10a)$$

$$W_i^{(k+1)} = \frac{W_i^{(k)} \cdot W_i^{(k+\frac{1}{2})}}{\langle W_i^{(k)}, W_i^{(k+\frac{1}{2})}\rangle}, \qquad i \in [m]. \qquad (2.10b)$$

We explain in the subsequent section how global labeling constraints can be taken into account within this framework.

3 Label Assignment with Global Constraints

This section details the class of global constraints that we impose on label assignments (Sect. 3.1), how linear filters defining these constraints are learned offline using basic techniques of numerical linear algebra (Sect. 3.2), and finally, in Sect. 3.3, how these constraints are taken into account using the assignment approach of Sect. 2.

3.1   Global Constraints

In order to incorporate some prior knowledge about the labelings, we consider linear $p \times p$ filters $h \in \mathcal{H} \subset \mathbb{R}^{p^2 \times n}$ operating on assignment matrices $W \in \overline{\mathcal{W}}$: For each label $j \in [n]$, we have a $p \times p$ filter $h^j \in \mathbb{R}^{p^2}$ in the usual sense, and the filter operation is given by

$$h * W := \sum_{j \in [n]} h^j * W^j, \tag{3.1}$$

with the common convolution of the 'label images' W^j, $j \in [n]$ with a $p \times p$ filter on the right-hand side. The space of filters $\mathcal{H}$ will be specified in Sect. 3.2. To avoid complications at and close to the boundary of the image region, we only take into account filter results $(h * W)_i$ at interior pixels i where the $p \times p$ filter support (centered at i) does not overlap with the boundary.

The filter result $(h * W)_i$ at a pixel $i \in [m]$ depends on the assignment within a $p \times p$ neighborhood of i and hence reflects the local spatial relation of the labels. Our objective is to control label assignments by constraining the filter results for a set

$$\{h^{(k)} \in \mathcal{H} \colon k = 1, \ldots, K\} \tag{3.2}$$

of K filters in order to take into account statistical prior information about the local geometry of labelings. Motivated by [3], where the ℓ^1-norm of the filter results of a grayscale image was considered in connection with non-smooth sparse regularization, we consider here the ℓ^2-norm of filter results which conforms to our *smooth* geometric label assignment scheme of Sect. 2.

Specifically, we consider *global convex constraints* of the form

$$c_{\text{low}} \leq \tfrac{1}{m} W^\top \mathbb{1}_m \leq c_{\text{up}}, \tag{3.3a}$$

$$\|h^{(k)} * W\|_{\ell^2} \leq d^{(k)}, \qquad k = 1, \ldots, K, \tag{3.3b}$$

where the parameter vectors $c_{\text{low}}, c_{\text{up}} \in \mathbb{R}^n$ impose lower and upper cardinality bounds for the assignment of each label of $\mathcal{L} = \{l^{(1)}, \ldots, l^{(n)}\}$ to the range of pixels $[m]$, whereas the parameters $d^{(k)}$ of (3.3b) constrain the output energy of each filter $h^{(k)}$, $k \in [K]$. To ensure that the region of feasible assignments W has a non-empty interior, we require

$$c_{\text{low}} < c_{\text{up}}, \qquad \langle c_{\text{low}}, \mathbb{1}_n \rangle < 1 < \langle c_{\text{up}}, \mathbb{1}_n \rangle \qquad \text{and} \qquad d^{(k)} > 0, \ k \in [K]. \tag{3.4}$$

As alternative to the ℓ^2-norm defining (3.3b), we also used a smooth approximation of the ℓ^1-norm denoted by

$$\|x\|_{\ell^1_\varepsilon} = \sum_{i \in [m]} |x_i|_\varepsilon, \qquad |x_i|_\varepsilon = \sqrt{x_i^2 + \varepsilon^2} - \varepsilon. \tag{3.5}$$

3.2   Learning Filters

We discuss how to choose filters for given classes of labelings and corresponding example data. First of all, we restrict the space of all possible $p \times p$ filters in order to eliminate some redundant degrees of freedom. To this end, we consider the decomposition of the space of all filters

$$\mathbb{R}^{p^2 \times n} = \mathcal{H}_0 \oplus \mathcal{H}_1 \oplus \mathcal{H}_2 \tag{3.6}$$

into subspaces given by[1]

$$\mathcal{H}_0 = \{h \in \mathbb{R}^{p^2 \times n} : h^i = h^j, \ \mathrm{mean}(h^j) = 0, \ \forall i, j \in [n]\},$$
$$\mathcal{H}_1 = \{h \in \mathbb{R}^{p^2 \times n} : \langle h_i, \mathbb{1}_n \rangle = 0, \ \forall i \in [p^2], \ \mathrm{mean}(h^j) = 0, \ \forall j \in [n]\}, \tag{3.7}$$
$$\mathcal{H}_2 = \{h \in \mathbb{R}^{p^2 \times n} : h_i = h_j, \ \forall i, j \in [p^2]\}.$$

These spaces are orthogonal to each other with respect to the Euclidean inner product. The space $\mathcal{H}_2$ consists of all filters h that are constant for each label, i.e. $h^j = c_j \cdot \mathbb{1}_{p^2}$ with $c_j \in \mathbb{R}$ for each $j \in [n]$. The space $\mathcal{H}_0$ consists of all zero-mean filters, which do not distinguish between the labels. For any filter $h \in \mathcal{H}_0$, we have $h * W = h^1 * \sum_j W^j = h^1 * \mathbb{1}_m = 0$ for all $W \in \mathcal{W}$, i.e. the subspace $\mathcal{H}_0$ does not represent any useful information for our purpose.

Thus, we can choose either $\mathcal{H} = \mathcal{H}_1$ or $\mathcal{H} = \mathcal{H}_1 \oplus \mathcal{H}_2$ as the actual space of filters. Our choice is

$$\mathcal{H} = \mathcal{H}_1, \qquad \dim \mathcal{H} = \dim \mathcal{H}_1 = (p^2 - 1)(n - 1) \tag{3.8}$$

for two reasons. Firstly, we use this framework for segmentation, where larger homogenous regions occur (e.g., background). Filters which return a small ℓ^2-norm for such labelings have (approximately) a zero-mean and therefore belong to $\mathcal{H}_1$. Secondly, we will use an inner point method for optimization, which requires a feasible initialization. In case of zero-mean filters, we can simply use homogenous assignments as initial assignment. As a result, we do not need an additional initialization process on which the final result might depend. This conforms to the philosophy to start the assignment process without any bias at the barycenter $C \in \mathcal{W}$ – cf. (2.5).

For learning the filters, we assume that sets $\mathcal{I}^+, \mathcal{I}^- \subset \overline{\mathcal{W}}^*$ for favorable and unfavorable label assignments are given. We are looking for filters $h \in \mathcal{H}$ such that $\|h * W\|_{\ell^2}$ is smaller for $W \in \mathcal{I}^+$ than for $W \in \mathcal{I}^-$. For simplicity, we choose

$$\frac{\mathrm{mean}_{W \in \mathcal{I}^+} \|h * W\|_{\ell^2}^2}{\mathrm{mean}_{W \in \mathcal{I}^-} \|h * W\|_{\ell^2}^2} < 1 \tag{3.9}$$

as criterion for filters $h \in \mathcal{H}$, which leads to a generalized eigenvalue problem. Specifically, let $\{e_{\mathcal{H}}^{(i)} : i = 1, \ldots, \dim \mathcal{H}\}$ be an orthonormal basis of $\mathcal{H}$ and consider the map

---

[1] Notation: Filters h are matrix-valued (image vectors $\times$ labels) with rows h_i and columns h^j. Superscripts in brackets $h^{(k)}$ index members of a collection of filters.

$$M : \overline{\mathcal{W}} \to \mathbb{R}^{\dim \mathcal{H} \times \dim \mathcal{H}}, \quad M(W)_{ij} = \langle e_{\mathcal{H}}^{(i)} * W, e_{\mathcal{H}}^{(j)} * W \rangle_{\ell^2}. \tag{3.10}$$

Then we have

$$\frac{\mathrm{mean}_{W \in \mathcal{I}^+} \|h * W\|_{\ell^2}^2}{\mathrm{mean}_{W \in \mathcal{I}^-} \|h * W\|_{\ell^2}^2} = \frac{x^\top A^+ x}{x^\top A^- x} \tag{3.11}$$

with $A^\pm = \mathrm{mean}_{W \in \mathcal{I}^\pm} M(W)$ and $h = \sum_i x_i e_{\mathcal{H}}^{(i)}$. As a consequence, a set of linearly independent filters satisfying the criterion (3.9) is given by the generalized eigenvectors of the matrix pencil (A^+, A^-) corresponding to eigenvalues less than 1. The filters corresponding to eigenvalues greater than 1 might also focus on useful features as can be seen, for example, in Fig. 7(d), where the last 16 filters correspond to eigenvalues greater than 1. These filters can be used additionally, since they further restrict the assignment and therefore may prevent some assignments, which were not taken into account by $\mathcal{I}^-$.

Having determined a set of filters as generalized eigenvectors, we normalize them in a post-processing step so as to meet the condition $h * W \in [-1, 1]^m$ for all $W \in \overline{\mathcal{W}}$, i.e.

$$\|h\|_{\mathcal{H}} = 1, \qquad \|h\|_{\mathcal{H}} := \max \Big\{ -\sum_{i \in [p^2]} \min_{j \in [n]} h_{ij}, \ \sum_{i \in [p^2]} \max_{j \in [n]} h_{ij} \Big\}. \tag{3.12}$$

3.3   Optimization

In order to take into account the constraints (3.3a) and (3.3b), we use log-barrier functions [4,9] that have been widely applied (e. g., in [1]). Given the parameters in (3.4) and filters $h^{(k)}$, $k \in [K]$, these functions read

$$\begin{aligned}
B_{\mathrm{low}}(W) &= -\big\langle \mathbb{1}_n, \log(\tfrac{1}{m} W^\top \mathbb{1}_m - c_{\mathrm{low}}) \big\rangle, \\
B_{\mathrm{up}}(W) &= -\big\langle \mathbb{1}_n, \log(c_{\mathrm{up}} - \tfrac{1}{m} W^\top \mathbb{1}_m) \big\rangle, \\
B_{\mathrm{filter}}(W) &= -\sum_{k=1}^{K} \log \big((d^{(k)})^2 - \|h^{(k)} * W\|_{\ell^2}^2 \big).
\end{aligned} \tag{3.13}$$

Summing up these functions yields the overall barrier function

$$B(W) = B_{\mathrm{low}}(W) + B_{\mathrm{up}}(W) + B_{\mathrm{filter}}(W) \tag{3.14}$$

for the constraints (3.3a) and (3.3b). Complementing the objective function

$$J_\tau(W) = J(W) + \tau B(W) \tag{3.15}$$

with a barrier parameter $\tau > 0$, modifies the flow (2.9) and the sequence (2.5) accordingly. Within the flow, τ is handled as a monotonously decreasing function $\tau : \mathbb{R}_{\geq 0} \to \mathbb{R}_{> 0}$ with $\lim_{t \to \infty} \tau(t) = 0$. The flow is initialized at

$$W_i(0) = c_{\mathrm{low}} + \frac{1 - \langle c_{\mathrm{low}}, \mathbb{1}_n \rangle}{\langle c_{\mathrm{up}}, \mathbb{1}_n \rangle - \langle c_{\mathrm{low}}, \mathbb{1}_n \rangle} (c_{\mathrm{up}} - c_{\mathrm{low}}), \qquad \forall i \in [m]. \tag{3.16}$$

Since we use zero-mean filters due to (3.8) and (3.7), the initialization (3.16) is strictly feasible for (3.3a) and (3.3b).

It remains to specify the gradients of the barrier functions that are required to evaluate the vector field, which defines the flow by the first equation of (2.9), with $J(W)$ replaced by $J_\tau(W)$ due to (3.15). The gradient $\nabla B(W) \in \mathbb{R}^{m \times n}$ of the barrier function (3.14) is given by

$$\nabla B_{\text{low}}(W)_i = -\frac{1}{m} \frac{1}{\frac{1}{m} W^\top \mathbb{1}_m - c_{\text{low}}}, \quad \nabla B_{\text{up}}(W)_i = \frac{1}{m} \frac{1}{c_{\text{up}} - \frac{1}{m} W^\top \mathbb{1}_m} \quad (3.17)$$

for each pixel $i \in [m]$, and by

$$\nabla B_{\text{filter}}(W)^j = 2 \sum_{k=1}^{K} \frac{h^{(k),j} \star (h^{(k)} * W)}{(d^{(k)})^2 - \|h^{(k)} * W\|_{\ell^2}^2} \quad (3.18)$$

for any label $j \in [n]$ with $h^{(k),j} \in \mathbb{R}^{p^2}$ being the j-th layer (column) of the k-th filter. Here, $\star$ denotes the cross-correlation operation, i.e. convolution with the mirrored filter. This convolution is performed on the whole image with zero-padding.

If the approximated ℓ^1-norm (3.5) is used instead of the ℓ^2-norm, the barrier function takes the form

$$B_{\text{filter}}(W) = -\sum_{k=1}^{K} \log \left(d^{(k)} - \|h^{(k)} * W\|_{\ell_\varepsilon^1} \right) \quad (3.19)$$

with the Euclidean gradient given by

$$\nabla B_{\text{filter}}(W)^j = \sum_{k=1}^{K} \frac{h^{(k),j} \star \nabla \|h^{(k)} * W\|_{\ell_\varepsilon^1}}{d^{(k)} - \|h^{(k)} * W\|_{\ell_\varepsilon^1}}, \quad \nabla \|x\|_{\ell_\varepsilon^1} = \frac{x}{\sqrt{x \cdot x + \varepsilon^2 \mathbb{1}}}, \quad (3.20)$$

where the operations of the latter right-hand side apply componentwise.

4   Experiments

In this section, we investigate the influence of the filter constraints on the labeling result. We test the new approach on several academic labeling scenarios and compare the results with those obtained without using these constraints.

Setup. We represent assignments by choosing for each label $j \in [n]$ some color $c^{(j)} \in [0,1]^3$ in the RGB color space. Then an assignment $W \in \overline{\mathcal{W}} \subset \mathbb{R}^{m \times n}$ is represented by the color image $I \in \mathbb{R}^{m \times 3}$ given by $I_i = \sum_{j \in [n]} W_{ij} c^{(j)} \in [0,1]^3$ for each pixel $i \in [m]$.

We consider *three different data sets* in order to check the effect of constraints on **(a)** primitive shape information, on **(b)** spatial relations (inclusion of regions), and **(c)** on the separation of fore- and background each defined by *several* labels.

The first data set **(a)** contains binary rectangles and ellipses. Filters of size 3×3 were trained for rectangles against ellipses, i.e. all assignments in a set $\mathcal{I}^+$ represent rectangles, while assignments in the complement set $\mathcal{I}^-$ represent ellipses. The second data set **(b)** comprises three labels (white, orange, black) forming white ellipses overlapped by orange ellipses on black background (see Fig. 2 for illustration). All ellipses have varying radii, orientation and position. Filters of size 5×5 were used to separate the positive class $\mathcal{I}^+$ defined by inclusions of regions, whereas these topological relations are violated in the negative class $\mathcal{I}^-$. The third data set **(c)** consists of Voronoi diagrams, with each polygon labeled by either one of three foreground labels (red, green, blue) and likewise for the background (black, gray, white). Both foreground and background are connected and the foreground is located in the center of the image domain. The negative class $\mathcal{I}^-$ is defined by randomly labelled Voronoi diagrams.

Implementation details. We solved the gradient flow by the explicit Euler method. A fixed step size Δt was used as long as $W(t_k)$ fulfilled the constraints. Otherwise the step size was reduced by backtracking line search. We did no analysis of the step size but rather used $\Delta t = 100$, which produced satisfying results. As usual for interior point methods, we used $\tau(t_k) = \alpha^{-k}\tau_0$ with $\alpha > 1$ for the barrier parameter. In our experiments, $\alpha = 1.03$ and $\tau_0 = 100$ turned out

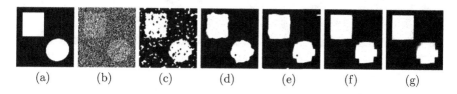

| (a) | (b) | (c) | (d) | (e) | (f) | (g) |

Fig. 1. Representing and enforcing rectangular structure. (a) Shows the original gray-scale image and (b) shows a noisy version of it, which was used as input data. (c) and (d) Show the labeling results *without imposing constraints* obtained through Riemannian averaging over neighborhoods of sizes 3×3 and 7×7 respectively. (e) and (f) Show the results of the new approach (without cardinality constraints) with neighborhood size 3×3 and four filters of size 3×3, which were trained for rectangles against ellipses. These 4 filters prefer horizontal and vertical edges. For (e), the ℓ^2-norm was used for filter constraints. For (f) and (g), $\| \cdot \|_{\ell^1_\varepsilon}$ with $\varepsilon = 0.1$ and $\varepsilon = 0.01$ was used.

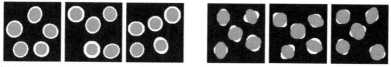

(a) *Positive* labeling examples in $\mathcal{I}^+$.   (b) *Negative* labeling examples in $\mathcal{I}^-$.

Fig. 2. Illustration of the training sets $\mathcal{I}^+, \mathcal{I}^- \subset \overline{\mathcal{W}}^*$. Positive examples (a) are defined by topological relations: orange ellipses are completely contained in the white ones. Negative examples (b) are labelings where this topological relation is violated. (Color figure online)

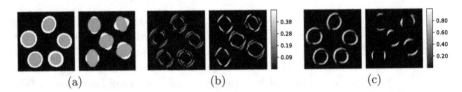

Fig. 3. Illustration of the generalized eigenvalue filters. (a) Shows one label assignment in $\mathcal{I}^+$ and one assignment in $\mathcal{I}^-$, used to illustrate the filter outputs in (b), (c). The computed filters of size 5×5 can be subdivided into two groups. The first 24 filters $h^{(1)}, \ldots, h^{(24)}$ respond to the boundary of the orange and black regions, and they have a large response at the border between the orange and black regions. This is illustrated by (b) which shows the absolute value of $h^{(1)} * W$. The last 24 filters $h^{(25)}, \ldots, h^{(48)}$ mainly respond to the boundary of the white regions. (c) shows the absolute value of $h^{(48)} * W$. (Color figure online)

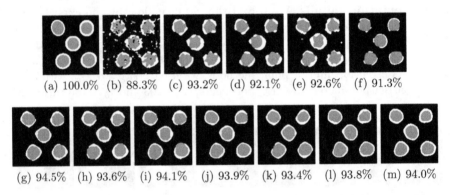

(a) 100.0% (b) 88.3% (c) 93.2% (d) 92.1% (e) 92.6% (f) 91.3%

(g) 94.5% (h) 93.6% (i) 94.1% (j) 93.9% (k) 93.4% (l) 93.8% (m) 94.0%

Fig. 4. Experimental results obtained with and without constraints. (a) Shows the ground truth assignment. A noisy version of this assignment was used as input data. The percentages of correctly labeled pixels are shown below the images. The results obtained *without constraints* are shown in (b)–(d) for neighborhood sizes 3×3, 5×5 and 7×7 respectively. For the results (e)–(m) of the new approach, 3×3 neighborhoods were used for spatial regularization. For (e), cardinality constraints were only used. For (f), filter constraints were only used (48 filters of size 5×5). For (g)–(m), *both cardinality constraints and filter constraints* were used. (g) and (h) were obtained with 24 filters using the ℓ^2-norm and the approximated ℓ^1-norm $\| \cdot \|_{\ell^1_\varepsilon}$ with $\varepsilon = 0.01$ respectively. For (i) and (j), 48 filters were used. For (k)–(m), the distance matrix was rescaled by a factor 0.01, and 48 filters were used with ℓ^2-norm as well as $\| \cdot \|_{\ell^1_\varepsilon}$ with $\varepsilon = 0.1$ and $\varepsilon = 0.01$ respectively.

to be a reliable choice. We terminated the iteration either after 1500 steps or when both the average entropy $-\frac{1}{m} \sum_{i \in [m]} \sum_{j \in [n]} W_{ij}^{(k)} \log W_{ij}^{(k)}$ dropped below a threshold $\epsilon_{\text{entropy}} > 0$ and $\tau(t_k)$ dropped below $\tau_{\min} > 0$. As in [2], we used $\epsilon_{\text{entropy}} = 10^{-3}$. For $\tau_{\min}$, we set 10^{-10}.

Results. (a) For the binary data set (rectangles/ellipses), the filter space $\mathcal{H}$ has dimension 8. We used the four filters corresponding to the eigenvalues less

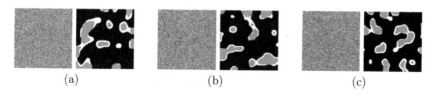

(a)   (b)   (c)

Fig. 5. Spatial pattern formation induced by pure noise and convex label constraints. All experiments were done using a 3×3 neighborhood for the spatial regularization, and using both cardinality constraints and filter constraints based on 48 filters and $\|\cdot\|_{\ell_\varepsilon^1}$ with $\varepsilon = 0.01$. Panels (a)–(c) show on the right random spatial labeling patterns induced by the random noise images on the left. These results demonstrate how filter constraints favor local shape and topological spatial structure on image labelings within our geometric approach to label assignments.

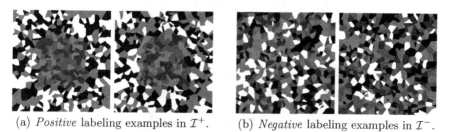

(a) *Positive* labeling examples in $\mathcal{I}^+$.   (b) *Negative* labeling examples in $\mathcal{I}^-$.

Fig. 6. Best viewed in color. Illustration of the training sets $\mathcal{I}^+, \mathcal{I}^- \subset \overline{\mathcal{W}}^*$. Both the foreground region and the background region of these Voronoi tilings are defined by three labels: red, green, blue and black, gray, white, respectively. Positive examples in $\mathcal{I}^+$ are defined by approximately square-shaped foreground regions that are simply connected and centered in the middle of the image domain. Negative examples $\mathcal{I}^-$ contain polygons that are randomly labeled and distributed over the image domain.

than 1. Inspecting these filters reveals discrete versions of the partial derivatives $\partial_{xy}, \partial_{xxy}, \partial_{xyy}$ and ∂_{xxyy}. The upper bounds for the filter constraints were set to $d = 2\|h * W\|$, where W is an assignment representing two rectangles. Figure 1 illustrates that using the filter constraints enables to remove noise, to regularize the rectangle, and to rectify the ellipse by *imposing local shape constraints*.

(b) For the second data set, we used all 48 filters obtained by the generalized eigenvalue problem. The first 24 filters corresponding to eigenvalues less than 1 contribute to separating the orange region from the background. The remaining 24 filters regularize the boundary of the white region (see Fig. 3). The upper bounds for the filter constraints were set to $d = \max_{W \in \mathcal{I}^+} \|h * W\|$. The bounds c_{low} and c_{up} were set in a similar way. We used the distance matrix $D_{ij} = \frac{1}{n\rho}\|\tilde{W}_i - e^{(j)}\|_2$, where the matrix $\tilde{W} \in \mathbb{R}^{m \times n}$ was obtained by adding white noise ($\sigma^2 = 4$) to the ground truth assignment. Figure 4 demonstrates that the constraints improve notably the results, and that in addition to the filter constraints, cardinality constraints are essential to reinforce *topological structure*. In

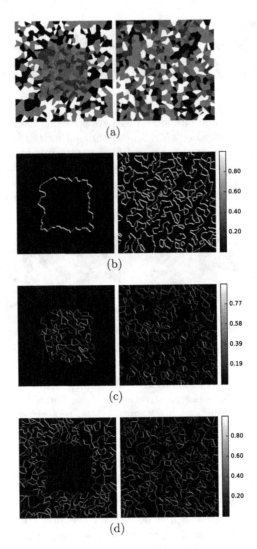

Fig. 7. Best viewed in color. Illustration of the generalized eigenvalue filters. (a) Shows one assignment in $\mathcal{I}^+$ and $\mathcal{I}^-$, respectively, used to illustrate the filter outputs. The computed filters of size 3×3 can be subdivided into three groups. The first eight filters $h^{(1)}, \ldots, h^{(8)}$ regularize the boundary between foreground (red, green, blue) and background (black, gray, white). (b) shows the absolute value of the filter result $h^{(4)} * W$ as example. Filters $h^{(9)}, \ldots, h^{(24)}$ regularize the boundaries *within the foreground* as illustrated by (c), which shows the absolute value of $h^{(10)} * W$. Eventually, filters $h^{(25)}, \ldots, h^{(40)}$ regularize the boundaries *within the background* as illustrated by (d).

order to demonstrate the potential of the constraints for *spatial pattern formation*, we repeated the experiments with *pure noise* as input data. Figure 5 demonstrated the strong regularizing effect of the constraints (Fig. 6).

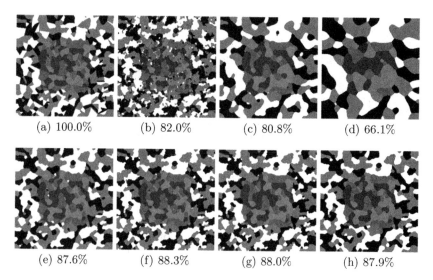

(a) 100.0%   (b) 82.0%   (c) 80.8%   (d) 66.1%

(e) 87.6%   (f) 88.3%   (g) 88.0%   (h) 87.9%

Fig. 8. Best viewed in color. Experimental results obtained with and without constraints. (a) shows the ground truth assignment. A noisy version of this assignment was used as input data. The percentages of correctly labeled pixels compared to the ground truth (a) are shown below the images. Panels (b)–(d) show the results obtained *without* constraints using neighborhood sizes $3 \times 3, 5 \times 5$ and 7×7 respectively. (e) is the result for neighborhood size 5×5, but with a rescaled (factor 100) distance matrix. (f)–(h) show the results *with* constraints using first 8 filters, 24 filters and 40 filters, respectively. These results were computed using 3×3 neighborhoods for spatial regularization and the ℓ^2-norm for the filter constraints.

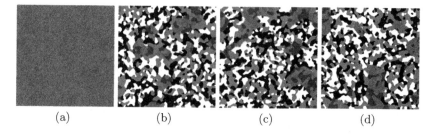

(a)   (b)   (c)   (d)

Fig. 9. Best viewed in color. Spatial pattern formation induced by pure noise and convex label constraints. (a) shows a random assignment W using the same color coding as for the Voronoi polygons. Each panel (b)–(d) shows the result of labeling a different random input image of type (a). All experiments were done using 3×3 neighborhoods, without cardinality constraints, and with filter constraints based on 40 filters and $\| \cdot \|_{\ell^1_\varepsilon}$ with $\varepsilon = 0.1$. The results demonstrate how the filter constraints enforce both the scale and the spatial structure of fore- and background regions that are randomly located due to the pure noise data.

(c) For the Voronoi data set, the filters determined by the eigenvalue problem can be subdivided into three groups: The first 8 filters (eigenvalues ≈ 0.12) contribute to separating the three foreground labels from three background labels. The next 16 filters (eigenvalues ≈ 0.52) regularize the foreground. The last 16 filters (eigenvalues $\approx 1.46 > 1$) regularize the background (Fig. 7). The distance matrix and the parameters for the constraints were set as described above for case (b). The results shown by Fig. 8(b), (f), (g) and (h) demonstrate the effect of the three groups of filters. Repeating the experiments with pure noise as input data illustrates how *spatial patterns* are induced by the constraints (Fig. 9).

5   Conclusion

We extended the smooth geometric image labeling approach of [2,11] in order to incorporate global convex constraints on the labeling result using linear filters in the label space. This extension was mathematically formulated so as to preserve smoothness of the overall approach. We showed how filters can be determined by a generalized eigenvalue problem in order to represent statistical prior knowledge about local shape and spatial relation. Experimental results demonstrate the potential of the approach for imposing these constraints onto labelings of noisy image data.

Our future work will focus on numerical aspects in order to make the approach more efficient for various applications. This includes, in particular, the investigation of how filters of small support can be used to represent and enforce the structure of labelings at multiple spatial scales.

Acknowledgments. We gratefully acknowledge support by the German Science Foundation, grant GRK 1653.

References

1. Alvarez, F., López, J.: Convergence to the optimal value for barrier methods combined with Hessian Riemannian gradient flows and generalized proximal algorithms. J. Convex Anal. **17**(3&4), 701–720 (2010)
2. Åström, F., Petra, S., Schmitzer, B., Schnörr, C.: Image labeling by assignment. J. Math. Imag. Vis. **58**(2), 211–238 (2017)
3. Benning, M., Gilboa, G., Grah, J.S., Schönlieb, C.-B.: Learning filter functions in regularisers by minimising quotients. In: Lauze, F., Dong, Y., Dahl, A.B. (eds.) SSVM 2017. LNCS, vol. 10302, pp. 511–523. Springer, Cham (2017). https://doi.org/10.1007/978-3-319-58771-4_41
4. Boyd, S., Vandenberghe, L.: Convex Optimization. Cambridge University Press, Cambridge (2004)
5. Cover, T., Thomas, J.: Elements of Information Theory, 2nd edn. Wiley, Hoboken (2006)
6. Kappes, J., Andres, B., Hamprecht, F., Schnörr, C., Nowozin, S., Batra, D., Kim, S., Kausler, B., Kröger, T., Lellmann, J., Komodakis, N., Savchynskyy, B., Rother, C.: A comparative study of modern inference techniques for structured discrete energy minimization problems. Int. J. Comp. Vis. **115**(2), 155–184 (2015)

7. Lowe, D.: Distinctive image features from scale-invariant keypoints. Int. J. Comp. Vis. **60**(2), 91–110 (2004)
8. Morel, J.M., Yu, G.: ASIFT: a new framework for fully affine invariant image comparison. SIAM J. Imag. Sci. **2**(2), 438–469 (2009)
9. Nesterov, Y., Nemirovskii, A.: Interior Point Polynomial Algorithms in Convex Programming. SIAM, Philadelphia (1994)
10. Portilla, J., Simoncelli, E.: A parametric texture model based on joint statistics of complex wavelet coefficients. Int. J. Comput. Vis. **40**(1), 49–70 (2000)
11. Savarino, F., Hühnerbein, R., Åström, F., Recknagel, J., Schnörr, C.: Numerical integration of Riemannian gradient flows for image labeling. In: Lauze, F., Dong, Y., Dahl, A.B. (eds.) SSVM 2017. LNCS, vol. 10302, pp. 361–372. Springer, Cham (2017). https://doi.org/10.1007/978-3-319-58771-4_29
12. Werner, T.: A linear programming approach to max-sum problem: a review. IEEE Trans. Patt. Anal. Mach. Intell. **29**(7), 1165–1179 (2007)
13. Xie, J., Lu, Y., Zhu, S.C., Wu, Y.: A theory of generative ConvNet. In: Proceedings of the ICML (2016)
14. Zhu, S., Mumford, D.: Prior learning and Gibbs reaction-diffusion. IEEE Trans. Patt. Anal. Mach. Intell. **19**(11), 1236–1250 (1997)

Discretized Convex Relaxations for the Piecewise Smooth Mumford-Shah Model

Christopher Zach[1(✉)] and Christian Häne[2]

[1] Toshiba Research Europe, Cambridge, UK
christopher.m.zach@gmail.com
[2] University of California, Berkeley, USA

Abstract. The Mumford-Shah model for image formation is an important, but also difficult energy functional. In this work we focus on several approaches based on convex relaxation operating on a discretized image domain. Existing methods typically use discretized intensity labels, but in this work we propose to retain the continuous label structure. To this end we employ a recently proposed framework for a new convex relaxation of the Mumford-Shah functional. Numerical results illustrate the performance of the various approaches.

Keywords: Mumford-Shah energy · Convex relaxations

1 Introduction

In this work we consider several approaches to approximately solve the continuous Mumford-Shah energy [12], i.e. to determine a minimizer of

$$E_{\text{Mumford-Shah}}(f, \Gamma) = \frac{\lambda}{2} \int_\Omega (f - g)^2 dx + \int_{\Omega \setminus \Gamma} \|\nabla f\|_2^2 dx + \mu \text{Length}(\Gamma). \quad (1)$$

The Mumford-Shah energy plays a central role in image analysis and low level computer vision and is very challenging to minimize. Note that the particular energy used here allows open boundaries as opposed to Chan-Vese-type models [5]. Since numerical algorithms require a finite representation of the problem, we focus on the following discretized version of the above energy,

$$E_{\text{Grid-MS}}(f) = \frac{\lambda}{2} \sum_{s \in \mathcal{V}} (f_s - g_s)^2 + \sum_{(s,t) \in \mathcal{E}} \min\{(f_s - f_t)^2, \mu\}, \quad (2)$$

where the node set $\mathcal{V}$ and edge set $\mathcal{E}$ are induced by a rectangular pixel grid, and the subscripts s/t etc. index the respective pixel. For simplicity we assume $\mathcal{E}$ is induced by a standard 4-neighborhood and utilize toroidal (wrap-around) boundary conditions. $E_{\text{Grid-MS}}$ on the 4-connected grid constitutes our baseline energy, which will be used to evaluate the results of the tested algorithms.

© Springer International Publishing AG, part of Springer Nature 2018
M. Pelillo and E. Hancock (Eds.): EMMCVPR 2017, LNCS 10746, pp. 548–563, 2018.
https://doi.org/10.1007/978-3-319-78199-0_36

In this work we restrict our attention to convex relaxations of the Mumford-Shah model aiming on approximately minimizing $E_{\text{Mumford-Shah}}$ (or $E_{\text{Grid-MS}}$). The main advantages of convex relaxations of difficult energies are (i) that the returned solution is independent of a starting point in contrast to local methods, and (ii) are usually much more efficient than global optimization methods such as simulated annealing. Hence, we do not further discuss non-convex approaches such as continuation methods (e.g. [3]), stochastic methods (e.g. [9]) or alternation-based optimization using line processes [8].

The contributions of this work can be summarized as follows: (i) we present an intuitive formulation for a discretized convex relaxation for the continuous Mumford-Shah energy (Sect. 4), which also leads to faster determination of minimizers. (ii) We propose a convex relaxation for $E_{\text{Grid-MS}}$ explicitly targeted at discrete domains, but–unlike fully discrete labeling approaches–assigns continuous labels (intensity values) to pixels (Sect. 5). We further show how an initially intractable model can be reduced to an equivalent more efficient one. (iii) Finally we numerically compare the results of these two formulations with baseline methods rooted in approximate discrete inference.

2 Background

Notations: As mentioned before we represent image domains as finite rectangular lattices with an edge set induced by a 4-neighborhood connectivity. Larger neighborhoods can be helpful to reduce the grid bias [4]. Thus, the degree of a node s, $\deg(s)$, is always 4.

The desired minimizer of the Mumford-Shah energy is a mapping $f : \mathcal{V} \to [0,1]$ from pixels to real-valued intensities. We assume normalized image intensities lying in $[0,1]$. Several of the methods described below require a discretization of the continuous label space, and using evenly spaced samples is a standard approach. Thus, the interval $[0,1]$ is subdivided into $N \in \mathbb{N}$ equal subranges. For a given discrete label state $i \in \{0,\ldots,N-1\}$ we associate a continuous value $\gamma^i \stackrel{\text{def}}{=} (i+1/2)/N$ (i.e. the midpoint of the respective subrange $[i/N, (i+1)/N]$). This choice of a uniform sampling is arbitrary, but also very convenient.

We denote the convex conjugate of an extended real-valued function ψ by ψ^*, and its biconjugate by ψ^{**}. Further, we use the notations $\iota_C(x)$ and $\iota\{x \in C\}$ to write a constraint $x \in C$ in functional form, i.e. $\iota_C(x) = 0$ iff $x \in C$ and ∞ otherwise. For a convex function ψ we denote the l.s.c. extension of its perspective $(x,y) \mapsto x\psi(y/x)$ to $x = 0$ by $\psi_\oslash$. $\psi_\oslash$ can be computed as the biconjugate of the standard perspective. We refer to [6] for a recent review of perspective functions.

The convex discrete-continuous model: In this section we briefly review the convex discrete-continuous formulation for inference proposed in [19], which generalizes approximate discrete inference (discrete state spaces and domains) to continuous-valued labels spaces by replacing the standard constant potentials with convex potentials functions: for given families of convex functions $\{\psi_s^i\}_{s \in \mathcal{V}}$

and $\{\psi_{st}^{ij}\}_{(s,t)\in\mathcal{E}}$ (with $i,j \in \{1,\dots,L\}$) the discrete-continuous formulation reads as

$$E_{\text{DC-MRF}}(\mathbf{x}, \mathbf{y}) = \sum_{(s,t)\in\mathcal{E}} \sum_{i,j} (\psi_{st}^{ij}) \oslash (x_{st}^{ij}, y_{st\to s}^{ij}, y_{st\to t}^{ij}) \tag{3}$$

subject to the following marginalization and "decomposition" constraints

$$x_s^i = \sum_j x_{st}^{ij} \qquad x_t^j = \sum_i x_{st}^{ij} \qquad y_s^i = \sum_j y_{st\to s}^{ij} \qquad y_t^j = \sum_i y_{st\to t}^{ij} \tag{4}$$

and simplex constraints $x_s \in \Delta^L$, $x_{st} \in \Delta^{L^2}$. The unknown vector $\mathbf{x}$ collects the "pseudo-marginals" (i.e. x_{st} indicates one-hot encoding of the active potential function ψ_{st}^{ij} state at edge (s,t)), and the unknowns $\mathbf{y}$ represent the assigned continuous labels in the solution. The DC-MRF model is an extension of the standard local-polytope relaxation for discrete labeling problems by allowing the unary and pairwise potentials now to be arbitrary piecewise convex functions.

The formulation Eq. 3 is used in [19] to model convex relaxations of non-convex continuous labeling tasks. In particular, the data term for a continuous labeling problem is allowed to be piecewise convex instead of globally convex. The discrete state in the final discrete-continuous label assignment encodes which case in the piecewise convex definition of the unary costs is active at the minimizer.

Figure 1 illustrates the smoothness term in $E_{\text{Grid-MS}}$ (Eq. 2) and the underlying concept of subranges used to restrict a function to smaller domains.

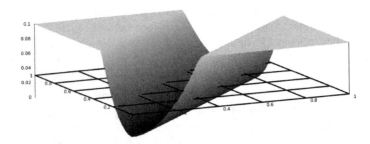

Fig. 1. Visualization of the smoothness term $\phi_{st}(f_s, f_t) = \min\{(f_s-f_t)^2, \mu\}$ in $E_{\text{Grid-MS}}$ (Eq. 2, with $\mu = 1/10$). The 2D function plot is shown together with a grid, that indicates the $N \times N$ subranges used to restrict ϕ_{st} to obtain $\phi_{st}^{ij} : [i/N, (i+1)/N] \times [j/N, (j+1)/N] \to \mathbb{R}$ (with $i,j \in \{0,\dots,N-1\}$). N is set to 5 in this example.

3   Baseline Model: Discrete Labeling on Discrete Grids

In this section we describe a formulation based on discrete label inference on a discrete pixel grid with 4-connectivity. We convert the continuous label space $[0,1]$ into a discrete label space by evenly sampled labels (recall Sect. 2). By using a standard local polytope relaxation (see e.g. [16]) we obtain the following convex program,

$$E_{\mathrm{MRF}}(\mathbf{x}) = \sum_{s,i} \theta_s^i x_s^i + \sum_{s \sim t} \sum_{ij} \theta^{ij} x_{st}^{ij} \tag{5}$$

$$\text{s.t. } x_s^i = \sum_j x_{st}^{ij} \qquad x_t^j = \sum_i x_{st}^{ij} \qquad \sum_i x_s^i = 1 \qquad x_s^i \ge 0, \, x_{st}^{ij} \ge 0,$$

where $\theta_s^i = \frac{\lambda}{2}(\gamma^i - g_s)^2$ and $\theta^{ij} = \min\{\mu, (\gamma^i - \gamma^j)^2\}$ (independent of the edge (s,t)). This choice for θ_s^i and θ^{ij} corresponds to point sampling of the continuous costs. In order to obtain a faithful approximation the number of states N should be large (e.g. $N = 256$ for 8-bit grayscale source images g). We implemented two methods for approximate inference in this setting: the first one is based on a dual block-coordinate (message-passing) method [10,11] utilizing the particular structure of the pairwise potentials for acceleration, and the second approach is a proximal method using a compressed but equivalent representation of marginals $\mathbf{x}$ [18] for efficiency.

Dual block coordinate method: A family of efficient algorithms for approximate inference in discrete MRFs is based on optimizing the dual using a block-coordinate approach [10,11]. We employ a slight variant of MPLP [10] by using the following dual,

$$E_{\mathrm{MRF\text{-}dual}}(\mathbf{p}) = \sum_{s \sim t} \min_{i,j} \left\{ p_{st \to s}^i + p_{st \to t}^j - \hat{\theta}_{st}^{ij} \right\} \qquad \text{s.t.} \sum_{t \in N(s)} p_{st \to s}^i = 0, \tag{6}$$

where $\hat{\theta}_{st}^{ij} = \theta_{st}^{ij} + \theta_s^i / \deg(s) + \theta_t^i / \deg(t)$ are reparametrized potentials such that the unary terms vanish. If we maximize over only $\{p_{st \to s}^i\}_{i \in \{1,\dots,L\}}$ for all edges (s,t) adjacent to s, it can be shown that a block-coordinate method yields the following updates,

$$\hat{p}_{st \to s}^i \leftarrow \min_j \left\{ \hat{\theta}_{st}^{ij} + p_{ts \to s}^j \right\} \qquad p_{st \to s}^i \leftarrow \frac{1}{\deg(s)} \sum_{r \in N(s)} \hat{p}_{sr \to s}^i - \hat{p}_{st \to s}^i \tag{7}$$

for all $t \in N(s)$ and for all i. Note that after the update, the messages $p_{st \to s}^i$ still satisfy the "flow conservation" constraint, $\sum_{t \in N(s)} p_{st \to s}^i = 0$. We accelerate the computation of $\hat{p}_{st \to s}^i$ significantly by exploiting the particular structure of θ_{st}^{ij} as truncated quadratic [7] if the number of states N is large. We use a fixed update schedule iterating from the top-left to the bottom-right pixel.

   Block coordinate methods monotonically increase the (dual) objective, but may stop too early at a fix-point without reaching a global maximizer. In our experiments we essentially did not experience this problem.

Compressed pairwise marginals: Since the pairwise potentials θ^{ij} are nothing else than truncated quadratic costs, we split the states j for a given state i into two groups: the "smooth" states

$$S^i \overset{\mathrm{def}}{=} \{j : (\gamma^i - \gamma^j)^2 < \mu\} = \{j : (i-j)^2 < N^2 \mu\} = \{j : |i-j| < N\sqrt{\mu}\}$$

and the complementary set of "jump" states

$$J^i \overset{\text{def}}{=} \{j : (\gamma^i - \gamma^j)^2 \geq \mu\} = \{j : |i - j| \geq N\sqrt{\mu}\}.$$

We can replace the pairwise marginals x_{st}^{ij}, $j \in J^i$, by two jump ("wildcard") marginals x_{st}^{i*} and x_{st}^{*i} such that

$$x_{st}^{i*} = \sum_{j \in J^i} x_{st}^{ij} \quad \text{and} \quad x_{st}^{*i} = \sum_{j \in J^i} x_{st}^{ji}$$

without affecting the optimal solution [18]. The reduced program reads as

$$E_{\text{reduced-MRF}}(\mathbf{x}) = \sum_{s,i} \theta_s^i x_s^i + \sum_{s \sim t} \sum_i \left(\sum_{j \in S^i} \theta^{ij} x_{st}^{ij} + \frac{\mu}{2} \left(x_{st}^{i*} + x_{st}^{*i} \right) \right) \quad (8)$$

$$\text{s.t. } x_s^i = \sum_{j \in S^i} x_{st}^{ij} + x_{st}^{i*} \qquad x_t^j = \sum_{i \in S^j} x_{st}^{ij} + x_{st}^{*j} \qquad \sum_i x_s^i = 1 \qquad \mathbf{x} \geq 0.$$

The even distribution of the jump cost μ to x_{st}^{i*} and x_{st}^{*i} is arbitrary and can be replaced by any two pairwise potentials such that their sum is μ. After introducing Lagrange multipliers $p_{st \to s}^i$ and $p_{st \to t}^j$ for the respective marginalization constraint, and q_s for the normalization constraint, we arrive at the following Lagrangian,

$$\min_{\mathbf{x} \geq 0} \max_{\mathbf{p}, \mathbf{q}} \sum_{s,i} \theta_s^i x_s^i + \sum_{s \sim t, i} \left(\sum_{j \in S^i} \theta^{ij} x_{st}^{ij} + \frac{\mu}{2} \left(x_{st}^{i*} + x_{st}^{*i} \right) \right) + \sum_s q_s \left(\sum_i x_s^i - 1 \right)$$

$$+ \sum_{s \sim t, i} \left(p_{st \to s}^i \left(x_s^i - \sum_{j \in S^i} x_{st}^{ij} - x_{st}^{i*} \right) + p_{st \to t}^i \left(x_t^i - \sum_{j \in S^i} x_{st}^{ji} - x_{st}^{*i} \right) \right).$$

In our implementation we eliminate the unknowns x_{st}^{i*} and x_{st}^{*i}, respectively by using the fact that e.g. $\min_{x_{st}^{i*} \geq 0} x_{st}^{i*} (\mu/2 - p_{st \to s}^i) = -\imath\{p_{st \to s}^i \leq \mu/2\}$, leading to additional but simple bounds constraints $p_{st \to s}^i \leq \mu/2$ and $p_{st \to t}^i \leq \mu/2$ in the saddle-point energy. We determine an optimal primal-dual pair using the preconditioned primal-dual algorithm [13]. Since this approach returns a global minimizer of the convexified labeling task, we use it to verify the result of the (much faster, but potentially stuck) dual coordinate method.

4   Review of a Discretized Continuous Convex Relaxation

In [14] a discretized formulation the Mumford-Shah model is proposed, which is based on a saddle-point energy derived for continuous non-convex and non-smooth problems [1]. Below we state the already discretized saddle-point energy using the notation of [15],

$$E_{\text{Saddle-point}}(\mathbf{x}; \mathbf{p}, \mathbf{b}) = \sum_{s,i} \theta_s^i x_s^i + \sum_{s,i} \left((p_s^i)^T \nabla x_s^i - b_s^i x_s^i \right) \quad (9)$$

$$\text{s.t. } \left\| p_s^i - p_s^j \right\|_* \leq \theta^{ij} \qquad -b_s^i \geq h^*(p_s^i - p_s^{i-1}) \qquad x_s \in \Delta.$$

The gradient ∇ is understood as finite difference approximation on regular pixel grids, and h^* is the conjugate of a convex function h. θ^{ij} are the cost associated with discontinuous transitions from a state i to state j. In the case of the Mumford-Shah energy one has $\theta_s^i = \lambda(i - g_s)^2/2$, $\theta^{ii} = 0$, $\theta^{ij} = \mu$ if $i \neq j$, and $h(\xi) = \|\xi\|^2$. We explicitly indicate the dual norm in the capacity constraints $\|p_s^i - p_s^j\|_* \leq \theta^{ij}$, since this will be helpful later.

It is possible to derive the primal program of the above saddle-point energy in a similar way to [18]. The following fact is helpful here: the conjugate of $\psi^*(b, r) \overset{\text{def}}{=} \iota\{-b \geq h^*(r)\}$ is given by $uh\left(\frac{v}{u}\right) + \iota\{u \geq 0\} = h_\oslash(u, v)$ for primal arguments u and v (see e.g. [19]). Due to space restriction we omit the otherwise non-interesting calculations and just state the corresponding primal program,

$$E_{\text{Discretized-MS}}(\mathbf{x}, \mathbf{y}, \mathbf{z}) = \sum_{s,i} \theta_s^i x_s^i + \sum_{s,i,j:i<j} \theta^{ij} \|y_s^{ij}\|_{**} + \sum_{s,i} \underbrace{x_s^i \, h\left(\frac{z_s^i}{x_s^i}\right)}_{=h_\oslash(x_s^i, z_s^i)} \quad (10)$$

$$\text{s.t. } \nabla_x x_s^i = \sum_{j:j<i} y_s^{ji} - \sum_{j:j>i} y_s^{ij} + z_s^i - z_s^{i+1} \quad x_s \in \Delta.$$

The bidual norm $\|\cdot\|_{**}$ equals to $\|\cdot\|_2$ if $\|\cdot\|_* = \|\cdot\|_2$, and $\|\cdot\|_{**} = \|\cdot\|_1$ for $\|\cdot\|_* = \|\cdot\|_\infty$. In order to be able to compare the results with labeling methods designed for discrete lattices, we use $\|\cdot\|_{**} = \|\cdot\|_1$ in our implementations. The utilized boundary conditions for $p_s^i - p_s^{i-1}$ (i.e. $p_s^0 - p_s^{-1} = 0$) imply $z_s^0 = z_s^N = 0$.

The gradient constraints are more intuitive, if we express them in terms of the sublevel function $u_s^i = \sum_{k=i}^{N-1} x_s^i$. Note that $u_s^0 = 1$ since $\sum_i x_s^i = 1$, and we can naturally introduce $u_s^N = 0$. One can show that the constraints on ∇x_s^i are equivalent to

$$\nabla u_s^i = \sum_{k=i}^{N-1} \nabla x_s^k = \sum_{k=i}^{N-1} \sum_{j=0}^{i-1} y_s^{jk} + z_s^i. \quad (11)$$

The first term on the right hand side, $\sum_{k=i}^{N-1} \sum_{j=0}^{i-1} y_s^{jk}$, has actually a simple interpretation: it counts all discontinuities from a label j below i to a label k above or equal i i.e. all jumps crossing label i. This constitutes the non-smooth parts of ∇u_i. z_s^i represents the smooth fraction of ∇u_s^i, which should be only non-zero if the jump part is zero. Note that $z_s^0 = z_s^N = 0$: since $u_s^0 = 1$ implies $\nabla u_s^0 = 0$ and the sum in the r.h.s. is empty, we consistently obtain $z_s^0 = 0$. Similarly we deduce $z_s^N = 0$.

The perspective $h_\oslash(x_s^i, z_s^i)$ essentially limits the evaluation of the smooth penalizer h to the graph of the function f (where $\partial_i u_s^i = x_s^i = 1$): if $\partial_i u_s^i = x_s^i = 0$, then the smooth part of the gradient, z_s^i, is forced to be 0 by $h_\oslash$. By noticing that y_s^{jk} appears $|k - j|$ times in $\sum_i \nabla u_s^i$ it can be shown, that the total (finite-difference) gradient of the sublevel function is given by

$$\nabla u_s = \sum_i \nabla u_s^i = \sum_{i=0}^{N-1} \sum_{k=i+1}^{N-1} |k - i| y_s^{ik} + \sum_i z_s^i. \quad (12)$$

Note that if u_s is a proper sublevel function, $u_s^i = \mathbf{1}(f_s \le (i+1)/N)$, then at most one y_s^{ik} is non-zero. Therefore the spatial (and finite-difference) gradient ∇u_s of the sublevel function can be decomposed into a non-smooth part (whose corresponding jumps have implied costs according to θ^{ij}) and a smooth component penalized according to a function $h(\cdot)$. This is visualized in Fig. 2.

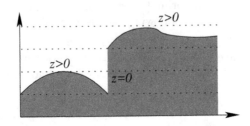

Fig. 2. Illustration of Eq. 12 using a 1-D image domain. Dotted lines indicated label subrange boundaries. If u is a proper sublevel function, $u_s^i = \mathbf{1}(f_s \le (i+1)/N)$, then the model $E_{\text{Discretized-MS}}$ (Eq. 10) forces the smooth gradient contribution z_s^i to 0 at discontinuities. The (finite-difference) spatial gradient of u at location s is a sum of smooth intra-subrange gradient z_s^i and a jump contribution. The latter one is zero where u is smooth and otherwise the jump height.

Remark 1. One of the puzzling aspects of $E_{\text{Discretized-MS}}$ is that *smooth* interactions via $h_\oslash$ between distant labels are not explicitly modeled. If $\mathbf{x}$ is purely integral (and therefore also the corresponding sublevel function $\mathbf{u}$), $\sum_i h_\oslash(x_s^i, z_s^i)$ has only one non-zero summand, regardless of the actual label difference. The fact that this formulation returns faithful results for N sufficient large is explained by the observation that minimizers of $E_{\text{Discretized-MS}}$ are generally fractional. Dropping the integrality constraint $x_s^i \in \{0, 1\}$ is actually necessary to produce faithful results. This can be seen in Fig. 3(b) that illustrates the fractional behaviour of the indicator values x_s^i and sublevel ones u_s^i using Tikhonov regularization (i.e. $\mu \ge 1$, $\lambda = 1/2$). The blurred result can be seen in Fig. 3(a). We selected $\mu \ge 1$ in order to have a fully convex problem in the first place, such that a subsequent convex relaxations should be tight (since it reduces to a continuous max-flow problem). Note that the minimizer of $E_{\text{Discretized-MS}}$ is an accurate approximation of the true solution in this case despite the fractional nature of $\mathbf{x}$. The necessity of allowing fractional solution in $E_{\text{Discretized-MS}}$ for faithful approximation is in strong contrast to the approach described in Sect. 5 below, which provides a globally optimal answer for $E_{\text{Grid-MS}}$ if the unknowns corresponding to $\mathbf{x}$ are all integral.

Implementation: We have two implementations: the first one follows the saddle-point formulation [14], but similar to [15] we avoid using Dykstra's algorithm to enforce $\|p_s^i - p_s^j\|_* \le \mu$ by introducing auxiliary variables (and therefore trading increased memory consumption to avoid nested iterations). Further, we use a depressed cubic approach [19] to project onto the parabola constraint

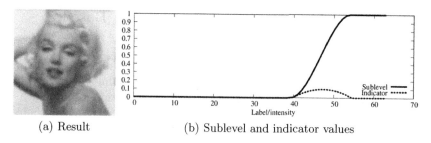

(a) Result   (b) Sublevel and indicator values

Fig. 3. Input image (a) and final values of u_s^i and x_s^i, respectively, for the center pixel for parameters $\lambda = 1/2$ and $\mu > 1$ (i.e. pure Tikhonov regularization).

$b_s^i \geq h^*(p_s^i - p_s^{i-1})$ in closed form (thus avoiding the Newton iterations utilized in [14]).

The second implementation replaces the capacity constraints $\{\|p_s^i - p_s^j\|_* \leq \mu\}_{i<j}$ (with $O(N^2)$ elements) by a weaker set of constraints $\{\|p_s^i\|_* \leq \mu/2\}_i$ with $O(N)$ inequalities.

$$E_{\text{reduced-saddle-point}}(\mathbf{x}; \mathbf{p}, \mathbf{b}) = \sum_{s,i} \theta_s^i x_s^i + \sum_{s,i} \left((p_s^i)^T \nabla x_s^i - b_s^i x_s^i \right) \tag{13}$$

$$\text{s.t. } \left\|p_s^i\right\|_* \leq \mu/2 \qquad -b_s^i \geq h^*(p_s^i - p_s^{i-1}) \qquad x_s \in \Delta.$$

If $\|\cdot\|_* = \|\cdot\|_2$ this is a (slightly) weaker relaxation than using the full set, since $\|p_s^i\|_* \leq \mu/2$ implies $\|p_s^i - p_s^j\|_2 \leq \mu$ but not the reverse. Fortunately, for the choice $\|\cdot\|_* = \|\cdot\|_\infty$ both formulations are equivalent. One way to prove that fact is using similar arguments as in [18] (see also Sects. 3 and 5), but below we present a different proof. The first observation is, that $E_{\text{Saddle-point}}$ is invariant to adding an edge-bias to the dual variables $\mathbf{p}$, i.e.

$$E_{\text{Saddle-point}}(\mathbf{x}; \mathbf{p}, \mathbf{b}) = E_{\text{Saddle-point}}(\mathbf{x}; \bar{\mathbf{p}}, \mathbf{b})$$

with $\bar{p}_s^i = p_t^i + \sigma_s$ for some $\sigma_s \in \mathbb{R}^2$. It is clear that the constraints in Eq. 9 are still satisfied after this transformation. The objective value is also unaffected, since

$$\sum_i (\bar{p}_s^i)^T \nabla x_s^i = \sum_i (p_s^i)^T \nabla x_s^i + \sigma_s^T \sum_i \nabla x_s^i = \sum_i (p_s^i)^T \nabla x_s^i,$$

since $\sum_i \nabla x_s^i = \nabla \sum_i x_s^i$ and $\sum_i x_s^i$ is the constant 1 from the normalization constraint. This observation corresponds to the one used in message passing for approximate discrete inference in order to center the messages (i.e. $\mathbf{p}$) after each update. The second observation is the following: If real numbers $\{p^i\}_i$ satisfy $|p^i - p^j| \leq \mu$ for all $i < j$, then we have $|p^i - \sigma| \leq \mu/2$ for all i if we set $\sigma \overset{\text{def}}{=} (\max_i p^i - \min_i p^i)/2$. Both facts together mean, that we can "center" the dual variables $\mathbf{p}$ in $E_{\text{Saddle-point}}$ such that $\|p_s^i - p_s^j\|_\infty \leq \mu$ implies $\|p_s^i - \sigma\|_\infty \leq \mu/2$, and a primal-dual optimal pair for $E_{\text{Saddle-point}}$ is also optimal for $E_{\text{reduced-saddle-point}}$.

5   Discretized Mumford-Shah Formulation Based on Discrete-Continuous MRFs

In this section we directly aim for a convex relaxation of $E_{\text{Grid-MS}}$ (Eq. 2). Since in $E_{\text{Grid-MS}}$ the data term is convex and the smoothness term is piecewise convex, we can immediately employ the discrete-continuous approach for MRFs [19] to obtain a respective convex relaxation. If the "switching variables" $\mathbf{x}$ in $E_{\text{DC-MRF}}$ (Eq. 3) are integral, we have a global optimal solution to the original non-convex, continuous labeling task. One way to strengthen the relaxation is to artificially subdivide the label space into subranges. Thus, the discrete state associated with each edge jointly selects one of the convex branches of the truncated quadratic $\xi \mapsto \min\{\mu, \xi^2\}$ and one out of N intensity subranges for its nodes. We start by presenting the full energy first and discuss its meaning and intuition subsequently:

$$E_{\text{DC-MS}}(\mathbf{x}, \mathbf{y}) = \sum_{s \sim t} \sum_{i,j,k} (\psi_{st}^{ijk}) \oslash (x_{st}^{ijk}, y_{st \to s}^{ijk}, y_{st \to t}^{ijk}) \tag{14}$$

$$\text{s.t. } x_s^i = \sum_{j,k} x_{st}^{ijk} \quad x_t^j = \sum_{i,k} x_{st}^{ijk} \quad y_s^i = \sum_{j,k} y_{st \to s}^{ijk} \quad y_t^j = \sum_{i,k} y_{st \to t}^{ijk}$$

$$y_{st \to s}^{ijk} \in \frac{x_{st}^{ijk}}{N} [i, i+1], \ y_{st \to t}^{ijk} \in \frac{x_{st}^{ijk}}{N} [j, j+1], \ x_s \in \Delta, \ x_{st}^{ijk} \geq 0,$$

with

$$\psi_{st}^{ij0}(z_s, z_t) \stackrel{\text{def}}{=} \frac{\lambda}{8}(z_s - g_s)^2 + \frac{\lambda}{8}(z_t - g_t)^2 + (z_s - z_t)^2 + \imath\{|z_s - z_t| \leq \sqrt{\mu}\}$$

$$\psi_{st}^{ij1}(z_s, z_t) \stackrel{\text{def}}{=} \frac{\lambda}{8}(z_s - g_s)^2 + \frac{\lambda}{8}(z_t - g_t)^2 + \mu + \imath\{z_s \leq z_t - \sqrt{\mu}\}$$

$$\psi_{st}^{ij2}(z_s, z_t) \stackrel{\text{def}}{=} \frac{\lambda}{8}(z_s - g_s)^2 + \frac{\lambda}{8}(z_t - g_t)^2 + \mu + \imath\{z_s \geq z_t + \sqrt{\mu}\}.$$

Each cost at edge (s, t) is piecewise convex with $N \times N \times 3$ branches: superscripts i and $j \in \{0, \dots, N-1\}$ indicate the value range the desired continuous label at node s (or t, respectively) is contained in. We use a uniform subdivision of the normalized intensity range $[0, 1]$ as indicated in the bounds constraints, i.e. the subrange associated with state i is $[i/N, (i+1)/N]$. This joint state to restrict the intensities at s and t is extended by indicating whether we have a smooth transition between nodes ($k = 0$) or an "increasing" ($k = 1$) or "decreasing" discontinuity ($k = 2$) along the edge (s, t). We also use here the observation that moving the unary costs to the pairwise ones is improving the relaxation [17], and we evenly distribute the node-wise data cost, $\lambda(z_s - g_s)^2/2$, to the 4 adjacent edges. Note that the actual term appearing in Eq. 14 is not ψ_{st}^{ijk}, but its corresponding perspective, leading essentially to a sum of quadratic-over-linear terms. This convex program is expensive to solve, but fortunately we can reduce the number of unknowns substantially to obtain a more tractable *equivalent* convex program. Since the pairwise cost in the Mumford-Shah energy is a truncated cost, we can use a discrete-continuous extension of "compressing the marginals" (recall Sect. 3).

Step 1: Dominance: The first simple observation is, that certain branches are dominated by other ones and consequently these branches will never be active in a minimizer. Essentially, the constraints $z_s \in [i, i+1]/N$ and $z_t \in [j, j+1]/N$ imply that certain x_{st}^{ijk} can be set to 0 without affecting the minimizer. We group the state pairs (i, j) into the following sets:

1. The set $\mathcal{S} \overset{\text{def}}{=} \{(i,j) : |i - j| \leq N\sqrt{\mu} - 1\}$ represents all subranges such that $z_s \in [i/N, (i+1)/N]$ and $z_t \in [j/N, (j+1)/N]$ are guaranteed to satisfy $(z_s - z_t)^2 \leq \mu$. Thus, we have $x_{st}^{ij1} = x_{st}^{ij2} = 0$ for $(i,j) \in \mathcal{S}$.

2. The set $\mathcal{J}_{\nearrow} \overset{\text{def}}{=} \{(i,j) : i \leq j - N\sqrt{\mu} - 1\}$. For $(i,j) \in \mathcal{J}_{\nearrow}$ the closest pair $z_s \in [i, i+1]/N$ and $z_t \in [j, j+1]/N$ has at least $\sqrt{\mu}$ distance, hence ψ_{st}^{ij0} is dominated in this case by ψ_{st}^{ij1} and $x_{st}^{ij0} = 0$. Further, x_{st}^{ij2} has to be 0 as well, since $z_s \geq z_t + \sqrt{\mu}$ is impossible to satisfy for any $(z_s, z_t) \in [i, i+1]/N \times [j, j+1]/N$ in this case.

3. The symmetrical set is $\mathcal{J}_{\searrow} \overset{\text{def}}{=} \{(i,j) : j \leq i - N\sqrt{\mu} - 1\}$ which w.l.o.g. leads to $x_{st}^{ij0} = x_{st}^{ij1} = 0$ in a minimizer. We further define $\mathcal{J} \overset{\text{def}}{=} \mathcal{J}_{\nearrow} \cup \mathcal{J}_{\searrow}$.

4. Finally we have the set of state pairs (i, j) which cannot be decided in advance ("undecided" state pairs), $\mathcal{U} \overset{\text{def}}{=} \{0, \ldots, N-1\}^2 \setminus (\mathcal{S} \cup \mathcal{J})$.

Note that none of the three branches ψ_{st}^{ijk}, $k = 0, 1, 2$, can be ruled out by dominance for a particular (i, j) if e.g. $N = 1$. For large N w.l.o.g. we can fix many unknowns x_{st}^{ijk} to 0 (and consequently also eliminate the respective $y_{st \to s}^{ijk}$ and $y_{st \to t}^{ijk}$ from the set of unknowns), we reduce the problem size substantially from $O(3N^2)$ to approximately $O(N^2)$, leading to a smaller memory footprint and faster convergence.

For $(i, j) \notin \mathcal{U}$ only one of the choices $k \in \{0, 1, 2\}$ can be active, and we can drop the state k for these pairs (i, j). Using dominance, the new cost functions are

$$\psi_{st}^{ij}(z_s, z_t) \overset{\text{def}}{=} \begin{cases} \frac{\lambda}{8}(z_s - g_s)^2 + \frac{\lambda}{8}(z_t - g_t)^2 + \mu & (i,j) \in \mathcal{J} \\ \frac{\lambda}{8}(z_s - g_s)^2 + \frac{\lambda}{8}(z_t - g_t)^2 + (z_s - z_t)^2 & (i,j) \in \mathcal{S} \end{cases} \tag{15}$$

If $(i,j) \in \mathcal{U}$ we have ψ_{st}^{ijk} ($k \in \{0, 1, 2\}$) as before. The reduced set of potentials is also reflected in a smaller set of unknowns x_{st}^{ij} for $(i,j) \in \mathcal{S} \cup \mathcal{J}$ (instead of having a full representation x_{st}^{ijk}). The same reduction applies to y_{st}^{ijk} as well. The simplified objective now reads as

$$E_{\text{DC-MS-I}}(\mathbf{x}, \mathbf{y}) = \sum_{(s,t) \in \mathcal{E}} \sum_{(i,j) \in \mathcal{U}, k} (\psi_{st}^{ijk})_{\oslash}(x_{st}^{ijk}, y_{st \to s}^{ijk}, y_{st \to t}^{ijk})$$

$$+ \sum_{(s,t) \in \mathcal{E}} \sum_{(i,j) \in \mathcal{S} \cup \mathcal{J}} (\psi_{st}^{ij})_{\oslash}(x_{st}^{ij}, y_{st \to s}^{ij}, y_{st \to t}^{ij}) \tag{16}$$

subject to the analogous marginalization and bounds constraints from Eq. 14.

Step 2: Introducing wildcard variables: The analysis in the previous section reduces the number of unknowns to approximately one third for large N. The still quadratic number of unknowns *per pixel* is a major obstacle for efficient implementations. Fortunately, the pairwise unknowns can be compressed in a similar way as sketched in Sect. 3. We introduce wildcard variables

$$x_{st}^{i*} \stackrel{\text{def}}{=} \sum_{j:(i,j)\in\mathcal{J}} x_{st}^{ij} \qquad\qquad y_{st\to s}^{i*} \stackrel{\text{def}}{=} \sum_{j:(i,j)\in\mathcal{J}} y_{st\to s}^{ij}$$

$$x_{st}^{*j} \stackrel{\text{def}}{=} \sum_{i:(i,j)\in\mathcal{J}} x_{st}^{ij} \qquad\qquad y_{st\to s}^{*j} \stackrel{\text{def}}{=} \sum_{i:(i,j)\in\mathcal{J}} y_{st\to s}^{ij}.$$

Consequently, the further reduced objective is

$$
\begin{aligned}
E_{\text{DC-MS-II}}(\mathbf{x},\mathbf{y}) = &\sum_{(s,t)\in\mathcal{E}} \sum_{(i,j)\in\mathcal{U},k} (\psi_{st}^{ijk})_{\oslash}(x_{st}^{ijk}, y_{st\to s}^{ijk}, y_{st\to t}^{ijk}) \\
&+ \sum_{(s,t)\in\mathcal{E}} \sum_{(i,j)\in\mathcal{S}} (\psi_{st}^{ij})_{\oslash}(x_{st}^{ij}, y_{st\to s}^{ij}, y_{st\to t}^{ij}) \\
&+ \sum_{(s,t)\in\mathcal{E}} \sum_{i} \left((\psi_{st}^{i*})_{\oslash}(x_{st}^{i*}, y_{st\to s}^{i*}) + (\psi_{st}^{*i})_{\oslash}(x_{st}^{*i}, y_{st\to t}^{*i})\right) \quad (17)
\end{aligned}
$$

subject to

$$x_s^i = \sum_{j:(i,j)\in\mathcal{S}} x_{st}^{ij} + \sum_{j:(i,j)\in\mathcal{U},k} x_{st}^{ijk} + x_{st}^{i*} \qquad x_t^j = \sum_{i:(i,j)\in\mathcal{S}} x_{st}^{ij} + \sum_{i:(i,j)\in\mathcal{U},k} x_{st}^{ijk} + x_{st}^{*j}$$

$$y_s^i = \sum_{j:(i,j)\in\mathcal{S}} y_{st\to s}^{ij} + \sum_{j:(i,j)\in\mathcal{U},k} y_{st\to s}^{ijk} + y_{st\to s}^{i*} \qquad y_t^j = \sum_{i:(i,j)\in\mathcal{S}} y_{st\to t}^{ij} + \sum_{i:(i,j)\in\mathcal{U},k} y_{st\to t}^{ijk} + y_{st\to t}^{*j},$$

and the respective non-negativity and bounds constraints. We also introduced

$$\psi_{st}^{i*}(z_s) \stackrel{\text{def}}{=} \frac{\lambda}{8}(z_s - g_s)^2 + \frac{\mu}{2} \qquad\qquad \psi_{st}^{*i}(z_t) \stackrel{\text{def}}{=} \frac{\lambda}{8}(z_t - g_t)^2 + \frac{\mu}{2}.$$

Proposition 1. *All formulations $E_{\text{DC-MS}}$, $E_{\text{DC-MS-I}}$ and $E_{\text{DC-MS-II}}$ are equivalent, i.e.*

$$\min_{\mathbf{x},\mathbf{y}} E_{\text{DC-MS}} = \min_{\mathbf{x},\mathbf{y}} E_{\text{DC-MS-I}} = \min_{\mathbf{x},\mathbf{y}} E_{\text{DC-MS-II}}.$$

Proof. (Sketch) The first equality follows from exploiting dominance relations as described in the previous section.

From Jensen's inequality it follows that $E_{\text{DC-MS-II}}(\mathbf{x},\mathbf{y}) \leq E_{\text{DC-MS-I}}(\mathbf{x},\mathbf{y})$, and consequently that $\min E_{\text{DC-MS-II}} \leq \min E_{\text{DC-MS-I}}$. The converse is also true: if we have a minimizer $(\mathbf{x},\mathbf{y})$ of $E_{\text{DC-MS-II}}$, then we can construct x_{st}^{ij} for all $(i,j) \in \mathcal{J}$ satisfying the marginalization and non-negativity constraints from the determined values for x_{st}^{i*} and x_{st}^{*i} (using the same procedure as in [18]), leaving the already existing x_{st}^{ij} (and y_{st}^{ij} etc.) untouched. From y_{st}^{i*} and y_{st}^{*j} we can also construct (for $(i,j) \in \mathcal{J}$)

$$y_{st \to s}^{ij} \leftarrow \begin{cases} \frac{x_{st}^{ij}}{x_{st}^{i*}} y_{st \to s}^{i*} & \text{if } x_{st}^{i*} > 0 \\ 0 & \text{otherwise} \end{cases} \qquad y_{st \to t}^{ij} \leftarrow \begin{cases} \frac{x_{st}^{ij}}{x_{st}^{*j}} y_{st \to s}^{*j} & \text{if } x_{st}^{*j} > 0 \\ 0 & \text{otherwise.} \end{cases}$$

Observe that this construction implies that $y_{st \to s}^{ij}/x_{st}^{ij} = y_{st \to s}^{i*}/x_{st}^{i*}$ (and $y_{st \to t}^{ij}/x_{st}^{ij} = y_{st \to t}^{*i}/x_{st}^{*i}$, respectively) for $(i,j) \in \mathcal{J}$, i.e. the ratios are constant. It is easy to verify that this assignment of y_{st}^{ij} also satisfies all the constraints in $E_{\text{DC-MS-I}}$. Further, we have (assuming $x_{st}^{i*} > 0$ and $x_{st}^{*i} > 0$ for now)

$$\sum_{(i,j) \in \mathcal{J}} (\psi_{st}^{ij}) \oslash (x_{st}^{ij}, y_{st \to s}^{ij}, y_{st \to t}^{ij}) = \sum_{(i,j) \in \mathcal{J}} x_{st}^{ij} \psi_{st}^{ij} \left(\frac{y_{st \to s}^{ij}}{x_{st}^{ij}}, \frac{y_{st \to t}^{ij}}{x_{st}^{ij}} \right)$$

$$= \sum_i x_{st}^{i*} \left(\frac{\mu}{2} + \frac{\lambda}{8} \left(\frac{y_{st \to s}^{i*}}{x_{st}^{i*}} - g_s \right)^2 \right) + \sum_i x_{st}^{*i} \left(\frac{\mu}{2} + \frac{\lambda}{8} \left(\frac{y_{st \to t}^{*i}}{x_{st}^{*i}} - g_t \right)^2 \right)$$

$$= \sum_i \left((\psi_{st}^{i*}) \oslash (x_{st}^{i*}, y_{st \to s}^{i*}) + (\psi_{st}^{*i}) \oslash (x_{st}^{*i}, y_{st \to t}^{*i}) \right).$$

Thus, the overall minimal objective value for $E_{\text{DC-MS-II}}$ and for $E_{\text{DC-MS-I}}$ are the same, since all other unknowns (and terms) remain the same. Overall, we also have $\min_{\mathbf{x},\mathbf{y}} E_{\text{DC-MS-I}} = \min_{\mathbf{x},\mathbf{y}} E_{\text{DC-MS-II}}$.

Using this construction we have reduced the number of unknowns (in terms of x_{st}) from approximately $N^2 + 4N$ per pixel to about $LN + 6N$, where $L \stackrel{\text{def}}{=} \max_i |\{j : (i,j) \in \mathcal{S}\}|$ is the "bandwidth" of the smooth states.

Remark 2. It seems surprising that this construction leads to less efficient representation with increasing μ. In the case $\mu > 1$, $L = N$ and we obtain again a $O(N^2)$ representation for *a fully convex original problem*. It is beyond the scope of this manuscript, but we conjecture that the fine-grained representation for states $\in \mathcal{S}$ can be coarsened, leading to a reduction of the unknowns. Since we are mostly interested in cases with $\mu \ll 1$ (i.e. edge-preservation), solving $E_{\text{DC-MS-II}}$ has sufficient efficiency.

Variations of $E_{DC-MS-II}$: Even with the reductions described above the formulation $E_{\text{DC-MS-II}}$ is still expensive to solve for N large (e.g. $N = 8$). Reasons are the non-linearity of the objective and the excessive number of invocations of the (slow) cubic root routine. We further observed, that setting N larger than 1 does not really improve the visual result to compensate for the increased computational cost. We modify $E_{\text{DC-MS-II}}$ by moving the (quadratic) data term from the pairwise potential functions to the unary terms, which leads to a weaker but more efficient to optimize formulation $E_{\text{DC-MS-III}}$.

6   Numerical Results

We implemented the different energy formulations using the optimization method as described in the respective section (block-coordinate method for

$E_{\text{dual-MRF}}$ and a first-order method for all the other energies). The energy values reported below and in the figures are computed with respect to $E_{\text{Grid-MS}}$. Computation of this energy requires conversion of the unknowns in the actual formulation into final intensities, which is done by computing the 1/2-isolevel of u_s^i for $E_{\text{Discretized-MS}}$, and by computing the "expected" intensity label for the other models, i.e.

$$f_s = \begin{cases} \sum_i \gamma^i x_s^i & \text{for } E_{\text{dual-MRF}} \text{ and } E_{\text{reduced-MRF}} \\ \sum_i y_s^i & \text{for } E_{\text{DC-MS}} \text{ and its variants} \end{cases}$$

given the minimizers $\mathbf{x}$ and $(\mathbf{x}, \mathbf{y})$, respectively.

Input $E = 24.58$ $E = 26.52$ $E = 29.92$

Input $E = 54.41$ $E = 60.15$ $E = 68.62$

Input $E = 53.82$ $E = 58.48$ $E = 70.16$

Fig. 4. Results of different relaxations for the choice $\lambda = 1/2$, $\mu = 0.005$. 1st column: input image. 2nd column: result of dual block coordinate method. 3rd column: result of $E_{\text{Discretized-MS}}$ with $N = 32$. 4th column: result of $E_{\text{DC-MS}}$ with $N = 1$. The given energy values are the ones for $E_{\text{Grid-MS}}$.

$N = 4, E = 91.31$ $\quad N = 4, E = 33.54$ $\quad N = 8, E = 42.90$ $\quad N = 8, E = 30.01$

$N = 4, E = 134.51$ $\quad N = 4, E = 80.21$ $\quad N = 8, E = 77.87$ $\quad N = 8, E = 68.55$

$N = 4, E = 131.95$ $\quad N = 4, E = 77.11$ $\quad N = 8, E = 82.15$ $\quad N = 8, E = 70.43$

Fig. 5. A comparison of $E_{\text{Discretized-MS}}$ (1st and 3rd column) and $E_{\text{DC-MS-III}}$ (2nd and 4th image) for small values of N.

Among the many possible settings for the discretization levels N a few ones are of particular interest: for standard discrete inference setting $N = 128$ leads to a reasonable tradeoff between memory consumption/run-time and visual quality of the solution. We focus on $N = 32$ for $E_{\text{Discretized-MS}}$ as suggested in [14]. Finally, the preferred setting is $N = 1$ for $E_{\text{DC-MS}}$, since the increase in run-time and memory consumption was not significantly compensated by improving the solution. $E_{\text{DC-MS}}$ with $N = 1$ had clearly the smallest memory footprint of all methods and settings for N which produced an at least visually convincing solution (see Fig. 4). Interestingly, the "Lena" image turned out to be a much more difficult input image than "Marilyn" for all formulations (with the exception of $E_{\text{dual-MRF}}$, whose performance is largely unaffected by the data set).

We were also interested in how $E_{\text{Discretized-MS}}$ and $E_{\text{DC-MS}}$ behave for rather low values of N (Fig. 5). Since $E_{\text{DC-MS}}$ with $N = 1$ already provides a relatively faithful result, we utilized $E_{\text{DC-MS-III}}$ in order to simultaneously judge how much moving the data term to the nodes weakens the relaxation. It does weaken the relaxation, but explicitly modeling continuous labels is able to avoid banding artefacts compared to $E_{\text{Discretized-MS}}$. In contrast to $E_{\text{DC-MS-III}}$, the model $E_{\text{DC-MS-II}}$ exhibits very slow information flow between intensity subranges $[i/N, (i+1)/N]$ whenever $N > 1$, leading to insistent banding artefacts in non-converged iterates.

A notorious problem of convex multi-labeling formulations is the slow convergence rate of first-order methods. In particular, iterates of $E_{\text{Discretized-MS}}$ for the "Marilyn" image are much closer to a low-energy baseline solution than the finally converged result. For instance, with $N = 16$ after 20.000 iterations evaluating $E_{\text{Grid-MS}}$ returns a value of 28.04, but after 50.000 iteration the energy increases to 29.03 and banding artefacts appear. Similar observations hold for $E_{\text{reduced-MRF}}$. One has to be careful in assessing the quality of any convex formulation, since for this particular problem proper convergence of the result is difficult to achieve in any reasonable time whenever $N \geq 16$.

The runtime of our OpenMP accelerated C++ implementations varies from a few seconds ($E_{\text{DC-MS}}$ with $N = 1$) to several minutes (the other methods) for 256×256 images. Each iteration to minimize $E_{\text{Discretized-MS}}$ or any variant of $E_{\text{DC-MS}}$ requires only fractions of a second, but visual and numerical convergence may easily require 10 000 iterations if $N \gg 1$.

In summary our recommendation for reasonable fast and faithful results for the Mumford-Shah energy are as follows: use a dual block-coordinate method with 64–128 labels (or, if memory consumption is a strong concern, use $E_{\text{DC-MS}}$ with $N = 1$) to obtain a good initialization for a subsequent non-convex refinement (e.g. in the spirit of [2] or by alternating minimization).

7   Conclusion

This work is a first step towards a better understanding of continuously inspired convex models for Mumford-Shah-type energies, and MRF-type relaxations retaining continuous label spaces. In this work we focus on anisotropic regularization, but many of the results will carry over to the isotropic setting. Future work will address e.g. a better understanding, why $E_{\text{Discretized-MS}}$ works quite well in discrete domains, and how the run-time performance of $E_{\text{DC-MS}}$ can be improved for $N \gg 1$.

References

1. Alberti, G., Bouchitté, G., Maso, G.D.: The calibration method for the Mumford-Shah functional and free-discontinuity problems. Calc. Var. Partial Differ. Eqn. **16**(3), 299–333 (2003)
2. Ambrosio, L., Tortorelli, V.: Approximation of functionals depending on jumps by elliptic functionals via Γ-convergence. Commun. Pure Appl. Math. **43**, 999–1036 (1990)
3. Blake, A., Zisserman, A.: Visual Reconstruction. MIT Press, Cambridge (1987)

4. Boykov, Y., Kolmogorov, V.: Computing geodesics and minimal surfaces via graph cuts. In: Proceedings of ICCV, pp. 26–33 (2003)
5. Chan, T.F., Vese, L.: Active contours without edges. IEEE Trans. Image Process. 10(2), 266–277 (2001)
6. Combettes, P.L.: Perspective functions: properties, constructions, and examples. Set-Valued Variational Anal. 1–18 (2016)
7. Felzenszwalb, P.F., Huttenlocher, D.P.: Efficient belief propagation for early vision. In: Proceedings of CVPR, pp. 261–268 (2004)
8. Geman, D., Reynolds, G.: Constrained restoration and the recovery of discontinuities. IEEE Trans. Pattern Anal. Mach. Intell. 14(3), 367–383 (1992)
9. Geman, S., Geman, D.: Stochastic relaxation, Gibbs distributions, and the Bayesian restoration of images. IEEE Trans. Pattern Anal. Mach. Intell. 6, 721–741 (1984)
10. Globerson, A., Jaakkola, T.: Fixing max-product: convergent message passing algorithms for MAP LP-relaxations. In: NIPS (2007)
11. Hazan, T., Shashua, A.: Norm-prodcut belief propagtion: primal-dual message-passing for LP-relaxation and approximate-inference. IEEE Trans. Inf. Theory 56(12), 6294–6316 (2010)
12. Mumford, D., Shah, J.: Optimal approximation by piecewise smooth functions and associated variational problems. Commun. Pure Appl. Math. 42, 577–685 (1989)
13. Pock, T., Chambolle, A.: Diagonal preconditioning for first order primal-dual algorithms in convex optimization. In: Proceedings of ICCV, pp. 1762–1769 (2011)
14. Pock, T., Cremers, D., Bischof, H., Chambolle, A.: An algorithm for minimizing the piecewise smooth Mumford-Shah functional. In: Proceedings of ICCV (2009)
15. Strekalovskiy, E., Goldluecke, B., Cremers, D.: Tight convex relaxations for vector-valued labeling problems. In: Proceedings of ICCV (2011)
16. Werner, T.: A linear programming approach to max-sum problem: a review. IEEE Trans. Pattern Anal. Mach. Intell. 29(7), 1165–1179 (2007)
17. Zach, C.: Dual decomposition for joint discrete-continuous optimization. In: AISTATS (2013)
18. Zach, C., Häne, C., Pollefeys, M.: What is optimized in convex relaxations for multi-label problems: connecting discrete and continuously-inspired MAP inference. IEEE Trans. Pattern Anal. Mach. Intell. (2013)
19. Zach, C., Kohli, P.: A convex discrete-continuous approach for Markov random fields. In: Proceedings of ECCV (2012)

A Projected Gradient Descent Method for CRF Inference Allowing End-to-End Training of Arbitrary Pairwise Potentials

Måns Larsson[1](✉) (iD), Anurag Arnab[2], Fredrik Kahl[1,3], Shuai Zheng[2], and Philip Torr[2]

[1] Chalmers University of Technology, Gothenburg, Sweden
mans.larsson@chalmers.se
[2] University of Oxford, Oxford, England
[3] Centre for Mathematical Sciences, Lund University, Lund, Sweden

Abstract. Are we using the right potential functions in the Conditional Random Field models that are popular in the Vision community? Semantic segmentation and other pixel-level labelling tasks have made significant progress recently due to the deep learning paradigm. However, most state-of-the-art structured prediction methods also include a random field model with a hand-crafted Gaussian potential to model spatial priors, label consistencies and feature-based image conditioning.

In this paper, we challenge this view by developing a new inference and learning framework which can learn pairwise CRF potentials restricted only by their dependence on the image pixel values and the size of the support. Both standard spatial and high-dimensional bilateral kernels are considered. Our framework is based on the observation that CRF inference can be achieved via projected gradient descent and consequently, can easily be integrated in deep neural networks to allow for end-to-end training. It is empirically demonstrated that such learned potentials can improve segmentation accuracy and that certain label class interactions are indeed better modelled by a non-Gaussian potential. In addition, we compare our inference method to the commonly used mean-field algorithm. Our framework is evaluated on several public benchmarks for semantic segmentation with improved performance compared to previous state-of-the-art CNN+CRF models.

Keywords: Conditional Random Fields · Segmentation
Convolutional Neural Networks

1 Introduction

Markov Random Fields (MRFs), Conditional Random Fields (CRFs) and more generally, probabilistic graphical models are a ubiquitous tool used in a variety

Electronic supplementary material The online version of this chapter (https://doi.org/10.1007/978-3-319-78199-0_37) contains supplementary material, which is available to authorized users.

M. Pelillo and E. Hancock (Eds.): EMMCVPR 2017, LNCS 10746, pp. 564–579, 2018.
https://doi.org/10.1007/978-3-319-78199-0_37

of domains spanning Computer Vision, Computer Graphics and Image Processing [4,23]. In this paper, we focus on the application of CRFs for Computer Vision problems involving per-pixel labelling such as image segmentation. There are many successful approaches in this line of research, such as the interactive segmentation of [31] using graph cuts and the semantic segmentation works of [25,35] where the parallel mean-field inference algorithm was applied for fast inference. Recently, Convolutional Neural Networks (CNNs) have dominated the field in a variety of recognition tasks [18,30,33]. However, we observe that leading segmentation approaches still include CRFs, either as a post-processing step [9–11,16], or as part of the deep neural network itself [2,22,26,27,38].

We leverage on the idea of embedding inference of graphical models into a neural network. An early example of this idea was presented in [7] where the authors back propagated through the Viterbi algorithm when designing a document recognition system. Similar to [2,3,37,38], we use a recurrent neural network to unroll the iterative inference steps of a CRF. This was first used in [32,38] to imitate mean-field inference and to train a fully convolutional network [10,28] along with a CRF end-to-end via back propagation. In contrast to mean-field, we do not optimize the KL-divergence. Instead, we use a gradient descent approach for the inference that directly minimises the Gibbs energy of the random field and hence avoids the approximations of mean-field. A similar framework was recently suggested in [3] for multi-label classification problems in machine learning with impressive results. However, [3] uses a Structured SVM approach for training whereas we do back propagation through the actual steps of the gradient descent method. Moreover, [15] have recently shown that one can obtain lower energies compared to mean-field inference using gradient descent based optimization schemes. Still, we lack formal algorithmic guarantees of the solution quality compared to, e.g., graph cuts [8].

In many works, the pairwise potentials consist of parametrized Gaussians [2,24,38] and it is only the parameters of this Gaussian which are learned. Our framework can learn arbitrary pairwise potentials which need not be Gaussian, cf. Fig. 1. An early work which learned potentials of a linear chain CRF for sequence modelling is [29]. In [12], a general framework for learning arbitrary potentials in deep structured models was proposed based on approximate Maximum Likelihood learning. One of the advantages with that framework is that data likelihood is maximized in the learning process. However, this involves approximating the partition function which is otherwise intractable. This hinders the handling of large structured output spaces like in our case. The experiments of [3,12] are limited to multi-label classification where there is no spatial relationship between different labels as in pixel-labelling tasks. Morever, [22] have also observed that the memory requirements of the method of [12] renders it infeasible for the large datasets common in Computer Vision. Another approach to learning arbitrary pairwise potentials was presented in [22] which uses Gibbs sampling. Again they struggle with the difficulty of computing the partition function. In the end, only experiments on synthetic data restricted to learned 2D potentials are presented.

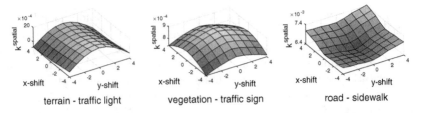

Fig. 1. Learned spatially invariant CRF filters for CITYSCAPES. These filters model contextual relationships between classes and their values can be understood as the energy added when setting one pixel to the first class (e.g., vegetation) and the other pixel with relative position (x-shift, y-shift) to the second class (e.g., traffic sign). Note how the terrain-traffic light filter favours vertical edges. In addition, the model can learn filters of different shapes which is shown by the road-sidewalk filter.

The authors of [9, 26] also learn arbitrary pairwise potentials to model contextual relations between parts of the image. However, their approaches still perform post-processing with a CRF model with parametric Gaussian potentials. In [20], a pairwise potential is learned based on sparse bilateral filtering. Applying such a filter can be regarded as one iteration of mean field CRF inference. Contrasting [20], we not only learn sparse high-dimensional bilateral filters, but also learn arbitrary spatial filters. Such spatial 2D potentials are computationally much more efficient and easier to analyze and interpret compared to their high-dimensional counterparts.

In summary, our contributions are as follows.

- Our main contribution is a new framework for non-parametric CRF inference and learning which is integrated with standard CNNs. During inference, we directly minimize the CRF energy using gradient descent and during training, we back propagate through the gradient descent steps for end-to-end learning.
- We analyze the learned filter kernels empirically and demonstrate that in many cases it is advantageous with non-Gaussian potentials.
- We experimentally compare our approach to several leading methodologies, e.g., [16, 28, 38] and improve on state of the art on two public benchmarks: NYU V2 [36] and CITYSCAPES [14].

Our framework has been implemented in both CAFFE [21] and MATCONVNET [34], and all source code will be made publicly available to facilitate further research.

2   CRF Formulation

Consider a Conditional Random Field over N discrete random variables $\mathcal{X} = \{X_1, \ldots, X_N\}$ conditioned on an observation I and let $\mathcal{G} = \{\mathcal{V}, \mathcal{E}\}$ be an undirected graph whose vertices are the random variables $\{X_1, \ldots, X_N\}$. Each random variable corresponds to a pixel in the image and takes values from a predefined set of L labels $\mathcal{L} = \{0, \ldots, L-1\}$. The pair $(\mathcal{X}, \boldsymbol{I})$ is modelled as a CRF characterized by the Gibbs distribution

$$P(\mathcal{X} = \boldsymbol{x}|\boldsymbol{I}) = \frac{1}{Z(\boldsymbol{I})} \exp(-E(\boldsymbol{x}|\boldsymbol{I})), \tag{1}$$

where $E(\boldsymbol{x}|\boldsymbol{I})$ denotes the Gibbs energy function with respect to the labeling $\boldsymbol{x} \in \mathcal{L}^N$ and $Z(\boldsymbol{I})$ is the partition function. To simplify notation the conditioning on $\boldsymbol{I}$ will from now on be dropped. The MAP inference problem for the CRF model is equivalent to the problem of minimizing the energy $E(\boldsymbol{x})$. In this paper, we only consider energies containing unary and pairwise terms. The energy function can hence be written as

$$E(\boldsymbol{x}) = \sum_{i \in \mathcal{V}} \psi_i(x_i) + \sum_{(i,j) \in \mathcal{E}} \psi_{ij}(x_i, x_j) \tag{2}$$

where $\psi_i : \mathcal{L} \to \mathbb{R}$ and $\psi_{ij} : \mathcal{L} \times \mathcal{L} \to \mathbb{R}$ are the unary and pairwise potentials, respectively.

2.1 Potentials

The unary potential $\psi_i(x_i)$ specifies the energy cost of assigning label x_i to pixel i. In this work we obtain our unary potentials from a CNN. Roughly speaking, the CNN outputs a probability estimate of each pixel containing each class. Denoting the output of the CNN for pixel i and class x_i as $z_{i:x_i}$, the unary potential is

$$\psi_i(x_i) = -w_u \log(z_{i:x_i} + \epsilon) \tag{3}$$

where w_u is a parameter controlling the impact of the unary potentials, and ϵ is introduced to avoid numerical problems.

The pairwise potential $\psi_{ij}(x_i, x_j)$ specifies the energy cost of assigning label x_i to pixel i while pixel j is assigned label x_j. Introducing pairwise terms in our model enables us to take dependencies between output data into account. We consider the following set of pairwise potentials

$$\psi_{ij}(x_i, x_j) = k_{x_i,x_j}^{spatial}(\boldsymbol{p}_i - \boldsymbol{p}_j) + k_{x_i,x_j}^{bilateral}(\boldsymbol{f}_i - \boldsymbol{f}_j) \tag{4}$$

Here $k_{x_i,x_j}^{spatial}$ denotes a spatial kernel with compact support. Its value depends on the relative position coordinates $\boldsymbol{p}_i - \boldsymbol{p}_j$ between pixels i and j. We do not restrict these spatial terms to any specific shape, cf. Fig. 1. However, we restrict the support of the potential meaning that if pixels i and j are far apart, then the value of $k_{x_i,x_j}^{spatial}(\boldsymbol{p}_i - \boldsymbol{p}_j)$ will be zero. CRFs with Gaussian potentials do not in theory have compact support, and therefore, they are often referred to as dense. However, in practice, the exponential function in the kernel drops off quickly and effectively. The interactions between pixels far apart are negligible and are commonly truncated to 0 after two standard deviations.

The term $k_{x_i,x_j}^{bilateral}$ is a bilateral kernel which depends on the feature vectors $\boldsymbol{f}_i$ and $\boldsymbol{f}_j$ for pixels i and j, respectively. Following several previous works [11, 24,

38], we let the vector depend on pixel coordinates p_i and RGB values associated to the pixel, hence f_i is a 5-dimensional vector. Note that for both the spatial and the bilateral kernels, there is one kernel for each label-to-label (x_i and x_j) interaction to enable the model to learn differently shaped kernels for each of these interactions.

2.2 Multi-label Graph Expansion and Relaxation

To facilitate a continuous relaxation of the energy minimisation problem we start off by expanding our original graph in the following manner. Each vertex in the original graph $\mathcal{G}$ will now be represented by L vertices $X_{i:\lambda}$, $\lambda \in \mathcal{L}$. In this way, an assignment of labels in $\mathcal{L}$ to each variable X_i is equivalent to an assignment of boolean labels 0 or 1 to each node $X_{i:\lambda}$, whereby an assignment of label 1 to $X_{i:\lambda}$ means that in the multi-label assignment, X_i receives label λ. As a next step, we relax the integer program by allowing real values on the unit interval $[0, 1]$ instead of booleans only. We denote the relaxed variables $q_{i:\lambda} \in [0, 1]$. We can now write our problem as a quadratic program

$$
\begin{aligned}
\min \quad & \sum_{i \in \mathcal{V}, \lambda \in \mathcal{L}} \psi_i(\lambda) q_{i:\lambda} + \sum_{\substack{(i,j) \in \mathcal{E} \\ \lambda, \mu \in \mathcal{L}}} \psi_{ij}(\lambda, \mu) q_{i:\lambda} q_{j:\mu} \\
\text{s.t.} \quad & q_{i:\lambda} \geq 0 && \forall i \in \mathcal{V}, \lambda \in \mathcal{L} \\
& \sum_{\lambda \in \mathcal{L}} q_{i:\lambda} = 1 && \forall i \in \mathcal{V}.
\end{aligned}
\tag{5}
$$

Note that the added constraints ensure that our solution lies on the probability simplex. A natural question is what happens when the domain is enlarged, allowing real values instead of booleans. Somewhat surprisingly, the relaxation is tight [6].

Proposition 1. *Let $E(x^*)$ and $E(q^*)$ denote the optimal values of (5), where x^* is restricted to boolean values. Then,*

$$ E(x^*) = E(q^*). $$

In the supplementary material, we show that for *any* real q, one can obtain a binary x such that $E(x) \leq E(q)$. In particular, it will be true for x^* and q^*, which implies $E(x^*) = E(q^*)$. Note that the proof is constructive.

In summary, it has been shown that to minimize the energy function $E(x)$ over $x \in \mathcal{L}^N$, one may work in the continuous domain, minimize over q, and then replace any solution q by a discrete solution x which has lower or equal energy. It will only be possible to find a local solution q, but still the discrete solution x will be no worse than q.

3   Minimization with Gradient Descent

To solve the program stated in (5) we propose an optimization scheme based on projected gradient descent, see Algorithm 1. It was designed with an extra condition in mind, that all operations should be differentiable to enable back propagation during training.

Initialize q^0
for t *from* 0 *to* $T - 1$ **do**
  Compute the gradient $\nabla_q E(q^t)$.
  Take a step in the negative direction, $\tilde{q}^{t+1} = \mathbf{q^t} - \gamma \, \nabla_q E$.
  Project $\tilde{q}_{i:\lambda}^{t+1}$ to the simplex $\triangle^L$ satisfying $\sum_{\lambda \in \mathcal{L}} \tilde{q}_{i:\lambda} = 1$ and $0 \leq \tilde{q}_{i:\lambda} \leq 1$,
  $q^{t+1} = \text{Proj}_{\triangle^L}(\tilde{q})$.
end
Output: q^{T-1}

Algorithm 1. Projected gradient descent algorithm.

The gradient $\nabla_q E$ of the objective function $E(q)$ in (5) has the following elements

$$\frac{\partial E}{\partial q_{i:\lambda}} = \psi_i(\lambda) + \sum_{\substack{j:(i,j)\in\mathcal{E} \\ \mu\in\mathcal{L}}} \psi_{ij}(\lambda,\mu)q_{j:\mu}. \tag{6}$$

The contribution from the spatial kernel in ψ_{ij}, cf. (4), can be written as

$$v_{i:\lambda}^{spatial} = \sum_{\substack{j:(i,j)\in\mathcal{E} \\ \mu\in\mathcal{L}}} k_{\lambda,\mu}^{spatial}(\boldsymbol{p}_i - \boldsymbol{p}_j)q_{j:\mu}. \tag{7}$$

Since the value of the kernel $v_{i:\lambda}^{spatial}$ only depends on the relative position of pixels i and j, the contribution for all pixels and classes can be calculated by passing $q_{j:\mu}$ through a standard convolution layer consisting of $L \times L$ filters of size $(2s + 1) \times (2s + 1)$ where L is the number of labels and s the number of neighbours each pixel interacts with in each dimension.

The contribution from the bilateral term is

$$v_{i:\lambda}^{bilateral} = \sum_{\substack{j:(i,j)\in\mathcal{E} \\ \mu\in\mathcal{L}}} k_{\lambda,\mu}^{bilateral}(\boldsymbol{f}_i - \boldsymbol{f}_j)q_{j:\mu}. \tag{8}$$

For this computation we utilize the method presented by Jampani et al. [20] which is based on the permutohedral lattice introduced by Adams et al. [1]. Efficient computations are obtained by using the fact that the feature space is generally sparsely populated. Similar to the spatial filter we get $L \times L$ filters, each having size of $(s + 1)^{d+1} - s^{d+1}$ where s is the number of neighbours each pixel interacts with in each dimension in the sparse feature space.

Next, we simply take a step in the negative direction of the gradient according to

$$\tilde{q}^{t+1} = \mathbf{q^t} - \gamma \, \nabla_q E, \tag{9}$$

where γ is the step size.

Finally, we want to project our values onto the simplex $\triangle^L$ satisfying $\sum_{\lambda \in \mathcal{L}} q_{i:\lambda} = 1$ and $0 \leq q_{i:\lambda} \leq 1$. This is done following the method by Chen and Ye [13] that efficiently calculates the euclidean projection on the probability simplex $\triangle^L$, for details see the supplementary materials. Note that this projection is done individually for each pixel i.

Comparison to Mean-Field. In recent years, a popular choice for CRF inference is to apply the mean-field algorithm. One reason is that the kernel evaluations can be computed with fast bilateral filtering [25]. As we have seen in this section, it can be accomplished with our framework as well. The main difference is that our framework directly optimizes the Gibbs energy which corresponds to the MAP solution while mean-field optimizes KL-divergence which does not.

4   Integration in a Deep Learning Framework

In this section we will describe how the gradient descent steps of Algorithm 1 can be formulated as layers in a neural network. We need to be able to calculate error derivatives with respect to the input given error derivatives with respect to the output. In addition we need to be able to calculate the error derivatives with respect to the network parameters. This will enable us to unroll the entire gradient descent process as a Recurrent Neural Network (RNN). A schematic of the data flow for one step is shown in Fig. 2. In the supplementary material, all derivative formulae are given.

Initialization. The variables q^0 are set as the output of the CNN, which has been pre-trained to estimate the probability of each pixel containing each class and has a softmax layers at the end to ensure that the variables lies within zero and one as well as sum to one for each pixel.

Gradient Computations. We have previously explained the gradient computations in Sect. 3 for the forward pass. To describe the calculation of the error

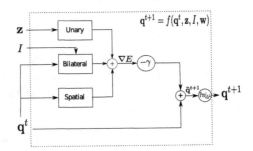

Fig. 2. The data flow of one iteration of the projected gradient descent algorithm. Each rectangle or circle represent an operation that can be performed within a deep learning framework.

derivatives we first notice that the gradient is calculated by summing three terms, the unary, spatial and bilateral pairwise terms. We can hence treat these three terms separately and combine them using element-wise summing.

The unary term in (3) is an element-wise operation with the CNN output as input and the unary weight w_u as parameter. The operation is obviously differentiable with respect to both the layer input as well as its parameter. Note that for w_u we get a summation over all class and pixel indexes for the error derivatives while for the input the error derivatives are calculated element-wise. The spatial pairwise term of the gradient can be calculated efficiently using standard 2D convolution. In addition to giving us an efficient way of performing the forward pass we can also utilize the 2D convolution layer to perform the backward pass, calculating the error derivatives with respect to the input and parameters. Similar to the spatial term, the bilateral term is also calculated utilizing a bilateral filtering technique. Jampani et al. [20] also presented a way to calculate the error derivatives with respect to the parameters for an arbitrary shaped bilateral filter.

Gradient Step. Taking a step in the negative direction of the gradient is easily incorporated in a deep learning framework by using an element-wise summing layer. The layer takes the variables q^t as the first input and the gradient (scaled by $-\gamma$) as the second input.

Simplex Projection. As a final step, the variables from the gradient step $\tilde{q}^t$ are projected onto the simplex $\triangle^L$. In reality we use a leaky version of the last step of the projection algorithm to avoid error derivatives becoming zero during back propagation. Since the projection is done individually for each pixel it can be described as a function $f(\tilde{q}) : \mathbb{R}^L \to \mathbb{R}^L$ of which we can calculate the Jacobian, see supplementary materials. Knowing the Jacobian, the error derivatives with respect to the input can be computed during back propagation.

5   Recurrent Formulation as a Deep Structured Model

Our iterative solution to the CRF energy minimisation problem by projected gradient descent, as described in the previous sections, is formulated as a Recurrent Neural Network (RNN). The input to the RNN is the image, and the outputs of a CNN, as shown in Fig. 3. The CNN's output, $\mathbf{z}$, are the unary potentials and obtained after the final softmax layer (since the CNN is initially trained for classification).

Each iteration of the RNN performs one projected gradient descent step to approximately solve (5). Thus, one update step can be represented by:

$$q^{t+1} = f(q^t, z, I, w). \tag{10}$$

As illustrated in Fig. 3, the gating function G_1 sets q^t to $\mathbf{z}$ at the first time step, and to q^{t-1} at all other time steps. In our iterative energy minimisation

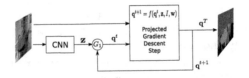

Fig. 3. The data flow of the deep structured model. Each rectangle or circle represent an operation that can be performed within a deep learning framework.

formulated as an RNN, the output of one step is the input to the next step. We initialise at $t = 0$ with the output of the unary CNN.

The output of the RNN can be read off $\boldsymbol{q}^T$ where T is the total number of steps taken. In practice, we perform a set number of T steps where T is a hyperparameter. It is possible to run the RNN until convergence for each image (thus a variable number of iterations per image), but we observed minimal benefit in the final Intersection over Union (IoU) from doing so, as opposed to $T = 5$ iterations.

The parameters of the RNN are the filter weights for the spatial and bilateral kernels, and also the weight for the unary terms. Since we are able to compute error derivatives with respect to the parameters, and input of the RNN, we can back propagate error derivates through our RNN to the preceding CNN and train our entire network end-to-end. Furthermore, since the operations of the RNN are formulated as filtering, training and inference can be performed efficiently in a fully-convolutional manner.

Implementation Details. Our proposed CRF model has been implemented in the CAFFE [21] library, and also has a Matlab wrapper allowing it to be used in MATCONVNET [34]. The Unary CNN part of our model is initialised from a pre-trained segmentation network. As we show in our experiments, different pre-trained networks for different applications can be used.

The CRF model has several tunable parameters. The step size γ and the number of iterations T specify the properties of the gradient decent algorithm. Too high a step size γ might make the algorithm not end up in a minimum while setting a low step size and a low number of iterations might not give the algorithm a chance to converge. The kernel sizes for the spatial and bilateral kernels also need to be set. Choosing the value of these parameters gives a trade-off between model expression ability and number of parameters, which may cause (or hinder) over-fitting.

The spatial weights of the CRF model are all initialized as zero with the motivation that we did not want to impose a shape for these filters, but instead see what was learned during training. The bilateral filters were initialized as Gaussians with the common Potts class interaction (the filters corresponding to interactions between the same class were set to zero) as done in [10,25,38]. Note that unlike [10,25] we are not limited to only Potts class interactions.

6 Experiments

We evaluate the proposed approach on three datasets: WEIZMANN HORSE [5], NYU V2 [36] and CITYSCAPES [14]. In these experiments, we show that the proposed approach, denoted CRF-Grad, has advantages over baseline approaches such as CRFasRNN [38] and complement other networks such as FCN-8s [28] and LRR [16]. In addition we compare our inference method to mean-field inference. Hyperparameter values not specified in this section can be found in the supplementary materials.

6.1 Weizmann Horse

The WEIZMANN HORSE dataset contains 328 images of horses in different environments. We divide these images into a training set of 150 images, a validation set of 50 images and a test set of 128 images. Our purpose is to verify our ability to learn reasonable kernels and study the effects of different settings on a relatively small dataset.

The CNN part of our model was initialized as an FCN-8s network [28] pretrained without the CRF layer. We then compare several variants of our model. We start off by training a variant of our CRF model only using the 2D spatial kernel. We compare these results to using a Gaussian spatial filter, where the parameters for the Gaussian kernel were evaluated using cross-validation. In addition we train the full model with both the spatial and bilateral kernels, once keeping the filters fixed as Gaussians and once allowing arbitrarily shaped filters.

The mean Intersection over Union (IoU) on the test set, for different configurations of our model, is shown in Table 1. Our results show that allowing arbitrarily shaped filters gives better performance than keeping the filters fixed as Gaussian. This is the case for both the spatial and bilateral kernels. Also,

Table 1. Quantitative results comparing our method to baselines as well as state-of-the-art methods. Mean intersection over union for the test set is shown for WEIZMANN HORSE and CITYSCAPES, for NYU V2 there is no test set and validation score is presented. In the entry denoted "full" the complete CRF-Grad layer was used, while in "spatial" no bilateral kernel was used. A (G) means that the filters were restricted to a Gaussian shape and (MF) means that the inference method was switched to mean field. For the entries denoted CRF-Grad the full model was used.

WEIZMANN HORSE		NYU V2		CITYSCAPES	
Method	IoU (%)	Method	IoU (%)	Method	IoU (%)
FCN-8s (only)	80.0	R-CNN [17]	40.3	CRFasRNN* [38]	62.5
FCN-8s + spatial (G)	81.3	Joint HCRF [36]	44.2	Deeplabv2-CRF [11]	70.4
FCN-8s + spatial	82.0	Modular CNN [19]	54.3	Adelaide_context [26]	71.6
FCN-8s + full (G)	82.9	CRFasRNN [38]	54.4	LRR-4x [16]	71.8
FCN-8s + full	**84.0**	CRF-Grad (Ours)	**55.0**	CRF-Grad (Ours)	**71.9**
FCN-8s + full (MF)	83.3				

*Note that CRFasRNN uses a different CNN model than ours on the CITYSCAPES dataset.

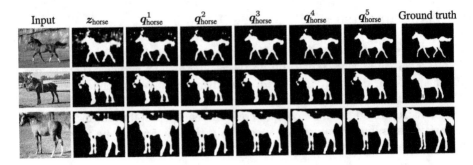

Fig. 4. Visualization of intermediate states of the CRF-Grad layer for the WEIZMANN HORSE dataset. Note how each step of the gradient descent algorithm refines the segmentation slightly, removing spurious outlier pixels classified as horse. z_{horse} is the CNN output.

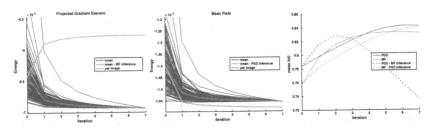

Fig. 5. Left and Middle: CRF energy, as defined in (5), as a function of inference iterations. The thin blue lines show the different instances, the thicker red line show the mean while the green line shows the mean when the inference method has been switched. The model has been trained with the inference method shown in respective title. For these calculations the leak factor was set to zero, meaning that the solutions satisfy the constraints of (5). Note that, for presentation purposes, all energies have been normalized to have the same final energy. Right: Mean IoU as a function of iterations for the two different models and inference methods. The number of iterations were set to five during training. All results are from the test set of the WEIZMANN HORSE data set. (Color figure online)

adding bilateral filters improves the results compared to using spatial filters only. In addition the model trained with our inference method performs better than the model (with the same unary and pairwise potentials) trained with mean-field inference.

In Fig. 4 the intermediate states of the layer (q^t for each gradient descent step) are shown. Figure 5 presents a convergence analysis for the full version of our layer as well as a comparison to mean-field inference. The results show that the CRF energy converges in only a few iteration steps and that increasing the number of iterations barely affects performance after $T = 6$ iterations. It also shows that our inference method achieves lower Gibbs energy as well as higher IoU results compared to mean-field, even for models trained with mean-field inference. Looking at using mean field inference for the model trained with projected gradient descent we see that the energy increases for each iteration. The reason for this is that the mean-field method doesn't actually minimize the energy but the Kullback-Liebler divergence between the Gibbs distribution and the solution. This is, as shown in the supplementary materials, equivalent to minimizing the energy minus the entropy. The extra entropy term favors "uncertain" solutions, i.e. solution with values not close to zero or one, which increases the energy for some models.

6.2   NYU V2

The NYU V2 dataset contains images taken by Microsoft Kinect V-1 camera in 464 indoor scenes. We use the official training and validation splits consisting of 795 and 654 images, respectively. Following the setting described in Wang *et al.* [36], we also include additional images for training. These are the images from the NYU V1 dataset that do not overlap with the images in the official validation set. This gives a total of 894 images with semantic label annotations for training. As in [36] we consider 5 classes conveying strong geometric properties: ground, vertical, ceiling, furniture and objects. The CNN part of our model was initialized as the fully convolutional network FCN-8s [28] pre-trained on the data. Afterwards we added our CRF-Grad layer and trained the model end-to-end.

As shown in Table 1, we achieved superior results for semantic image segmentation on the NYU V2 dataset. Some example segmentations are shown in Fig. 6. Additional examples are included in the supplementary material.

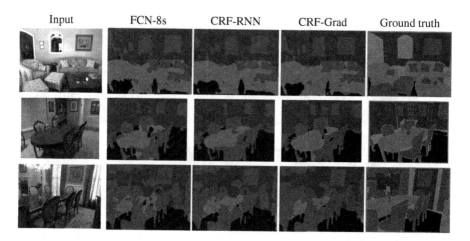

Fig. 6. Qualitative results on the NYU V2 dataset. Note that the CRF-Grad captures the shape of the object instances better compared to the baselines. This effect is perhaps most pronounced for the paintings hanging on the walls.

6.3 Cityscapes

The CITYSCAPES dataset [14] consists of a set of images of street scenes collected from 50 different cities. The images are high resolution (1024 × 2048) and are paired with pixel-level annotations of 19 classes including road, sidewalk, traffic sign, pole, building, vegetation and sky. The training, validation and test sets consist of 2975, 500 and 1525 images, respectively. In addition there are 20000 coarsely annotated images that can be used for training. The CNN part of our model was initialized as an LRR network [16] pre-trained on both the fine and the coarse annotations. We then added our CRF-Grad layer and trained the model end-to-end on the finely annotated images only.

In Table 1 the results of evaluating our model on the test set are compared to current state of the art. As can be seen, our model is on par although the improvement upon LRR is minor. An interesting aspect of CITYSCAPES is that it contains classes of thin and vertical objects, e.g., traffic light and pole. What we noticed is that the spatial filters for these classes usually get a more oblong shape. This type of pairwise filters does not add as much energy for switching classes going in the horizontal direction, favoring vertically elongated segmentations. This can be seen in the spatial filter for the class interaction between "terrain" and "traffic light" in Fig. 1. Some example segmentations are shown in Fig. 7. Additional examples as well as class-wise results are included in the supplementary material. Our method got better, or equal, results for 14 of the 19 classes compared to the baseline.

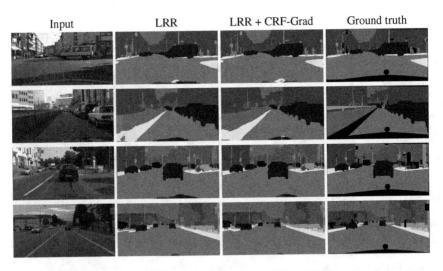

Fig. 7. Qualitative results on the CITYSCAPES validation set. Black regions in the ground truth are ignored during evaluation. Our CRF models contextual relationships between classes, hence unlike LRR, it does not label "road" as being on top of "sidewalk" (Row 2). Note that the traffic lights are better segmented with the additional CRF-Grad layer. Adding the CRF-Grad layer increased the IoU of the class traffic lights from 66.8 to 68.1.

7   Concluding Remarks

In this paper, we have introduced a new framework capable of learning arbitrarily shaped pairwise potentials in random fields models. In a number of experiments, we have empirically demonstrated that our developed framework can improve state-of-the-art CNNs by adding a CRF layer. We have also seen that the learned filters are not necessarily Gaussian, and may capture other kinds of interactions between labels. In addition, we have shown advantages of our inference method compared to the commonly used mean-field method.

A key factor for the success of deep learning and by now a well-established paradigm is that the power of convolutions should be used, especially for the first layers in a CNN. Our work supports that repeated usage of convolutions in the final layers is also beneficial. We also note that our gradient descent steps resemble the highly successful RESNET [18], as one step in gradient descent is, in principle, an identity transformation plus a correction term.

There are several future research avenues that we intend to explore. In our model, many free variables are introduced and this may lead to over-fitting. One way to compensate would be to collect larger datasets and consider data augmentation. An alternative approach would be to directly encode geometric shape priors into the random fields and thereby reducing the required amount of data.

Acknowledgements. This work has been funded by the Swedish Research Council (grant no. 2016-04445), the Swedish Foundation for Strategic Research (Semantic Mapping and Visual Navigation for Smart Robots), Vinnova/FFI (Perceptron, grant no. 2017-01942), ERC (grant ERC-2012-AdG 321162-HELIOS) and EPSRC (grant Seebibyte EP/M013774/1 and EP/N019474/1).

References

1. Adams, A., Baek, J., Davis, M.A.: Fast high-dimensional filtering using the permutohedral lattice. In: Computer Graphics Forum (2010)
2. Arnab, A., Jayasumana, S., Zheng, S., Torr, P.H.S.: Higher order conditional random fields in deep neural networks. In: Leibe, B., Matas, J., Sebe, N., Welling, M. (eds.) ECCV 2016. LNCS, vol. 9906, pp. 524–540. Springer, Cham (2016). https://doi.org/10.1007/978-3-319-46475-6_33
3. Belanger, D., McCallum, A.: Structured prediction energy networks. In: International Conference on Machine Learning (2016)
4. Blake, A., Kohli, P., Rother, C.: Markov Random Fields for Vision and Image Processing. MIT Press, Cambridge (2011)
5. Borenstein, E., Ullman, S.: Class-specific, top-down segmentation. In: Heyden, A., Sparr, G., Nielsen, M., Johansen, P. (eds.) ECCV 2002. LNCS, vol. 2351, pp. 109–122. Springer, Heidelberg (2002). https://doi.org/10.1007/3-540-47967-8_8
6. Boros, E., Hammer, P.L.: Pseudo-boolean optimization. Discret. Appl. Math. **123**, 155–225 (2002)
7. Bottou, L., Bengio, Y., Le Cun, Y.: Global training of document processing systems using graph transformer networks. In: IEEE Conference on Computer Vision and Pattern Recognition, pp. 489–494. IEEE (1997)

8. Boykov, Y., Veksler, O., Zabih, R.: Fast approximate energy minimization via graph cuts. IEEE Trans. Pattern Anal. Mach. Intell. **23**(11), 1222–1239 (2001)

9. Chandra, S., Kokkinos, I.: Fast, exact and multi-scale inference for semantic image segmentation with deep Gaussian CRFs. In: Leibe, B., Matas, J., Sebe, N., Welling, M. (eds.) ECCV 2016. LNCS, vol. 9911, pp. 402–418. Springer, Cham (2016). https://doi.org/10.1007/978-3-319-46478-7_25

10. Chen, L., Papandreou, G., Kokkinos, I., Murphy, K., Yuille, A.L.: Semantic image segmentation with deep convolutional nets and fully connected CRFs. In: International Conference on Learning Representations (2015)

11. Chen, L., Papandreou, G., Kokkinos, I., Murphy, K., Yuille, A.L.: DeepLab: semantic image segmentation with deep convolutional nets, atrous convolution, and fully connected CRFs. arXiv preprint arXiv:1606.00915 (2016)

12. Chen, L.C., Schwing, A.G., Yuille, A.L., Urtasun, R.: Learning deep structured models. In: International Conference Machine Learning, Lille, France (2015)

13. Chen, Y., Ye, X.: Projection onto a simplex. arXiv preprint arXiv:1101.6081 (2011)

14. Cordts, M., Omran, M., Ramos, S., Rehfeld, T., Enzweiler, M., Benenson, R., Franke, U., Roth, S., Schiele, B.: The cityscapes dataset for semantic urban scene understanding. In: IEEE Conference on Computer Vision and Pattern Recognition (2016)

15. Desmaison, A., Bunel, R., Kohli, P., Torr, P.H.S., Kumar, M.P.: Efficient continuous relaxations for dense CRF. In: Leibe, B., Matas, J., Sebe, N., Welling, M. (eds.) ECCV 2016. LNCS, vol. 9906, pp. 818–833. Springer, Cham (2016). https://doi.org/10.1007/978-3-319-46475-6_50

16. Ghiasi, G., Fowlkes, C.C.: Laplacian pyramid reconstruction and refinement for semantic segmentation. In: Leibe, B., Matas, J., Sebe, N., Welling, M. (eds.) ECCV 2016. LNCS, vol. 9907, pp. 519–534. Springer, Cham (2016). https://doi.org/10.1007/978-3-319-46487-9_32

17. Girshick, R., Donahue, J., Darrell, T., Malik, J.: Rich feature hierarchies for accurate object detection and semantic segmentation. In: IEEE Conference on Computer Vision and Pattern Recognition (2014)

18. He, K., Zhang, X., Ren, S., Sun, J.: Deep residual learning for image recognition. In: IEEE Conference on Computer Vision and Pattern Recognition (2016)

19. Jafari, O.H., Groth, O., Kirillov, A., Yang, M.Y., Rother, C.: Analyzing modular CNN architectures for joint depth prediction and semantic segmentation. In: International Conference on Robotics and Automation (2017)

20. Jampani, V., Kiefel, M., Gehler, P.V.: Learning sparse high dimensional filters: image filtering, dense CRFs and bilateral neural networks. In: IEEE Conference on Computer Vision and Pattern Recognition, June 2016

21. Jia, Y., Shelhamer, E., Donahue, J., Karayev, S., Long, J., Girshick, R., Guadarrama, S., Darrell, T.: Caffe: convolutional architecture for fast feature embedding. arXiv preprint arXiv:1408.5093 (2014)

22. Kirillov, A., Schlesinger, D., Zheng, S., Savchynskyy, B., Torr, P.H.S., Rother, C.: Joint training of generic CNN-CRF models with stochastic optimization. In: Lai, S.-H., Lepetit, V., Nishino, K., Sato, Y. (eds.) ACCV 2016. LNCS, vol. 10112, pp. 221–236. Springer, Cham (2017). https://doi.org/10.1007/978-3-319-54184-6_14

23. Koller, D., Friedman, N.: Probabilistic Graphical Models. MIT Press, Cambridge (2009)

24. Kraehenbuehl, P., Koltun, V.: Parameter learning and convergent inference for dense random fields. In: Proceedings of the 30th International Conference on Machine Learning, pp. 513–521 (2013)

25. Krähenbühl, P., Koltun, V.: Efficient inference in fully connected CRFs with Gaussian edge potentials. In: Neural Information Processing Systems (2011)
26. Lin, G., Shen, C., Hengel, A., Reid, I.: Efficient piecewise training of deep structured models for semantic segmentation. In: IEEE Conference on Computer Vision and Pattern Recognition, June 2016
27. Liu, Z., Li, X., Luo, P., Loy, C.C., Tang, X.: Semantic image segmentation via deep parsing network. In: International Conference on Computer Vision (2015)
28. Long, J., Shelhamer, E., Darrell, T.: Fully convolutional networks for semantic segmentation. In: IEEE Conference on Computer Vision and Pattern Recognition (2015)
29. Peng, J., Bo, L., Xu, J.: Conditional neural fields. In: Advances in Neural Information Processing Systems, pp. 1419–1427 (2009)
30. Ren, S., He, K., Girshick, R., Sun, J.: Faster R-CNN: towards real-time object detection with region proposal networks. In: Neural Information Processing Systems (2015)
31. Rother, C., Kolmogorov, V., Blake, A.: "GrabCut": interactive foreground extraction using iterated graph cuts. In: ACM Transactions on Graphics, pp. 309–314 (2004)
32. Schwing, A., Urtasun, R.: Fully connected deep structured networks. arXiv preprint arXiv:1503.02351 (2015)
33. Simonyan, K., Zisserman, A.: Very deep convolutional networks for large-scale image recognition. In: International Conference on Learning Representations (2015)
34. Vedaldi, A., Lenc, K.: MatConvNet - convolutional neural networks for MATLAB. In: Proceeding of the ACM International Conference on Multimedia (2015)
35. Vineet, V., Warrell, J., Torr, P.H.S.: Filter-based mean-field inference for random fields with higher-order terms and product label-spaces. In: Fitzgibbon, A., Lazebnik, S., Perona, P., Sato, Y., Schmid, C. (eds.) ECCV 2012. LNCS, vol. 7576, pp. 31–44. Springer, Heidelberg (2012). https://doi.org/10.1007/978-3-642-33715-4_3
36. Wang, P., Shen, X., Lin, Z., Cohen, S., Price, B., Yuille, A.: Towards unified depth and semantic prediction from a single image. In: IEEE Conference on Computer Vision and Pattern Recognition (2014)
37. Wang, W., Fidler, S., Urtasun, R.: Proximal deep structured models. In: Neural Information Processing Systems (2016)
38. Zheng, S., Jayasumana, S., Romera-Paredes, B., Vineet, V., Su, Z., Du, D., Huang, C., Torr, P.: Conditional random fields as recurrent neural networks. In: International Conference on Computer Vision (2015)

Author Index

Printed in the United States
By Bookmasters